Contributions to the

THEORY OF TRANSCENDENTAL NUMBERS

MATHEMATICAL SURVEYS AND MONOGRAPHS · *Number 19*

Contributions to the
THEORY OF TRANSCENDENTAL NUMBERS

BY

GREGORY V. CHUDNOVSKY

AMERICAN MATHEMATICAL SOCIETY
PROVIDENCE, RHODE ISLAND

Portions of this volume were translated
from the Russian by G. A. Kandall

1980 *Mathematics Subject Classification.* Primary 10F35, 10F37.

Library of Congress Cataloging in Publication Data

Chudnovsky, G. (Gregory), 1952–

Contributions to the theory of transcendental numbers.

(Mathematical surveys and monographs; no. 19)

Bibliography: p.

Includes index.

1. Numbers, Transcendental. I. Title. II. Series.

QA247.5.C48 1984 512′.73 83-15728

ISBN 0-8218-1500-8

TABLE OF CONTENTS

To My Parents

PREFACE

Transcendental number theory, which began its development in the time of Euler, is particularly attracted to classical constants of analysis, especially to those connected with the exponential function. The irrationality and transcendence proofs of π and e, achieved by the successive efforts of Euler, Lambert, Hermite and Lindemann mark the spectacular beginning of transcendental number theory. Hilbert's seventh problem (or, as Gel′fond [**1**] put it, the Euler-Hilbert problem) of the transcendence of α^β for algebraic $\alpha \neq 0, 1$ and algebraic irrational β, was solved partially by Gel′fond and Kuzmin (1929) and completely by Gel′fond and Schneider (1934). These results mark a mature state of transcendental number theory, with the Gel′fond-Schneider method as its main instrument for studying the algebraic independence of values of functions satisfying algebraic differential (and functional) equations. In modern form, transcendental number theory and related studies of diophantine approximations exploit a combination of various auxiliary algebraic techniques with analytic approximation and interpolation methods traced back to the works of Hermite, Siegel, Gel′fond and Schneider and with important multidimensional generalizations introduced by Baker. We refer the reader to the excellent book of Baker [**2**] for an overall picture of the field. Progress in this field over the last decade has been connected with the introduction of new functional and algebraic techniques from different fields of mathematics, including mathematical physics. This interaction with the outside world has not yet fully developed and applications of transcendental number theory are still scarce; though one amazing example of the absence of particle production for a particular σ-model that is conditional on the transcendence result of Gel′fond [**1**] was presented by Lüscher [**3**].

This volume consists of a collection of papers devoted primarily to transcendental number theory and diophantine approximations written by the author since 1970. Most of the materials included in this volume are English translations of Russian manuscripts of the author, written in 1970 (Chapter 6), 1974 (Chapters 1 and 2 and parts of Chapter 5), 1974–1977 (Chapters 3, 9, Appendix and parts

0076-5376/84 $1.00 + $.25 per page

of Chapter 7). These papers and other papers included in this volume were available to specialists in manuscript form, but this is the first time that they have been collected together and published.

Though the earlier papers have been preserved in the form in which they were prepared initially, the volume is organized in such a way as to reflect recent progress and to allow readers to follow recent developments in the field. As a guide to the volume we have included an expanded and updated text of the author's address to the 1978 International Mathematical Congress as an Introductory paper. We also made an effort to interrelate different directions of the research, and some papers in the volume were updated to include cross-references and to unify the exposition, though for the sake of completeness some repetitions still remain.

The Introductory paper surveys the progress in transcendental number theory with particular emphasis on transcendence, algebraic independence and measures of the algebraic independence of numbers connected with exponential and elliptic functions. Special sections of the Introduction are devoted to period relations for Abelian varieties of CM-type, the author's results on the Schanuel conjecture, measures of algebraic independence of periods and quasi-periods of elliptic curves, etc. Proofs of the appropriate results are presented in different papers (chapters) of this volume.

Chapter 1 deals with the proofs of the author's 1974 results on the Schanuel conjecture, which show, essentially, that the number of algebraically independent elements in the set S for complex numbers of the form $\{e^{\alpha_i\beta_j}\}$, $\{\alpha_i, e^{\alpha_i\beta_j}\}$, $\{\alpha_i, \beta_j, e^{\alpha_i\beta_j}\}$ $(1 \leqslant i \leqslant N, 1 \leqslant j \leqslant M)$ grows logarithmically with $\min(N, M)$. The author has improved his results since then (now we know that in a set S of the form above there are not less than $|S|/(M + N)$ algebraically independent elements). However, the methods of Chapter 1 are still novel and introduce a variety of useful auxiliary algebraic techniques to deal with specializations of algebraic varieties over fields of finite degrees of transcendence, especially in the cases, when the degrees and sizes of the coefficients of equations defining algebraic varieties grow independently. The techniques of Chapter 1 also serve as a prelude to a different elementary technique used to prove more powerful results (see, especially, Chapters 2, 4, 7 and 8).

While the analytic methods of Chapter 1 are based on the Gel′fond-Schneider method, Chapter 2 applies Baker's analytic methods to the problem of algebraic independence. The results proved in Chapter 2 were first announced in 1974 [**4**]. Baker's method, developed in Chapter 4, is useful later (cf. Chapters 5 and 7). To suggest the complexity of the problems under consideration, we mention one result of Chapter 2, that there are at least two algebraically independent numbers among π, $\pi^{\sqrt{D}}$, $e^{i\pi\sqrt{D}}$. Still, we do not know whether π and e^{π} are algebraically independent, or even, whether $\pi/\ln\pi$ is rational.

Chapter 3 on "colored sequences" is based on two 1975 papers of the author which seem to have vanished in one of the East European magazines. Chapter 3 and Chapter 4, which immediately follow, deal with the criterion of the algebraic independence of two or more numbers. The starting point of all these studies is the apparent impossibility of extending the well-known Gel′fond [**1**] criterion of the transcendence of a single number θ to the case of the algebraic independence of numbers $\theta_1, \ldots, \theta_n$ for $n > 1$. Gel′fond's criterion (or its subsequent modifications) states that it is impossible for a transcendental number θ to have a "dense" sequence of algebraic approximations that are good: say, $|P(\theta)| < H(P)^{-Cd(P)}$ for infinitely many $P(x) \in \mathbf{Z}[x]$ and a sufficiently large $C > 0$. Though a straightforward generalization of this criterion to $n > 1$ numbers is impossible, various results that are important for applications are presented in Chapters 3 and 4. Chapter 3 deals mainly with the case, when the algebraic independence of several numbers depends on the measure of transcendence of each of them. Chapter 4 presents an attractive and simple criterion of the algebraic independence of two numbers in terms of a "dense" system of polynomials in two variables with rational integer coefficients, assuming small values at these numbers. This criterion and its n-dimensional generalization are the best possible, and constitute the necessary algebraic part in the proof of the algebraic independence of various numbers connected with exponential, elliptic and Abelian functions. See particularly the discussion of §6, Chapter 4 on analytic methods in the algebraic independence proofs.

Chapter 5 is devoted to diophantine approximations. It is in this chapter that we use Siegel's 1929 [**5**] method in the theory of diophantine approximations, which is based on the study of linear differential equations. In Siegel's method one recognizes important features of Hermite's 1873 construction [**6**] of approximations to exponential functions. Siegel's method was later further developed in the theory of E-functions. At the same time Siegel's (and Hermite's) work was generalized in the theory of Hermite-Padé approximations initiated by Mahler in 1935 (see his paper [**7**], published in 1968). It turns out that the combination of the author's conjugate auxiliary function method [**8**], explained in detail in Chapter 9, and Siegel's method gives a new best possible result on diophantine approximations of numbers generated by values of E-functions at rational points. Chapter 5 is devoted to an exposition of such results. First, in §1 of Chapter 5, we present a different approach to the Thue-Siegel theorem on (noneffective) rational approximations to algebraic numbers. This exposition follows the author's 1974 preprint [**9**] and uses the Gel′fond-Baker method. We also show how an alternative use of linear differential equations gives much better bounds on constants than those known from Baker's theory of linear forms in logarithms of algebraic numbers. Our results are applied to bound the number of solutions of Thue's diophantine equations. In §2 we describe the essence of Siegel's method [**5**] and in §§3, 4 we prove our main results on the best possible bounds of the measures of diophantine approximations of arbitrary numbers from the fields generated over $\mathbf{Q}$ by values of exponential and other E-functions, satisfying linear differential equations over $\mathbf{Q}(x)$, at rational points. The results proved here are based on the author's technique of G-invariance explained in Chapter 9, and we recommend the reader to get acquainted with Chapter 9,

§§0–4, before reading §§3, 4 of Chapter 5. The results of Chapter 5 were recently significantly extended by the author to a larger class of linear forms in values of E-functions and G-functions (including linear forms in algebraic numbers) in the direction of the effectivization of analogs of Roth's and Schmidt's theorems.

Chapter 6 is a translation of the author's paper of 1970 on diophantine sets, and represents an expanded version of his 1970 article [**13**].

Chapter 7 is devoted mainly to the transcendence and algebraic independence of numbers connected with elliptic and exponential functions. It contains a survey of the author's results prior to 1981 (in §1) and various methods of proofs of these results in §§2–5. Among these results are the algebraic independence of π/ω, η/ω for a period ω and a quasi-period $\eta = 2\zeta(\omega/2)$ of an elliptic curve defined over $\overline{\mathbf{Q}}$; the algebraic independence of π and ω, and of π/ω and $e^{i\pi\sqrt{D}}$ for a period ω of an elliptic curve with a complex multiplication by $\sqrt{D}$. In Chapter 7 we also present a discussion of properties of L-functions of elliptic curves and the Birch-Swinnerton-Dyer conjecture used to bound heights of integer points on elliptic curves (see §1, Introductory paper). We discuss proofs of the algebraic independence for values of elliptic (and exponential) functions in §§7–9. §§10 and 11 of Chapter 7 are devoted to proofs of the author's generalization of the Lindemann-Weierstrass theorem for arbitrary Abelian CM-varieties. We also present analogs of the Lindemann-Weierstrass theorem for arbitrary systems of functions satisfying differential equations and the laws of addition. These results were first announced in [**10**] and in the elliptic case were presented in [**11**].

In Chapter 8 we give the bound on the measure of the algebraic independence of π/ω and η/ω, which is close to the best possible. Another important feature of this result is in its dependence of height of "H^{-C}" form. This result implies, e.g. an "H^{-C}" form of the measure of diophantine approximations for numbers such as $\Gamma(1/3)$, $\Gamma(1/4)$, etc. The proof of Theorem 1.1 in Chapter 8 is obviously extended to other cases of algebraically independent numbers connected with elliptic functions studied by the author (see §4 of the Introductory paper).

We leave beyond the scope of this book various results of the author on the measures of diophantine approximations of classical constants of analysis based on Padé-type approximation techniques. Some of these results are touched upon in [**12**].

Chapter 9 presents a group of the author's results on a generalization of the Straus-Schneider theorem on the number of complex z such that a given meromorphic function f assumes, together with all its derivatives, rational integer values at z. Our exposition is based on unpublished manuscripts of 1977–1978 discussing various one and multidimensional generalizations of these theorems and intersects with [**8**]. Chapter 9 could serve as a good introduction to Chapter 5.

As an appendix we included a paper on the extremality of certain multidimensional manifolds prepared by A. I. Vinogradov and the author in 1976.*

---

* This paper was initially prepared for a Festschrift for the 85th birthday of I. M. Vinogradov.

This book was prepared through the generous effort of the American Mathematical Society and with the help of D. V. Chudnovsky. The author is deeply thankful to the Society and to its Editorial Staff for their patience during the preparation of this volume which lasted several long years. Unfortunately the author's state of health did not contribute positively towards the completion of this project. The author specially thanks Professor W. J. LeVeque for his kindness and encouragement, Dr. B. Silver and Mr. M. Carcieri for their care and help and Dr. G. Kandall for his translations. The author is indebted to Professor L. Bers for his help. This book is dedicated to the author's parents.

REFERENCES

1. A. O. Gel'fond, *Transcendental and algebraic numbers* (Moscow, 1952); Dover, New York, 1960.
2. A. Baker, *Transcendental number theory*, 2nd ed., Cambridge Univ. Press, Cambridge, 1979.
3. M. Lüscher, Nuclear Physics, **B135** (1978), 1–19.
4. G. V. Chudnovsky, Uspekhi Math. Nauk **31**, No. 4, (1976), 281–282 (Russian).
5. C. L. Siegel, Abh. Preuss. Akad. Wiss., No. 1 (1929).
6. Ch. Hermite, C. R. Acad. Sci. Paris **77** (1873), 18–24, 74–79, 226–233, 285–293.
7. K. Mahler, Composito Math. **19** (1968), 95–166.
8. G. V. Chudovsky, Ann. of Math. (2) **109** (1979), 353–377.
9. ______, Preprint IM-74-9, Institute of Mathematics, Kiev, USSR (Russian); in Topics of Number Theory, Colloquia Math. Societatis J. Bolyai, Vol. 13, North-Holland, Amsterdam, 1974, pp. 19–30.
10. ______, C. R. Acad. Sci. Paris Sér. A **288** (1979), 439–440.
11. ______, Invent. Math. **61** (1980), 267–280.
12. ______, *Bifurcation phenomena in mathematical physics and related topics*, D. Reidel Publishing Company, Boston, 1980, 448–510; Lecture Notes Phys., vol. 120, Springer, New York, 1980, pp. 103–150; Arithmetic and Geometry, v. I, Birkhäuser, Boston, Mass., 1983, 61–105.
13. ______, Uspekhi Math. Nauk **25** No. 4, (1970), 171, 185–186.

INTRODUCTION

ALGEBRAIC INDEPENDENCE OF VALUES OF EXPONENTIAL AND ELLIPTIC FUNCTIONS

0. We present a review of the theory of transcendence and algebraic independence of numbers connected with the exponential, elliptic and Abelian functions. The present review was prepared in August, 1978 as an extended version of the report [**C11**] for the International Congress of Mathematicians in Helsinki. It has been updated in 1981 to cover some recent progress. Some new developments (including e.g. irrationality of numbers and G-function method (cf. [**C14**]), zeroes of auxiliary functions [**B17**, **M8**], etc.) are not covered here as they deserve separate exposition.

Let $\wp(z)$ denote the Weierstrass elliptic function with algebraic invariants g_2, g_3 and $\zeta(z)$ the ζ-function, $\zeta'(z) = -\wp(z)$. Let ω, η denote any pair of periods and quasi-periods of $\wp(z)$: $\zeta(z+\omega) = \zeta(z) + \eta$, and let ω_i, η_i denote fundamental periods and quasi-periods of $\wp(z)$. We call the point u algebraic for $\wp(z)$ if $\wp(u) \in \overline{\mathbf{Q}}$. For a finite set $S \subset \mathbf{C}$, $\#S$ denotes the maximal number of algebraically independent (a.i.) elements in S.

1. Linear independence of algebraic points of elliptic and Abelian functions. Linear independence is often the first step of a proof of algebraic independence. Even when this step does not lead to a proof of algebraic independence (e.g. for logarithms of algebraic numbers where we do not know anything reasonable about algebraic independence), it is often very useful for applications to know linear independence and a measure of linear independence. This becomes especially clear from the works of A. Baker [**B1**, **B2**, **B3**] and H. Stark [**S8**] on estimates of linear forms in logarithms of algebraic numbers and applications of these estimates to diophantine equations and the class-number problem. The report of R. Tijdeman [**T2**] contains a survey of such results.

It was realized long ago (practically in the first of C. Siegel's works [**S6**]) that estimates for linear forms in algebraic points of elliptic and Abelian functions are

0076-5376/84 $1.00 + $.25 per page

extremely important for effective determination of integer points on curves of positive genus (see Lang **[L2, L3]**).

Only in 1974 did D. Masser **[M3]** prove a linear independence result for the algebraic points of an elliptic curve with complex multiplication. Later S. Lang **[L2]** and D. Masser **[M4]** obtained more general linear independence results for algebraic points of Abelian varieties of CM-type. More important are the lower bounds for linear combinations of algebraic points of Abelian varieties with algebraic coefficients, that were obtained simultaneously in **[M3, M4]**. The proofs of D. Masser use analytical arguments. J. Coates-S. Lang **[C22]** used Kummer theory and the examination of l-adic representations of Tate modules in the case of complex multiplication (K. Ribet's theorem **[R4]**, see also **[R5]**).

Let $\mathbf{A}$ be an Abelian variety of dimension d, defined over $\overline{\mathbf{Q}}$ with Abelian functions $A_1(\bar{z}),\ldots,A_d(\bar{z})$, $B(\bar{z})$. Assume that $A_1(\bar{z}),\ldots,A_d(\bar{z})$ are algebraically independent and $B(\bar{z})$ is algebraic over $\mathbf{Q}[A_1,\ldots,A_d]$, and assume that the partial derivatives $\partial/\partial z_i$ map $\overline{\mathbf{Q}}[A_1,\ldots,A_d, B]$ into itself. Let $U \subset \mathbf{C}^d$ be the set of algebraic points of $\mathbf{A}$, i.e. such that $\vec{u} \in \mathbf{C}^d$ and $A_1(\vec{u}),\ldots,A_d(\vec{u})$ are algebraic.

Let $\mathbf{A}$ have CM-type; this means that the endomorphism ring End($\mathbf{A}$) is isomorphic to an order E in a CM-field of degree $2d$ (that is, a quadratic imaginary extension of a totally real field). Without loss of generality we assume that the matrices representing the elements of End($\mathbf{A}$) are in diagonal form. Let $\vec{u}_1,\ldots,\vec{u}_m$ be algebraic points of $\mathbf{A}$ (from U), linearly independent over End($\mathbf{A}$), and $\vec{u}_0$ be the column vector all of whose components are unity.

The following result was proved for elliptic curves with complex multiplications ($d = 1$) by Anderson **[A1]** and in the general case by D. Masser **[M4]**.

THEOREM 1. *For any $K > dm + 1$ and any integer $N \geqslant 1$ there exists $C_1 > 0$ depending only on K, N, $\vec{u}_1,\ldots,\vec{u}_m$ and $\mathbf{A}$ such that*

$$|B_0\vec{u}_0 + B_1\vec{u}_1 + \cdots + B_m\vec{u}_m| > C_1 H^{-(\log\log H)^K} \tag{1}$$

for any diagonal matrices B_0, $B_1,\ldots,B_m$ with algebraic entries of degrees at most N and heights at most H, such that $B_0,\ldots,B_m$ are not all singular.

This result covers previous results of Lang, Coates, Masser and can be partially generalized to the case of singular matrices $B_0,\ldots,B_m$. Define

$$\overline{\Lambda} = B_0\vec{u}_0 + B_1\vec{u}_1 + \cdots + B_m\vec{u}_m, \tag{2}$$

where B_0, $B_1,\ldots,B_m$ are diagonal matrices with algebraic entries of degrees $\leqslant N$ and heights $\leqslant H$, not all zero. Then D. Masser's bounds **[M3, M4]** give us

$$|\overline{\Lambda}| > C_2 e^{-(\log H)^K} \tag{3}$$

for any $K > 2d^2m + d + 1$ and $K > (d^2 - d)m + d$ if $B_0\vec{u}_0 \neq 0$ and for $C_2 > 0$ depending on K, $\vec{u}_1,\ldots,\vec{u}_m$, $\mathbf{A}$, N.

The results (1) and (3) are very important for applications (see below).

One very attractive problem is the generalization of results (1) and (3) to elliptic curves or Abelian varieties defined over $\overline{\mathbf{Q}}$ and not of CM-type. Recently,

D. Bertrand and D. Masser proved [**B9**] a linear independence result for algebraic points on an arbitrary elliptic curve (Abelian variety **A**, $\dim \mathbf{A} = d$, with d real multiplications). This result covers A. Baker's theorem [**B2**] on the linear independence of logarithms of algebraic numbers; the proof uses only the Schneider-Lang theorem [**L1**] and so far does not yield quantitative results of the form (1) or (3).

Recently, bounds (1) and (3) were improved in another aspect, known for linear forms in logarithms as "H^{-C}" bounds [**B3**]. For instance, using arithmetic analysis of linear differential equations satisfied by Abelian integrals, we showed [**C14**] that the linear form $\overline{\Lambda}$ in (2) can be estimated below as

$$|\overline{\Lambda}| > H^{-C}$$

for B_i with algebraic entries of heights $\leqslant H$ and $C > 0$ depending on $\vec{u}_1, \ldots, \vec{u}_m$, **A**, N.

The natural field for applications of bounds of linear forms in algebraic points on Abelian varieties, is, of course, diophantine equations. First of all, there is the

Problem. To effectivize Siegel's theorem on the finiteness of the number of integer points on an algebraic curve $\mathcal{C}$ of genus $g > 1$.

All the results obtained in this direction (archimedian as well as p-adic) we call "conditional" because they are implicitly ineffective: they depend on knowledge of a basis of the Mordell-Weil group of the Jacobian $J(\mathcal{C})$ of $\mathcal{C}$.

We call the curve $\mathcal{C}$ of genus $g \geqslant 1$ a CM-curve if its Jacobian $J(\mathcal{C})$ is a CM-variety or weaker, if $\mathcal{C}$ admits a nonconstant rational map (defined over $\overline{\mathbf{Q}}$) into a CM-variety. As an application of the bounds (1), (3) to CM-curves, we have D. Masser's [**M4**] result.

THEOREM 2. *Let $\mathcal{C}$ be CM-curve defined over the number field* **K**. *If P is the point from $\mathcal{C}(\mathbf{K})$ of height $H(P)$ and denominator $D(P)$, then*

$$H(P) < C_3 \exp(\log^{\varepsilon} D(P)) \tag{4}$$

for $C_3 > 0$ depending only on **K**, $\mathcal{C}$, $\varepsilon > 0$.

In this direction the p-adic results give stronger corollaries. Since 1976, D. Bertrand [**B4**] has developed methods to estimate linear forms in algebraic points of elliptic curves and Abelian varieties of CM-type in the p-adic domain. He and Y. Flicker [**B8**, **F1**] have proved precise results of the form (1), (3) in the p-adic case. E.g.:

THEOREM 3 [**B1**]. *Let $\mathcal{C}$ be an elliptic CM-curve defined over* **K**. *If P is a point from $\mathcal{C}(\mathbf{K})$, $H(P)$ is the height of P and* pr $D(P)$ *is the greatest prime factor of the denominator $D(P)$ of P, then*

$$\operatorname{pr} D(P) > C_4(\log H(P))^{C_5} \tag{5}$$

for some $C_4 > 0$, $C_5 > 0$ depending on $\mathcal{C}$ and **K** *(more precisely C_5 depends only on the rank of $J(\mathcal{C})(\mathbf{K})$).*

Such results would be extremely useful if it were possible to make them completely effective in terms of $\mathcal{C}$ and **K**. There are two obstacles on the way: (a) the absence of an effective bound for generators of the Mordell-Weil group of $J(\mathcal{C})(\mathbf{K})$ in terms of $\mathcal{C}$, **K**; (b) the need for a lower bound for $|\overline{\Lambda}|$ in (2) in terms of the heights of $\vec{u}_1,\ldots,\vec{u}_m$.

In the direction (b) first results were already obtained by Coates-Lang **[C22]** but the bound we really need has not been proved in general. We will formulate it as a

CONJECTURE. *In the previous notation, let $\vec{u}_i$ be an algebraic point of* **A** *of degree $\leqslant M$ and height $\leqslant U_i$: $i = 1,\ldots,m$. Then for any $\varepsilon > 0$*

$$|\overline{\Lambda}| > \exp\left(-C_6 \log H \cdot \prod_{i=1}^{m} \log U_i^{1+\varepsilon}\right), \tag{6}$$

where $C_6 > 0$ depends on M, N and $\varepsilon > 0$ for matrices $B_0, B_1,\ldots,B_m$, not all zero, with algebraic entries of degrees $\leqslant N$ and heights $\leqslant H$.

This conjecture is true for $m = 2$ for an arbitrary CM-variety **A**. Recently an answer to this conjecture was established by the author for an arbitrary $m \geqslant 2$ and an elliptic curve **E** with complex multiplication, with slightly different dependence on H in (6), cf. **[C15b]**.

In direction (a), there are no unconditional positive results (and there is even some suspicion about the existence of effective bounds for generators of the Mordell-Weil group **[M2]**). The Birch-Swinnerton-Dyer conjecture enables us to solve (a) **[M2, L3]**. For this one should estimate $\prod_{i=1}^{m} \log U_i$, where $\vec{u}_1,\ldots,\vec{u}_m$ are generators of the Mordell-Weil group of **A(K)**. According to the general form of the Birch-Swinnerton-Dyer conjecture (see report of J. Coates **[C23]**), we need the bound for $|L_{\mathbf{A}}^{(m)}(1)|$, where $L_{\mathbf{A}}(s) = \sum a_n n^{-s}$ is a properly defined L-function of the Abelian variety **A** defined over **K** and m is the rank of Mordell-Weil group of **A(K)**. For general **A**, it is unclear what are good upper bounds for $|L_{\mathbf{A}}^{(m)}(1)|$. However it becomes possible to do this in case $L_{\mathbf{A}}(s)$ satisfies a functional equation of predicted form (e.g. in the CM-case). Thus for an elliptic curve $\mathbf{A} = \mathbf{E}$ defined over **Q** of conductor N we have,

$$|L_{\mathbf{E}}^{(m)}(1)| \underset{m}{\ll} N^{1/4} \tag{7}$$

assuming Weil conjecture for **E**, and

$$|L_{\mathbf{E}}^{(m)}(1)| \underset{m,\varepsilon}{\ll} N^{\varepsilon} \tag{8}$$

assuming also the Riemann hypothesis for $L_{\mathbf{E}}(s)$.

In the CM-case bounds (7)–(8) together with estimate (6) for $m = 2$ enable us to obtain bounds for integer points on some curves of positive genus.

THEOREM 4. *Let k be an integer such that $|k|$ is the power of one prime p and* **E** *be the curve*

$$y^2 = x^3 + k. \tag{9}$$

If $L_{\mathbf{E}}(s)$ satisfies the Riemann hypothesis and the Birch-Swinnerton-Dyer conjecture, then for integer solutions x, y of (9) *we have*

$$\max(|x|, |y|) \leqslant \exp\left(C_7 p^{1/6+\varepsilon}\right) \tag{10}$$

for $C_7 > 0$ depending only on $\varepsilon > 0$.

(This is a considerable improvement of H. Stark's general bound [**S9**].)

Using the remark after the Conjecture, one can generalize the result (10) to an arbitrary elliptic curve **E** defined over $\overline{\mathbf{Q}}$ and having complex multiplication. For example, if a CM elliptic curve **E** is defined over **Q**, having the conductor N, then the bound for the height $H(P)$ of an integer point P of $\mathbf{E}(\mathbf{Q})$, is of the form $H(P) \leqslant \exp(c(\varepsilon)N^{1/12+\varepsilon})$ for arbitrary $\varepsilon > 0$ and $c(\varepsilon)$ depending only on ε and the rank of $\mathbf{E}(\mathbf{Q})$.

Recently a paper appeared [**G2**], in which the author deduced bounds weaker than (10) based on the scheme above, but with bounds weaker than (6) or (8).

Now, we shall deal with the "transcendental" part of linear independence results.

The first such results were obtained by A. Baker [**B1, B2**], and then by J. Coates [**C21**] and D. Masser [**M3**] for the product of two elliptic curves.

Let **A** be an Abelian variety of dimension d defined over $\overline{\mathbf{Q}}$; with Abelian functions $A_1(\bar{z}),\ldots,A_d(\bar{z})$, $B(\bar{z})$. Let $H_1(\bar{z}),\ldots,H_d(\bar{z})$ be algebraically independent (over $\mathbf{C}(z)$) quasi-periodic functions of **A**, i.e. such that for any period $\vec{\omega}$ of **A**,

$$H_i(\vec{x} + \vec{\omega}) = H_i(\vec{x}) + \eta_i(\vec{\omega}): \qquad i = 1,\ldots,d,$$

and $\partial/\partial z_j$ maps $\mathbf{Q}[A_1,\ldots,A_d, B, H_1,\ldots,H_d]$ into itself: $j = 1,\ldots,d$.

THEOREM 5 (D. MASSER [**M6**]). *Let **A** be an abelian variety of dimension* 2 (*and A_1, A_2, B, H_1, H_2 be its periodic and quasi-periodic functions*). *If $\vec{\omega} = (\omega_1, \omega_2)$ is a nonzero period of **A** then the linear combination*

$$l = \alpha_1\omega_1 + \alpha_2\omega_2 + \beta_1\eta_1(\vec{\omega}) + \beta_2\eta_2(\vec{\omega})$$

with algebraic α_1, α_2, β_1, β_2 is either zero (*then **A** is not a simple variety*) *or transcendental for $|\alpha_1| + |\alpha_2| + |\beta_1| + |\beta_2| > 0$.*

These and other results applied to the CM-case show that

(i) the dimension of the set

$$\{B(m/5, n/5): m \not\equiv -n \pmod 5\}$$

over $\overline{\mathbf{Q}}$ is exactly five and

(ii) the dimension of the set

$$\{B(m/5, n/5): m, n \in \mathbf{Z}\}$$

over **Q** is exactly six [**M7**].

We propose here a new unsolved problem: to prove that there are no new *linear* relations between π, periods and quasi-periods of an Abelian variety **A** over $\overline{\mathbf{Q}}$, other than the Riemann (Legendre) relations and different complex multiplications of **A**.

However if we have a product of an Abelian variety with a degenerate one, the situation becomes better. E.g. let $u_1, \ldots, u_m$ be algebraic points of an elliptic curve $\mathbf{E}$ defined over $\overline{\mathbf{Q}}$ and $\log \alpha_1, \ldots, \log \alpha_n$ be logarithms of algebraic numbers, linearly independent with πi over $\mathbf{Q}$. Then

THEOREM 6. *For algebraic numbers $\beta_1, \ldots, \beta_m, \gamma_1, \ldots, \gamma_n$ such that $|\gamma_1| + \cdots + |\gamma_n| > 0$ we have*

$$l' = \beta_1 u_1 + \cdots + \beta_m u_m + \gamma_1 \log \alpha_1 + \cdots + \gamma_n \log \alpha_n \neq 0. \tag{11}$$

Moreover we have a lower bound for $|l'|$ in terms of heights and degrees of u_i, α_j, β_i, γ_j.

As results in transcendence theory, obtained by refinement of the Gelfond-Schneider method, we must mention very important generalizations of the Blanksby-Montgomery theorem **[B11]** by C. Stewart **[S10]** and Dobrowolsky **[D2]**.

2. Algebraic independence in the elliptic function case. The situation with algebraic independence for numbers connected with exponential and elliptic functions was very poor until very recently. Besides the Lindemann-Weierstrass theorem (1882) and Gelfond's example of algebraically independent α^β and α^{β^2} for algebraic $\alpha \neq 0, 1$ and β cubic irrationality (**[G1]**, 1949), there were *no* concrete examples of algebraically independent numbers (see, however, **[R1, B16]**).

We can say that the lack of knowledge in this field is determined by the methods used for the analysis of algebraic independence. The introduction of nontrivial algebraic (algebraic-geometrical) and multidimensional considerations gave us some ideas of how to develop Transcendence Theory and also supplied us with many (at least some) new examples of algebraically independent numbers.

In the elliptic case since 1975 we have constructed several pairs of algebraically independent numbers connected with algebraic points of $\wp(z)$ or periods of $\wp(z)$ **[C3, C11]**. Historically the first result was the following.

THEOREM 1. *If ω_1, ω_2 are fundamental periods of $\wp(z)$ and η_1, η_2 are corresponding quasi-periods, then*

$$\# \{\omega_1, \omega_2, \eta_1, \eta_2\} \geqslant 2. \tag{1}$$

The most interesting case is that of complex multiplication.

COROLLARY 2. *If ω is a the period of $\wp(z)$ with complex multiplication, then ω and π are a.i.*

We can see exactly what ω is, thanks to the Selberg-Chowla formula **[S11, W7]**. We get from this formula and Corollary 2 that for the discriminant m of a quadratic imaginary field and $\chi(n) = (n/m)$,

$$\omega_m = \prod_{i=1}^{m-1} \Gamma(i/m)^{\chi(i)} \text{ and } \pi \text{ are algebraically independent.} \tag{2}$$

Since that we have improved (1) considerably **[C11, C19]**.

THEOREM 3. *For any pair* ω, η *of period, quasi-period of* $\wp(z)$ *(i.e.* $\zeta(z+\omega) = \zeta(z) + \eta$*) the two numbers*

(3) $$\pi/\omega,\ \eta/\omega \text{ are algebraically independent.}$$

Moreover, we have Theorem 3 in the case when $\wp(z)$ does not necessarily have algebraic invariants. In this case, when $\wp(z)$ is arbitrary, we have

(4) $$\#\{g_2, g_3, \pi/\omega, \eta/\omega\} \geqslant 2.$$

From (4) it can be deduced e.g. that for the modular invariant $J(q)$, $q = e^{2\pi i z}$, that if $J(q)$ is algebraic and $\theta = q(d/dq)$, then $\theta J(q)$ and $\theta^2 J(q)$ are a.i. The last result as well as (3) was generalized to the p-adic case by D. Bertrand **[B7]**.

We can present an interesting generalization of (3) for the case of algebraic nonperiodic points of $\wp(z)$ **[C11, C19]**.

THEOREM 4. *Let* u *be an algebraic point of* $\wp(z)$ *linearly independent with* ω *over* $\mathbf{Q}$. *Then*

(5) $$\zeta(u) - \eta u/\omega \text{ and } \eta/\omega \text{ are a.i.}$$

E.g. from this follows

COROLLARY 5. *For* u *algebraic for* $\wp(z)$, *linearly independent with* ω *over* $\mathbf{Q}$, $\zeta(u) - \eta u/\omega$ *is transcendental.*

This transcendence result was provable only by using the algebraic independence result. The result (5) also was proved in the p-adic case by D. Bertrand **[B7]**.

We have two more results of the author on algebraic independence for the case of complex multiplication.

THEOREM 6. *Let* $\wp(z)$ *have complex multiplication and* u *be an algebraic point of* $\wp(z)$; *then*

(6) $$u \text{ and } \zeta(u) \text{ are a.i.}$$

This is the natural generalization of Corollary 2.

THEOREM 7. *Let* $\wp(z)$ *have complex multiplication by* β. *Then for period* ω *of* $\wp(z)$

(7) $$\pi/\omega \text{ and } e^{\pi i \beta} \text{ are a.i.}$$

Also we have elliptic (and Abelian) analogues of the Lindemann-Weierstrass theorem (see below).

Let us see how our results can be interpreted in terms of some problems of analysis and partial differential equations. We will present one application. Let $u(x)$ be a periodic (with period T) potential that is n-band, or, in other words, a solution of a stationary nth order Korteweg-de Vries equation (S. Novikov **[N1]**, H. P. McKean **[M9]** and see **[D3]** for early results) $\Sigma_{i=0}^{n} c_i R_i[u] = 0$.

Then the spectrum for the Schrödinger operator $L = -d^2/dx^2 + u(x)$ has n-band structure.

$$\lambda_0 < \lambda_1 \leqslant \lambda_2 < \lambda_3 \leqslant \lambda_4 < \cdots \leqslant \lambda_{2n} < +\infty, \tag{8}$$

where λ_i are the first $2n+1$ points of T-periodic or T-antiperiodic spectrum $L\psi = \lambda\psi$ and $(\lambda_{2i-1}, \lambda_{2i})$: $i = 1,\ldots,n$, are n forbidden zonas. All λ_j: $j = 0,\ldots,2n$, are determined algebraically by the system of differential equations satisfied by $u(x)$ and its derivatives. They are called "algebraic eigenvalues".

However there are infinitely many (degenerate) points λ_i: $i = 2n+2,\ldots$ of the periodic and antiperiodic spectrum (i.e. $\lambda_{2i-1} = \lambda_{2i}$: $i = n+1, n+2,\ldots$). They are sometimes called "transcendental", but it is possible to give precise sense to this word and to prove that such numbers are indeed transcendental. Let us take the best known example of an n-band potential, the Lamé potential, $u(x) = n(n+1)\wp(x)$, where $\wp(x)$ have algebraic invariants.

COROLLARY 8 [**C10**]. *For the Lamé potential $u(x) = n(n+1)\wp(x)$ the first $2n+1$ points of the periodic and antiperiodic spectrum are algebraic, while the others are transcendental. (E.g. all eigenvalues μ_i of $L\psi = \mu_i\psi$ with periodic boundary conditions $\psi(0) = \psi(T) = 0$ for $i = n+1, n+2,\ldots$ are transcendental.)*

Another problem arising from Abelian varieties associated with Fermat curves is the problem of the transcendence of values of the Γ-function in rational, but not integer points. Of course, $\Gamma(1/2) = \sqrt{\pi}$ is transcendental. From Corollary 2 we know also that

$$\Gamma(1/6),\quad \Gamma(1/4),\quad \Gamma(1/3),\quad \Gamma(2/3),\quad \Gamma(3/4),\quad \Gamma(5/6) \tag{9}$$

are transcendental (and each of the numbers in (9) is a.i. with π). These examples correspond to elliptic curves of CM-type, but in order to investigate general $\Gamma(m/n)$, we must investigate the arithmetic nature of the periods and quasi-periods of Abelian varieties of CM-type in the Shimura sense. This problem turns out to be the most important and difficult in Transcendence Theory. It is very interesting that analytical difficulties in transcendence proof in this situation are tied up with algebraic difficulties.

Let $\mathbf{A}$ be an Abelian variety defined over $\overline{\mathbf{Q}}$ of dimension d. Then we have Abelian functions $A_1(\bar{z}),\ldots,A_d(\bar{z})$, $B(\bar{z})$, where $A_1(\bar{z}),\ldots,A_d(\bar{z})$ are algebraically independent and $B(\bar{z})$ is algebraic over $\overline{\mathbf{Q}}[A_1,\ldots,A_d]$. The functions $A_1(\bar{z}),\ldots,A_d(\bar{z})$, $B(\bar{z})$ are assumed to be regular at $\bar{z} = 0$; these functions are $2d$-periodic with common periods $\vec{\omega}_1,\ldots,\vec{\omega}_{2d} \in \mathbf{C}^d$. There are also d quasi-periodic functions $H_1(\bar{z}),\ldots,H_d(\bar{z})$, algebraically independent over $\mathbf{C}(\bar{z})$, with quasi-periods η_{ij},

$$H_j(\vec{x} + \vec{\omega}_i) = H_j(\vec{x})\eta_{ij}: \qquad j = 1,\ldots,d;\ i = 1,\ldots,2d.$$

We put

$$\vec{\eta}_i = (\eta_{ij})_{j=1,\ldots,d}, \qquad \vec{\omega}_i = (\omega_{ij})_{j=1,\ldots,d}: \qquad i = 1,\ldots,2d. \tag{10}$$

The main problem for Abelian functions can be formulated as follows: to determine the number D of algebraically independent elements among the entries of $\Omega \cup H = \{\vec{\omega}_j, \vec{\eta}_j: j = 1,\dots,2d\}$.

For general Abelian varieties there is not even a conjecture expressing this number $D = D(\mathbf{A})$ in terms of $\mathbf{A}$ (definitely for the same dimension d, D vary). Moreover, even the CM-case is extremely nontrivial.

For a general Abelian variety we have only results on the algebraic independence of two numbers [**C19**, Ch. 7].

THEOREM 9. *Let* $\mathbf{A}$ *be defined over* $\overline{\mathbf{Q}}$. *Then* $\deg\operatorname{tr} Q(\Omega \cup H) \geqslant 2$. *Moreover for numbers* (10) *we have*

$$(11)\qquad \#\left[\{\omega_{ij}: i = 1,\dots,2d, j = 1,\dots,d\} \cup \{\eta_{ij_0}: i = 1,\dots,2d\}\right] \geqslant 2$$

for any $j_0 \leqslant d$. *Now if* $\{j_1,\dots,j_K\}$, $\{j_{K+1},\dots,j_{d+1}\}$ *are two subsets of* $\{1,\dots,d\}$ *of cardinality* K *and* $d+1-K$, *then*

$$(12)\qquad \#\left[\{\omega_{ij_s}: i = 1,\dots,2d, s = 1,\dots,K\} \cup \{\eta_{ij_s}: i = 1,\dots,2d,\ s = K+1,\dots,d+1\}\right] \geqslant 2.$$

It is also easy to obtain results analogous to Theorem 3 and involving fewer numbers than in (11), (12), but in any case it will be results on the existence of 2 a.i. numbers.

Since absolutely unclear difficulties appear in the analytic proofs it becomes obvious that there may exist some algebraic mechanism responsible for this. Approximately at the same time (1976–1978) people working with Abelian varieties of CM-type (Deligne, Shimura, Gross) found new algebraic relations between periods. Now we can formulate the problem precisely.

Let $\mathbf{A}$ be a CM-variety in the sense that for the endomorphism ring $\operatorname{End}(\mathbf{A})$ we have $\operatorname{End}(\mathbf{A}) \otimes \mathbf{Q} = \mathbf{K}$ for a CM-field $\mathbf{K}$ (quadratic imaginary extension of a totally real field) of degree $2d$, $[\mathbf{K}:\mathbf{Q}] = 2d$. We may assume that $\mathbf{A}$ is normalized in the sense that $\operatorname{End}(\mathbf{A})$ is represented in $\mathbf{A}$ by diagonal matrices, and we have a representation of $\mathbf{K}$ in $\mathbf{C}^d$ by the direct sum $\Sigma_{i=1}^n \tau_i$, where τ_i runs through a set of representatives of pairs of the complex conjugate embedding of $\mathbf{K}$ in $\mathbf{C}$. Thus $\mathbf{K}$ acts on $(z_1,\dots,z_d)$ by $\gamma(z_1,\dots,z_d) = (\tau_1(\gamma)z_1,\dots,\tau_d(\gamma)z_d)$ for $\gamma \in \mathbf{K}$.

Then $(\mathbf{K},\varphi)$ is called the CM-type of $\mathbf{A}$ for $\varphi = \Sigma_{n=1}^n \tau_i$ [**S2**]. According to Shimura, Taniyama and T. Kubota we introduce the notion of the rank of the CM-type $(\mathbf{K},\varphi)$. Let $I_{\mathbf{K}}^0(\mathbf{Q})$ be the module of all formal linear combinations $\psi = \Sigma_\tau c_\tau \tau$ for $c_\tau \in \mathbf{Q}$ of the injections τ of $\mathbf{K}$ into $\mathbf{C}$ such that $c_\tau + c_{\tau\rho}$ does not depend on τ (ρ is the complex conjugation). If $\mathbf{M}$ is the Galois closure of $\mathbf{K}$ over $\mathbf{Q}$, then by the rank $r(\mathbf{K},\varphi)$ of the CM-type $(\mathbf{K},\varphi)$ we mean the dimension of the subspace of $I_{\mathbf{K}}^0(\mathbf{Q})$ generated over $\mathbf{Q}$ by $\varphi\gamma$ for all $\gamma \in \operatorname{Gal}(\mathbf{M}/\mathbf{Q})$.

Then $r(\mathbf{K},\varphi) \leqslant d+1$ and rank $r(\mathbf{K},\varphi)$ is the main invariant connected with the division field $\mathbf{A}_l$. The rank is also the main invariant for the determination of $\deg\operatorname{tr} Q(\Omega \cup H)$. Results of Shimura [**S3–S5**] and Deligne [**D1**] show that for a

CM-variety **A** (which is simple) and for the corresponding CM-type $(\mathbf{K}, \varphi)$ (which is then primitive)

(13) the number $D(\mathbf{A})$ of algebraically independent elements in $\Omega \cup H$ is at most rank $r(\mathbf{K}, \varphi)$ of $(\mathbf{K}, \varphi)$.

E.g. Shimura [**S4**] gives examples of primitive CM-types $(\mathbf{L}, \psi)$ such that $[\mathbf{L} : \mathbf{Q}] = 2^g$, $(\mathbf{L}, \psi)$ is primitive and for a corresponding CM-variety **B**, $\dim(\mathbf{B}) = 2^{g-1}$, there are at most $g + 1$ algebraically independent periods and quasi-periods.

CONJECTURE 10. *For any* CM-*variety* **A** *and its* CM-*type* $(\mathbf{K}, \varphi)$ *the number* $D(\mathbf{A})$ *is exactly the rank of* $(\mathbf{K}, \varphi)$

$$\deg \operatorname{tr} Q(\Omega \cup H) = r(\mathbf{K}, \varphi). \tag{14}$$

As a particular case of (14) we get that for prime $l \geqslant 3$ the numbers π, $\Gamma(1/l), \ldots, \Gamma(n/l)$ for $n = (l-1)/2$ are a.i.

The main question for transcendence results in the case of "degenerate" CM-varieties with $r(\mathbf{K}, \varphi) < d + 1$ is the following: are there some functional relations of θ-functions responsible for new algebraic relations between periods?

Because of the complicated nature of even CM-varieties, it probably would be too early to formulate any general conjecture on algebraic values of Abelian and quasi-periodic functions at arbitrary points.

In 1978 another type of algebraic independence results for values of elliptic and Abelian functions was obtained by the author using an algebraic-geometric approach. These results are analogs of the Lindemann-Weierstrass theorem (cf. [**L1**]) for elliptic and Abelian functions. We formulate the corresponding results in the CM-case [**C7**, **C12**]:

THEOREM 11. *Let* $\wp(z)$ *have complex multiplication in* $\mathbf{Q}(\tau)$. *If* $\alpha_1, \ldots, \alpha_n$ *are algebraic numbers, linearly independent over* $\mathbf{Q}(\tau)$, *then* n *numbers* $\wp(\alpha_1), \ldots, \wp(\alpha_n)$ *are algebraically independent.*

The analytic part of the proof of the theorem is presented in [**C7**] together with the detailed proof for $n = 2$. For the proof for $n = 3$ see [**C19**, Ch. 4] and in general [**C12**], [**C19**, Ch. 7].

Let **A** be an Abelian variety of CM-type of dimension d defined over $\overline{\mathbf{Q}}$ with Abelian functions $A_1(\bar{z}), \ldots, A_d(\bar{z})$, $B(\bar{z})$ as defined above in §1. The algebraic independence results of the Lindemann-Weierstrass type for values of $A_i(\bar{z})$ at algebraic points can be summarized as follows:

THEOREM 12. *Let* **A** *be a CM-variety of dimension* d. *If* $\vec{\alpha}_1, \ldots, \vec{\alpha}_k$ *are algebraic vectors* (*from* $\overline{\mathbf{Q}}^d$) *linearly independent over* End(**A**), *then among*

$$A_i(\vec{\alpha}_j): \qquad i = 1, \ldots, d,\ j = 1, \ldots, k$$

there are $\geqslant k$ *a.i. numbers.*

When many coordinates of $\vec{\alpha}_i$ are zeroes we have a better result.

THEOREM 13. *Let* **A** *be a CM-variety. If* $\vec{\alpha}_1,\ldots,\vec{\alpha}_k \in \overline{\mathbf{Q}}^d$ *are linearly independent over* End(**A**) *and all coordinates of* $\vec{\alpha}_i$ *but the* k_0*th are zeroes* ($k_0 \leqslant d$), $i = 1,\ldots,k$, *then the numbers*

These results are particular cases of more general statements on the algebraic independence of values of functions satisfying algebraic laws of addition [**C19**]. Proof for two and three algebraically independent numbers is given in [**C19**, Ch. 4]; in general, cf. [**C19**, Ch. 7]. These results can be applied in non-CM-cases too. E.g.

THEOREM 14. *Let* $\alpha_1,\ldots,\alpha_n$ *be algebraic numbers, linearly independent over* **Q**. *Then*

$$\#\{\wp(\alpha_i): i = 1,\ldots,n\} \geqslant [(n+1)/2],$$

i.e. among $\wp(\alpha_i)$: $i = 1,\ldots,n$, *there are at least* $(n+1)/2$ *algebraically independent numbers.*

In general, one has algebraic independence results for values of functions that parametrize an arbitrary group variety defined over $\overline{\mathbf{Q}}$. We refer the reader to e.g. [**C12**] for the formulation of some of these results. In the next section we consider the most interesting case of the exponential function.

We complete this section with a short exposition of a new result on the transcendence of values of elliptic functions. Let u and u_0 be algebraic points of $\wp(x)$ and let $u - u_0$ be a nonperiod of $\wp(x)$. It was then proved in [**W5**] that the number

$$\frac{\sigma(u-u_0)}{\sigma(u)\sigma(u_0)} \cdot e^{u\zeta(u_0)}$$

is transcendental. From this it follows that for an algebraic number β, for an algebraic point u of $\wp(x)$, and for a pair ω, η of period, quasi-period of $\wp(x)$, the number $\exp\{\omega\zeta(u) - \eta u + \beta\omega\}$ is transcendental [**W5**]. M. Laurent proved the transcendence of periods of elliptic integrals of the third kind [**L4**]. Then E. Reyssat obtained some results on the algebraic independence of two (or three) numbers among several values of

$$\wp(x), \qquad \zeta(x) - \eta x/\omega, \qquad \frac{\sigma(x-u)}{\sigma(x)\sigma(u)} e^{x\zeta(u)}$$

[**R3**]. Recently D. Bertrand-M. Laurent [**B10**] obtained results on the transcendence of a Jacobi θ-function in the CM-case.

3. Algebraic independence in the exponential function case. In the exponential case the ultimate aim is to prove the

SCHANUEL CONJECTURE. *Let* $\alpha_1,\ldots,\alpha_n$ *be complex numbers linearly independent over* **Q**. *Then among*

$$\alpha_1,\ldots,\alpha_n, e^{\alpha_1},\ldots,e^{\alpha_n}$$

there are at least n algebraically independent numbers.

It is now a common opinion (first formulated by S. Lang) that the Schanuel Conjecture implies all reasonable statements on the transcendence and algebraic independence of values of an exponential function. We have no idea how to prove this conjecture for general $\alpha_1,\ldots,\alpha_n$ even if $n = 2$.

For many years the most important case of the Conjecture has been the case of numbers of the form $\alpha^{\beta},\ldots,\alpha^{\beta^{d-1}}$, where α and β are algebraic, $\alpha \neq 0, 1$ and β is of degree d. Assuming the Schanuel Conjecture, these numbers are a.i.

This set of numbers is a particular subclass of the following sets of numbers connected with the exponential function and has been the major target for investigation since 1949 **[G1]**. Let $\alpha_1,\ldots,\alpha_N$ and $\beta_1,\ldots,\beta_M$ be two sequences of complex numbers linearly independent over **Q**. We define three sets of complex numbers:

$$(1) \qquad S_i = \{e^{\alpha_i\beta_j}\}, \qquad S_2 = \left\{\beta_j, e^{\alpha_i\beta_j}\right\}, \qquad S_3 = \left\{\alpha_i, \beta_j, e^{\alpha_i\beta_j}\right\}$$
$$(i = 1,\ldots,N; j = 1,\ldots,M).$$

Additional conditions imposed on $\alpha_1,\ldots,\alpha_N$ and $\beta_1,\ldots,\beta_M$ bound their measure of linear independence,

$$(2) \qquad \left|\sum_{i=1}^{N} u_i\alpha_i\right| > \exp\left(-\tau \sum_{i=1}^{N} |u_i|\right), \qquad \left|\sum_{j=1}^{M} v_j\beta_j\right| > \exp\left(-\tau \sum_{j=1}^{M} |v_j|\right),$$

for integers $u_1,\ldots,u_N$ and $v_1,\ldots,v_M$ if $\sum_{i=1}^{N} |u_i| > 0$, $\sum_{j=1}^{M} |v_j| > 0$.

In **[G1]** Gelfond proposed a method that enabled researchers to examine the case $\#S_i \geqslant 1$ or $\#S_i \geqslant 2$ for $i = 1, 2, 3$. The investigation of this case (one or two algebraically independent numbers) was completed after Gelfond **[G1]** in the papers of Brownawell, Tijdeman and Waldschmidt **[B13, T1, W1]**.

The situation with three or more algebraically independent numbers is more difficult. The main obstacles, as usual, are algebraic-geometric. More precisely, in the course of the analytic part of the proof of the existence of $n + 1$ algebraically independent elements of S_i, one obtains a system of polynomials $P_i(x_1,\ldots,x_z) \in \mathbf{Z}[x_1,\ldots,x_n]$ of the height $\leqslant H_i$ and degree $\leqslant d_i$, $i = 1, 2, 3,\ldots$ that satisfy

$$(3) \qquad |P_i(\theta_1,\ldots,\theta_n)| < H_i^{-d_i^{\nu}}: \qquad i = 1, 2, 3,\ldots,$$

for fixed $(\theta_1,\ldots,\theta_n)$ and $\nu > n$. The system of polynomials $P_i(x_1,\ldots,x_n)$ typically satisfies some additional conditions, see **[C19]**. The first general results on $\#S_i$ were obtained in 1973/1974 by the author **[C1, C2a, C2b]** using resultants (elimination of variables according to Kronecker), when $\nu > 2^n - 1$. The corresponding result was

THEOREM 1 **[C1]**. *If* $|S_i|/(M + N) \geqslant 2^n$, *then* $\#S_i \geqslant n + 1$ (*if* $i = 3$, *then one requires that* $|S_3|/(M + N) > 2^n$).

The complete proof of Theorem 1 can be found in **[C19]**. The additional assumptions (2) on α_i, β_j in **[C1, C19]** can be removed.

Since then we have found new methods for dealing with problems (3). These methods are based on the investigation of the singularities of intersections of hypersurfaces $P_i = 0$, $P_{i+1} = 0, \ldots, P_{i+k} = 0$ or of ideals $(P_i, P_{i+1}, \ldots, P_{i+k})$ in $\mathbf{C}[x_1, \ldots, x_n]$. The analysis of singularities based upon the consideration of invariants connected with the order of multiplicity of the intersection of hypersurfaces (we call these invariants higher Jacobians) allows us to replace 2^n by $n+1$ in Theorem 1 [**C11**].

Later we found an elementary method to treat inequalities (3). This elementary method is based on the observation that if near $\bar{\theta} = (\theta_1, \ldots, \theta_n)$ there is more than one singular point of intersection $(P_i, \ldots, P_{i+k})$, it is possible to prove that the number of singular points near $\bar{\theta}$ is not very large, and that they cannot be too close to each other [**C8**, **C9**, **C19**]. The case of three algebraically independent numbers is the simplest and we have a completely elementary proof [**C9**, **C19**] of

THEOREM 2. *Let* $|S_i|/(M+N) \geq 3$ (*and* $|S_3|/(M+N) > 3$ *for* $i = 3$), *then there are at least three algebraically independent numbers among the elements in* S_i.

The proof follows from the criterion (presented in Ch. 4, Theorem 4.1 in this volume), of the algebraic independence of two numbers. We prove also the corresponding general result.

THEOREM 3. *Let* S_i *be as before. If* $d \geq 2$ *and*

$$(4) \qquad |S_i|/(M+N) \geq d \qquad (\text{and } |S_3|/(M+N) > d \text{ for } i = 3),$$

then there are d *algebraically independent numbers among the elements in* S_i.

The assumptions (2) for $d \geq 4$ can be removed from the proof of Theorem 3, cf. [**C19**, Ch. 7].

For $d = 2$ this corresponds to the results of Gelfond, Brownawell, Tijdeman and Waldschmidt.

The results closest to Schanuel's Conjecture are proved for example for algebraic powers of algebraic numbers [**C15a**]:

COROLLARY 4. *Let* $\alpha \neq 0, 1$ *be algebraic numbers and* β *be an algebraic number of degree* $d \geq 2$. *Then among* α^{β^j}: $j = 1, \ldots, d-1$, *there are at least* $(d+1)/2$ *algebraically independent numbers.*

This is "half" of the Schanuel Conjecture. For $d = 3$ we get Gelfond's result. If d is of 5th degree, then among $\alpha^{\beta}, \ldots, \alpha^{\beta^4}$ there are 3 a.i. numbers (and if β is of the 7th degree, then $\#\{\alpha^{\beta}, \ldots, \alpha^{\beta^6}\} \geq 4$). Similarly,

$$\#\{e, e^e, \ldots, e^{e^7}\} \geq 3, \qquad \#\{e, e^e, \ldots, e^{e^{11}}\} \geq 4.$$

We see that our results give only some parts of the Schanuel Conjecture. However even these partial results give us hope of one day proving the whole conjecture (if nobody finds a counterexample first). In any case, analytical methods of Siegel-Gelfond-Schneider type should be considerably modified for this.

The number of problems connected with the simplest cases of the Schanuel Conjecture is innumerable. We cite the simplest example.

PROBLEM 5. *Let β be a quadratic irrationality and* $\log \alpha_1$, $\log \alpha_2$ *be logarithms of algebraic numbers, linearly independent over* $\mathbf{Q}$. *One should try to prove that* $\log \alpha_1$ *and* α_1^β *are a.i. and that* α_1^β, α_2^β *are a.i.*

We do not know the solution of this problem, which is the simplest from the point of view of existing analytic methods (hint: in (3) $n = 1$, $\nu = 1$).

Present techniques give us the possibility of only adding to the numbers θ_1, θ_2 a third number θ_3 (connected with an exponential or elliptic function) and then showing $\#\{\theta_1, \theta_2, \theta_3\} \geqslant 2$. There are several results in this case for θ_1, θ_2 from Problem 5 and we present some of them.

THEOREM 6. *In the notation of Problem* 5 *we have*

$$\#\left\{\frac{\log \alpha_1}{\log \alpha_2}, \alpha_1^\beta, \alpha_2^\beta\right\} \geqslant 2, \tag{5}$$

$$\#\{\log \alpha_1, \alpha_1^\beta, \alpha_2^\beta\} \geqslant 2, \tag{6}$$

$$\#\{\pi, \pi^{\beta i}, e^{\pi\beta}\} \geqslant 2, \qquad \beta \notin Q(i). \tag{7}$$

Only the result (5) follows from Gelfond's **[G1]** method. Results (6) and (7) are proved in this volume **[C19]**, cf. **[C4]** using Baker's method.

THEOREM 7. *Let $\wp(z)$ have complex multiplication by β, and let ω be a nonzero period of $\wp(z)$ and u be an algebraic point of $\wp(z)$. In the notation of Problem* 5 *we have*

$$\#\{\omega, \alpha_1^\beta, \alpha_2^\beta\} \geqslant 2, \tag{8}$$

$$\#\left\{\frac{\pi}{\omega}, \alpha_1^\beta, \alpha_2^\beta\right\} \geqslant 2, \tag{9}$$

$$\#\left\{\frac{\pi}{\omega}, \frac{\log \alpha_1}{\pi}, \alpha_1^\beta\right\} \geqslant 2, \tag{10}$$

$$\#\left\{\frac{\pi}{\omega}, \log \alpha_1, \alpha_1^\beta\right\} \geqslant 2, \tag{11}$$

$$\#\{u, \log \alpha_1, \alpha_1^\beta\} \geqslant 2. \tag{12}$$

Results (8), (9) were obtained in 1975–1977 using a combination of exponential and elliptic functions; see **[C19**, Ch. 7].

There are similar results on the existence of two a.i. among three numbers connected with exponential and elliptic functions. For the proof of Theorem 7 cf. **[C3]** and **[C19]**.

Multidimensional considerations are very important for further progress in Transcendence Theory. Lately they have become a subject of particular attention and have led to considerable progress in studies of the zeroes of auxiliary functions (Brownawell-Masser-Wustholz) **[B17, M8]** and transcendence proofs **[B9, B10, M8, W5]**.

In the multidimensional case one still needs the general form of the Schwarz lemma. Previous results [**B12, C6, W2, W5**] show that the Schwarz lemma depends on properties of singularities of hypersurfaces in $\mathbf{C}^n$. We propose one algebraic-geometric problem. Let

$$\Omega(S; K) = \min\Big\{\deg P\colon P(\bar{x}) \in \mathbf{C}[\bar{x}],\ \partial_{\bar{x}}^{\bar{k}} P(\vec{w}) = 0 \quad \text{for all } \bar{k} \in \mathbf{N}^n,\ |\bar{k}| < K,\ \vec{w} \in S\Big\}$$

for finite $S \subset \mathbf{C}^n$.

QUESTION 8. For a given $|S| < \infty$ to describe all possible values of $\Omega_0(S) = \lim_{K\to\infty} \Omega(S; K)/K$.

Using the properties of $\Omega_0(S)$ [**C6**] we find e.g. that for the meromorphic transcendental function $f(\bar{z})$ in $\mathbf{C}^n$ of order of growth $\leqslant \rho$, the set $S(f)$ of $\vec{w} \in \overline{\mathbf{Q}}^n$ with $\partial_{\bar{x}}^{\bar{k}} f(\vec{w}) \in \mathbf{Z}$ for all $\bar{k} \in \mathbf{N}^n$ is contained in a hypersurface of degree $\leqslant n\rho$ [**C5, C6**]. For further results see [**C19**].

4. Measures of transcendence and algebraic independence. Soon after the proof of transcendence of the first concrete numbers (Liouville numbers, e, π) it became clear that the question of transcendence (or algebraic independence) is closely connected with the problem of diophantine approximation or, as people sometimes said, the "arithmetical nature" of transcendental numbers.

The problem of the "arithmetic nature" of a given number (or sequence of numbers) is, roughly speaking, the problem of how this number is approximated by algebraic (rational, algebraic of bounded degree, etc.) numbers. All the well-known classifications of transcendental numbers are based on the property of approximations of numbers by algebraic ones.

First of all, in order to distinguish which numbers are well approximated and which are badly approximated, we have the following corollary of Dirichlet's box principle.

LEMMA 1. *If the numbers $\theta_1, \ldots, \theta_n$ are algebraically independent numbers, then for any N, H there exists a polynomial $P(x_1, \ldots, x_n) \in \mathbf{Z}[x_1, \ldots, x_n]$ of degree $\leqslant N$ and height $\leqslant H$ with*

$$(1) \qquad 0 < |P(\theta_1, \ldots, \theta_n)| < e^{c_1 N^n} H^{-N^n}$$

for $c_1 = c_1(\theta_1, \ldots, \theta_n) > 0$.

It is conjectured that for almost all $(\theta_1, \ldots, \theta_n) \in \mathbf{R}^n$ (up to measure zero) the upper bound for $|P(\theta_1, \ldots, \theta_n)|$ (if $P(x_1, \ldots, x_n) \in \mathbf{Z}[x_1, \ldots, x_n]$) is not considerably different from the right side in (1). At least this conjecture is true for $n = 1$ (see [**B2**]).

However to prove that bounds for measure of transcendence ($n = 1$) or algebraic independence ($n \geqslant 2$) of concrete numbers are of Dirichlet type is extremely difficult.

Work in this directin started in 1899, when E. Borel considered the case $n = 1$, $\theta_1 = e$.

During the last 80 years this direction has developed into a new branch of Transcendence Theory: Measures of transcendence. For numbers associated with functions of classical analysis we have only two general analytic methods to treat measures of transcendence.

(A) Siegel's method [**S7**] for E-functions,

(B) The Gelfond-Schneider method and its variants: Baker's method, etc.

Using (A) K. Mahler [**M1**] obtained very good bounds for the measure of algebraic independence of $e^{\alpha_1},\dots,e^{\alpha_n}$ for algebraic numbers $\alpha_1,\dots,\alpha_n$ linearly independent over $\mathbf{Q}$. For $P(x_1,\dots,x_n)\in\mathbf{Z}[x_1,\dots,x_n]$, $H(P)\leqslant H$, $d(P)\leqslant d$,

$$|P(e^{\alpha_1},\dots,e^{\alpha_n})|> H^{-c_2d^n} \tag{2}$$

for $c_2=c_2(\alpha_1,\dots,\alpha_n)$. However this bound holds only for $\log\log H\geqslant c_3d^{2n}$. Analogous bounds hold for other E-functions [**S7**] $f(z)$ instead of e^z.

See the general results of [**C19**, Ch. 5].

Method (B) was applied with great success to estimating the measure of transcendence of a single number connected with exponentials: e^α, $\log\alpha$, α^β, $\log\alpha/\log\beta,\dots$ for algebraic α, β. These measures of transcendence were obtained for independent degree and height (or even for the case when the degree is more essential than the height [**C2b**]).

For the best possible results as of 1972 see the book of Cijsouw [**C20**]; for the situation as of 1978 see [**W6**], based on estimates of linear forms in logarithms of algebraic numbers, by Baker's method [**B2**].

For the case of a single number connected with the elliptic function $\wp(z)$ (such as ω, η, π/ω, η/ω, $\zeta(u)$, e^u, $u,\dots$ for an algebraic point of $\wp(z)$) we refer to papers of E. Reyssat [**R2**]. The work of [**R2**] contains all known bounds for the measure of transcendence of one number connected with an elliptic function. The main method of the proof is just the Schneider-Gelfond method together with knowledge of the number of zeroes of meromorphic functions of the form $F(z)=P(f_1(z),f_2(z))$, where $f_1(z)$, $f_2(z)$ have the form $\alpha z+\beta\zeta(z)$ or $\wp(\gamma z)$. The bound for the number of zeroes of $F(z)$ in a given disc $|z|\leqslant R$ is a corollary of results of D. Masser [**M3**] and D. Brownawell-D. Masser [**B17**].

Thus, we can say that for now we have used all existing analytical methods for the investigation of the measure of transcendence of *one* number, connected with the exponential and elliptic functions. Some minor problems still remain, but the existing analytical methods give no hope of considerably improving the bounds for the measure of transcendence.

For the case of two and more numbers the main difficulties are of an algebraic character. With the same sort of analytic arguments (construction of the auxiliary function $F(z)=P(f_1(z),\dots,f_n(z))$ and investigation of its zeroes or small values) we now have very deep algebraic problems caused by the singularities of algebraic hypersurfaces.

This explains why for a long time there were practically no known bounds for the measure of algebraic independence of two numbers (as well as no new examples of two algebraically independent numbers connected with exponential and elliptic functions).

In order to discuss measures of transcendence it is useful to introduce S. Lang's notion of type of transcendence [**L1**]. This notion corresponds to the situation when the degree and logarithmic height of the polynomial are of the same order.

DEFINITION 2. Let $\theta_1,\ldots,\theta_n$ be algebraically independent numbers. We say that $(\theta_1,\ldots,\theta_n)$ are of type of transcendence $\leqslant \tau$ if there exists a constant $C > 0$ such that for every $P(x_1,\ldots,x_n) \in \mathbf{Z}[x_1,\ldots,x_n]$, $H(P) \leqslant H$, $d(P) \leqslant d$, $P \not\equiv 0$, we have

$$(3) \qquad |P(\theta_1,\ldots,\theta_n)| > \exp\left(-C(\log H(P) + d(P))^{\tau}\right).$$

It is clear from Lemma 1 that $\tau \geqslant n + 1$ for any $(\theta_1,\ldots,\theta_n)$. We know that for $n = 1$ almost all numbers θ have type of transcendence $\leqslant 2 + \varepsilon$ for any $\varepsilon > 0$. It seems very natural that the following classical conjecture is true.

CONJECTURE. *For almost all* $(\theta_1,\ldots,\theta_n) \in \mathbf{C}^n$, *the type of transcendence* $\tau \leqslant n + 1 + \varepsilon$ *for any* $\varepsilon > 0$.

We know the proof of the part of the conjecture for $n = 2$ [**V1**].

From the results of [**C20**, **R2**] previously mentioned it follows that numbers

$$\pi, \pi/\omega \text{ are of type } \leqslant 2 + \varepsilon \text{ for any } \varepsilon > 0,$$

while

$$e^{\alpha}, \eta/\omega \text{ are of type } \leqslant 3 + \varepsilon \text{ for any } \varepsilon > 0, \text{ etc.} \ldots$$

However up to now there are no examples of a number with type of transcendence exactly 2. Probably π and π/ω are such examples but we have no idea how to prove this.

For the case of two or more numbers, up to 1975 we simply had *no* examples of several numbers and *finite* type of transcendence! Only in 1975 did the author prove that pairs of numbers (π and ω) and (π/ω and $e^{\pi i \beta}$), for ω the period of $\wp(z)$ with complex multiplications by β, are two pairs of numbers with finite type of transcendence.

Since 1975 we found a few more pairs of algebraically independent numbers with finite type of transcendence. However, as if mentioned before the main difficulties in the proof of measure of algebraic independence are algebraic geometrical. These difficulties occur as explained in §3, when, in the course of the proof of algebraic independence, one analyzes the system $P_i(x_1,\ldots,x_n) \in \mathbf{Z}[x_1,\ldots,x_n]$: $i = 1, 2, 3, \ldots$, of polynomials (see (3) of §3), having small value's at the given point $(\theta_1,\ldots,\theta_n)$, for which we try to obtain the measure of algebraic independence. Both nonelementary and elementary methods give good bounds of the measure of algebraic independence of numbers, whose algebraic independence is established in §2. We present the list of results obtained by our methods. Also (for comparison) we present the previous estimates.

Probably the most important result in this direction is the proof that the numbers

$$(4) \qquad \pi/\omega, \eta/\omega \text{ are of the type of transcendence } \leqslant 3 + \varepsilon \text{ for any } \varepsilon > 0.$$

The result (4) is "almost" nonimprovable according to Lemma 1. However it would be important to show that the type of transcendence is exactly 3.

Here, as before $\wp(z)$ is the Weierstrass elliptic function with algebraic invariants g_2, g_3. By ω and η we denote a period and quasi-period associated with $\wp(z)$ and $\zeta(z)$:

$$2\zeta(\omega/2) = \eta.$$

For simplicity, everywhere below we denote by $P(x, y)$ a nonzero polynomial from $\mathbf{Z}[x, y]$ of height $\leqslant H$ and of degree $\leqslant d$. We also put $h = \log H$, $t = h + d$.

THEOREM 3. *The type of transcendence of π/ω, η/ω is at most $3 + \varepsilon$ for any $\varepsilon > 0$. Moreover, if $P(x, y) \in \mathbf{Z}[x, y]$, then*

$$(5) \qquad |P(\pi/\omega, \eta/\omega)| > \exp\big(-c_4(\log H + d\log d)\, d^2 \log^2(d + 1)\big)$$

for $c_4 > 0$.

Theorem 3 in its complete form **[C14]** is proved below **[C19]**. The history of Theorem 3 is as follows. After the proof in 1976 of the algebraic independence of π/ω, η/ω, in 1977 the author found that π/ω, η/ω have type of transcendence $\leqslant 6 + \varepsilon$ for any $\varepsilon > 0$. Earlier, in 1975, the bound for type of transcendence $\leqslant 6 + \varepsilon$ was obtained by the author in the case of complex multiplication (Corollary 2, §12). A little later the same result was obtained by D. Masser. In 1977 the author also proved the following bound for the measure of algebraic independence of π/ω, η/ω **[C3]**:

$$(6) \quad |P(\pi/\omega, \eta/\omega)| > \exp\big(-c_5\big[\log Hd^5 \log^8(\log Hd) + d^6 \log^9(\log Hd)\big]\big)$$

in the notation of Theorem 3. By nonelementary methods it was proved also that the type of transcendence of π/ω, η/ω is $\leqslant 5 + \varepsilon$ for any $\varepsilon > 0$. Then in March, 1978 in a series of lectures at the University of Maryland, College Park, the author gave the following bound based on an elementary method:

$$(7) \quad |P(\pi/\omega, \eta/\omega)| > \exp\big(-c_6\big[\log Hd^{3.5} \log^8(\log Hd) + d^{4.5} \log^9(\log Hd)\big]\big).$$

In **[C8]** we presented a completely elementary proof that the type of transcendence is $\leqslant 4.2 + \varepsilon$ for any $\varepsilon > 0$, and the same kind of technique gives us Theorem 3. Finally, in **[C11]** we presented a bound

$$(8) \qquad |P(\pi/\omega, \eta/\omega)| > \exp\big(-c_4'(\log H + d)d^2 \log^3(d\log H)\big).$$

This bound was subsequently improved in **[C14]** to its present "H^{-c}" form (5).

The results that we obtained are almost the best possible from the point of view of the analytic part of the proof.

As before in §2, we consider the following pairs of numbers that were previously proved by the author to be algebraically independent:

1) π/ω, η/ω;

2) $\zeta(u) - \eta u/\omega$, η/ω for u algebraic for $\wp(z)$;

3) π/ω, $e^{\pi i\beta}$ for ω a period of $\wp(z)$ with complex multiplication by β.

For the numbers $\zeta(u) - \eta/\omega$, η/ω we have

THEOREM 4. *Let* (ω, η) *be a pair, a period and a quasi-period of* $\wp(z)$, *and let* u *be an algebraic point of* $\wp(z)$. *Then*

$$(9) \qquad |P(\zeta(u) - \eta u/\omega, \eta/\omega)| > H^{-c_7 d^8 \log^9(d+1)}$$

if $\log H \geqslant c_8 d$, *and* c_7, $c_8 > 0$ *depending on* $\wp(z), u$. *In particular, the numbers* $\zeta(u) - \eta u/\omega$, η/ω *have type of transcendence* $\leqslant 9 + \varepsilon$ *for any* $\varepsilon > 0$.

We present one earlier result (1978, cf. **[C11]**).

$$(10) \qquad |P(\zeta(u) - \eta u/\omega, \eta/\omega)| > \exp\left(-c_9[h^{1+\varepsilon}d^{9.5+\varepsilon} + d^{13.5+\varepsilon}]\right)$$

for $c_9 > 0$ depending on $\wp(z)$, u and ε for every $\varepsilon > 0$.

In **[C11]** the bound of type (9) was presented, but with H replaced by He^d and $\log(d+1)$ by $\log(t+1)$; hence the bound was not of the "H^{-c}" form.

THEOREM 5. *Let* $\wp(z)$ *have complex multiplication by a quadratic imaginary number* β *and let* ω *be a nonzero period of* $\wp(z)$. *Then*

$$(11) \qquad |P(\pi/\omega, e^{\pi i\beta})| > \exp\left(-c_{10}(h + d^3)^{7/3}\log^7 t\right)$$

for $c_{10} > 0$ *depending on* $\wp(z)$. *In particular, the type of transcendence of* π/ω, $e^{\pi i\beta}$ *is* $\leqslant 7 + \varepsilon$ *for any* $\varepsilon > 0$.

The previous result for these numbers **[C3]** in 1977 was the following:

$$(12) \qquad |P(\pi/\omega, e^{\pi i\beta})| > \exp\left(-c_{11}(d^2\log d + dh)^5\right) \quad \text{for } c_{11} > 0.$$

For numbers u and $\zeta(u)$ we do not know whether the type of transcendence is finite or not. For this case we have one rather old bound **[C11]** for the measure of algebraic independence.

THEOREM 6. *Let* $\wp(z)$ *have complex multiplications and* u *be an algebraic point of* $\wp(z)$. *Then*

$$(13) \qquad |P(u, \zeta(u))| > \exp\left(-c_{12}\left[he^{c_{13}d^2(\log h)^{1/2}}\right]\right) = \exp\left(-c_{12}\left[\log H(P)\cdot e^{c_{13}d^2(\log\log H(P))^{1/2}}\right]\right)$$

for $\log\log H(P) = \log h \geqslant c_{14}d^4$ *where* c_{12}, c_{13}, $c_{14} > 0$ *depend only on* $\wp(z)$, u.

This is all we know about estimates for the type of transcendence of pairs of numbers connected with elliptic and exponential functions. However there are a few more situations when we know some bounds for the measure of transcendence of two algebraically independent numbers. First of all we have the classical Gelfond example of numbers α^β, α^{β^2} for a cubic irrationality β. In this case we have the following modern estimates (D. Brownawell **[B14]** and the author **[C12, B15]**):

THEOREM 7 **[B14]**. *Let* $\alpha \neq 0, 1$ *be an algebraic number*, β *a cubic irrationality. Then*

$$|P(\alpha^{\beta}, \alpha^{\beta^2})| > \exp(-\exp[c_{15}d^3(d+h)]) \quad \textit{for } c_{15} > 0. \tag{14}$$

This estimate was subsequently improved, cf. **[C8]**, to

$$|P(\alpha^{\beta}, \alpha^{\beta^2})| > \exp(-\exp[c_{16}(d+h)^{2+\varepsilon}]) \tag{15}$$

for $c_{16} > 0$ depending on α, β and ε, for every $\varepsilon > 0$.

Estimate (15) can be improved to

$$|P(\alpha^{\beta}, \alpha^{\beta^2})| > \exp(-c'_{17}H^{c_{17}d^{1+\varepsilon}}) \tag{16}$$

for $c'_{17}, c_{17} > 0$ depending on α, β and ε, for every $\varepsilon > 0$; see **[C14, C12]**.

Similar lower bounds for algebraically independent numbers connected with Abelian varieties of CM-type can be presented. E.g. let **A** be an Abelian variety of CM-type of dimension 3 and let $A_i(x_1, x_2, x_3)$: $i = 1, 2, 3$, be its Abelian functions normalized as in §1. Let the number $A_1(u, 0, 0)$ be algebraic for $u \neq 0$. Then $A_2(u, 0, 0)$, $A_3(u, 0, 0)$ are algebraically independent and, moreover,

$$|P(A_2(u,0,0), A_3(u,0,0))| > \exp(-\exp t^{\lambda}) \tag{17}$$

for some $\lambda > 0$ and $t \geqslant t_0$.

The bound of form (17) is also true for two numbers $\wp(u)$, $\wp(u\beta)$ **[C12]**, where u is an algebraic point of $\wp(x)$ with c.m., and β is a cubic irrationality, whose a.i. was proved by D. Masser-K. Wüstholz **[M8]**.

Now we consider measures of algebraic independence, when h is sufficiently large with respect to d. In this case we are interested, not in (3), but in the so called "normal" bounds of K. Mahler's type (2):

$$|P(\theta_1, \ldots, \theta_n)| > H(P)^{-c(d(P))} \tag{18}$$

for $P(x_1, \ldots, x_n) \in \mathbf{Z}[x_1, \ldots, x_n]$, $P \not\equiv 0$, and $H(P)$ sufficiently large with respect to $d(P)$. The set of numbers $\theta_1, \ldots, \theta_n$ satisfying (18) is called normal. While the results of Siegel-Mahler show that numbers $e^{\alpha_1}, \ldots, e^{\alpha_n}$ are normal for algebraic $\alpha_1, \ldots, \alpha_n$, Baker's method **[B2]** shows that a number $\Sigma\,\beta_i \log \alpha_i / \Sigma\,\delta_j \log \gamma_j$, if transcendental, is normal for algebraic α_i, β_i, γ_j, δ_j.

Recently we discovered a new method for proving the normality of numbers associated with elliptic and Abelian functions **[C14]**. Among these newly discovered normal numbers are:

$$\frac{\Sigma\,\beta_i u_i}{\Sigma\,\gamma_j u'_j}, \qquad u, \qquad \alpha_1 u + \alpha_2 \zeta(u), \qquad \zeta(u) - \frac{\eta}{\omega} u,$$

for u, u_i, u'_j that are algebraic points of $\wp(x)$,

$$\Sigma\,\beta_i u_i + \Sigma\,\gamma_i \zeta(u_i)$$

in the c.m. case;

u/v for algebraic points u on $\wp_1(x)$ and v on $\wp_2(x)$ with a.i. $\wp_1(x)$, $\wp_2(x)$; $\wp(\alpha_1), \ldots, \wp(\alpha_n)$ in the c.m. case, where α_i, β_i, γ_j are algebraic numbers.

We start the presentation of these results with a statement about the normality of periods of Abelian varieties.

THEOREM 8. *Periods (or quasi-periods) of Abelian varieties with real (complex) multiplications are normal transcendental numbers.*

Transcendence results in Theorem 8 belong to [**S1, B6**].

COROLLARY 9. *Periods of Abelian varieties of CM-type are normal transcendental numbers. In particular, for rational (noninteger) a and b the value of Euler's B-function is a normal transcendental number,*

$$|P(B(a,b))| > H(P)^{-c(d,a,b)}. \tag{19}$$

Similarly, lower bounds for linear forms $\beta_1 u_1 + \cdots + \beta_m u_m + \beta_0$ in algebraic points $u_1, \ldots, u_m$ of an elliptic curve **E** have normal or "H^{-c}" form

$$|\beta_1 u_1 + \cdots + \beta_m u_m + \beta_0| > H^{-c}$$

for $H(\beta_0) \leqslant H$, as noted in §1.

A large class of normal measures of algebraic independence is provided by a generalization of the Lindemann-Weierstrass theorem for elliptic and Abelian functions (§2, [**C7, C19**, Ch. 7]).

Similarly, low bounds for linear forms $\beta_1 u_1 + \cdots + \beta_m u_m + \beta_0$ in algebraic points $u_1, \ldots, u_m$ of an elliptic curve **E** have normal or "H^{-c}" form as noted in §1.

A large class of normal measures of algebraic independence is provided by a generalization of the Lindemann-Weierstrass theorem for elliptic and Abelian functions (§2, [**C7, C19**]).

THEOREM 10. *Let $\wp(x)$ have complex multiplication and let $\alpha \neq 0$ be an algebraic number. If $P(x) \in \mathbf{Z}[x]$, $P \not\equiv 0$, $d(P) \leqslant d$, $H(P) \leqslant H$, then*

$$|P(\wp(\alpha))| > H^{-c_{18} d}, \tag{20}$$

if $\log\log H \geqslant c_{19} d^3$ *for* $c_{18}, c_{19} > 0$.

THEOREM 11. *Let $\wp(z)$ have complex multiplication by $\mathbf{Q}(\tau)$ and let α, β be algebraic numbers, linearly independent over $\mathbf{Q}(\tau)$. If $P(x, y) \in \mathbf{Z}[x, y]$, $P \not\equiv 0$, $d(P) \leqslant d$, $H(P) \leqslant H$, then*

$$|P(\wp(\alpha), \wp(\beta))| > H^{-c_{20} d^2} \tag{21}$$

for $\log\log H \geqslant c_{21} d^5$ *and* $c_{20}, c_{21} > 0$.

For numbers $\wp(\alpha)$, in the case of $\wp(z)$ without complex multiplication, we have

THEOREM 12. *Let $\alpha \neq 0$ be an algebraic number and let $\wp(z)$ have algebraic invariants. Then for $P(X) \in \mathbf{Z}[x]$, $P(x) \not\equiv 0$, $d(P) \leqslant d$, $H(P) \leqslant H$ we have*

$$|P(\wp(\alpha))| > H^{-c_{22} d^2} \tag{22}$$

when $\log\log H \geqslant c_{23} d^3$ *for* $c_{22}, c_{23} > 0$.

One can find the proof of Theorems 10–12 in **[C7]**. In general, we have the following elliptic generalization of Mahler's bound (2), see **[C19**, Ch. 5,7]. Let $\wp(x)$ have complex multiplication in an imaginary quadratic field **K** and let $\alpha_1, \ldots, \alpha_n$ be algebraic numbers linearly independent over **K**. Then for $P(x_1, \ldots, x_n) \in \mathbf{Z}[x_1, \ldots, x_n]$, $P \not\equiv 0$, we have

$$(23) \qquad |P(\wp(\alpha_1), \ldots, \wp(\alpha_n))| > H(P)^{-c_{24}(\bar{\alpha})d(P)^n}$$

where $c_{24}(\bar{\alpha}) > 0$ depends only on $[\mathbf{K}(g_2, g_3, \alpha_1, \ldots, \alpha_n):\mathbf{Q}]$ provided that $\log\log H(P) > c_{25}d(P)^{2n+1}$.

Similar results are true for Abelian varieties of CM-type of an arbitrary dimension, cf. **[C19]**. We present one such result in dimension two.

Let **A** be an Abelian variety of CM-type and dimension two, defined over $\overline{\mathbf{Q}}$ with Abelian functions $A_1(\bar{z})$, $A_2(\bar{z})$, $B(\bar{z})$, where $A_1(\bar{z})$, $A_2(\bar{z})$ are Abelian functions of **A**, algebraically independent over **C** and regular with $B(\bar{z})$ at $\bar{z} = 0$ **[M5, C22]**.

THEOREM 13. *Let* **A**, $A_1(\bar{z})$, $A_2(\bar{z})$, $B(\bar{z})$ *be as above. If* α *is algebraic* $\alpha \neq 0$, $P(x, y) \in \mathbf{Z}[x, y]$, $P \not\equiv 0$, $d(P) \leqslant d$, $H(P) \leqslant H$, *then*

$$(24) \qquad |P(A_1(\alpha, 0), A_2(\alpha, 0))| > H^{-c_{26}d^2}$$

for $\log\log H \geqslant c_{27}d^5$.

Theorem 6 can be supplied with an additional normality statement (cf. **[C14]**).

THEOREM 14. *If* $\wp(z)$ *has complex multiplication and* u *is an algebraic point of* $\wp(z)$, *then*

$$|P(u, \zeta(u))| > H^{-c_{28}(d)}$$

with $c_{28}(d) > 0$ *depending on* d *and* u, $\wp(z)$ *only.*

Theorem 14 for example implies that any number $P(u, \zeta(u))$ is normal. As a corollary (or, equally, as a corollary of Theorem 3), values of the Γ-function $\Gamma(1/3)$, $\Gamma(1/4)$, etc. listed in §2 are normal transcendental numbers. E.g. for $\Gamma(1/3)$,

$$|\Gamma(1/3) - p/q| > |q|^{-\gamma}$$

for some $\gamma > 0$. The best value of γ is unknown although a natural conjecture would be $\gamma \leqslant 2 + \varepsilon$, if $|q| \geqslant q_0(\varepsilon)$. For a simpler number $\Gamma(1/2) = \sqrt{\pi}$, the corresponding measure of irrationality is that of π studied in **[C17]**. The bounds from **[C17]** give for this γ the value $\gamma \leqslant 39.7799\ldots$. New methods based on Padé approximations **[C16]** give for $\Gamma(1/2)$ the new value of the exponent γ only as $\gamma \leqslant 29$.

We conclude this section with some conjectures.

CONJECTURE 15. *If* l *is prime*, $l \geqslant 3$, *then the numbers*

$$\pi, \quad \Gamma(1/l), \ldots, \Gamma(n/l) \quad \text{for } n = (l-1)/2$$

have type of transcendence exactly $(l+1)/2$ (*or at least* $\leqslant (l+1)/2 + \varepsilon$ *for any* $\varepsilon > 0$).

CONJECTURE 16. *For* $\wp(z)$ *of simple nature e.g.* $g_2 = 0$ *or* $g_3 = 0$,

$$|\wp(1) - p/q| > |q|^{-2-\varepsilon}$$

for integers p, q *with* $|q| > c_{28}(\varepsilon)$.

REFERENCES

A1. M. Anderson, *Inhomogeneous linear forms in algebraic points of an elliptic function*, Transcendence Theory: Advances and Applications, Academic Press, London and New York, 1977, Chapter 7, pp. 121–143.

B1. A. Baker, *On the quasi-periods of the Weierstrass ζ-function*, Göttingen Nachr. **N16** (1969), 145–157.

B2. ______, *Transcendental number theory*, Cambridge Univ. Press, Cambridge, 1975.

B3. ______, *The theory of linear forms in logarithms*, Transcendence Theory: Advances and Applications, Academic Press, London and New York, 1977, Chapter 1.

B4. D. Bertrand, *Approximations diophantiennes p-adique et courbes élliptiques*, Thèse, Paris, 1977 (see also, Compositio Math. **37** (1978), 21–50).

B5. ______, *Transcendence de valeurs de la fonction gamma d'après G. V. Chudnovsky*, Séminaire D-P-P, **17** (1975/1976), G8.

B6. ______, *Sur les périodes de formes modulaires*, C. R. Acad. Sci. Paris Sér. A **288** (1979), 531–534.

B7. ______, *Fonctions modulaires, courbes de Tate et indépendence algébrique*, Séminaire D-P-P, **19** (1977/78), no. 36.

B8. D. Bertrand and Y. Flicker, *Linear forms on abelian varieties over local fields*, Acta Arith. **38** (1979).

B9. D. Bertrand and D. Masser, *Linear forms in elliptic integrals*, Invent. Math. **58** (1980), 283–288.

B10. D. Bertrand and M. Laurent, *Propriétés de transcendence de nombres liés aux fonctions thêta*, C. R. Acad. Sci. Paris Sér. A **292** (1981), 747–749.

B11. P. E. Blanksby and H. L. Montgomery, *Algebraic integers near the unit circle*, Acta Arith. **18** (1971), 355–369.

B12. E. Bombieri and S. Lang, *Analytic subgroups of group varieties*, Invent. Math. **11** (1970), 1–14.
E. Bombieri, *Algebraic values of meromorphic maps*, Invent. Math. **10** (1970), 267–287.

B13. D. Brownawell, *Gelfond's method for algebraic independence*, Trans. Amer. Math. Soc. **210** (1975), 1–26.

B14. ______, *Algebraic independence of cubic powers of certain Liouville numbers*, Compositio Math. **38** (1979), 355–368.

B15. ______, *Some remarks on semi-resultants*, Transcendence Theory: Advances and Applications, Academic Press, London and New York, 1977, Chapter 14.

B16. D. Brownawell and K. Kubota, *The algebraic independence of Weierstrass functions and some related numbers*, Acta Arith. **39** (1977), 111–149.

B17. D. W. Brownawell and D. W. Masser, *Multiplicity estimates for analytic functions*, I, J. Reine Angew. Math. **314** (1980), 200–216; II, Duke Math. J. **47** (1980), 273–295.

C1. G. V. Chudnovsky, *A mutual transcendence measure for some classes of numbers*, Soviet. Math. Dokl. **15** (1974), 1424–1428.

C2a. ______, *Analytical methods in diophantine approximations*, Preprint IM-74-9, Inst. Math., Kiev, 1974. (and see this volume, Chapter 1).

C2b. ______, *Some analytical methods in the theory of transcendental numbers*, Preprint IM-74-8, Inst. Math., Kiev, 1974 (and see this volume, Chapter 1).

C3. ______, *Algebraic independence of constants connected with exponential and elliptic functions*, Dokl. Akad. Nauk Ukrain. SSR Ser. A **1976**, 698–701, 767. (Russian. English summary)

C4. ______, *Baker's method in the theory of transcendental numbers*, Uspekhi Math. Sci. **31** (1976), 281–282 (and see this volume, Chapter 2).

C5. ______, *A new method for the investigation of arithmetical properties of analytic functions*, Ann. of Math. (2) **109** (1979), 353–376.

C6a. ______, *Arithmetical properties of values of analytical functions*, Conf. on Diophantine Approximations (Oberwolfach, 1977).

C6b. ______, *Values of meromorphic functions of order* 2, Séminaire D-P-P (1977/78), Novembre 1977, 19e année, University Publishing, Paris, 1978, pp. 4501–4518.

C6c. ______, *Singular points on complex hypersurfaces and multidimensional Schwarz lemma*, Séminaire D-P-P (1977/78), January 1977, 19e année, University Publishing, Paris, pp. 1–40. Reprinted in Progress in Math., Birkhauser-Verlag, vol. 12, 1980, pp. 29–69.

C7. ______, *Algebraic independence of the values of elliptic function at algebraic point. Elliptic analogue of Lindemann-Weierstrass theorem*, Invent. Math. **61** (1980), 267–280. (and see this volume, Chapter 7).

C8. ______, *Algebraic grounds for the proof of algebraic independence.* I. *Elementary algebra*, Comm. Pure Appl. Math. **34** (1981), 1–28.

C9. ______, *Criteria of algebraic independence of several numbers*, Lecture Notes in Math., vol. 925, Springer-Verlag, Berlin and New York, 1982, pp. 323–368.

C10. G. V. Chudnovsky and D. V. Chudnovsky, *Remark on the nature of spectrum of Lamé equation*, *Problems from Transcendence Theory*, Letters Nuovo Cimento **29** (1980), 545–550.

C11. G. V. Chudnovsky, *Algebraic independence of values of exponential and elliptic functions*, Proc. Internat. Congr. Math. (Helsinki, 1978), vol. 1, Acad. Sci. Fenn, Helsinki, 1980, pp. 339–350.

C12. ______, *Independénce algébrique dans la méthode de Gelfond-Schneider*, C. R. Acad. Sci. Paris Sér. A **291** (1980), A365–A368.

C13. ______, *Sur la mesure de la transcendence de nombres dependent d'équations differentielles de type fuchsien*, C. R. Acad. Sci. Paris Sér. A **291** (1980), A485–A487.

C14. ______, *Measures of irrationality, transcendence and algebraic independence. Recent progress*, Lectures at Exeter Conf. Number Theory (Exeter, April 1980), in Journees Arithmetiques, 1980, ed. by J. V. Armitage, Cambridge Univ. Press, Cambridge, 1982, pp. 11–83.

C15a. ______, *On Schanuel's hypothesis. Three algebraically independent numbers.* I, II, Notices Amer. Math. Soc. **23** (1976), A-272; Preliminary report, A-422.

C15b. ______, *The bound for linear forms in elliptic logarithms for the complex multiplication case*, Notices Amer. Math. Soc. **26** (1979), A-39.

C16. ______, *Padé approximation and the Riemann monodromy problem* (Cargesé Lectures, June 1979), Bifurcation Phenomena in Mathematical Physics and Related Topics, Reidel, Boston, 1980, pp. 448–510.

C17. ______, *Formules d'Hermité pour les approximants de Padé de logarithmes et de fonctions binomes, et mesures d'irrationalité*, C. R. Acad. Sci. Paris Sér. A **288A** (1979), A965–A967.

C18. ______, *Algebraic independence of constants connected with exponential and elliptic functions*, Meeting of AMS, Pullman, Washington, June 1975, Notices Amer. Math. Soc. **22** (1975), A-486.

C19. ______, This volume, Chapters 1, 4, 7, 8.

C20. P. L. Cijsouw, *Transcendence measures*, Thesis, 1972.

C21. J. Coates, *Linear relations between $2\pi i$ and the periods of two elliptic curves*, Diophantine Approximation and its Applications, Academic Press, London, 1973, pp. 77–99.

C22. J. Coates and S. Lang, *Diophantine approximation on Abelian varieties with complex multiplication*, Invent. Math. **34** (1976), 129–133.

C23. J. Coates, *The arithmetic of elliptic curves with complex multiplication*, Proc. Internat. Congr. Math. (Helsinki, 1978), vol. 1, Acad. Sci. Fenn., Helsinki, 1980, pp. 351–355.

D1. P. Deligne, *Cycles de Hodge absolus et périodes des intégrales des variétés abéliennes*, Bull. Soc. Math. France Mém. No. 2 (1980), 23–33.

D2. E. Dobrowolski, *On a question of Lehmer*, Bull. Soc. Math. France Mém. No. 2 (1980), 35–39.

D3. J. Drach, *Sur l'intégration par quadratures de l'équation $d^2y/dx^2 = [\varphi(x) + h]y$*, C. R. Acad. Sci. Paris **168** (1919), 47–50, 337–340.

F1. Y. Z. Flicker, *Linear forms on arithmetic Abelian varieties: ineffective bound*, Bull. Soc. Math. France Mém. No. 2 (1980), 41–47.

G1. A. O. Gelfond, *Transcendental and algebraic numbers*, Dover, New York, 1960.

G2. H. Groscot, *Points entiers sur les courbes élliptiques*, Thèse de troisième cycle (Paris VI), 1979.

L1. S. Lang, *Introduction to transcendental numbers*, Addison-Wesley, Reading, Mass., 1966.

L2. ______, *Diophantine approximation on Abelian varieties with complex multiplication*, Adv. in Math. **17** (1975), 281–336.

L3. ______, *Elliptic curves, diophantine analysis*, Springer-Verlag, Berlin and New York, 1978.

L4. M. Laurent, *Transcendence de périodes d'intégrales élliptiques*, J. Reine Angew. Math. **316** (1980), 123–139.

L5. F. Lindemann, *Uber die Zahl π*, Math. Ann. **20** (1882), 213–225.

M1. K. Mahler, *Zur approximation der Exponentialfunktion und des Logarithmus*, J. Reine Angew. Math. **166** (1932), 118–150.

M2. Yu. I. Manin, *Cyclotomic fields and modular curves*, Uspekhi Math. Nauk **25** (1971), 7–71.

M3. D. Masser, *Elliptic functions and transcendence*, Lecture Notes in Math., vol. 437, Springer-Verlag, Berlin and New York, 1975.

M4. ______, *Linear forms in algebraic points of Abelian functions*, Parts I, II, Math. Proc. Cambridge Philos. Soc. **77** (1975), 499–513; **79** (1976), 55–70. Part III, Proc. London Math. Soc. **33** (1976), 549–564.

M5. ______, *The transcendence of certain quasi-periods associated with Abelian functions in two variables*, Compositio Math. **35** (1977), 239–258.

M6. ______, *Diophantine approximations and lattices with complex multiplication*, Invent. Math. **45** (1978), 61–82.

M7. ______, *On quasi-periods of Abelian functions with complex multiplication*, Bull. Soc. Math. France Mém. No. 2 (1980), 55–68.

M8. D. Masser and G. Wüstholz, *Zero estimates on group varieties* (to appear).

M9. H. P. McKean, *Algebraic curves of infinite genus arising in the theory of nonlinear waves*, Proc. Internat. Congr. Math. (Helsinki, 1978), vol. 2, Acad Sci. Fenn., Helsinki, 1980, pp. 777–783.

N1. S. P. Novikov, *Linear operators and integrable Hamiltonian systems*, Proc. Internat. Congr. Math. (Helsinki, 1978), vol. 1, Acad. Sci. Fenn., Helsinki, 1980, p. 187.

R1. K. Ramachandra, *Contributions to the theory of transcendental numbers*, Acta Arith. **14** (1968), 65–88.

R2. E. Reyssat, *Approximation algébrique de nombres liés aux fonctions élliptiques et exponentielles*, Bull. Soc. Math. France, **108** (1980), 47–79.

R3. ______, C. R. Acad. Sci. Paris, Sér. A290 (1980), 439–441.

R4. K. A. Ribet, *Dividing rational points on Abelian varieties of CM-type*, Compositio Math. **33** (1976), 69–74.

R5. ______, *Division fields of Abelian varieties with complex multiplication*, Bull. Soc. Math. France Mém. No. 2 (1980), 75–94.

S1. Th. Schneider, *Introduction aux nombres transcendants*, Gauthier-Villars, Paris, 1959.

S2. G. Shimura and Y. Taniyama, *Complex multiplication of abelian varieties*, Publ. Math. Soc. Japan, N6, 1961.

S3. G. Shimura, *On some problems of algebraicity*, Proc. Internat. Congr. Math. (Helsinki, 1978), vol. 1, Acad. Sci. Fenn., Helsinki, 1980, pp. 373–379.

S4. ______, *Automorphic forms and the periods of Abelian varieties*, J. Math. Soc. Japan **31** (1979), 561–592.

S5. ______, *The arithmetic of certain zeta functions and automorphic forms on orthogonal groups*, Ann. of Math. (2) **110** (1980).

S6. C. L. Siegel, *Über einige Anwendungen diophantischer Approximationen*, Abh. Preuss. Akad. Wiss. Phys-Math. Kl. No. 1, 1929.

S7. ______, *Transcendental numbers*, Ann. of Math. Studies, Princeton Univ. Press, Princeton, N.J., 1949.

S8. H. M. Stark, *Further advances in the theory of linear forms in logarithms*, Diophantine Approximation and its Applications, Academic Press, London, 1973, pp. 255–293.

S9. ______, *Effective estimates of solutions of some Diophantine equations*, Acta Arith. **24** (1973), 251–259.

S10. C. L. Stewart, *On a theorem of Kronecker and related question of Lehmer*, Séminaire de Théorie des Nombres (Bordeaux, 1977–78), No. 7, 11pp.

S11. A. Selberg and S. D. Chowla, *On Epstein's zeta-function*, J. Reine Angew. Math. **227** (1967), 86–110.

T1. R. Tijdeman, *On the number of zeroes of general exponential polynomials*, Indag. Math. **33** (1971), 1–7.

T2. ______, *Exponential diophantine equations*, Proc. Internat. Congr. Math. (Helsinki, 1978), vol. 1, Acad. Sci. Fenn., Helsinki, 1980, pp. 381–387.

T3. *Transcendence theory: Advances and applications*, Academic Press, New York and London, 1977.

V1. A. I. Vinogradov and G. V. Chudnovsky, *Extremality of certain manifolds*, this volume, Appendix.

W1. M. Waldschmidt, *Nombres transcendants*, Lecture Notes in Math., vol. 402, Springer-Verlag, Berlin and New York, 1974.

W2. ______, *Propriétés arithmétiques de fonctions de plusieurs variables*. I, Séminaire P. Lelong (Analyse), 15e année (1974/1975), Lecture Notes in Math., vol. 518, Springer-Verlag, Berlin and New York, 1976; II, ibid, 16e année (1975/1976), vol. 524, 1977.

W3. ______, *Les travaux de G. V. Choodnovsky sur les nombres transcendants*, Séminaire Bourbaki (1975/1976), no. 488, Lecture Notes in Math., vol. 567, Springer-Verlag, Berlin and New York, 1977, pp. 274–292.

W4. ______, *On functions of several complex variables having algebraic Taylor coefficients*, Transcendence Theory: Advances and Applications, Academic Press, London and New York, 1977, Chapter 11.

W5. ______, *Nombres transcendants et groupes algébriques*, Astérisque No. 69–70 (1980).

W6. ______, *Transcendence measures for exponentials and logarithms*, J. Austral. Math. Soc. **25** (1978), 445–465.

W7. A. Weil, *Elliptic functions according to Eisenstein and Kronecker*, Springer, New York, 1976.

CHAPTER 1

SOME ANALYTIC METHODS IN THE THEORY OF TRANSCENDENTAL NUMBERS

0. The first part of this chapter is devoted mainly to an exposition of a method for investigating the algebraic independence of numbers connected with analytic functions. We first consider the classical problem of the arithmetic nature of values of the exponential function. The machinery we propose is very useful for studying the mutual transcendence measure of a set of numbers. The analytic basis of this approach is the method of A. O. Gel′fond (more precisely, Gel′fond's third method [**1**]). The main difficulties in extending this method to the case of several algebraically independent numbers are algebraic in nature, i.e. they involve properties of fields of finite transcendence degree over **Q**. This investigation can be regarded as the first step towards a proof of Schanuel's well-known conjecture to the effect that there are n algebraically independent numbers among

$$\alpha_1, \ldots, \alpha_n, e^{\alpha_1}, \ldots, e^{\alpha_n}$$

if $\alpha_1, \ldots, \alpha_n$ are linearly independent over **Q**.

This conjecture has been proved in the case of fields of formal power series (Ax [**2**]), but in the complex case, except for $n = 1$ (the Hermite-Lindemann theorem), the answer is not known.

The author considers one of the main results of the present paper to be the procedure affording a partial answer to complex versions of Schanuel's conjecture. It is shown that among a set of N values of the exponential function, only $O(\log_2 N)$ are algebraically independent. In view of this, we can find for the first time a set of numbers in which the number of algebraically independent ones tends to infinity along with the size of the set. Finally, refinements in the analytic part of the method in combination with algebraic considerations enable us to improve this result.

0076-5376/84 $1.00 + $.25 per page

A procedure for investigating several algebraically independent numbers was first suggested by S. Lang [**3, 4**]. His hypothetical method consisted of the following. First, one establishes good lower bounds for polynomials in one variable at a transcendental point. Transcendental numbers θ for which there exists a good estimate $|P(\theta)| > \exp(-ct(P)^\tau)$, where $c, \tau > 0$ and $t(P) = \ln H(P) + \deg P$, are called numbers of finite type τ. Similarly, one can define the concept of finite type for n-tuples of numbers $(\theta_1,\ldots,\theta_n)$ by considering polynomials in n variables with rational integer coefficients. Lang's idea was that by choosing at the outset n numbers of finite type, one obtains $n+1$, $n+2,\ldots$ algebraically independent numbers. Exactly as in the usual method of Gel′fond, by starting with an algebraic number one arrives at a transcendental number, two algebraically independent numbers.

The methods of the author set forth below are close to those used by Gel′fond. However, they also have an inductive aspect reminiscent of Lang's scheme. An essential feature of the proposed scheme is the absence of a priori assumptions about fields of finite type, and the calculations are carried out simultaneously in fields of different transcendence degrees.

Along with the proof of algebraic independence we estimate the mutual transcendence measure of numbers connected with the exponential function. Present analytic methods do not allow us, in general, to improve this estimate. Note that the estimate of the transcendence measure depends on the height and the degree of the polynomial, and, although good estimates with respect to the height have long been known, here we obtain for the first time a good estimate with respect to the degree of the polynomial. Good estimates with respect to $d = \deg(P)$ are established for the numbers e^α, α^β (α, β algebraic), and e^π. These estimates, which have the form $|P(\theta)| > \exp(-cd\ln^2 d)$, are close to unimprovable trivial estimates.

We use the standard notation and concepts of [**1, 5**]. The symbols $\mathbf{Z}, \mathbf{R}, \mathbf{C}$ have the usual meaning. We denote by $\mathbf{A}$ the field of all algebraic numbers. If $\mathbf{K}$ is an algebraic number field, then $\mathbf{I_K}$ denotes the ring of integers of this field. If $\mathbf{B}$ is a ring, then $\mathbf{B}[x_1,\ldots,x_n]$ is the ring of polynomials in $x_1,\ldots,x_n$ with coefficients in $\mathbf{B}$. If $P(x_1,\ldots,x_n)$ is a polynomial, then $\deg_{x_i}(P)$ denotes the degree of occurrence of x_i in $P(x_1,\ldots,x_n)$, and $\deg(P) = d(P)$ denotes the degree of $P(x_1,\ldots,x_n)$, i.e. $\max_i \deg_{x_i}(P)$.

The elementary facts concerning discriminants and resultants used below can be found in [**5**].

1. We present in a somewhat new form a series of lemmas on resultants, mainly due to A. O. Gel′fond. The need for these auxiliary lemmas can be explained as follows. In most of the present paper it is possible to manage without resultants and operations with polynomials. In order to do this it is necessary to use Gel′fond's method and auxiliary functions for direct approximations of transcendental numbers by algebraic ones.

We begin with a refinement of the concepts of height and length of an algebraic number. By the height $H(\alpha)$ of an algebraic number α we usually mean the maximum of the moduli of the coefficients of the minimal polynomial $\mu_\alpha(z) \in \mathbf{Z}[z]$ for α. The minimal polynomial $\mu_\alpha(z)$ is the primitive irreducible polynomial with rational integer coefficients of which α is a root.

In addition to $H(\alpha)$ in the case of the bounded degree of α, i.e. the degree of $\mu_\alpha(z)$, we use the concepts of the size of α, the type of α, and so on [**3**]. Along with the height $H(\alpha)$ and the logarithmic height $h(\alpha)$ (see below) it will be convenient to use the concept of the length of an algebraic number.

We mention one well-known lemma:

LEMMA 1.1. *Suppose we have a polynomial of degree $n \geqslant 1$ with integral coefficients*:

$$P(z) = p_n z^n + \cdots + p_1 z + p_0 = p_n(z - \alpha_1) \cdots (z - \alpha_n).$$

Then

$$|p_n| \prod_{i=1}^{n} \max(1, |\alpha_i|) \leqslant |p_0| + \cdots + |p_n|. \tag{1.1}$$

This lemma has been proved often (see [**6**]). In this exact form it is due to K. Mahler.

We now give two definitions closely related to Lemma 1.1.

DEFINITION 1.2. (i) Suppose $P(z) \in \mathbf{Z}[z]$ and $P(\alpha) = 0$. If $P(z) = a(z - \alpha_1) \cdots (z - \alpha_d)$, where $d = \deg(P)$, we put $H_P(\alpha) = \prod_{i=1}^{d} \max(1, |\alpha_i|)$; $\hat{H}_P(\alpha)$ is the height of $P(z)$, the maximum of the moduli of the coefficients of $P(z)$; $\hat{L}_P(\alpha)$ is the length of $P(z)$, the sum of the moduli of the coefficients of $P(z)$; $L_P(\alpha) = (d + 1)\hat{H}_P(\alpha)$.

(ii) $\hat{H}_P$ is the unreduced height relative to $P(z)$; $\hat{L}_P$ is the unreduced length relative to $P(z)$. Finally,

$$L(\alpha) = (\deg(\alpha) + 1)H(\alpha) = (\deg(\mu_\alpha(z)) + 1)\hat{H}_{\mu_\alpha}(\alpha),$$

where $H(\alpha)$ is the height of α.

LEMMA 1.3. *Suppose $P(z)$, $R(z) \in \mathbf{Z}[z]$, α is an algebraic number of degree $n \geqslant 1$, and $P(\alpha) = R(\alpha) = 0$. If a is the smallest natural number such that $a \cdot \alpha$ is an algebraic integer, and p, r are the leading coefficients of $P(z)$, $R(z)$, then*:

$$H_P(\alpha) \leqslant \hat{L}_P(\alpha)/|p|, \qquad H_P(\alpha) \leqslant (\deg(P) + 1)\hat{H}_P(\alpha)/|p|. \tag{1.2}$$

$$H_{\mu_\alpha}(\alpha) \leqslant H_P(\alpha). \tag{1.3}$$

$$H_{\mu_\alpha}(\alpha) \leqslant (n + 1)H(\alpha)/a. \tag{1.4}$$

$$a \leqslant |p|. \tag{1.5}$$

Also, if $H(P)$ is the height of $P(z)$ and $L(P)$ is the length of $P(z)$, then:

$$H(P) = \hat{H}_P(\alpha). \tag{1.6}$$

$$L(P) = \hat{L}_P(\alpha). \tag{1.7}$$

$$H_P(\alpha) \leqslant L(P)/|p| \leqslant (\deg(P) + 1)H(P)/|p|. \tag{1.8}$$

(1.9) $$L(P) \leqslant (\deg(P)+1)H(P).$$

(1.10) $$L_P(\alpha) = (\deg(P)+1)H(P).$$

Finally, if $P(z)$ divides $R(z)$, i.e. if there exists $Q(z) \in \mathbf{Z}[z]$ such that $R(z) = P(z)\cdot Q(z)$, then:

(1.11) $$\deg(P(z)) \leqslant \deg(R(z)).$$

(1.12) $$H_P(\alpha) \leqslant H_R(\alpha).$$

In view of Lemma 1.3, we will denote $\hat{H}_P(\alpha)$ and $\hat{L}_P(\alpha)$ simply by $H(P)$ and $L(P)$. Do not confuse $L_P(\alpha)$ with $L(P)$. The difference between them can be clearly seen from (1.9) and (1.10), or by considering $P_n(z) = z^n - 1$ and $\alpha_n = e^{2\pi i/n}$, in which case $L(P_n) = 2$ and $L_{P_n}(\alpha_n) = n$.

The function H_P was chosen because of property (1.12). This last property can be conveniently expressed as follows:

COROLLARY 1.3.1. *Suppose α is an algebraic number and $\mu_\alpha(z)$ is its minimal polynomial, $\mu_\alpha(z) = a(z-\alpha_1)\cdots(z-\alpha_n)$. Denote $H_{\mu_\alpha}(\alpha)$ by $\mathcal{H}(\alpha)$, i.e.*

(1.13) $$\mathcal{H}(\alpha) = \prod_{i=1}^{n} \max(1, |\alpha_i|).$$

Then $\mathcal{H}(\alpha)$ is the smallest number in the set $\{H_P(\alpha)\colon P(z) \in \mathbf{Z}[z]$ and $P(\alpha) = 0\}$.

PROOF. If $P(z) \in \mathbf{Z}[z]$ and $P(\alpha) = 0$, then, as is well known, $\mu_\alpha(z)$ divides $P(z)$ and Corollary 1.3.1 follows from (1.12).

Clearly, $\mathcal{H}(\alpha)$ calls to mind the true height. Properly speaking, $\mathcal{H}(\alpha)$ differs by only a factor (depending on $a = a(\alpha)$) from the height $\mathcal{H}_K(\alpha)$ described in detail by S. Lang [**4**].

It would be preferable not to use the height $\mathcal{H}_K$, since this would unnecessarily complicate the exposition.

In view of the definition of $\mathcal{H}(\alpha)$ and (1.8), we have

(1.14) $$\mathcal{H}(\alpha)\cdot a(\alpha) \leqslant L(\alpha) \leqslant (\deg(\alpha)+1)H(\alpha),$$

where $a(\alpha)$ is the leading coefficient of $\mu_a(z)$ (the so-called *denominator* of α) and $L(\alpha)$ is the length of α, i.e. the length of $\mu_\alpha(z)$.

It would be completely trivial to estimate $\mathcal{H}(\alpha)$ from below in terms of $H(\alpha)$ by means of Viète's theorem. We now explain why the height $H(\alpha)$ is unsuitable. This is suggested by the following. Corollary 1.3.1 for $H(\alpha)$ instead of $\mathcal{H}(\alpha)$ is not true in general. In other words, if for an algebraic number α we have $P(\alpha) = 0$, $P(z) \in \mathbf{Z}[z]$, and estimate $H(P) \leqslant H$, then it does not follow that $H(\alpha) \leqslant H$ as happens when $H_P(\alpha) \leqslant H$ for $\mathcal{H}(\alpha)$. Very interesting examples of such "bad" behavior of the height $H(P)$ can be found in Gel'fond [**1, 7**]. A completely trivial example is the case $P(z) = (z^2+z+1)^2$ and $R(z) = (z^3-1)^2$, where $P(z)$ divides $R(z)$, but $H(P) = 3$ and $H(R) = 2$, or, for example, $\alpha_0 = (-3+\sqrt{5})/2$, with $\mu_{\alpha_0}(z) = z^2+3z+1$ and $H(\alpha_0) = 3$. However, $P_0(\alpha_0) = 0$, where $P_0(z) = (z^2+3z+1)(z-1)$, although $H(P_0) = 2 < H(\alpha_0)$. These examples are not spectacular and were chosen at random, but they do show why $H(\alpha)$ is unsuitable. The same examples suggest that in passing from $H(P)$ to $H(R)$, where $R(z)$ is divisible by $P(z)$, we cannot essentially increase the height. That an essential increase is impossible is corroborated by the following well-known lemmas of Gel′fond.

LEMMA 1.4. *Suppose $P_1(z_1,\ldots,z_s),\ldots,P_m(z_1,\ldots,z_s)$ are arbitrary polynomials in s variables with complex coefficients and with heights $H_1,\ldots,H_m$. If we denote the height of the polynomial $P(z_1,\ldots,z_s) = P_1(z_1,\ldots,z_s)\cdots P_m(z_1,\ldots,z_s)$ by H and the degrees of $P(z_1,\ldots,z_s)$ with respect to $z_1,\ldots,z_s$ by $n_1,\ldots,n_s$, we have*

$$H \geqslant e^{-n}H_1\cdots H_m, \qquad n = \sum_{i=1}^{s} n_i. \tag{1.15}$$

LEMMA 1.5. *If $P(z_1,\ldots,z_s)$ is a polynomial of degrees $n_1,\ldots,n_s$ with respect to $z_1,\ldots,z_s$ and of height H', and if $R(z_1,\ldots,z_s) = (P(z_1,\ldots,z_s))^m$ is of height H, then*

$$H \geqslant \prod_{i=1}^{s}(1+2mn_i)^{-1}H'^m. \tag{1.16}$$

Weaker variants of these lemmas have been proved earlier. Lemma 1.4 shows that for bounded n or for n small in comparison with H it is of no particular value to introduce $\mathcal{H}(\alpha)$; however, our scheme involves an investigation so detailed that without $\mathcal{H}$ we would not be able to manage. Let us illustrate the use of $\mathcal{H}$ by the example of a very frequently used lemma concerning the estimation from below of a polynomial in algebraic numbers.

LEMMA 1.6. *Suppose that α_i is an algebraic number of degree d_i, height H_i, and length $L(\alpha_i)$: $i = 1,\ldots,m$. Let d denote the degree of $\mathbf{Q}(\alpha_1,\ldots,\alpha_m)$ and let*

$$P(z_1,\ldots,z_m) = \sum_{i_1=0}^{N_1}\cdots\sum_{i_m=0}^{N_m} C_{i_1,\ldots,i_m} z_1^{i_1}\cdots z_m^{i_m}$$

be a polynomial in $\mathbf{Z}[z_1,\ldots,z_m]$ of height B, i.e. $B = \max_{i_1,\ldots,i_m}|C_{i_1,\ldots,i_m}|$. Then either $P(\alpha_1,\ldots,\alpha_m) = 0$, or else

(1.17)

$$|P(\alpha_1,\ldots,\alpha_m)| \geqslant \{(N_1+1)\cdots(N_m+1)B\}^{-d+1}\prod_{i=1}^{m}\{(d_i+1)H_i\}^{-N_id/d_i}$$

i.e. more precisely, in terms of lengths,

$$|P(\alpha_1,\ldots,\alpha_m)| \geqslant (L(P))^{-d+1}\prod_{i=1}^{m}\{L(\alpha_i)\}^{-N_id/d_i} \tag{1.18}$$

where $L(P)$ is the length of $P(z_1,\ldots,z_m)$ and $L(\alpha_i)$ is the length of α_i.

Actually, this lemma admits an improvement: the length $L(\alpha_i)$ can be replaced by the height $\mathcal{H}(\alpha_i)$.

LEMMA 1.7. *Suppose* α_i *is an algebraic number of degree* d_i *and 2-height* $\mathcal{H}_i = \mathcal{H}(\alpha_i)$: $i = 1,\ldots,m$, *and* $d = \deg(\mathbf{Q}(\alpha_1,\ldots,\alpha_m))$. *If* $P(z_1,\ldots,z_m)$ *is a polynomial in* $\mathbf{Z}[z_1,\ldots,z_m]$ *of height* $H(P)$, *length* $L(P)$, *and degrees* $N_1,\ldots,N_m$ *with respect to* $z_1,\ldots,z_m$, *then either*

$$|P(\alpha_1,\ldots,\alpha_m)| = 0,$$

or else

(1.19)

$$|P(\alpha_1,\ldots,\alpha_m)| \geqslant \{(N_1+1)\cdots(N_m+1)H(P)\}^{-d+1}\prod_{i=1}^{m}\{\mathcal{H}(\alpha_i)\}^{-N_i d/d_i}$$

that is to say,

$$|P(\alpha_1,\ldots,\alpha_m)| \geqslant \{L(P)\}^{-d+1}\prod_{i=1}^{m}\{\mathcal{H}(\alpha_i)\}^{-N_i d/d_i}. \tag{1.20}$$

PROOF. Since the proof of (1.20) (and therefore (1.19)) differs only slightly from that of Lemma 1.6, we give only an outline.

If $\alpha_i = \alpha_{i1},\ldots,\alpha_{id}$ are the conjugates of α_i in $\mathbf{K} = \mathbf{Q}(\alpha_1,\ldots,\alpha_m)$ and $a_i = a(\alpha_i)$ is the denominator of α_i, then the number

$$\prod_{i=1}^{m} a_i^{N_i d/d_i}\cdot\prod_{j=1}^{d}P(\alpha_{1j},\ldots,\alpha_{mj})$$

is a rational integer. In addition, obviously,

$$\prod_{j=2}^{d}|P(\alpha_{1j},\ldots,\alpha_{mj})| \leqslant (L(P))^{d-1}\prod_{i=1}^{m}\prod_{j=1}^{d}\left(\max\{1,|\alpha_{ij}|\}\right)^{N_i}. \tag{1.21}$$

By Lemma 1.1 and the definition of $\mathcal{H}$, $\mathcal{H}(\alpha_i) = a_i H_{\mu_{\alpha_i}}(\alpha_i)$ and

$$\prod_{j=2}^{d}|P(\alpha_{1j},\ldots,\alpha_{mj})| \leqslant \{L(P)\}^{d-1}\prod_{i=1}^{m}\{a_i^{-1}\mathcal{H}(\alpha_i)\}^{N_i d/d_i}. \tag{1.22}$$

From this we directly obtain (1.20) and (1.19).

Then (1.19) implies (1.17), and (1.20) implies (1.18), according to Lemma 1.3.

We now consider polynomials and their roots and resultants. We first present several lemmas which are essential for Gel′fond's method. Two of them are due to Gel′fond himself [**1**, 7], and one is in [**8**].

LEMMA 1.8. *Suppose* $P(z)$, $Q(z) \in \mathbf{Z}[z]$ *are nontrivial polynomials of degrees* $d(P)$, $d(Q)$ *and heights* $H(P)$, $H(Q)$. *Assume that for some* θ *we have*

$$\begin{aligned}(1.23)\quad &\max\{|P(\theta)|,|Q(\theta)|\}\\ &< \left\{H(P)^{d(Q)}H(Q)^{d(P)}\cdot(d(P)+2)^{d(Q)/2}\cdot(d(Q)+2)^{d(P)/2}\right\}^{-1}.\end{aligned}$$

Then $P(z)$ *and* $Q(z)$ *have a common root, i.e. a common nontrivial polynomial divisor.*

LEMMA 1.9. *Suppose θ is a complex number and $P(z) \in \mathbf{Z}[z]$ is a polynomial of degree $n > 1$ and height H. If*

$$|P(\theta)| < \exp\{-\lambda n(n + \ln H)\}$$

where $\lambda > 4$, then there exists an irreducible divisor $Q(z)$ of $P(z)$ such that

$$|Q(\theta)| \leqslant \exp\left\{-\frac{\lambda - 4}{S} n(n + \ln H)\right\}$$

where

$$\deg(Q(z)) \leqslant n/S \quad \textit{and} \quad \deg(Q(z)) + \ln H(Q) \leqslant 2(n + \ln H)/S.$$

The proof of Lemma 1.9 is effected by means of Lemma 1.8 and can be found in [**1**, **7**, **6**].

LEMMA 1.10. *Suppose $P(z) \in \mathbf{Z}[z]$ is a polynomial without multiple roots (say, irreducible), of degree n and height H, and let $\xi_1, \dots, \xi_n$ be its roots. Then for any number θ,*

(1.24)

$$\min_{i=1,\dots,n} |\theta - \xi_i| \cdot e^{-n^2} H^{-n} \leqslant |P(Q)|$$

$$\leqslant \min_{i=1,\dots,n} |\theta - \xi_i| \cdot n^2 H (H + 2)^{n-1} \max\{1, |\theta|^n\}.$$

The proof of Lemma 1.10 is given in [**6**]; it makes use of properties of the discriminant $D = D(P)$ of the polynomial $P(z)$.

Lemma 1.10 is unsatisfactory for two reasons. First, it applies only to irreducible polynomials. It is true that Lemma 1.9 already contains the means to avoid this obstacle. But the estimate in Lemma 1.9 is of the form $H^{-\lambda^* n}$ when $\ln H \geqslant n$. Secondly, in Lemma 1.9 and (1.24) there are factors of the form e^{-cn^2}, which is at once unsatisfactory: it is too far from the natural upper bound of Dirichlet's theorem. Since the upper bound in (1.24) is entirely satisfactory, we want to find a good lower bound. We first consider irreducible polynomials.

LEMMA 1.11. *Suppose $P(z) \in \mathbf{Z}[z]$ is an irreducible polynomial of degree n, height $H = H(P)$, length $L = L(P)$, and 2-height $\mathcal{H}_P$, and let $\xi_1, \dots, \xi_n$ be its roots. Then for any θ,*

$$\min_{i=1,\dots,n} |\theta - \xi_i| \cdot 2^{-n+1} \cdot n^{-n+1} L(P)^{-2n+2} \leqslant |P(\theta)| \tag{1.25}$$

or

$$\min_{i=1,\dots,n} |\theta - \xi_i| \cdot n^{-3n} \cdot H^{-2n} \cdot 2^{-3n} \leqslant |P(\theta)|. \tag{1.26}$$

PROOF. Let $\min_{i=1,\dots,n} |\theta - \xi_i| = |\theta - \xi_{i_0}|$. As is well known, the discriminant $D = D(P)$ of the polynomial $P(z)$ can be written in the form

$$D = a^{2n-1} \prod_{i \neq j} (\xi_i - \xi_j) = a^{n-1} \prod_{i=1}^{n} P'(\xi_i),$$

where a is the leading coefficient of $P(z)$ and $P'(z)$ is the derivative. Since D is a rational integer, then $|D| \geq 1$. Also,

$$(1.27)\quad |a|^{n-1}\left|\prod_{i=1, i\neq i_0}^{n} P'(\xi_i)\right| \leq |a^{n-1}| \cdot \prod_{i=1, i\neq i_0}^{n} n \cdot L(P)\max\{1, |\xi_i|^{n-1}\}$$
$$= |a|^{n-1} n^{n-1} L(P)^{n-1} \cdot \prod_{i=1, i\neq i_0}^{n} (\max\{1, |\xi_i|\})^{n-1}.$$

By definition of H_P (see Definition 1.2),

$$H_P \geq \prod_{i=1, i\neq i_0}^{n} \max\{1, |\xi_i|\}.$$

Furthermore, by Lemma 1.1,

$$(1.28)\qquad aH_P \leq L(P).$$

Thus, in view of (1.27),

$$\left|a^{n-1} \prod_{i=1, i\neq i_0}^{n} P'(\xi_i)\right| \leq (|a|\, n\, L(P) H_P)^{n-1}$$

or, in view of (1.28),

$$(1.29)\qquad \left|a^{n-1} \prod_{i=1, i\neq i_0}^{n} P'(\xi_i)\right| \leq \left(n(L(P))^2\right)^{n-1}.$$

Let us now estimate $P'(\xi_{i_0})$ from above. We have for $i = 1, \ldots, n$, $P'(\xi_{i_0}) = a\prod_{i=1, i\neq i_0}^{n}(\xi_{i_0} - \xi_i)$. Thus, for $i \neq i_0$,

$$|\xi_{i_0} - \xi_i| \leq |\theta - \xi_{i_0}| + |\theta - \xi_i| \leq 2|\theta - \xi_i|,$$

by definition of i_0, and

$$(1.30)\qquad \left|P'(\xi_{i_0})\right| \leq a \prod_{i=1, i\neq i_0}^{n} 2|\theta - \xi_i|$$
$$= 2^{n-1} a \prod_{i=1, i\neq i_0}^{n} |\theta - \xi_i| = 2^{n-1} \frac{|P(\theta)|}{|\theta - \xi_{i_0}|}.$$

Here, of course, we assume that $\theta \neq \xi_{i_0}$, otherwise Lemma 1.11 is trivial. Taking into account (1.29), (1.30), and the fact that $|D| \geq 1$, we obtain

$$|\theta - \xi_{i_0}| \leq 2^{n-1}|P(\theta)| n^{n-1}(L(P))^{2n-2}$$

and Lemma 1.11 is proved.

Inequality (1.25)–(1.26) can be rewritten in terms of $L_P = (n+1)H$:

$$(1.31)\qquad \min_{i=1,\ldots,n} |\theta - \xi_i| \cdot L_P^{-4n} \leq |P(\theta)|.$$

Let us not be confused by the fact that in (1.26), in contrast to Lemma 1.10, H^{-2n} occurs instead of H^{-n}. This is inessential. The main thing is that instead of

e^{-3n^2} we have $e^{-3n\ln n}$. This is close to the estimate obtained by means of Dirichlet's principle.

Not having at present an improved variant of Lemma 1.9, we consider polynomials with roots of arbitrary multiplicity. Instead of $|\theta - \xi_i|$ there will now occur $|\theta - \xi_i|^{s_i}$, where s_i is the multiplicity of ξ_i in $P(z)$. To prove the corresponding assertion it is necessary to introduce the concept of the so-called semidiscriminant of a polynomial.

Suppose $P(z) \in \mathbf{Z}[z]$ is of degree n, $P(z) = a(z - t_1)^{s_1} \cdots (z - t_k)^{s_k}$, where $t_1, \dots, t_k$ are distinct and $\sum_{i=1}^k s_i = n$. The natural form of the semidiscriminant of $P(z)$ is

$$SD(P) = a^{n-1} \prod_{i=1}^{k} \left(P^{(s_i)}(t_i)/s_i!\right). \tag{1.32}$$

It can be verified directly that $SD(P)$ is a nonzero rational integer.

We have (say, by Taylor's formula)

$$\frac{P^{(s_i)}(t_i)}{s_i!} = a \prod_{j=1, j\neq i}^{k} (t_i - t_j)^{s_j}. \tag{1.33}$$

Or, if $P = P_1^{\tau_1} \cdots P_l^{\tau_l}$ is a factorization into irreducibles and t_i is a root of the polynomial $P_{f(i)}(z)$, we obtain

$$\frac{P^{(s_i)}(t_i)}{s_i!} = \left(P'_{f(i)}(t_i)\right)^{\tau_{f(i)}} \prod_{m=1, m\neq f(i)}^{l} \left(P_m(t_i)\right)^{\tau_m}.$$

Now $\{t_1, \dots, t_k\} = \bigcup_{m=1}^l T_m$, where T_m consists of the roots of $P_m(z)$. Thus,

$$\begin{aligned} \prod_{i=1}^{k} \frac{\left(P^{(s_i)}(t_i)\right)}{s_i!} &= \prod_{m=1}^{l} \prod_{t_i \in T_m} \frac{\left(P^{(s_i)}(t_i)\right)}{s_i!} \\ &= \prod_{m=1}^{l} \prod_{P_m(t_i)=0} \left\{ (P'_m(t_i))^{\tau_m} \prod_{m'\neq m} (P_{m'}(t_i))^{\tau_{m'}} \right\} \\ &= \prod_{m=1}^{l} \left\{ D(P_m)^{\tau_m} \prod_{m'\neq m} (\operatorname{res}(P_m, P_{m'}))^{\tau_{m'}} \right\} \times A. \end{aligned} \tag{1.34}$$

Let a_m denote the leading coefficient of $P_m(z)$. Then $a = a_1^{\tau_1} \cdots a_l^{\tau_l}$. We will determine A (expressed in terms of the a_m) so as to involve only a. We have

$$A = \prod_{m=1}^{l} a_m^{-\tau_m(\deg(P_m)-1)} \cdot \prod_{m=1}^{l} \left\{ \prod_{m'\neq m} a_m^{-\tau_m/\deg(P_{m'})} \right\}.$$

We find the power of a_m ($m = 1, \dots, l$) occurring in A:

$$A = a_1^{-\gamma_1} \cdots a_l^{-\gamma_l}$$

and

$$a_m^{\gamma_m} = a_m^{\tau_m(\deg(P_m)-1)} \cdot \prod_{m'\neq m} a_m^{\tau_{m'}\deg(P_{m'})} = a_m^{\sum_{k=1}^l \tau_k \deg(P_k) - \tau_m}.$$

It is known that $\deg(P(z)) = n = \sum_{k=1}^{l} \tau_k \deg(P_k)$, hence

$$A = a \prod_{m=1}^{l} a_m^{-n}.$$

Since $\prod_{m=1}^{l} a_m^n$ divides $\prod_{m=1}^{l} a_m^{\tau_m n} = a^n$, it follows from (1.32) and (1.34) that

$$SD(P) = a^{n-1} \prod_{i=1}^{k} \frac{P^{(s_i)}(t_i)}{s_i!}$$

is a nonzero rational integer.

We now estimate the separate components of $SD(P)$. Consider in the definition of $SD(P)$ all factors except the one corresponding to $i = i_0$. We obtain

$$SD_{i_0}(P) = a^{n-1} \prod_{i=1, i\neq i_0}^{k} \frac{P^{(s_i)}(t_i)}{s_i!}.$$

But we have

$$\left| \frac{P^{(s_i)}(t_i)}{s_i!} \right| \leqslant (\max\{1, |t_i|\})^{n-s_i} (H(P)) \cdot (C_n^{s_i} + \cdots + 1).$$

Therefore,

$$|SD_{i_0}(P)| \leqslant |a|^{n-1} H(P)^{k-1} \prod_{i\neq i_0} (C_n^{s_i} + \cdots + 1) \prod_{\substack{i=1 \\ i\neq i_0}}^{k} \max\{1, |t_i|\}^{n-s_i}.$$

By Lemma 1.1,

$$|a| \cdot \prod_{i=1}^{k} \max\{1, |t_i|\}^{s_i} \leqslant L(P).$$

We obtain $(n-1)s_i = ns_i - s_i \geqslant n - s_i$ for $s_i \geqslant 1$. Therefore,

$$\begin{aligned} |SD_{i_0}(P)| &\leqslant H(P)^{k-1} L(P)^{n-1} \prod_{i\neq i_0} (C_n^{s_i} + C_{n-1}^{s_i} + \cdots + 1) \\ &\leqslant n^{n-1} H(P)^{n-1} L(P)^{n-1} \prod_{i\neq i_0} C_n^{s_i}. \end{aligned}$$

But $C_n^{s_i} \leqslant n^{s_i}$ and $\sum_{i\neq i_0} s_i \leqslant n-1$. Consequently,

$$|SD_{i_0}(P)| \leqslant n^{2(n-1)} H(P)^{n-1} L(P)^{n-1}. \tag{1.35}$$

Let us estimate the second factor in $SD(P)$:

$$\frac{P^{(s_{i_0})}(t_{i_0})}{s_{i_0}!}.$$

According to (1.33),

$$\frac{P^{(s_{i_0})}(t_{i_0})}{s_{i_0}!} = a \prod_{j\neq i_0} (t_{i_0} - t_j)^{s_j}.$$

Now suppose θ is a complex number and $|\theta - t_{i_0}| = \min_{i=1,\dots,n}|\theta - t_i|$. Then $|t_{i_0} - t_i| \leq 2|\theta - t_i|$. Therefore,

$$\left|\frac{P^{(s_{i_0})}(t_{i_0})}{s_{i_0}!}\right| \leq 2^{n-1}a\prod_{j\neq i_0}|\theta - t_j|^{s_j}. \tag{1.36}$$

As a consequence of (1.35) and (1.36) we obtain

LEMMA 1.12. *Suppose $P(z)\in\mathbf{Z}[z]$ is a polynomial of degree n, height $H(P)$, and length $L(P)$, and let $t_1,\dots,t_k$ be its distinct roots of multiplicities $s_1,\dots,s_k$; $\sum_{i=1}^k s_i = n$. If θ is any number, then for $|\theta - t_{i_0}| = \min_{i=1,\dots,k}|\theta - t_i|$ we have*

$$|\theta - t_{i_0}|^{s_{i_0}}2^{-n+1}n^{-2(n-1)}H(P)^{-n+1}L(P)^{-n+1} \leq |P(\theta)| \leq |\theta - t_{i_0}|^{s_{i_0}}\cdot 2^nL(P)\max\{1,|\theta|^n\}. \tag{1.37}$$

PROOF. If θ is a root of $P(z)$, then (1.37) holds. If $\theta \neq t_{i_0}$, then, in view of (1.36) and (1.35), we have

$$1 \leq |SD(P)| \leq 2^{n-1}n^{2(n-1)}H(P)^{n-1}L(P)^{n-1}\frac{|P(\theta)|}{|\theta - t_{i_0}|^{s_{i_0}}}, \tag{1.38}$$

which establishes the left-hand inequality in (1.37). To prove the other part of (1.37) we use Taylor's formula:

$$P(\theta) = \frac{P^{(s_{i_0})}(\sigma)}{s_{i_0}!}(\theta - t_{i_0})^{s_{i_0}}. \tag{1.39}$$

Consider two polynomials R_1 and R_2 with known heights and two corresponding divisors P_1 and P_2 such that $|P_1(\theta)|$ and $|P_2(\theta)|$ are small. Suppose P_1 and P_2 are coprime. What lower bound can we obtain for $\max\{|P_1(\theta)|, |P_2(\theta)|\}$? If we use Gel'fond's Lemmas 1.9 and 1.4–1.5, then for $\ln H \geq n$ we obtain a very good estimate. Otherwise, because of the factor e^n attached to $H(P_i)$, we obtain an unsatisfactory estimate. How can we improve it? It is necessary to do the following: instead of $H(P_i)$ we use H_{P_i}, since we know H_{R_i} and, by Lemma 1.3, $H_{P_i} \leq H_{R_i}$.

We now present several lemmas which are completely sufficient for our purposes, but which are not the best possible results.

LEMMA 1.13. *Suppose that $P(x), Q(x)\in\mathbf{Z}[x]$ and $P(x)$ is irreducible. Denote by $H(P)$ the height of $P(x)$ and by $H(T)$ and $d(T)$ the height and degree of the polynomial $T(x)$, where $T(x) = (P(x))^fQ(x)$. If $P(x)$ does not divide $Q(x)$, then a lower bound for* $\max\{|P(\theta)|, |Q(\theta)|\}$ *for any θ is given by*

$$\max\{|P(\theta)|, |Q(\theta)|\} > L(P)^{-100d(T)}\cdot L(T)^{-100d(P)}\cdot d(T)^{-100d(T)}. \tag{1.40}$$

PROOF. Since $P(x)$ is irreducible, then $P(x)$ and $Q(x)$ are coprime, i.e. the resultant $\operatorname{Res}(P,Q)$ of $P(x)$ and $Q(x)$ is nonzero. Assume that (1.40) does not hold. We will obtain an upper bound for a quantity closely connected with $\operatorname{Res}(P,Q)$.

We have $T = Q \cdot P^f$, hence for any root α of the polynomial $P(z)$ we obtain

$$(1.41) \qquad T^{(f)}(\alpha) = f!Q(\alpha) \cdot (P'(\alpha))^f.$$

Now for any root α of $P(z)$ we have $Q(\alpha) \neq 0$ and $P'(\alpha) \neq 0$, since $P(x)$ is irreducible. Therefore, if q and p are the leading coefficients of $Q(x)$ and $P(x)$, it follows that

$$(1.42) \qquad p^{d(Q)} \cdot p^{f(d(P)-1)} \prod_{P(\alpha)=0} \frac{T^{(f)}(\alpha)}{f!}$$

is a nonzero number. Here the product extends over all roots of $P(z)$. It follows from (1.41) that (1.42) is the product of two rational integers: $\mathrm{Res}(P, Q)$ and $(D(P))^f$. Therefore, the number in (1.42) is a rational integer, of modulus $\geqslant 1$. In particular,

$$(1.43) \qquad \left| p^{d(T)-f} \prod_{P(\alpha)=0} \frac{T^{(f)}(\alpha)}{f!} \right| \geqslant 1$$

since $d(T) - f = d(Q) + fd(P) - f = d(Q) + f(d(P) - 1)$.

Since (1.40) does not hold, we can use Lemma 1.11. Let $\alpha_1, \ldots, \alpha_{d(P)}$ be all of the roots of $P(z)$, and let $|\theta - \alpha_{i_0}| = \min\{|\theta - \alpha_i| : i = 1, \ldots, d(P)\}$. Then, according to (1.25),

$$|\theta - \alpha_{i_0}| \leqslant |P(\theta)| \cdot (2n)^{n-1} L(P)^{2n-2}, \qquad n = d(P).$$

Since (1.40) does not hold, we obtain

$$(1.44) \qquad |\theta - \alpha_{i_0}| \leqslant L(P)^{-80d(T)} \cdot L(T)^{-80d(P)} \cdot d(T)^{-80d(T)}.$$

We will estimate from above the quantity appearing on the left in (1.43). As usual, we first estimate one factor

$$A = \left| p^{d(T)-f} \cdot \prod_{i=1, i\neq i_0}^{d(P)} \frac{T^{(f)}(\alpha_i)}{f!} \right|.$$

By Lemma 1.1, we have

$$(1.45) \qquad \begin{aligned} A &\leqslant |p|^{d(T)-f} \cdot \prod_{i=1, i\neq i_0}^{d(P)} \max\{1, |\alpha_i|\}^{d(T)-f} \\ &\quad \times H(T)^{d(P)-1} \left(C_{d(T)}^f + \cdots + 1\right)^{d(P)-1} \\ &\leqslant H(T)^{d(P)-1} d(T)^{(f+1)(d(P)-1)} \cdot L(P)^{d(T)-f}. \end{aligned}$$

We will simplify (1.45) later. In the meantime, let us estimate the second factor in (1.43), taking into account that (1.40) does not hold; the second factor is

$$B = \frac{T^{(f)}(\alpha_{i_0})}{f!}.$$

It follows from Lemma 1.4 that $H(P)^f \cdot H(Q) \leqslant e^{d(T)} \cdot H(T)$. It is easy to see that

$$(1.46) \qquad H(P')^f \cdot H(Q) \leqslant d(P)^f \cdot e^{d(T)} \cdot H(T).$$

We next consider that $|x^k - y^k| \leqslant |x - y| \cdot k \cdot (\max(|x|, |y|))^{k-1}$. Therefore,

$$|Q(\alpha_{i_0}) - Q(\theta)| \leqslant |\alpha_{i_0} - \theta| \cdot H(Q) \cdot d(Q)(|\theta| + 1)^{d(Q)}.$$

In view of (1.44) and the negation of (1.40), we obtain

$$(1.47) \quad |Q(\alpha_{i_0})| \leqslant H(Q)d(Q)(|\theta| + 1)^{d(Q)} \times (ed(P))^{d(P)} \cdot L(P)^{-80d(T)} \cdot L(T)^{-80d(P)} \cdot d(T)^{-80d(T)}.$$

Using (1.41), (1.46), and (1.47), we have

$$(1.48) \qquad |B| \leqslant (|\theta| + 1)^{d(Q)+f(d(P)-1)} \cdot d(P)^f e^{d(T)} \cdot H(T) \cdot d(Q) \times L(P)^{-10d(T)} \cdot L(T)^{-10d(P)} \cdot d(T)^{-10d(T)}.$$

Finally, in view of (1.45) and (1.48), we obtain the upper bound:

$$\left| p^{d(T)-f} \prod_{P(\alpha)=0} \frac{T^{(f)}(\alpha)}{f!} \right| \leqslant H(T)^{d(P)} \cdot (L(P))^{d(T)-f} \cdot d(T)^{(f+1)(d(P)-1)} \times d(P)^f d(Q)(|\theta| + 1)^{d(T)} e^{d(T)} (ed(P))^{d(P)} \cdot L(P)^{2d(P)} \leqslant \cdots \leqslant L(P)^{-d(T)} \cdot L(T)^{-d(P)} \cdot d(T)^{-d(T)}.$$

The last number is < 1, which contradicts (1.43), and Lemma 1.13 is proved.

2. The auxiliary results proved above enable us to investigate the transcendence measure of certain numbers. Pay particular attention to $\deg(P) = d$. We succeed in our estimates only because in the auxiliary lemmas d occurs in the best way. We will first indicate what type of estimate is most desirable for our purposes. Let $\mathrm{Pol}(d, H) = \{P(x) \in \mathbf{Z}[x]\colon P(x)$ is irreducible, $\deg(P) \leqslant d$, $H(P) \leqslant H$, $P(x) \not\equiv 0\}$. For a transcendental number α we are interested in the quantity

$$(2.1) \qquad m(\alpha, d, H) = \sup\{-\ln|P(\alpha)| : P(x) \in \mathrm{Pol}(d, H)\}.$$

Dirichlet's principle yields the estimate $m(\alpha, d, H) \geqslant d \ln H - c(\alpha) \cdot d$. Moreover, in view of §1, for almost all α (i.e. for a set of $\alpha \in \mathbf{C}^1$ of measure 1) we have $m(\alpha, d, H) \leqslant c_1 d \ln(dH)$, where c_1 is some absolute constant. It is not clear whether we can show for almost all α that $m(\alpha, d, H) \leqslant c_2 d \ln H$.

We are mainly interested in estimating $m(\alpha, d, H)$ in the case where H is bounded, since estimates in the case where d is bounded have already been rather thoroughly studied. There is another reason why the case where H is bounded is interesting. In this case, for transcendental numbers α, only Gel'fond's method is effective; the method of Siegel or Mahler (for $e, \pi, \ln\beta\colon \beta \in \mathbf{A}$) gives a completely unsatisfactory estimate of the type $m(\alpha, d, O(1)) \leqslant e^{O(d)}$. At the same time, Gel'fond's method furnishes an estimate in which d occurs in a polynomial. It is the aim of the present section to improve this estimate.

An estimate of $|P(e)|$ for $P(x) \in \mathbf{Z}[x]$ with bounded height was first obtained by N. I. Fel′dman **[9]**; this estimate, however, can be improved.

For simplicity, the method of proof of the main theorems of this section will be illustrated by the example of the number e.

We first give several lemmas of an analytic nature which are usually used in Gel′fond's method.

LEMMA 2.1. *Suppose F is an entire function, and the natural numbers P, T and the numbers $A, R \in \mathbf{R}$ are such that $R \geqslant 2P$ and $A > 2$. Put*

$$M_r = \max_{|z| \leqslant r} |F(z)| \qquad (r > 0)$$

and

$$E = \max_{\substack{t=0,\ldots,T-1 \\ p=0,\ldots,P-1}} |F^{(t)}(p)| \cdot \frac{1}{t!}.$$

Then

$$M_R \leqslant 2M_{AR}(2/A)^{PT} + (9R/P)^{PT}E. \tag{2.2}$$

This lemma, a variant of the well-known Schwarz lemma, is proved in **[6]**.

The following auxiliary result is due to Tijdeman **[10]**.

LEMMA 2.2. *Suppose*

$$F(z) = \sum_{k=0}^{K-1} \sum_{m=0}^{M-1} C_{k,m} z^k e^{\omega_k z}$$

where the $C_{k,m} \in \mathbf{C}^1$ and the $\omega_k \in \mathbf{C}^1$ are such that $\omega_k \neq \omega_l$ for $k \neq l$. Put

$$\Omega = \max\Big(1, \max_{k=0,\ldots,K-1} |\omega_k|\Big)$$

and

$$\omega = \min\Big(1, \min_{k,l=0,\ldots,K-1;\, k\neq l} |\omega_k - \omega_l|\Big).$$

Also, for natural numbers T', P' put

$$E = \max_{\substack{t=0,\ldots,T'-1 \\ p=0,\ldots,P'-1}} |F^{(t)}(p)|.$$

If

$$T'P' \geqslant 2KM + 13\Omega P', \tag{2.3}$$

then

$$|C_{k,m}| \leqslant P' \left\{ \frac{6}{\sqrt{K}} \cdot \frac{\Omega}{\omega} \max\left\{6, \frac{KM}{\max(1, P'-1)}\right\}\right\}^{KM} \cdot 72^{T'P'} \cdot E \tag{2.4}$$

for $k = 0,\ldots,K-1$ and $m = 0,\ldots,M-1$.

If, in addition, $\omega_k = k\theta$, *where* $\theta \in \mathbf{C}^1$ $(k = 0, 1, \ldots, K-1)$, *then* (2.4) *can be replaced by*

$$(2.5) \qquad |C_{k,m}| \leqslant P'\left\{\frac{6}{K} \cdot \frac{\Omega}{\omega} \max\left\{6, \frac{KM}{\max(1, P'-1)}\right\}\right\}^{KM} \cdot 72^{T'P'} \cdot E.$$

The proof uses Dirichlet's principle:

LEMMA 2.3. *Any set of r linear forms in s variables*

$$\sum_{\sigma=1}^{s} a_{\rho,\sigma} x_\sigma: \qquad \rho = 1, \ldots, r,$$

$$s > 2r, \quad a_{\rho,\sigma} \in \mathbf{C}^1, \quad |a_{\rho,\sigma}| \leqslant A: \qquad \rho = 1, \ldots, r; \sigma = 1, \ldots, s,$$

has the following property. For any natural number C there exist rational integers $C_1, \ldots, C_s$ *not all zero such that* $|C_\sigma| \leqslant C$: $\sigma = 1, \ldots, s$, *and*

$$\left|\sum_{\sigma=1}^{s} a_{\rho,\sigma} C_\sigma\right| \leqslant \sqrt{2} \cdot s \cdot A \cdot C^{1-s/2r}: \qquad \rho = 1, \ldots, r.$$

We now give an estimate of the transcendence measure of e.

THEOREM 2.4. *Suppose* $P(z) \in \mathbf{Z}[z]$ *is an irreducible polynomial,* $P(z) \not\equiv 0$, *of degree* $\leqslant d$ *and height* $\leqslant H$. *Then there exists an absolute constant* $c_0 > 0$ *such that*

$$(2.6) \qquad |P(e)| > \exp\left(-c_0 d^2 \ln(Hd) \ln^2 d\right).$$

PROOF. In the course of the proof, the quantities $c_i > 0$ denote effectively computable absolute constants.

Assume that

$$(2.7) \qquad |P(e)| < \exp\left(-c_0 d^2 \ln(Hd) \ln^2 d\right)$$

and $d + \ln H \geqslant c_1$, where c_0, $c_1 > 0$ are sufficiently large constants.

We choose the following positive integers:

$$K = [d \ln d], \quad M = \left[\frac{d \cdot \ln(Hd) \ln d}{\ln \ln(Hd)}\right], \quad C = \left[\exp\{c_2 d \ln(Hd) \ln^2 d\}\right]$$

$$P = [c_3 d \ln d], \quad T = \left[c_4 \frac{d \ln(Hd) \ln d}{\ln \ln(Hd)}\right], \qquad T' = \left[c_5 \frac{d \ln(Hd) \ln d}{\ln \ln(Hd)}\right].$$

Consider the following auxiliary function:

$$(2.8) \qquad F(z) = \sum_{k=0}^{K-1} \sum_{m=0}^{M-1} C_{k,m} z^m e^{kz}.$$

We will indicate how to choose the numbers $C_{k,m} \in \mathbf{C}^1$. As in the usual method of Gel′fond, we seek them in the form of a polynomial in e:

$$(2.9) \qquad C_{k,m} = \sum_{l=0}^{L-1} C_{k,m,l} \cdot e^l, \qquad L = d^2 \ln^2 d.$$

The same symbol $C_{k,m}(z)$ will denote the polynomial in z obtained by replacing e in $C_{k,m}$ by z. The numbers $C_{k,m,l}$ are rational integers with $|C_{k,m,l}| \leqslant C$ for $k = 0,\dots,K-1$; $m = 0,\dots,M-1$; $l = 0,\dots,L-1$. We choose these numbers below with the aid of Dirichlet's principle.

For $t = 0, 1,\dots$ and $p = 0, 1,\dots$ we have

$$F^{(t)}(p) = \sum_{k=0}^{K-1} \sum_{m=0}^{M-1} \sum_{l=0}^{L-1} C_{k,m,l}e^{l} \times \sum_{\tau=0}^{m} \binom{t}{\tau} \frac{m!}{(m-\tau)!} p^{m-\tau}k^{t-\tau} \cdot e^{kp}. \tag{2.10}$$

We choose rational integers $C_{k,m,l}$ by means of Lemma 2.3 so that $F^{(t)}(p) = 0$ for $t = 0, 1,\dots, T-1$ and $p = 0, 1,\dots, P-1$. For this we rewrite (2.10) in the form

$$F^{(t)}(p) = \sum_{k=0}^{K-1} \sum_{m=0}^{M-1} \sum_{l=0}^{L-1} C_{k,m,l}e^{l+kp} \cdot P_{k,m,t,p}$$

with rational integral $P_{k,m,t,p}$, where

$$|P_{k,m,t,p}| \leqslant M \cdot 2^{T} \cdot M!\, e^{M \ln P} \cdot e^{T \ln k} \leqslant \exp\{c_6 \cdot d \ln(Hd) \cdot \ln^2 d\}. \tag{2.11}$$

Applying Lemma 2.3 and taking into account that the number of equations needed to determine the $C_{k,m,l}$ is $TP(L+KP)$ and the number of unknowns is KML, we obtain the following.

(2.12) There exist $C_{k,m,l} \in \mathbf{Z}$: $k = 0,1,\dots,K-1$; $m = 0, 1,\dots, M-1$; $l = 0, 1,\dots, L-1$, not all zero, such that $F^{(t)}(p) = 0$ for all $t = 0, 1,\dots, T-1$; $p = 0, 1,\dots, P-1$.

The following remark concerning the polynomials $C_{k,m}(z)$ is important for what is to come.

Let $\tau \geqslant 0$ be the largest natural number such that $(P(z))^{\tau}$ divides all of the polynomials $C_{k,m}(z)$. Now $C_{k,m}(z) = (P(z))^{\tau} \cdot C'_{k,m}(z)$, where $C'_{k,m}(z)$ is a polynomial with rational integral coefficients, and at least one of the $C'_{k,m}(z)$ for some $k = 0, 1,\dots, K-1$; $m = 0, 1,\dots, M-1$, is coprime to $P(z)$.

Let

$$C'_{k,m}(z) = \sum_{l=0}^{L'-1} C'_{k,m,l}\, z^{l},$$

where $L' \leqslant L$: $k = 0, 1,\dots, K-1$; $m = 0, 1,\dots, M-1$. We take

$$F_1(z) = \sum_{k=0}^{K-1} \sum_{m=0}^{M-1} C'_{k,m} z^{m} e^{kz},$$

where

$$C'_{k,m} = \sum_{l=0}^{L'-1} C'_{k,m,l}e^{l}$$

is obtained from $C'_{k,m}(z)$ by replacing z by e. Since e is a transcendental number, we have

$$F^{(t)}(z) = (P(e))^{\tau} \cdot F_1^{(t)}(z). \tag{2.13}$$

In particular, it follows from (2.12) that

$$F_1^{(t)}(p) = 0: \qquad t = 0, 1, \ldots, T-1; \quad p = 0, 1, \ldots, P-1. \tag{2.14}$$

In the sequel we will need a rough estimate of $\max |C'_{k,m,l}|$. This estimate can be obtained from Lemma 1.4 and has the form

$$\begin{gathered} |C'_{k,m,l}| \leqslant \exp\big(L + c_7 d \ln(Hd) \ln^2 d\big), \\ k = 0, 1, \ldots, K-1; \quad m = 0, 1, \ldots, M-1; \quad l = 0, 1, \ldots, L'-1. \end{gathered} \tag{2.15}$$

For $|z| \leqslant 2P \ln(Hd)$ we have

$$|F(z)| < \exp\big(c_8 d^2 \ln(Hd) \ln^2 d\big) \tag{2.16}$$

and

$$|F_1(z)| < \exp\big(c_9 d^2 \ln(Hd) \ln^2 d\big). \tag{2.17}$$

Applying Lemma 2.1, we obtain

$$\max_{|z| \leqslant 2P} |F(z)| \leqslant \exp\big(-c_{10} d^2 \ln(Hd) \ln^2 d\big) \tag{2.18}$$

and

$$\max_{|z| \leqslant 2P} |F_1(z)| \leqslant \exp\big(-c_{10} d^2 \ln(Hd) \ln^2 d\big).$$

In particular, if c_3 is sufficiently large in comparison with c_4, we have

$$\begin{aligned} |F^{(t)}(p)| &= \left| \frac{t!}{2\pi i} \cdot \int_{|z|=2P} \frac{F(z)}{(z-p)^{t+1}} dz \right| \\ &\leqslant T'^{T'} \cdot 2P \cdot \max_{|z| \leqslant 2P} |F(z)| \leqslant \exp\big(-c_{11} d^2 \ln(Hd) \ln^2 d\big); \end{aligned} \tag{2.19}$$

for $t = 0, 1, \ldots, T'-1$; $p = 0, 1, \ldots, P-1$.

It is entirely obvious that, analogously,

$$|F_1^{(t)}(p)| \leqslant \exp\big(-c_n d^2 \ln(Hd) \ln^2 d\big). \tag{2.20}$$

We carry out the final stage of the proof in the spirit of [**1**]. Consider $F^{(t)}(p)$ and $F_1^{(t)}(p)$. They can be represented in the form

$$F^{(t)}(p) = \sum_{k=0}^{K-1} \sum_{m=0}^{M-1} C_{k,m}(e) \cdot e^{kp} \cdot P_{k,m,p,t}$$

and

$$F_1^{(t)}(p) = \sum_{k=0}^{K-1} \sum_{m=0}^{M-1} C'_{k,m}(e) \cdot e^{kp} \cdot P_{k,m,p,t}$$

with rational integral $P_{k,m,p,t} \neq 0$. Replacing e by z, we obtain a polynomial $F_{t,p}(z)$, where $F_{t,p}(e) = F^{(t)}(p)$, and a polynomial $F^1_{t,p}(z)$, where $F^1_{t,p}(e) = F_1^{(t)}(p)$. If $\tau = 0$, then at least one of the polynomials $C_{k,m}(z)$ is coprime to $P(z)$. If there exist $t = 0, 1, \ldots, T' - 1$ and $p = 0, 1, \ldots, P - 1$ such that $F_{t,p}(z)$ is coprime to $P(z)$, then, in view of (2.7), (2.19), and Lemma 1.8, we arrive at a contradiction. Here it is essential to use the fact that

$$(2.21)\quad H(F_{t,p}) \leqslant c_{12} d \ln(Hd) \ln^2 d, \qquad \deg(F_{t,p}) \leqslant c_{12} d^2 \ln^2 d: \quad t = 0, 1, \ldots, T' - 1; \quad p = 0, 1, \ldots, P - 1.$$

We may therefore assume that all of the polynomials $F_{t,p}(z)$ for t, p within the above-mentioned limits are divisible by $P(z)$. It then follows from (2.7) and (2.21) that

$$(2.22)\qquad |F^{(t)}(p)| < \exp(-(c_0 - 1) d^2 \ln(Hd) \cdot \ln^2 d).$$

Choosing $c_0 > 0$ to be large in comparison with $c_1, \ldots, c_{10}$ and applying Lemma 2.2, we obtain

$$(2.23)\qquad |C_{k,m}(e)| < \exp(-c_{13} d^2 \ln(Hd) \cdot \ln^2 d),$$

for all $k = 0, 1, \ldots, K - 1$; $m = 0, 1, \ldots, M - 1$, and for an arbitrarily large $c_{13} > 0$. But one of the polynomials $C_{k,m}(z)$ is coprime to $P(z)$ (since $\tau = 0$). For this polynomial and $P(z)$ we again apply Lemma 1.8, taking into account that $\deg(C_{k,m}) \leqslant L$, $H(C_{k,m}) \leqslant C$. Estimates (1.23) and (2.23) lead to a contradiction.

Thus, $\tau > 0$. We now work with the function $F_1(z)$. Consider first the case where one of the polynomials $F^1_{t,p}(z)$ with $t = 0, 1, \ldots, T' - 1$; $p = 0, 1, \ldots, P - 1$, is coprime to $P(z)$. According to (2.13),

$$F_{t,p}(z) = (P(z))^\tau \cdot F^1_{t,p}(z).$$

We now use Lemma 1.13 and estimate (1.40). Taking for $H(F_{t,p})$ and $\deg(F_{t,p})$ the values in (2.21) and considering that $P(x) \in \mathrm{Pol}(d, H)$, we see that for large $c_1, \ldots, c_{10} > 0$ estimates (1.40) and (2.20) lead to a contradiction.

Consider the second case, where all of the polynomials $F^1_{t,p}(z)$ are divisible by $P(z)$. Apply the rough estimate of the height of $F^1_{t,p}(z)$ given in (2.15):

$$(2.24)\quad H(F^1_{t,p}) \leqslant c_{14} d(\ln(Hd) + d) \ln^2 d, \qquad \deg(F^1_{t,p}) \leqslant c_{14} d^2 \ln^2 d; \quad t = 0, 1, \ldots, T' - 1; \quad p = 0, 1, \ldots, P - 1.$$

This in conjunction with Lemma 1.4 and (2.7) yields an estimate better than (2.20):

$$(2.25)\qquad |F^{(t)}(p)| < \exp(-c_{15} d^2 \ln(Hd) \ln^2 d),$$

and if c_0 is sufficiently large in comparison with $c_1, \ldots, c_{14}$, then c_{15} can be arbitrarily large in comparison with $c_1, \ldots, c_{14}$. Using Lemma 2.2, we obtain

$$(2.26)\qquad |C'_{k,m}| < \exp(-c_{16} d^2 \ln(Hd) \ln^2 d),$$

where c_{16} is arbitrarily large in comparison with $c_1, \ldots, c_{14}$. Consider $C_{k,m}(z) = C'_{k,m}(z) \cdot (P(z))^\tau$. In view of (2.7), (2.26), Lemma 1.13, and (1.40), we obtain that

$C'_{k,m}(z)$ is divisible by $P(z)$. Then $C_{k,m}(z) = (P(z))^{\tau+1}Q_{k,m}(z)$ for all $k = 0, 1, \ldots, K-1$; $m = 0, 1, \ldots, M-1$. However, we have assumed that τ is the smallest power to which $P(z)$ occurs in all $C_{k,m}(z)$. This contradiction refutes (2.7) for sufficiently large $c_0 > 0$. The theorem is proved.

Analogously, using 1.12 and (1.37), we can prove

THEOREM 2.5. *Suppose $P(x) \in \mathbf{Z}[x]$, $P(x) \not\equiv 0$, is a polynomial of degree $\leqslant d$ and height $\leqslant H$. Then there exists an absolute constant $c'_0 > 0$ such that*

$$|P(e)| > \exp\left(-c'_0 d^2 \ln(Hd) \ln^2 d\right). \tag{2.27}$$

Using Lemma 1.12, we also obtain

COROLLARY 2.6. *There exists an effectively computable constant $c''_0 > 0$ such that*

$$|e - \xi| > \exp\left(-c''_0 d^2 \ln(Hd) \ln^2 d\right) \tag{2.28}$$

for all algebraic numbers ξ of degree $\leqslant d$ and height $\leqslant H$.

The method employed in the proof of Theorem 2.4 can also be used absolutely in the same lines to improve estimates of the transcendence measure of other numbers. In this regard, the analytic part of the proof remains the same.

THEOREM 2.7. *There exists an absolute constant $c_1 > 0$ such that*

$$|e^\pi - \xi| > \exp\left(-c_1 d^2 \ln H (\ln((\ln H) d))^2\right)$$

for all $\xi \in \mathbf{A}$ of degree $\leqslant d$ and height $\leqslant H$, and

$$|P(e^\pi)| > \exp\left(-c_1 d^2 \ln H (\ln((\ln H) d))^2\right)$$

for all $P(x) \in \mathbf{Z}[x]$, $P(x) \not\equiv 0$, of degree $\leqslant d$ and height $\leqslant H$.

The proof makes use of the auxiliary function

$$F(z) = \sum_{k=0}^{K-1} \sum_{l=0}^{L-1} C_{k,l} e^{(ik+l)\pi z},$$

$$K = \left[c_2 d \ln H (\ln((\ln H) d))\right], \qquad L = \left[c_3 d \ln(d \ln H)\right],$$

and $C_{k,e}$ is chosen as a polynomial in e^π.

By another choice of auxiliary function with zeros at points of the form $a + bi$ we can even prove

THEOREM 2.8. *There exists an effectively computable constant $c'_1 > 0$ such that*

$$|P(e^\pi)| > \exp\left(-c'_1 \frac{d \ln^2(Hd) \ln(d \ln H)}{(\ln \ln(Hd))^2}\right)$$

for all $P(x) \in \mathbf{Z}[x]$, $P(x) \not\equiv 0$, of degree $\leqslant d$ and height $\leqslant H$, and

$$|e^\pi - \xi| > \exp\left(-c'_1 \frac{d \ln^2(Hd) \ln(d \ln H)}{(\ln \ln(Hd))^2}\right)$$

for all $\xi \in \mathbf{A}$ of degree $\leqslant d$ and height $\leqslant H$.

PROOF OF THEOREM 2.8. The proof is essentially the same as the proof of Theorem 2.4. The only difference is in the choice of the auxiliary function. We choose as an auxiliary function the following one:

$$F(z)=\sum_{\lambda_1=0}^{L-1}\sum_{\lambda_2=0}^{L-1}C_{\lambda_1,\lambda_2}e^{\pi(\lambda_1+\lambda_2 i)z}.$$

The coefficients C_{λ_1,λ_2}, being polynomials in e^π, are determined by Siegel's lemma from the following system of linear equations:

$$F^{(\sigma)}(z_1+z_2 i)=0:\qquad \sigma=0,\dots,S_0;\quad z_1,z_2=0,1,\dots,X_0.$$

The parameters L, X_0 and S_0 are determined in terms of d and H in the following way:

$$L=\left[d^{1/2}\frac{\ln^{1/2}(d\ln H)\ln(dH)}{(\ln\ln(dH))^{3/2}}\right],\qquad X_0=\left[c_1''d^{1/2}\frac{\ln^{1/2}(d\ln H)}{(\ln\ln(dH))^{1/2}}\right];$$

$$S_0=\left[c_1'''\frac{\ln^2(dH)}{(\ln\ln(dH))^2}\right].$$

From Theorem 2.8 we obtain the curious

COROLLARY 2.9. *If* $P(x)\in\mathbf{Z}[x]$, $P(x)\not\equiv 0$, *and* $P(x)$ *has bounded coefficients,* $H(P)\leqslant c$, *and degree* $\leqslant d$, *then*

$$|P(e^\pi)|>\exp(-\tilde{c}d\ln^3 d), \tag{2.29}$$

where the constant $\tilde{c}$ *depends, in general, on* c.

The lower bound (2.29) is, in fact, very close to a trivial upper bound. The bound (2.29) can be also proved for numbers of the form $e^{\pi\alpha}$, where $\alpha\in\mathbf{A}$, $\alpha\notin i\mathbf{Q}$.

We give two more examples in which the measure $m(\alpha, d, H)$ is determined in the best way with respect to d.

THEOREM 2.9. *Suppose* $\alpha\in\mathbf{A}$, $\alpha\neq 0$. *There exists a constant* $c_2(\alpha)>0$ *such that*

$$|e^\alpha-\xi|>\exp(-c_2(\alpha)d^2\ln(Hd)\ln^2 d)$$

for all $\xi\in\mathbf{A}$ *of degree* $\leqslant d$ *and height* $\leqslant H$, *and*

$$|P(e^\alpha)|>\exp(-c_2(\alpha)d^2\ln(Hd)\ln^2 d)$$

for all $P(x)\in\mathbf{Z}[x]$, $P(x)\not\equiv 0$, *such that* $H(P)\leqslant H$ *and* $\deg(P)\leqslant d$.

The proof of Theorem 2.9 is almost identical to the proof of Theorem 2.4, but with another auxiliary function of the form

$$F(z)=\sum_{k=0}^{K-1}\sum_{m=0}^{M-1}C_{k,m}z^m e^{\alpha k z},$$

$$K=\left[c_3(\alpha)d\ln d\right],\qquad M=\left[c_4(\alpha)\cdot\frac{d\ln(Hd)\ln d}{\ln\ln(Hd)}\right].$$

THEOREM 2.10. *Suppose* $\alpha, \beta \in \mathbf{A}$, $\alpha \neq 0, 1$, $\beta \notin \mathbf{Q}$ *and* $\ln \alpha$ *is a fixed value of the logarithm of* α. *Then there exists an effectively computable constant* $c_5(\alpha, \beta) > 0$ *such that*

$$|\alpha^\beta - \xi| > \exp\left(-c_5 d^3 \ln H(1+\ln d)^{-3}(\ln(\ln Hd))^2\right)$$

for all $\xi \in \mathbf{A}$ *of degree* $\leqslant d$ *and height* $\leqslant H$, *and*

$$|P(\alpha^\beta)| > \exp\left(-c_5 d^3 \ln H(1+\ln d)^{-3}(\ln(\ln Hd))^2\right)$$

for all $P(x) \in \mathbf{Z}[x]$, $P(x) \not\equiv 0$, *such that* $H(P) \leqslant H$ *and* $\deg(P) \leqslant d$.

3. The symbols M, N, and n denote arbitrary natural numbers ($\geqslant 1$), and $\alpha_1,\dots,\alpha_N$ and $\beta_1,\dots,\beta_M$ are any sequences of complex numbers that are linearly independent over $\mathbf{Q}$. We are mainly interested in studying algebraically independent numbers in the set $\{\alpha_i, \beta_j, e^{\alpha_i\beta_j}: 1 \leqslant i \leqslant N,\ 1 \leqslant j \leqslant M\}$. As in the papers of Gel′fond, Lang, and others, we distinguish three subsystems of this set.

(i) $e^{\alpha_i\beta_j}$: $1 \leqslant i \leqslant N$, $1 \leqslant j \leqslant M$,

(ii) $\beta_1,\dots,\beta_M$, $e^{\alpha_i\beta_j}$: $1 \leqslant i \leqslant N$, $1 \leqslant j \leqslant M$,

and, finally,

(iii) $\alpha_1,\dots,\alpha_N$, $\beta_1,\dots,\beta_M$, $e^{\alpha_i\beta_j}$: $1 \leqslant i \leqslant N$, $1 \leqslant j \leqslant M$.

We denote these sets by S_1, S_2, and S_3, respectively.

Let us turn to one of the main problems, which we consider in the general case. We can state it as follows:

to investigate approximations of the numbers in S_1, S_2, S_3 by elements of fields of finite transcendence degree. In particular, to investigate approximations of these numbers by algebraic numbers, and so on.

It would be very desirable to find the transcendence measure and the mutual transcendence measure of numbers connected with the exponential function. We begin the discussion with several definitions and explain what is meant by the mutual transcendence measure.

DEFINITION 3.1.A. Suppose α is an arbitrary number and, as usual, $w(\alpha, n, H) = \min|P(\alpha)|$, where min extends over all polynomials $P(x) \in \mathbf{Z}[x]$ of degree $\leqslant n$ and height $\leqslant H$. If φ: $\mathbf{N}^2 \to \mathbf{R}$ is a nonnegative function, then α has *transcendence measure* $\leqslant \varphi$ if $\ln w(\alpha, n, H) \geqslant -\varphi(n, H)$ for $n + \ln H \geqslant c_0$, where c_0 depends only on α. Analogously, α has *transcendence type* $\leqslant \varphi$ if φ: $\mathbf{N} \to [0, \infty)$ and $\ln w(\alpha, t, e^t) \geqslant -\varphi(t)$ for $t \geqslant c_0(\alpha)$.

We are interested in both the transcendence measure and the transcendence type. In addition, we will need even more precise estimates. However, for the general propositions which will be given the transcendence type must be used.

Analogously, we state

DEFINITION 3.1.B. For numbers $\alpha_1,\dots,\alpha_k$ we put

$$w(\alpha_1,\dots,\alpha_k, n, H) = \min\{|P(\alpha_1,\dots,\alpha_k)|: P(x_1,\dots,x_n) \in \mathbf{Z}[x_1,\dots,x_k],$$
$$\deg_{x_1}(P) + \cdots + \deg_{x_k}(P) \leqslant n, H(P) \leqslant H\}.$$

We say that $(\alpha_1,\dots,\alpha_k)$ has *mutual transcendence measure* $\leqslant \varphi$ if $\varphi\colon \mathbf{N}^2 \to [0, \infty)$ and $\ln w(\alpha_1,\dots,\alpha_k, n, H) \geqslant -\varphi(n, H)$ for $n + \ln H \geqslant c_0(\alpha_1,\dots,\alpha_k)$. The *mutual transcendence type* is defined analogously.

Of course, instead of the transcendence measure of a number α we could define the measure of approximation of α by algebraic numbers. These measures are equivalent (see §1 above).

As is well known, rather good transcendence measures have been obtained for the numbers $\ln\alpha$, e^α, α^β, $\ln\alpha/\ln\beta$ $(\alpha, \beta \in \mathbf{A})$ connected with the exponential function. Something is also known about the mutual transcendence measure of α^β and α^{β^2}, where $\alpha, \beta \in \mathbf{A}$ and β is of degree 3 **[11]**. We will see below that all of these results are consequences of certain general theorems on the approximation of numbers in S_i by elements of fields of finite transcendence degree.

Let us elaborate upon the meaning of these theorems. For simplicity, we consider the family $S_1 = \{e^{\alpha_i\beta_j}\}$. Let $\mathfrak{D}(S_1) = \mathfrak{D}_1$ denote the transcendence degree of $\mathbf{Q}(S_1)$, i.e. the largest number of algebraically independent quantities among the $e^{\alpha_i\beta_j}$. The number $\mathfrak{D}_1$ plays a prominent role in the formulation of our results. Recall that $\mathfrak{D}_1$ is closely connected with a second "invariant" $\kappa_1 = MN/(M+N)$. This connection is given by

PROPOSITION 3.2. (1) (*Lang, Waldschmidt and Brownawell* **[12–14]**). *If* $\kappa_1 = MN/(M+N) > 1 = 2^0$, *then* $\mathfrak{D}_1 \geqslant 1$. *If* $\kappa_1 = MN/(M+N) \geqslant 2^1$, *then* $\mathfrak{D}_1 \geqslant 2$.
(2) *If* $n \geqslant 1$, *then* $\kappa_1 = MN/(M+N) \geqslant 2^n$ *implies* $\mathfrak{D}_1 \geqslant n+1$.

Assertion (2) of Proposition 3.2 is actually Theorem 4.1. Note that according to a conjecture of Lang **[3]**, $\kappa_1 > 1$ in assertion (1) can be replaced by $\kappa_1 \geqslant 1$. In this event, the whole proposition assumes the unified form: if $n \geqslant 0$, then $\kappa_1 \geqslant 2^n$ implies $\mathfrak{D}_1 \geqslant n+1$. At present there is no proof of Lang's conjecture in the general case.

We now explain the role played here by transcendence measures. The fact is that in the proof of the proposition there are unused reserves. Therefore, this proof enables us to show not only that $\kappa_1 \geqslant 2^n$ implies $\mathfrak{D}_1 \geqslant n+1$ (i.e. $\mathfrak{D}_1 \geqslant [\log_2\kappa_1] + 1$, or more precisely, $\mathfrak{D}_1 \geqslant [\log_2\kappa_1] + 1$ when $\kappa_1 > 1$), but also that the numbers in S_1 cannot be very well approximated by elements of fields of transcendence degree n. In short, this is one possibility for determining the mutual transcendence measure in the set S_1. But there is another, more interesting, possibility. Suppose, for example, $\kappa_1 \geqslant 2^n$ (> 1) and the transcendence degree $\mathfrak{D}_1$ of $\mathbf{Q}(S_1)$ is exactly $n+1$. We then choose in S_1 $n+1$ algebraically independent numbers $\vartheta_1, \ldots, \vartheta_{n+1}$, in terms of which we express the others. Then it turns out to be possible to determine the mutual transcendence measure of $\vartheta_1, \ldots, \vartheta_{n+1}$.

Let us now pass from S_1 to S_2 and S_3, where there is no doubt of the possibility of obtaining good results. We again denote by $\mathfrak{D}_2$ and $\mathfrak{D}_3$ the transcendence degrees of $\mathbf{Q}(S_2)$ and $\mathbf{Q}(S_3)$, and by κ_2 and κ_3 the following invariants: for S_2 the number $M(N+1)/(M+N)$, and for S_3 the number $(MN+M+N)/(M+N)$. Then we obtain as a consequence of well-known theorems and Theorems 4.2 and 4.3 the following

PROPOSITION 3.3. (1) *If* $\kappa_2 = M(N+1)/(M+N) > 1$, *then* $\mathfrak{D}_2 \geqslant 1$; *if* $\kappa_3 = (MN+M+N)/(M+N) > 1$, *then* $\mathfrak{D}_3 \geqslant 1$.

(2) *If* $n \geqslant 1$, *then* $\kappa_2 = M(N+1)/(M+N) \geqslant 2^n$ *implies* $\mathfrak{D}_2 \geqslant n+1$, *and* $\kappa_3 > 2^n$ *implies* $\mathfrak{D}_3 \geqslant n+1$.

The assertions in (1) are theorems of Gel′fond (1934) [**2**] and Lindemann.

Indeed, suppose $\kappa_2 > 1$ and $\mathfrak{D}_2 = 1$. The simplest case is the following: $N = 1$, $M = 2$. Then S_2 has the form $S_2 = \{\beta_1, \beta_2, e^{\beta_1\alpha}, e^{\beta_2\alpha}\}$ where, say, $\beta_1, \beta_2 \in \mathbf{A}$. Assertion (1) of Proposition 3.2 asserts that for $\beta \in \mathbf{A}$ and any $\alpha \neq 0$, one of the numbers $e^{\alpha}, e^{\alpha\beta}$ is transcendental. The condition $\mathfrak{D}_2 = 1$ means that precisely one of these numbers is transcendental. Then, clearly, some number of the form $\alpha_0^{\beta_0}$, where $\alpha_0 \in \mathbf{A}$, $\beta_0 \in \mathbf{A}$, $\alpha_0 \neq 0, 1$, $\beta_0 \neq 0$, is a transcendental generator of $\mathbf{Q}(S_2)$. The transcendence type of this number is $\leqslant O(t^4)$, which is in complete agreement with Gel′fond's theorem [**1**] concerning numbers of the form $\alpha_0^{\beta_0}$.

Regarding S_3, note that the simplest possibility with $\kappa_3 > 1$ and $\mathfrak{D}_3 = 1$ is $S_3 = \{\alpha, e^{\alpha}\}$. Now a transcendental generator of $\mathbf{Q}(S_3)$ is either $\ln \alpha$, where $\alpha \in \mathbf{A}$, $\alpha \neq 0, 1$, or e^{α}, where $\alpha \in \mathbf{A}$, $\alpha \neq 0$.

The theorem below shows that the transcendence type of these numbers is $\leqslant O(t^3 \ln t)$.

Below will be proved for the first time general (and in some sense unimprovable) theorems on approximations of values of the exponential function.

We describe separately the algebraic and analytic parts of the sequel in the general case. The analytic part of our exposition is Gel′fond's method in its classical form (Gel′fond's third method). The algebraic part differs substantially from the usual approach. It is necessary to carry out the calculations for the "functional case", i.e. for fields of arbitrary transcendence degree. As always in such cases, we make extensive use of the connection between function and number fields. It seems useful to preface a detailed proof with an explanation of its principal features, as Gel′fond did. We concentrate more on the algebraic aspects, since the analytic part is thoroughly analyzed in [**1, 7**].

To the assumptions on the S_i already mentioned we add the following:

(H) There exist constants τ_0 and τ_1, depending only on $\alpha_1,\ldots,\alpha_N$ and $\beta_1,\ldots,\beta_M$, such that for any rational integers $u_1,\ldots,u_N$ and $v_1,\ldots,v_M$ with $|u|=\sum_{i=1}^N|u_i|>\tau_0$, $|v|=\sum_{j=1}^M|v_j|>\tau_0$ we have

$$|u_1\alpha_1+\cdots+u_N\alpha_N|>\exp(-\tau_1|u|),\qquad (3.1)$$
$$|v_1\beta_1+\cdots+v_M\beta_M|>\exp(-\tau_1|v|).$$

It will be shown below that (3.1) can be removed. We mention in passing that assumption (H), first used in [**1**, **7**], is satisfied for most of the specific sets $\{\alpha_1,\ldots,\alpha_N\}$ and $\{\beta_1,\ldots,\beta_M\}$ in which we are interested. Actually, assumption (H) can be completely eliminated from all proofs. This elimination is accomplished by purely algebraic methods, and it would be very interesting to obtain an analytic proof.

Let us pause briefly to indicate some terminology and notation. If $\mathcal{K}$ is a field of transcendence degree k, there exist numbers $\vartheta_1,\ldots,\vartheta_k$ which are algebraically independent over $\mathbf{Q}$ (a so-called transcendence basis of $\mathcal{K}$) and a number ω which is algebraic over $\mathbf{Q}(\vartheta_1,\ldots,\vartheta_k)$ such that $\mathcal{K}=\mathbf{Q}(\vartheta_1,\ldots,\vartheta_k,\omega)$. Purely transcendental extensions $\mathbf{Q}(\vartheta_1,\ldots,\vartheta_k)$ will be called fields of the form R_k, and their finite algebraic extensions $\mathcal{K}=\mathbf{Q}(\vartheta_1,\ldots,\vartheta_k,\omega)$ will be called fields of the form R_k^*. Thus, $\operatorname{tr\,deg}\mathcal{K}=k$ if and only if $\mathcal{K}$ is a field of the form R_k^*. By the integers of the field $\mathcal{K}=\mathbf{Q}(\vartheta_1,\ldots,\vartheta_k)$ of the form R_k we mean the set of all $R(\vartheta_1,\ldots,\vartheta_k)$, where $R(x_1,\ldots,x_k)\in\mathbf{Z}[x_1,\ldots,x_k]$. Let $I_{\mathcal{K}}$ denote the ring of integers of $\mathcal{K}$. If $\mathcal{K}'=\mathcal{K}(\omega)$ is a field of the form R_k^*, $\mathcal{K}$ being a field of the form R_k, then the ring of integers $I_{\mathcal{K}'}$ of $\mathcal{K}'$ is the integral closure of $I_{\mathcal{K}}$ in $\mathcal{K}'$ [**5**]. In other words, the elements of $I_{\mathcal{K}'}$ are those elements of $\mathcal{K}'$ which are solutions of equations in one variable with integral coefficients from $I_{\mathcal{K}}$ and with leading coefficient unity. Clearly, if $\mathcal{K}_1=\mathcal{K}_0(\omega)$ is a field of the form R_k^* and $\mathcal{K}_0$ is a field of the form R_k, then there exists an integer ω_1 over $\mathcal{K}_0$ such that $\mathcal{K}_1=\mathcal{K}_0(\omega_1)$, and each element of $\mathcal{K}_1$ has the form a/b, where $a=P(\vartheta_1,\ldots,\vartheta_k,\omega_1)$, $P(x_1,\ldots,x_{k+1})\in\mathbf{Z}[x_1,\ldots,x_{k+1}]$, and $b\in I_{\mathcal{K}_0}$.

So in the sequel, when we speak of a field $\mathcal{K}$ of transcendence degree k, we have a representation in the form $\mathcal{K}=\mathbf{Q}(\vartheta_1,\ldots,\vartheta_k,\omega_1)$, where ω_1 is integral over $\mathbf{Q}(\vartheta_1,\ldots,\vartheta_k)$ and any element of $\mathcal{K}$ can be written as a/b, where $a=R(\vartheta_1,\ldots,\vartheta_k,\omega_1)$, $b=P(\vartheta_1,\ldots,\vartheta_k)$, with $P(x_1,\ldots,x_k)\in\mathbf{Z}[x_1,\ldots,x_k]$, $R(x_1,\ldots,x_k,x_{k+1})\in\mathbf{Z}[x_1,\ldots,x_k,x_{k+1}]$.

In accordance with the above notation, for a field $\mathcal{K}=\mathbf{Q}(\bar{S}_1)$, $\mathbf{Q}(\bar{S}_2)$, $\mathbf{Q}(\bar{S}_3)$ of transcendence degree $\mathfrak{D}=\mathfrak{D}_1,\ \mathfrak{D}_2,\ \mathfrak{D}_3$ we have a representation $\mathcal{K}=\mathbf{Q}(\vartheta_1,\ldots,\vartheta_{\mathfrak{D}},\omega_1)$, where the integer ω_1 over $\mathbf{Q}(\vartheta_1,\ldots,\vartheta_{\mathfrak{D}})$ has degree $\nu\geqslant 1$ and $\omega_2,\ldots,\omega_\nu$ are the conjugates of ω_1. Now each element of S_1, S_2, or S_3 has the form a/b, where a is a polynomial in $\vartheta_1,\ldots,\vartheta_{\mathfrak{D}},\omega_1$ and b is a polynomial in $\vartheta_1,\ldots,\vartheta_{\mathfrak{D}}$ with rational integral coefficients.

We now introduce auxiliary functions for the families S_1, S_2, and S_3. In the sequel, in this part of the paper, the symbols $c_i>0$ will denote effectively defined constants which depend only on $\alpha_1,\ldots,\alpha_N$, $\beta_1,\ldots,\beta_M$, M, N, $\vartheta_1,\ldots,\vartheta_{\mathfrak{D}}$, ω_1. We emphasize that $\vartheta_1,\ldots,\vartheta_{\mathfrak{D}}$, ω_1 are not to be considered fixed once and for all; in our arguments we only require a representation of $\mathcal{K}=\mathbf{Q}(S_i)$ in the form $\mathcal{K}=\mathbf{Q}(\vartheta_1,\ldots,\vartheta_{\mathfrak{D}},\omega_1)$.

The symbols $X, Y, Z, W, V,\ldots$ denote sufficiently large (i.e. with respect to certain constants of the form c_i) natural numbers.

For each X we define three functions f_1^X, f_2^X, f_3^X of a complex variable z, which correspond to the three families S_1, S_2, S_3, as follows:

$$
\begin{aligned}
f_1^X(z) &= \sum_{\lambda_1=0}^{L_1}\cdots\sum_{\lambda_M=0}^{L_1} C^1_{\lambda_1,\ldots,\lambda_M} e^{(\lambda_1\beta_1+\cdots+\lambda_M\beta_M)z},\\
f_2^X(z) &= \sum_{\lambda_1=0}^{X}\cdots\sum_{\lambda_M=0}^{X} C^2_{\lambda_1,\ldots,\lambda_M} e^{(\lambda_1\beta_1+\cdots+\lambda_M\beta_M)z}, \qquad (3.2)\\
f_3^X(z) &= \sum_{\lambda_0=0}^{L_0}\sum_{\lambda_1=0}^{X}\cdots\sum_{\lambda_M=0}^{X} C^3_{\lambda_0,\ldots,\lambda_M} z^{\lambda_0} e^{(\lambda_1\beta_1+\cdots+\lambda_M\beta_M)z}.
\end{aligned}
$$

Here,

$$L_1=[X^{N/(M+N)}], \qquad L_0=[X^{(M+N)/N}\ln^{-1}X].$$

The numbers $C^1_{\lambda_1,\ldots,\lambda_M}$, $C^2_{\lambda_1,\ldots,\lambda_M}$, and $C^3_{\lambda_0,\lambda_1,\ldots,\lambda_M}$ are chosen to be elements of the ring of integers of the field $\mathbf{Q}(\vartheta_1,\ldots,\vartheta_{\mathfrak{D}})$. More precisely,

$$
\begin{aligned}
C^1_{\lambda_1,\ldots,\lambda_M} &= \sum_{l_1=0}^{X}\cdots\sum_{l_{\mathfrak{D}}=0}^{X} C^1_{l_1,\ldots,l_{\mathfrak{D}},\lambda_1,\ldots,\lambda_M}\vartheta_1^{l_1}\cdots\vartheta_{\mathfrak{D}}^{l_{\mathfrak{D}}},\\
C^2_{\lambda_1,\ldots,\lambda_M} &= \sum_{l_1=0}^{L_2}\cdots\sum_{l_{\mathfrak{D}}=0}^{L_2} C^2_{l_1,\ldots,l_{\mathfrak{D}},\lambda_1,\ldots,\lambda_M}\vartheta_1^{l_1}\cdots\vartheta_{\mathfrak{D}}^{l_{\mathfrak{D}}}, \qquad (3.3)\\
C^3_{\lambda_0,\lambda_1,\ldots,\lambda_M} &= \sum_{l_1=0}^{L_3}\cdots\sum_{l_{\mathfrak{D}}=0}^{L_3} C^3_{l_1,\ldots,l_{\mathfrak{D}},\lambda_1,\ldots,\lambda_M}\vartheta_1^{l_1}\cdots\vartheta_{\mathfrak{D}}^{l_{\mathfrak{D}}},
\end{aligned}
$$

where

$$L_2=[X^{(M+N)/(N+1)}\ln^{1/(N+1)}X], \qquad L_3=[X^{(M+N)/N}].$$

In the sequel, the rational integers $C^i_{l_1,\ldots,l_{\mathfrak{D}},\lambda_1,\ldots,\lambda_M}$ will be briefly denoted by $C^i_{l,\lambda}$. These $C^i_{l,\lambda}$ are determined by Dirichlet's principle (in accordance with the Siegel Lemma 2.3) from the following conditions:

$$
\begin{aligned}
&(3.4)\qquad f_1^X(x_1\alpha_1+\cdots+x_N\alpha_N)=0 \quad \text{for integers } x_i\colon i=1,\ldots,N,\\
&\qquad 0\le x_i\le[c_1X^{M/(N+M)}], \qquad \max|C^1_{l_1,\ldots,l_{\mathfrak{D}},\lambda_1,\ldots,\lambda_M}|<\exp(c_2X);
\end{aligned}
$$

$$
\begin{aligned}
&(3.5)\quad \left(f_2^X(y_1\alpha_1+\cdots+y_N\alpha_N)\right)^{(k)}=0 \quad \text{for integers } y_i,k\colon i=1,\ldots,N,\\
&0\le y_i\le[c_3X^{(M-1)/(N+1)}\ln^{1/(N+1)}X], \quad 0\le k\le[X^{(M+N)/(N+1)}\ln^{-N/(N+1)}X],\\
&\qquad \max|C^2_{l_1,\ldots,l_{\mathfrak{D}},\lambda_1,\ldots,\lambda_M}|<\exp(C_4X^{(M+N)/(N+1)}\ln^{1/(N+1)}X);
\end{aligned}
$$

(3.6) $\left(f_3^X(y_1\alpha_1+\cdots+y_N\alpha_N)\right)^{(k)}=0$ for integers $y_i,\ k$: $i=1,\ldots,N$,

$$0\leqslant y_i\leqslant\left[c_5X^{M/N}\right],\qquad 0\leqslant k\leqslant\left[X^{(N+M)/N}\ln^{-1}X\right],$$

$$\max\left|C^3_{l_1,\ldots,l_{\mathfrak{D}},\lambda_0,\ldots,\lambda_M}\right|<\exp\left(c_6X^{(N+M)/N}\right).$$

Such a choice of rational integers $C^i_{l_1,\ldots,l_{\mathfrak{D}},\ldots}$: $i=1,2,3$, not all zero, is possible in view of the fact that the numbers of the form $e^{\alpha_i\beta_j}$, α_i, and β_j belong to the approximate set S_1, S_2, or S_3. Indeed, there exist polynomials $T^i_X(x_1,\ldots,x_{\mathfrak{D}_i})$, $S^i_X(x_1,\ldots,x_{\mathfrak{D}_i},y)$ such that for any $i=1,2,3$ we obtain for $S_i=\{a_{X,i}\}$ a representation of the form

$$a_{X,i}=\frac{S^i_X(\vartheta_1,\ldots,\vartheta_{\mathfrak{D}_i},\omega_1)}{T^i_X(\vartheta_1,\ldots,\vartheta_{\mathfrak{D}_i})}.$$

Therefore, Lemma 2.3 can be applied to the functions f_1^X, f_2^X, and f_3^X in the standard way: see, e.g., the detailed exposition for the case $\mathfrak{D}_i=1$ in A. O. Gel'fond [**1**, 7].

If for such a choice of the numbers $C^i_{l,\lambda}$ it turns out that all of the polynomials

$$C^i_\lambda(x_1,\ldots,x_{\mathfrak{D}_i})=\sum_{l_1}\cdots\sum_{l_{\mathfrak{D}_i}}C^i_{l_1,\ldots,l_{\mathfrak{D}_i},\lambda_1,\ldots}x_1^{l_1}\cdots x_{\mathfrak{D}_i}^{l_{\mathfrak{D}_i}}$$

for all $\lambda=(\lambda_1,\ldots,\lambda_M)$ have a common divisor $R_0(x_1,\ldots,x_{\mathfrak{D}_i})$, then it is possible to divide the function $f_i(z)$ by $R_0(\vartheta_1,\ldots,\vartheta_{\mathfrak{D}_i})$, where we have put $x_1=\vartheta_1,\ldots,x_{\mathfrak{D}_i}=\vartheta_{\mathfrak{D}_i}$. The nontriviality of this procedure is explained by the fact that $\mathbf{Q}(\vartheta_1,\ldots,\vartheta_{\mathfrak{D}_i},\omega_1)$ has transcendence degree $\mathfrak{D}_i$, i.e. $\vartheta_1,\ldots,\vartheta_{\mathfrak{D}_i}$ are algebraically independent numbers.

As a result, we obtain a function $f_i^X(z)$ which also has the form (3.2) and which satisfies (3.3) with other rational integers $C^i_{l,\lambda}$. In addition, (3.4)–(3.6) are satisfied, but, in view of Lemma 1.4, the constants c_2, c_4, c_6 must be replaced by others.

Thus, we may assume (cf. [**1**, 7]) that the polynomials $C^i_\lambda(x_1,\ldots,x_{\mathfrak{D}_i})$: $\lambda=(\lambda_1,\ldots,\lambda_M)$, generated by the functions f_1^X, f_2^X, and f_3^X have no common divisor.

To obtain an upper bound for the values of the functions $f_i^X(z)$ we use a formula of Hermite.

In view of (3.4), we can write

$$|f_1^X(z)|$$
$$\leqslant\left|\frac{1}{2\pi i}\cdot\int_\Gamma\prod_{x_1=0}^{[c_1X^{M/(N+M)}]}\cdots\prod_{x_N=0}^{[c_1X^{M/(N+M)}]}\frac{z-x_1\alpha_1-\cdots-x_N\alpha_N}{\zeta-x_1\alpha_1-\cdots-x_N\alpha_N}\cdot\frac{f(\zeta)}{\zeta-z}\cdot d\zeta\right|,$$

where Γ is the circle $|\zeta|-X^{(M+1)/(N+M)}$ and $|z|\leqslant X^{M/(N+M)}\ln X$. As usual, for $\kappa_1=MN/(M+N)$ we have

(3.7) $$|f_1^X(z)|<\exp(-c_7X^{\kappa_1}\ln X)$$

for $|z| \leqslant X^{M/(N+M)} \ln X$, when $\kappa_1 > 1$. Similarly, for all of the functions $f_2(z)$ and $f_3(z)$ we obtain the estimates

$$(3.8) \qquad |(f_2^X(z))^{(k)}| < \exp(-c_7 X^M \ln X)$$

for $|z| \leqslant [X^{(M-1)/(N+1)} \ln^{N/(N+1)} X]$, $k \leqslant [X^{(M+N)/(N+1)}]$ as $\kappa_2 > 1$,

$$(3.9) \qquad |(f_3^X(z))^{(k)}| < \exp(-c_7 X^{(MN+M+N)/N})$$

for $|z| \leqslant X^{M/N} \ln X$ and $k \leqslant X^{(N+M)/N}$ as $\kappa_3 > 1$.

We next use Tijdeman's auxiliary lemma. Here assumption (H) comes into play. It will be eliminated later, but for the moment we use it. From Tijdeman's lemma [**10**] we directly obtain

LEMMA 3.4. *There exist constants $c_8, c_9 > 0$ such that*

(i) *if for all integers $0 \leqslant x_i \leqslant [c_8 X^{M/(N+M)}]$, $1 \leqslant i \leqslant N$, we have*

$$|f_1(x_1\alpha_1 + \cdots + x_N\alpha_N)| < \exp(-c_9 X^{\kappa_1} \ln X),$$

then for any $\lambda_1, \ldots, \lambda_M$ it follows that

$$|C^1_{\lambda_1,\ldots,\lambda_M}| < \exp(-c_{10} X^{\kappa_1} \ln X);$$

(ii) *if for all integers $0 \leqslant x_i \leqslant 9[c_8 X^{(M-1)/(N+1)} \ln^{1/(N+1)} X]$, $0 \leqslant k \leqslant [c_{11} X^{(M+N)/(N+1)} \ln^{-N/(N+1)} X]$, we have*

$$|f_2^{(k)}(x_1\alpha_1 + \cdots + x_N\alpha_N)| < \exp(-c_9 X^M \ln X),$$

then for any $\lambda_1, \ldots, \lambda_M$ it follows that

$$|C^2_{\lambda_1,\ldots,\lambda_M}| < \exp(-c_{10} X^M \ln X);$$

(iii) *if for all integers $0 \leqslant x_i \leqslant [c_8 X^{M/N}]$, $1 \leqslant i \leqslant N$, $0 \leqslant k \leqslant [c_{12} X^{(M+N)/N} \ln^{-1} X]$, we have*

$$|f_3^{(k)}(x_1\alpha_1 + \cdots + x_N\alpha_N)| < \exp(-c_9 X^{(MN+M+N)/N}),$$

then for any $\lambda_0, \lambda_1, \ldots, \lambda_M$ it follows that

$$|C^3_{\lambda_0,\lambda_1,\ldots,\lambda_M}| < \exp(-c_{10} X^{(MN+M+N)/N}).$$

Let $T^i(x_1, \ldots, x_{\mathfrak{D}_i}) = \prod_{X, a_{X,i} \in S_i} T^i_X(x_1, \ldots, x_{\mathfrak{D}_i})$. Then there exists a constant c_{13} such that $(T^1(\vartheta_1, \ldots, \vartheta_{\mathfrak{D}_1}))^{[c_{13}X]}(f_1^X(z))$ is an integer of the field $\mathcal{K}_1 = \mathbf{Q}(\vartheta_1, \ldots, \vartheta_{\mathfrak{D}_1}, \omega_1)$ for $z = x_1\alpha_1 + \cdots + x_N\alpha_N$, where the x_i are within the limits indicated in Lemma 3.4(i). Similar assertions hold for $f_2^X(z)$ and $f_3^X(z)$. Taking into account the explicit form of the functions $f_i^X(z)$ and making the obvious estimates of the polynomials therein, we obtain, for example,

$$(T^1(\vartheta_1, \ldots, \vartheta_{\mathfrak{D}_1}))^{[c_{13}X]}(f_1^X(x_1\alpha_1 + \cdots + x_N\alpha_N)) = P_X(\vartheta_1, \ldots, \vartheta_{\mathfrak{D}_1}, \omega_1)$$

and $t(P_X) \leqslant c_{11} X$ for $0 \leqslant x_i \leqslant [c_8 X^{M/(N+M)}]$.

Thus, applying Lemma 3.4, we obtain for $i = 1, 2, 3$

PROPOSITION 3.5. *For any sufficiently large* $X > c_{15}$, *either for some* $P_X(x_1, \ldots, x_{\mathfrak{D}_i}, y) \in \mathbf{Z}[x_1, \ldots, x_{\mathfrak{D}_i}, y]$ *we have*

$$\exp(-c_{16}X^{\kappa_i}\ln^{\gamma_i} X) < |P_X(\vartheta_1, \ldots, \vartheta_{\mathfrak{D}_i}, \omega_1)| < \exp(-c_{17}X^{\kappa_i}\ln^{\gamma_i} X), \qquad t(P_X) \leq c_{18}X, \tag{3.10}$$

or else for some family $C_{X,l}(x_1, \ldots, x_{\mathfrak{D}_i})$: $l \in \mathfrak{L}_{X,i}$, *of polynomials in* $\mathbf{Z}[x_1, \ldots, x_{\mathfrak{D}_i}]$ *we have the estimates*

$$|C_{X,l}(\vartheta_1, \ldots, \vartheta_{\mathfrak{D}_i})| < \exp(-c_{19}X^{\kappa_i}\ln^{\gamma_i} X), \qquad t(C_{X,l}) \leq c_{20}X, \tag{3.11}$$

and the polynomials $C_{X,l}(x_1, \ldots, x_{\mathfrak{D}_i})$: $l \in \mathfrak{L}_{X,i}$, *have no common divisor.*

The proposition follows directly from Lemma 3.4, on replacing $[X^{(M+N)/(N+1)}\ln^{1/(N+1)} X]$ by X when $i = 2$, and $[X^{(M+N)/N}]$ by X when $i = 3$.

REMARK 3.6. In the statement of the proposition

$$\kappa_1 = \frac{MN}{M+N}, \qquad \kappa_2 = \frac{M(N+1)}{M+N},$$

$$\kappa_3 = \kappa_1 + 1, \qquad \gamma_1 = 1, \qquad \gamma_2 = N/(M+N), \qquad \gamma_3 = 0.$$

The form of κ_i and γ_i is clear from (3.2) and (3.3).

To put the proposition in an acceptable form it is only necessary to rid P_X of ω_1. We use the argument of **[15]**. For any $i = 1, 2, 3$ we have the following

PROPOSITION 3.7. *For any sufficiently large* $X > c_{21}$, *either for some* $P_X(x_1, \ldots, x_{\mathfrak{D}_i}) \in \mathbf{Z}[x_1, \ldots, x_{\mathfrak{D}_i}]$ *we have*

$$\exp(-c_{22}X^{\kappa_i}\ln^{\gamma_i} X) < |P_X(\vartheta_1, \ldots, \vartheta_{\mathfrak{D}_i})| < \exp(-c_{23}X^{\kappa_i}\ln^{\gamma_i} X), \qquad t(P_X) \leq c_{24}X, \tag{3.12}$$

or else for some family $C_{X,l}(x_1, \ldots, x_{\mathfrak{D}_i})$: $l \in \mathfrak{L}_{X,i}$, *of polynomials in* $\mathbf{Z}[x_1, \ldots, x_{\mathfrak{D}_i}]$ *we have the estimates*

$$|C_{X,l}(\vartheta_1, \ldots, \vartheta_{\mathfrak{D}_i})| < \exp(-c_{25}X^{\kappa_i}\ln^{\gamma_i} X), \qquad t(C_{X,l}) \leq c_{26}X, \tag{3.13}$$

and the polynomials $C_{X,l}(x_1, \ldots, x_{\mathfrak{D}_i})$: $l \in \mathfrak{L}_{X,i}$, *have no common divisor.*

PROOF. Clearly, (3.13) and (3.11) are one and the same. Assume that (3.13) does not hold. Then we obtain (3.10). Denote $P_X(\vartheta_1, \ldots, \vartheta_{\mathfrak{D}_i}, \omega_1)$ by $R_1(\vartheta_1, \ldots, \vartheta_{\mathfrak{D}_i}) + \cdots + R_\nu(\vartheta_1, \ldots, \vartheta_{\mathfrak{D}_i})\omega_1^{\nu-1}$, or briefly by $R_1(\theta) + \cdots + R_\nu(\theta)\omega_1^{\nu-1}$.

We will show that for any i, $1 \leq i \leq \nu$, there exist polynomials $Q_i^i, \ldots, Q_\nu^i$ in $\mathbf{Z}[x_1, \ldots, x_\nu]$ such that for $w = Q_i^i(\omega_1, \ldots, \omega_\nu)R_i(\theta) + \cdots + Q_\nu^i(\omega_1, \ldots, \omega_\nu)R_\nu(\theta)$ we have

$$\exp(-c_{27}X^{\kappa_i}\ln^{\gamma_i} X) < |w| < \exp(-c_{28}X^{\kappa_i}\ln^{\gamma_i} X), \tag{3.14}$$

where the type $t(Q_j^i)$ is bounded by an absolute constant. We do this by induction. For $i - 1$, (3.14) follows from (3.10). Assume that (3.14) has been proved for a fixed i $(1 \leq i \leq \nu)$. Let $w^{(2)}, \ldots, w^{(d)}$, where $d = [\mathbf{Q}(\omega_1, \ldots, \omega_\nu, \theta) : \mathbf{Q}(\theta)]$, denote the conjugates of w.

Then $w_0 = ww^{(2)} \cdots w^{(d)}$ is a polynomial in $\vartheta_1, \ldots, \vartheta_{\mathfrak{D}_i}$ such that $|w_0| < \exp(-c_{29} X^{\kappa_i} \ln^{\gamma_i} X)$. If (3.12) holds for w_0, the proposition is proved. Otherwise, by the induction assumption, for some l, $2 \leqslant l \leqslant d$, we have

$$|w^{(l)}| < \exp(-c_{30} X^{\kappa_i} \ln^{\gamma_i} X) \quad \text{for an arbitrary large } c_{30} > 0.$$

If $w^{(l)} = \tilde{Q}_i^i(\omega_1, \ldots, \omega_\nu) R_i(\theta) + \cdots + \tilde{Q}_\nu^i(\omega_1, \ldots, \omega_\nu) R_\nu(\theta)$, then for $v = w\tilde{Q}_i^i(\omega_1, \ldots, \omega_\nu) - w^{(l)} Q_i^i(\omega_1, \ldots, \omega_\nu)$ we obtain

$$\exp(-c_{31} X^{\kappa_i} \ln^{\gamma_i} X) > |v| > \exp(-c_{32} X^{\kappa_i} \ln^{\gamma_i} X).$$

Here, $v = Q_{i+1}^{i+1}(\omega_1, \ldots, \omega_\nu) R_{i+1}(\theta) + \cdots + Q_\nu^{i+1}(\omega_1, \ldots, \omega_\nu) R_\nu(\theta)$, i.e. (3.14) is proved for $i + 1$.

From (3.14) with $i = \nu$ we directly obtain (3.12), and the proposition is proved.

Proposition 3.7 is essentially the analytic part of the proof of the main results [**16**]:

THEOREM I. *If*

$$\kappa_1 = MN/(M + N) \geqslant 2^n,$$

then in the set S_1 *of numbers* (i) *there are* $n + 1$ *which are algebraically independent.*

THEOREM II. *If*

$$\kappa_2 = (MN + M)/(M + N) \geqslant 2^n,$$

then in the set S_2 *of numbers* (ii) *there are* $n + 1$ *which are algebraically independent.*

THEOREM III. *If*

$$\kappa_3 = (MN + M + N)/(M + N) > 2^n,$$

then in the set S_3 *of numbers* (iii) *there are* $n + 1$ *which are algebraically independent.*

The proof of Theorems I–III is carried out by contradiction in §5 using a single scheme. It is essential to the proof to establish by induction, for any j, $1 \leqslant j \leqslant n$, the following assertion with fixed $\theta_1, \ldots, \theta_n$:

(A_j) for any $X > c_1(j)$ there exists a polynomial $P(x_1, \ldots, x_j) \in \mathbf{Z}[x_1, \ldots, x_j]$, $P(x_1, \ldots, x_j) \not\equiv 0$, such that

$$t(P) = \max\{\ln H(P), \deg(P)\} \leqslant c_2 X^{2^{n-j}}$$

and

$$|P(\theta_1, \ldots, \theta_j)| < \exp(-c_3 X^{\kappa_i} \ln^{\gamma_i} X).$$

For given c_2 and c_3 we denote this condition by $\langle c_2 X^{2^{n-j}}, c_3 X^{\kappa_i} \ln^{\gamma_i} X \rangle_j$. Here $i = 1, 2, 3$, and the meaning of $\kappa_i \geqslant 2^n$ and γ_i is explained in 3.6. The algebraically independent numbers $\{\theta_1, \ldots, \theta_n\}$ contain a transcendence basis of the field $\mathbf{Q}(\bar{S}_i)$ when $\operatorname{tr} \deg \mathbf{Q}(\bar{S}_i) \leqslant n$.

Note that (A_n) holds by virtue of 3.7. In §5 all of the (A_j) will be proved by induction on $n-j$. In particular, (A_1) holds. However, according to the assumptions of Theorems I–III and 3.6, either $\kappa_i > 2^n$ or else $\kappa_i = 2^n$ and $\gamma_i > 0$. Therefore, (A_1) and the Gel'fond lemma (Lemma VII of Chapter III of [**1**]) imply that the number θ_1 is algebraic, which is impossible.

The proofs of Theorems I–III given in §5 are based on lemmas proved in §4. The notation for $\theta_1,\dots,\theta_n$ introduced above is preserved in §5. More precisely, we will assume that if one of the Theorems I–III is not true, then for some $i = 1,2,3$ we have $\mathbf{Q}(S_i) \subset \mathbf{Q}(\theta_1,\dots,\theta_n,\omega)$, where $\theta_1,\dots,\theta_n$ are algebraically independent and ω is algebraic over $\mathbf{Q}(\theta_1,\dots,\theta_n)$.

We give a brief illustration of the algebraic part of the proof on the example of Theorem I.

According to 3.7, (A_n) holds. Assume that (A_m) holds for all m such that $n-l<m\leqslant n$. We will verify (A_l), where $l<m$. Suppose $\mathcal{K}_0 = \mathbf{Z}[\alpha_1,\dots,\alpha_k]$ is a finitely generated extension of $\mathbf{Z}$. For elements of $\mathcal{K}_0$ we introduce the concept of the type $t_{\mathcal{K}_0}$: for $x\in\mathcal{K}_0$ we put $t_{\mathcal{K}_0}(x)\leqslant t$ if $x = P(\alpha_1,\dots,\alpha_k)$ and $t(P)\leqslant t$ for $P(x_1,\dots,x_k)\in\mathbf{Z}[x_1,\dots,x_k]$. If ζ is algebraic over $\mathcal{K}_0$ and is a root of the ($\mathcal{K}_0$-irreducible) equation $p_dx^d + p_{d-1}x^{d-1} + \cdots + p_0 = 0$, then we put $t_{\mathcal{K}_0}(\zeta)\leqslant t$ if $d\leqslant t$ and $\max_{0\leqslant i\leqslant d} t_{\mathcal{K}_0}(p_i)\leqslant t$ for $p_i\in\mathcal{K}_0$.

Applying (A_{l+1}) and replacing the condition of "smallness" of the polynomial $P(\theta_1,\dots,\theta_{l+1})$ by that of "nearness" of θ_{l+1} to a root of the algebraic equation $P(\theta_1,\dots,\theta_l,z)=0$ over $\mathbf{Q}(\theta_1,\dots,\theta_l)$ (in the spirit of §1), we obtain either the condition (A_l) for a given X: $\langle c_9X^{2^{n-l}}, c_{10}X^{2^n}\ln X\rangle_l$ or else the existence of $s_1\leqslant c_{11}X^{2^{n-l-1}}$ such that

$$|\theta_{l+1}-\xi_1| < \exp\left(-c_{12}X^{2^n}\ln X\cdot s_1^{-1}\right)$$

for a number ξ_1, algebraic over $\mathcal{K}_l = \mathbf{Q}(\theta_1,\dots,\theta_l)$ and of type $\leqslant c_{13}X^{2^{n-l-1}}s_1^{-1}$ over $\mathcal{K}_l$.

Using all of the conditions (A_{l+r}): $1\leqslant r\leqslant n-l$, we obtain inductively for all k, $l+1\leqslant k\leqslant n$, numbers $s_{k-l}\leqslant c_{14}X^{2^{n-k}}$ and numbers $\xi_1,\dots,\xi_{n-l}$ such that $|\theta_k-\xi_{k-l}| < \exp(-c_{15}X^{2^n}\ln Xs_1^{-1}\cdots s_{k-l}^{-1})$, and also a number ξ_{k-l}, algebraic over the field $\mathcal{K}_{k-1} = \mathbf{Q}(\theta_1,\dots,\theta_l,\xi_1,\dots,\xi_{k-l-1})$, having type $t_{\mathcal{K}_{k-1}}(\xi_{k-l})\leqslant c_{16}X^{2^{n-k}}s_1^{-1}\cdots s_{k-l}^{-1}$ over $\mathcal{K}_{k-1}$.

The numbers $C^1_{\lambda_1,\dots,\lambda_M}$ in the auxiliary function $f(z)$ are now chosen as follows:

$$C^1_{\lambda_1,\dots,\lambda_M} = \sum_{i_1=0}^{Z}\cdots\sum_{i_l=0}^{Z}\sum_{i_{l+1}=0}^{W_1}\cdots\sum_{i_n=0}^{W_{n-l}} C^1_{\lambda_1,\dots,\lambda_M,i_1,\dots,i_n} \times\theta_1^{i_1}\cdots\theta_l^{i_l}\cdot\xi_1^{i_{l+1}}\cdots\xi_{n-l}^{i_n},$$

where $Z=[c_{17}X]$, $W_1 = \deg_{\mathcal{K}_l}(\xi_1),\dots,W_{n-l} = \deg_{\mathcal{K}_{n-1}}(\xi_{n-l})$. Here, if ζ is algebraic over the field $\mathcal{K}_0$, $\deg_{\mathcal{K}_0}(\zeta)$ denotes its degree. Using Dirichlet's principle, it is easy to verify that there exist rational integers $C^1_{\lambda_1,\dots,\lambda_M,i_1,\dots,i_n}$, not all zero, such that (3.9) holds. The rest of the proof amounts only to calculating the norm of the number $f_1(z_0)$ (for a suitable $z_0 = x_1\alpha_1+\cdots+x_N\alpha_N$) in an algebraic extension of the field $\mathcal{K}_l = \mathbf{Q}(\theta_1,\ldots,\theta_l)$ and the type of an element $g(\theta_1,\ldots,\theta_l) \in \mathcal{K}_l$ equal to this norm.

To conclude §3 we mention some applications of Theorems I–III. The general algebraic scheme of the arguments enables us to obtain all of the necessary auxiliary results for fields of finite transcendence degree. Since the analytic part of the proof remains unchanged, the combination of §§4–5 and any variant of Gel'fond's method enables us to generalize all of the general theorems established in this area. For example, the theorems of Schneider, Lang [**17**, **3**], Ramachandra [**18**], and [**19**, **3**] pertaining to the transcendence of values of arbitrary meromorphic functions can be generalized to the case of n algebraically independent numbers.

We give below some corollaries of Theorems I–III.

COROLLARY 3.8. *Suppose $\alpha \neq 0, 1$ is any algebraic number and β is an irrational algebraic number of degree $\geqslant 2^n - 1$. Then among the numbers $\alpha^\beta,\ldots,\alpha^{\beta^m}$, where $m \geqslant 2^{n+1} - 4$, there are n that are algebraically independent. In particular, if $d \geqslant 2^n - 1$, where d is the degree of β, then among the numbers $\alpha^\beta,\ldots,\alpha^{\beta^{d-1}}$ there are n that are algebraically independent.*

Here $n \geqslant 0$. In the case $n = 1$, 3.8 becomes Gel'fond's theorem of 1934 [**20**], and in the case $n = 2$, Gel'fond's theorem of 1949 [**1**, **7**]. The case $n = 3$ was considered by the author in [**15**].

In 3.8 we establish the existence of three algebraically independent numbers among $\alpha^\beta,\ldots,\alpha^{\beta^6}$ for algebraic numbers $\alpha \neq 0, 1$ and β of degree 7, and the existence of four algebraically independent numbers among $\alpha^\beta,\ldots,\alpha^{\beta^{14}}$ for β of degree 15. This is weaker than Gel'fond's conjecture that among $\alpha^\beta,\ldots,\alpha^{\beta^{d-1}}$, where d is the degree of β, all of the numbers are algebraically independent. However, this is the first positive result in the direction of Gel'fond's conjecture.

COROLLARY 3.9. *If $n > 1$, then among the numbers $e, e^e,\ldots,e^{e^{2^{n+1}-5}}$ there are n that are algebraically independent.*

We can formulate still other corollaries of Theorems I–III. Note that in all concrete cases the conjecture (H) is true, even with a more precise estimate for linear forms in the α_i or having the form $-\exp(-\tau_1 \ln x)$ but not $\exp(-\tau_1 x)$.

In the proofs of §§4–5 the main tool is the calculation of types or heights of elements of certain algebraic or function fields. Much more general functions of the height are considered in [**21**]. From the point of view of explicit quantitative estimates it is more convenient to use somewhat cruder concepts.

ALGEBRAIC INDEPENDENCE OF NUMBERS CONNECTED WITH THE EXPONENTIAL FUNCTION

4. Properties of height and type in finitely generated extensions of Z. In this part of the chapter we present auxiliary statements related to polynomials, their roots and resultants. Coefficients of polynomials are taken from rings $\mathcal{L}$ which are of finite type over **Z**. The core of the statements we prove lies in the process of successive reduction of type (degree) of transcendence of the rings under consideration. The procedure we propose, though generating worse estimates, permits us in the end to reduce everything to polynomials over **Z**. The key in the proof is the elimination theory used as an analog of Sylvester resultants. In this process in order to achieve the sharpest estimates the use of determinants should be avoided.

Besides we, in fact, use not resultants but their natural generalizations—semiresultants to take care of the case when polynomials might have a common divisor. A notion of semidiscriminant is introduced which corresponds locally to the set of those primes p for which, after reduction (mod p), the number of irreducible components together with multiplicities is preserved.

Unfortunately, explicit use of resultants and their generalizations, necessary for getting precise estimates, makes all considerations more complicated. For this reason we present the proof of existence of resultants from the general viewpoint. To prove the existence of resultants we use certain forms of compactness theorems which were applied in similar situations in the papers of A. Robinson [**22a, 22b**].

Let us consider arbitrary polynomials $P(x)$ and $Q(x)$ with coefficients from the field **C** of complex numbers. As in the ordinary resultant theory it is necessary to show the existence of such a complex number r, expressed by a predicate formula depending on parameters—the coefficients of $P(x)$ and $Q(x)$, and such that there exist polynomials $S(x)$, $T(x)$ from $\mathbf{C}[x]$ satisfying the equality

$$P(x)S(x) + Q(x)T(x) = r.$$

This in particular implies that if $P(x)$ and $Q(x)$ have a common root, then $r = 0$. Therefore the second statement is necessary to show that $r \neq 0$, if $P(x)$ and $Q(x)$ do not have common roots. Finally, for estimates it is important to prove that the degrees of polynomials depend only on degrees of $P(x)$ and $Q(x)$. All these statements are proved using model completeness, completeness and the existence of elimination quantifiers for the elementary theory of all algebraically closed fields of characteristic zero, a model of which is the field **C**.

A description of the signature and the system of axioms of the theory T_{ac}—of algebraically closed fields of characteristic zero is given in the books [**22**]. The statement on the existence of the elimination of quantifiers for the theory T_{ac} (Tarski Theorem [**22**]) is formulated in the following way.

TARSKI THEOREM. *For any formula $\varphi(v_0,\dots,v_{n-1})$ having the signature of the theory T_{ac} with free variables $v_0,\dots,v_{n-1}$ there exists a formula, $T_\varphi(v_0,\dots,v_{n-1})$ of signature T_{ac} with free variables $v_0,\dots,v_{n-1}$, which is quantifier free and for which the following conditions are satisfied: for any model $\mathfrak{A}$ of the theory T_{ac} and arbitrary elements $a_0,\dots,a_{n-1}$ of the basic set of $\mathfrak{A}$ we have $\mathfrak{A} \models \varphi(a_0,\dots,a_{n-1})$ if and only if $\mathfrak{A} \models T_\varphi(a_0,\dots,a_{n-1})$.*

This theorem in particular contains the Seidenberg-Tarski theorem [**22, 23**] and the existence of resultant r is an immediate consequence of it. Indeed, let us apply the theorem to the formula $\varphi = (\exists x)(P(x) = 0 \,\&\, Q(x) = 0)$. Then T_φ is a predicate formula defining r. It is evident that r is determined only by coefficients of $P(x)$ and $Q(x)$. To prove this it suffices to substitute all the undetermined coefficients of $P(x)$ and $Q(x)$ by pairwise different symbols of free variables. Then T_φ is written as a system of equalities: $r_1 = 0, \dots, r_n = 0$, where $r_i = r_i(v_0, \dots, v_{n-1})$ is a polynomial in variables $v_0, \dots, v_{n-1}$. Thus the properties of the theory T_{ac} imply the existence of the resultant.

The existence of the necessary expressions for the degree (height) of r as a function of the coefficients of $P(x)$, $Q(x)$ can also be deduced starting from the compactness theorem, cf. [**22**].

We introduce notations and definitions to be used everywhere in this section. The symbol $\mathfrak{L}$ denotes a finitely generated extension of $\mathbf{Z}$, where $\mathfrak{L} = \mathbf{Z}[\alpha_1, \dots, \alpha_l]$ stands for an extension generated by complex numbers $\alpha_1, \dots, \alpha_l$. We assume l to be bounded, in other words, all the constants under consideration depend on l. By $\mathcal{K} = \mathcal{F}(\mathfrak{L})$ we denote the factorial field of $\mathfrak{L}$, i.e. $\mathcal{K} = \mathbf{Q}(\alpha_1, \dots, \alpha_l)$.

The function of $\mathfrak{L}$-type $t_\mathfrak{L}(\cdot)$ in the ring $\mathfrak{L}$ is introduced in the following way: for $p \in \mathfrak{L}$ we put $t_\mathfrak{L}(p) = \inf t(P)$, where the inf is taken over all polynomials $P(x_1, \dots, x_l) \in \mathbf{Z}[x_1, \dots, x_l]$ such that $p = P(\alpha_1, \dots, \alpha_l)$ and $t(P)$ is the type of $P(x_1, \dots, x_l)$ in the sense of §1. Analogously for a polynomial $Q(x_1, \dots, x_k) \in \mathfrak{L}[x_1, \dots, x_k]$ we define the $\mathfrak{L}$-type $t_\mathfrak{L}(Q)$ as a maximum of $\mathfrak{L}$-types $t_\mathfrak{L}(\cdot)$ of coefficients of $Q(x_1, \dots, x_k)$ and degree $d(Q)$ of polynomial $Q(x_1, \dots, x_k)$, where $d(Q) = d_{x_1}(Q) + \cdots + d_{x_k}(Q)$.

For the sake of simplicity we assume $\max |\alpha_i|$ to be bounded. This assumption can be eliminated if we modify the definition of $\mathfrak{L}$-type for elements of $\mathfrak{L}$. Instead of $t_\mathfrak{L}(p) = \inf t(P)$ it is necessary to put $t_\mathfrak{L}(P) = \inf t^*(P)$. Here for a polynomial $P(x_1, \dots, x_l) \in \mathbf{Z}[x_1, \dots, x_l]$ and given $\alpha_1, \dots, \alpha_l$ we put

$$t^*(P) = \max\left\{\ln H(P),\, d(P) \cdot \max_{1 \leqslant i \leqslant l} (|\alpha_i|, 1)\right\}.$$

In other words, $t^*(P) \geqslant t(P)$ and if $\max_{1 \leqslant i \leqslant l} |\alpha_i|$ is bounded then $t^*(P)$ differs from $t(P)$ by a multiplicative constant. Hence the new notion of $\mathfrak{L}$-type differs from the old one by a multiplicative constant $\max_{1 \leqslant i \leqslant l}(|\alpha_i|, 1)$. Since everywhere below we consider only the case of bounded generators of $\mathfrak{L}$, there is no particular reason to introduce the new notion of $\mathfrak{L}$-type. It should be noted, however, that the new notion of $\mathfrak{L}$-type, which it is convenient to denote as $t^*_\mathfrak{L}(\cdot)$, gives a simple estimate $|p|$ for $p \in \mathfrak{L}$,

$$|p| \leqslant \exp 2t^*_\mathfrak{L}(p) \quad \text{for any } p \in \mathfrak{L}, p \neq 0.$$

Similarly, the height $H(Q)$ of a polynomial $Q(x_1, \dots, x_k) \in \mathfrak{L}[x_1, \dots, x_k]$ in the ordinary sense, i.e. the maximum of moduli of coefficients, is estimated in terms of $t^*_\mathfrak{L}(Q)$.

To simplify the computations we introduce the logarithmic height. For $P(x_1, \dots, x_k) \in \mathfrak{L}[x_1, \dots, x_k]$ we denote by $h_\mathfrak{L}(P)$ the maximum of $\mathfrak{L}$-types of coefficients of $P(x_1, \dots, x_k)$. This notion we name the $\mathfrak{L}$-height, which should not be confused with the ordinary height, so $h_\mathfrak{L}(P)$ is the logarithmic height.

We turn now to a theorem on symmetric functions. Consider a set of variables $t_1,\dots,t_n$ and the elementary symmetric polynomials $f_1,\dots,f_n$ in $t_1,\dots,t_n$:

$$f_1 = t_1 + \cdots + t_n, \dots, f_n = t_1 \cdots t_n.$$

By the degree of a polynomial $P(x_1,\dots,x_n)$ with respect to x_i we mean the highest power to which x_i occurs in P; by the degree of a monomial $ax_1^{\nu_1}\cdots x_n^{\nu_n}$ we mean $\nu_1 + \cdots + \nu_n$.

For a set of natural numbers $(\alpha_1,\dots,\alpha_k)$ we denote by $S_{\alpha_1,\dots,\alpha_k}$ the sum $\Sigma_\kappa t_{\kappa(1)}^{\alpha_1}\cdots t_{\kappa(k)}^{\alpha_k}$, where Σ_κ extends over all one-to-one mappings κ: $\{1,\dots,k\} \to \{1,\dots,n\}$. The expression $\Sigma_\kappa t_{\kappa(1)}^{\alpha_1}\cdots t_{\kappa(k)}^{\alpha_k}$ differs by a multiplicative constant from the traditional expression $\Sigma\, t_1^{\alpha_1}\cdots t_k^{\alpha_k}$:

$$\sum t_1^{\alpha_1}\cdots t_k^{\alpha_k} = \frac{1}{j_1!\cdots j_s!} S_{\alpha_1,\dots,\alpha_k},$$

where $j_1 + \cdots + j_s = k$, s is the number of distinct $\alpha_1',\dots,\alpha_s'$ among $\alpha_1,\dots,\alpha_k$, and $j_t = |\{i \leqslant n\colon\ \alpha_t' = \alpha_i\}|$ for $t = 1,\dots,s$. For any symmetric polynomial $S(t_1,\dots,t_n)$ we denote by $L_f(S)$ the length of $S(\cdots)$ as a polynomial in $f_1,\dots,f_n$. By means of Newton's formulas we can show that $L_f(S_\alpha) \leqslant e^\alpha$ for any α. Expressing $S_{\alpha_1,\dots,\alpha_k}$ in terms of S_β, we obtain

$$S_{\alpha_1,\dots,\alpha_k} = \sum \pm B(\alpha_1,\dots,\alpha_k),$$

where the sum extends over the $k!$ blocks $B(\alpha_1,\dots,\alpha_k)$ having the form $S_{\beta_1}\cdots S_{\beta_l}$, where $\beta_1 + \cdots + \beta_l = \alpha_1 + \cdots + \alpha_k$ and each β_i has the form $\beta_i = \alpha_{f_1} + \cdots + \alpha_{f_t}$. From this and the estimate for $L_f(S_\alpha)$ we obtain

$$L_f\left(S_{\alpha_1,\dots,\alpha_k}\right) \leqslant k!e^{\alpha_1+\cdots+\alpha_k}.$$

Consequently, for the sums $\Sigma\, t_1^{\alpha_1}\cdots t_k^{\alpha_k}$, which are also symmetric polynomials in $t_1,\dots,t_s$ with rational integral coefficients, we have

$$L\left(\sum t_1^{\alpha_1}\cdots t_k^{\alpha_k}\right) \leqslant d^k e^{\alpha_1+\cdots+\alpha_k} \leqslant d^k e^{dk}$$

with $d \geqslant \max\{\alpha_1,\dots,\alpha_k\}$. This estimate enables us to prove the following

THEOREM ON SYMMETRIC FUNCTIONS (CF. **[5]**). *Suppose $P(t_1,\dots,t_n) \in \mathfrak{L}[t_1,\dots,t_n]$ is a symmetric polynomial in $t_1,\dots,t_n$. Then*

$$P(t_1,\dots,t_n) = P_1(f_1,\dots,f_n)$$

for some $P_1(z_1,\dots,z_n) \in \mathfrak{L}[z_1,\dots,z_n]$. If $P(t_1,\dots,t_n)$ has $\mathfrak{L}$-height $H_\mathfrak{L}(P) \leqslant H$ and degree $\leqslant d$ with respect to each of the variables t_i, then each monomial occurring in $P_1(z_1,\dots,z_n)$ has degree $\leqslant d$, and the $\mathfrak{L}$-height $H_\mathfrak{L}(P_1) \leqslant H + 2dn$.

In the inductive scheme of §5 we will use a number of simple lemmas on polynomials in several variables and their resultants. The two lemmas given below are similar to Lemmas 2 and 3 of **[24]** and general results **[25]**.

LEMMA 4.1. *Suppose $\theta_1,\ldots,\theta_n$, θ are arbitrary complex numbers and $P(x_1,\ldots,x_n, y)$, $S(x_1,\ldots,x_n, y) \in \mathbf{Z}[x_1,\ldots,x_n, y]$, where*

$$t(P) = \max\{\ln H(P), d_{x_1}(P) + \cdots + d_y(P)\} \leqslant T_1,$$
$$t(S) = \max\{\ln H(S), d_{x_1}(S) + \cdots + dy(S)\} \leqslant T_2.$$

If

$$(4.1)\qquad |P(\theta_1,\ldots,\theta_n,\theta)| + |S(\theta_1,\ldots,\theta_n,\theta)| < \delta\{(1+|\theta|)\cdots(1+|\theta_n|)\}^{-2nT_1T_2}(T_1+T_2)^{-(T_1+T_2)}e^{-2T_1T_2}.$$

then either there exists $R(x_1,\ldots,x_n) \in \mathbf{Z}[x_1,\ldots,x_n]$, $R(\theta_1,\ldots,\theta_n) \neq 0$, such that

$$(4.2)\quad t(R) \leqslant 4nT_1T_2 + (T_1+T_2)\ln(T_1+T_2) + 2T_1T_2, \quad |R(\theta_1,\ldots,\theta_n)| \leqslant \delta,$$

or else the polynomials $P(\theta_1,\ldots,\theta_n, y)$ and $S(\theta_1,\ldots,\theta_n, y)$ have a common root.

PROOF. Let

$$P(x) = \sum_{r=0}^{q} P_r x^r, \qquad P_r = \sum_{i_1=0}^{k_1}\cdots\sum_{i_n=0}^{k_n} P_{r,i_1,\ldots,i_n}\theta_1^{i_1}\cdots\theta_n^{i_n},$$
$$S(x) = \sum_{s=0}^{t} S_s x^r, \qquad S_s = \sum_{j_1=0}^{m_1}\cdots\sum_{j_n=0}^{m_n} S_{s,j_1,\ldots,j_n}\theta_1^{j_1}\cdots\theta_n^{j_n},$$

where $P_{r,i_1,\ldots,i_n}$, $S_{s,j_1,\ldots,j_n} \in \mathbf{Z}$. Assume that the polynomials $P(\theta_1,\ldots,\theta_n, y)$ and $S(\theta_1,\ldots,\theta_n, y)$ have no common root. Then their y-resultant

$$R(P,S) = \begin{vmatrix} P_0, & P_1, & \cdots, & P_q, & 0, & \cdots, & 0 \\ \vdots & \vdots & & \vdots & \vdots & & \vdots \\ 0, & 0, & \cdots, & P_{q-t}, & \cdots & \cdots, & P_q \\ \vdots & \vdots & & \vdots & & & \vdots \\ 0, & 0, & \cdots, & S_0, & \cdots & \cdots, & S_t \end{vmatrix}$$

is different from zero. Therefore $R(P,S) = R(\theta_1,\ldots,\theta_n)$, where $R(x_1,\ldots,x_n) \in \mathbf{Z}[x_1,\ldots,x_n]$, and a direct calculation proves the first of the inequalities (4.2). At the same time, we have a representation

$$R(P,S) = \begin{vmatrix} P(\theta), & P_1, & \cdots, & P_q, & \cdots, & 0 \\ \vdots & \vdots & & \vdots & & \vdots \\ \theta^{t-1}P(\theta), & 0, & \cdots, & P_{q-t}, & \cdots, & P_q \\ \vdots & \vdots & & \vdots & & \vdots \\ \theta^{q-1}S(\theta), & 0, & \cdots, & S_0, & \cdots, & S_t \end{vmatrix}.$$

Applying Hadamard's inequality for determinants, we obtain from (4.1) the second inequality in (4.2).

The following lemma is an n-dimensional variant of a lemma of Gel'fond (in which $n = 0$). In the case $n = 1$, this lemma was proved in [**15**] (see also [**24** and **25**]).

LEMMA 4.2. *Suppose $\theta_1,\ldots,\theta_n$, θ are algebraically independent complex numbers, $P(x_1,\ldots,x_n, y) \in \mathbf{Z}[x_1,\ldots,x_n, y]$, and $t(P) \leq T$, where $T \geq \gamma_0(\theta_1,\ldots,\theta_n,\theta)$. If*

$$(4.3) \qquad |P(\theta_1,\ldots,\theta_n,\theta)| < \exp(-\lambda T^2)$$

for $\lambda > \gamma_1 = \gamma_1(\theta_1,\ldots,\theta_n,\theta)$, then either there exists $P_1(x_1,\ldots,x_n, y) \in \mathbf{Z}[x_1,\ldots,x_n, y]$ such that $P_1(\theta_1,\ldots,\theta_n, y)$ is a divisor of $P(\theta_1,\ldots,\theta_n, y)$, $P_1(\theta_1,\ldots,\theta_n, y)$ is a power of a $\mathbf{Q}(\theta_1,\ldots,\theta_n)$-irreducible polynomial in $\mathbf{Z}[\theta_1,\ldots,\theta_n, y]$, and

$$(4.4) \qquad t(P_1) \leq 3T, \qquad |P_1(\theta_1,\ldots,\theta_n,\theta)| < \exp(-\lambda T^2/4),$$

or else for some $R(x_1,\ldots,x_n) \in \mathbf{Z}[x_1,\ldots,x_n]$, $R(\theta_1,\ldots,\theta_n) \neq 0$, we have $t(R) \leq \gamma_2 T^2$, $\gamma_2 = \gamma_2(n)$, and

$$(4.5) \qquad |R(\theta_1,\ldots,\theta_n)| < \exp(-\lambda T^2/8),$$

PROOF. Assume there exists no $R(x_1,\ldots,x_n)$, $R(\theta_1,\ldots,\theta_n) \neq 0$, satisfying (4.5) with $\gamma_2 = 64n$. Then $P(\theta_1,\ldots,\theta_n, y) = R_1(\theta_1,\ldots,\theta_n, y) \cdot R_2(\theta_1,\ldots,\theta_n, y)$, where $R_i(x_1,\ldots,x_n, y) \in \mathbf{Z}[x_1,\ldots,x_n, y]$: $i = 1, 2$, and $R_1(\theta_1,\ldots,\theta_n, y)$, $R_2(\theta_1,\ldots,\theta_n, y)$ are coprime, implies

$$(4.6) \qquad \max\{|R_1(\theta_1,\ldots,\theta_n,\theta)|, |R_2(\theta_1,\ldots,\theta_n,\theta)|\} > \exp(\lambda T^2/3).$$

Indeed, if (4.6) does not hold, we can apply Lemma 4.1. We obtain a polynomial $R(x_1,\ldots,x_n) \in \mathbf{Z}[x_1,\ldots,x_n]$, $R(\theta_1,\ldots,\theta_n) \neq 0$, satisfying (4.2). According to 1.4, we have for the types $t(R_i)$ of the polynomials $R_i(x_1,\ldots,x_n, y)$ the estimate

$$t(R_1) + t(R_2) \leq 3t(P).$$

Thus, for the type of $R(x_1,\ldots,x_n)$ we have $t(R) \leq (4n+3)T^2$ when $T \geq \gamma_0(n)$. From (4.1) and (4.2) we obtain

$$|R(\theta_1,\ldots,\theta_n)| < \exp(-\lambda T^2/4),$$

when $T \geq \gamma_0'(n, \theta_1,\ldots,\theta)$ and $\lambda \geq \gamma_1'(\theta_1,\ldots,\theta)$, i.e. (4.5) holds. Estimate (4.6) is proved. We have a representation

$$P(\vec{\theta}, y) = AP_1(\vec{\theta}, y) \cdots P_m(\vec{\theta}, y),$$

where $A \in \mathbf{Z}[\vec{\theta}]$, the $P_i(\vec{\theta}, y)$ $(i = 1,\ldots,m)$ are powers of distinct irreducible polynomials in $\mathbf{Z}[\vec{\theta}, y]$, and $|P_1(\vec{\theta},\theta)| \leq \cdots \leq |P_m(\vec{\theta},\theta)|$. Let $s \leq m$ be such that $|P_s(\vec{\theta},\theta)| < 1$, $|P_{s+1}(\vec{\theta},\theta)| \geq 1$. Since (4.5) does not hold, we obtain for the polynomial $P'(\vec{\theta}, y) = P_1(\vec{\theta}, y) \cdots P_s(\vec{\theta}, y)$ an estimate of the form (4.3)

$$|P'(\vec{\theta}, y)| \leq \exp(-7\lambda T^2/8).$$

There exists $l \leq [\frac{1}{2}s] + 1$ such that

$$|P_1(\vec{\theta},\theta)| \cdots |P_{l-1}(\vec{\theta},\theta)| \geq |P_l(\vec{\theta},\theta)| \cdots |P_s(\vec{\theta},\theta)|,$$
$$|P_1(\vec{\theta},\theta)| \cdots |P_l(\vec{\theta},\theta)| \leq |P_{l+1}(\vec{\theta},\theta)| \cdots |P_s(\vec{\theta},\theta)|.$$

From this and (4.6) we obtain

$$|P_{l+1}(\vec{\theta},\theta)|\cdots|P_s(\vec{\theta},\theta)|>\exp(-\lambda T^2/3)$$

and

$$|P_l(\vec{\theta},\theta)|\cdots|P_s(\vec{\theta},\theta)|<\exp(-3\lambda T^2/4).$$

Finally,

$$|P_1(\vec{\theta},\theta)|\leqslant|P_l(\vec{\theta},\theta)|\leqslant\exp(-5\lambda T^2/12),$$

i.e. the second inequality in (4.4) is proved.

We establish a series of auxiliary algebraic statements related to polynomials from n variables over rings $\mathfrak{L}$ of finite type over **Z**. Formally they resemble lemmas of Gel′fond type [**1**] (for $\mathfrak{L}=\mathbf{Z}$) or that of Brownawell [**25**] (for $\mathfrak{L}$ having "finite type of transcendence over Q") but are different in essence. It is no longer possible, as before, to single out irreducible divisors of polynomials and to examine them only. The reason for this is that it is impossible to obtain a good estimate of the $\mathfrak{L}$-height ($\mathfrak{L}$-type) of divisors of polynomials in terms of the $\mathfrak{L}$-type of a polynomial itself even when $\mathfrak{L}$ is a ring of algebraic numbers. As is shown in [**1**, **7**], even in this case the estimates inevitably contain the discriminant of $\mathfrak{L}$, regulator of $\mathfrak{L}$, etc. In order to avoid unnecessary change for the worse in the estimates, which is caused by growth of degrees of polynomials as a logarithm of height, we have to examine polynomials with multiple roots, etc. It is natural to introduce instead of discriminants, resultants and so on, more subtle (nondegenerating) notions of semidiscriminant, semiresultant. These notions have a simple arithmetic sense. While the discriminant corresponds to the set of primes p, for which a polynomial (manifold) has nondegenerate reduction [**26**], the semidiscriminant corresponds to the set of those p, for which the number of irreducible factors (components) and their multiplicities are preserved under reduction (mod p). Semidiscriminants and semiresultants allow us to connect arithmetics with analysis. In the examination of rings of finite type over **Z** there appear other peculiarities which are explained at the end of §4 and which occur in the proofs of Theorems I–III of §5.

We introduce one of the equivalent definitions of semidiscriminant (cf. [**27**, **28**, **29**]). Most naturally, preserving the analogy with the definition of a discriminant, would be to define $SD(P)$ in the following way:

$$SD(P)=a^{n-1}\prod_{u_i\neq u_j}(u_i-u_j),$$

where u_i, u_j are arbitrary roots of polynomial $P(x)$ and are counted with their respective multiplicities. Hence we obtain another definition of semidiscriminant, modifying (1.32).

$$SD(P)=a^{n-1}\prod_{i=1}^{k}\left(\frac{P^{(s_i)}(t_i)}{s_i!}\right)^{s_i}.$$

As t_i is a root of multiplicity s_i for $P(x)$, then

$$\frac{P^{(s_i)}(t_i)}{s_i!} = a \prod_{\substack{j=1 \\ j\neq i}}^{k} (t_i - t_j)^{s_j}. \tag{1.33}$$

Therefore, $SD(P) \neq 0$.

Let us present a general result concerning common roots of polynomials. As in [**1**, **7**], this statement plays the central role in all the auxiliary lemmas.

LEMMA 4.3. *Suppose $P(x) \in \mathfrak{L}[x]$ is any polynomial of degree n and $\mathfrak{L}$-type $t_{\mathfrak{L}}(P) \leqslant T$, with distinct roots $t_1, \ldots, t_k$ of multiplicities $s_1, \ldots, s_k$, where $\sum_{i=1}^{k} s_i = n$. If θ is any number and $|\theta - t_{i_0}| = \min_{i=1,\ldots,k} |\theta - t_i|$, we have for some $u \in \mathfrak{L}$, $u \neq 0$, $t_{\mathfrak{L}}(u) \leqslant c_1 nT$, the estimate*

$$|P(\sigma)| \geqslant |\sigma - t_{i_0}|^{s_{i_0}} \exp(-c_2 nT)\,|u|\,. \tag{4.7}$$

PROOF. We estimate $|SD(P)|$ from above. We have

$$\left|\frac{P^{(s_i)}(t_i)}{s_i!}\right| \leqslant (\max\{1, |t_i|\})^{n-s_i} H(P)(C_n^{s_i} + C_{n-1}^{s_i} + \cdots + 1).$$

Therefore, if

$$SD_{i_0}(P) = a^{n-1} \prod_{\substack{i=1 \\ i\neq i_0}}^{k} \left(\frac{P^{(s_i)}(t_i)}{s_i!}\right)^{s_i} \cdot \left(\frac{P^{(s_{i_0})}(t_{i_0})}{s_{i_0}!}\right)^{s_{i_0}-1},$$

we obtain

$$|SD_{i_0}(P)| \leqslant |a|^{n-1} H(P)^{\sum_{i=1}^{k} s_i - 1} \prod_{i=1}^{k} (C_n^{s_i} + \cdots + 1)^{s_i}$$
$$\times \prod_{i=1}^{k} \max\{1, |t_i|\}^{(n-s_i)s_i}.$$

According to Lemma 1.1,

$$|SD_{i_0}(P)| \leqslant H(P)^{n-1} L(P)^{n-1} \prod_{i=1}^{k} (C_n^{s_i} + \cdots + 1)^{s_i}$$
$$\leqslant n^{2n} H(P)^{2n} \prod_{i=1}^{k} 2^{ns_i} \leqslant n^{2n} H(P)^{2n} \cdot 2^{n^2}.$$

Expressing $H(P)$ in terms of T, we see that $H(P) < \exp(c_1 T)$, i.e.

$$|SD_{i_0}(P)| \leqslant \exp(c_2' Tn), \tag{4.8}$$

since $T \geqslant n$. Furthermore, in view of (1.13),

$$P^{(s_{i_0})}(t_{i_0})/s_{i_0}! = a \prod_{j\neq i_0} (t_{i_0} - t_j)^{s_j}.$$

By definition of i_0, we have $|t_{i_0} - t_i| \leqslant |\theta - t_{i_0}| + |\theta - t_i| \leqslant 2|\theta - t_i|$. Thus,

$$|P^{(s_{i_0})}(t_{i_0})/s_{i_0}!| \leqslant 2^{n-1}|a| \prod_{j \neq i_0} |\theta - t_j|^{s_j} = 2^{n-i}\frac{|P(\theta)|}{|\theta - t_{i_0}|^{s_{i_0}}}$$

if $\theta \neq t_{i_0}$. It follows from this inequality and (4.8) that

$$|SD(P)| \leqslant \exp(c_3' nT)\frac{|P(\theta)|}{|\theta - t_{i_0}|^{s_{i_0}}}. \tag{4.9}$$

It remains to show only that $SD(P) \in \mathcal{K}$. To do this we use the theorem on symmetric functions. We first introduce the following notation: let $\xi_1, \ldots, \xi_n$ be all roots of $P(x)$, i.e.

$$\xi_1 = \cdots = \xi_{s_1} = t_1, \quad \xi_{s_1+1} = \cdots = \xi_{s_1+s_2} = t_2, \quad \xi_{s_1+\cdots+s_{k-1}+1} = \cdots = \xi_n = t_k.$$

Then any elementary symmetric function f_i in $\xi_1, \ldots, \xi_n$ has the form $\pm p_i/a$, where p_i is some coefficient of $P(x)$, i.e. $af_i \in \mathfrak{L}$.

Unfortunately, $SD(P)$ is not immediately an elementary symmetric function either of the t_i or the ξ_j, but $SD(P)$ can be transformed into one. This can be done as follows.

For $i = 1, \ldots, k$ let T_i denote the set of those $j \leqslant n$ such that $\xi_j = t_i$, i.e. $T_i = \{s_1 + \cdots + s_{i-1} + 1, \ldots, s_1 + \cdots + s_i\}$. For any permutation κ of the set $\{1, \ldots, n\}$ we consider the expression

$$SD^{\kappa}(P) = a^{n-1}\prod_{i=1}^{k}\prod_{j \in T_i}\frac{P^{(s_i)}(\xi_{\kappa(j)})}{s_i!}. \tag{4.10}$$

We will show that, for any permutation κ, either $SD^{\kappa}(P) = SD(P)$ or else $SD^{\kappa}(P) = 0$. Indeed, for any $j \in \{1, \ldots, n\}$, let s^j be the multiplicity of the root ξ_j in $P(x)$. Then $s^j = s_i$ for $j \in T_i$. If $s^j = s^{\kappa(j)}$ for each $j \in \{1, \ldots, n\}$, then (4.10) implies that $SD^{\kappa}(P) = SD(P)$. Assume that for some $j_0 \in \{1, \ldots, n\}$ we have $s^{j_0} \neq s^{\kappa(j_0)}$. Then there exists $j' \in \{1, \ldots, n\}$ such that $s^{\kappa(j')} < s^{j'}$. For if this were not so, we would have $\sum_{j=1}^n s^j \leqslant \sum_{j=1}^n s^{\kappa(j)}$ and therefore, since $s^{\kappa(j_0)} \neq s^{j_0}$, we would have $\sum_{j=1}^n s^j < \sum_{j=1}^n s^{\kappa(j)}$. This contradicts the fact that κ is one-to-one. Thus, $s^{\kappa(j')} < s^{j'}$, and then $\xi_{\kappa(j')}$ is a root of $P(x)$ of multiplicity $s^{\kappa(j')} < s^{j'}$, i.e. $SD^{\kappa}(P) = 0$ in view of (4.10). In addition to (4.10) consider the following sum, which can be viewed as a symmetric polynomial in the variables $\xi_1, \ldots, \xi_n$:

$$SD(P) = \sum_{\kappa} SD^{\kappa}(P). \tag{4.11}$$

Here $\sum_{\kappa}$ means that the summation extends over all permutations κ of the set $\{1, \ldots, n\}$.

It follows from the above argument that $SD_1(P) = mSD(P)$, where m is a natural number, $m \leqslant n!$.

To prove (4.7) it now suffices to show that $u = SD_1(P)$ is an element of $\mathfrak{L}$ and $t_{\mathfrak{L}}(u) \leqslant c_4' nT$. We can then apply (4.9) with $SD(P) = u/m$, $m \leqslant n!$.

We make use of (4.11) and (4.10). According to (4.11), $SD_1(P)$ is symmetric as a polynomial in $\xi_1,\ldots,\xi_n$. We can now use the theorem on symmetric functions, taking into account the observation that the elementary symmetric functions f_i in $\xi_1,\ldots,\xi_n$ have the form $f_i = \pm P_i/a$, where $P_i \in \mathfrak{L}$, $t_{\mathfrak{L}}(P_i) \leqslant T$.

Indeed, $SD_1(P)$ has the form $a^{n-1}S$, where S is a symmetric polynomial in $\xi_1,\ldots,\xi_n$. The degree of this polynomial $S = S(\xi_1,\ldots,\xi_n)$ with respect to each of the variables ξ_j is at most $n-1$, and the $\mathfrak{L}$-height $H_{\mathfrak{L}}(S)$ of $S(x_1,\ldots,x_n)$ is at most $c_5'n(T+\ln n) \leqslant 2c_5'nT$. By the theorem on symmetric functions, we have a representation $S = S_1(f_1,\ldots,f_n)$ for some $S_1(x_1,\ldots,x_n) \in \mathfrak{L}[x_1,\ldots,x_n]$, where the degree of S_1 with respect to all variables $f_1,\ldots,f_n$ (i.e. the maximum of the degrees of the monomials occurring in S_1) is at most $n-1$ and the $\mathfrak{L}$-height $H_{\mathfrak{L}}(S_1)$ of $S_1(x_1,\ldots,x_n)$ is at most $c_6'nT$, $T \geqslant n$. But then $a^{n-1}S = a^{n-1}S_1(\pm P_1/a,\ldots,\pm P_n/a)$ is an element of $\mathfrak{L}$ having $\mathfrak{L}$-type at most $2c_6'nT$. This proves Lemma 4.3.

It is necessary to prove analogues of Lemmas 1.6 and 1.7 on estimates of $|P(\zeta)|$, where $P(x) \in \mathfrak{L}[x]$, for the case where ζ is a root of multiplicity s of a polynomial in $\mathfrak{L}[x]$ with known type. So suppose $Q(x) \in \mathfrak{L}[x]$ is of degree d, $t_{\mathfrak{L}}(Q) \leqslant t$, and ζ is a root of $Q(x)$ of multiplicity s. Let $\zeta = \zeta_1,\ldots,\zeta_d$ be all roots of $Q(x)$, enumerated so that $\zeta_1 = \cdots = \zeta_s,\ldots$.

Consider an arbitrary polynomial $P(x) \in \mathfrak{L}[x]$ of degree N and of $\mathfrak{L}$-type $t_{\mathfrak{L}}(P) \leqslant T$. Let us assume that $P(\zeta) \neq 0$ and find a lower bound for $|P(\zeta)|$ in terms of T and t. The simplest way of doing this is to use, instead of the norm of $P(\zeta)$ in $\mathcal{K}(\zeta)/\mathcal{K}$, the following expression:

$$a^N \prod_{j=1}^{d} P(\zeta_j), \tag{4.12}$$

where a is the leading coefficient of $Q(x)$. The expression (4.12), unfortunately, cannot be applied directly, since for some j it can happen that $P(\zeta_j) = 0$ even though $P(\zeta) \neq 0$, since the ζ_j are not all conjugate to ζ. Therefore, instead of (4.12) we consider the similar expression of seminorm

$$N(P) = a^N \prod_{\substack{j=1 \\ P(\zeta_j)\neq 0}}^{d} P(\zeta_j),$$

and denote by $\Lambda \subseteq \{1,\ldots,d\}$ the set of those j for which $P(\zeta_j) \neq 0$. For each subset $S \subseteq \{1,\ldots,d\}$ of cardinality $|S| = |\Lambda|$ we define

$$N_S(P) = a^N \prod_{\substack{j=1 \\ j\in S}}^{d} P(\zeta_j). \tag{4.13}$$

Then for any $S \subseteq \{1,\ldots,d\}$, $|S| = |\Lambda|$, we have either $S = \Lambda$ and $N_S(P) = N(P)$ or else $N_S(P) = 0$. Indeed, if $S \neq \Lambda$, then $j \notin \Lambda$ for some $j \in S$. In this case, $P(\zeta_j) = 0$ and $N_S(P) = 0$.

We denote by $P_\Lambda(n)$ the collection of all subsets of $\{1,\ldots,d\}$ of cardinality $|\Lambda|$. In the spirit of (4.11) we consider the formal expression

$$N_1(P) = \sum_{S\in P_\Lambda(n)} N_S(P). \tag{4.14}$$

By what we have proved, $N_1(P) = N(P)$, since $N_S(P) = 0$ for $S \in P_\Lambda(n)$, $S \neq \Lambda$. The essential point is that $N_1(P)$ is a symmetric function in the variables $\zeta_1,\ldots,\zeta_d$. Indeed, if κ is a permutation of the set $\{1,\ldots,d\}$, then $\{\kappa(S)\colon S \in P_\Lambda(n)\} = P_\Lambda(n)$ and $\kappa(S) \neq \kappa(S')$ when $S \neq S'$. Moreover,

$$a^N \prod_{\substack{j=1\\ j\in S}} P(\zeta_{\kappa(j)}) \equiv N_{\kappa(S)}(P),$$

i.e. $N_1(P)$ is a symmetric function in the variables $\zeta_1,\ldots,\zeta_d$. According to (4.13) and (4.14), $N_1(P)$ can be written in the form $N_1(P) = a^N S$, where $S = S(\zeta_1,\ldots,\zeta_d)$ is a symmetric polynomial in $\zeta_1,\ldots,\zeta_d$. Let us estimate the height $H_\mathfrak{L}(S)$ and degree $d_{x_i}(S)$ of the polynomial

$$S(x_1,\ldots,x_d) \in \mathfrak{L}[x_1,\ldots,x_d], \qquad S = S(\zeta_1,\ldots,\zeta_d).$$

First of all, it follows from (4.13) and (4.14) that the degree $d_{x_i}(S) \leqslant N$ and the $\mathfrak{L}$-height $H_\mathfrak{L}(S)$ satisfies $H_\mathfrak{L}(S) \leqslant c_7' dT \leqslant c_7' tT$. Therefore, if we apply the theorem on symmetric functions to the polynomial $S(x_1,\ldots,x_d)$, we obtain $S = S_1(f_1,\ldots,f_d)$, where the f_i are the elementary symmetric polynomials in $\zeta_1,\ldots,\zeta_d$, and $S_1(x_1,\ldots,x_d) \in \mathfrak{L}[x_1,\ldots,x_d]$, the degree of each monomial occurring in S_1 is at most N, and $t_\mathfrak{L}(S_1) \leqslant c_8' dT \leqslant c_8' tT$. Since $f_i = P_i/a$, where $P_i \in \mathfrak{L}$, $t_\mathfrak{L}(P_i) \leqslant t$, it follows that $N_1(P) \in \mathfrak{L}$ and we have an estimate of the type, of the form $t_\mathfrak{L}(N_1(P)) \leqslant c_9'(tN + dT) \leqslant c_{10}' tT$, since $T \geqslant N$, $t \geqslant d$. We make separate mention of the last conclusion and another consequence of our calculations:

$$N(P) \neq 0,\ N(P) \in \mathfrak{L}, \qquad t_\mathfrak{L}(N(P)) \leqslant c_{10}'(tN + dt) \leqslant 2c_{10}' Tt. \tag{4.15}$$

We can use seminorm $N(P)$ instead of the norm in the proof of

LEMMA 4.4. *Suppose $Q(x) \in \mathfrak{L}[x]$ is a polynomial of degree d and type $t_\mathfrak{L}(Q) \leqslant t$, and ζ is a root of $Q(x)$ of multiplicity s. If $P(x) \in \mathfrak{L}[x]$ has degree N and $\mathfrak{L}$-type $t_\mathfrak{L}(P) \leqslant T$, then either $P(\zeta) = 0$ or else for some $x \in \mathfrak{L}$, $x \neq 0$, $t_\mathfrak{L}(x) \leqslant c_3(dt + Nt) \leqslant 2c_3 Tt$, we have*

$$|P(\zeta)| \geqslant \exp(-c_4 Tt/s)\,|x|^{1/s}. \tag{4.16}$$

PROOF. We use the notation and assertions given above.

$$N(P) = a^N \prod_{i=1,\ P(\zeta_i)\neq 0}^{d} P(\zeta_i), \tag{4.17}$$

where $\zeta_1 = \zeta, \ldots, \zeta_d$ are all of the roots of $P(x)$. We have
(4.18)

$$\left| a^N \prod_{i=s+1, P(\zeta_i)\neq 0}^{d} P(\zeta_i) \right| \leq |a|^N (N+1)^{d-s} H(P)^{d-s} \prod_{i=s+1}^{d} (\max\{1, |\zeta_i|\})^N$$
$$\leq (N+1)^d H(P)^d H(Q)^N \leq \exp(c'_{11}(dT + Nt)).$$

Moreover, since ζ is a root of $Q(x)$ of multiplicity s, it follows that $\zeta = \zeta_1 = \cdots = \zeta_s$ and

$$\left| \prod_{i=1}^{s} P(\zeta_i) \right| = |P(\zeta)|^s.$$

Comparing this expression with (4.18) and (4.17), we obtain

$$|N(P)| \leq \exp(c'_{11}(dT + Nt)) \, |P(\zeta)|^s.$$

Putting $x = N(P)$, we at once obtain (4.16).

LEMMA 4.5. *Suppose $\zeta_1, \ldots, \zeta_k$ are algebraic numbers over $\mathfrak{L}$, $\mathfrak{L}_i = \mathfrak{L}[\zeta_1, \ldots, \zeta_{i-1}]$, and ζ_i is a root of a polynomial $Q_i(x) \in L_i[x]$ of multiplicity s_i with $t_{\mathfrak{L}_i}(Q_i) \leq t_i$. If $P(x_1, \ldots, x_k) \in \mathfrak{L}[x_1, \ldots, x_k]$, $t_{\mathfrak{L}}(P) \leq T$, then either*

$$P(\zeta_1, \ldots, \zeta_k) = 0,$$

or else for some $v \in \mathfrak{L}$, $v \neq 0$, $t_{\mathfrak{L}}(v) \leq c_5 T t_1 \cdots t_k$, we have

$$|P(\zeta_1, \ldots, \zeta_\alpha)| \geq \exp(-c_6 T t_1 \cdots t_k / s_1 \cdots s_k) \, |v|^{1/s_1 \cdots s_k}.$$

PROOF. Assume that Lemma 4.5 has been proved for polynomials in $k-1$ variables and for any $\mathfrak{L}'$. We will prove Lemma 4.5 for k and $\mathfrak{L}$. Consider the ring $\mathfrak{L}' = \mathfrak{L}[\zeta_1]$, the number $\zeta'_1 = \zeta_2, \ldots, \zeta'_{k-1} = \zeta_k$ and the polynomial $P'(x_1, \ldots, x_{k-1}) = P(\zeta_1, x_1, \ldots, x_{k-1})$. By our assumption concerning a polynomial in $k-1$ variables, if $P(\zeta_1, \ldots, \zeta_k) = P'(\zeta'_1, \ldots, \zeta'_{k-1}) \neq 0$, then

$$|P(\zeta_1, \ldots, \zeta_k)| > \exp(-c'_{13} T' t_2 \cdots t_k / s_2 \cdots s_k) \, |u|^{1/s_2 \cdots s_k}$$

for $T' = t_{\mathfrak{L}'}(P') \leq t_{\mathfrak{L}}(P) \leq T$ and $u \in \mathfrak{L}'$, $u \neq 0$, $t_{\mathfrak{L}'}(u) \leq c'_{14} T t_2 \cdots t_k$. Using Lemma 4.4 we estimate $|u| = |\mathfrak{U}(\zeta_1)|$, where $\mathfrak{U}(x) \in \mathfrak{L}[x]$ and $t_{\mathfrak{L}}(\mathfrak{U}) \leq c'_{15} T t_2 \cdots t_k$.

We obtain

$$|u| > \exp(-c'_{16} T t_1 \cdots t_k / s_1) \, |v|^{1/s_1}$$

with $v \in \mathfrak{L}$, $v \neq 0$, $t_{\mathfrak{L}}(v) \leq c'_{17} T t_1 \cdots t_k$. Since $s_1 \leq \deg(Q_1(x)) \leq t_1$, we obtain Lemma 4.5.

Lemma 4.5 will often be used in the proofs of §§4–6.

In addition to Lemma 4.5 we need similar assertions, in the spirit of §1, for resultants of polynomials and their common divisors.

LEMMA 4.6. *Suppose $\zeta_1,\dots,\zeta_k$ are algebraic numbers over $\mathfrak{L}$, $\mathfrak{L}_i = \mathfrak{L}[\zeta_1,\dots,\zeta_{i-1}]$, and ζ_i is a root of a polynomial $Q_i(x) \in \mathfrak{L}_i[x]$ of multiplicity s_i with $t_{\mathfrak{L}_i}(Q_i) \leq t_i$. Assume that $P(x_1,\dots,x_k, x) \in \mathfrak{L}[x_1,\dots,x_k, x]$, $t_{\mathfrak{L}}(P) \leq T$, and $\xi_1,\dots,\xi_l$ are all of the distinct roots of the polynomial $P(x) = P(\zeta_1,\dots,\zeta_k, x)$ of multiplicities $r_1,\dots,r_l$, with $n = \sum_{i=1}^{l} r_i$. Then there exists $w \in \mathfrak{L}$, $w \neq 0$, $t_{\mathfrak{L}}(w) \leq c_7 nTt_1 \cdots t_k$, such that for any θ we have*

$$|P(\theta)| > \min_{i=1,\dots,l} |\theta - \xi_i|^{r_i} \exp(-c_8 nTt_1 \cdots t_k/s_1 \cdots s_k)\,|w|^{1/s_1\cdots s_k}.$$

PROOF. Lemma 4.6 follows from Lemmas 4.3 and 4.5.

LEMMA 4.7. *Suppose $P(x)$, $Q(x) \in \mathfrak{L}[x]$ are polynomials of degrees d_P, d_Q and $\mathfrak{L}$-types t_P, t_Q, and $P_1(x)$, $Q_1(x) \in \mathfrak{K}[x]$ are polynomials with leading coefficients of unity such that $P_1(x)^s$ divides $P(x)$ and $Q_1(x)$ divides $Q(x)$ in $\mathfrak{K}[x]$. Then there exists $r = r(P, Q) \in \mathfrak{L}$ such that $r \neq 0$, $t_{\mathfrak{L}}(r) \leq c_{10}(d_P t_Q + d_Q t_P) \leq 2c_{10} t_P t_Q$, and if*

$$\max\{|P_1(\theta)|, |Q_1(\theta)|\} < \exp\bigl(-c_9(d_P t_Q + d_Q t_P)/s\bigr)\,|r|^{1/s} \tag{4.19}$$

for each θ, then $P_1(x)$, $Q_1(x)$ have common roots.

PROOF. We begin with the definition of the number $r = r(P, Q)$, the semiresultant of $P(x)$ and $Q(x)$. Let p and q denote the leading coefficients of $P(x)$ and $Q(x)$, and $t_1,\dots,t_{d_P}$ and $u_1,\dots,u_{d_Q}$ denote the roots of $P(x)$ and $Q(x)$; put

$$r = p^{d_Q} \cdot q^{d_P} \cdot \prod_{i=1}^{d_P} \prod_{\substack{j=1 \\ t_i \neq u_j}}^{d_Q} (t_i - u_j). \tag{4.20}$$

Then $r \neq 0$, and it remains to show that $r \in \mathfrak{L}$ and to estimate $t_{\mathfrak{L}}(r)$. Let $\mathfrak{M}$ denote the set of all $(i, j) \in \{1,\dots,d_P\} \times \{1,\dots,d_Q\}$ such that $t_i = u_j$, and let $s = |\mathfrak{M}|$. For each subset $F \subseteq \{1,\dots,d_P\} \times \{1,\dots,d_Q\}$ of cardinality s we put

$$r_F = p^{d_Q} q^{d_P} \prod_{i=1}^{d_P} \prod_{\substack{j=1 \\ (i,j) \notin F}}^{d_Q} (t_i - u_j). \tag{4.21}$$

Consider the formal sum

$$r_0 = \sum_F r_F, \tag{4.22}$$

where the sum $\sum_F$ extends over all subsets of $\{1,\dots,d_P\} \times \{1,\dots,d_Q\}$ of cardinality s. Then r_0 is a symmetric function relative to the t_i and to the u_j. Suppose, for example, that κ is a permutation of the set $\{1,\dots,d_P\}$; put $\kappa(F) = \{(\kappa(i), j)$: $(i, j) \in F\}$. Since κ is one-to-one, $|\kappa(F)| = s$ and it follows from (4.21) that $(r_F)^\kappa$, the expression obtained from r_F by replacing t_i by $t_{\kappa(i)}$, agrees with $r_{\kappa(F)}$. Consequently, r_0 is a symmetric polynomial relative to the t_i and to the u_j. In particular, $r_0 = p^{d_Q} \cdot q^{d_P} \cdot R$, where $R = R(t_1, \ldots; u_1, \ldots)$ is a symmetric polynomial relative to the t_i and to the u_j with rational integral coefficients. Applying the theorem on symmetric functions to $R(t_1, \ldots; u_1, \ldots)$, we obtain a polynomial $R_1(f_1, \ldots, f_{d_P}; g_1, \ldots, g_{d_Q})$ with rational integral coefficients such that $R = R_1$, where $f_1, \ldots, f_{d_P}$ are the elementary symmetric polynomials in $t_1, \ldots, t_{d_P}$ and $g_1, \ldots, g_{d_Q}$ are the elementary symmetric polynomials in $u_1, \ldots, u_{d_Q}$.

The polynomial $R(\cdots)$ has degree $\leqslant d_Q$ in each of the variables t_i and has degree $\leqslant d_P$ in each of the variables u_j. Thus, the degree of each monomial in the f_i occurring in $R_1(\cdots)$ is $\leqslant d_Q$ and the degree of each monomial in the g_j is $\leqslant d_P$. The height of $R_1(\cdots)$ can be crudely estimated by $\exp(4d_P d_Q)$. By the definitions of f_i, g_j and the types of $P(x)$, $Q(x)$ we have $f_i = p_i/p$, $g_j = q_j/q$, where $t_{\mathfrak{L}}(p_i) \leqslant t_P$, $t_{\mathfrak{L}}(q_j) \leqslant t_Q$. Thus, $r_0 = p^{d_Q} \cdot q^{d_P} \cdot R$ is an element of $\mathfrak{L}$ and $t_{\mathfrak{L}}(r_0) \leqslant c'_{18}(t_P d_Q + t_Q d_P)$.

We will show, however, that $r = r_0$, where r is defined in (4.20). Suppose $F \subset \{1, \ldots, d_P\} \times \{1, \ldots, d_Q\}$, $F \neq \mathfrak{M}$, $|F| = s$. Then $(i, j) \notin F$ for some $(i, j) \in \mathfrak{M}$. It then follows from (4.21) that $r_F = 0$. According to (4.22), $r_0 = r$ and, by the above,

(4.23) $$r \neq 0, \qquad t_{\mathfrak{L}}(r) \leqslant c'_{18}(t_P d_Q + t_Q d_P).$$

We use r to prove (4.19). Suppose t_j is any root of $P_1(x)$, occurring with multiplicity f_j in $Q(x)$. We denote the set of all such j by $\mathfrak{P}$. Assume that $P_1(x)$ and $Q_1(x)$ have no common roots. Then $q^{-1}Q(x) = (P_{1j}(x))^{f_j} Q_{2j}(x)$, where $P_{1j}(x)$, $Q_{2j}(x) \in \mathfrak{K}[x]$ are polynomials with leading coefficients unity, t_j is a root of an irreducible polynomial $P_{1j}(x)$ dividing $P(x)$, and $Q_{2j}(t_j) \neq 0$. In this case, according to (1.21),

(4.24) $$q^{-1}Q^{(f_j)}(t_j) = f_j! Q_{2j}(t_j) \cdot \left(P'_{1j}(t_j)\right)^{f_j}.$$

We can also express (4.24) as follows:

(4.25) $$q^{-1}Q^{(f_j)}(t_j) = f_j! \prod_{\substack{i=1 \\ u_i \neq t_j}}^{d_Q} (t_j - u_i).$$

Moreover, $Q_1(x)$ and $P_1(x)$ have no common roots. Thus, $Q_1(x)$ and $P_{1j}(x)$ have no common roots and $Q_{2j}(x)$ can be written in the from $Q_{2j}(x) = Q_1(x) Q_{3j}(x)$, where $Q_{3j}(x) \in \mathfrak{K}[x]$ has leading coefficient unity. In view of (4.24) and (4.25), for $j \in \mathfrak{P}$ we have

$$\prod_{i=1,\, u_i \neq t_j}^{d_Q} (t_j - u_i) = Q_1(t_j) Q_{3j}(t_j) \cdot \left(P'_{1j}(t_j)\right)^{f_j}.$$

We apply the semiresultant r, making use of (4.20):

(4.26)
$$|r| \leqslant \left| p^{d_Q} q^{d_P} \prod_{j=1,\, j \notin \mathfrak{P}}^{d_P} \prod_{i=1,\, u_i \neq t_j}^{d_Q} (t_j - u_i) \prod_{j \in \mathfrak{P}} \left(Q_1(t_j) Q_{3j}(t_j) \left(P'_{1j}(t_j)\right)^{f_j}\right) \right|.$$

We estimate separately the factors in (4.26). We have

$$\left|\prod_{j=1,\, j\notin\mathcal{P}}^{d_P}\ \prod_{i=1,\, u_i\neq t_j}^{d_Q}(u_i-t_j)\right|\leqslant 2^{d_Pd_Q}\prod_{j\in\mathcal{P}}\prod_{i=1}^{d_Q}\left(\max\{|t_j|,1\}\max\{|u_i|,1\}\right).$$

To estimate the second factor in (4.26) we use the representation $q^{-1}Q(x)=Q_1(x)Q_{3j}(x)(P_{1j}(x))^{f_j}$. Since $H(P'_{1j})^{f_j}\leqslant d(P_{1j})^{f_j}H(P_{1j})^{f_j}$, it follows from Lemma 1.4 that

$$H(Q_1)\cdot H(Q_{3j})\cdot H(P'_{1j})^{f_j}\leqslant|q|^{-1}\exp(c'_{19}t_Q)d(P_{1j})^{f_j}, \tag{4.27}$$

since $\ln H(Q)\leqslant c'_{20}t_Q$. Thus, we can estimate the other factors in (4.26):
(4.28)

$$\begin{aligned}\left|\prod_{j\in\mathcal{P}}Q_{3j}(t_j)\cdot\left(P'_{1j}(t_j)\right)^{f_j}\right|&\leqslant\prod_{j\in\mathcal{P}}\Big(d(Q)\cdot d(P_{1j})^{f_j}\cdot H(Q_{3j})\\&\qquad\times H(P'_{1j})^{f_j}\left(\max\{1,|t_j|\}\right)^{d(Q_{3j})+f_j(d(P_{1j})-1)}\Big)\\&\leqslant e^{2d_Pd_Q}\prod_{j\in\mathcal{P}}\left(H(Q_{3j})H(P'_{1j})^{f_j}\max\{1,|t_j|^{d_Q-d(Q_1)}\}\right),\end{aligned}$$

since $d(Q_{3j})+f_j(d(P_{1j})-1)+d(Q_1)\leqslant d_Q$ and $d(P_{1j})\leqslant d_P$. Also,

$$\begin{aligned}|Q_1(t_j)|&\leqslant|Q(\theta)-Q_1(t_j)|+|Q_1(\theta)|\\&\leqslant d_QH(Q_1)\max\{|\theta^{d(Q_1)}-t_j^{d(Q_1)}|,\dots,|\theta-t_j|\}+|Q_1(\theta)|\\&\leqslant\exp(c'_{20}t_Q)H(Q_1)|\theta-t_j|\max\{1,|t_j|^{d(Q_1)}\}+|Q_1(\theta)|.\end{aligned}$$

We can assume that $\max\{|P_1(\theta)|,|Q_1(\theta)|\}<1$. Otherwise, Lemma 4.7 can be obtained from a trivial estimate of $|r|$ using (4.20) and 1.1:

$$\begin{aligned}|r|&\leqslant|p|^{d_Q}|q|^{d_P}\prod_{i=1}^{d_P}\prod_{j=1}^{d_Q}2\max\{|t_i|,|u_j|,1\}\\&\leqslant|p|^{d_Q}|q|^{d_P}\prod_{i=1}^{d_P}\prod_{j=1}^{d_Q}\left(\max\{|t_i|,1\}\max\{|u_j|,1\}\right)\\&\leqslant L(P)^{d_Q}L(Q)^{d_P}\leqslant\exp\left(c'_{21}(d_Qt_P+d_Pt_Q)\right).\end{aligned}$$

Thus, $|P_1(\theta)|$, $|Q_1(\theta)|<1$ and it follows from (4.28) that

$$\begin{aligned}&\left|\prod_{j\in\mathcal{P}}Q_1(t_j)Q_{3j}(t_j)\left(P'_{1j}(t_j)\right)^{f_j}\right|\\&\qquad\leqslant\exp\left(c'_{22}(d_Pt_Q+d_Qt_P)\right)\prod_{j\in\mathcal{P}}\max\{1,|t_j|^{d_Q}\}\\&\qquad\quad\times\prod_{j\in\mathcal{P}}H(Q_1)H(Q_{3j})H(P'_{1j})^{f_j}\Big(\prod_{j\in\mathcal{P}}|\theta-t_j|+|Q_1(\theta)|\Big).\end{aligned}$$

Therefore, we obtain from (4.27) the estimate

$$\left|\prod_{j\in\mathcal{P}} Q_1(t_j)Q_{3j}(t_j)\left(P'_{1j}(t_j)\right)^{f_j}\right|$$
$$\leqslant \exp\left(c'_{23}(d_P t_Q + d_Q t_P)\right)|q|^{-|\mathcal{P}|}\times e^{d_P d_Q}\times\prod_{j\in\mathcal{P}}\max\left\{1,|t_j|^{d_Q}\right\}$$
$$\times\prod_{j\in\mathcal{P}}\left(|\theta-t_j|+|Q_1(\theta)|\right)\leqslant\exp\left(c'_{24}(d_P t_Q+d_Q t_P)\right)$$
$$\times\prod_{j\in\mathcal{P}}\max\left\{1,|t_j|^{d_Q}\right\}\times|q|^{-|\mathcal{P}|}\left(\prod_{j\in\mathcal{P}}|\theta-t_j|+|Q_1(\theta)|\right).$$

Moreover, $\prod_{j\in\mathcal{P}}|\theta-t_j|\leqslant|P_1(\theta)|^s$, since $P_1(x)^s$ divides $P(x)$ and the t_j: $j\in\mathcal{P}$, are those roots of $P(x)$ which are roots of $P_1(x)$. Therefore,

$$\left|\prod_{j\in\mathcal{P}} Q_1(t_j)Q_{3j}(t_j)\left(P'_{1j}(t_j)\right)^{f_j}\right| \tag{4.29}$$
$$\leqslant\exp\left(c'_{25}(d_P t_Q+d_Q t_P)\right)\prod_{j\in\mathcal{P}}\max\left\{1,|t_j|^{d_Q}\right\}$$
$$\times|q|^{-|\mathcal{P}|}\left(|P_1(\theta)|^s+|Q_1(\theta)|\right).$$

Finally, using (4.26) and (4.29), we obtain from 1.1 the estimate
(4.30)

$$|r|\leqslant\exp\left(c'_{26}(d_P t_Q+d_Q t_P)\right)|p^{d_Q}q^{d_P}|$$
$$\times\prod_{j\notin\mathcal{P}}\prod_{i=1}^{d_Q}\left(\max\{|t_j|,1\}\max\{|u_i|,1\}\right)$$
$$\times\prod_{j\in\mathcal{P}}\left(\max\{|t_j|,1\}\right)^{d_Q}\cdot|q|^{-|\mathcal{P}|}\left(|P_1(\theta)|^s+|Q_1(\theta)|\right)$$
$$\leqslant\exp\left(c'_{27}(d_P t_Q+d_Q t_P)\right)L(P)^{d_Q}\cdot L(Q)^{d_P-|\mathcal{P}|}\left(|P_1(\theta)|^s+|Q_1(\theta)|\right)$$
$$\leqslant\exp\left(c'_{28}(d_P t_Q+d_Q t_P)\right)\max\{|P_1(\theta)|^s,|Q_1(\theta)|\}.$$

From (4.30) and estimates of the $\mathfrak{L}$-type $t_{\mathfrak{L}}(r)$ we directly obtain Lemma 4.10.

By analogy with Lemma 4.5 we can establish a generalization of Lemma 4.7. We use Lemma 4.4 and the scheme outlined in Lemma 4.5.

LEMMA 4.8. *Suppose $\zeta_1,\dots,\zeta_k$ are algebraic numbers over $\mathfrak{L}$, $\mathfrak{L}_i=\mathfrak{L}[\zeta_1,\dots,\zeta_{i-1}]$, and ζ_i is a root of a polynomial $Q_i(x)\in\mathfrak{L}_i[x]$ of multiplicity s_i with $t_{\mathfrak{L}_i}(Q_i)\leqslant t_i$. Assume that $P(x_1,\dots,x_k,y)$, $Q(x_1,\dots,x_k,y)\in\mathfrak{L}[x_1,\dots,x_k,y]$ are polynomials* *of $\mathfrak{L}$-types t_P, t_Q and $P_1(y)$, $Q_1(y) \in \mathcal{K}_{k+1}[y]$ are polynomials with leading coefficients unity such that $P_1(y)^{s_{k+1}}$ divides $P(\zeta_1,\ldots,\zeta_k, y)$ and $Q_1(y)$ divides $Q(\zeta_1,\ldots,\zeta_k, y)$. Then there exists $r = r(P, Q) \in \mathfrak{L}$, $r \neq 0$, $t_{\mathfrak{L}}(r) \leqslant c_{12}t_Pt_Qt_1 \cdots t_k$, such that if*

$$(4.31) \qquad \max\{|P_1(\theta)|, |Q_1(\theta)|\} < \exp\left(-c_{11}t_Pt_Qt_1 \cdots t_k/s_1 \cdots s_{k+1}\right) \times |r|^{1/s_1\cdots s_{k+1}}$$

for each θ (with bounded $|\theta|$), then $P_1(x)$ and $Q_1(x)$ have common roots.

The proof of Lemma 4.8 follows immediately from Lemmas 4.7 and 4.5.

In the proofs of §§5–6 we use auxiliary results more complicated than Lemmas 4.7 and 4.8. Let us begin with the simpler

LEMMA 4.9. *Suppose ζ_1, ζ_2 are numbers such that ζ_i is a root of multiplicity s_i of $Q_i(x) \in \mathfrak{L}_{i-1}[x]$, where $\mathfrak{L}_0 = \mathfrak{L}$, $\mathfrak{L}_1 = \mathfrak{L}[\zeta_1]$, and $t_{\mathfrak{L}_i}(Q_i) \leqslant t_i$. Assume that $P_i(x) \in \mathcal{K}_i[x]$ is a $\mathcal{K}_i$-irreducible polynomial with leading coefficient unity which divides $Q_i(x)$, $P_i(\zeta_i) = 0$, and $T_0(x_1, x_2) \in \mathfrak{L}[x_1, x_2]$, $t_{\mathfrak{L}}(T_0) \leqslant T$. If $T_1(x_1, x_2) \in \mathcal{K}_0[x_1, x_2]$, $T_2(x_2) \in \mathcal{K}_1[x_2]$ are such that*

$$T_1(x_1, x_2)(P_1(x_1))^{\sigma_1} = T_0(x_1, x_2), \qquad T_1(\zeta_1, x_2) = (P_2(x_2))^{\sigma_2}T_2(x_2),$$

then there exists $u \in \mathfrak{L}$, $u \neq 0$, $t_{\mathfrak{L}}(u) \leqslant c_{13}Tt_1t_2$, such that either

$$(4.32) \qquad |T_2(\zeta_2)| \geqslant \exp(-c_{14}Tt_1t_2/s_1s_2)\,|u|^{1/s_1s_2}$$

or else $T_2(\zeta_2) = 0$.

PROOF. Assume that $T_2(\zeta_2) \neq 0$. Then $T_1(\zeta_1, x_2) \not\equiv 0$. Denote by $A_0(x_1)$ the leading coefficient of $T_0(x_1, x_2)$ with respect to x_2, and by $A_1(x_1)$ the leading coefficient of $T_1(x_1, x_2)$. Then

$$(4.33) \qquad A_0(x_1) = A_1(x_1)(P_1(x_1))^{\sigma_1}$$

and, according to 1.4, $H(A_1)H(P_1)^{\sigma_1} \leqslant e^{d(A_0)}H(A_0)$. Since the leading coefficient of $P_1(x_1)$ is unity, we have $H(P_1) \geqslant 1$, and it follows from $t_{\mathfrak{L}}(A_0) \leqslant T$ that $H(A_1) \leqslant e^{O(T)}$. Similarly, from the equality $T_1(x_1, x_2)(P_1(x_1))^{\sigma_1} = T_0(x_1, x_2)$ we obtain an estimate of $H(T_1)$, the maximum of the moduli of the coefficients of $T_1(x_1, x_2)$:

$$(4.34) \qquad H(T_1) \leqslant \exp(c'_{29}T).$$

We use the semiresultant r_1 of the polynomials $Q_2(x_2) \in \mathfrak{L}_1[x_2]$ and $T_1(\zeta_1, x_2) \in \mathcal{K}_1[x_2]$, which was constructed in the proof of 4.7 (see (4.20)).

From the representation of r_1 in the form

$$(4.35) \qquad r_1 = q^{d(T_1)}A_1^{d(Q_2)} \prod_{i=1}^{d(T_1)} \prod_{j=1,\, t_i \neq u_j}^{d(Q_2)} (t_i - u_j),$$

where q, $A_1 = A_1(\zeta_1)$ are the leading coefficients of $Q_2(x)$, $T_1(\zeta_1, x)$ and $d(Q_2) = d_Q$, $d(T_1) = d_T$ are their degrees with respect to x, it follows that $r_1 \neq 0$. As in the proof of Lemma 4.7, we conclude that $r_1 = q^{d(T_1)}A_1^{d(Q_2)}R_1(f_1,\dots,f_{d_T}; g_1,\dots,g_{d_Q})$, where $R_1(\cdots)$ is a polynomial with rational integral coefficients, of height $\leqslant \exp(4d_Qd_T)$ and of degree $\leqslant d_Q$ with respect to the whole set $\{f_1,\dots,f_{d_T}\}$ and of degree $\leqslant d_T$ with respect to $\{g_1,\dots,g_{d_Q}\}$. Here, $f_1,\dots,f_{d_T}$ and $g_1,\dots,g_{d_Q}$ are the elementary symmetric functions in $t_1,\dots,t_{d_T}$, the roots of $T_1(\zeta_1, x)$, and in $u_1,\dots,u_{d_Q}$, the roots of $Q_2(x)$. It is clear that $f_i = p_i/A_1$ and $g_j = q_j/q$, where the p_i are coefficients of $T_1(\zeta_1, x)$ and the q_j are coefficients of $Q_2(x)$. Writing $Q_2(x)$ in the form $Q_2(x_2) = Q_2(\zeta_1, x_2)$, $Q_2(x_1, x_2) \in \mathfrak{L}[x_1, x_2]$, we see that $q_j \in \mathfrak{L}[\zeta_1]$ and $q_j = \bar{q}_j(\zeta_1)$, where $\bar{q}_j(x) \in \mathfrak{L}[x]$. Using the equality $T_0(x_1, x_2) = (P_1(x_1))^{\sigma_1}T_1(x_1, x_2)$, we conclude that $p_i = \bar{p}_i(\zeta_1)$, where $\bar{p}_i(x_1) \in \mathcal{K}[x_1]$ and $\bar{p}_j(x_1)(P_1(x_1))^{\sigma_1}$ are coefficients of $T_0(x_1, x_2)$. Moreover, $A_1(x_1)(P_1(x_1))^{\sigma_1} = A_0(x_1)$. Therefore, f_i can be written in the form $f_i = \bar{f}_i(\zeta_1)$, where $\bar{f}_i(x_1) \in \mathcal{K}[x_1]$, $\bar{f}_i(x_1) = p_i^0(x_1)/A_0(x_1)$, and $t_{\mathfrak{L}}(p_i^0)$, $t_{\mathfrak{L}}(A_0) \leqslant c_{30}'T$. Thus, in $R(f_1,\dots; g_1,\dots)$ we replace the numbers $f_i = \bar{f}_i(\zeta_1)$, $g_j = \bar{g}_j(\zeta_1)$ by $\bar{f}_i(x_1)$, $\bar{g}_j(x_1)$ and consider the following polynomial in x_1:

$$r_1(x_1) = \bar{q}(x_1)^{d_T}A_1(x_1)^{d_Q}R(\bar{f}_1(x_1),\dots; \bar{g}_1(x_1),\dots).$$

According to the above observations, $r_1(x_1) \in \mathcal{K}[x_1]$, and

$$R(\bar{f}_1(x_1),\dots;\bar{g}_1(x_1),\dots)(A_0(x_1))^{d_Q}(\bar{q}(x_1))^{d_T}$$

is an element of $\mathfrak{L}[x_1]$ having $\mathfrak{L}$-type $\leqslant c_{31}'(Td_Q + t_2d_T)$. Consequently, $(P_1(x_1))^{\sigma_1d_Q}r_1(x_1)$ is a polynomial in $\mathfrak{L}[x_1]$. Let us denote it by $r_2(x_1)$ and estimate its type. Since $t_{\mathfrak{L}}(A_0) \leqslant T$, it follows that

$$t_{\mathfrak{L}}(r_2) \leqslant c_{32}'(Td_Q + td_T) \leqslant 2c_{32}'Tt_2, \tag{4.36}$$

where $r_2(x_1) = (P_1(x_1))^{\sigma_1d_Q}r_1(x_1)$. We apply Lemma 4.7 to the polynomials $r_2(x_1)$ and $Q_1(x_1)$. Consider their divisors $P_1(x_1) \in \mathcal{K}[x_1]$ and $r_1(x_1) \in \mathcal{K}[x_1]$. Denote $r_1(x_1)$ by $br_1'(x_1)$, where $r_1'(x_1) \in \mathcal{K}[x_1]$ has leading coefficient unity. Since $P_1(\zeta_1) = 0$, $r_1(\zeta_1) = r_1 \neq 0$, and $P_1(x_1)$ is $\mathcal{K}$-irreducible, it follows that $P_1(x_1)$ and $r_1'(x_1)$ have no common roots. Therefore, for some $v_1 \in \mathfrak{L}$, $v_1 \neq 0$, $t_{\mathfrak{L}}(v_1) \leqslant c_{33}'Tt_1t_2$, we obtain from Lemma 4.7 that

$$|r_1'(\zeta_1)| > \exp(-c_{34}'Tt_1t_2/s_1)|v_1|^{1/s_1}. \tag{4.37}$$

If $a_0 = 1$ is the leading coefficient of $A_0(x_1)$, then, by definition of $r_2(x)$, b is the leading coefficient of $r_2(x)$, i.e. $b \in \mathfrak{L}$ and $t_{\mathfrak{L}}(b) \leqslant 2c_{32}'Tt_2$ by virtue of (4.36). Multiplying (4.37) by b, we obtain

$$|r_1(\zeta_1)| > \exp(-c_{34}'Tt_2t_1/s_1)|v_1|^{1/s_1}|b|.$$

Put $v_2 = v_1(b)^{s_1}$; then $t_{\mathfrak{L}}(v_2) \leqslant c_{35}'(Tt_2t_1 + Tt_2s_1) \leqslant 2c_{35}'Tt_2t_1$. It follows that

$$|a_0r_1| > \exp(-c_{34}'Tt_2t_1/s_1)|v_2|^{1/s_1}. \tag{4.38}$$

It remains to estimate $|r_1|$ from above in terms of $T_2(\zeta_2)$, in order to obtain (4.32). First, since $t_{\mathfrak{L}}(A_0) \leqslant T$, we have $|a_0| \leqslant \exp(c'_{36}T)$. Therefore, it follows from (4.38) that

$$(4.39) \qquad |r_1| > \exp(-c'_{37}Tt_2t_1/s_1)\,|v_2|^{1/s_1}.$$

To estimate r_1, the semiresultant of the polynomials $Q_2(x_2) \in \mathcal{K}_1[x_2]$ and $T_1(\zeta_1, x_2) \in \mathcal{K}_1[x_2]$, it suffices to use the argument of the proof of Lemma 4.7. Indeed, let $T_1(x) = T_1(\zeta_1, x) \in \mathcal{K}[x]$. According to (4.34), $H(T_1) \leqslant \exp(c'_{38}T)$. It follows from the equality $T_1(x) = P_2(x)^{\sigma_2}T_2(x)$ (since $H(P_2) \geqslant 1$) that $H(T_2) \leqslant \exp(c'_{39}T)$.

Let $\mathcal{P}$ denote the set of those j for which u_j is a root of $P_2(x)$. Since $P_2(x)$ is $\mathcal{K}_1$-irreducible, $P_2(\zeta_2) = 0$, and $T_2(\zeta_2) \neq 0$, it follows, as we have noted, that $r_1 \neq 0$ and that $P_2(x)$, $T_2(x)$ have no common roots. Since $T_1(x) = P_2(x)^{\sigma_2}T_2(x)$, we can write for any $u_j, j \in \mathcal{P}$: $T_1^{(\sigma_2)}(u_j) = \sigma_2!T_2(u_j)(P_2'(u_j))^{\sigma_2}$. From (4.24)–(4.26) we obtain

$$(4.40) \quad |r_1| \leqslant \Big| q_2^{d(T_1)}A_1^{d(Q_2)} \prod_{j=1,\, j\notin\mathcal{P}}^{d(Q_2)} \prod_{i=1,\, t_i\neq u_j}^{d(T_1)} (t_i - u_j) \prod_{j\in\mathcal{P}} T_2(u_j)(P_2'(u_j))^{\sigma_2} \Big|.$$

However, $P_2'(u_j) = \prod_{u'}(u_j - u')$, where $\prod_{u'}$ extends over all roots u' of $P_2(x)$ different from u_j. Therefore,

$$(4.41) \qquad \prod_{j\in\mathcal{P}} |P_2'(u_j)| \leqslant \exp\left(c'_{40}d(P_2)^2\right) \cdot \prod_{j\in\mathcal{P}} \left(\max\{1, |u_j|\}\right)^{d(P_2)}.$$

But $\sigma_2 d(P_2) \leqslant d(T_1) \leqslant T$ and $d(P_2) \leqslant d(Q_2) \leqslant t_2$. Therefore, it follows from (4.41) that

$$(4.42) \qquad \prod_{j\in\mathcal{P}} |P_2'(u_j)|^{\sigma_2} \leqslant \exp(c'_{40}Tt_2) \cdot \prod_{j\in\mathcal{P}} \left(\max\{1, |u_j|\}\right)^{\sigma_2 d(P_2)}.$$

Moreover, by definition of s_2 we have $u_j = \zeta_2$ for $j \in \mathcal{P}_2 \subset \mathcal{P}$, $|\mathcal{P}_2| = s_2$. Therefore,

$$(4.43) \qquad \left|\prod_{j\in\mathcal{P}} T_2(u_j)\right| \leqslant |T_2(\zeta_2)|^{s_2} \prod_{j\in\mathcal{P}\setminus\mathcal{P}_2} |T_2(u_j)|$$
$$\leqslant |T_2(\zeta_2)|^{s_2}\exp(c'_{39}t_2T) \cdot \prod_{j\in\mathcal{P}\setminus\mathcal{P}_2} \max\left\{1, |u_j|^{d(T_2)}\right\}.$$

Note that $d(P_2)\sigma_2 + d(T_2) = d(T_1)$. Thus, (4.42) and (4.43) imply that
(4.44)

$$\prod_{j\in\mathcal{P}} |T_2(u_j)(P_2'(u_j))^{\sigma_2}| \leqslant \exp(c'_{41}Tt_2)\,|T_2(\zeta_2)^{s_2}| \prod_{j\in\mathcal{P}} \left(\max\{1, |u_j|\}\right)^{d(T_1)}.$$

In view of (4.40), (4.44), and Lemma 1.1, we conclude that

$$(4.45)\qquad |r_1| \le |q_2|^{d(T_1)} |A_1|^{d(Q_2)} \left| \prod_{j\notin\mathcal{P}} \prod_{i=1,\, t_i\ne u_j}^{d(T_1)} (t_i - u_j) \right|$$

$$\times \exp(c'_{41}Tt_2) |T_2(\zeta_2)|^{s_2} \prod_{j\in\mathcal{P}} (\max\{1, |u_j|\})^{d(T_1)}$$

$$\le \exp(c'_{41}Tt_2) 2^{d(T_1)d(Q_2)} |T_2(\zeta_2)|^{s_2} |q_2|^{d(T_1)}$$

$$\times |A_1|^{d(Q_2)} \prod_{j\notin\mathcal{P}} \prod_{i=1}^{d(T_1)} (\max\{1, |t_i|\}\max\{1, |u_j|\})$$

$$\times \prod_{j\in\mathcal{P}} (\max\{1, |u_j|\})^{d(T_1)}$$

$$\le \exp(c'_{42}Tt_2) |T_2(\zeta_2)|^{s_2} L(Q_2)^{d(T_1)} L(T_2)^{d(Q_2)}$$

$$\le \exp(c'_{43}Tt_2) |T_2(\zeta_2)|^{s_2}.$$

Note that $s_1 \le t_1$, i.e. $\exp(c'_{43}Tt_2) \le \exp(c'_{43}Tt_2t_1/s_1)$. Finally, from (4.45) we obtain

$$(4.46)\qquad |r_1| \le \exp(c'_{43}Tt_2t_1/s_1) |T_2(\zeta_2)|^{s_2}.$$

Combining (4.39) and (4.46), we see that

$$|T_2(\zeta_2)|^{s_2} \ge \exp(-c'_{44}Tt_2t_1/s_1) |v_2|^{1/s_1}.$$

Since $t_{\mathcal{L}}(v_2) \le 2c'_{45}Tt_2t_1$ and $v_2 \ne 0$, we obtain (4.32) and Lemma 4.9 is proved.

In a similar way, using Lemmas 4.5 and 4.6, we can prove a much stronger result.

LEMMA 4.10. *Suppose $\zeta_1, \ldots, \zeta_k$ are algebraic numbers over $\mathcal{L}$, $\mathcal{L}_i = \mathcal{L}[\zeta_1, \ldots, \zeta_{i-1}]$, and ζ_i is a root of a polynomial $Q_i(x) \in \mathcal{L}_i[x]$ of multiplicity s_i with $t_{\mathcal{L}_i}(Q_i) \le t_i$: $i = 1, \ldots, k$. Let $R_i(x) \in \mathcal{K}_i[x]$ be a $\mathcal{K}_i$-irreducible polynomial with leading coefficient unity which divides $Q_i(x)$ and is such that $R_i(\zeta_i) = 0$. Assume that*

$$T_1(x_1, \ldots, x_k) \in \mathcal{L}[x_1, \ldots, x_k], \qquad t_{\mathcal{L}}(T_1) \le T,$$

$$T_2(x_1, \ldots, x_k) \in \mathcal{K}[x_1, \ldots, x_k], \ldots, T_{k+1}(x_k) \in \mathcal{K}_{k-1}[x_k],$$

and $T_{k+1}(\zeta_k) \not\equiv 0$. If

$$T_1(x_1, \ldots, x_k) = (R_1(x_1))^{\sigma_1} T_2(x_1, \ldots, x_k),$$

$$T_2(\zeta_1, x_2, \ldots, x_k) = (R_2(x_2))^{\sigma_2} T_3(x_2, \ldots, x_k), \ldots, T_k(\zeta_{k-1}, x_k)$$

$$= (R_k(x_k))^{\sigma_k} T_{k+1}(x_k),$$

then for some $v \in \mathcal{L}$, $v \ne 0$, $t_{\mathcal{L}}(v) \le c_{15}Tt_1 \cdots t_k$, we have

$$(4.47)\qquad |T_{k+1}(\xi_k)| \ge \exp(-c_{16}Tt_1 \cdots t_k/s_1 \cdots s_k) |v|^{1/s_1 \cdots s_k}.$$

We make an observation which is important for what follows. First, Lemma 4.4 and its generalizations are consequences of Lemma 4.7. Indeed, it suffices to put $\theta = \zeta$ in Lemma 4.7. The assumption that the leading coefficients of $P_1(x)$ and $Q_1(x)$ are unity is inessential: if we obtain a good lower bound for $|p^{-1}P(\zeta)|$, where p is the leading coefficient of $P(x)$;

$$|p^{-1}P(\zeta)| \geqslant \exp(-c'_{46}t_P t_Q/s)\,|u|^{1/s},$$

then for $u' = p^s u$ we have

$$|P(\zeta)| \geqslant \exp(-c'_{47}t_P t_Q/s)\,|u'|^{1/s},$$

where $t_{\mathfrak{L}}(u') \leqslant c'_{48}(t_{\mathfrak{L}}(u) + st_{\mathfrak{L}}(P)) \leqslant c'_{48}(t_{\mathfrak{L}}(u) + d_Q t_P)$, since $s \leqslant d_Q$. Therefore, to prove *all* of the lemmas of §4 it suffices to use the concept of the semiresultant $r = r(P, Q)$ of the polynomials $P(x)$ and $Q(x)$, which was constructed in Lemma 4.7; Lemmas 4.3 and 4.6 are also obtained by means of the semiresultant, since the semidiscriminant $SD(P)$ constructed in the proof of Lemma 4.3 is $r(P, P)$. A further observation: to obtain all of the estimates in Lemmas 4.3–4.10 it is not necessary to use estimates of the $\mathfrak{L}$-*types* of the corresponding polynomials. As the calculations show, we are only interested in the maximum of the absolute value of the coefficients of the polynomials $P(x)$, $Q(x),\ldots$, i.e. the *types* of the polynomials $P(x)$, $Q(x),\ldots$; an estimate of the $\mathfrak{L}$-types of $P(x)$ and $Q(x)$ is only needed to obtain an estimate of the $\mathfrak{L}$-type of $r = r(P, Q)$. Therefore, in all of the estimates (4.17), (4.16), and (4.19), we can replace $t_P, t_Q,\ldots$ by $t(P)$, $t(Q),\ldots$, the types of the polynomials $P(x)$, $Q(x),\ldots$ in the usual sense, i.e. $t(P) = \max\{\ln H(P), d(P)\}$.

Suppose $\mathfrak{L}_0$ is a fixed finitely generated extension of $\mathbf{Z}$ with extensions $\mathfrak{L}_j = \mathfrak{L}_0[\alpha_1,\ldots,\alpha_j]$, and let $\mathfrak{L} = \mathfrak{L}_n$. We denote by $\mathcal{K}_j$ the corresponding fields: $\mathcal{K}_j = \mathcal{K}_0(\alpha_1,\ldots,\alpha_j)$. Consider a polynomial $P(x) \in \mathcal{K}[x]$; we write $P(x)$ in the form $P(x) = P(\alpha_1,\ldots,\alpha_n, x)$ even though $P(x_1,\ldots,x_n, x) \notin \mathcal{K}_0[x_1,\ldots,x_n, x]$, since in the case of a transcendental α_n the coefficients of $P(x)$ are rational combinations of $\alpha_1,\ldots,\alpha_n$ but not necessarily polynomials. Consider the following situation, taken from §5 and reflected in 4.9:

There exist polynomials $R_1(x_1),\ldots,R_n(x_n)$: $R_i(x_i) = R_i(\alpha_1,\ldots,\alpha_{i-1}, x_i) \in \mathcal{K}_{i-1}[x_i]$ and polynomials $P_1(x_1,\ldots,x_n, x) \in \mathfrak{L}_0[x_1,\ldots,x_n, x]$,

$$\begin{aligned} P_2(x_2,\ldots,x_n, x) &= P_2(\alpha_1, x_2,\ldots,x) \in \mathcal{K}_1[x_2,\ldots,x],\ldots,P_n(x_n, x) \\ &= P_n(\alpha_1,\ldots,\alpha_{n-1}, x_n, x) \in \mathcal{K}_{n-1}[x_n, x] \end{aligned}$$

such that

(4.48)

$$\begin{aligned} P(\alpha_1,\ldots,x_n, x)R_n(x_n) = P_n(\alpha_1,\ldots,\alpha_{n-1}, x_n, x),\ldots,P_3(\alpha_1, x_2,\ldots,x_n, x)R_2(x_2) \\ = P_2(\alpha_1, x_2,\ldots,x_n, x), \\ P_2(x_1,\ldots,x_n, x)R_1(x_1) = P_1(x_1,\ldots,x_n, x). \end{aligned}$$

If (4.48) holds, we denote this briefly by

$$[P(x)\cdot R_n(x_n)\cdots R_1(x_1)]_{\alpha_1,\ldots,\alpha_n}=P_1(x_1,\ldots,x_n,x).$$

The main problem is to establish analogues of Lemmas 4.3–4.10 for $P(x)$ when instead of the $\mathcal{L}$-type of $P(x)$ (which is very difficult to estimate) we use the $\mathcal{L}_0$-type of $P_1(x_1,\ldots,x_n,x)$ (which is actually the $\mathcal{L}$-type of $P_1(\alpha_1,\ldots,\alpha_n,x)$).

If some α_i is algebraic over $\mathcal{L}_{i-1}=\mathcal{L}_0[\alpha_1,\ldots,\alpha_{i-1}]$, then we denote by $R_i(x_i)\in\mathcal{K}_{i-1}[x_i]$ the $\mathcal{K}_{i-1}$-irreducible polynomial which has α_i as a root and leading coefficient unity. If α_i is transcendental over $\mathcal{L}_{i-1}$, we put $R_i(x_i)\equiv 1$. Along with ξ, a number algebraic over $\mathcal{L}_n=\mathcal{L}$, we consider a polynomial $Q(x)=Q(\alpha_1,\ldots,\alpha_n,x)\in\mathcal{K}_n[x]$ such that

$$(4.49)\quad [Q(x)\cdot(R_n(x_n))^{\sigma_n}\cdots(R_1(x_1))^{\sigma_1}]_{\alpha_1,\ldots,\alpha_n}=Q_0(x_1,\ldots,x_n,x)$$

for certain $\sigma_1\geqslant 0,\ldots,\sigma_n\geqslant 0$ and $Q_0(x_1,\ldots,x_n,x)\in\mathcal{L}_0[x_1,\ldots,x_n,x]$. As the type $t_{\mathcal{L}_n}(\xi)$ it turns out that we can use $t_{\mathcal{L}}(Q_0)$.

As already mentioned, we need an estimate of the type $t(Q)$ of $Q(x)$, in the usual sense. We obtain it from the following: first, $t_{\mathcal{L}}(Q_0)\leqslant ct(Q_0)$ and secondly, applying in succession Lemma 1.4 to the equalities (4.48), which represent the explicit form of (4.49), we obtain $t(Q)\leqslant c't(Q_0)\leqslant c''T$, since $H(R_i)\geqslant 1$ in view of the fact that the leading coefficients of the $R_i(x_i)$ are unity. Thus, it follows from (4.49) that $t(Q)\leqslant c'''t_{\mathcal{L}_0}(Q_0)$.

LEMMA 4.11. *Let $\mathcal{L}_n=\mathcal{L}_0[\alpha_1,\ldots,\alpha_n]$ and, as above, let $R_i(x_i)$ denote the $\mathcal{K}_{i-1}$-irreducible polynomial in $\mathcal{K}_{i-1}[x_i]$ with leading coefficient unity such that $R_i(\alpha_i)=0$, if α_i is algebraic over $\mathcal{L}_{i-1}=\mathcal{L}_0[\alpha_1,\ldots,\alpha_{i-1}]$, and let $R_i(x_i)\equiv 1$ otherwise. Assume that ζ is an algebraic number, $Q(x)\in\mathcal{K}_n[x]$, $Q(\zeta)=0$, and for the analogue of the $\mathcal{L}_n$-type of ζ we have*

$$(4.50)\qquad [Q(x)R_n(x_n)^{\sigma_n}\cdots R_1(x_1)^{\sigma_1}]_{\alpha_1,\ldots,\alpha_n}=Q_0(x_1,\ldots,x_n,x)$$

with $t_{\mathcal{L}_0}(Q_0)\leqslant t_Q$. Let $R(x)\in\mathcal{K}_n[x]$ denote the $\mathcal{K}_n$-irreducible polynomial with leading coefficient unity of which ζ is a root, and suppose $R(x)^s$ divides $Q(x)$, i.e. ζ is a root of $Q(x)$ of multiplicity s.

If $S(x)\in\mathcal{K}_n[x]$ is any polynomial such that

$$(4.51)\qquad S(x)\cdot R(x)^{\sigma_0}=P(x)$$

where $P(x)\in\mathcal{K}_n[x]$ and

$$(4.52)\qquad [P(x)\cdot R_n(x_n)^{\tau_n}\cdots R_1(x_1)^{\tau_1}]_{\alpha_1,\ldots,\alpha_n}=P_0(x_1,\ldots,x_n,x)$$

with $t_{\mathcal{L}_0}(P_0)\leqslant t_P$, then there exist $S_1(x)$ and $P'(x)$ such that $S_1(\alpha_n)\neq 0$ and

$$(4.53)\qquad S_1(x_n)R_n(x_n)^{\sigma_0'}=P'(x_n)$$

where $P'(x_n)\in\mathcal{K}_{n-1}[x_n]$ and

$$(4.54)\quad [P'(x_n)\cdot R_{n-1}(x_{n-1})^{\kappa_{n-1}}\cdots R_1(x_1)^{\kappa_1}]_{\alpha_1,\ldots,\alpha_n}=P_0'(x_1,\ldots,x_n)$$

with $t_{\mathfrak{L}_0}(P_0') \leqslant c_{49}' t_P t_Q$, *and either*

$$|S(\zeta)| \geqslant \exp(-c_{50}' t_P t_Q / s) |S_1(\alpha_n)|^{1/s} \tag{4.55}$$

or else $S(\zeta) = 0$.

PROOF. We use in its entirety the scheme of the proof of Lemmas 4.9 and 4.10. As usual, as $S_1(\alpha_n)$ we take the resultant of the polynomials $Q(x)$ and $P(x)$, where, in view of (4.52),

$$\begin{aligned} P(x) &= P(\alpha_n, x), & P(x_n, x) R_n(x_n)^{\sigma_n} &= P_n(x_n, x), \\ Q(x) &= Q(\alpha_n, x), & Q(x_n, x) R_n(x_n)^{\tau_n} &= Q_n(x_n, x). \end{aligned} \tag{4.56}$$

Inequality (4.55) with the definition $S_1(\alpha_n) = r(P, Q)$ is a simple consequence of relations (4.40)–(4.46) in the proof of Lemma 4.9, since, as we have mentioned, the proof involves only the types $t(Q)$ and $t(P)$, which (see above) can be estimated in terms of $t_{\mathfrak{L}_0}(Q_0) = t_Q$ and $t_{\mathfrak{L}_0}(P_0) = t_P$:

$$t(P) \leqslant c_{51}' t_P, \qquad t(Q) \leqslant c_{52}' t_Q.$$

We must prove (4.53) and (4.54) and obtain an estimate for $t_{\mathfrak{L}_0}(P_0')$. It follows from Lemma 4.7 that

$$S_1(\alpha_n) = p^{d_Q} q^{d_P} S(f_1, \dots, f_{d_P}; g_1, \dots, g_{d_Q}),$$

where $d_Q = d_x Q_n(x_n, x)$ is the degree of $Q(x)$, $d_P = d_x P(x_n, x)$ is the degree of $P(x)$, and $f_1, \dots, f_{d_P}$ have the form p_i/p_n and $g_1, \dots, g_{d_Q}$ the form q_i/q_n. As usual, p_n, q_n are the leading coefficients of $P(x)$, $Q(x)$, and p_i, q_i are the other coefficients of $P(x)$, $Q(x)$. Now $p_i = \bar{p}_i(\alpha_n)$, $p_n = \bar{p}_n(\alpha_n)$, $q_i = \bar{q}_i(\alpha_n)$, $q_n = \bar{q}_n(\alpha_n)$, and $\bar{p}_i(x_n), \dots, \bar{q}_n(x_n)$ are elements of $\mathcal{K}_{n-1}[x_n]$ (if α_n is transcendental over $\mathcal{K}_{n-1}$, then, in view of (4.56), $\bar{p}_i$ and $\bar{q}_i$ are polynomials in α_n, since $R_n(x_n) \equiv 1$; if α_n is algebraic over $\mathcal{K}_{n-1}$, then $\bar{p}_i$ and $\bar{q}_i$ can be represented as elements of $\mathcal{K}_{n-1}[\alpha_n]$).

Using (4.56), we obtain $\bar{p}_i(x_n)(R_n(x_n))^{\tau_n} = \bar{k}_i^n(x_n)$, $\bar{q}_i(x_n)(R_n(x_n))^{\sigma_n} = \bar{l}_i^n(x_n)$, where $\bar{k}_i^n(x_n)$, $\bar{l}_i^n(x_n)$ are coefficients of the polynomial $P_n(x_n, x)$ with respect to x. Replacing α_n by x_n in $S_1(\alpha_n)$, we have

$$S_1(x_n) = \bar{p}_n(x_n)^{d_Q} \bar{q}_n(x_n)^{d_P} S(\bar{f}_1^n, \dots, \bar{f}_{d_P}^n; \bar{g}_1^n, \dots, \bar{g}_{d_Q}^n),$$

where the $\bar{f}_i^n$ have the form $\bar{k}_i^n(x_n)/\bar{k}_n^n(x_n)$ and the $\bar{g}_j^n$ the form $\bar{l}_j^n(x_n)/\bar{l}_n^n(x_n)$. Since the degree of $S(\cdots;\cdots)$ with respect to the set of variables $\{f_1, \dots, f_{d_P}\}$ is $\leqslant d_Q$ and with respect to $\{g_1, \dots, g_{d_Q}\}$ is $\leqslant d_P$, and since $R_n(x_n)^{\tau_n} \bar{p}_n(x_n) = \bar{k}_n^n(x_n)$, $R_n(x_n)^{\sigma_n} \bar{q}_n(x_n) = \bar{l}_n^n(x_n)$, it follows that

$$R_n(x_n)^{\tau_n d_Q + \sigma_n d_P} S_1(x_n) = S_2(x_n), \tag{4.57}$$

where $S_2(x_n) \in \mathcal{K}_{n-1}[x_n]$ and

$$S_2(x_n) = \bar{k}_n^n(x_n)^{d_Q} \cdot \bar{l}_n^n(x_n)^{d_P} S(\bar{f}_1^n, \dots). \tag{4.58}$$

Similarly, using (4.50) and (4.52),

$$P_n(x_{n-1}, x_n, x)R_{n-1}(x_{n-1})^{\tau_{n-1}} = P_{n-1}(x_{n-1}, x_n, x),$$

$$Q_n(x_{n-1}, x_n, x)R_{n-1}(x_{n-1})^{\sigma_{n-1}} = Q_{n-1}(x_{n-1}, x_n, x),$$

where $P_{n-1}(x_{n-1},\ldots,x)$, $Q_{n-1}(x_{n-1},\ldots,x) \in \mathcal{K}_{n-2}[x_{n-1},\ldots,x]$. Then $\bar{k}_i^n(x_{n-1}, x_n)R_{n-1}(x_{n-1})^{\tau_{n-1}} = \bar{k}_i^{n-1}(x_{n-1}, x_n)$, etc., and substituting into (4.58) we obtain, by analogy with (4.57),

$$R_{n-1}(x_{n-1})^{\tau_{n-1}d_Q+\sigma_{n-1}d_P}S_2(x_{n-1}, x_n) = S_3(x_{n-1}, x_n), \tag{4.59}$$

where $S_3(x_{n-1}, x_n) \in \mathcal{K}_{n-2}[x_{n-1}, x_n]$ and

$$S_3(x_{n-1}, x_n) = \bar{k}_n^{n-1}(x_{n-1}, x_n)^{d_Q}\bar{l}_n^{n-1}(x_{n-1}, x_n)^{d_P}S(\bar{f}_1^{n-1},\ldots). \tag{4.60}$$

Continuing this procedure, using (4.50) and (4.52), we obtain, in view of (4.57), (4.59), and analogous equalities,

$$\left[S_2(x_n)\cdot R_{n-1}(x_{n-1})^{\tau_{n-1}d_Q+\sigma_{n-1}d_P}\cdots R_1(x_1)^{\tau_1 d_Q+\sigma_1 d_P}\right]_{\alpha_1,\ldots,\alpha_{n-1}} = S_{n+1}(x_1,\ldots,x_n), \tag{4.61}$$

where $S_{n+1}(x_1,\ldots,x_n) \in \mathcal{K}_0[x_1,\ldots,x_n]$ and, as in (4.58) and (4.60),

$$S_{n+1}(x_1,\ldots,x_n) = k(x_1,\ldots,x_n)^{d_Q}\cdot l(x_1,\ldots,x_n)^{d_P} \times S\left(f_1^0,\ldots,f_{d_P}^0; g_1^0,\ldots,g_{d_Q}^0\right), \tag{4.62}$$

where $k(x_1,\ldots,x_n)$, $l(x_1,\ldots,x_n)$ are the leading coefficients of $P_0(x_1,\ldots,x_n, x)$, $Q_0(x_1,\ldots,x_n, x)$ with respect to x, and f_i^0, g_j^0 have the form $k_i(x_1,\ldots,x_n)/k(x_1,\ldots,x_n)$, $l_j(x_1,\ldots,x_n)/l(x_1,\ldots,x_n)$, where $k_i(x_1,\ldots,x_n)$, $l_j(x_1,\ldots,x_n)$ are coefficients of P_0, Q_0 with respect to x. It follows from properties of $S(\cdots;\cdots)$ and (4.62) that $S_{n+1}(x_1,\ldots,x_n)$ can be expressed as a polynomial in the $k_i(x_1,\ldots,x_n)$, $l_j(x_1,\ldots,x_n)$ with rational integral coefficients which are $\leqslant \exp(2d_Pd_Q)$ in absolute value, i.e. $S_{n+1}(x_1,\ldots,x_n) \in \mathcal{L}_0[x_1,\ldots,x_n]$. Since $t_{\mathcal{L}_0}(P_0) \leqslant t_P$ and $t_{\mathcal{L}_0}(Q_0) \leqslant t_Q$, we obtain $t_{\mathcal{L}_0}(S_{n+1}) \leqslant c'_{53}(d_Pt_Q + d_Qt_P)$. This proves (4.53) and (4.54).

Estimate (4.55) is, as we have mentioned, a consequence of (4.46) and the fact that $t(P) \leqslant c'_{54}t_P$, $t(Q) \leqslant c'_{55}t_Q$.

A corollary of Lemma 4.11 can be obtained when α_n is algebraic over $\mathcal{K}_{n-1}$, i.e. we apply Lemma 4.11 several times.

COROLLARY 4.12. *We keep the notation of Lemma* 4.11 *for* α_i *and* ζ. *Assume that* $\alpha_i,\ldots,\alpha_n$ *are algebraic over* $\mathcal{K}_{i-1}$ *and* α_{i-1} *is transcendental over* $\mathcal{K}_{i-2}$ (*i.e.* $R_{i-1}(x_{i-1}) \equiv 1$), *and for* $i \leqslant j \leqslant n$ *there is a* $Q_j(x_j) \in \mathcal{K}_{j-1}[x_j]$ *such that*

$$\left[Q_j(x_j)\cdot R_{j-1}(x_{j-1})^{\lambda_{j-1}}\cdots R_1(x_1)^{\lambda_1}\right]_{\alpha_1,\ldots,\alpha_{j-1}} = Q_j^0(x_1,\ldots,x_j)$$

where $Q_j^0(x_1,\ldots,x_j) \in \mathcal{L}_0[x_1,\ldots,x_j]$, $t_{\mathcal{L}_0}(Q_j^0) \leqslant t_j$: $1 \leqslant j \leqslant n$, *and* α_j *is a root of* $Q_j(x_j)$ *of multiplicity* s_j. *Then* (4.50)–(4.52) *imply the existence of* $P_1(x_{i-1}) \in \mathcal{K}_{i-2}[x_{i-1}]$ *such that* $P_1(\alpha_{i-1}) \neq 0$,

$$\left[P_1(x_{i-1})\cdot R_{i-2}(x_{i-2})^{\mu_i-2}\cdots R_1(x_1)^{\mu_1}\right]_{\alpha_1,\ldots,\alpha_{i-2}} = R_0(x_1,\ldots,x_{i-1})$$

where $R_0(x_1,\dots,x_{i-1}) \in \mathfrak{L}_0[x_1,\dots,x_{i-1}]$, $t_{\mathfrak{L}_0}(R_0) \leqslant c'_{56} t_P t_Q t_i \cdots t_n$, *and either* $S(\zeta) = 0$ *or else*

$$|S(\zeta)| \geqslant \exp(-c'_{57} t_P t_Q t_i \cdots t_n / s_i \cdots s_n) \, |P_1(\alpha_{i-1})|^{1/s_i \cdots s_n}.$$

We give an analogue of Lemma 4.3. For the proof one uses (4.6)–(4.9), which involve not the $\mathfrak{L}$-type of $P(x)$, but simply $t(P)$. Regarding the representation of $SD(P)$ in the form (4.49), we have $SD(P) = r(P, P)$ and we apply (4.57)–(4.62) to $r(P, P)$ (where $P(x) \equiv Q(x)$).

LEMMA 4.13. *Suppose* $\mathfrak{L}_n = \mathfrak{L}_0[\alpha_1,\dots,\alpha_n]$; *put* $R_i(x_i) \equiv 1$ *if* α_i *is transcendental over* $\mathfrak{L}_{i-1} = \mathfrak{L}_0[\alpha_1,\dots,\alpha_{i-1}]$ *and let* $R_i(x_i)$ *be the* $\mathfrak{K}_{i-1}$*-irreducible polynomial in* $\mathfrak{K}_{i-1}[x_i]$ *with leading coefficient unity having* α_i *as a root if* α_i *is algebraic over* $\mathfrak{L}_{i-1}$. *If* $P(x) \in \mathfrak{K}_n[x]$ *is of degree* $N \geqslant 1$ *and*

$$\left[P(x) \cdot R_n(x_n)^{\sigma_n} \cdots R_1(x_1)^{\sigma_1}\right]_{\alpha_1,\dots,\alpha_n} = P_0(x_1,\dots,x_n, x)$$

with $t_{\mathfrak{L}_0}(P_0) \leqslant T$, *then there exists a root* ξ *of* $P(x)$ *of multiplicity* s *and* $R(x_n) \in \mathfrak{K}_{n-1}[x]$, $R(\alpha_n) \neq 0$, *such that*

$$|P(\theta)| \geqslant |\xi - \theta|^s \cdot \exp(-c'_{58} \cdot T^2) \, |R(\alpha_n)| \,, \tag{4.63}$$

where $[R(x_n) R_{n-1}(x_{n-1})^{\tau_{n-1}} \cdots R_1(x_1)^{\tau_1}]_{\alpha_1,\dots,\alpha_{n-1}} = R_0(x_1,\dots,x_n)$ *and* $t_{\mathfrak{L}_0}(R_0) \leqslant c'_{59} T^2$; *the relation* (4.63) *holds for any* θ, *with* c'_{58} *depending on* $|\theta|$.

Let us show how the situation described above and examined in Lemmas 4.11–4.13 may arise in practice. Typical (cf. below §5) is the case when ζ and ξ are algebraic numbers over $\mathfrak{L}$, with $\mathfrak{L}_1 = \mathfrak{L}[\zeta]$ and ξ is a root of the polynomial $Q(y) \in \mathfrak{L}_1[y]$. As usual, $Q(y) = Q(\zeta, y)$ for $Q(x, y) \in \mathfrak{L}[x, y]$, but even for the $\mathfrak{L}$-type of $Q(x, y)$ a good estimate is not available. Nevertheless it is often possible to obtain a good estimate of the $\mathfrak{L}$-type of a polynomial $Q_1(x, y)$ having the form $Q_1(x, y) = Q(x, y)P(x)^\tau$. It would be natural to consider ξ simply as a root of $Q_1(\zeta, y)$. However, in the case $P(\zeta) = 0$ it is impossible, as $Q_1(\zeta, y)$ is identically zero as a polynomial in y, i.e. it does not have roots. Lemmas 4.11–4.13 compensate for the possibility of degeneration while passing from polynomials in $n - 1$ variables over a field with higher degree of transcendence, to polynomials in n variables over a field with a lower degree of transcendence.

5. Proof of theorems on algebraically independent values of the exponential function. In this section we completely prove Theorems I–III, which were formulated in §3. For this purpose we prove by induction the auxiliary assertions (A_k): $1 \leqslant k \leqslant n$; when $k = n$, (A_n) is a consequence of 3.7. We also establish below a criterion for the algebraic independence of several numbers, a generalization of the transcendence criterion of Gel'fond [7].

We use once again the notation of §3.

To illustrate the methods used we give the proof of Theorems I–III in the first nontrivial case, $n = 2$. This proof is taken from the author's paper [**15**].

THEOREM 5.1. *Suppose* $\kappa_i \geqslant 4$ *if* $i = 1, 2$ *and* $\kappa_3 > 4$ *if* $i = 3$. *Then the field* $\mathbf{Q}(S_i)$ *has transcendence degree* > 2, *i.e. the set* S_i *contains three algebraically independent numbers.*

In particular, Theorem 5.1 implies (cf. §3) this result: if $\alpha \neq 0, 1$ is an algebraic number and β is algebraic of degree 7, then among the six numbers $\alpha^{\beta}, \ldots, \alpha^{\beta^6}$ there are three that are algebraically independent.

To prove 5.1 we need two lemmas, which are special cases of Lemmas 4.1 and 4.2. For the sake of completeness we give these lemmas, 5.2 and 5.3, separately.

The proof of Theorem 5.1 is presented for $i = 1$, $\kappa_1 = \kappa$. In the case $i = 2, 3$ we need only replace everywhere below the function $f_1(z) = f(z)$ by the $f_i(z)$ from §3 and κ_1 by κ_i: $i = 2, 3$.

LEMMA 5.2 [**30**]. *Suppose* $P_1(x, y)$, $P_2(x, y)$ *are polynomials in* $\mathbf{Z}[x, y]$ *having no common divisors and* $R(x)$ *is the resultant of these polynomials with respect to* y. *Then* $R(x)$ *is a polynomial in* $\mathbf{Z}[x]$, $R(x) \not\equiv 0$, *such that*

$$t(R) \leqslant 16t(P_1)t(P_2) + \big(t(P_1) + t(P_2)\big)\ln\big(t(P_1) + t(P_2)\big).$$

Also, if the numbers θ_1, θ_2 *are algebraically independent over* $\mathbf{Q}$, *then*

$$|R(\theta_1)| \leqslant \max\{|P_1(\theta_1, \theta_2)|, |P_2(\theta_1, \theta_2)|\} \cdot e^{\gamma_1(t(P_1)+t(P_2))^2}.$$

Here γ_1 *is a constant depending on* θ_1, θ_2: $\gamma_1 = 2\ln\{(1 + |\theta_1|)(1 + |\theta_2|) + 2\}$.

LEMMA 5.3. *Suppose* $P(x, y) \in \mathbf{Z}[x, y]$ *and* θ_1, θ_2 *are algebraically independent numbers. Suppose in addition that* $|P(\theta_1, \theta_2)| < e^{-\lambda t^4}$, *where* $t \geqslant t(P) + \gamma_2(\theta_1, \theta_2)$, $\lambda > 1$. *Then either there exists a divisor* $P_1(x, y)$ *of* $P(x, y)$ *which is a power of a* $\mathbf{Q}$-*irreducible polynomial such that*

$$|P_1(\theta_1, \theta_2)| < e^{-\lambda t^4/3} \quad \textit{and} \quad t(P_1) \leqslant 3t(P) \leqslant 3t,$$

or else there exists a polynomial $R(x) \not\equiv 0$ *such that*

$$|R(\theta_1)| < e^{-\lambda t^4/4} \quad \textit{and} \quad t(R) \leqslant (8t(P))^2 \leqslant 3^2t^2. \tag{5.1}$$

PROOF. We use the argument of [**1**], Lemma VI of Chapter III, §4. Assume there exists no $R(x) \not\equiv 0$ satisfying (5.1). Then from $P(x, y) = R_1(x, y) \cdot R_2(x, y)$, where R_1, R_2 are coprime polynomials in $\mathbf{Z}[x, y]$ and $|R_1(\theta_1, \theta_2)| \geqslant |R_2(\theta_1, \theta_2)|$, it follows that

$$e^{-\lambda t^4/3} < |R_1(\theta_1, \theta_2)|. \tag{5.2}$$

Indeed, if (5.2) does not hold, we construct the resultant $R(x)$ of R_1 and R_2. According to Lemma 1.4 and [**1**], Lemma II of Chapter III, §4, we have $t(R_1) + t(R_2) \leqslant 3t(P)$. Therefore, by Lemma 5.2,

$$|R(\theta_1)| < e^{-\lambda t^4/4} \quad \text{and} \quad t(R) \leqslant 8^2(t(P))^2,$$

and condition (5.1) holds, which is impossible.

We now represent $P(x, y)$ as a product of powers of distinct irreducible polynomials with integral coefficients:

$$P(x, y) = P_1(x, y) \cdots P_s(x, y), \qquad |P_1(\theta_1, \theta_2)| \leqslant \cdots \leqslant |P_s(\theta_1, \theta_2)|.$$

Then there exists $r \leqslant s/2$ such that $|P_1(\theta_1, \theta_2)| \cdots |P_{r-1}(\theta_1, \theta_2)| \geqslant |P_r(\theta_1, \theta_2)| \cdots |P_s(\theta_1, \theta_2)|$ and $|P_1(\theta_1, \theta_2)| \cdots |P_r(\theta_1, \theta_2)| \leqslant |P_{r+1}(\theta_1, \theta_2)| \cdots |P_s(\theta_1, \theta_2)|$. In view of (5.2), we obtain

$$|P_r(\theta_1, \theta_2)| \cdots |P_s(\theta_1, \theta_2)| < e^{-2\lambda t^4/3}$$

and

$$|P_{r+1}(\theta_1, \theta_2)| \cdots |P_s(\theta_1, \theta_2)| > e^{-\lambda t^4/3}$$

which implies that $|P_1(\theta_1, \theta_2)| \leqslant |P_r(\theta_1, \theta_2)| < e^{-\lambda 2t^4/3}$, and we obtain the remaining estimates in Lemma 5.3 since, by Lemma 1.4 or [**1**], $H(P_1) \leqslant H(P)e^{2d(P)}$ and, obviously, $d(P_1) \leqslant d(P)$, i.e. $t(P_1) \leqslant 3t(P)$.

PROOF OF THEOREM 5.1. The analytic scheme of the proof follows the method of [**1**]. The argument of the first step of the proof is the same as in [**1**].

Assume that by adjoining the MN numbers $e^{\alpha_i\beta_j}$ $(1 \leqslant i \leqslant N, 1 \leqslant j \leqslant M)$ to the field **Q** we obtain a field **K** of transcendence degree $\leqslant 2$. However, in view of a theorem of [**31**], among the numbers $e^{\alpha_i\beta_j}$ $(1 \leqslant i \leqslant N, 1 \leqslant j \leqslant M)$ there are two that are algebraically independent. Therefore, the field $\mathbf{K} = \mathbf{Q}(e^{\alpha_i\beta_j}\colon 1 \leqslant i \leqslant N, 1 \leqslant j \leqslant M)$ is of type Q_2 and is generated by numbers θ_1, θ_2, and ω_1. Let $\omega_2, \ldots, \omega_\nu$ denote the conjugates of ω_1 relative to the field $\mathbf{Q}(\theta_1, \theta_2)$. By assumption, the MN numbers

$$e^{\alpha_i\beta_j} \qquad (1 \leqslant i \leqslant N, 1 \leqslant j \leqslant M)$$

are equal to MN numbers of the field $\mathbf{K} = \mathbf{Q}(\theta_1, \theta_2, \omega_1)$:

$$S_i/T_i \quad (i = 1, \ldots, MN), \qquad T_0 = T_1 \cdots T_{MN},$$

where S_i, T_i are integers of **K** and $T_i \in \mathbf{Q}(\theta_1, \theta_2)$.

In what follows, X denotes any (sufficiently large) natural number and the $c_i > 0$ are constants which do not depend on X (but depend only on $\alpha_1, \ldots, \alpha_N$ and $\beta_1, \ldots, \beta_M$). Consider the function $f(z)$ $(= f_1^X(z))$,

$$f(z) = \sum_{k_1=0}^{L} \cdots \sum_{k_M=0}^{L} C_{k_1,\ldots,k_M} e^{(k_1\beta_1 + \cdots + k_M\beta_M)z}, \tag{5.3}$$

$$C_{k_1,\ldots,k_M} = \sum_{l_0=0}^{X} \sum_{l_1=0}^{X} C_{l_0,l_1,k_1,\ldots,k_M} \theta_1^{l_0}\theta_2^{l_1}, \qquad L = [X^{N/(M+N)}].$$

Using the Dirichlet principle (see Lemma 2.3), we choose rational integers $C_{l_0,l_1,k_1,\ldots,k_M}$, not all zero, such that

$$f(x_1\alpha_1 + \cdots + x_N\alpha_N) = 0, \qquad 0 \leqslant x_i \leqslant [c_1 X^{M/(M+N)}]\colon 1 \leqslant i \leqslant N, \tag{5.4}$$

$$\max |C_{l_0,l_1,k_1,\ldots,k_M}| < \exp(c_2 X). \tag{5.5}$$

If it happens that all of the polynomials

$$C_{k_1,\ldots,k_M}(x, y) = \sum_{l_0=0}^{X} \sum_{l_1=0}^{X} C_{l_0,l_1,k_1,\ldots,k_M} x^{l_0} y^{l_1}$$

have a common divisor $R_0(x, y)$, then we can divide the function $f(z)$ by this divisor with $x = \theta_1$, $y = \theta_2$. The resulting function, which we also denote by $f(z)$, has the form (5.3) and satisfies conditions (5.4) and (5.5), perhaps with a different c_2. We therefore assume (cf. **[1]**) that the polynomials $C_{k_1,\ldots,k_M}(x, y) = C^X_{k_1,\ldots,k_M}(x, y)$ have no common divisor.

According to **[1]**, Lemma III of Chapter III, or a lemma of **[10]**, there exists

$$z_0 = x_1^0\alpha_1 + \cdots + x_N^0\alpha_N, \qquad 0 \leqslant x_i^0 \leqslant \left[c_3 X^{M/(M+N)}\right]: 1 \leqslant i \leqslant N,$$

such that

$$f_0(z_0) = T_0^{\tau L} f\left(x_1^0\alpha_1 + \cdots + x_N^0\alpha_N\right) \neq 0,$$

where $\tau = \max(x_i^0 + 1: 1 \leqslant i \leqslant N) \leqslant (c_3 + 1)X^{M/(M+N)}$.

We estimate the magnitude of $f(z)$ using (5.3)–(5.5) as

(5.6)

$$|f(z)| \leqslant \left| \frac{1}{2\pi i} \int_\Gamma \prod_{x_1=0}^{[c_1 X^{M/(M+N)}]} \cdots \prod_{x_N=0}^{[c_1 X^{M/(M+N)}]} \frac{z - x_1\alpha_1 - \cdots - x_N\alpha_N}{\zeta - x_1\alpha_1 - \cdots - x_N\alpha_N} \cdot \frac{f(\zeta)}{\zeta - z} d\zeta \right|,$$

where Γ is the circle $|\zeta| = X^{(M+1)/(N+M)}$ and $|z| \leqslant X^{M/(N+M)} \ln X$.

Therefore, for $X > c_4$ we obtain, putting $\kappa = MN/(M+N)$,

$$(5.7) \qquad |f_0(z_0)| < \exp(-c_5 X^\kappa \ln X).$$

By assumption, $f_0(z_0)$ is an integer of the field $\mathbf{K}$, i.e. $f_0(z_0) = P(\theta_1, \theta_2, \omega_1)$, where $d(P) + \ln H(P) \leqslant c_6 X$ in accordance with (5.3) and (5.5). Consider the numbers $f_{i0} = P(\theta_1, \theta_2, \omega_i)$ for $i = 2, \ldots, \nu$.

Since $f_{10} = f_0(z_0) \neq 0$, it follows that

$$P_X(\theta_1, \theta_2) = \prod_{i=1}^{\nu} f_{i0} \neq 0,$$

and $P_X(\theta_1, \theta_2)$ is a polynomial in θ_1 and θ_2 with rational integral coefficients such that

$$(5.8) \quad t(P_X) = d(P_X) + \ln H(P_X) \leqslant c_7 X, \qquad |P_X(\theta_1, \theta_2)| < \exp(-c_8 X^\kappa \ln X)$$

for $\kappa = MN/(M+N)$, $X > c_9$.

In the sequel we will find useful the following

LEMMA 5.4. *There exist two constants $c_{10}, c_{11} > 0$ such that if for all $0 \leqslant x_i \leqslant [c_{10} X^{M/(N+M)}]$: $1 \leqslant i \leqslant N$, we have*

$$(5.9) \qquad |f(x_1\alpha_1 + \cdots + x_N\alpha_N)| \leqslant \exp(-c_{11} X^\kappa \ln X),$$

then for any $k_1, \ldots, k_M$ we obtain

$$(5.10) \qquad |C_{k_1,\ldots,k_M}| < \exp(-c_{12} X^\kappa \ln X).$$

PROOF OF LEMMA 5.4. Lemma 5.4 is obtained from a lemma of Tijdeman (see [**10**] and §3). According to a lemma of [**10**] or §3, it follows from assumption (H), (5.3)–(5.5), and (5.9) that for a $c_{10} \geqslant 16$ we have

$$\max |C_{k_1,\ldots,k_M}| \leqslant \exp\big(c_{10}X^\kappa \ln X + \tau' N/(M+N)X^\kappa \ln X + c_{10}^N \tau' M/(M+N)X^\kappa \ln X - c_{11}X^\kappa \ln X\big),$$

where τ' is the constant in (H). Choosing $c_{11} > c_{10}^N(1+\tau')$, we obtain (5.10) for any $k_1,\ldots,k_M$ assuming the values $0,\ldots,L$ and for $c_{12} = c_{11} - c_{10}^N(1+\tau')$. Lemma 5.4 is proved.

Let the symbol $\langle Y, Z\rangle$ signify that there exists a polynomial $P(x) \not\equiv 0$ such that $t(P) \leqslant Y$ and $|P(\theta_1)| < \exp(-Z)$.

According to [**1**], Lemma VII of Chapter III, to complete the proof it suffices to show that $\langle X^2 \ln^{1/4} X, X^4 \ln^{5/8} X\rangle$ for any sufficiently large X. Then θ_1 will be an algebraic number, contrary to our assumption, and the field **K** will have transcendence degree > 2.

We next apply Lemma 5.3. Since $\kappa = MN/(M+N) \geqslant 4$, it follows from (5.8) that $|P_X(\theta_1, \theta_2)| < \exp(-c_8 X^4 \ln X)$ and $t(P_X) \leqslant c_7 X$.

If (5.1) holds for any sufficiently large X with $t = c_7X$, we obtain $\langle X^2 \ln^{1/4} X, X^4 \ln^{5/8} X\rangle$. In this case, by a lemma of Gel'fond, Theorem 5.1 is proved. We therefore consider the second case, where, by Lemma 5.3, for an infinite sequence of $X > 0$ there exists a divisor $Q_X(x, y)$ of $P_X(x, y)$ having the form $Q_X(x, y) = (R_X(x, y))^{s_X}$, where $R_X(x, y)$ is **Q**-irreducible and

$$|Q_X(\theta_1, \theta_2)| < \exp(-c_{13}X^\kappa \ln X), \qquad t(Q_X) \leqslant c_{14}X. \tag{5.11}$$

Let $t_X = t(R_X)$ and $\delta_X = -\ln |R_X(\theta_1, \theta_2)|/\ln X$. Lemma 1.4 or Lemma II of [**1**], Chapter III, shows that

$$\tfrac{1}{3} t_X s_X \leqslant t(Q_X) \leqslant c_{14}X. \tag{5.12}$$

In view of (5.11) and (5.12), we obtain $\delta_X \geqslant c_{13}X^\kappa (s_X)^{-1} \geqslant \frac{1}{3}c_{13}X^\kappa t_X (t(Q_X))^{-1} \geqslant c_{15}X^{\kappa-1}t_X$. Thus,

$$\delta_X \geqslant c_{15}X^{\kappa-1}t_X. \tag{5.13}$$

Here, in (5.11)–(5.13), and later on, the (sufficiently large) $X > 0$ belongs to an infinite sequence $X_1,\ldots,X_n,\ldots$ of natural numbers for which nonfulfillment of $\langle X^2 \ln^{1/4} X, X^4 \ln^{5/8} X\rangle$ is assumed.

Case 1. Suppose $\delta_X > X^\kappa \ln^{1/8} X$. Put $Y = [X \ln^{-1/16} X]$.

LEMMA 5.5. *For any* $T(x, y) \in \mathbf{Z}[x, y]$ *with* $|T(\theta_1, \theta_2)| < \exp(-c_{16}Y^\kappa \ln Y)$ *and* $t(T) \leqslant c_{17}Y$ *we have* $|T(\theta_1, \theta_2)| < \exp(-Y^\kappa \ln^{9/8} Y)$.

PROOF OF LEMMA 5.5. If $R_X(x, y)$ divides $T(x, y)$, then, since $\delta_X > X^\kappa \ln^{1/8} X$ and $|R_X(\theta_1, \theta_2)| \leqslant \exp(-\delta_X \ln X)$, we obtain $|T(\theta_1, \theta_2)| < \exp(-Y^\kappa \ln^{9/8} Y)$. If $R_X(x, y)$ and $T(x, y)$ are coprime, then, considering their resultant, we obtain from Lemma 5.2 the relation $\langle X^2 \ln^{1/4} X, X^4 \ln^{5/8} X\rangle$, which is not so, hence Lemma 5.5 is proved.

It follows from Lemma 5.4 that either for certain $y_1,\ldots,y_N$: $0 \leqslant y_i \leqslant [c_{10}Y^{M/(M+N)}] = Y_0$, we have

$$|f^Y(y_1\alpha_1 + \cdots + y_N\alpha_N)| > \exp(-c_{11}Y^\kappa \ln Y), \tag{5.14}$$

or else for all $k_1,\ldots,k_M$: $0 \leqslant k_j \leqslant [Y^{N/(M+N)}]$, we have

$$|C^Y_{k_1,\ldots,k_M}| < \exp(-c_{12}Y^\kappa \ln Y). \tag{5.15}$$

Since the $C^Y_{k_1,\ldots,k_M}(x, y)$ have no common divisor, there exists among them a polynomial $C(x, y)$ which is coprime to $R_X(x, y)$. In the case (5.15), if we consider the resultant of $C(x, y)$ and $R_X(x, y)$ and if we use (5.3), (5.5), (5.15), and Lemma 5.2, we obtain $\langle X^2 \ln^{1/4} X, X^4 \ln^{5/8} X\rangle$. We may therefore assume that for certain $y_1,\ldots,y_N$: $0 \leqslant y_i \leqslant Y_0$, condition (5.14) holds.

We denote $T_0^{[c_{18}Y]} f^Y(z)$ with $z = y_1\alpha_1 + \cdots + y_N\alpha_N$ by $T_1(\theta_1, \theta_2) + \cdots + T_\nu(\theta_1, \theta_2)\omega_1^{\nu-1}$, where $t(T_i) \leqslant c_{18}Y$.

We will show that for any i, $1 \leqslant i \leqslant \nu$, there exist polynomials $Q^i_i,\ldots,Q^i_\nu$ in $\mathbf{Z}[x_1,\ldots,x_\nu]$ such that for

$$w = Q^i_i(\omega_1,\ldots,\omega_\nu)T_i(\theta_1, \theta_2) + \cdots + Q^i_\nu(\omega_1,\ldots,\omega_\nu)T_\nu(\theta_1, \theta_2)$$

we have

$$\exp(-c_{20}Y^\kappa \ln Y) < |w| < \exp(-c_{19}Y^\kappa \ln Y), \tag{5.16}$$

where the type $t(Q^i_j)$ is bounded by an absolute constant. For $i = 1$ this follows from (5.4) and (5.14). Assume that (5.16) has been proved for i, $1 \leqslant i < \nu$. Let $w^{(2)},\ldots,w^{(d)}$, where $d = [\mathbf{Q}(\omega_1,\ldots,\omega_\nu, \theta_1, \theta_2)\colon \mathbf{Q}(\theta_1, \theta_2)]$, denote the conjugates of w. Then $w_0 = ww^{(2)} \cdots w^{(d)}$ is (on replacing θ_1, θ_2 by x, y) a polynomial in x, y such that $|w_0(\theta_1, \theta_2)| < \exp(-c_{21}Y^\kappa \ln Y)$. It follows from Lemma 5.5 that $|w_0(\theta_1, \theta_2)| < \exp(-Y^\kappa \ln^{9/8} Y)$ or $|w^{(l)}| < \exp(-Y^\kappa \ln^{9/8} Y/d)$ for $1 \leqslant l \leqslant d$. If $w^{(l)} = \tilde{Q}^i_i(\omega_1,\ldots,\omega_\nu)T_i(\theta_1, \theta_2) + \cdots + \tilde{Q}^i_\nu(\omega_1,\ldots,\omega_\nu)T_\nu(\theta_1, \theta_2)$, then for $v = w\tilde{Q}^i_i(\omega_1,\ldots,\omega_\nu) - w^{(l)}Q^i_i(\omega_1,\ldots,\omega_\nu)$ we obtain

$$\exp(-c_{23}Y^\kappa \ln Y) > |v| > \exp(-c_{22}Y^\kappa \ln Y).$$

Here $v = Q^{i+1}_{i+1}(\omega_1,\ldots,\omega_\nu)T_{i+1}(\theta_1, \theta_2) + \cdots + Q^{i+1}_\nu(\omega_1,\ldots,\omega_\nu)T_\nu(\theta_1, \theta_2)$, i.e. (5.16) is proved for $i + 1$.

From (5.16) with $i = \nu$ we obtain

$$\exp(-c_{24}Y^\kappa \ln Y) > |T_\nu(\theta_1, \theta_2)| > \exp(-c_{25}Y^\kappa \ln Y),$$

which contradicts Lemma 5.5, thus completing Case 1.

Case 2. Suppose $\delta_X \leqslant X^\kappa \ln^{1/8} X$. Put $Y = [\delta_X^{1/(\kappa-1)} t_X^{-1/(\kappa-1)} \ln^{1/8} X]$. According to (5.13), $\ln X \leqslant \ln Y \leqslant 2 \ln X$. Putting $p = \max(1, [X^\kappa/\delta_X])$, we obtain, in view of our assumptions and (5.13),

$$\delta_X p \geqslant \tfrac{1}{2}X^4, \qquad Y^\kappa \ln Y \geqslant X^4 \ln X, \tag{5.17}$$

$$t_X pY \leqslant \delta_X^{1/(\kappa-1)} \cdot t_X^{(\kappa-2)/(\kappa-1)} \cdot \ln^{1/8} X \cdot p \leqslant X^2 \ln^{3/16} X. \tag{5.18}$$

Let i be the largest power ($i \geq 0$) of $R_X(x, y)$ which divides $P_Y(x, y)$. Then obviously $it_X \leq 3t(P_Y)$ and

$$\tfrac{1}{3} i\delta_X \ln X \leq \delta_X t(P_Y) t_X^{-1} \ln X \leq c_7 \cdot \delta_X \cdot Y \cdot t_X^{-1} \ln X$$
$$\leq c_{26} \delta_X^{\kappa/(\kappa-1)} \cdot t_X^{-\kappa/(\kappa-1)} \cdot \ln^{9/8} X.$$

It follows that

$$(5.19) \qquad i\delta_X \ln X \leq Y^\kappa \ln^{3/4} Y.$$

Write $P_Y(x, y)$ in the form $(R_X(x, y))^i S(x, y)$, where the polynomial $S(x, y)$ is coprime to $R_X(x, y)$. In view of (5.19) and the inequality $|P_Y(\theta_1, \theta_2)| < \exp(-c_8 Y^\kappa \ln Y)$, we obtain $|S(\theta_1, \theta_2)| < \exp(-\frac{1}{2} c_8 Y^\kappa \ln Y)$. By considering the resultant of $S(x, y)$ and $(R_X(x, y))^p$ with respect to y, we obtain (see [30] and Lemma 5.2) $\langle X^2 \ln^{1/4} X, X^4 \ln^{5/8} X\rangle$ and the analysis of Case 2 is complete. Theorem 5.1 is proved.

The statement (A_n) was already established in Proposition 3.7. Now, as a preparation to the proofs of Theorems I–III, we prove the statement (A_{n-1}) using the method employed in the proof of Theorem 5.1.

THEOREM 5.6. *Suppose* $\mathbf{Q}(S_i) = \mathbf{Q}(\theta_1, \ldots, \theta_d, \omega)$, *where* $\operatorname{tr}\deg \mathbf{Q}(S_i) = d$ *and* $\kappa_i \geq 4$. *Then for any* $X > c_{2i}$ *there exists* $P_X^1(x_1, \ldots, x_{d-1}) \in \mathbf{Z}[x_1, \ldots, x_{d-1}]$ *such that* $P_X^1(x_1, \ldots, x_{d-1}) \not\equiv 0$ *and*

$$(5.20) \qquad |P_X^1(\theta_1, \ldots, \theta_{d-1})| < \exp(-c_{3i} X^{\kappa_i} \ln^{\gamma_i} X),$$

$$(5.21) \qquad t(P_X^1) \leq c_{4i} X^2.$$

PROOF. By the symbol $\langle Y, Z\rangle$ we mean that for some $P(x_1, \ldots, x_{d-1}) \in \mathbf{Z}[x_1, \ldots, x_{d-1}]$, $P(x_1, \ldots, x_{d-1}) \not\equiv 0$, we have $t(P) \leq Y$ and $|P(\theta_1, \ldots, \theta_{d-1})| < \exp(-Z)$. We will show that for any $X > c_2$ we have $\langle c_5 X^2, c_6 X^\kappa \ln^\gamma X\rangle$. Using 3.7, we obtain $P_X(x_1, \ldots, x_d) \in \mathbf{Z}[x_1, \ldots, x_d]$, $P_X(x_1, \ldots, x_d) \not\equiv 0$, such that

$$(5.22) \qquad |P_X(\theta_1, \ldots, \theta_d)| < \exp(-c_7 X^\kappa \ln^\gamma X),$$

$$(5.23) \qquad t(P_X) \leq c_8 X.$$

Applying Lemma 4.2 we obtain either $R(x_1, \ldots, x_{d-1}) \in \mathbf{Z}[x_1, \ldots, x_{d-1}]$, $R(x_1, \ldots, x_{d-1}) \not\equiv C$, such that

$$(5.24) \quad t(R) \leq \gamma_1 c_8^2 X^2, \qquad |R(\theta_1, \ldots, \theta_{d-1})| < \exp(-c_7/8 \cdot X^{\kappa_i} \ln^{\gamma_i} X),$$

or else a divisor $Q_X(x_1, \ldots, x_d)$ of $P_X(x_1, \ldots, x_d)$ of the form $(R_X(x_1, \ldots, x_d))^{s_X}$, where $R_X(x_1, \ldots, x_d)$ is irreducible over $\mathbf{Q}$ and

$$(5.25) \qquad |Q_X(\theta_1, \ldots, \theta_d)| < \exp(-c_7/4 \cdot X^{\kappa_i} \ln^{\gamma_i} X),$$

$$(5.26) \qquad t(Q_X) \leq 3c_8 X.$$

If (5.24) holds, we have $\langle \gamma_1 c_8^2 X^2, c_4 X^{\kappa_i} \ln^{\gamma_i} X/8\rangle$. Let $t_X = t(R_X) \leq 3t(Q_X)/s_X$ and $\delta_X = -\ln|R_X(\theta_1, \ldots, \theta_d)|$. We have

$$(5.27) \qquad \delta_X \geq c_7/4 X^{\kappa_i} \ln^{\gamma_i} X (s_X)^{-1}.$$

We apply 3.7 to $f_i^Y(z)$ with $Y=[Xs_X^{-1/\kappa_i}/c_9]$ for a sufficiently large constant $c_9>0$. We obtain either $T_Y(x_1,\ldots,x_d)\in\mathbf{Z}[x_1,\ldots,x_d]$ and the estimates

$$\exp(-\tilde{c}_{22}Y^{\kappa_i}\ln^{\gamma_i}Y)<|T_Y(\theta_1,\ldots,\theta_d)|<\exp(-\tilde{c}_{23}Y^{\kappa_i}\ln^{\gamma_i}Y), \tag{5.28}$$

$$t(T_Y)\leqslant\tilde{c}_{24}Y, \tag{5.29}$$

or else for some family $C_l(x_1,\ldots,x_d)$: $l\in\mathfrak{L}_Y$, of polynomials in $\mathbf{Z}[x_1,\ldots,x_d]$ without a common divisor we have the estimates

$$|C_l(\theta_1,\ldots,\theta_d)|<\exp(-\tilde{c}_{25}Y^{\kappa_i}\ln^{\gamma_i}Y), \tag{5.30}$$

$$t(C_l)\leqslant\tilde{c}_{26}Y. \tag{5.31}$$

Suppose the second alternative holds. Then there exists $l\in\mathfrak{L}_Y$ such that $R_X(x_1,\ldots,x_d)$ and $C_l(x_1,\ldots,x_d)$ are coprime. In view of Lemma 4.1 there exists $R^1(x_1,\ldots,x_{d-1})\in\mathbf{Z}[x_1,\ldots,x_{d-1}]$, $R^1(x_1,\ldots,x_{d-1})\not\equiv 0$, such that

$$|R^1(\theta_1,\ldots,\theta_{d-1})|<\exp(-c_{10}X^{\kappa_i}\ln^{\gamma_i}X/s_X), \tag{5.32}$$

$$t(R^1)\leqslant c_{11}X^2/s_X^{1+1/\kappa_i}. \tag{5.33}$$

Raising $R^1(x_1,\ldots,x_{d-1})$ to the power s_X, we obtain $\langle c_{12}X^2/s_X^{1/\kappa_i}, c_{13}X^{\kappa_i}\ln^{\gamma_i}X\rangle$, as required. Similarly for (5.18) and (5.19). Assume that $R_X(x_1,\ldots,x_d)$ and $T_Y(x_1,\ldots,x_d)$ are coprime. In view of (5.17) and (5.18), we have

$$\max\{|R_X(\theta_1,\ldots,\theta_d)|,|T_Y(\theta_1,\ldots,\theta_d)|\}<\exp(-c_{14}X^{\kappa_i}\ln^{\gamma_i}X/s_X), \tag{5.34}$$

$$t(R_X)\cdot t(T_Y)\leqslant c_{15}X^2/s_X^{1+1/\kappa_i}, \tag{5.35}$$

which, in view of 4.1, implies $\langle c_{16}X^2/s_X^{1/\kappa_i}, c_{17}X^{\kappa_i}\ln^{\gamma_i}X\rangle$. If $T_Y(x_1,\ldots,x_d)$ is divisible by $R_X(x_1,\ldots,x_d)$, then from 1.4 we obtain

$$|T_Y(\theta_1,\ldots,\theta_d)|\leqslant|R_X(\theta_1,\ldots,\theta_d)|e^{c_{18}X} \tag{5.36}$$

$$|T_Y(\theta_1,\ldots,\theta_d)|<\exp(-c_7/8X^{\kappa_i}\ln^{\gamma_i}X/s_X). \tag{5.37}$$

On the other hand, by (5.28) and the definition of Y,

$$|T_Y(\theta_1,\ldots,\theta_d)|>\exp(-1/c_{19}X^{\kappa_i}\ln^{\gamma_i}X/s_X) \tag{5.38}$$

for an arbitrarily large c_{19}, which contradicts (5.37). Theorem 5.20 is proved.

Let us turn to the proof of a criterion for the algebraic independence of several numbers. The existing criterion of Gel′fond **[1]** for the transcendence of one number is the best. As is now known (E. Bombieri; see **[3]**), it is impossible to prove an analogous result for two or more numbers. The theorems proved in **[32]** show that for any function ψ: $N\to N$ there exist algebraically independent numbers θ_1, θ_2 such that

$$|a\theta_1+b\theta_2+c|<\exp(-\psi(X))$$

for $0<\max(|a|,|b|,|c|)<X$ and any sufficiently large natural number X. However, it is possible to obtain an analogue of Gel′fond's criterion if we consider a dense (in the sense of **[1]**) sequence of polynomials assuming small,

"but not too small", values. The precise formulation of this principle is given by

LEMMA 5.7 (CONCERNING ALGEBRAIC INDEPENDENCE). *Suppose* $\theta_1,\ldots,\theta_n$ *are arbitrary complex numbers and* $\sigma(x)$, $\theta(x)$ *are nonnegative, monotone increasing functions such that* $\sigma(N+1)\leqslant a_0\sigma(N)$, $\theta(N+1)\leqslant a_0\theta(N)$ *and* $\lim_{x\to\infty}\theta(x)=\infty$, $\lim_{x\to\infty}\sigma(x)=\infty$. *If for any* $N>N_0$ *there exists* $P_N(x_1,\ldots,x_n)\in\mathbf{Z}[x_1,\ldots,x_n]$ *such that* $t(P_N)=\max\{\ln H(P_N), d(P_N)\}\leqslant\sigma(N)$ *and*

$$(5.39)\quad \exp\left(-c_1\sigma^{2^n}(N)\theta(N)\right)<|P_N(\theta_1,\ldots,\theta_n)|<\exp\left(-c_2\sigma^{2^n}(N)\theta(N)\right)$$

for certain constants $c_1>c_2>0$, *then the numbers* $\theta_1,\ldots,\theta_n$ *are algebraically dependent.*

REMARK 5.8. The exponent 2^n in (5.39) is certainly not the best; it is possible that it can be lowered to $n+1$ (best from the point of view of Dirichlet's principle).

PROOF. For $n=1$ the lemma follows from Gel'fond's lemma.

Let $\sigma_0^{-1}(N)$ denote the natural number M such that $\sigma(x)=N$ and $[x]=M$. Then $x-1\leqslant\sigma_0^{-1}(N)\leqslant x$. Therefore, $N/a_0\leqslant\sigma(\sigma_0^{-1}(N))\leqslant N$. Let $\theta_1(N)=\theta(\sigma_0^{-1}(N))$. Clearly, $\theta_1(N)$ is an unbounded, monotone increasing function. Since for $N>N_1$ we have $\sigma_0^{-1}(N)>N_0$, it follows that for $\sigma_1(N)=\sigma(\sigma_0^{-1}(N))$ we obtain from (5.39) the existence of $P_N^n(x_1,\ldots,x_n)\in\mathbf{Z}[x_1,\ldots,x_n]$ such that $t(P_N^n)\leqslant\sigma_1(N)$ and

$$(5.40)\quad \exp\left(-c_1\sigma_1^{2^n}(N)\theta_1(N)\right)<|P_N^n(\theta_1,\ldots,\theta_n)|<\exp\left(-c_2\sigma_1^{2^n}(N)\theta_1(N)\right).$$

By definition of $\sigma_1(N)$,

$$(5.41)\quad N/a_0\leqslant\sigma_1(N)\leqslant N.$$

We will prove that for any k, $1\leqslant k\leqslant n$, the following assertion is true:

(B_k) For any $N>c_{3k}$ there exists $P_N^k(x_1,\ldots,x_k)\in\mathbf{Z}[x_1,\ldots,x_k]$ such that $t(P_N^k)\leqslant c_{4k}N^{2^{n-k}}$, $P_N^k(x_1,\ldots,x_k)\not\equiv 0$, and

$$(5.42)\quad |P_N^k(\theta_1,\ldots,\theta_k)|<\exp\left(-c_{5k}N^{2^n}\theta_1(N)\right).$$

Assertion (B_n) is true by virtue of (5.40) and (5.41). Assume that (B_i) holds for all i, $k+1\leqslant i\leqslant n$, $k<n$. We will prove (B_k).

We fix a sufficiently large $N>c_6$ and will construct for this N a polynomial $P_N^k(x_1,\ldots,x_k)\in\mathbf{Z}[x_1,\ldots,x_k]$, $P_N^k(x_1,\ldots,x_k)\not\equiv 0$, $t(P_N^k)\leqslant c_7N^{2^{n-k}}$ such that (5.42) holds. As usual, we write $\langle Y, X\rangle_k$ if there exists $P(x_1,\ldots,x_k)\in\mathbf{Z}[x_1,\ldots,x_k]$, $P(x_1,\ldots,x_k)\not\equiv 0$, such that $t(P)\leqslant Y$ and $|P(\theta_1,\ldots,\theta_k)|<\exp(-X)$.

Using (B_{k+1}), we obtain

$$P_N^{k+1}(x_1,\ldots,x_{k+1})\in\mathbf{Z}[x_1,\ldots,x_{k+1}],\qquad P_N^{k+1}(x_1,\ldots,x_{k+1})\not\equiv 0,$$

such that $t(P_N^{k+1})\leqslant c_8N^{2^{n-k-1}}$ and

$$(5.43)\quad |P_N^{k+1}(\theta_1,\ldots,\theta_{k+1})|<\exp\left(-c_9N^{2^n}\theta_1(N)\right).$$

We now use Lemma 4.2. It follows from (5.43) that either

$$\langle c_{10}N^{2^{n-k}}, c_{11}N^{2^n}\theta_1(N)\rangle_k$$

or else there exists a polynomial $P'_N(x_1,\ldots,x_{k+1}) \in \mathbf{Z}[x_1,\ldots,x_{k+1}]$ which divides $P_N^{k+1}(x_1,\ldots,x_{k+1})$ and is a power of a **Q**-irreducible polynomial such that

$$(5.44)\qquad |P'_N(\theta_1,\ldots,\theta_{k+1})| < \exp\left(-c_{12}N^{2^n}\cdot\theta_1(N)\right).$$

Using Lemma 4.3, we represent $P'_N(x_1,\ldots,x_{k+1})$ in the form $P'_N(x_1,\ldots,x_{k+1}) = (R_N(x_1,\ldots,x_{k+1}))^{s_1}$. Then, by Lemma 1.4, $t(P'_N) \leqslant 3t(P_N^{k+1})$. Again using Lemma 1.4, or 1.5, we obtain $t(R_N) \leqslant c_{13}N^{2^{n-k-1}}/s_1$. Therefore, in view of (5.44), we can write

$$(5.45)\quad |R_N(\theta_1,\ldots,\theta_{k+1})| < \exp\left(-c_{12}N^{2^n}\theta_1(N)/s_1\right),\qquad t(R_N) \leqslant c_{13}N^{2^{n-k-1}}/s_1.$$

Assume in what follows that $\theta_1,\ldots,\theta_n$ are algebraically independent.

Let $\mathfrak{L}$ denote the ring $\mathbf{Z}[\theta_1,\ldots,\theta_k]$. Since $R_N(x_1,\ldots,x_{k+1})$ is irreducible over **Q**, it follows that $R_N(\theta_1,\ldots,\theta_k, x)$ is irreducible over $\mathbf{K} = \mathbf{Q}(\theta_1,\ldots,\theta_k)$. This is a consequence of the algebraic independence of $\theta_1,\ldots,\theta_k$. By Lemma 4.3, there exists a root ξ_{k+1} of the polynomial $R_N(\theta_1,\ldots,\theta_k, x)$ such that

$$(5.46)\quad |R_N(\theta_1,\ldots,\theta_k,\theta_{k+1})| > |\theta_{k+1} - \xi_{k+1}| \times |u| \times \exp\left(-c_{14}N^{2^{n-k}}/s_1\right),$$

for $u \in \mathfrak{L}$, $u \neq 0$, $t_{\mathfrak{L}}(u) \leqslant c_{15}N^{2^{n-k}}/s_1$. We write u in the form $P_0(\theta_1,\ldots,\theta_k)$, where $P_0(x_1,\ldots,x_k) \in \mathbf{Z}[x_1,\ldots,x_k]$. If we do not have

$$\langle 2c_{15}N^{2^{n-k}}, c_{16}N^{2^n}\theta_1(N)\rangle_k,\qquad c_{16} < c_{12},$$

then for the polynomial $u^{s_1} = (P_0(x_1,\ldots,x_k))^{s_1}$, which is of type $\leqslant 2c_{15}N^{2^{n-k}}$, we have the estimate

$$(5.47)\qquad |u^{s_1}| \geqslant \exp\left(-c_{16}N^{2^n}\theta_1(N)\right).$$

Combining (5.46) and (5.47), we obtain

$$(5.48)\quad |R_N(\theta_1,\ldots,\theta_{k+1})| > |\theta_{k+1} - \xi_{k+1}| \exp\left(-c_{14}N^{2^k}/s_1 - c_{16}N^{2^n}\theta_1(N)/s_1\right).$$

Since $c_{16} < c_{12}$, relations (5.45) and (5.48) yield

$$(5.49)\qquad |\theta_{k+1} - \xi_{k+1}| < \exp\left(-c_{17}N^{2^n}\theta_1(N)/s_1\right).$$

Here ξ_{k+1} is a root of the polynomial $R_N(\theta_1,\ldots,\theta_k, x)$, i.e. it is an algebraic number over $\mathfrak{L}$ of $\mathfrak{L}$-type $t_{\mathfrak{L}}(\xi_{k+1}) \leqslant c_{13}N^{2^{n-k-1}}/s_1$ (see (5.45)).

It is somewhat more complicated to establish the existence of approximations ξ_i to the numbers θ_i. The difficulty is that for polynomials $P_N^i(x_1,\ldots,x_i) \not\equiv 0$ satisfying (B_i) it is possible to have

$$(5.50)\qquad \frac{\partial P_N^i(x_1,\ldots,x_i)}{\partial x_i} \equiv 0,$$

i.e. it is impossible to select an approximation to θ_i by means of $P_N^i(x_1,\ldots,x_i)$. We will show how to get around this obstacle.

For this let us consider the smallest $i \leqslant n$ such that a number θ_i can be approximated by numbers ξ_i of type $\exp(-O(N^{2^n}\theta_1(N)/s_1 s_2))$, where ξ_i is a root of multiplicity s_2 of a polynomial from $\mathfrak{L}_i[x]$,

$$\mathfrak{L}_i = \mathbf{Z}[\theta_1,\ldots,\theta_k, \xi_{k+1}, \theta_{k+2},\ldots,\theta_{i+1}].$$

The concrete form of the absolute constant in front of the symbol $O(\cdot)$ is of no importance for sufficiently large N. Here it should be pointed out that $i = i(N) \leqslant n$, i.e. the choice of i is not unique, but depends on N. In what follows, in order not to repeat this in every case, we take N to be an integer, which is considerably larger than all the constants that are used. Instead of repeating all the time "sufficiently large N and some constant c", we could consider a function $c = c(N)$ and require only $\overline{\lim}_{N\to\infty} c(N) < \infty$. The validity of the latter condition can always be easily checked. Nevertheless we are going to speak everywhere about certain constants without giving in each case their explicit form though in each case it can be done very easily.

Let l_1 be the smallest natural number, $k + 1 < l_1 \leqslant n$, such that

(5.51) there exists a polynomial $R_N^{l_1}(x_1,\ldots,x_{l_1}) \in \mathbf{Z}[x_1,\ldots,x_{l_1}]$, $t(R_N^{l_1}) \leqslant c_{18}N^{2^{n-l_1}}$, such that

$$|R_N^{l_1}(\theta_1,\ldots,\theta_k, \xi_{k+1}, \theta_{k+2},\ldots,\theta_{l_1})| < \exp(-c_{19}N^{2^n}\theta_1(N)/s_1), \tag{5.52}$$

$$\partial R_N^{l_1}/\partial x_{l_1} \not\equiv 0 \tag{5.53}$$

for certain constants c_{18}, c_{19}.

We will first show that $l_1 \leqslant n$ exists. If it does not, we can apply (5.39) and (5.40). We choose a real $x_0 > 0$ so that $x_0^{2^n}\theta_1(x_0) = N^{2^n}\theta_1(N)/s_1$ and put $X = [c_{20}x_0]$, where c_{20} is chosen so that

$$\frac{c_{17}}{2}N^{2^n}\theta_1(N)/s_1 > c_1\sigma^{2^n}(X)\cdot\theta_1(X). \tag{5.54}$$

Since $l_1 \leqslant n$ does not exist, (5.52) or (5.53) is not satisfied for the polynomial $P_X^n(x_1,\ldots,x_n) \not\equiv 0$ satisfying (5.40), $t(P_X^n) \leqslant X \leqslant c_{20}N$. If (5.53) is not satisfied, then, using the nonexistence of l_1, $k + 1 \leqslant l_1 \leqslant n$, we obtain

$$\frac{\partial P_X^n}{\partial x_{k+2}} \equiv \cdots \equiv \frac{\partial P_X^n}{\partial x_n} \equiv 0: \qquad x_1 = \theta_1,\ldots,x_k = \theta_k, x_{k+1} = \xi_{k+1}. \tag{5.55}$$

Therefore, putting $P_X^n(\theta_1,\ldots,x_{k+1}) \equiv P_X^n(\theta_1,\ldots,x_n)$, we obtain from (5.40) and (5.49) that

$$|P_X^n(\theta_1,\ldots,\theta_k, \xi_{k+1})| < \exp(-c_{21}N^{2^n}\theta_1(N)/s_1).$$

Using Lemmas 4.4 and 4.5, we obtain $u \in \mathbf{Z}[\theta_1,\ldots,\theta_k]$, $u = P(\theta_1,\ldots,\theta_k)$, $P(x_1,\ldots,x_k) \in \mathbf{Z}[x_1,\ldots,x_k]$, such that $u \neq 0$, $t(P) \leqslant c_{22}XN^{2^{n-k-1}}/s_1 \leqslant c_{23}N^{2^{n-k}}/s_1$, and

$$|P_X^n(\theta_1,\ldots,\theta_k, \xi_{k+1})| > |u|\exp(-c_{24}N^{2^{n-k}}/s_1) \tag{5.56}$$

when $P_X^n(\theta_1,\dots,\theta_k,\xi_{k+1}) \neq 0$. If (5.56) is satisfied, then for the polynomial $P(x_1,\dots,x_k) \in \mathbf{Z}[x_1,\dots,x_k]$, $P(x_1,\dots,x_k) \not\equiv 0$, we have $\langle c_{23}N^{2^{n-k}}/s_1, \frac{1}{2}c_{24}N^{2^n}\theta_1(N)/s_1\rangle_k$. Raising this polynomial to the power s_1, we obtain by virtue of 1.4 the desired assertion $\langle c_{25}N^{2^{n-k}}, c_{26}N^{2^n}\theta_1(N)\rangle_k$. If (5.56) is not satisfied, then, according to Lemma 4.4,

$$P_X^n(\theta_1,\dots,\theta_k,\xi_{k+1}) = 0.$$

In view of (5.49), it follows that

$$|P_X^n(\theta_1,\dots,\theta_k,\theta_{k+1})| < \exp\left(-c_{17}/2N^{2^n}\theta_1(N)/s_1\right). \tag{5.57}$$

According to (5.54), it follows from (5.57) that

$$|P_X^n(\theta_1,\dots,\theta_{k+1})| < \exp\left(-c_1\sigma_1^{2^n}(X)\theta_1(X)\right),$$

which contradicts (5.40). Thus, $l_1 \leqslant n$ satisfying (5.51) exists.

Then, in view of the minimality of l_1, we obtain by means of Lemmas 4.3 and 4.6 an approximation to θ_{l_1} in $\mathbf{Q}(\theta_1,\dots,\theta_{l_1-1})$, and so on. We state the general assertion as a lemma:

LEMMA 5.9. *Assume there have been constructed natural numbers* $l_0,\dots,l_s$: $l_0 = k+1 < l_1 < \cdots < l_s < n$, *and numbers* $\zeta_0,\dots,\zeta_s$ *satisfying the following conditions*:

(a) ζ_i *is algebraic over the ring*

$$\mathfrak{L}_i = \mathbf{Z}\left[\theta_1,\dots,\theta_k,\zeta_0,\theta_{k+1},\dots,\theta_{l_1-1},\zeta_1,\theta_{l_1+1},\dots,\theta_{l_i-1}\right],$$

and ζ_i *is a root of multiplicity* s_{l_i} *of a polynomial* $Q_i(x) \in \mathfrak{L}_i[x]$, $t_{\mathfrak{L}_i}(Q_i) \leqslant c_{27}N^{2^{n-l_i}}$, *such that*

$$|\zeta_i - \theta_{l_i}| < \exp\left(-c_{28}N^{2^n}\theta_1(N)/s_{l_0}\cdots s_{l_i}\right) \tag{5.58}$$

for all $0 \leqslant i \leqslant s$;

(b) l_j *is the smallest natural number* l, $l_{j-1} < l \leqslant n$, *for which there exists* $P(x) \in \mathfrak{L}'[x]$, $\mathfrak{L}' = \mathbf{Z}[\theta_1,\dots,\theta_k,\zeta_0,\dots,\theta_{l_{j-1}},\dots,\theta_{l-1}]$, $t_{\mathfrak{L}'}(P) \leqslant c_{29}N^{2^{n-l}}$, *such that*

$$\begin{gathered} |P(\theta_l)| < \exp\left(-c_{30}N^{2^n}\theta_1(N)/s_{l_0}\cdots s_{l_{j-1}}\right), \\ P(x) \not\equiv 0: \qquad 1 \leqslant j \leqslant s. \end{gathered} \tag{5.59}$$

Then there exists an l_{s+1}, $l_s < l_{s+1} \leqslant n$, *and a polynomial*

$$Q_{s+1} \in \mathfrak{L}_{s+1}[x], \qquad \mathfrak{L}_{s+1} = \mathbf{Z}\left[\theta_1,\dots,\theta_k,\zeta_0,\dots,\zeta_s,\theta_{l_s+1},\dots,\theta_{l_{s+1}-1}\right],$$

such that

$$\begin{gathered} Q_{s+1}(x) \not\equiv 0, \qquad t_{\mathfrak{L}_{s+1}}(Q_{s+1}) \leqslant c_{31}N^{2^{n-l_{s+1}}}, \\ |Q_{s+1}(\theta_{l_{s+1}})| < \exp\left(-c_{32}N^{2^n}\theta_1(N)/s_{l_0}\cdots s_{l_s}\right) \end{gathered} \tag{5.60}$$

and there exists ζ_{s+1}, *a root of* $Q_{s+1}(x)$ *of multiplicity* $s_{l_{s+1}}$, *such that*

$$|\theta_{l_{s+1}} - \zeta_{s+1}| < \exp\left(-c_{33}N^{2^n}\theta_1(N)/s_{l_0}\cdots s_{l_{s+1}}\right). \tag{5.61}$$

PROOF OF 5.9. Note that for $s = 0$ the lemma follows from (5.51) and Lemma 4.3. Assume, in the general case, that the desired $l_{s+1} \leqslant n$ does not exist. Consider the ring $\mathfrak{L}_s = \mathbf{Z}[\theta_1, \ldots, \theta_k, \zeta_0, \ldots, \theta_{l_s - 1}]$. Let x be a real number such that

$$x^{2^n}\theta_1(x) = N^{2^n}\theta_1(N)/s_{l_0} \cdots s_{l_s}. \tag{5.62}$$

The existence of a number satisfying (5.62) follows from the monotonicity of $\theta_1(x)$. We next define $X_1 = [c_{34}x] \leqslant c_{34}N$ so that, as above, we have

$$c_{28}/2 \cdot N^{2^n}\theta_1(N)/s_1 > c_1\sigma_1^{2^n}(X_1)\theta_1(X_1)$$

(cf. (5.40) and (5.58)). This is possible by virtue of (5.41). We use the polynomial $P_X^n(x_1, \ldots, x_n)$ in (5.40) for the given $X_1 = X$ constructed for N. We denote simply by t the smallest t, $1 \leqslant t \leqslant n$, such that

$$\frac{\partial P_X^n}{\partial x_{t+1}} \equiv \frac{\partial P_X^n}{\partial x_{t+2}} \equiv \cdots \equiv 0,$$

where we have put in P_X^n: $x_1 = \theta_1, \ldots, x_k = \theta_k$, $x_{k+1} = \xi_{k+1}, \ldots, x_{l_s - 1} = \theta_{l_s - 1}$, $x_{l_s} = \xi_s$, $x_{l_s + 1} = \theta_{l_s + 1}, \ldots, x_t = \theta_t$.

Consider first the case where $t < l_s$. We then obtain a number $P_X^n(\theta_1, \ldots, \theta_k, \xi_{k+1}, \ldots, \vartheta_t)$, where we have put $\vartheta_t = \theta_t$ for $t \neq l_i$ and $\vartheta_{l_i} = \zeta_i$, and, according to (5.40), (5.58), and (5.62), we have

$$|P_X^n(\theta_1, \ldots, \theta_k, \xi_{k+1}, \ldots, \vartheta_t)| < \exp\left(-c_{35}N^{2^n}\theta_1(N)/s_{l_0} \cdots s_{l_s}\right). \tag{5.63}$$

Consider the case $\vartheta_t = \theta_t$ for $l_i < t < l_{i+1}$ and a polynomial $Q_2(x) \in \mathfrak{L}^2[x]$, $\mathfrak{L}^2 = \mathbf{Z}[\theta_1, \ldots, \theta_k, \zeta_0, \ldots, \theta_{l_i + 1}, \ldots, \theta_{t-1}]$, such that $Q_2(\theta_i) = P_X^n(\theta_1, \ldots, \theta_k, \xi_{k+1}, \ldots, \theta_t)$. We will show that $Q_2(x)$ satisfies inequalities of the form (5.59), despite our assumption concerning the minimality of l_{i+1} and the fact that $l_i < t < l_{i+1}$. Indeed, by the choice of t, $\partial P_X^n/\partial x_t \not\equiv 0$, i.e. $Q_2(x) \not\equiv 0$. Let us estimate the $\mathfrak{L}^2$-type of $Q_2(x)$: by definition of P_X^n we have

$$t_{\mathfrak{L}^2}(Q_2) \leqslant X \leqslant c_{34}N.$$

Raise $Q_2(x)$ to the power $N^{2^{n-t}-1}$; denote the resulting polynomial by $P_2(x) \in \mathfrak{L}^2[x]$.

We have $t_{\mathfrak{L}^2}(P_2) \leqslant c_{34}N^{2^{n-t}}$ and, by definition of the numbers s_i, we obtain ($s_j = 1$ for $j \neq l_{i'}$)

$$s_{l_{i+1}} \cdots s_{l_s} \leqslant c_{36}N^{2^{n-l_{i+1}}} \cdots N^{2^{n-l_s}}$$

$$= c_{36}N^{2^{n-l_{i+1}+1} - 2^{n-l_s}} < c_{36}N^{2^{n-t}-1},$$

since, by definition of i and t, $t \leqslant l_{i+1} - 1$. Therefore, raising (5.63) to the power $N^{2^{n-t}-1}$, we have

$$|P_2(\theta_t)| < \exp\left(-c_{37}N^{2^n}\theta_1(N)/s_{l_0} \cdots s_{l_i}\right).$$

It follows from this and the estimate of the $\mathfrak{L}^2$-type of $P_2(x)$ that (5.59) holds, even though $l_i < t < l_{i+1}$, which contradicts (b).

The case $\vartheta_t = \zeta_i$ for $t = l_i$ is handled analogously. We need only use Lemmas 4.4 and 4.5, progressively discarding $\zeta_i, \ldots$ to arrive at the first transcendental $\theta_{t'}$, $t' < t$, where $\vartheta_{t'+1} = \zeta_{i'}, \ldots, \vartheta_t = \zeta_i$. For the polynomial in θ_t, we obtain, by virtue of (5.63), Lemmas 4.4 and 4.5, an estimate contradicting (5.59).

Let us turn to the case where $l_s < t$. In view of (5.63) and the assumption of the nonexistence of $l_{s+1} > l_s$, we conclude from the falsity of (5.61) that for the polynomial $P(x_t) = P_X^n(\theta_1, \ldots, \theta_k, \xi_{k+1}, \ldots, \theta_{t-1}, x_t)$ we have $P(x_t) \equiv 0$; this contradicts the definition of t. Therefore, $l_s = t$. We denote by t' the largest integer $< t$ such that $t' \neq l_j$, i.e. $\vartheta_{t'} \neq \zeta_j$. Consider the ring $\mathfrak{L}^* = \mathbf{Z}[\theta_1, \ldots, \theta_k, \zeta_0, \ldots, \theta_{t'}]$ and its algebraic extension $\mathfrak{L}_I = \mathfrak{L}^*[\vartheta_{t'+1}, \ldots, \vartheta_t]$, where, according to (a), the numbers ϑ_j are roots of multiplicities s_j of polynomials $Q_j(x) \in \mathfrak{L}^*[\vartheta_{t'+1}, \ldots, \vartheta_{j-1}, x]$ satisfying inequalities of the form (5.58).

We write (5.63) in the form

$$(5.64)\quad |P_X^n(\theta_1, \ldots, \theta_k, \ldots, \theta_{t'}, \vartheta_{t'+1}, \ldots, \vartheta_t)| < \exp\left(-c_{35} N^{2^n} \theta_1(N)/s_{l_0} \cdots s_{l_s}\right).$$

Using Lemma 4.5, we obtain $w \in \mathfrak{L}^*$, $w \neq 0$, $t_{\mathfrak{L}^*}(w) \leqslant c_{38} X N^{2^{n-t}} \cdots N^{2^{n-t'-1}} \leqslant c_{38} N^{2^{n-t'}}$, such that either

$$(5.65)\quad |P_X^n(\theta_1, \ldots, \theta_k, \ldots, \theta_{t'}, \vartheta_{t'+1}, \ldots, \vartheta_t)| \geqslant \exp\left(-c_{39} N^{2^{n-t'}}/s_{t'+1} \cdots s_t\right) \times |w|^{1/s_{t'+1} \cdots s_t}$$

or else $P_X^n(\theta_1, \ldots, \theta_{t'}, \vartheta_{t'+1}, \ldots, \vartheta_t) = 0$. It is clear that $s_j \leqslant N^{2^{n-j}}$ i.e. $s_{l_0} \cdots s_{t'} \leqslant c_{40} N^{2^{n-1} - 2^{n-t'}}$. Therefore, combining (5.64) and (5.65), we obtain

$$\exp\left(-c_{41} N^{2^n} \theta_1(N)/s_{l_0} \cdots s_{l_s}\right) > |w|^{1/s_{t'+1} \cdots s_t}.$$

Writing $w \in \mathfrak{L}^*$ in the form $w = R(\theta_1, \ldots, \theta_k, \zeta_0, \ldots, \theta_{t'})$, we have the inequality

$$(5.66)\quad \exp\left(-c_{41} N^{2^n} \theta_1(N)/s_{l_0} \cdots s_{t'-1}\right) > |R(\theta_1, \ldots, \zeta_0, \ldots, \theta_{t'})|.$$

If $t' \leqslant k$, then (5.66) becomes the assertion $\langle c_{38} N^{2^{n-k}}, c_{41} N^{2^n} \theta_1(N) \rangle_k$. If $k + 1 < t'$, then $l_i < t' < l_{i+1}$ for some i. From the estimate of the $\mathfrak{L}^*$-type of w and (5.66) we obtain for the polynomial $R(x_{t'}) = R(\theta_1, \ldots, \theta_k, \zeta_0, \ldots, x_{t'})$ the estimate (5.59) of condition (b), even though $l_i < t' < l_{i+1}$. This contradicts the minimum property of $l_{i+1} > l_i$.

Thus we have proved that inequality (5.65) does not hold. Therefore, by Lemma 4.5,

$$P_X^n(\theta_1, \ldots, \theta_{t'}, \vartheta_{t'+1}, \ldots, \vartheta_t) = 0.$$

We use condition (a) and the estimates (5.58) for replacing the numbers ζ_{l_i} with θ_{l_i}. We obtain the inequality

$$|P_X^n(\theta_1, \ldots, \theta_t)| < \exp\left(-c_{28}/2 N^{2^n} \theta_1(N)/s_{l_0} \cdots s_{l_s}\right).$$

On the other hand, according to (5.40),

$$|P_X^n(\theta_1, \ldots, \theta_t)| > \exp\left(-c_1 \sigma_1^{2^n}(X) \theta_1(X)\right),$$

since $P_X^n(\theta_1,\ldots,x_t) \equiv P_X^n(\theta_1,\ldots,x_n)$. The last two inequalities contradict each other for the choice $X = X_1$. Therefore, in all cases we have reached a contradiction to the nonexistence of l_{s+1}. The desired l_{s+1}, $l_s < l_{s+1} \leqslant n$, exists, and l_{s+1} is the smallest integer for which (5.60) holds. From (5.60) we also obtain (5.61). Indeed, apply Lemma 5.6 to some polynomial $Q(x) \in \mathfrak{L}_{s+1}[x]$, $\mathfrak{L}_{s+1} = \mathbf{Z}[\theta_1,\ldots,\theta_k,\zeta_0,\ldots,\theta_{l_{s+1}-1}]$, satisfying (5.60). First write $\mathfrak{L}_{s+1}$ in the form $\mathfrak{L}_{t'}$, $[\vartheta_{t'},\ldots,\vartheta_{l_{s+1}-1}]$, where $\vartheta_{t'} = \zeta_{i'},\ldots,\vartheta_{l_{s+1}-1} = \zeta_s$ are algebraic over $\mathfrak{L}_{t'} = \mathbf{Z}[\theta_1,\ldots,\theta_k,\ldots,\theta_{t'-1}]$. By Lemma 4.6 applied to the ring $\mathfrak{L}_{t'}$, there exists a root ζ_{s+1} of $Q(x)$ of multiplicity s_{l_s+1} such that

$$|Q(\theta_{l_{s+1}})| > |\theta_{l_{s+1}} - \zeta_{s+1}|^{s_{l_{s+1}}} \cdot |v|^{1/s_{t'}\cdots s_{l_s}}$$
$$\times \exp\left(-c_{42}N^{2^{n-l_{s+1}+1}} \cdot N^{2^{n-t'}} \cdots N^{2^{n-l_s}}/s_t' \cdots s_{l_s}\right)$$

for $v \in \mathfrak{L}_{t'} = \mathbf{Z}[\theta_1,\ldots,\theta_k,\ldots,\theta_{t'-1}]$, $v \neq 0$,

$$t_{\mathfrak{L}_t'}(v) \leqslant c_{43}N^{2^{n-l_{s+1}+1}} \cdot N^{2^{n-t'}} \cdots N^{2^{n-l_s}}$$
$$\leqslant c_{43}N^{2^{n-l_s}} \cdot N^{2^{n-t'}} \cdots N^{2^{n-l_s}} \leqslant c_{43}N^{2^{n-t'+1}}.$$

Using (5.60), we obtain the estimate

$$\exp\left(-c_{44}N^{2^n}\theta_1(N)/s_{l_0}\cdots s_{t'-1}\right) > |v|, \tag{5.67}$$

or $\exp(-c_{45}N^{2^n}\theta_1(N)/s_{l_0}\cdots s_{l_s}) > |\theta_{l_{s+1}} - \zeta_{s+1}|^{s_{l_{s+1}}}$.

Assume first that (5.67) holds.

We write v in the form $v = R(\theta_1,\ldots,\theta_{t'-1}) \in \mathfrak{L}_{t'} = \mathbf{Z}[\theta_1,\ldots,\theta_k,\zeta_0,\ldots,\theta_{t'-1}]$; $R(x_1,\ldots,x_{t'-1}) \in \mathbf{Z}[x_1,\ldots,x_{t'-1}]$. If $t'-1 = k$, then (5.67) together with $t_{\mathfrak{L}_{t'}}(v) \leqslant c_{43}N^{2^{n-(t'-1)}} = c_{43}N^{2^{n-k}}$ yields $\langle c_{43}N^{2^{n-k}}, c_{44}N^{2^n}\theta_1(N)\rangle_k$. The case $t'-1<k$ is impossible, since then, by definition of t', $\vartheta_{t'} = \theta_t$ is algebraic over $\mathbf{Q}(\theta_1,\ldots,\theta_{t-1})$, which is not so. If $t'-1>k$, then $l_i < t' < l_{i+1}$. In this case, (5.67) leads to a contradiction of the minimum property of l_{i+1}, $0 \leqslant i \leqslant s$.

Finally, (5.67) does not hold, and for some root ζ_{s+1} of $Q(x) \in \mathfrak{L}_{s+1}[x]$ of multiplicity $s_{l_{s+1}}$ we have

$$|\theta_{l_{s+1}} - \zeta_{s+1}| < \exp\left(-c_{45}N^{2^n}\theta_1(N)/s_{l_0}\cdots s_{l_{s+1}}\right).$$

Therefore, estimates (5.60) and (5.61) are completely proved, hence so is Lemma 5.9.

Since Lemma 5.9 is proved, we can continue the construction of the numbers l_i satisfying conditions (a) and (b) of Lemma 5.9. Note that $l_0 = k+1$. Continuing this procedure and taking into account the finiteness of n, we obtain that $l_m = n$ for some m (where $m \leqslant n-2$, since $k \geqslant 1$).

We will show that this is impossible. Consider a constant c_{46} such that

$$|\theta_{l_i} - \zeta_i| < \exp\left(-c_{46}N^{2^n}\theta_1(N)/s_{l_0}\cdots s_{l_i}\right)$$

for all $i = 0,\ldots,m$. Choose $X = X_2$ such that

$$c_{46}/2N^{2^n}\theta_1(N)/s_{l_0}\cdots s_{l_m} > c_1\sigma_1^{2^n}(X)\theta_1(X)$$

and $X \leqslant c_{47}N$. Replacing θ_{l_i} by ζ_i in (5.40), we obtain

$$(5.68) \qquad |P^n_{X_2}(\theta_1,\dots,\theta_k,\zeta_0,\dots,\zeta_m)| < \exp\left(-c_{47}N^{2^n}\theta_1(N)/s_{l_0}\cdots s_{l_m}\right).$$

We apply Lemma 4.5 to the ring $\mathfrak{L}^* = \mathbf{Z}[\theta_1,\dots,\theta_k,\zeta_0,\dots,\theta_t]$ such that $t' = l_i$, for all t', $t < t' \leqslant n$, i.e. all of the numbers $\theta_{t'}$ can be approximated by numbers that are algebraic over $\mathfrak{L}^*$. In view of the estimates of the types of the numbers ζ_i, (see Lemma 5.9), we obtain from (5.68) and Lemma 4.5 the inequality

$$\exp\left(-c_{48}N^{2^n}\theta_1(N)/s_{l_0}\cdots s_{l_m}\right) > |w|^{1/s_{t+1}\cdots s_{l_m}}$$

for $w \in \mathfrak{L}^*$, $w \neq 0$, $t_{\mathfrak{L}^*}(w) \leqslant c_{49}N\cdot N^{2^{n-t-1}}\cdots N = c_{49}N^{2^{n-t}}$. As before, the case $t < k$ is impossible, $t = k$ implies $\langle c_{49}N^{2^{n-k}}, c_{48}N^{2^n}\theta_1(N)\rangle_k$, and $t > k$ leads to a contradiction of the minimum property of l_{i+1}, $l_i < t < l_{i+1}$.

As a result, for any sufficiently large N we have $\langle O(N^{2^{n-k}}), O(N^{2^n}\theta_1(N))\rangle_k$, i.e. (B_k) is proved.

In the case $k = 1$, assertion (B_1) implies, in view of Lemma VII of [**1**], Chapter III, that θ is algebraic, which contradicts our assumption.

Remark 5.10. As the proof of Lemma 5.9 shows, we need not assume that the function $\theta(x)$ is unbounded. We need only assume that $\lim_{x\to\infty}\theta(x) > C(a_0, c_1, c_2, \theta_1,\dots,\theta_n)$.

6. This section is solely devoted to the complete proof of Theorems I–III. We start first with the exposition of the proof in the so called "nondegenerate" case [**16, 28**]. In this case all the polynomials $P^l_X(x_1,\dots,x_l)$ from (A_l) for all l, $k+1 \leqslant l \leqslant n$ contain the variable x_l with a nonzero coefficient. We present in this case the proofs of Theorems I–III.

Lemma 6.1. *Suppose $\theta_1,\dots,\theta_n$ are algebraically independent and for any l, $k+1 \leqslant l \leqslant n$, we have proved the following assertion*:

(6.1) (A'_l) *for every $X > c_5$ there exists $P^l_X(z_1,\dots,z_l) \in \mathbf{Z}[z_1,\dots,z_l]$, $t(P^l_X) \leqslant c_6X^{2^{n-l}}$, $P^l_X(z_1,\dots,z_l) \not\equiv 0$, such that*

$$(6.2) \qquad |P^l_X(\theta_1,\dots,\theta_l)| < \exp\left(-c_7X^{2^n}\theta(X)\right)$$

for a given monotone, unboundedly increasing function $\theta(X+1) \leqslant c_8\theta(X)$.

Then for any fixed natural number $X > c_9$, either (A'_k) and (6.2) hold, or else for the same X and each i, $k+1 \leqslant i \leqslant n$, there exist algebraic numbers ξ_i over $\mathfrak{L} = \mathbf{Z}[\theta_1,\dots,\theta_k]$ and $s_i \leqslant c_{11}X^{2^{n-i}}$ such that

$$(6.3) \qquad |\xi_i - \theta_i| < \exp\left(-c_{10}X^{2^n}\theta(X)/s_{k+1}\cdots s_i\right).$$

Here ξ_i and s_i: $k+1 \leqslant l \leqslant n$, are numbers such that for the ring $\mathfrak{L}_i = \mathfrak{L}[\xi_{k+1},\dots,\xi_{i-1}]$, $\mathfrak{L}_{k+1} = \mathfrak{L}$, there exists a polynomial $P_i(x) \in \mathfrak{L}_i[x]$ of $\mathfrak{L}_i$-type $t_{\mathfrak{L}_i}(P_i) \leqslant c_{11}X^{2^{n-i}}$ of which ξ_i is a root of multiplicity s_i.

Proof. Fix a sufficiently large X. As usual, we denote by $\langle Y, Z\rangle_k$ the fact that there exists $P(x_1,\dots,x_k) \in \mathbf{Z}[x_1,\dots,x_k]$, $P(x_1,\dots,x_k) \not\equiv 0$, such that $t(P) \leqslant Y$

and $|P(\theta_1,\ldots,\theta_k)|<\exp(-Z)$. Using (A'_{k+1}), we obtain $P_X^{k+1}(x_1,\ldots,x_{k+1})\in\mathbf{Z}[x_1,\ldots,x_{k+1}]$, $P_X^{k+1}(x_1,\ldots,x_k)\not\equiv 0$, such that $t(P_X^{k+1})\leqslant X^{2^{n-k-1}}$ and

$$(6.4)\qquad |P_X^{k+1}(\theta_1,\ldots,\theta_{k+1})|<\exp\bigl(-c_{12}X^{2^n}\theta(X)\bigr).$$

We now use Lemma 4.2. It follows from (6.4) that either $\langle c_{13}X^{2^{n-k}}, c_{14}X^{2^n}\theta(X)\rangle_k$, or else there exists a polynomial $P_X^1(x_1,\ldots,x_{k+1})\in\mathbf{Z}[x_1,\ldots,x_{k+1}]$, which divides $P_X^{k+1}(x_1,\ldots,x_{k+1})$ and is a power of a $\mathbf{Q}$-irreducible polynomial, such that

$$(6.5)\qquad |P_X^1(\theta_1,\ldots,\theta_{k+1})|<\exp\bigl(-c_{15}X^{2^n}\theta(X)\bigr).$$

Represent $P_X^1(x_1,\ldots,x_{k+1})$ in the form

$$P_X^1(x_1,\ldots,x_{k+1})=\bigl(R_X(x_1,\ldots,x_{k+1})\bigr)^{s_{k+1}}.$$

Then, by Lemma 1.4, $t(R_X)\leqslant c_{16}X^{2^{n-k-1}}/s_{k+1}$. Therefore, in view of (6.5), we can write

$$(6.6)\qquad \begin{aligned}|R_X(\theta_1,\ldots,\theta_{k+1})|&<\exp\bigl(-c_{17}X^{2^n}\theta(X)/s_{k+1}\bigr),\\ t(R_X)&\leqslant c_{18}X^{2^{n-k-1}}/s_{k+1}.\end{aligned}$$

Since $R_X(x_1,\ldots,x_{k+1})$ is irreducible over $\mathbf{Q}$, it follows that $R_X(\theta_1,\ldots,\theta_k,x)$ is irreducible over $\mathcal{K}=\mathbf{Q}(\theta_1,\ldots,\theta_k)$. According to Lemma 4.6, there exists a root ξ_{k+1} of the polynomial $R_X(\theta_1,\ldots,\theta_k,x)$ such that

$$(6.7)\qquad |R_X(\theta_1,\ldots,\theta_{k+1})|>|\theta_{k+1}-\xi_{k+1}|\cdot|u|\cdot\exp\bigl(-c_{19}X^{2^{n-k}}/s_{k+1}\bigr)$$

for $u\in\mathfrak{L}$, $u\neq 0$, $t_{\mathfrak{L}}(u)\leqslant c_{20}X^{2^{n-k}}/s_{k+1}$. Write u in the form $P(\theta_1,\ldots,\theta_k)$, where $P(x_1,\ldots,x_k)\in\mathbf{Z}[x_1,\ldots,x_k]$. If we do not have $\langle c_{21}X^{2^{n-k}}, c_{22}X^{2^n}\theta(X)\rangle_k$, $c_{22}<c_{17}$, then for the polynomial $u^{s_{k+1}}=(P(x_1,\ldots,x_k))^{s_{k+1}}$ of type $\leqslant c_{21}X^{2^{n-k}}$ we obtain the estimate

$$(6.8)\qquad |u^{s_{k+1}}|>\exp\bigl(-c_{22}X^{2^n}\theta(X)\bigr).$$

Combining (6.7) and (6.8), we obtain

$$(6.9)\qquad \begin{aligned}|R_X(\theta_1,\ldots,\theta_{k+1})|>&|\theta_{k+1}-\xi_{k+1}|\\ &\times\exp\bigl(\bigl(-c_{19}X^{2^{n-k}}+c_{22}X^{2^n}\theta(X)\bigr)/s_{k+1}\bigr).\end{aligned}$$

Since $c_{22}<c_{17}$, it follows from (6.9) and (6.6) that

$$(6.10)\qquad |\theta_{k+1}-\xi_{k+1}|<\exp\bigl(-c_{23}X^{2^n}\theta(X)/s_{k+1}\bigr).$$

Thus, (6.3) holds for $i=k+1$. Let us assume that (6.3) has been proved for any i, $k+1\leqslant i\leqslant l$, $l<n$. We will prove (6.3) for $i=l+1$.

Applying (A'_{l+1}), $k<l+1\leqslant n$, we obtain for any $X>c_{24}$ the existence of a polynomial $P_X^{l+1}(z_1,\ldots,z_{l+1})\in\mathbf{Z}[z_1,\ldots,z_{l+1}]$, $P_X^{l+1}(z_1,\ldots,z_{l+1})\not\equiv 0$, $t(P_X^{l+1})\leqslant c_6X^{2^{n-l-1}}$, such that

$$(6.11)\qquad |P_X^{l+1}(\theta_1,\ldots,\theta_{l+1})|<\exp\bigl(-c_7X^{2^n}\theta(X)\bigr).$$

Choose a real x_0 such that $x_0^{2^n}\theta(x_0) = X^{2^n}\theta(X)/s_{k+1}\cdots s_l$ and put $X_0 = [x_0]$. Then from (6.2) we have for X_0

$$(6.12)\qquad |P_{X_0}^{l+1}(\theta_1,\ldots,\theta_{l+1})| < \exp\left(-c_{25}X^{2^n}\theta(X)/s_{k+1}\cdots s_l\right),$$
$$t\left(P_{X_0}^{l+1}\right) \leqslant x_7 X_0^{2^{n-l-1}}.$$

Since $\theta(x)$ is a monotone function, we have $X_0 \leqslant X$ and we obtain $t(P_{X_0}^{l+1}) \leqslant c_7 X^{2^{n-l-1}}$. Using the inequalities (6.3) for $i = k+1,\ldots,l$, we deduce from (6.12) that

$$|P_{X_0}^{l+1}(\theta_1,\ldots,\theta_k, \xi_{k+1}\cdots\xi_l, \theta_{l+1})| < \exp\left(-c_{26}X^{2^n}\theta(X)/s_{k+1}\cdots s_l\right).$$

But it follows from the definition of X_0 that for $X > c_9$ we have $X_0^{2^{n-l-1}} < X^{2^{n-l-1}}\theta(X)/s_{k+1}\cdots s_l$, and since $l \geqslant 1$, we have

$$(6.13)\quad |P_{X_0}^{l+1}(\theta_1,\ldots,\theta_k, \xi_{k+1},\ldots,\xi_l, \theta_{l+1})| < \exp\left(-c_{27}X^{2^n}\theta(X)/s_{k+1}\cdots s_l\right).$$

Put $\mathfrak{L}_{l+1} = \mathfrak{L}[\xi_{k+1},\ldots,\xi_l]$. From Lemma 4.6 we obtain the existence of a root ξ_{l+1} of multiplicity s_{l+1} of the polynomial $P_{X_0}^{l+1}(\theta_1,\ldots,\theta_k,\ldots,\xi_l, x) \in \mathfrak{L}_{l+1}[x]$ such that

$$(6.14)\qquad |P_{X_0}^{l+1}(\theta_1,\ldots,\theta_k,\ldots,\xi_l, \theta_{l+1})| \geqslant |\theta_{l+1} - \xi_{l+1}|^{s_{l+1}}$$
$$\times\exp\left(-c_{28}X^{2^{n-k}}/s_{k+1}\cdots s_l\right)|w|^{1/s_{k+1}\cdots s_2}$$

for $w \in \mathfrak{L}$, $w \neq 0$, $t_{\mathfrak{L}}(w) \leqslant c_{29}X^{2^{n-k}}$.

If we do not have $\langle 3c_{29}X^{2^{n-k}}, c_{30}X^{2^n}\theta(X)\rangle_k$ with $c_{30} < c_{27}$, then for $w \in \mathfrak{L}$, $t_{\mathfrak{L}}(w) \leqslant c_{29}X^{2^{n-k}}$, we have the estimate

$$(6.15)\qquad |w| \geqslant \exp\left(-c_{30}X^{2^n}\theta(X)\right).$$

Comparing (6.15), (6.14), and (6.13), we obtain

$$(6.16)\qquad \exp\left(-c_{31}X^{2^n}\theta(X)/s_{k+1}\cdots s_l\right) > |\theta_{l+1} - \xi_{l+1}|^{s_{l+1}}.$$

Since ξ_{l+1} is a root of multiplicity s_{l+1} of the polynomial

$$P_{X_0}^{l+1}(\theta_1,\ldots,\theta_k, \xi_{k+1},\ldots,\xi_l, x) \in \mathfrak{L}_{l+1}[x]$$

of $\mathfrak{L}_{l+1}$-type $\leqslant c_6 X^{2^{n-l-1}}$, it follows from (6.16) that (6.3) is proved also for $i = l+1$. Therefore, either (6.3) is proved for all i, $k+1 \leqslant i \leqslant n$, or else we have $\langle O(X^{2^{n-k}}), O(X^{2^n}\theta(X))\rangle_k$.

We will now complete the proof of Theorems I–III in the "nondegenerate" case. We will show that (A_k) follows from (A_m) for all m, $k+1 \leqslant m \leqslant n$, $k \geqslant 1$. Fix a sufficiently large natural number X. We will prove $\langle O(X^{2^{n-k}}), O(X^{\kappa_i}\ln^{\gamma_i} X)\rangle_k$. In view of Lemma 6.1, we may assume that we are given $\mathfrak{L} = \mathbf{Z}[\theta_1,\ldots,\theta_k]$, $\mathfrak{L}_i = \mathfrak{L}[\xi_{k+1},\ldots,\xi_{i-1}]$, $\mathfrak{L}_{k+1} = \mathfrak{L}$, with algebraic numbers ξ_i over $\mathfrak{L}_i$ which are roots of multiplicities s_i of polynomials $Q_i(x) \in \mathfrak{L}_i[x]$ with $t_{\mathfrak{L}_i}(Q_i) \leqslant c_2' X^{2^{n-i}}$ and

$$(6.17)\qquad |\theta_i - \xi_i| < \exp\left(-c_3' X^{\kappa}\ln^{\gamma} X/s_{k+1}\cdots s_i\right)$$

for all $i = k+1,\ldots,n$. Consider some equation defining ω: $S(\theta_1,\ldots,\theta_n,\omega)=0$, where $S(x_1,\ldots,x_n,y)\in\mathbf{Z}[x_1,\ldots,x_n,y]$ is a polynomial with bounded type. Then we obtain for some root ω^* of the polynomial $S(\theta_1,\ldots,\theta_k,\ldots,\xi_n,x)$

$$(6.18)\qquad |\omega-\omega^*|<\exp(-c_4' X^{\kappa_i}\ln^{\gamma_i} X/s_{k+1}\cdots s_n).$$

For any $a\in S_i$, it follows from (6.17) and (6.18) that

$$(6.19)\qquad |a-S_a/T_a|<\exp(-c_5' X^{\kappa_i}\ln^{\gamma_i} X/s_{k+1}\cdots s_n),$$

where $T_a\in\mathfrak{L}_{n+1}=\mathbf{Z}[\theta_1,\ldots,\theta_k,\ldots,\xi_n]$, $s_a\in\mathfrak{L}_{n+1}[\omega^*]=\mathfrak{L}_{n+2}$, and the types $t_{\mathfrak{L}_{n+1}}(T_a)$, $t_{\mathfrak{L}_{n+2}}(S_a)$ are bounded.

For the rest of the proof we fix some $i=1,2,3$, say, for definiteness, $i=1$. Choose for X a natural number Y such that $(\gamma_1=1)$

$$(6.20)\qquad Y^{\kappa_1}\ln Y = 1/c_6' X^{\kappa_1}\ln X/s_{k+1}\cdots s_n$$

for a sufficiently large constant $c_6'>0$. Consider the function $f_1^Y(z)$.

Now instead of (3.3) we choose the coefficients $C^1_{\lambda_1,\ldots,\lambda_M}$ in a new form. Consider a finitely generated extension $\mathbf{Z}[z_1,\ldots,z_n]$ of $\mathbf{Z}$; let y denote a root of the equation $S(z_1,\ldots,z_n,y)=0$, and let $S_{i,j}$ and $T_{i,j}$ denote polynomials in $\mathbf{Z}[z_1,\ldots,z_n,y]$ and $\mathbf{Z}[z_1,\ldots,z_n]$: $i=1,\ldots,N$; $j=1,\ldots,M$. We consider coefficients C_λ in the form

$$(6.21)\qquad C_{\lambda_1,\ldots,\lambda_M}=\sum_{l_1=0}^{Y}\cdots\sum_{l_n=0}^{Y} C^Y_{\lambda_1,\ldots,\lambda_M,l_1,\ldots,l_n}\cdot z_1^{l_1}\cdots z_n^{l_n}.$$

Instead of the system of equations (3.4) there will be a new system:

$$(6.22)\qquad \sum_{\lambda_1=0}^{L_1}\cdots\sum_{\lambda_M=0}^{L_1} C^1_{\lambda_1,\ldots,\lambda_M}\prod_{i=1}^{N}\prod_{j=1}^{M}\left(\frac{S_{i,j}}{T_{i,j}}\right)^{x_i\lambda_j}=0$$

$$L_1=[Y^{M/(M+N)}]$$

for all rational integers x_i, $0\leqslant x_i\leqslant[c_7'Y^{M/(M+N)}]$: $1\leqslant i\leqslant N$. Each of the equations (6.22) can be rewritten as follows:

$$(6.23)\qquad \sum_{j_0=0}^{\nu-1}\sum_{j_1=0}^{[c_8'Y]}\cdots\sum_{j_n=0}^{[c_8'Y]} y^{j_0}z_1^{j_1}\cdots z_n^{j_n}$$
$$\times\sum_{\lambda_1=0}^{L_1}\cdots\sum_{\lambda_M=0}^{L_M}\sum_{l_1=0}^{Y}\cdots\sum_{l_n=0}^{Y} C^Y_{\lambda_1,\ldots,\lambda_M,l_1,\ldots,l_n}$$
$$\times F_{\lambda_1,\ldots,\lambda_M,l_1,\ldots,l_n,j_1,\ldots,j_n,x_1,\ldots,x_n}=0$$

where the $F_{\bar\lambda,\bar l,\bar j,\bar x}$ are rational integers of absolute value $\leqslant\exp(c_8'Y)$ for any rational integers x_i, $0\leqslant x_i\leqslant[c_7'Y^{M/(M+N)}]$: $1\leqslant i\leqslant N$. According to Lemma 2.3, there exist rational integers $C^Y_{\lambda_1,\ldots,\lambda_M,l_1,\ldots,l_n}$, not all zero, such that

$$(6.24)\qquad \max|C^Y_{\lambda_1,\ldots,\lambda_M,l_1,\ldots,l_n}|\exp(c_9'Y)$$

and the system (6.22) is solvable for all $x_i\leqslant[c_7'Y^{M/(M+N)}]$: $1\leqslant i\leqslant N$.

Consider the function $f_{11}(z)$, whose coefficients are the numbers $C^1_{\lambda_1,\ldots,\lambda_M}$ determined by means of (5.40) with $z_1 = \theta_1,\ldots,z_k = \theta_k$, $z_{k+1} = \xi_{k+1},\ldots,z_n = \xi_n$ and

(6.25) $$f_n(z) = \sum_{\lambda_1=0}^{L_1} \cdots \sum_{\lambda_M=0}^{L_1} C^1_{\lambda_1,\ldots,\lambda_M} e^{(\lambda_1\beta_1+\cdots+\lambda_M\beta_M)z}.$$

Taking into account (6.22) and (6.24), we obtain from (6.19) and (6.20) the inequalities

(6.26) $$|f^Y_{11}(x_1\alpha_1 + \cdots + x_n\alpha_n)| < \exp(-c'_{10}Y^{\kappa_1}\ln Y)$$

for all x_i, $0 \leqslant x_i \leqslant [c'_7Y^{M/(M+N)}]$: $1 \leqslant i \leqslant N$, and an arbitrarily large constant $c'_{10}\ (= c'_{11}\cdot c'_6)$. Using (6.26), consider the Hermite interpolation formula

(6.27) $$f_n(z) = \frac{1}{2\pi i}\int_{\Gamma_0} \frac{f_{11}(\zeta)}{\zeta - z}\prod_{i=1}^{p}\left(\frac{z-t_i}{\zeta-t_i}\right)d\zeta$$
$$-\frac{1}{2\pi i}\sum_{i=1}^{p} f_{11}(t_i)\int_{\Gamma_i}\prod_{j=1}^{p}\left(\frac{z-t_j}{\zeta-t_j}\right)\frac{(\zeta-t_i)\,d\zeta}{(\zeta-z)},$$

where $p = ([c'_7Y^{M/(M+N)}]+1)^N$, and Γ_0 and Γ_i $(1 \leqslant i \leqslant p)$ are the circles $|\zeta| = Y^{(1+M)/(M+N)}\ln Y$ and $|\zeta - t_i| = \frac{1}{2}\exp(-\tau_1Y^{M/(M+N)})$, where $t_1,\ldots,t_p$ are all of the numbers $z = x_1\alpha_1 + \cdots + x_N\alpha_N$ with $0 \leqslant x_i \leqslant [c'_7Y^{M/(M+N)}]$: $1 \leqslant i \leqslant N$. We choose the constant $\tau_1 > 0$ starting from assumption (H). If

$$|z| \leqslant Y^M/(M+N)\mathrm{In}Y,$$

then

(6.28) $$|f_{11}(z)| \leqslant \exp(-c'_{12}Y^{\kappa_1}\ln Y)$$

for some $c'_{12} > 0$.

Thus, for any x_i, $|x_i| \leqslant [c'_{13}Y^{M/(M+N)}]$: $1 \leqslant i \leqslant N$, we obtain

(6.29) $$|f_{11}(x_1\alpha_1 + \cdots + x_N\alpha_N)| < \exp(-c'_{14}Y^{\kappa_i}\ln Y).$$

According to [**10**], there are two alternatives:

(a) there exist rational integers x^0_i, $0 \leqslant x^0_i \leqslant [c_{13}Y^{M/(M+N)}]$: $1 \leqslant i \leqslant N$, such that

(6.30) $$\exp(-c'_{15}Y^{\kappa_1}\ln Y) < |f_{11}(x^0_1\alpha_1 + \cdots + x^0_N\alpha_N)|,$$

or else

(b) for all $\lambda_1,\ldots,\lambda_M$ taking the values $0,\ldots,L_1 = [Y^{N/(M+N)}]$ we have

(6.31) $$|C^1_{\lambda_1,\ldots,\lambda_M}| < \exp(-c'_{16}Y^{\kappa_1}\ln Y).$$

Assume that (a) holds. Let $f^*_{11}(x^0_1\alpha_1 + \cdots + x^0_N\alpha_N)$ denote the quantity obtained from $f_{11}(x^0_1\alpha_1 + \cdots + x^0_N\alpha_N)$ by replacing $e^{\alpha_i\beta_j}$ by $S_{i,j}/T_{i,j}$ (see (6.19)).

By definition of $T_{i,j}$, $\varphi = (\Pi_{i,j}T_{i,j})^{[c'_{13}Y]}f^*_{11}(x^0_1\alpha_1 + \cdots + x^0_N\alpha_N)$ is an element of the ring $\mathfrak{L}_{n+2} = \mathbf{Z}[\theta_1,\ldots,\xi_n,\omega^*]$. In view of (6.29), (6.30), and (6.21)–(6.24), we have

(6.32) $$t_{\mathfrak{L}_{n+2}}(\varphi) \leqslant c'_{19}Y.$$

Applying 4.5, we conclude that, since $\varphi \neq 0$, for some $w \in \mathfrak{L}$, $w \neq 0$, $t_{\mathfrak{L}}(w) \leqslant c'_{20} Y X^{2^{n-k-1}} \cdots X \leqslant c'_{20} X^{2^{n-k}}$, we have

$$|\varphi| > \exp\left(-c'_{21} X^{2^{n-k}} / s_{k+1} \cdots s_n\right) |w|^{1/s_{k+1} \cdots s_n}. \tag{6.33}$$

Comparing (6.32) and (6.33), and taking into account (6.20), we obtain $\langle c'_{22} X^{2^{n-k}}, c'_{23} X^{\kappa_1} \ln X \rangle_k$. Therefore, we may assume that the inequalities (6.30) are impossible for x_i, $0 \leqslant x_i \leqslant [c'_{13} Y^{M/(M+N)}]$: $1 \leqslant i \leqslant N$. Consequently, (6.31) holds for any $\lambda_1, \ldots, \lambda_M$ taking the values $0, \ldots, L_1$, where $L_1 = [Y^{N/(M+N)}]$. By definition, $C^1_{\lambda_1, \ldots, \lambda_M} \in \mathfrak{L}_{n+1}$ and $C^1_{\lambda_1, \ldots, \lambda_M} = \Gamma(\xi_{k+1}, \ldots, \xi_n)$, where $t_{\mathfrak{L}}(\Gamma) \leqslant c'_{24} Y$. According to 4.10, either

$$C^1_{\lambda_1, \ldots, \lambda_M} = 0 \tag{6.34}$$

or else for some $v \in \mathfrak{L}$, $v \neq 0$, $t_{\mathfrak{L}}(v) \leqslant c'_{25} Y X^{2^{n-k-1}} \cdots X = c'_{25} Y X^{2^{n-k}-1} \leqslant c'_{25} X^{2^{n-k}}$, we have

$$|C^1_{\lambda_1, \ldots, \lambda_M}| > \exp\left(-c'_{26} X^{2^{n-k}} / s_{k+1} \cdots s_n\right) |v|^{1/s_{k+1} \cdots s_n}. \tag{6.35}$$

If $\langle c'_{25} X^{2^{n-k}}, c'_{27} X^{\kappa_1} \ln X \rangle_k$ does not hold for a c'_{27} such that

$$c'_{27} X^{\kappa_1} \ln X / s_{k+1} \cdots s_n < \tfrac{1}{2} c'_{16} Y^{\kappa_1} \ln Y,$$

then, by definition of v,

$$|v| > \exp(-c'_{27} X^{\kappa_1} \ln X).$$

It then follows from (6.35) that

$$|C^1_{\lambda_1, \ldots, \lambda_M}| > \exp(-2 c'_{28} X^{\kappa_1} \ln X / s_{k+1} \cdots s_n). \tag{6.36}$$

Since $c'_{16} Y^{\kappa_1} \ln Y > 2 c'_{27} X^{\kappa_1} \ln X / s_{k+1} \cdots s_n$, inequalities (6.35) and (6.36) are contradictory. Therefore, (6.34) holds for any $\lambda_1, \ldots, \lambda_M$.

If $C^1_{\lambda_1, \ldots, \lambda_M} = 0$ for all $\bar{\lambda} = (\lambda_1, \ldots, \lambda_M)$, then, according to (6.21),

$$\sum_{l_1=0}^{Y} \cdots \sum_{l_n=0}^{Y} C^Y_{\bar{\lambda}, l_1, \ldots, l_{n-1}, l_n} \theta_1^{l_1} \cdots \theta_k^{l_k} \xi_{k+1}^{l_{k+1}} \cdots \xi_n^{l_n} = 0$$

for all $\bar{\lambda}$. Let us denote by j, $j \leqslant n$, the smallest integer such that

$$\sum_{l_1=0}^{Y} \cdots \sum_{l_j=0}^{Y} C^Y_{\bar{\lambda}, l_1, \ldots, l_n} \theta_1^{l_1} \cdots \theta_k^{l_k} \xi_{k+1}^{l_{k+1}} \cdots \xi_j^{l_j} = 0 \tag{6.37}$$

for all $\bar{\lambda}_0$ and $l_{j+1}, \ldots, l_n$ taking the values $0, \ldots, Y$. Since $\theta_1, \ldots, \theta_k$ are algebraically independent, $k + 1 \leqslant j$. We write the equalities (6.37) in the form

$$P^j_{\bar{\lambda}, l_{j+1}, \ldots, l_n}(\theta_1, \ldots, \theta_k, \xi_{k+1}, \ldots, \xi_j) = 0 \tag{6.38}$$

for all $\bar{\lambda}$ and $l_{j+1}, \ldots, l_n$ taking the values $0, \ldots, Y$, where $P^j_{\bar{\lambda}, L_{j+1}, \ldots, l_n}(x_1, \ldots, x_j) \in \mathbf{Z}[x_1, \ldots, x_j]$ and, by (6.24) and (6.21),

$$t\left(P_{\bar{\lambda}, l_{j+1}, \ldots, l_n}\right) \leqslant c'_{29} Y. \tag{6.39}$$

Moreover, by definition of j, for certain $\bar{\lambda}_0$, $l_j^0, \ldots, l_n^0$ we have

$$P^{j-1}_{\bar{\lambda}_0, l_j^0, \ldots, l_n^0}(\theta_1, \ldots, \theta_k, \xi_{k+1}, \ldots, \xi_{j-1}) \neq 0. \tag{6.40}$$

For any i, $k+1 \leqslant i \leqslant n$, let $P_i(x)$ be a $\mathcal{K}_i$-irreducible polynomial in $\mathcal{K}_i[x]$ with leading coefficient unity which divides $Q_i(x)$ and be such that $P_i(\xi_i) = 0$.

Consider, starting from (6.37) and (6.38), polynomials $P^1_{\bar\lambda, l_{j+1},\ldots,l_n}(x) = P^j_{\bar\lambda, l_{j+1},\ldots,l_n}(\theta_1,\ldots,\theta_k, \xi_{k+1},\ldots,\xi_{j-1}, x) \in \mathcal{L}_j[x]$. According to (6.40), $P^1_{\bar\lambda, l_{j+1},\ldots,l_n}(x) \not\equiv 0$ for certain $\bar\lambda$, $l_{j+1},\ldots,l_n$. By (6.38), the polynomials $P^1_{\bar\lambda, l_{j+1},\ldots,l_n}(x)$ are divisible by $P_j(x)$ in $\mathcal{K}_j[x]$. Let $\sigma_j > 0$ be the largest σ such that all of the $P^1_{\bar\lambda, l_{j+1},\ldots,l_n}(x)$ are divisible by $(P_j(x))^\sigma$. Then we can write

$$(6.41) \qquad P^1_{\bar\lambda, l_{j+1},\ldots,l_n}(x) = \left(P_j(x)\right)^{\sigma_j} Q^1_{\bar\lambda, l_{j+1},\ldots,l_n}(x),$$

where $Q^1_{\bar\lambda,\bar l}(x) \in \mathcal{K}_j[x]$. Let us turn to the definition of the $C_{\lambda_1,\ldots,\lambda_M}$ in (6.21). We obtain

$$(6.42) \qquad C_{\lambda_1,\ldots,\lambda_M} = \sum_{l_{j+1}=0}^{Y} \cdots \sum_{l_n=0}^{Y} P^j_{\bar\lambda, l_{j+1},\ldots,l_n}(x_1,\ldots,x_j) \cdot x_{j+1}^{l_{j+1}} \cdots x_n^{l_n}.$$

Using (6.41), we obtain

$$(6.43) \qquad \tilde C_{\lambda_1,\ldots,\lambda_M} = \sum_{l_{j+1}=0}^{Y} \cdots \sum_{l_n=0}^{Y} \left(P_j(x_j)\right)^{\sigma_j} Q^1_{\bar\lambda, l_{j+1},\ldots,l_n}(x_j) \times x_{j+1}^{l_{j+1}} \cdots x_n^{l_n} \in \mathcal{L}_j\left[x_j,\ldots,x_n\right].$$

Put $C^{(1)}_{\lambda_1,\ldots,\lambda_M} = \sum_{l_{j+1}=0}^{Y} \cdots \sum_{l_n=0}^{Y} Q^1_{\bar\lambda, l_{j+1},\ldots,l_n}(x_j) x_{j+1}^{l_{j+1}} \cdots x_n^{l_n}$. Then, $C^{(1)}_{\bar\lambda} \in \mathcal{K}_j[x_j,\ldots,x_n]$ and

$$(6.44) \qquad \tilde C_{\bar\lambda} = \left(P_j(x_j)\right)^{\sigma_j} C^{(1)}_{\bar\lambda}.$$

Note that by definition of the $Q^1_{\bar\lambda,\bar l}$ in (6.41) and σ_j, some $Q^1_{\bar\lambda_0,\bar l_0}(x)$ is not divisible by $P_j(x)$ and, since $P_j(x)$ is irreducible over $\mathcal{K}_j$, then $Q^1_{\bar\lambda_0, l_0}(\xi_j) \neq 0$. Therefore, the polynomials

$$P^2_{\bar\lambda, l_{j+2},\ldots,l_n}(x_{j+1}) = \sum_{l_{j+1}=0}^{Y} Q^1_{\bar\lambda, l_{j+1},\ldots,l_n}(\xi_j) x_{j+1}^{l_{j+1}}$$

in $\mathcal{K}_{j+1}[x]$ are not all identically zero. We again denote by $\sigma_{j+1} \geqslant 0$ the largest $\sigma \geqslant 0$ such that $P^2_{\bar\lambda, l_{j+2},\ldots,l_n}(x_{j+1})$ is divisible by $(P_{j+1}(x_{j+1}))^\sigma$ in $\mathcal{K}_{j+1}[x_{j+1}]$ for any $\bar\lambda$, l. Then

$$(6.45) \qquad P^2_{\bar\lambda, l_{j+2},\ldots,l_n}(x_{j+1}) = \left(P_{j+1}(x_{j+1})\right)^{\sigma_{j+1}} Q^2_{\bar\lambda, l_{j+2},\ldots,l_n}(x_{j+1}),$$

where $Q^2_{\bar\lambda,\bar l}(x_{j+1}) \in \mathcal{K}_{j+1}[x_{j+1}]$ and, for certain $\bar\lambda$, $\bar l$, we have $Q^2_{\bar\lambda,\bar l}(\xi_{j+1}) \neq 0$. Continuing this procedure up to the variable x_n, we arrive at the following.

For any i, $1 \leqslant i \leqslant n-j+1$, there exist polynomials

$$P^i_{\bar\lambda, l_{j+i},\ldots,l_n}(x_{j+i-1}) \in \mathcal{K}_{j+i-1}[x_{j+i-1}], \qquad Q^i_{\bar\lambda, l_{j+i},\ldots,l_n}(x_{j+i-1}) \in \mathcal{K}_{j+i-1}[x_{j+i-1}]$$

such that

$$(6.46) \quad P^i_{\bar\lambda, l_{j+i},\ldots,l_n}(x_{j+i-1}) = \left(P_{j+i-1}(x_{j+i-1})\right)^{\sigma_{j+i-1}} Q^i_{\bar\lambda, l_{j+i},\ldots,l_n}(x_{j+i-1}),$$

and, for certain $\bar{\lambda}$ and $\bar{l}$, $Q^i_{\bar{\lambda},\bar{l}}(\xi_{j+i-1}) \neq 0$ and

$$P^{i+1}_{\bar{\lambda},l_{j+i+1},\ldots}(x_{j+1}) = \sum_{l_{j+i}=0}^{Y} Q^i_{\bar{\lambda},l_{j+i},\ldots}(\xi_{j+i-1}) x^{l_{j+i}}_{j+i}.$$

We also put

(6.47)

$$C^i_{\bar{\lambda}} = \sum_{l_{j+i}=0}^{Y} \cdots \sum_{l_n=0}^{Y} Q^i_{\bar{\lambda},l_{j+i},\ldots}(x_{j+i-1}) \cdot x^{l_{j+i}}_{j+1} \cdots x^{l_n}_n \in \mathcal{K}_{j+i-1}[x_{j+i-1},\ldots,x_n].$$

It is now easy to see that the following equalities hold:

$$\tilde{C}_{\bar{\lambda}} = C_{\bar{\lambda}}(\theta_1,\ldots,\theta_k, \xi_{k+1},\ldots,\xi_{j-1}, x_j,\ldots,x_n),$$

$$\begin{aligned} (6.48)\quad \tilde{C}_\lambda &= (P_j(x_j))^{\sigma_j} C^1_{\bar{\lambda}},\ldots,C^i_{\bar{\lambda}}(\xi_{j+i-1}, x_{j+i},\ldots,x_n) \\ &= (P_{j+i}(x_{j+i}))^{\sigma_{j+i}} \cdot C^{i+1}_{\bar{\lambda}}(x_{j+1},\ldots,x_n),\ldots,C^{n-j}_{\bar{\lambda}}(\xi_{n-1}, x_n) \\ &= (P_n(x_n))^{\sigma_n} \cdot C^{n-j+1}_{\bar{\lambda}}(x_n). \end{aligned}$$

In view of (6.47) and (6.48), from the fact that $Q^{n-j+1}_{\bar{\lambda}}(\xi_n) \neq 0$ for some $\bar{\lambda}$ we obtain

$$C^{n-j+1}_{\bar{\lambda}}(\xi_n) \neq 0 \tag{6.49}$$

for some $\bar{\lambda}$.

We choose the coefficients of $f_1(z)$ in the form

$$C^K_{\lambda_1,\ldots,\lambda_M} = C_{\lambda_1,\ldots,\lambda_M}(\xi_n). \tag{6.50}$$

Let us estimate $\max|C^K_{\lambda_1,\ldots,\lambda_M}|$. According to (6.24), $H(\tilde{C}_{\bar{\lambda}}) \leqslant \exp(c'_{30}Y)$ for $\tilde{C}_{\bar{\lambda}} = \tilde{C}_{\bar{\lambda}}(x_j,\ldots,x_n)$. Since $P_i(x)$ has leading coefficient unity, the equalities (6.48) together with 1.4 yield in succession $H(C^i_{\bar{\lambda}}) \leqslant \exp(c'_{31}Y)$ for all i, $1 \leqslant i \leqslant n-j+1$. In particular, $H(C^{n-j+1}_{\bar{\lambda}}) \leqslant c'_{32}Y$. Therefore, in view of (6.50),

$$\max|C^K_{\lambda_1,\ldots,\lambda_M}| \leqslant \exp(c'_{32}Y). \tag{6.51}$$

We construct a function $f^K_1(z)$ of the form (6.25) with coefficients $C^K_{\lambda_1,\ldots,\lambda_M}$. Dividing (6.22) by $P_j(x_j)^{\sigma_j}$, substituting $x_j = \xi_j$, dividing by $P_{j+1}(x_{j+1})^{\sigma_{j+1}}$, and so on, we obtain

$$f^{K*}_1(x_1\alpha_1 + \cdots + x_N\alpha_N) = 0 \tag{6.52}$$

for $0 \leqslant x_i \leqslant [c'_{33}Y^{M/(M+N)}]$: $1 \leqslant i \leqslant N$, where $f^{K*}_1(x_1\alpha_1 + \cdots + x_N\alpha_N)$ is obtained from $f^K_1(x_1\alpha_1 + \cdots + x_N\alpha_N)$ by replacing $e^{\alpha_i\beta_j}$ by $S_{i,j}/T_{i,j}$ in accordance with (6.19). Taking into account (6.51) for the function f^{K*}_1, we have (6.26). As above, one of the following two alternatives holds:

(a″) for certain x^1_i, $0 \leqslant x^1_i \leqslant [c'_{33}Y^{M/(M+N)}]$: $1 \leqslant i \leqslant N$, we have

$$\begin{aligned} (6.53)\quad \exp(-c'_{34}Y^{\kappa_1}\ln Y) &< |f^{K*}_1(x^1_1\alpha_1 + \cdots + x^1_N\alpha_N)| \\ &< \exp(-c'_{35}Y^{\kappa_1}\ln Y), \end{aligned}$$

(b″) for all $\lambda_1,\ldots,\lambda_M$ taking the values $0,\ldots,L_1$ we have

$$|C^K_{\lambda_1,\ldots,\lambda_M}| < \exp(-c'_{36}Y^{\kappa_1}\ln Y). \tag{6.54}$$

Case (b″) follows trivially from (6.48) and Lemma 4.10. Indeed, in the notation of this lemma we put $\mathfrak{L}' = \mathfrak{L}_j$ $(j \geqslant k+1)$, $T_1(x_j,\ldots,x_n) \equiv \tilde{C}_{\bar\lambda}(x_j,\ldots,x_n)$, and so on. Applying (6.48) and the estimate $t_{\mathfrak{L}_j}(\tilde{C}_{\bar\lambda}) \leqslant c'_{37}Y$, we obtain either $C^{n-j+1}_{\bar\lambda}(\xi_n) = 0$ (which is impossible for all $\bar\lambda$ in view of (6.49), or else for $v \in \mathfrak{L}_j$, $v \neq 0$, $t_{\mathfrak{L}}(v) \leqslant c'_{38}YX\cdots X^{2^{n-j}} = c'_{38}YX^{2^{n-j+1}-1}$, the estimate

$$|C^K_{\bar\lambda}| \geqslant \exp\left(-c'_{39}YX^{2^{n-j+1}-1}/s_j\cdots s_n\right)|v|^{1/s_j\cdots s_n}. \tag{6.55}$$

Comparing (6.55) with (6.54) and taking into account (6.20), we have

$$\exp(-c'_{40}X^{\kappa_1}\ln X/s_{k+1}\cdots s_n) \geqslant |v|^{1/s_j\cdots s_n}. \tag{6.56}$$

From Lemma 4.5 we then obtain $\langle c'_{41}X^{2^{n-k}}, c'_{42}X^{\kappa_1}\ln X\rangle_k$. Consider case (a″). Multiplying (6.53) by $T_1 = (\Pi_{i,j}T^t_{i,j})^{[c'_{43}Y]}$, we obtain

$$\exp(-c'_{44}Y^{\kappa_1}\ln Y) < |T_1f_1^{K^*}(x_1^1\alpha_1 + \cdots + x_N^1\alpha_N)| < \exp(-c'_{45}Y^{\kappa_1}\ln Y). \tag{6.57}$$

Using the definition (6.25) and the definition of $C^K_{\lambda_1,\ldots,\lambda_M}$, we conclude that $T_1f_1^{K^*}(x_1^1\alpha_1 + \cdots + x_N^1\alpha_N) = \psi_1$ is an element of $\mathfrak{K}_n[\xi_n,\omega^*]$, $\psi_1 = \psi_1(\xi_n,\omega^*)$ with $\psi_1(x_n) \in \mathfrak{K}_n[x_n]$. We apply (6.48). Multiplying $\psi_1(x_n)$ by $(R_n(x_n))^{\sigma_n}$, we obtain, according to (6.48), ψ_2 in $\mathfrak{K}_{n-1}[\xi_{n-1},x_n]$, $\psi_2 = \psi_2(\xi_{n-1},x_n)$, $\psi_2(x_{n-1},x_n) \in \mathfrak{K}_{n-1}[x_{n-1},x_n]$. Continuing this procedure, we finally obtain $\psi_{n-j+2}(x_j,\ldots,x_n) \in \mathfrak{L}_j[x_j,\ldots,x_n]$ of type $t_{\mathfrak{L}_j}(\psi_{n-j+2}) \leqslant c'_{46}Y$ such that

$$\begin{aligned}\psi_{n-j+2}(x_j,\ldots,x_n) &= (R_j(x_j))^{\sigma_j}\psi_{n-j+1}(x_j,\ldots,x_n),\ldots,\psi_2(\xi_{n-1},x_n)\\ &= (R_n(x_n))^{\sigma_n}\psi_1(x_n),\end{aligned}$$

where, according to (6.57), $\psi_1(\xi_n) \neq 0$. In view of Lemma 4.10 and (5.57), we arrive at the inequality

$$\exp(-c'_{47}Y^{\kappa_1}\ln Y) \geqslant |v_2|^{1/s_j\cdots s_n}$$

for $v_2 \in \mathfrak{L}_j$, $v_2 \neq 0$, $t_{\mathfrak{L}_j}(v_2) \leqslant c_{48}YX^{2^{n-j+1}}$. In view of (6.20) and 4.10, we obtain $\langle c'_{49}X^{2^{n-k}}, c'_{50}X^{\kappa_1}\ln X\rangle_k$. Thus, assertion (A_k) is completely proved for any $k \geqslant 1$, which is impossible for $k = 1$ by Lemma VII of Chapter III of [**1**].

We come now to the complete proof of Theorems I–III. For this, as above, we are going to prove all the statements (A_k). We do not use now an additional assumption $\partial P^i_X/\partial x_i \not\equiv 0$. The previous proof was based on this assumption. As in the case of §5, when we say "some constant" and for a "sufficiently large integers X" we mean that there exists some absolute constant c, satisfying certain conditions for all integers X much larger than c. For example, one can put $X > \exp\cdots\exp(c + \max_i c_i)$, where the exp is taken $n+1$ times and where c_i are all constants (and there is only a finite number of them) encountered before. There is also another way, as described above. Instead of fixing c we can look at c

as a function, $c = c(X)$ and require $\overline{\lim}_{X\to\infty} c(X) < \infty$. To check this condition in each case is not difficult. However, there is no need of formalization along these lines because all the main elementary proofs have been made transparent in the "nondegenerate" case. The present exposition is simply a natural addition to them.

Fix a sufficiently large natural number X. We will prove for X the assertion $\langle O(X^{2^{n-k}}), O(X^{\kappa_i}\ln^{\gamma_i} X)\rangle_k$ with certain absolute constants. We use the argument of 5.7.

In view of (A_{k+1}) and Lemma 4.3, we at once obtain the existence of $\zeta_0 = \xi_{k+1}$, algebraic over $\mathfrak{L}_{k+1} = \mathbf{Z}[\theta_1,\ldots,\theta_k]$, such that ξ_{k+1} is a root of multiplicity s_{k+1} of a polynomial $Q_{k+1}(x) \in \mathfrak{L}_{k+1}[x]$ with $t_{\mathfrak{L}_{k+1}}(Q_{k+1}) \leqslant c_0 X^{2^{n-k-1}}$ and

$$(6.58)\qquad |\theta_{k+1} - \xi_{k+1}| < \exp(-c_1 X^{\kappa}\ln^{\gamma} X/s_{k+1}),$$

where $i = 1, 2, 3$ is fixed and $\kappa = \kappa_i$, $\gamma = \gamma_i$. Indeed, in the contrary case we obtain from (A_{k+1}) and 4.3 a $u \in \mathfrak{L}_{k+1}$, $u \neq 0$, $t_{\mathfrak{L}_{k+1}}(u) \leqslant c_2 X^{2^{n-k}}$, such that

$$|u| < \exp(-c_3 X^{\kappa}\ln^{\gamma} X),\quad \text{i.e. } \langle c_2 X^{2^{n-k}}, c_3 X^{\kappa}\ln^{\gamma} X\rangle_k.$$

As in Lemma 4.3 we can generalize (6.58). Assume we have constructed numbers $l_0,\ldots,l_r$ such that $l_0 = k+1 < \cdots < l_r \leqslant n$, and numbers $\zeta_0,\ldots,\zeta_r$ satisfying conditions of type (a) and (b) of Lemma 5.9. In what follows we put $\xi_i = \theta_i$ if $i \notin \{l_0,\ldots,l_r\}$ and $\xi_{l_i} = \zeta_i$.

LEMMA 6.2. *Suppose* $\mathfrak{L}_j = \mathbf{Z}[\xi_i,\ldots,\xi_{j-1}]$ *and let* $R_j(x_j) \in \mathcal{K}_j[x_j]$ *denote a* $\mathcal{K}_j$*-irreducible polynomial with leading coefficient unity such that* $R_j(\xi_j) = 0$ *if* $j = l_i$ *and* $\xi_{l_i} = \zeta_i$ *is algebraic over* $\mathfrak{L}_{l_i}$, *or such that* $R_j(x_j) \equiv 1$ *if* $j \notin \{l_0,\ldots,l_r\}$, *i.e. if* $\xi_j = \theta_j$ *is transcendental over* $\mathfrak{L}_{j-1}$.

Also, let $Q_{l_i}(x) \in \mathcal{K}_{l_i}[x]$ *denote a polynomial such that, in the notation of* §4,

$$(6.59)\qquad \left[Q_{l_i}(x_{l_i}) \cdot R_{l_i-1}(x_{l_i-1})^{\sigma'} \cdots R_1(x_1)^{\sigma''}\right]_{\xi_1,\ldots,\xi_{l_i-1}} = Q_{l_i}(x_1,\ldots,x_{l_i}),$$

where $t_{\mathbf{Z}}(Q_{l_i}) \leqslant c_4 X^{2^{n-l_i}}$ *and* $R_{l_i}(x)^{s_{l_i}}$ *divides* $Q_{l_i}(x)$, *i.e.* ξ_{l_i} *is a root of* $Q_{l_i}(x)$ *of multiplicity* s_{l_i}.

(a) *Suppose*

$$(6.60)\qquad |\theta_{l_i} - \xi_{l_i}| < \exp\left(-c_5 X^{\kappa}\ln^{\gamma} X/s_{l_0}\cdots s_{l_i}\right)$$

and

(b) *for any* $i, 0 \leqslant i < r$, l_{i+1} *is the smallest natural number* l, $l_i < l \leqslant n$, *such that there exists* $Q(x_l) \in \mathcal{K}_l[x_l]$, $Q(x_l) \not\equiv 0$, *such that*

$$(6.61)\qquad \left[Q(x_l) \cdot R_{l-1}(x_{l-1})^{\sigma_{l-1}} \cdots R_1(x_1)^{\sigma_1}\right]_{\xi_1,\ldots,\xi_{l-1}} = Q_l(x_1,\ldots,x_l)$$

where $t_{\mathbf{Z}}(Q_l) \leqslant c_6 X^{2^{n-l}}$ *and*

$$(6.62)\qquad |Q(\theta_l)| < \exp\left(-c_7 X^{\kappa}\ln^{\gamma} X/s_{l_0}\cdots s_{l_i}\right).$$

Then for any $P(x_1,\ldots,x_{l_r}) \in \mathbf{Z}[x_1,\ldots,x_{l_r}]$, *either* $P(\xi_1,\ldots,\xi_{l_r}) = 0$ *or for* $t(P) \leqslant c_8 X$ *we have*

$$(6.63)\qquad |P(\xi_1,\ldots,\xi_{l_r})| > \exp(-c_9 X^{\kappa} \ln^{\gamma} X / s_{l_0} \cdots s_{l_r}),$$

or $\langle c_{10} X^{2^{n-k}}, c_{11} X^{\kappa} \ln^{\gamma} X\rangle_k$ *for any constants* c_8, c_9, *where* c_{10}, c_{11} *are effectively determined by* c_8, c_9.

PROOF. Assume that $P(\xi_1,\ldots,\xi_{l_r}) \neq 0$. Let t denote the smallest natural number, $1 < t < l_r$, such that all of the numbers $\xi_{t+1},\ldots,\xi_{l_r}$ are algebraic over $\mathfrak{L}_{t+1} = \mathbf{Z}[\xi_1,\ldots,\xi_t]$, i.e. have the form ζ_i: $0 \leqslant i \leqslant r$. Note that $k \leqslant t$, since $\xi_1 = \theta_1,\ldots,\xi_k = \theta_k$ are algebraically independent. Assume that $t = k$. We apply Lemma 4.12 to $\mathfrak{L}_0 = \mathbf{Z}[\theta_1,\ldots,\theta_k] = \mathfrak{L}_{k+1}$, taking into account that $R_i(x_i) \equiv 1$ for $1 \leqslant i \leqslant k$ and $P(\theta_1,\ldots,\theta_k, x_{k+1},\ldots,x_n) \in \mathfrak{L}_{k+1}[x_{k+1},\ldots,x_n]$. We obtain $u \in \mathfrak{L}_{k+1}$, $u \neq 0$, $t_{\mathfrak{L}_{k+1}}(u) \leqslant c_{12} X^{2^{n-k}-1} t(P)$, such that

$$|P(\xi_1,\ldots,\xi_{l_r})| > \exp\left(-c_{13} t(P) X^{2^{n-k}-1} / s_{k+1} \cdots s_{l_r}\right) |u|^{1/s_{k+1}\cdots s_{l_r}}.$$

Assume that $\langle c_{12} X^{2^{n-k}}, c_{14} X^{\kappa} \ln^{\gamma} X\rangle_k$ does not hold. Then

$$|u| > \exp(-c_{14} X^{\kappa} \ln^{\gamma} X).$$

Therefore,

$$|P(\xi_1,\ldots,\xi_{l_r})| > \exp(-c_{15} X^{\kappa} \ln^{\gamma} X / s_{k+1} \cdots s_{l_r}).$$

Suppose $k < t$, i.e. $l_i < t < l_{i+1}$ for some i. We apply Lemma 4.12 to $\mathfrak{L}_0 = \mathbf{Z}[\theta_1,\ldots,\theta_t] = \mathfrak{L}_{t+1}$. We obtain $u \in \mathfrak{L}_{t+1}$, $u \neq 0$, $t_{\mathfrak{L}_{t+1}}(u) < c_{16} X^{2^{n-t}}$, such that

$$(6.64)\qquad |P(\xi_1,\ldots,\xi_{l_r})| > \exp\left(-c_{17} X^{2^{n-t}} / s_{t+1} \cdots s_{l_r}\right) |u|^{1/s_{t+1}\cdots s_{l_r}}.$$

Since l_{i+1} is the smallest natural number $> l_i$ satisfying (6.59), it follows from the falsity of (6.63) and (6.64) that $u \in \mathfrak{L}_t$ (since, according to (b), θ_t cannot occur in u nontrivially), and

$$|u| < \exp(-c_{18} X^{\kappa} \ln^{\gamma} X / s_{l_0} \cdots s_{t-1}).$$

Similarly, according to (b), θ_{t-1} cannot occur in u nontrivially, and so on. If $\xi_{t-1} = \zeta_i$, we apply Lemma 4.12 to $\mathfrak{L}_0 = \mathfrak{L}_{t'+1}$, where $\xi_{t'} = \theta_t$, and where $\xi_{t'+1},\ldots,\xi_{t-1}$ are algebraic over $\mathfrak{L}_{t'+1}$. We again obtain $u' \in \mathfrak{L}_{t'+1}$, $t_{\mathfrak{L}_{t'+1}}(u') \leqslant c_{19} X^{2^{n-t}} X^{2^{n-t+1}} \cdots X^{2^{n-t'-1}} \leqslant X^{2^{n-t'}}$, such that

$$|u'| < \exp(-c_{20} X^{\kappa} \ln^{\gamma} X / s_{l_0} \cdots s_{t'}).$$

Continuing this procedure, we obtain either (6.63) or $\langle c_{22} X^{2^{n-k}}, c_{21} X^{\kappa} \ln^{\gamma} X\rangle_k$. Lemma 6.2 is proved.

LEMMA 6.3. *We preserve all the notation of Lemma* 6.2 *and assume that* $l_0,\ldots,l_r$; $\zeta_0,\ldots,\zeta_r$ *satisfy* (a) *and* (b) *of Lemma* 6.2. *If there exist* l, $l_r < l \leqslant n$, *and* $Q_1(x_l) \in \mathcal{K}_l[x_l]$, $Q_1(x_l) \not\equiv 0$, *such that* $t_{\mathbf{Z}}(Q_1) \leqslant c_{23} X^{2^{n-l}}$, *where*

$$(6.65)\qquad \begin{gathered} [Q_1(x_l) \cdot R_{l-1}(x_{l-1})^{\sigma_{l-1}} \cdots R_1(x_1)^{\sigma_1}]_{\xi_1,\ldots,\xi_{l-1}} = Q_1(x_1,\ldots,x_l), \\ |Q_1(\theta_l)| < \exp(-c_{24} X^{\kappa} \ln^{\gamma} X / s_{l_0} \cdots s_{l_r}), \end{gathered}$$

then either there exist l_{r+1}, $l_r < l_{r+1} \leqslant n$, *and* $\xi_{l_{r+1}} = \zeta_{r+1}$, *a root of* $Q_{l_{r+1}}(x) \in \mathcal{K}_{l_{r+1}}[x]$ *of multiplicity* $s_{l_{r+1}}$, *such that* (a) *and* (b) *of Lemma* 6.2 *hold for* $l_0,\ldots,l_{r+1}$ *and* $\zeta_0,\ldots,\zeta_{r+1}$, *or else* $\langle c_{25}X^{2^{n-k}}, c_{26}X^{\kappa}\ln^{\gamma} X\rangle_k$, *where* c_{25}, c_{26} *are determined by* c_{23}, c_{24}.

PROOF. Denote by $l_{r+1} \leqslant n$ the smallest natural number such that $l_r < l_{r+1}$ and (6.63) holds with $Q_1(x) \in \mathcal{K}_{l_{r+1}}[x]$, $Q_1(x) \not\equiv 0$. Therefore, (b) holds. We will prove the existence of ζ_{r+1}. From Lemma 4.13 we obtain the estimate

$$(6.66)\qquad |Q_1(\theta_{l_{r+1}})| > |\theta_{l_{r+1}} - \zeta_{r+1}|^{s_{l_{r+1}}} \exp\left(-c_{28}X^{2^{n-l_{r+1}+1}}\right)|w|,$$

where ζ_{r+1} is a root of $Q_1(x)$ of multiplicity $s_{l_{r+1}}$, $|w| \neq 0$, and $|w| = |R(\xi_{l_{r+1}-1})|$, with $R(x) \in \mathcal{K}_{l_{r+1}-1}(x)$ and

$$\left[R(x_{l_{r+1}-1}) \cdot R_{l_{r+1}-2}(x_{l_{r+1}-2})^{\sigma'} \cdots R_1(x_1)^{\sigma''}\right]_{\xi_{l_{r+1}-2},\ldots,\xi_1}$$
$$= R_0(x_1,\ldots,x_{l_{r+1}-1}), \qquad t_Z(R_0) \leqslant c_{27}X^{2^{n-l_{r+1}+1}}.$$

If $\xi_{l_{r+1}-1} = \theta_{l_{r+1}-1}$, then $R(x)$ is constant, and in view of (6.65), (6.66), and the fact that l_{r+1} was chosen to satisfy minimum conditions, etc., we obtain that w does not depend on $\xi_{l_r+1},\ldots$. If $w = w(\theta_1,\ldots,\theta_k, \xi_{k+1},\ldots,\xi_{l_r})$, we can apply Lemmas 4.12 and 6.2. Since $w \neq 0$, it follows that either $\langle O(X^{2^{n-k}}), O(X^{\kappa}\ln^{\gamma} X)\rangle_k$ or else

$$|w| \geqslant \exp\left(-c_{29}X^{\kappa}\ln^{\gamma} X/s_{l_0}\cdots s_{l_r}\right).$$

It follows from this last inequality, (6.65), and (6.66) that

$$(6.67)\qquad |\theta_{l_{r+1}} - \xi_{l_{r+1}}| < \exp\left(-c_{30}X^{\kappa}\ln^{\gamma} X/s_{l_0}\cdots s_{l_{r+1}}\right).$$

To complete the proof of Theorems I–III it remains to show that the sequences $l_0,\ldots,l_r$ and $\zeta_0,\ldots,\zeta_r$ in Lemma 6.2 can be extended to $k+1 = l_0 < \cdots < l_s = n$ and that the case $l_s = n$ is impossible.

We combine the proofs of these two assertions into one.

Assume we have constructed a sequence $l_0,\ldots,l_r$: $k+1 = l_0 < \cdots < l_r \leqslant n$ and $\zeta_0,\ldots,\zeta_r$ satisfying conditions (a) and (b) of Lemma 6.2 and admitting no continuation. We will show that (A_k) then follows.

Let $l_r = m \leqslant n$ and $A = \{l_0,\ldots,l_r\}$. Then the fact that $l_0,\ldots,l_r$ and $\zeta_0,\ldots,\zeta_r$ admit no continuation $l_{r+1} > l_r$ and ζ_{r+1} also satisfying (a) and (b) of Lemma 6.2 implies, in view of Lemma 6.3, the following:

(6.68) If $Q(x_{m+1},\ldots,x_n) \in \mathfrak{L}_{m+1}[x_{m+1},\ldots,x_n]$ and $Q(x_{m+1},\ldots,x_n) \not\equiv 0$, then from

$$|Q(\xi_{m+1},\ldots,\xi_n)| < \exp(-c_1'X^{\kappa}\ln^{\gamma} X/s_1\cdots s_m)$$

with $t_{\mathfrak{L}_{m+1}}(Q) \leqslant c_2'X$ follows $\langle O(X^{2^{n-k}}), O(X^{\kappa}\ln^{\gamma} X)\rangle_k$, where the constants implied by $O(\cdot)$ and $O(\cdot)$ depend on c_1', c_2'.

We also introduce the following notation: $\xi_i = \theta_i$ for $i \notin A$, $\xi_{l_i} = \zeta_i$ for $l_i \in A$; $\mathfrak{L}_j = \mathbf{Z}[\theta_1,\ldots,\theta_k, \xi_{k+1},\ldots,\xi_{j-1}]$. According to Lemma 6.2, for $i \in A$ there exists

$Q_i(x) \in \mathcal{K}_i[x]$ satisfying (6.59) such that ξ_i is a root of $Q_i(x)$ of multiplicity s_i and

$$|\theta_i - \xi_i| < \exp(-c_3' X^\kappa \ln^\gamma X / s_{k+1} \cdots s_i), \tag{6.69}$$

where we have put $s_j = 1$ for $j \notin A$.

In addition to (6.69) we must also consider an analogous inequality for a number ω such that $\mathbf{Q}(S_i) \subset \mathbf{Q}(\theta_1, \ldots, \theta_n, \omega)$.

Consider some equation defining ω:

$$S(\theta_1, \ldots, \theta_n, \omega) = 0, \tag{6.70}$$

where $S(x_1, \ldots, x_n, y) \in \mathbf{Z}[x_1, \ldots, x_n, y]$ is a polynomial with type bounded by a constant.

In view of (6.69), we obtain

$$|S(\theta_1, \ldots, \theta_k, \xi_{k+1}, \ldots, \xi_n, \omega)| < \exp(-c_4' X^\kappa \ln^\gamma X / s_{k+1} \cdots s_n).$$

Let $b(x_1, \ldots, x_n)$ denote the leading coefficient of the polynomial $S(y) = S(x_1, \ldots, x_n, y)$. In view of (6.69) and (6.70), we obtain

$$|b(\theta_1, \ldots, \theta_k, \ldots, \xi_n)| \prod_{\omega^*} |\omega - \omega^*| \leqslant \exp(-c_5' X^\kappa \ln^\gamma X / s_{k+1} \cdots s_n). \tag{6.71}$$

Since $b(x_1, \ldots, x_n)$ has bounded type, the inequality

$$|b(\theta_1, \ldots, \theta_k, \xi_{k+1}, \ldots, \xi_n)| < \exp(-c_6' X) \tag{6.72}$$

is impossible for $X > c_7'$. Indeed, (6.69) and (6.70) imply that $|b(\theta_1, \ldots, \theta_n)| < \exp(-c_8' X)$. This inequality cannot be satisfied for infinitely many natural numbers X, since then $b(\theta_1, \ldots, \theta_n) = 0$, which contradicts the fact that $\theta_1, \ldots, \theta_n$ are algebraically independent.

Therefore, (6.71) and (6.72) yield, for some root ω^* of the equation $S(\theta_1, \ldots, \theta_k, \xi_{k+1}, \ldots, \xi_n, x) = 0$, the inequality

$$|\omega - \omega^*| < \exp(-c_9' X^\kappa \ln^\gamma X / s_{k+1}, \ldots, s_n). \tag{6.73}$$

For any element $a \in S_i$, i.e. a number of the form α_i, β_j, or $e^{\alpha_i \beta_j}$, we obtain approximate representations by elements of the field $\mathbf{Q}(\theta_1, \ldots, \theta_k, \xi_{k+1}, \ldots, \xi_n, \omega^*)$. For any element $a \in S_i$ we obtain from (6.69), (6.73), and §3 the inequality

$$|a - S_a / T_a| < \exp(-c_{10}' X^\kappa \ln^\gamma X / s_{k+1} \cdots s_n), \tag{6.74}$$

where $T_a \in \mathfrak{L}_{n+1} = \mathbf{Z}[\theta_1, \ldots, \theta_k, \xi_{k+1}, \ldots, \xi_n]$, $S_a \in \mathfrak{L}_{n+1}[\omega^*] = \mathfrak{L}_{n+2}$, and the types $t_{\mathfrak{L}_{n+1}}(T_a)$ and $t_{\mathfrak{L}_{n+2}}(S_a)$ are bounded by an absolute constant $\leqslant c_{11}'$. This follows from the representation of $a \in S_i$ as an element of the field $\mathbf{Q}(\theta_1, \ldots, \theta_n, \omega)$ by a number of type $\leqslant c_{12}'$ and the fact that $\max\{|\theta_i|, |\omega|\} \leqslant c_{13}'$. The expressions S_a, T_a with $a = e^{\alpha_i \beta_j}$ are denoted by $S_{i,j}$, $T_{i,j}$.

For the rest of the proof we fix some $i = 1, 2, 3$; say, for definiteness, $i = 1$. As in §3, we introduce an auxiliary function of the form $f_1(z)$ in (3.2). We choose for a given X a natural number Y such that ($\gamma_1 = 1$)

$$[Y^{\kappa_1} \ln Y] = [1/c_{14}' \cdot X^{\kappa_1} \ln X / s_{k+1} \cdots s_n] \tag{6.75}$$

for a sufficiently large constant $c'_{14} > 0$. Consider the function $f_1^Y(z)$ defined in §3, (3.2).

Let us consider the choice of coefficients $C^1_{\lambda_1,\ldots,\lambda_M}$ in a more general case. For this let us consider an arbitrary finitely generated extension of $\mathbf{Z}$, in particular, this may be a ring of polynomials in n variables. We fix some integer Y and the corresponding function $f_1^Y(z)$ given by (3.2).

Consider a finitely generated extension $\mathbf{Z}[x_1,\ldots,x_n]$ of $\mathbf{Z}$; let y denote a root of the equation $S(z_1,\ldots,z_n, y) = 0$ and let $S^t_{i,j}$ and $T^t_{i,j}$ denote polynomials in $\mathbf{Z}[z_1,\ldots,z_n, y]$ and $\mathbf{Z}[z_1,\ldots,z_n]$: $i = 1,\ldots,N$; $j = 1,\ldots,M$. We represent the coefficients $C^1_{\bar\lambda}$ in the form

$$(6.76) \qquad C^1_{\lambda_1,\ldots,\lambda_M} = \sum_{l_1=0}^{Y} \cdots \sum_{l_n=0}^{Y} C^Y_{\lambda_1,\ldots,\lambda_M,l_1,\ldots,l_n} z_1^{l_1} \cdots z_n^{l_n}.$$

Instead of the system of equations (3.4) we have the new system

$$(6.77) \qquad \sum_{\lambda_1=0}^{L_1} \cdots \sum_{\lambda_M=0}^{L_1} C^1_{\lambda_1,\ldots,\lambda_M} \prod_{i=1}^{N} \prod_{j=1}^{M} \left(\frac{S^t_{i,j}}{T^t_{i,j}} \right)^{x_i\lambda_j} = 0, \qquad L_1 = [Y^{M/(M+N)}],$$

for all rational integers x_i, $0 \leqslant x_i \leqslant [c'_{15} Y^{M/(M+N)}]$: $1 \leqslant i \leqslant N$. Each of the equations in (6.77) can be rewritten, after multiplying by $(\Pi_{i,j} T^t_{i,j})^{[c'_{15}Y]}$, as follows (cf. [**5**]):

$$(6.78) \qquad \sum_{j_0=0}^{\nu-1} \sum_{j_1=0}^{[c'_{16}Y]} \cdots \sum_{j_n=0}^{[c'_{16}Y]} y^{j_0} z_1^{j_1} \cdots z_n^{j_n}$$
$$\times \sum_{\lambda_1=0}^{L_1} \cdots \sum_{\lambda_M=0}^{L_M} \sum_{l_1=0}^{Y} \cdots \sum_{l_n=0}^{Y} C^Y_{\lambda_1,\ldots,\lambda_M,l_1,\ldots,l_n}$$
$$\times F_{\lambda_1,\ldots,\lambda_M,l_1,\ldots,l_n,j_1,\ldots,j_n,x_1,\ldots,x_N} = 0,$$

where the $F_{\bar\lambda,\bar l,\bar j,\bar x}$ are rational integers $\leqslant \exp(c'_{16}Y)$ in absolute value for any rational integers x_i, $0 \leqslant x_i \leqslant [c'_{15} Y^{M/(M+N)}]$: $1 \leqslant i \leqslant N$. The representability of the equations (6.77) in the form (6.78) follows immediately from (6.76) and the fact that

$$S^t_{i,j} \in \mathbf{Z}[z_1,\ldots,z_n, y], \qquad T^t_{i,j} \in \mathbf{Z}[z_1,\ldots,z_n].$$

The number of equations (6.78) is equal to $([c'_{15}Y^{M/(M+N)}] + 1)^N$. The equations (6.78) can be rewritten in the form of $\nu([c'_{16}Y] + 1)^n([c'_{15}Y^{M/(M+N)}] + 1)^N$ equations in $(Y + 1)^n[Y^{\kappa_1}]$ unknowns $C^Y_{\lambda_1,\ldots,\lambda_M,l_1,\ldots,l_n}$ with rational integral coefficients not exceeding $\exp(c'_{16}Y)$ in absolute value. Since $c'_{16} \leqslant c'_{17}c'_{15}$ for some absolute constant c'_{17} not depending on the choice of c'_{15}, we can apply Lemma 2.3. Choosing c'_{15} so that $\nu 2^n {c'_{17}}^n 2^N {c'_{15}}^{n+N} < \frac{1}{2}$, we obtain from 2.3 that the equations (6.78) are solvable in rational integers $C^Y_{\lambda_1,\ldots,l_n}$ not all zero. In addition, according to 2.3,

$$(6.79) \qquad \max |C^Y_{\lambda_1,\ldots,\lambda_M,l_1,\ldots,l_n}| \leqslant \exp(c'_{18}Y).$$

Thus, the system (6.77) is satisfied for all $x_i \leqslant [c'_{15}Y^{M/(M+N)}]$: $1 \leqslant i \leqslant N$, in such a way that (6.79) holds. Consider the function $f_{11}(z)$ whose coefficients are the numbers $C^1_{\lambda_1,\ldots,\lambda_M}$ defined by means of (6.76) with $z_1 = \theta_1,\ldots,z_k = \theta_k$, $z_{k+1} = \xi_{k+1},\ldots,z_{n-1} = \xi_{n-1}$, $z_n = \xi_n$, i.e.

(6.80) $$f_{11}(z) = \sum_{\lambda_1=0}^{L_1} \cdots \sum_{\lambda_M=0}^{L_1} C^1_{\lambda_1,\ldots,\lambda_M} e^{(\lambda_1\beta_1+\cdots+\lambda_M\beta_M)z}.$$

Let us consider that case, when $z_1,\ldots,z_n$ are complex numbers, i.e. y is also a complex number and the numbers $e^{\alpha_i\beta_j}$ are approximated by the elements of the field $\mathbf{Q}(z_1,\ldots,z_n, y)$:

$$\left|\frac{S_{ij}(z_1,\ldots,z_n, y)}{T_{ij}(z_1,\ldots,z_n)} - e^{\alpha_i\beta_j}\right| < \varepsilon,$$

for all $i = 1,\ldots,N, j = 1,\ldots,M$. Then as shown in (6.76) and (6.79) we obtain

$$|f^Y_{11}(x_1\alpha_1 + \cdots + x_N\alpha_N)| < \varepsilon \exp(c'Y)$$

for $x_i \leqslant [c'_{15}Y^{M/(M+N)}]$.

In view of (6.77) and (6.79), we obtain from (6.74) the inequalities

$$|f^Y_{11}(x_1\alpha_1 + \cdots + x_N\alpha_N)| < \exp(-c'_{19}X^{\kappa_1} \ln X/s_{k+1} \cdots s_n)$$

when $x_i \leqslant [c'_{15}Y^{M/(M+N)}]$. In view of (6.75), we obtain

(6.81) $$|f^Y_{11}(x_1\alpha_1 + \cdots + x_N\alpha_N)| < \exp(-c'_{20}Y^{\kappa_1} \ln Y)$$

for all x_i, $0 \leqslant x_i \leqslant [c'_{15}Y^{M/(M+N)}]$: $1 \leqslant i \leqslant N$, and an arbitrarily large constant c'_{20} $(= c'_{19}c'_{14})$. Consider, using (6.81), Hermite's interpolation formula (see [**1**, **15**]):

(6.82) $$f_{11}(z) = \frac{1}{2\pi i}\int_{\Gamma_0} \frac{f_{11}(\zeta)}{\zeta - z} \prod_{i=1}^{p} \left(\frac{z - t_i}{\zeta - t_i}\right) d\zeta$$
$$- \frac{1}{2\pi i}\sum_{i=1}^{p} f_{11}(t_i) \int_{\Gamma_i} \prod_{j=1}^{p} \left(\frac{z - t_j}{\zeta - t_j}\right) \frac{(\zeta - t_i)\, d\zeta}{(\zeta - z)},$$

where $p = ([c'_{15}Y^{M/(M+N)}] + 1)^N$, Γ_0 and Γ_i $(1 \leqslant i \leqslant p)$ are the circles $|\zeta| = Y^{1+M/(M+N)} \ln Y$ and $|\zeta - t_i| = \frac{1}{2}\exp(-\tau_1 Y^{M/(M+N)})$, and $t_1,\ldots,t_p$ are all of the numbers $z = x_1\alpha_1 + \cdots + x_N\alpha_N$ for $0 \leqslant x_i \leqslant [c'_{15}Y^{M/(M+N)}]$: $1 \leqslant i \leqslant N$. The constant $\tau_1 > 0$ is chosen starting from assumption (H). If $|z| \leqslant Y^{M/(M+N)} \ln Y$, then, estimating the modulus of the right-hand side of (6.82) and using (6.81) and (H), we obtain in the standard way (see [**1**, **11**, **15**]) that

(6.83) $$|f_{11}(z)| \leqslant \exp(-c'_{21}Y^{\kappa_1} \ln Y)$$

for some constant $c'_{21} > 0$. Since the inequality (6.83) holds for $|z| \leqslant Y^{M/(M+N)} \ln Y$, it can be applied to $z = x_1\alpha_1 + \cdots + x_N\alpha_N$ with $0 \leqslant x_i \leqslant [c'_{22}Y^{M/(M+N)}]$ for an arbitrarily large $c'_{22} > 0$.

Thus, for any x_i, $|x_i| \leqslant [c'_{22}Y^{M/(M+N)}]$, $1 \leqslant i \leqslant N$, it follows from (6.83) that

(6.84) $$|f_{11}(z)| < \exp(-c_{21}Y^{\kappa_1} \ln Y).$$

To complete the investigation of the function $f_{11}(z)$ we employ a lemma of **[10]**; we can use precisely the argument given by the author in **[15]**, or in **[1]**. For the sake of completeness we state

LEMMA 6.4 **[10]**. *Suppose* $\alpha_0,\ldots,\alpha_{n-1}$ *are n distinct numbers*, $\beta_0,\ldots,\beta_{s-1}$ *are s distinct numbers, and*

$$a = \max_{\nu}(|\alpha_\nu|,1),\ a_1 = \min_{\mu} \prod_{\nu=0,\,\nu\neq\mu}^{n-1} |\alpha_\mu - \alpha_\nu|^{1/n},\ a_2 = \min_{\mu\neq\nu}\left(|\alpha_\mu - \alpha_\nu|^{1/n},1\right);$$

$$b = \max_{\sigma}(|\beta_\sigma|,1),\ b_1 = \min_{\rho} \prod_{\sigma=0,\,\sigma\neq\rho}^{s-1} |\beta_\rho - \beta_\sigma|^{1/s},\ b_2 = \min_{\rho\neq\sigma}\left(|\beta_\rho - \beta_\sigma|^{1/s},1\right).$$

If A_ν $(\nu = 0,1,\ldots,n-1)$ *are any numbers*, $A = \max_\nu|A_\nu|$, *and*

$$F(z) = \sum_{\nu=0}^{n-1} A_\nu e^{\alpha_\nu z},$$

then for $rs \geqslant 2n + 13ab$ *and* $E = \max_{0\leqslant\rho\leqslant r-1,\,0\leqslant\sigma\leqslant s-1}|F^{(\rho)}(\beta_\sigma)|$ *we have*

$$A \leqslant 4s\left(\frac{a}{a_1a_2}\max\left(b,\frac{n}{b}\right)\right)^n\left(\frac{12b}{b_1b_2}\right)^r E. \tag{6.85}$$

If we apply (6.85) to the function $f_{11}(z) = F(z)$, using (6.80) and (H) to estimate a_1, a_2, b_1, b_2 in the spirit of **[1]**, we obtain

$$\max|C^1_{\lambda_1,\ldots,\lambda_M}| \leqslant \exp(-c'_{23}Y^{\kappa_1}\ln Y)E, \tag{6.86}$$

where $E = \max\{|f_{11}(x_1\alpha_1 + \cdots + x_N\alpha_N)|:\ 0 \leqslant x_i \leqslant [c'_{22}Y^{M/(M+N)}],\ 1 \leqslant i \leqslant N\}$, $c'_{22} > 16$, and c'_{23} is some constant defined by the α_i and β_i. From (6.86) follows one of two possibilities: either

(a) there exist rational integers x_i^0, $0 \leqslant x_i^0 \leqslant [c'_{22}Y^{M/(M+N)}]$: $1 \leqslant i \leqslant N$, such that

$$\exp(-c'_{24}Y^{\kappa_1}\ln Y) < |f_{11}(x_1^0\alpha_1 + \cdots + x_N^0\alpha_N)|, \tag{6.87}$$

or else

(b) for all $\lambda_1,\ldots,\lambda_M$ assuming the values $0,\ldots,L_1 = [Y^{N/(M+N)}]$,

$$|C^1_{\lambda_1,\ldots,\lambda_M}| < \exp(-c'_{25}Y^{\kappa_1}\ln Y). \tag{6.88}$$

Assume that (a) holds. Let $f^*_{11}(x_1^0\alpha_1 + \cdots + x_N^0\alpha_N)$ denote the quantity obtained from $f_{11}(x_1^0\alpha_1 + \cdots + x_N^0\alpha_N)$ by replacing $e^{\alpha_i\beta_j}$ by $S^t_{i,j}/T^t_{i,j}$ (see (6.74) with $a = e^{\alpha_i\beta_j}$). In view of the definition of $f_{11}(z)$, (6.76), (6.79), and (6.80), we obtain from (6.74) and (6.84) with $x_i \leqslant [c'_{22}Y^{M/(M+N)}]$ that

$$|f^*_{11}(x_1^0\alpha_1 + \cdots + x_N^0\alpha_N)| < \exp(-c'_{21}Y^{\kappa_1}\ln Y/2). \tag{6.89}$$

By definition of $T^t_{i,j}$, $(\Pi_{i,j}T^t_{i,j})^{[c'_{22}Y]}f^*_{11}(x_1^0\alpha_1 + \cdots + x_N^0\alpha_N) = \varphi$ is an element of the ring $\mathfrak{L}_{n+2} = \mathfrak{L}_{n+1}[\omega^*]$.

According to (6.76), (6.79), and (6.89),

$$|\varphi| < \exp(-c'_{26}Y^{\kappa_1}\ln Y), \qquad t_{\mathfrak{L}_{n+2}}(\varphi) \leqslant c'_{27}Y. \tag{6.90}$$

It follows from (6.87) that $\exp(-c'_{28}Y^{\kappa_1}\ln Y) < |\varphi|$. Therefore, $\varphi \neq 0$ and for the number ψ, the $\mathfrak{L}_{n+2}/\mathfrak{L}_{n+1}$-norm of φ, we have $\psi \neq 0$, $\psi \in \mathfrak{L}_{n+1} = \mathbf{Z}[\theta_1,\ldots,\theta_k, \xi_{k+1},\ldots,\xi_n]$. In view of (6.90) and the definition of φ, we obtain

$$(6.91) \qquad |\psi| < \exp(-c'_{29}Y^{\kappa_1}\ln Y), \qquad t_{\mathfrak{L}_{n+2}}(\psi) \leq c'_{30}Y.$$

Applying (6.68), we deduce from (6.91) that $\langle c'_{31}X^{2^{n-k}}, c'_{32}X^{\kappa_1}\ln X\rangle_k$. We may therefore assume that (6.87) is impossible for any rational integers x_i, $0 \leq x_i \leq [c'_{22}Y^{M/(M+N)}]$: $1 \leq i \leq N$. Consequently, (6.88) holds for any $\lambda_1,\ldots,\lambda_M$ assuming the values $0,\ldots,L_1$.

The number $C^1_{\lambda_1,\ldots,\lambda_M}$ can be written in the form $C^1_{\lambda_1,\ldots,\lambda_M} = \Gamma(\theta_{m+1},\ldots,\theta_n)$, where $\Gamma(x_{m+1},\ldots,x_n) \in \mathfrak{L}_{m+1}[x_{m+1},\ldots,x_n]$. Using (6.68) or Lemma 6.2, we obtain, in view of (6.88) and (6.76), that either $\Gamma(x_{m+1},\ldots,x_n) \equiv 0$ or else $\langle c'_{33}X^{2^{n-k}}, c'_{34}X^{\kappa_1}\ln X\rangle_k$ for constants c'_{33}, c'_{34} determined by c'_{25}. It suffices to consider the case $\Gamma \equiv 0$. However, $C^1_{\lambda_1,\ldots,\lambda_M} = \Gamma(\theta_{m+1},\ldots,\theta_n)$. Therefore,

$$(6.92) \qquad C^1_{\lambda_1,\ldots,\lambda_M} = 0$$

for all $\lambda_1,\ldots,\lambda_M$ assuming the values $0,\ldots,L_1$ and, according to properties of $\Gamma(x_{m+1},\ldots,x_n)$ and the definition of $C^1_{\lambda_1,\ldots,\lambda_M}$, the polynomials $C^1_{\lambda_1,\ldots,\lambda_M}(x_1,\ldots,x_n)$ do not depend on $x_{m+1},\ldots,x_n$:

$$(6.93) \qquad C^1_{\lambda_1,\ldots,\lambda_M} = C^1_{\lambda_1,\ldots,\lambda_M}(x_1,\ldots,x_m)$$

and also, in view of (6.92), $C^1_{\lambda_1,\ldots,\lambda_M}(\theta_1,\ldots,\theta_k, \xi_{k+1},\ldots,\xi_m) = 0$. In the notation $C^1_{\lambda_1,\ldots,\lambda_M}$ the parameters $l_{m+1},\ldots,l_n$ will be omitted.

For all $\bar{\lambda} = (\lambda_1,\ldots,\lambda_M)$ we obtain

$$\sum_{l_1=0}^{Y} \cdots \sum_{l_m=0}^{Y} C^Y_{\bar{\lambda},l_1,\ldots,l_m}\theta_1^{l_1}\cdots\theta_k^{l_k}\xi_{k+1}^{l_{k+1}}\cdots\xi_m^{l_m} = 0.$$

We denote by j, $j \leq m$, the smallest natural number such that

$$(6.94) \qquad \sum_{l_1=0}^{Y} \cdots \sum_{l_j=0}^{Y} C^Y_{\bar{\lambda},l_1,\ldots,l_j}\theta_1^{l_1}\cdots\theta_k^{l_k}\xi_{k+1}^{l_{k+1}}\cdots\xi_j^{l_j} = 0$$

for all $\bar{\lambda}$ and all $l_{j+1},\ldots,l_m$ assuming the values $0,\ldots,Y$. Since $\theta_1,\ldots,\theta_k$ are algebraically independent, it follows that $k+1 \leq j$. We write the equalities (6.94) in the form

$$(6.95) \qquad \mathbf{P}^j_{\bar{\lambda},l_{j+1},\ldots,l_m}(\theta_1,\ldots,\theta_k, \xi_{k+1},\ldots,\xi_j) = 0$$

for all $\bar{\lambda}$ and all $l_{j+1},\ldots,l_m$ assuming the values $0,\ldots,Y$, where $\mathbf{P}^j_{\bar{\lambda},l_{j+1},\ldots,l_m}(x_1,\ldots,x_j) \in \mathbf{Z}[x_1,\ldots,x_j]$ and, according to (6.79) and (6.76),

$$(6.96) \qquad t\left(\mathbf{P}_{\bar{\lambda},l_{j+1},\ldots,l_m}\right) \leq c'_{35}Y.$$

Furthermore, by definition of j, for certain $\bar{\lambda}_0$, $l^0_j,\ldots,l^0_m$ we have

$$(6.97) \qquad \mathbf{P}^{j-1}_{\bar{\lambda}_0,l^0_j,\ldots,l^0_m}(\theta_1,\ldots,\theta_k, \xi_{k+1},\ldots,\xi_{j-1}) \neq 0.$$

Returning to the notation of Lemma 6.2, for any i, $j \leqslant i \leqslant m$, $i \in A$, we denote by $P_i(x)$ the $\mathcal{K}_i$-irreducible polynomial in $\mathcal{K}_i[x]$ with leading coefficient unity such that $P_i(\xi_i) = 0$.

Consider, starting from (6.97), the polynomials

$$P^1_{\bar{\lambda},l_{j+1},\ldots,l_m}(x) = \mathbf{P}^j_{\bar{\lambda},l_{j+1},\ldots,l_m}(\theta_1,\ldots,\theta_k,\xi_{k+1},\ldots,\xi_{j-1},x) \in \mathfrak{L}_j[x].$$

According to (6.97), $P^1_{\bar{\lambda},l_{j+1},\ldots,l_m}(x) \not\equiv 0$ for certain $\bar{\lambda}, l_{j+1},\ldots,l_m$. In view of (6.95), the polynomials $P^1_{\bar{\lambda},l_{j+1},\ldots,l_m}(x)$ are divisible by $P_j(x)$ in $\mathcal{K}_j[x]$. Let $\sigma_j > 0$ denote the largest σ for which all of the $P^1_{\bar{\lambda},l_{j+1},\ldots,l_m}(x)$ are divisible by $(P_j(x))^\sigma$. Then we can write

$$P^1_{\bar{\lambda},l_{j+1},\ldots,l_m}(x) = \left(P_j(x)\right)^{\sigma_j} Q^1_{\bar{\lambda},l_{j+1},\ldots,l_m}(x), \tag{6.98}$$

where $Q^1_{\bar{\lambda},\bar{l}}(x) \in \mathcal{K}_j[x]$. Returning to the definition of $C^1_{\lambda_1,\ldots,\lambda_M}$ in (6.76), we obtain

$$C^1_{\lambda_1,\ldots,\lambda_M} = \sum_{l_{j+1}=0}^{Y} \cdots \sum_{l_m=0}^{Y} \mathbf{P}^j_{\bar{\lambda},l_{j+1},\ldots,l_m}(x_1,\ldots,x_j) x_{j+1}^{l_{j+1}} \cdots x_m^{l_m}. \tag{6.99}$$

Using (6.98), we obtain

$$\tilde{C}_{\lambda_1,\ldots,\lambda_M} = \sum_{l_{j+1}=0}^{Y} \cdots \sum_{l_m=0}^{Y} \left(P_j(x_j)\right)^{\sigma_j} Q^1_{\bar{\lambda},l_{j+1},\ldots,l_m}(x_j) \tag{6.100}$$
$$\times x_{j+1}^{l_{j+1}} \cdots x_m^{l_m} \in \mathfrak{L}_j[x_j,\ldots,x_m].$$

Put

$$C^{(1)}_{\lambda_1,\ldots,\lambda_M} = \sum_{l_{j+1}=0}^{Y} \cdots \sum_{l_m=0}^{Y} Q^1_{\bar{\lambda},l_{j+1},\ldots,l_m}(x_j) x_{j+1}^{l_{j+1}} \cdots x_m^{l_m}.$$

Then $C^{(1)}_{\bar{\lambda}} \in \mathcal{K}_j[x_j,\ldots,x_m]$ and

$$\tilde{C}_{\bar{\lambda}} = \left(P_j(x_j)\right)^{\sigma_j} \cdot C^{(1)}_{\bar{\lambda}}. \tag{6.101}$$

Note that, by definition of $Q^1_{\bar{\lambda},\bar{l}}$ in (6.98) and σ_j, some $Q^1_{\bar{\lambda}_0,\bar{l}_0}(x)$ is not divisible by $P_j(x)$, and since $P_j(x)$ is $\mathcal{K}_j$-irreducible, $Q^1_{\bar{\lambda}_0,\bar{l}_0}(\xi_j) \neq 0$. Therefore, the polynomials

$$P^2_{\bar{\lambda},l_{j+2},\ldots,l_m}(x_{j+1}) = \sum_{l_{j+1}=0}^{Y} Q^1_{\bar{\lambda},l_{j+1},\ldots,l_m}(\xi_j) x_{j+1}^{l_{j+1}} \in \mathcal{K}_{j+1}[x_{j+1}]$$

are not all identically zero. We again denote by $\sigma_{j+1} \geqslant 0$ the largest $\sigma \geqslant 0$ such that $P^2_{\bar{\lambda},l_{j+2},\ldots,l_m}(x_{j+1})$ is divisible by $(P_{j+1}(x_{j+1}))^{\sigma_{j+1}}$ in $\mathcal{K}_{j+1}[x_{j+1}]$ for any $\bar{\lambda}, \bar{l}$. Then

$$P^2_{\bar{\lambda},l_{j+2},\ldots,l_m}(x_{j+1}) = \left(P_{j+1}(x_{j+1})\right)^{\sigma_{j+1}} \cdot Q^2_{\bar{\lambda},l_{j+2},\ldots}(x_{j+1}), \tag{6.102}$$

where $Q^2_{\bar{\lambda},\bar{l}}(x_{j+1}) \in \mathcal{K}_{j+1}[x_{j+1}]$ and, for certain $\bar{\lambda}$ and $\bar{l}$, $Q^2_{\bar{\lambda},\bar{l}}(\xi_{j+1}) \neq 0$. If we continue this procedure up to the variable x_m, we arrive at the following equalities in the notation of §4 (here $R_i(x_i) \equiv 1$ if ξ_i is transcendental over $\mathfrak{L}_i$):

$$(6.103) \qquad C_{\bar\lambda}^{m-j+1}(x_m)\cdot P_m(x_m)^{\sigma_m} = C_{\bar\lambda}^{m-j}(\xi_{m-1}, x_m),$$

$$\left[C_{\bar\lambda}^{m-j}(x_m)\cdot P_{m-1}(x_{m-1})^{\sigma_{m-1}}\cdots P_1(x_1)^{\sigma_1}\right]_{\xi_1,\dots,\xi_{m-1}} = \tilde C_{\bar\lambda}(x_1,\dots,x_m)$$

where $C_{\bar\lambda}^{m-j+1}(x_m) \in \mathcal{K}_m[x_m]$; also, by choice of j, we obtain $C_{\bar\lambda}^{m-j+1}(\xi_m) \neq 0$, since the σ_i are chosen so that $C_{\bar\lambda}^{m-j+1}(x_m)$ is not divisible by $P_m(x_m)$, and so on. Thus,

$$(6.104) \qquad C_{\bar\lambda}^{m-j+1}(\xi_m) \neq 0$$

for some $\bar\lambda$. This is the reason for a new choice of coefficients of $f_1(z)$ in the form

$$(6.105) \qquad C^K_{\lambda_1,\dots,\lambda_M} = C^{m-j+1}_{\lambda_1,\dots,\lambda_M}(\xi_m).$$

First, we obtain an estimate for $\max|C^K_{\lambda_1,\dots,\lambda_M}|$. This is very simple: according to (6.79), $H(\tilde C_{\bar\lambda}) \leqslant \exp(c'_{36}Y)$ for $\tilde C_{\bar\lambda} = \tilde C_{\bar\lambda}(x_j,\dots,x_m)$. Since all of the polynomials $P_i(x)$ have leading coefficient unity, the equalities (6.103) together with 1.4 yield

$$(6.106) \qquad \max|C^K_{\lambda_1,\dots,\lambda_M}| \leqslant \exp(c'_{37}Y).$$

We again obtain a function $f_1^K(z)$ of the form (6.80) with coefficients $C^K_{\lambda_1,\dots,\lambda_M}$. If in succession we divide the identities (6.77) by $P_j(x_j)^{\sigma_j}$, put $x_j = \xi_j$, divide by $P_{j+1}(x_{j+1})^{\sigma_{j+1}}$, and so on, we obtain

$$(6.107) \qquad f_1^{K*}(x_1\alpha_1 + \cdots + x_N\alpha_N) = 0$$

for $0 \leqslant x_i \leqslant [c'_{15}Y^{M/(M+N)}]$: $1 \leqslant i \leqslant N$, where $f_1^{K*}(x_1\alpha_1 + \cdots + x_N\alpha_N)$ is obtained from $f_1^K(x_1\alpha_1 + \cdots + x_N\alpha_N)$ by replacing $e^{\alpha_i\beta_j}$ by $S_{i,j}/T_{i,j}$ in accordance with (6.71). In view of (6.106), the function f_1^K satisfies (6.81). As above, one of two possibilities holds:

(a′) for certain x_i^1, $0 \leqslant x_i^1 \leqslant [c'_{33}Y^{M/(M+N)}]$: $1 \leqslant i \leqslant N$, we have

$$(6.108) \qquad \exp(-c'_{39}Y^{\kappa_1}\ln Y) < |f_1^{K*}(x_1^1\alpha_1 + \cdots + x_N^1\alpha_N)| < \exp(-c'_{40}Y^{\kappa_1}\ln Y);$$

(b′) for all $\lambda_1,\dots,\lambda_M$ assuming the values $0,\dots,L_1$,

$$(6.109) \qquad |C^K_{\lambda_1,\dots,\lambda_M}| < \exp(-c'_{41}Y^{\kappa_1}\ln Y).$$

Case (b′) is derived from (6.103) and Lemma 4.12.

Indeed, apply Lemma 4.12, taking into account that $t_{\mathfrak{L}_j}(\tilde C_{\bar\lambda}) \leqslant c'_{42}Y$. By means of Lemma 4.12 we successively pass to analogues of (6.103), eliminating those ξ_i which are algebraic over $\mathfrak{L}_i$. If we arrive at a ξ_i which is transcendental over $\mathfrak{L}_i$, we use condition (b) of Lemma 6.2. This will signify the impossibility of a good upper bound for $|Q(\xi_i)|$ with $Q(x) \in \mathcal{K}_i[x]$. Continuing this procedure up to $\mathfrak{L}_j = \mathbf{Z}[\theta_1,\dots,\xi_{j-1}]$, we obtain either $C_{\bar\lambda}^{m-j+1}(\xi_m) = 0$ (which is impossible for all $\bar\lambda$ in view of (6.104) or else $v \in \mathfrak{L}_j$, $v \neq 0$, such that

$$t_{\mathfrak{L}_j}(v) \leqslant c'_{43}Y\cdot X\cdots X^{2^{n-j}} = c'_{43}Y\cdot X^{2^{n-j+1}-1},$$

$$(6.110) \qquad |C^K_{\bar\lambda}| \geqslant \exp\left(-c'_{44}YX^{2^{n-j+1}-1}/s_j\cdots s_n\right)|v|^{1/s_j\cdots s_n}.$$

Comparing (6.110) with (6.109) and taking into account (6.75), we obtain

$$(6.111) \qquad \exp\left(-c'_{45}X^{\kappa_1}\ln X/s_{k+1}\cdots s_n\right) > |v|^{1/s_j\cdots s_n},$$

i.e., according to Lemma 4.5, $\langle c'_{46}X^{2^{n-k}}, c'_{47}X^{\kappa_1}\ln X\rangle_k$. Consider case (a′). If we multiply (6.110) by $T_1 = (\Pi_{i,j}T_{i,j})^{[c'_{48}Y]}$, we obtain

$$(6.112) \quad \exp\left(-c'_{49}Y^{\kappa_1}\ln Y\right) < \left|T_1 f_1^{K^*}\left(x_1^1\alpha_1 + \cdots + x_N^1\alpha_N\right)\right| < \exp\left(-c'_{50}Y^{\kappa_1}\ln Y\right).$$

Using the definition (6.80) and the definition of $C^K_{\lambda_1,\ldots,\lambda_M}$, we conclude that $T_1 f_1^{K^*}(x_1^1\alpha_1 + \cdots + x_N^1\alpha_N) = \psi_1$ is an element of $\mathcal{K}_{n+1}[\omega^*]$, $\psi_1 = \psi_1(\omega^*)$ for $\psi_1(x) \in \mathcal{K}_{n+1}[x]$. We use (6.103). Multiplying $\psi_1(x_m,\ldots)$ by $(P_m(x))^{\sigma_m}$, we arrive, according to (6.103), at ψ_2 in $\mathcal{K}_{m-1}[\xi_{m-1}, x_m,\ldots,x_n, x]$,

$$\psi_2(x_{m-1}, x_m,\ldots,x_n, x) \in \mathcal{K}_{m-1}[x_{m-1},\ldots,x].$$

Continuing this procedure, we finally obtain $\psi_{m-j+2}(x_j,\ldots,x) \in \mathfrak{L}_j[x_j,\ldots,x_n, x]$ having type $t_{\mathfrak{L}_j}(\psi_{m-j+2}) \leqslant c'_{51}Y$ and such that

$$\begin{aligned}
&\psi_{m-j+2}(x_j,\ldots,x_n, x) \\
&\quad = \left(P_j(x_j)\right)^{\sigma_j}\psi_{m-j+1}(x_j,\ldots,x_n, x),\ldots,\psi_2(\xi_{m-1}, x_m,\ldots,x_n, x) \\
&\quad = \left(P_m(x_m)\right)^{\sigma_m}\psi_1(x_m,\ldots,x_n, x),
\end{aligned}$$

where, according to (6.112), $\psi_1(\xi_m,\ldots,\omega^*) \neq 0$. In view of Lemmas 4.12 and 6.2, we arrive, using (6.112), at the inequality

$$\exp\left(-c'_{52}Y^{\kappa_1}\ln Y\right) > |v_2|^{1/s_j\cdots s_n}$$

for some $v_2 \in \mathfrak{L}_j$, $v_2 \neq 0$, $t_{\mathfrak{L}_j}(v_2) \leqslant c'_{53}YX^{2^{n-j+1}-1}$. In view of (6.75), we obtain, according to Lemma 4.5, $\langle c'_{54}X^{2^{n-k}}, c'_{55}X^{\kappa_1}\ln X\rangle_k$. Thus, we have completely proved assertion (A_k) for any $k \geqslant 1$. This is impossible for $k = 1$ in view of Lemma VII of **[1]**, Chapter III.

Our proof of Theorems I–III establishes

PROPOSITION 6.5. *Suppose $\kappa_1 > 2^k$, i.e. $\kappa_i \geqslant 2^k$ if $i = 1, 2$ and $\kappa_3 > 2^k$ if $i = 3$. If n is the transcendence degree of $\mathbf{Q}(S_i)$, then the assertion (A_i) is true for $n - k \leqslant i \leqslant n$.*

7. Mutual transcendence measure for certain classes of numbers. In this section we obtain estimates for the measure of approximation of numbers connected with the exponential function by elements of fields of finite transcendence degree. The simplest problem involving the approximation of such numbers by algebraic numbers arose in the first applications of the methods of Gel′fond and Siegel. The most complete results have been obtained for numbers connected with Siegel's E-functions **[8]**. However, they do not take into account the degrees of the corresponding fields. This prevents us from extending Siegel's method to investigate E-functions at nonalgebraic points. Gel′fond's method does not possess such a deficiency. It was, in fact, by considering numbers with increasing degrees that we established Theorems I–III. The proof of these theorems given in §§5, 6 contained a certain "surplus": owing to the fact that $\kappa_i + \gamma_i > 2^n$ it is possible to take into account increasing types and degrees of numbers of the form $e^{\alpha_i\beta_j}$. More precisely, instead of Theorems I–III we can prove theorems concerning lower bounds of the measure of approximation of numbers in S_i by elements of fields of transcendence degree $\leqslant n$. For $n = 0$ this kind of argument was actually used by A. O. Gel′fond in his 1949 method [**1**]. Apart from these results there existed only fragmentary information on estimating approximations of numbers in S_i by elements of fields of nonzero transcendence degree, cf. however [**11, 25, 19**].

Below we prove for the first time general theorems on the mutual transcendence measure of numbers connected with the set S_i. In the statements of these theorems we use the concept of the type of a polynomial and the estimates are given in terms of the type, not the height and degree, as is customary. This is done in order to give the estimates a uniform character and allow for the possibility of simultaneous growth of the height and degree. The lemmas of §4 enable us in each individual case to take into consideration the influence of d and H.

On the basis of §4, using the arguments of §§5–6, we will prove the following theorems.

Suppose the transcendence degree of $\mathbf{Q}(S_i)$, $i = 1, 2, 3$, is exactly equal to $n + 1$, $n \geqslant 0$, $\mathbf{Q}(S_i) = \mathbf{Q}(\theta_1^{(i)},\dots,\theta_{n+1}^{(i)}, \omega)$, where $\theta_1^{(i)},\dots,\theta_{n+1}^{(i)}$ are algebraically independent numbers. Let $P(x_1,\dots,x_k)$, $k \leqslant n+1$, be a polynomial in $\mathbf{Z}[x_1,\dots,x_k]$ of type $t(P) = \max\{\ln H(P), \deg(P)\} \leqslant T$ such that $P(x_1,\dots,x_k) \not\equiv 0$.

THEOREM 7.1. *When* $i = 1$, *if* $\kappa_1 \geqslant 2^n$, *then*

$$|P(\theta_1^{(1)},\dots,\theta_k^{(1)})| > \exp\left(-\exp\left(c_1 T^{2^{k-1}}\right)\right), \tag{7.1}$$

and if $\kappa_1 > 2^n$, *then*

$$|P(\theta_1^{(1)},\dots,\theta_k^{(1)})| > \exp\left(-c_1 T^{\kappa_1 2^{k-1}/(\kappa_1 - 2^n)} (\ln T)^{-2^n/(\kappa_1 - 2^n)}\right). \tag{7.2}$$

THEOREM 7.2. *When* $i = 2$, *if* $\kappa_2 \geqslant 2^n$, *then*

$$|P(\theta_1^{(2)},\dots,\theta_k^{(2)})| > \exp\left(-\exp\left(c_2 T^{(N+M)2^{k-1}/N}\right)\right), \tag{7.3}$$

and if $\kappa_2 > 2^n$, *then*

$$|P(\theta_1^{(2)},\dots,\theta_k^{(2)})| > \exp\left(-c_2 T^{\kappa_2 2^{k-1}/(\kappa_2 - 2^n)} (\ln T)^{-2^n N/(M+N)(\kappa_2 - 2^n)}\right). \tag{7.4}$$

THEOREM 7.3. *When* $i = 3$, *if* $\kappa_3 > 2^n$, *then*

$$|P(\theta_1^{(3)},\dots,\theta_k^{(3)})| > \exp\left(-c_3 T^{\kappa_3 2^{k-1}/(\kappa_3 - 2^n)}\right). \tag{7.5}$$

We also assume, to eliminate trivial cases, that $\kappa_i > 1$.

From Theorems 7.1–7.3 it is possible to obtain as corollaries the well-known results of [**1, 11, 6, 9, 38, 39**], pertaining to this question.

The scheme of the proofs of 7.1–7.3 is uniform and completely analogous, for example, to the proof of Theorem 7.1. We need only replace the auxiliary function $f_1(z)$ by the function $f_2(z)$ or $f_3(z)$ in (3.2). The algebraic part of the argument does not change in any way. In 7.1 we consider the more complicated case $\kappa_1 > 2^n$. If $\kappa_1 = 2^n$, then the estimates are much simpler and we can use the argument of [**11**].

Suppose $\kappa_1 > 2^n$ and $\mathbf{Q}(\theta_1,\ldots,\theta_{n+1},\omega) = \mathbf{Q}(S_1)$, where $\theta_1,\ldots,\theta_{n+1}$ are algebraically independent over $\mathbf{Q}$ and ω is algebraic over $\mathbf{Q}(\theta_1,\ldots,\theta_{n+1})$. Consider for $k \leqslant n+1$ a polynomial $P(x_1,\ldots,x_k) \in \mathbf{Z}[x_1,\ldots,x_k]$ of type $\leqslant T$ such that $P(x_1,\ldots,x_k) \not\equiv 0$. We will prove (7.2) by induction on k. Assume that the corresponding estimate has already been proved for all $k' < k$.

First of all, in view of §§3,6, for any l, $2 \leqslant l \leqslant n+1$, there exists $P_X^l(x_1,\ldots,x_l) \in \mathbf{Z}[x_1,\ldots,x_l]$ such that $t(P_X^l) \leqslant X^{2^{n+1-l}}$, $P_X^l(x_1,\ldots,x) \not\equiv 0$, and

$$(7.6) \qquad |P_X^l(\theta_1,\ldots,\theta_l)| < \exp(-c_1' X^{\kappa_1} \ln X).$$

Put $\mathfrak{T}_0 = T^{\kappa_1 2^{k-1}/(\kappa_1 - 2^n)}(\ln T)^{-2^n/(\kappa_1-2^n)}$, where

$$(7.7) \qquad |P(\theta_1,\ldots,\theta_k)| < \exp(-c_0'\mathfrak{T}_0),$$

for an arbitrarily large constant $c_0' > 0$. We may assume that $P(x_1,\ldots,x_k)$ is $\mathbf{Q}$-irreducible. Indeed, if $P(x_1,\ldots,x_k)$ is $\mathbf{Q}$-reducible, we apply Lemma 4.2, obtaining either a divisor $P_1(x_1,\ldots,x_k)$ of $P(x_1,\ldots,x_k)$ such that $P_1(x_1,\ldots,x_k)$ is a power of a $\mathbf{Q}$-irreducible polynomial and

$$(7.8) \qquad |P_1(\theta_1,\ldots,\theta_k)| < \exp(-\tfrac{1}{8}\mathfrak{T}_0),$$

or else a polynomial $R(x_1,\ldots,x_{k-1}) \in \mathbf{Z}[x_1,\ldots,x_{k-1}]$, $R(x_1,\ldots,x_{k-1}) \not\equiv 0$, such that

$$(7.9) \qquad |R(\theta_1,\ldots,\theta_{k-1})| < \exp(-\tfrac{1}{8}\mathfrak{T}_0)$$

and $t(R) \leqslant \gamma T^2$. From (7.9) there follows an inequality contradicting (7.2) for polynomials in $k-1$ variables. We write $P_1(x_1,\ldots,x_k)$ in the form $(P_2(x_1,\ldots,x_k))^s$, where $P_2(x_1,\ldots,x_k)$ is a $\mathbf{Q}$-irreducible polynomial. In view of (7.8), inequality (7.7) holds for $P_2(x_1,\ldots,x_k)$ if T is replaced by $3T/s$.

We apply the lemmas of §4 separately to polynomials in $i < k$ variables, to $P(x_1,\ldots,x_k)$, and to the polynomials $P_X^i(x_1,\ldots,x_i)$ with $i > k$. Put

(7.10)

$$\chi_i = 2^{n+k-i}/(\kappa_1 - 2^n), \quad \mu_i = 2^{n+1-i}/(\kappa_1 - 2^n): \quad k+1 \leqslant i \leqslant n+1,$$
$$\chi_k = 1, \quad \mu_k = 0, \quad \chi_j = 2^{k-1-j}\kappa_1/(\kappa_1 - 2^n),$$
$$\mu_j = 2^{n-j}/(\kappa_1 - 2^n): \quad 0 \leqslant j \leqslant k-1.$$

For any j, $0 \leqslant j < k$, and any i, $j < i \leqslant n+1$, we have

$$(7.11) \qquad \chi_j \geqslant \chi_{j+1} + \cdots + 2\chi_i, \qquad \mu_j \geqslant \mu_{j+1} + \cdots + 2\mu_i.$$

As in §6, we establish auxiliary assertions similar to the (A_k) in §§3 and 6.

LEMMA 7.4. *For any $j, 0 \leqslant j \leqslant k$, we have*

(C_j) *there exists $v \in \mathfrak{L}_j = \mathbf{Z}[\theta_1,\ldots,\theta_j]$, $v \neq 0$, such that $t_{\mathfrak{L}_j}(v) \leqslant c_2' T^{\chi_j}(\ln T)^{-\mu_j}$ and*

$$(7.12) \qquad | v | < \exp(-c_3' \mathfrak{T}_0)$$

where c_3' is sufficiently large in comparison with c_2'.

First of all, (C_k) holds by virtue of (7.7). To prove (C_{k-1}) we assume that $\partial P/\partial x_k \not\equiv 0$, otherwise we can apply (7.6) for $k-1$. According to Lemma 4.3, there exists ξ_k, algebraic over $\mathbf{Q}(\theta_1,\ldots,\theta_{k-1})$, such that ξ_k is a root (of multiplicity 1) of $P(x) \in \mathfrak{L}[x]$, $\mathfrak{L} = \mathbf{Z}[\theta_1,\ldots,\theta_{k-1}]$, $P(x) = P(\theta_1,\ldots,\theta_{k-1}, x)$, and $| \theta_k - \xi_k | < \exp(-\frac{1}{2}\mathfrak{T}_0)$. Here $t_{\mathfrak{L}}(P) \leqslant T$ and we can use the proof of Theorems I–III. We need only add to (6.69) the inequality for $| \theta_k - \xi_k |$. To define Y we use, instead of (6.75),

$$(7.13) \qquad Y = [x_0], \qquad x_0^{\kappa_1} \ln x_0 = 1/c_4' \cdot \mathfrak{T}_0/s_{k+1} \cdots s_{n+1}$$

for a sufficiently large c_4' (which does not depend on the choice of c_0'). We make a simple observation. Consider in the proof of Theorem I the numbers ξ_i: $l \leqslant i \leqslant n+1$, approximating the θ_i such that

$$(7.14) \qquad | \xi_i - \theta_i | < \exp(-c_5' Y^{\kappa_1} \ln Y)$$

for an arbitrarily large constant c_5'. Let the t_i are the "$\mathfrak{L}_{l_i}$-types" ($\mathfrak{L}_l = \mathbf{Z}[\theta_1,\ldots,\theta_{l-1}]$) of ξ_i in the sense of the definition at the end of §4.

In other words, if $P_i(x)$ is the $\mathcal{K}_i$-irreducible polynomial with leading coefficient unity of which ξ_i is a root and if $Q_i(x) \in \mathcal{K}_i[x]$, where $Q(\xi_i) = 0$ and

$$\left[Q_i(x_i) \cdot P_{i-1}(x_{i-1})^{\tau_{i-1}} \cdots P_1(x_1)^{\tau_1}\right]_{\xi_1,\ldots,\xi_{i-1}} = Q_i(x_1,\ldots,x_i),$$

then $t_{\mathbf{Z}}(Q_i) \leqslant t_i$. Assume that s_i is the multiplicity of ξ_i in $Q_i(x_i)$ (or that $s_i = 1$ if ξ_i is transcendental over $\mathfrak{L}_i$). If $Y\prod_{i=l}^{n+1} t_i \leqslant (1/c_6')Y^{\kappa_1} \ln Y \cdot s_l \cdots s_{n+1}$ for a sufficiently large constant, then using, as in §4, Lemmas 4.5, 4.6, 4.12, and 4.13, we obtain an element $v \in \mathfrak{L}_l$, $v \neq 0$, such that $t_{\mathfrak{L}_l}(v) \leqslant c_7' Y \prod_{i=l}^{n+1} t_i$ and

$$(7.15) \qquad | v | < \exp(-c_8' Y^{\kappa} \ln Y \cdot s_{k+1} \cdots s_{n+1})$$

for a sufficiently large $c_8' > 0$. From the choice of $\mathfrak{T}_0$ and the definition of t_i: $t_i \leqslant c_9' Y^{2^{n-i}}$ we at once obtain (C_{k-1}).

The inductive passage from (C_j) to (C_{j-1}) is carried out in the same way. Using (C_j), we obtain an approximation ξ_j to θ_j, and so on. We again arrive at the inequality (7.15). Thus we have established 7.1–7.3.

Let us discuss the question of eliminating assumption (H) from the proofs of §§6 and 7. Note at the outset that the only place where (H) is used is Proposition 3.7 or 6.4. Therefore, we say that (H) does not hold for a given Y if 3.7 (or 6.4) does not hold for the function $f_i^Y(z)$, i.e. if there exists an upper bound for the linear form in the α_i or β_j having the appearance $\exp(-Y \ln Y)$.

We will show, for example, how to eliminate (H) from Theorem III in the first nontrivial case: $n = 2$. For this purpose it is best to use the proof of Theorem 5.1. From it we can make a very important deduction:

(7.16) Either for a given sufficiently large X we have $\langle O(X^2), O(X^{\kappa_3})\rangle_1$ or else for some $X \geqslant Y = [X/c]$ assumption (H) does not hold.

Indeed, as can be seen from the proof of Theorem 5.1, the only place where (H) was used was Case 2, in which (H) was applied to the function $f_i^Y(z)$. The proof of (5.2) does not require (H), since it follows from Lemma 6.4 with $E = 0$.

Assuming, therefore, that $\langle O(X^2), O(X^{\kappa_3})\rangle_1$ does not hold for an infinite sequence of natural numbers $X_1, \ldots, X_n, \ldots$, we conclude that (H) does not hold. We write the latter assertion in the following form (for definiteness we consider the α_i): There exist rational integers $u_1, \ldots, u_N$ such that

$$|u_1\alpha_1 + \cdots + u_N\alpha_N| < \exp(-\tau_1|u|) \tag{7.17}$$

if $|u| = \sum_{i=1}^N |u_i| > \tau_0$, where τ_0, τ_1 are arbitrarily large constants. By definition of S_i, inequality (7.17) can be written in the form

$$|Q_0(\theta_1, \theta_2, \omega)| < \exp(-\tau' v), \tag{7.18}$$

where $Q_0(x, y, z) \in \mathbf{Z}[x, y, z]$, $t(Q_0) \leqslant \ln v$ for an arbitrarily large v, and $\deg(Q_0)$ is bounded.

Since $\deg(Q_0)$ is bounded, there exists a $\mathbf{Q}$-irreducible polynomial $Q_1(x, y)$, $|Q_1(\theta_1, \theta_2)| < \exp(-\tau_2 v)$, such that $t(Q_1) \leqslant \tau_2 \ln v$, $\deg(Q_1)$ is bounded, and

$$|Q_1(\theta_1, \theta_2)| < \exp(-\tau_3 v). \tag{7.19}$$

Let w denote the type of $Q_1(x, y)$; $w \leqslant \tau_2 \ln v$. Then, for an arbitrarily large v, w is sufficiently large. Consider the interval $[w^{1/200}, w^{1/2}]$. Assume first that (H) holds for any $X \in [w^{1/200}, w^{1/2}]$. Then, according to (7.16), $\langle \tau_4 X^2, \tau_5 X^{\kappa_3}\rangle_1$ for all $X \in [w^{1/200}, w^{1/2}]$. Let $X_0 = [w^{1/2}]$. There exists $P(x) \in \mathbf{Z}[x]$, $P(x) \not\equiv 0$, $t(P) \leqslant \tau_6 w$, such that

$$|P(\theta_1)| < \exp\left(-\tau_7 w^{\kappa_3/2}\right). \tag{7.20}$$

We apply Lemma 4.2 or 1.9, obtaining a $\mathbf{Q}$-irreducible polynomial $R(x) \in \mathbf{Z}[x]$ such that

$$|R(\theta_1)| < \exp\left(-\tau_8 w^{\kappa_3/2}/s\right) \tag{7.21}$$

and $t(R) = t \leqslant \tau_9 w/s$. Assume that $t^{1/2} \in [x^{1/100}, w^{1/2}]$. Consider some assertion $\langle \frac{1}{4}t, \tau_9 t^{\kappa_3/2}\rangle_1$, which holds by assumption. Then there exists $R_1(x) \in \mathbf{Z}[x]$, $R_1(x) \not\equiv 0$, such that $t(R_1) \leqslant \frac{1}{4}t$ and $|R_1(\theta)| < \exp(-\tau_9 t^{\kappa_3/2})$. Since $R_1(x)$ and $R(x)$ are coprime, $w^{\kappa_3/2}/s \geqslant t^{\kappa_3/2}$, and $\kappa_3 > 4$, it follows that 1.8 contradicts (7.21). Therefore, $t \leqslant w^{1/100}$. Since $s \leqslant \tau_9 w$, it follows from (7.21) that $|R(\theta_1)| < \exp(-\tau_{10} w^{\kappa_3/2-1})$ and, since $\kappa_3 > 4$,

$$|R(\theta_1)| < \exp(-\tau_{10} w) \tag{7.22}$$

if $t(R) \leqslant t$, $t \leqslant w^{1/100}$.

Now assume that (H) does not hold for the whole interval $[w^{1/200}, w^{1/2}]$. Then there exist rational integers $x_1,\dots,x_N$ (or $y_1,\dots,y_M$ for the β_j) such that $\max_{i=1,\dots,N}|x_i|\leqslant|x|$, where $|x|\in[w^{1/200}, w^{1/2}]$ and

$$|x_1\alpha_1+\cdots+x_N\alpha_N|<\exp(-\tau_{11}|x|).$$

As above, this inequality can be rewritten in the form

$$|Q_2(\theta_1,\theta_2)|<\exp(-\tau_{12}|x|)$$

for an irreducible $Q_2(x, y)\in\mathbf{Z}[x, y]$, $H(Q_2)\leqslant|x|$, $\deg(Q_2)$ bounded. But $Q_2(x, y)$ is not equal to $Q_1(x, y)$, since $|x|<w^{1/2}<H(Q_1)$. Therefore, according to Lemma 4.1, there exists $R_1(x)\in\mathbf{Z}[x]$, $R_1(x)\not\equiv 0$, $t(R_1)\leqslant\tau_{13}\ln|x|\ln w\leqslant(\ln w)^3$, such that

$$|R_1(\theta_1)|<\exp(-\tau_{14}|x|)\leqslant\exp(-\tau_{14}w^{1/200}). \tag{7.23}$$

In view of (7.19)–(7.22) or (7.19)–(7.23) and Lemma 1.9, we can draw the following conclusion:

There exist $\mathbf{Q}$-irreducible polynomials $Q_2(x, y)\in\mathbf{Z}[x, y]$ and $R_2(x)\in\mathbf{Z}[x]$ such that

$$\begin{aligned} &|R_2(\theta_1)|<\exp(-\tau_{15}T), \qquad t(R_2)\leqslant\tau_{16}T^{1/100},\\ &|Q_2(\theta_1,\theta_2)|<\exp(-\tau_{17}T), \qquad t(Q_2)\leqslant\tau_{18}\ln T. \end{aligned} \tag{7.24}$$

Using 1.10 and Lemma 4.6, we obtain from (7.24) algebraic numbers ξ, ζ such that

$$|\theta_1-\xi|<\exp(-\tau_{19}T), \qquad |\theta_2-\zeta|<\exp(-\tau_{20}T), \tag{7.25}$$

ξ has type $\leqslant\tau_{21}T^{1/100}$ over $\mathbf{Q}$, and ζ has type $\leqslant\tau_{22}\ln T$ over $\mathbf{Q}(\xi)$. Such inequalities, however, are impossible. To see this we need only consider the set $S'=\{\alpha_1,\beta_1,e^{\alpha_1\beta_1}\}$, which obviously satisfies (H), and apply the method expounded in the proof of 7.1. We can also use the argument of [**31**]. Thus, Theorem III (and likewise 7.3) does not depend on (H). In a similar way, (H) can be eliminated from Theorems II and 7.2. It is more complicated to eliminate (H) from Theorems I and 7.1, though it is also possible.

Note that earlier, in [**1**], A. O. Gel′fond used (H) with $n=1$; this assumption was subsequently eliminated in [**31** and **33**].

We indicated one metric result relevant to the methods of this present paper.

REMARK 7.5. For almost all complex numbers θ we have for $P(x)\in\mathbf{Z}[x]$ of height H and degree d, $Hd\geqslant c_0(\theta)$, the inequality

$$|P(\theta)|>\exp(-6d\ln(Hd)).$$

The proof is obtained in the usual way: using 1.12 we can find a root ξ of $P(x)$ of multiplicity $s\leqslant d$ such that

$$|\theta-\xi|^s\exp(-3d\ln(Hd))\leqslant|P(\theta)|.$$

Therefore, the set of numbers θ satisfying $|P(\theta)|<\exp(-6d\ln(Hd))$ for infinitely many $P(x)$ has measure zero.

In particular, if we consider bounded H, then we obtain a lower bound of the form $\exp(-cd\ln d)$. This provides an answer to a question of V. G. Sprindžuk [**36, 37**] as to the order, in terms of d, of approximations of transcendental numbers by algebraic ones.

We mention briefly the p-adic analogues of Theorems I–III. The difficulties concentrated in the analytic part have been successfully overcome: see [**34, 35, 19**]. The algebraic part of the arguments of §§4 and 5–6 carries over to non-Archimedean metrics. Therefore, the theorems of §6 carry over (as in [**19, 35**]) to p-adic numbers.

We preserve the notation of §3. Assume that $\alpha_1, \ldots, \alpha_N$ and $\beta_1, \ldots, \beta_M$ are numbers in an algebraically closed field which is complete in the metric $|\cdot|_p$ and that $\alpha_1, \ldots, \alpha_N$ and $\beta_1, \ldots, \beta_M$ are **Q**-linearly independent sets with $|\alpha_i\beta_j|_p < p^{-1/(p-1)}$: $1 \leqslant i \leqslant N$. Then

THEOREM 7.6. *If $\kappa_i > 2^n$, then there are $n+1$ algebraically independent numbers in the set S_i: $i = 1, 2, 3$.*

REFERENCES

1. A. O. Gel'fond, *Transcendental and algebraic numbers* (translated from the 1st Russian ed. by L. F. Boron), Dover, New York, 1960.

2. J. Ax, *On Schanuel's conjectures*, Ann. of Math. **93** (1971), 252–268.

3. S. Lang, *Transcendental numbers and diophantine approximations*, Bull. Amer. Math. Soc. **77** (1971), 635–677.

4. ______, *Introduction to transcendental numbers*, Addison-Wesley, Reading, Mass., 1966.

5. ______, *Algebra*, Addison-Wesley, Reading, Mass., 1965.

6. P. L. Cijsouw, *Transcendental numbers*, North-Holland, Amsterdam, 1972.

7. A. O. Gel'fond, *On the algebraic independence of transcendental numbers of certain classes*, Uspehi Mat. Nauk **4** (1949), 14–18 (Russian) = Amer. Math. Soc. Transl. (1) **2** (1962), 125–169.

______, *Selecta*, "Nauka", Moscow, 1973. (Russian)

8. N. I. Fel'dman and A. B. Shidlovsky, *The development and present state of the theory of transcendental numbers*, Russian Math. Surveys **22** (1967), 1–79.

9. N. I. Fel'dman, *On the problem of the measure of transcendence of e*, Uspehi Mat. Nauk. **18** (1963), 207–213. (Russian)

10. R. Tijdeman, *An auxiliary result in the theory of transcendental numbers*, J. Number Theory **5** (1973), 80–94.

11. A. O. Gel'fond and N. I. Fel'dman, *On the measure of mutual transcendence of certain numbers*, Izv. Akad. Sci. USSR **14** (1950), 493–500.

12. M. Waldschmidt, *Amélioration d'un théorème de Lang sur l'indépendance algébrique d'exponentielles*, C. R. Acad. Sci. Paris Sér. A **272** (1971), 413–415.

13. ______, *Indépendance algébrique des valeurs de la fonction exponentielle*, Bull. Soc. Math. France **99** (1971), 285–304.

14. W. D. Brownawell, *The algebraic independence of certain values of the exponential function*, J. Number Theory **6** (1974), 22–31.

15. G. V. Chudnovsky, *Algebraic independence of several values of the exponential function*, Mat. Zametki **15** (1974), 661–672.

16. ______, *A mutual transcendence measure for some classes of numbers*, Soviet Math. Dokl. **15** (1974), 1424–1428.

______, *Transcendence measure for some classes of numbers*, Proc. All Union Conf. Number Theory (Vilnius), Univ. Publ. House, 1974, pp. 304–305. (Russian)

17. Th. Schneider, *Einführung in die transzendenten zahlen*, Springer, New York, 1957.

18. K. Ramachandra, *Contributions to the theory of transcendental numbers*. I, II, Acta Arith. **14** (1968), 65–88.

19. M. Waldschmidt, *Propriétés arithmétiques des valeurs de fonctions méromorphes algébriquement indépendantes*, Acta Arith. **23** (1973), 19–88.

20. A. O. Gel'fond, *Sur le septieme problème de Hilbert*, Izv. Akad. Nauk SSSR **7** (1934), 623–634. (Russian)

21. Yu. I. Manin, *The Tate height of points on an Abelian variety, its variants and applications*, Amer. Math. Soc. Transl. (2) **59** (1966), 82–110.

22a. A. Robinson, *Introduction to model theory and the metamathematics of algebra*, North-Holland, Amsterdam, 1963.

b. ______, *On the metamathematics of algebra*, North-Holland, Amsterdam, 1951.

23. A. Seidenberg, *A new decision method for elementary algebra*, Ann. of Math. (2) **60** (1954), 365–374.

______, *An elimination theory for differential algebra*, Publ. in Math. (N. S.), Univ. of Calif. **3** (1956), 31–66.

24. A. A. Shmelev, *A criterion for algebraic dependence of transcendental numbers*, Math. Notes **16** (1974), 921–926.

25. W. D. Brownawell, *Gel'fond's method for algebraic independence*, Trans. Amer. Math. Soc. **210** (1975), 1–26.

26. S. Lang, *Diophantine geometry*, Interscience, New York, 1962.

27. G. V. Chudnovsky, *Some analytical methods in the theory of transcendental numbers*, Institute of Mathematics, Ukrainian SSR Academy of Sciences, Preprint IM-74-8 (Kiev, 1974), "Naukova Dumka", Kiev, USSR, pp. 1–47. (Russian)

28. ______, *Analytical methods in Diophantine approximations*, Institute of Mathematics, Ukrainian SSR Academy of Sciences, Preprint IM-74-9 (Kiev, 1974), "Naukova Dumka", Kiev, USSR, pp. 1–52. (Russian)

29. W. D. Brownawell, *Some remarks on semi-resultants*, Transcendence Theory: Advances and Applications, Academic Press, New York, 1977, pp. 205–210.

30. A. A. Shmelev, *On the question of algebraic independence of algebraic powers of algebraic numbers*, Math. Notes **11** (1972), 387–392.

31. ______, *A. O. Gel'fond's method in the theory of transcendental numbers*, Math. Notes **10** (1971), 672–678.

32. J. W. S. Cassels, *Introduction to diophantine approximations*, Cambridge Univ. Press, Cambridge, 1957.

33. R. Tijdeman, *On the algebraic independence of certain numbers*, Indag. Math. **33** (1971), 146–162.

34. W. W. Adams, *Transcendental numbers in the p-adic domain*, Amer. J. Math. **88** (1966), 279–308.

35. T. N. Shorey, *Algebraic independence of certain numbers in the p-adic domain*, Indag. Math. **34** (1972), 423–435, 436–442.

36. V. G. Sprindžuk, *Mahler's problem in metric number theory*, Transl. Math. Mono., vol. 25, Amer. Math. Soc., Providence, R. I., 1969.

37. A. Baker, *On a theorem of Sprindžuk*, Proc. Roy. Soc. London **A292** (1966), 92–104.

38. ______, *Transcendental number theory*, Cambridge Univ. Press, Cambridge, 1979.

39. N. I. Fel'dman, *Approximation of certain transcendental numbers. I. The approximation of logarithms of algebraic numbers*, Amer. Math. Soc. Transl. (2) **59** (1966), 224–245.

Translated by G. A. KANDALL

CHAPTER 2

BAKER'S METHOD IN THE THEORY OF TRANSCENDENTAL NUMBERS

The method of A. Baker was first employed in the theory of transcendental numbers in early papers [**1**] of its author. It was shown [**1**, **2**] that if the logarithms $\ln \alpha_1, \ldots, \ln \alpha_n$ of the algebraic numbers $\alpha_1, \ldots, \alpha_n$ are linearly independent over $\mathbf{Q}$ and the numbers $\beta_0, \beta_1, \ldots, \beta_n$ are algebraic, then the number

$$e^{\beta_0 + \beta_1 \ln \alpha_1 + \cdots + \beta_n \ln \alpha_n} = e^{\beta_0} \cdot \alpha_1^{\beta_1} \cdots \alpha_n^{\beta_n}$$

is transcendental. In [**10**] M. Waldschmidt announced without any proofs several results on the application of Baker's method to Gel′fond's scheme. His results are of the following form: if $\mathbf{K}$ is a field of transcendence type $\leqslant \tau$ and

$$MN \geqslant (\tau - 1)(M + N) + Nr_2,$$

where $r_2 + 1$ is the dimension of the vector space over $\mathbf{K}$ generated by $\{1, \beta_1, \ldots, \beta_M\}$ and $N \geqslant \tau$, $M > r_2$, then among the numbers α_i, $e^{\alpha_i \beta_j}$ $(1 \leqslant i \leqslant N,$ $1 \leqslant j \leqslant M)$ there is one which is transcendental over $\mathbf{K}$. The end of the proof with his method has, however, gaps.

The first interesting application of Baker's method is a proof of the existence of two algebraically independent numbers. In this case the best result must have the following form: if $\mathbf{K} \subseteq \mathbf{C}$, $\operatorname{tr} \deg \mathbf{K} \leqslant 1$, the dimension of $\{1, \beta_1, \ldots, \beta_M\}$ over $\mathbf{K}$ is $r_2 + 1$, and $MN \geqslant M + N + Nr_2$, then among the numbers α_i, $e^{\alpha_i \beta_j}$ $(1 \leqslant i \leqslant N,$ $1 \leqslant j \leqslant M)$ there is one which is transcendental over $\mathbf{K}$. Precisely this result is obtained in §§1 and 3.

For the final part of the proof, which is always crucial in Baker's method, it is possible to use only Stark's method [**4**]. This is the first application of this method to problems of transcendental number theory. It requires several lemmas concerning finitely generated extensions of $\mathbf{Q}$.

0076-5376/84 $1.00 + $.25 per page

1. THEOREM 1.1. *Suppose* $\{\alpha_1, \alpha_2\}$ *and* $\{\beta_1, \ldots, \beta_4\}$ *are sets of complex numbers which are linearly independent over* $\mathbf{Q}$, *and* $\mathbf{K}$ *is a subfield of* $\mathbf{C}$ *of transcendence degree at most* 1. *If at most two of the numbers* $\{1, \beta_1, \ldots, \beta_4\}$ *are linearly independent over* $\mathbf{K}$, *then one of the numbers* α_i, $e^{\alpha_i\beta_j}$: $i = 1, 2$; $j = 1, \ldots, 4$, *does not belong to* $\mathbf{K}$.

From this we can deduce an especially important result.

THEOREM 1.2. *Suppose* $\{\alpha_1, \alpha_2\}$ *and* $\{\beta_1, \ldots, \beta_4\}$ *are linearly independent over* $\mathbf{Q}$ (*i.e.* $N = 2$, $M = 4$). *Then among the numbers*

$$\alpha_1, \alpha_2, \beta_1, \beta_2, \beta_3/\beta_4, e^{\alpha_i\beta_j}: \qquad i = 1, 2; \quad j = 1, \ldots, 4,$$

there are two which are algebraically independent.

The method of proof of Theorem 1.1 is also used to prove the much more general Theorem 3.1 and is used in Chapter 6 to investigate diophantine problems. Several consequences of Theorem 1.2 will be established below in §2.

PROOF OF THEOREM 1.1. Suppose $\mathbf{K} = \mathbf{Q}(\theta, \omega)$, where ω is an algebraic integer over $\mathbf{Q}(\theta)$ of degree ν, θ is transcendental, and $\alpha_i, e^{\alpha_i\beta_j} \in \mathbf{K}$: $i = 1, 2$; $j = 1, \ldots, 4$. Then, by definition of the β_j: $j = 1, \ldots, 4$, we have

$$\beta_j = a_j + b_j\kappa: \qquad j = 1, \ldots, 4,$$

where $a_j, b_j \in \mathbf{K}$. Therefore, there exist polynomials T_j^1, T_j^2, T_i^3, $T_{i,j}^4$ in $\mathbf{Z}[x]$ and polynomials S_j^1, S_j^2, S_i^3, $S_{i,j}^4$ in $\mathbf{Z}[x, y]$ such that

$$\alpha_i = \frac{S_i^3(\theta, \omega)}{T_i^3(\theta)}, \quad a_j = \frac{S_j^1(\theta, \omega)}{T_j^1(\theta)}, \quad b_j = \frac{S_j^2(\theta, \omega)}{T_j^2(\theta)}, \quad e^{\alpha_i\beta_j} = \frac{S_{i,j}^4(\theta, \omega)}{T_{i,j}^4(\theta)}, \tag{1.1}$$

for $i = 1, 2$; $j = 1, \ldots, 4$. We put $T_0 = \prod_{i=1,2; j=1,\ldots,4} T_j^1(\theta)T_j^2(\theta)T_i^3(\theta)T_{i,j}^4(\theta)$. In the sequel, $c_i > 0$ will denote constants depending only on α_i, β_j, θ, and ω.

LEMMA 1.3. *For any sufficiently large* Y *there exists a polynomial* $P(x) \in \mathbf{Z}[x]$ *such that* $t(P) = \max\{\ln H(P), d(P)\} \leqslant Y$ *and*

$$|P(\theta)| < \exp\left(-c_1 Y^2 \ln^{1/3} Y\right). \tag{1.2}$$

PROOF OF 1.3. Consider an auxiliary function $F(z_1, z_2)$ connected with a given natural number $X > c_2$:

$$F(z_1, z_2) = \sum_{\lambda_0=0}^{L-1} \sum_{\lambda_1=0}^{L_0-1} \cdots \sum_{\lambda_4=0}^{L_0-1} C_{\lambda_0,\lambda_1,\ldots,\lambda_4} z_1^{\lambda_0} \times e^{(\lambda_1 a_1 + \lambda_2 a_2 + \lambda_3 a_3 + \lambda_4 a_4)z_1} \cdot e^{\kappa(\lambda_1 b_1 + \lambda_2 b_2 + \lambda_3 b_3 + \lambda_4 b_4)z_2}. \tag{1.3}$$

We use $F(z_1, z_2)$ to prove (1.2) for $X > c_2$ with $Y = [c_0 X \ln^{-1/3} X]$. For this purpose we choose parameters L, L_0, etc. in the form

$$\begin{aligned} L &= [X \ln^{-4/3} X], & L_0 &= [X^{1/2} \ln^{-1/3} X], \\ X_0 &= [c_3 X^{1/2}], & S_0 &= [c_4 X \ln^{-4/3} X]; \qquad c_3, c_4 < 1. \end{aligned} \tag{1.4}$$

For simplicity we introduce the notation

$$F_{\sigma_1,\sigma_2}(z) = \frac{\partial^{\sigma_1+\sigma_2}F(z_1, z_2)}{\partial z_1^{\sigma_1}\partial z_2^{\sigma_2}}\Big|_{z_1=z_2=z}.$$

The coefficients $C_{\lambda_0,\lambda_1,\ldots,\lambda_4}$ are chosen in the form

$$(1.5)\qquad C_{\lambda_0,\lambda_1,\ldots,\lambda_4} = \sum_{l=0}^{L_1-1} C_{\lambda_0,\ldots,\lambda_4,l}\,\theta^l, \qquad L_1 = \left[c_5 X \ln^{-1/3} X\right].$$

We choose rational integers $C_{\lambda_0,\ldots,\lambda_4,l}$ so that the system of equations

$$(1.6)\qquad F_{\sigma_1,\sigma_2}(x_1\alpha_1 + x_2\alpha_2) = 0$$

is satisfied for any natural numbers σ_1, σ_2, x_1, x_2 such that $0 \leqslant \sigma_1 + \sigma_2 \leqslant S_0$; $0 \leqslant x_1, x_2 \leqslant X_0$. To prove the existence of the coefficients $C_{\lambda_0,\ldots,\lambda_4,l}$ we use the Siegel Lemma. We state the necessary result as a lemma from [**11**]:

LEMMA 1.4. *Suppose* $\mathbf{K}_0$ *is an extension of* $\mathbf{Q}$ *of finite transcendence degree* q *and* $(\theta_1,\ldots,\theta_q,\omega)$ *is a system of generators of* $\mathbf{K}_0$ *over* $\mathbf{Q}$. *Then there exists a constant* $C(\mathbf{K}_0) > 0$ *such that if* n *and* r *are natural numbers,* $n \geqslant 2r > 0$, *and* $a_{i,j}$ $(1 \leqslant i \leqslant n, 1 \leqslant j \leqslant r)$ *are elements of the ring* $\mathbf{Z}[\theta_1,\ldots,\theta_q]$, *then there exist* $\xi_1,\ldots,\xi_n$ *in* $\mathbf{Z}[\theta_1,\ldots,\theta_q]$, *not all zero, such that*

$$\sum_{i=1}^{n} a_{ij}\xi_i = 0: \qquad j = 1,\ldots,r,$$

and

$$\max_{1\leqslant i\leqslant n} t(\xi_i) \leqslant C(\mathbf{K}_0)\left[\max_{i,j} t(a_{i,j}) + \ln n\right].$$

We apply this lemma to the system of equations (1.6). Moreover, taking the partial derivatives of $F(z_1, z_2)$, we obtain

$$(1.7)\quad F_{\sigma_1,\sigma_2}(z) = \sum_{\lambda_0=0}^{L-1}\sum_{\lambda_1=0}^{L_0-1}\cdots\sum_{\lambda_4=0}^{L_0-1} C_{\lambda_0,\lambda_1,\ldots,\lambda_4}$$
$$\times \sum_{m=0}^{\sigma_1} \frac{\sigma_1!\lambda_0!}{m!(\sigma_1-m)!(\lambda_0-m)!} z_1^{\lambda_0-m}(\lambda_1 a_1 + \cdots + \lambda_4 a_4)^{\sigma_1-m}$$
$$\times e^{(\lambda_1 a_1+\cdots+\lambda_4 a_4)z_1}\kappa^{\sigma_2}(\lambda_1 b_1 + \cdots + \lambda_4 b_4)^{\sigma_2} e^{\kappa(\lambda_1 b_1+\cdots+\lambda_4 b_4)z_2}.$$

Consider the following system of equations in the $C_{\lambda_0,\ldots,\lambda_4}$ with coefficients from the ring $\mathbf{Z}[\theta, \omega]$:

$$T_0^{[2X\ln^{-1/3}X]}\kappa^{-\sigma_2}F_{\sigma_1,\sigma_2}(x_1\alpha_1 + x_2\alpha_2) = 0: \qquad \sigma_1 + \sigma_2 \leqslant S_0;\quad x_1, x_2 \leqslant X_0,$$

which is equivalent to (1.6). In view of (1.1), (1.7), and Lemma 1.4, this system is solvable if $c_3^2 c_4 < 2$. Furthermore, we have an estimate of the type $t(C_{\lambda_0,\ldots,\lambda_4})$ as a polynomial in θ: $\max t(C_{\lambda_0,\dots,\lambda_4}) \leqslant C(\mathbf{K})(c_6 X \ln^{-1/3} X + 4\ln X)$. Therefore, system (1.6) is solvable for the $C_{\lambda_0,\dots,\lambda_4}$ of the form (1.5) and

$$(1.8) \qquad \max\left|C_{\lambda_0,\dots,\lambda_4,l}\right| \leqslant \exp\left(c_6' X \ln^{-1/3} X\right),$$

where $c_6' > C(\mathbf{K})(c_6 + 1)$.

In the sequel we will also use the identity

$$(1.9) \qquad \left(F_{\sigma_1,\sigma_2}(z)\right)^{(s)} = \sum_{\sigma_1'=0}^{s} \sum_{\sigma_2'=0,\sigma_1'+\sigma_2'=s}^{s} \frac{s!}{\sigma_1'!\sigma_2'!} F_{\sigma_1+\sigma_1',\sigma_2+\sigma_2'}(z).$$

For the rest of the proof of 1.1 we require a generalization of the Baker-Stark theory [**3**, **4**] to the case of fields of nonzero transcendence degree. First, we cite three lemmas from [**3**], which are valid, according to [**5**], for any subfields of $\mathbf{C}$. Note that 1.7 follows from 1.6, and 1.6 from 1.5. Below, the symbol $\mathbf{K}$ denotes a subfield of $\mathbf{C}$.

LEMMA 1.5. *If $\mathbf{K}$ contains a pth root of unity ζ_p, where p is a prime, and $\alpha \neq 0$ belongs to $\mathbf{K}$, then either $\alpha^{1/p}$ belongs to $\mathbf{K}$ or else $\mathbf{K}(\alpha^{1/p})$ is an extension of $\mathbf{K}$ of degree p.*

LEMMA 1.6. *Suppose α, β are nonzero elements of $\mathbf{K}$ and $\alpha^{1/p}, \beta^{1/p}$ are fixed pth roots for a given p. Then either $\mathbf{K}(\alpha^{1/p}, \beta^{1/p})$ is an extension of $\mathbf{K}(\alpha^{1/p})$ of degree p or else $\beta = \alpha^j\gamma^p$ for some γ in $\mathbf{K}$ and some integer j, $0 \leqslant j < p$.*

LEMMA 1.7. *Suppose $\alpha_1,\dots,\alpha_n$ are nonzero elements of $\mathbf{K}$ and $\alpha_1^{1/p},\dots,\alpha_n^{1/p}$ denote fixed pth roots for a prime p. If $\mathbf{K}' = \mathbf{K}(\alpha_1^{1/p},\dots,\alpha_{n-1}^{1/p})$, then either $\mathbf{K}'(\alpha_n^{1/p})$ is an extension of $\mathbf{K}'$ of degree p or else $\alpha_n = \alpha_1^{j_1}\cdots\alpha_{n-1}^{j_{n-1}}\gamma^p$ for some $\gamma \in \mathbf{K}$ and certain integers $j_1,\dots,j_{n-1}$ with $0 \leqslant j_l < p$.*

We now apply the above lemmas to the case $\mathbf{K} = \mathbf{Q}(\theta_1,\dots,\theta_n,\omega)$, where $\vec{\theta} = (\theta_1,\dots,\theta_n)$ is an n-tuple of algebraically independent complex numbers and ω is an algebraic integer over $\mathbf{Q}(\theta_1,\dots,\theta_n) = \mathbf{K}_0$ of degree D, $[\mathbf{K}:\mathbf{K}_0] = D$. Everywhere below, the fields $\mathbf{K}$ and $\mathbf{K}_0$ (i.e. the numbers $\theta_1,\dots,\theta_n$, D, and ω) will be viewed as fixed. Consider the algebraic numbers ξ of the field $\mathbf{K}_0$. We are interested in the heights $H_{\mathbf{K}}(\xi)$ and the degrees $d_{\mathbf{K}}(\xi)$ of these numbers. Suppose $\xi \in \mathbf{K}_0$. Then ξ satisfies an equation

$$R(\theta_1,\dots,\theta_n,z) = R(\vec{\theta},z) = 0$$

with $R(\vec{x},z) \in \mathbf{Z}[\vec{x},z] = \mathbf{Z}[x_1,\dots,x_n,z]$. From among all such $R(\vec{x},z) \in \mathbf{Z}[\vec{x},z]$ we choose a polynomial $R_0(\vec{x},z)$ which is irreducible in $\mathbf{Z}[\vec{x},z]$ and primitive (the greatest common divisor of its coefficients is 1). The polynomial $R_0(\vec{x},z)$ is called the minimal polynomial for ξ. We then put $H_{\mathbf{K}}(\xi) = H(R_0)$, $d_{\mathbf{K}}(\xi) = \deg_z(R_0)$, and $d(\xi) = d(R_0)$, where $H(R_0)$ is the height of $R_0(x_1,\dots,x_n,z)$, $\deg_z(R_0)$ is its degree with respect to z, and $\deg(R_0) = d(R_0)$ is the sum of the degrees over all variables. As usual, $t_{\mathbf{K}}(\xi)$ denotes the type of ξ, i.e. $\max\{\ln H_{\mathbf{K}}(\xi), d(\xi)\}$. Our main problem at the moment is to find a relation between the types of elements α, β of $\mathbf{K}$ such that $\alpha = \beta^m$.

So suppose $\alpha, \beta \in \mathbf{K}$ and, if $\omega^{(1)} = \omega, \ldots, \omega^{(D)}$ denote the conjugates of ω over $\mathbf{K}$, let $\alpha^{(1)} = \alpha, \ldots, \alpha^{(D)}$ and $\beta^{(1)} = \beta, \ldots, \beta^{(D)}$ be the corresponding elements of the conjugate fields $\mathbf{K}^{(1)} = \mathbf{K} = \mathbf{K}_0(\omega^{(1)}), \ldots, \mathbf{K}^{(D)} = \mathbf{K}_0(\omega^{(D)})$. Let $R_0(z) = R_0(\vec{\theta}, z)$ be the minimal polynomial for β and let $d_{\mathbf{K}}(\beta) = d$. Then $k = D/d$ is an integer and the polynomial $(R_0(\vec{\theta}, z))^k$ has the form $f(\vec{\theta}, z)$:

$$f(\vec{\theta}, z) = B(z - \beta^{(1)}) \cdots (z - \beta^{(D)}),$$

where $B = B(\theta_1, \ldots, \theta_n)$ with $B(x_1, \ldots, x_n) \in \mathbf{Z}[x_1, \ldots, x_n]$ and $f(\vec{x}, z) \in \mathbf{Z}[\vec{x}, z]$. Similarly, for $\alpha \in \mathbf{K}$, $d_{\mathbf{K}}(\alpha) = e$, $l = D/e$, and $P_0(\vec{\theta}, z)$ the minimal polynomial for α we put $g(\vec{\theta}, z) = (P_0(\vec{\theta}, z))^l$, where

$$g(\vec{\theta}, z) = A(z - \alpha^{(1)}) \cdots (z - \alpha^{(D)})$$

and $A = A(\theta_1, \ldots, \theta_n)$, $A(x_1, \ldots, x_n) \in \mathbf{Z}[x_1, \ldots, x_n]$, $g(\vec{x}, z) \in \mathbf{Z}[\vec{x}, z]$. We also put

$$G(\vec{\theta}, z) = A(z^m - \alpha^{(1)}) \cdots (z^m - \alpha^{(D)}) = g(\vec{\theta}, z^m)$$

and

$$h(\vec{\theta}, z) = \prod_{j=1}^{m} f(\vec{\theta}, x\zeta_m^j),$$

where m is a natural number. If $\alpha = \beta^m$, then it is easy to see that

$$B^m G(\vec{\theta}, z) = (-1)^{D(m+1)} A h(\vec{\theta}, z).$$

Using Gauss' lemma in $\mathbf{Z}[\vec{\theta}, z]$, we conclude that $h(z) = h(\vec{\theta}, z)$ and $G(z) = G(\vec{\theta}, z)$ are primitive polynomials in $\mathbf{Z}[\vec{\theta}, z]$. Therefore, $B^m = \pm A$.

We consider the connection between the types of the polynomials

$$h(x_1, \ldots, x_n, z) \quad \text{and} \quad G(x_1, \ldots, x_n, z).$$

By the above, $G(x_1, \ldots, x_n, z) = \pm h(x_1, \ldots, x_n, z)$ and $h(x_1, \ldots, x_n, z) = \prod_{j=1}^{m} f(x_1, \ldots, x_n, z\zeta_m^j)$. Furthermore, $G(x_1, \ldots, x_n, z) = g(x_1, \ldots, x_n, z^m)$. Since $H(f_j) = H(f)$, where $f_j = f(x_1, \ldots, x_n, z\zeta_m^j)$, we obtain the following inequalities for types:

$$m\frac{t(f)}{2} \leqslant t(h) \leqslant 2mt(f), \qquad t(G) = t(h); \qquad t(G) = \max\{t(g), mD\}.$$

By definition, $t(\alpha) = t(P_0)$ and $t(\beta) = t(R_0)$, where $P_0(\vec{\theta}, z)$ and $R_0(\vec{\theta}, z)$ are the minimal polynomials for α and β. Since

$$g(x_1, \ldots, x_n, z) = (P_0(x_1, \ldots, x_n, z))^l, \quad l = D/d_{\mathbf{K}}(\alpha),$$
$$f(x_1, \ldots, x_n, z) = (R_0(x_1, \ldots, x_n, z))^k, \quad k = D/d_{\mathbf{K}}(\beta),$$

it follows that

$$\frac{D}{2d_{\mathbf{K}}(\alpha)}t(\alpha) \leqslant t(g) \leqslant \frac{2D}{d_{\mathbf{K}}(\alpha)}t(\alpha); \qquad \frac{D}{2d_{\mathbf{K}}(\beta)}t(\beta) \leqslant t(f) \leqslant \frac{2D}{d_{\mathbf{K}}(\beta)}t(\beta).$$

Combining all these inequalities, we obtain

LEMMA 1.8. *Suppose* $\mathbf{K} = \mathbf{K}_0(\omega)$, $[\mathbf{K}:\mathbf{K}_0] = D$, *where* $\mathbf{K}_0 = \mathbf{Q}(\theta_1,\dots,\theta_n)$ *is a purely transcendental extension of* $\mathbf{Q}$. *If* α *and* β *are elements of* $\mathbf{K}$ *such that* $\alpha = \beta^m$ *for some natural number* m, *we have the following inequalities relating* $t(\alpha)$ *and* $t(\beta)$ $(d_{\mathbf{K}}(\alpha), d_{\mathbf{K}}(\beta) \leqslant D)$:

$$\frac{d_{\mathbf{K}}(\beta)}{8d_{\mathbf{K}}(\alpha)} t(\alpha) \leqslant t(\beta)m \leqslant \max\left\{\frac{8d_{\mathbf{K}}(\beta)}{d_{\mathbf{K}}(\alpha)} t(\alpha), 4d_{\mathbf{K}}(\beta)m\right\}.$$

From 1.8 we directly obtain

LEMMA 1.9. *Suppose* $\mathbf{K} = \mathbf{K}_0(\omega)$, *where* $[\mathbf{K}:\mathbf{K}_0] = D < \infty$ *and* $\mathbf{K}_0$ *is a purely transcendental extension of* $\mathbf{Q}$, $\mathbf{K}_0 = \mathbf{Q}(\theta_1,\dots,\theta_n)$. *Let* $\alpha_1,\dots,\alpha_k$ *be fixed numbers in* $\mathbf{K}$ (*i.e. having bounded types* $t_{\mathbf{K}}(\alpha_i)$). *If* $\alpha_1^{j_1}\cdots\alpha_k^{j_k} = \gamma^p$ *for a prime* p *and rational integers* $j_1,\dots,j_k$: $|j_i| < p$, *then the type* $t_{\mathbf{K}}(\gamma)$ *is also bounded. More precisely,* $t_{\mathbf{K}}(\gamma)$ *is less than some absolute constant depending only on* $\max t_{\mathbf{K}}(\alpha_i)$, $\mathbf{K}$, *and* k.

The above lemmas are used to complete the proof of 1.1. Consider a sufficiently large $c_7 > 0$ such that the set $\mathcal{P}$ of primes $p < c_7$ has cardinality $> \frac{1}{2}c_7/\ln c_7$. Fix $p > 11$, $p \in \mathcal{P}$. We use this p to reduce $L_0 (\operatorname{mod} p^i)$. For each $i \geqslant 0$ we put

$$(1.10) \qquad L_i = [L_0/p^i], \qquad X_i = [X_0 p^i], \qquad S_i = [S_0/4^i].$$

We now choose for the parameter i an upper bound $\mathcal{T}$: $i \leqslant \mathcal{T}$, so that $L_{\mathcal{T}} = 0$, i.e. $L_0 < p^{\mathcal{T}}$. It suffices to take

$$(1.11) \qquad \mathcal{T} = [\ln L_0/\ln p] + 1.$$

For $i \geqslant 0$ and a fixed sequence of rational integers $\vec{\lambda} = (\lambda_1^0,\dots,\lambda_1^{i-1};\dots;\lambda_4^0,\dots,\lambda_4^{i-1})$ in $[0, p)$ of length $4i$ we put

$$(1.12) \quad C^{(\vec{\lambda})}_{\lambda_0,m_1,\dots,m_4} = p^{-i\lambda_0} C_{\lambda_0, m_1p^i + \lambda_1^{i-1}p^{i-1} + \cdots + \lambda_1^0, \dots, m_4p^i + \lambda_4^{i-1}p^{i-1} + \cdots + \lambda_4^0}$$

where $m_1,\dots,m_4$ assume the values $0, 1,\dots,L_i$, if $m_jp^i + \lambda_j^{i-1}p^{i-1} + \cdots + \lambda_j^0 < L_0$ for all $j = 1,\dots,4$. If $m_jp^i + \lambda_j^{i-1}p^{i-1} + \cdots + \lambda_j^0 \geqslant L_0$ for some $j = 1,\dots,4$, then we put

$$C^{(\vec{\lambda})}_{\lambda_0,m_1,\dots,m_4} = 0.$$

We next define a function $F^i(z_1, z_2)$ as follows:

$$(1.13) \qquad F^i(z_1, z_2) = \sum_{\lambda_0=0}^{L-1} \sum_{m_1=0}^{L_i} \cdots \sum_{m_4=0}^{L_i} C^{(\vec{\lambda})}_{\lambda_0,m_1,\dots,m_4} \times z^{\lambda_0} e^{(m_1a_1 + \cdots + m_4a_4)z_1} e^{\kappa(m_1b_1 + \cdots + m_4b_4)z_2}.$$

LEMMA 1.10. *For each rational integer $i \leqslant \mathfrak{T}$ and any sequence of rational integers $\vec{\lambda} = (\lambda_1^0,\dots,\lambda_1^{i-1};\dots;\lambda_4^0,\dots,\lambda_4^{i-1})$ in $[0, p)$ and function $F^i(z_1, z_2) = F^{i,\vec{\lambda}}(z_1, z_2)$ we have*

$$F^{i,\vec{\lambda}}_{\sigma_1,\sigma_2}(x_1\alpha_1 + x_2\alpha_2) = 0 \tag{1.14}$$

for all $\sigma_1 + \sigma_2 \leqslant S_i$ and $x_1, x_2 \leqslant X_i$.

Let Ω_i denote the set of all quadruples of natural numbers $(\sigma_1, \sigma_2, x_1, x_2)$ such that $\sigma_1 + \sigma_2 \leqslant S_i$ and $x_1, x_2 \leqslant X_i$.

PROOF OF 1.10. As (1.6) shows, (1.14) holds for $i = 0$. Assume that (1.14) has been proved for all $i' < i$, $\vec{\lambda}'$ of length $4i'$, and $(\sigma_1, \sigma_2, x_1, x_2) \in \Omega_{i'}$. We will show that (1.14) holds for i.

Fix a sequence $\vec{\lambda} = (\lambda_1^0,\dots,\lambda_1^{i-1};\dots;\lambda_4^0,\dots,\lambda_4^{i-1})$ and the new sequence $\vec{\lambda}' = (\lambda_1^0,\dots,\lambda_1^{i-2};\dots;\lambda_4^0,\dots,\lambda_4^{i-2})$ (for $i = 1$ we put $\vec{\lambda}' = 0$). Consider the functions $\mathbf{F}(z_1, z_2) = F^{i-1,\vec{\lambda}'}(z_1, z_2)$ and $\mathbf{F}_{\sigma_1,\sigma_2}(z) = F^{i-1,\vec{\lambda}'}_{\sigma_1,\sigma_2}(z)$. According to (1.14), with the value $i - 1$, $\mathbf{F}_{\sigma_1,\sigma_2}(x_1\alpha_1 + x_2\alpha_2) = 0$ for $(\sigma_1, \sigma_2, x_1, x_2) \in \Omega_{i-1}$. In view of (1.9), it follows that

$$\mathbf{F}^{(s)}_{\sigma_1,\sigma_2}(x_1\alpha_1 + x_2\alpha_2) = 0, \tag{1.15}$$

if $\sigma_1 + \sigma_2 \leqslant \frac{1}{2}S_{i-1}$, $s \leqslant \frac{1}{2}S_{i-1}$, and $x_1, x_2 \leqslant X_{i-1}$.

For given fixed σ_1, σ_2 with $\sigma_1 + \sigma_2 \leqslant S_i$ we denote $\mathbf{F}_{\sigma_1,\sigma_2}(z)$ by $f_i(z)$. Then we obtain from (1.15) the integral representation

$$|f_i(z)| = \left| \frac{1}{2\pi i} \int_\Gamma \prod_{x_1=0}^{X_{i-1}} \prod_{x_2=0}^{X_{i-1}} \left\{ \frac{z - x_1\alpha_1 - x_2\alpha_2}{\zeta - x_1\alpha_1 - x_2\alpha_2} \right\}^{[S_{i-1}/2]} \frac{f_i(\zeta)\,d\zeta}{z - \zeta} \right| \tag{1.16}$$

where Γ is the circle $|\zeta| = X^{3/2}$ and $|z| < |\zeta|$.

Using (1.13) and a formula of the form (1.7), we obtain

$$\max_{|\zeta| = X^{3/2}} |f_i(\zeta)| \leqslant \exp\big(4\ln X + c_5 X \ln^{-1/3} X + 2S_{i-1}\ln X + 3L\ln X + c_8 L_0/p^{i-1} \cdot X^{3/2}\big). \tag{1.17}$$

Substituting (1.17) into (1.16) and taking into account that $f_i(z) = \mathbf{F}_{\sigma_1,\sigma_2}(z)$, we deduce that

$$\left|\mathbf{F}_{\sigma_1,\sigma_2}(z)\right| < \exp\big(-c_9 X_{i-1}^2 S_{i-1} \ln X\big), \tag{1.18}$$

for $|z| \leqslant X^{5/4}$. We apply this inequality to $z = (x_1/p)\alpha_1 + (x_2/p)\alpha_2$ for rational integers $x_1, x_2 \leqslant X_{i-1}p$. In view of (1.7), (1.12), and (1.13), the number $\varphi = \kappa^{-\sigma_2} T_0^{L+S_0} p^{L+(i-1)L} \mathbf{F}_{\sigma_1,\sigma_2}((x_1\alpha_1 + x_2\alpha_2)/p)$ has the form

$$\varphi = \sum_{j_{11}=0}^{W} \cdots \sum_{j_{41}=0}^{W} \sum_{j_{42}=0}^{W} H_{j_{11},\dots,j_{42}} \cdot e^{\alpha_1\beta_1 j_{11}/p} \times e^{\alpha_2\beta_1 j_{21}/p} \cdots e^{\alpha_1\beta_4 j_{41}/p} \cdot e^{\alpha_2\beta_4 j_{42}/p}, \tag{1.19}$$

where the $H_{j_{11},\dots,j_{42}}$ are numbers in the ring $\mathbf{Z}[\theta,\omega]$ having types

$$\leqslant c_9\big(X\ln^{-1/3}X + S_0\ln X + L\ln(pX_{i-1}) + L_{i-1}X_{i-1}\big) \leqslant c_{10}X\ln^{-1/3}X$$

and $W = [c_{11}X\ln^{-1/3}X]$. Consider the norm of φ in the field $\mathbf{K}_q/\mathbf{Q}(\theta)$, where $\mathbf{K}_q = \mathbf{Q}(\theta,\omega,e^{\alpha_1\beta_1/p},\dots,e^{\alpha_2\beta_4/p})$. If $\varphi_1 = \varphi,\dots,\varphi_\eta$, where $\eta = [\mathbf{K}_q:\mathbf{Q}(\theta)]$, are the conjugates of φ, we consider the number

$$\mathfrak{p} = T_0^{W\cdot p^8}\cdot\prod_{i=1}^{\eta}\varphi_i$$

in the ring $\mathbf{Z}[\theta]$. We write $\mathfrak{p}$ in the form $\mathfrak{p} = \mathbf{P}(\theta)$, where $\mathbf{P}(x)\in\mathbf{Z}[x]$. We estimate the type of $\mathbf{P}(x)$ by means of (1.19) and the theorem on symmetric functions. We obtain

$$t(\mathbf{P}) \leqslant c_{12}X\ln^{-1/3}Xp^8. \tag{1.20}$$

Using inequality (1.18) with $z = (x_1/p)\alpha_1 + (x_2/p)\alpha_2$, we obtain an estimate of $\mathfrak{p} = \mathbf{P}(\theta)$ in terms of $\varphi = \varphi_1$:

$$\begin{aligned} |\mathbf{P}(\theta)| &\leqslant |\varphi|\cdot\exp\big(c_{13}X\ln^{-1/3}Xp^8\big)\\ &\leqslant \big|\mathbf{F}_{\sigma_1,\sigma_2}\big((x_1/p)\cdot\alpha_1 + (x_2/p)\cdot\alpha_2\big)\big|\exp\big(c_{14}X\ln^{-1/3}Xp^8\big)\\ &\leqslant \exp\big(-c_{15}X_{i-1}^2S_{i-1}\ln X\big), \end{aligned} \tag{1.21}$$

since, according to (1.10),

$$X_{i-1}^2S_{i-1}\ln X \geqslant X_0^2S_0\ln X(p^2/4)^{i-1} \geqslant c_{16}X^2\ln^{-1/3}X(p/2)^{2i-2} \tag{1.22}$$

and $p > 11$. Since p is bounded, $p < c_7$, it follows directly from (1.20)–(1.22) that Lemma 1.3 is true with $Y = [c_{12}c_7^8X\ln^{-1/3}X]$ whenever $\mathbf{P}(x)\not\equiv 0$. Therefore, we may assume that for an infinite sequence of X's we have $\mathbf{P}(x)\equiv 0$. By definition of the norm, $\mathfrak{p} = 0$ and $\varphi = 0$. Thus,

$$\mathbf{F}_{\sigma_1,\sigma_2}\big((y_1/p)\cdot\alpha_1 + (y_2/p)\cdot\alpha_2\big) = 0 \tag{1.23}$$

for all $y_1, y_2 \leqslant X_{i-1}p$, $\sigma_1+\sigma_2 \leqslant [\frac{1}{2}S_{i-1}]$. This equality can already be used for the inductive step. We write out (1.23) explicitly, using (1.7):

$$\begin{aligned} &\sum_{\lambda_0=0}^{L-1}\sum_{m_1=0}^{L_{i-1}}\cdots\sum_{m_4=0}^{L_{i-1}} C^{(\vec{\lambda}')}_{\lambda_0,m_1,\dots,m_4}\sum_{s=0}^{\sigma_1}\frac{\sigma_1!\lambda_0!}{(\sigma_1-s)!s!(\lambda_0-s)!}\\ &\times\big((y_1/p)\alpha_1 + (y_2/p)\alpha_2\big)^{\lambda_0-s}(m_1a_1+\cdots+m_4a_4)^{\sigma_1-s}(m_1b_1+\cdots+m_4b_4)^{\sigma_2}\\ &\times\prod_{i=1}^{2}\prod_{j=1}^{4}\big(e^{\alpha_i\beta_j}\big)^{m_jy_i/p} = 0. \end{aligned} \tag{1.24}$$

In this situation we use Lemmas 1.4–1.9, which were established earlier. Consider first the case $(y_1,p) = 1$, $p\mid y_2$. Writing (1.24) in abbreviated form:

$$\sum_{\lambda_0=0}^{L-1}\sum_{m_1,\dots,m_4=0}^{L_{i-1}} D^{(\vec{\lambda}')}_{\lambda_0,m_1,\dots,m_4,y_1,y_2,\sigma_1,\sigma_2}\prod_{j=1}^{4}\big(e^{\alpha_i\beta_j}\big)^{m_jy_i/p} = 0, \tag{1.25}$$

where the $D_{\lambda_0,\ldots,y_2,\sigma_1,\sigma_2}$ are elements of the field $\mathbf{K} = \mathbf{Q}(\theta, \omega)$, we arrive at the field $\mathbf{K}((e^{\alpha_1\beta_1})^{1/p},\ldots,(e^{\alpha_1\beta_4})^{1/p})$.

Assume first that the degree of $\mathbf{K}((e^{\alpha_1\beta_1})^{1/p},\ldots,(e^{\alpha_1\beta_4})^{1/p})$ over $\mathbf{K}$ is equal to p^4. Analogously, it suffices to assume that the degree of

$$\mathbf{K}\Big(\big(e^{\alpha_2\beta_1}\big)^{1/p},\ldots,\big(e^{\alpha_2\beta_4}\big)^{1/p}\Big)$$

over $\mathbf{K}$ is equal to p^4. The possibility that this condition does not hold will be discussed after the completion of the proof of 1.10.

Since $(y_1, p) = 1$, it follows from our assumption concerning the degree of the field that for any fixed $\lambda_1^{i-1},\ldots,\lambda_4^{i-1}$ in $[0, p)$ we have

$$(1.26)\qquad \sum_{\lambda_0=0}^{L-1}\ \sum_{m_1=0,m_1\equiv\lambda_1^{i-1}}^{L_{i-1}}\cdots\sum_{m_4=0,m_4\equiv\lambda_4^{i-1}}^{L_{i-1}} D_{\lambda_0,m_1,\ldots,y_2,\sigma_1,\sigma_2} \times\prod_{j=1}^{4}\big(e^{\alpha_1\beta_j}\big)^{m_jy_1/p} = 0.$$

Here the congruences $\equiv$ are to be understood as (mod p).

We now denote $(y_1/p)\alpha_1 + y_2'\alpha_2$, where $y_2' = y_2/p$ is a natural number, by z_p. We write each $m_j = 0,\ldots,L_{i-1}$ such that $m_j \equiv \lambda_j^{i-1}$ (mod p) in the form $m_j = \lambda_j^{i-1} + m_j'p$. By the definition of L_i in (1.10), $m_j' \leqslant L_i$. In view of the definition (1.12), we obtain in this notation

$$(1.27)\qquad C^{\vec{\lambda},i}_{\lambda_0,m_1',\ldots,m_4'} = p^{-\lambda_0}C^{\vec{\lambda}',i-1}_{\lambda_0,m_1'p+\lambda_1^{i-1},\ldots,m_4'p+\lambda_4^{i-1}} = p^{-\lambda_0}C^{\vec{\lambda}',i-1}_{\lambda_0,m_1,\ldots,m_4}$$

according to the definition of $\vec{\lambda}$ and $\vec{\lambda}'$. Taking into account (1.27), (1.26), and the abbreviated form (1.25) of (1.24), we obtain

(1.28)

$$\sum_{\lambda_0=0}^{L-1}\sum_{m_1'=0}^{L_i}\cdots\sum_{m_4'=0}^{L_i} p^{\lambda_0}C^{(\vec{\lambda})}_{\lambda_0,m_1',\ldots,m_4'}\Big(z_p^{\lambda_0}e^{((m_1'p+\lambda_1^{i-1})a_i+\cdots+(m_4'p+\lambda_4^{i-1})a_4)z_p}\Big)^{(\sigma_1)}$$
$$\cdot\Big(e^{\kappa((m_1'p+\lambda_1^{i-1})b_1+\cdots+(m_4'p+\lambda_4^{i-1})b_4)z_p}\Big)^{(\sigma_2)} = 0$$

for any natural numbers $y_1 \leqslant X_{i-1}p$, $(y_1, p) = 1$, $y_2' \leqslant X_{i-1}$ and any $\sigma_1 + \sigma_2 \leqslant \frac{1}{2}S_{i-1}$. We also put

$$A_1 = \lambda_1^{i-1}a_1 + \cdots + \lambda_4^{i-1}a_4;\qquad B_1 = \kappa\big(\lambda_1^{i-1}b_1 + \cdots + \lambda_4^{i-1}b_4\big);$$
$$\gamma_1 = (m_1a_1 + \cdots + m_4a_4)p;\qquad \gamma_2 = \kappa(m_1b_1 + \cdots + m_4b_4)p.$$

It is easy to see that

$$(1.29)\qquad \begin{aligned} \big(e^{\gamma_2z_p}\big)^{(\sigma_2)} &= \left(\sum_{s=0}^{\sigma_2} B_{s,\sigma_2}\big(e^{(\gamma_2+B_1)z_p}\big)^{(s)}\right)e^{-B_1zp},\\ \big(ze^{\gamma_1z}\big)^{(\sigma_1)} &= \sum_{s_1=0}^{\sigma_1} C_{\sigma_1}^{s_1}(-A_1)^{\sigma_1-s_1}\big(z\cdot e^{(\gamma_1+A_1)z}\big)^{(s_1)}\cdot e^{A_1z},\end{aligned}$$

where the coefficients B_{s,σ_2} depend only on s, σ_2, and B_1. From (1.28) and (1.29) we obtain, for any $\sigma_1 + \sigma_2 \leq \frac{1}{2}S_{i-1}$,

$$(1.30) \qquad \sum_{\lambda_0=0}^{L-1} \sum_{m_1'=0}^{L_i} \cdots \sum_{m_4'=0}^{L_i} p^{\lambda_0} \cdot C^{(\vec{\lambda})}_{\lambda_0, m_1', \ldots, m_4'} \left(z_p^{\lambda_0} e^{\gamma_1 z_p}\right)^{(\sigma_1)} \times \left(e^{\gamma_2 z_p}\right)^{(\sigma_2)} = 0.$$

If we denote $pz_p = y_1\alpha_1 + py_2'\alpha_2$ by z_p^1, we obtain, by transforming (1.30),

$$(1.31) \quad \sum_{\lambda_0=0}^{L-1} \sum_{m_1'=0}^{L_i} \cdots \sum_{m_4'=0}^{L_i} C^{(\vec{\lambda})}_{\lambda_0, m_1', \ldots, m_4'} \sum_{s=0}^{\sigma_1} \frac{\sigma_1!\lambda_0!}{s!(\sigma_1 - s)!(\lambda_0 - s)!} \times \left(z_p^1\right)^{\lambda_0 - s} (m_1'a_1 + \cdots + m_4'a_4)^{\sigma_1 - s} (m_1'b_1 + \cdots + m_4'b_4)^{\sigma_2} \times e^{\kappa(m_1'\beta_1 + \cdots + m_4'\beta_4)z_p^1} = 0$$

for $z_p^1 = y_1\alpha_1 + y_2\alpha_2$, where $(y_1, p) = 1$, $p \mid y_2$, $\sigma_1 + \sigma_2 \leq \frac{1}{2}S_{i-1}$, and $y_1, y_2 \leq X_{i-1}p$. In the notation of (1.13), the equalities (1.31) mean that

$$(1.32) \qquad F^{\vec{\lambda}, i}_{\sigma_1, \sigma_2}(y_1\alpha_1 + y_2\alpha_2) = 0$$

for all $\sigma_1 + \sigma_2 \leq \frac{1}{2}S_{i-1}$ and $y_1, y_2 \leq X_{i-1}p$, $(y_1, p) = 1$, $p \mid y_2$. For fixed σ_1, σ_2 with $\sigma_1 + \sigma_2 \leq S_i \leq \frac{1}{4}S_{i-1}$, it follows from (1.32) and (1.9) that

$$(1.33) \quad \left|F^i_{\sigma_1,\sigma_2}(z)\right| = \left|\frac{1}{2\pi i}\int_\Gamma \prod_{y_1=0,(y_1,p)=1}^{X_{i-1}p} \prod_{y_2=0;\,p|y_2}^{X_{i-1}p} \left\{\frac{z - y_1\alpha_1 - y_2\alpha_2}{\zeta - y_1\alpha_1 - y_2\alpha_2}\right\}^{S_i} \frac{F^i_{\sigma_1,\sigma_2}(\zeta)}{z - \zeta} d\zeta\right|.$$

The number of pairs of natural numbers (y_1, y_2) such that $y_1, y_2 \leq X_{i-1}p$, $(y_1, p) = 1$, $p \mid y_2$ is at least X_{i-1}^2. Therefore, it follows from (1.33) with $\Gamma: |\zeta| = X^{3/2}$ that

$$(1.34) \qquad \left|F^i_{\sigma_1\sigma_2}(z)\right| < \exp\left(-c_{17}X_{i-1}^2 S_i \ln X\right).$$

It is easy to see that if $(\sigma_1, \sigma_2, x_1, x_2) \in \Omega_i$, then the number

$$\varphi = \kappa^{-\sigma_2} p^{iL} T_0^{[2X_iL_i + 2X\ln^{-1/3}X]} F^i_{\sigma_1,\sigma_2}(x_1\alpha_1 + x_2\alpha_2)$$

belongs to the ring $\mathbf{Z}[\theta, \omega]$. This follows from (1.13) as $p^{iL}C^{\vec{\lambda}}_{\lambda_0, m_1', \ldots, m_4'}$ belongs to $\mathbf{Z}[\theta, \omega]$. Since $p^i \leq p^{\mathfrak{T}} \leq L + 1$, it follows that the type

$$t(\varphi) \leq c_{18}\left(X_iL_i + X\,ln^{-1/3}X\right).$$

Consider the norm of φ in $\mathbf{Q}(\theta, \omega)/\mathbf{Q}(\theta)$. We have $\operatorname{Norm}\varphi = R(\theta)$, where $R(x) \in \mathbf{Z}[x]$. In view of (1.34) and the inequality for $t(\varphi)$, it follows from (1.22) that

$$(1.35) \qquad |R(\theta)| < \exp\left(-c_{19}X^2\ln^{-1/3}X\right), \qquad t(R) \leq c_{20}X\ln^{-1/3}X.$$

If $R(x) \not\equiv 0$, we obtain 1.3 with $Y = [c_{20} X \ln^{-1/3} X]$. If $R(x) \equiv 0$, then $\varphi = 0$ and

$$F^{i}_{\sigma_1,\sigma_2}(x_1\alpha_1 + x_2\alpha_2) = 0 \tag{1.36}$$

for all $(\sigma_1, \sigma_2, x_1, x_2) \in \Omega_i$. Equalities (1.14) and Lemma 1.10 are proved.

We apply 1.10 and (1.14) with $i = \mathfrak{T}$ and $\sigma_1 = \sigma_2 = 0$ (note that $S_{\mathfrak{T}} \geqslant S/L_0 > 0$). We obtain

$$\sum_{\lambda_0=0}^{L-1} C^{(\vec{\lambda})}_{\lambda_0,0,\ldots,0}(x_1\alpha_1 + x_2\alpha_2)^{\lambda_0} = 0 \tag{1.37}$$

for any natural numbers x_1, $x_2 \leqslant X_{\mathfrak{T}}$ and any sequence $\vec{\lambda}$ of length $4\mathfrak{T}$. But according to (1.11) and (1.4), $X^2_{\mathfrak{T}} \geqslant XL_0^2 \geqslant X^{1.5} > L$. Thus, the number of equations in (1.34) is greater than the number of unknowns. Since α_2/α_1 is irrational, (1.37) has a unique solution: $C^{\vec{\lambda}}_{\lambda_0,0,\ldots,0} = 0$ for all $\lambda_0 = 0, 1, \ldots, L$.

We take into account also that any $m_j = 0,\ldots,L_0$: $j = 1,\ldots,4$, can be represented in the form $m_j = \lambda^0_j + \lambda^1_j p + \cdots + \lambda^{\mathfrak{T}-1}_j p^{\mathfrak{T}-1}$, since $p^{\mathfrak{T}} > L_0$. Consequently, for $\vec{\lambda} = (\lambda^0_1,\ldots,\lambda^{\mathfrak{T}-1}_1; \cdots; \lambda^0_4,\ldots,\lambda^{\mathfrak{T}-1}_4)$, it follows from (1.12) that $C_{\lambda_0,\lambda_1,\ldots,\lambda_4} = 0$ for any $\lambda_0, \lambda_1,\ldots,\lambda_4$, which contradicts the choice of the C_λ. Thus, finally, Lemma 1.3 is proved with $Y = [c_{21} X \ln^{-1/3} X]$ and with it 1.1, since (1.2) implies that θ is algebraic by a lemma of Gel'fond **[8]**.

It remains to eliminate the assumption made in 1.10 that the degree of the field $\mathbf{K}_1 = \mathbf{K}(e^{\alpha_1\beta_1/p},\ldots,e^{\alpha_1\beta_4/p})$ or the field $\mathbf{K}_2 = \mathbf{K}(e^{\alpha_2\beta_1/p},\ldots,e^{\alpha_2\beta_4/p})$ is equal to p^4 for some $p \in \mathfrak{P}$. Assume this is not the case. Fix $i = 1, 2$. According to Lemma 1.7, we arrive, let us say, at the relation

$$(e^{\alpha_i\beta_1})^{j_1} \cdots (e^{\alpha_i\beta_3})^{j_3} e^{\alpha_i\beta_4} = \gamma^p_{i,p} \tag{1.38}$$

where $0 \leqslant j_1,\ldots,j_3 < p$ and $\gamma_{i,p} \in \mathbf{K} = \mathbf{Q}(\theta, \omega)$ for all $p \in \mathfrak{P}$. It follows from (1.38) and 1.9 that the type $t(\gamma_{i,p})$ is bounded: $t(\gamma_{i,p}) < c_{22}$, where c_{22} depends only on $\max t(e^{\alpha_i\beta_j})$, θ, and ω, i.e. does not depend on the choice of p. By construction, $|\mathfrak{P}| \geqslant \frac{1}{2}c_7/\ln c_7$ for an arbitrarily large c_7. Since the number of elements of $\mathbf{Q}(\theta, \omega)$ having bounded type is bounded, for $c_7 \gg c_{22}$ we obtain two distinct p_i, $q_i \in \mathfrak{P}$ with $\gamma_{i,p_i} = \gamma_{i,q_i}$. From (1.38) we obtain

$$(e^{\alpha_i\beta_1})^{j_1 q_i} \cdots e^{\alpha_i\beta_1 q_i} = (e^{\alpha_i\beta_1})^{j'_1 p_i} \cdots e^{\alpha_i\beta_4 p_i}. \tag{1.39}$$

From (1.39) with $i = 1, 2$ we arrive at nontrivial relations of the form

$$\alpha_1\beta_1 x_1 + \cdots + \alpha_1\beta_4 x_4 = 2\pi i s_1, \qquad \alpha_2\beta_1 y_1 + \cdots + \alpha_2\beta_4 y_4 = 2\pi i s_2. \tag{1.40}$$

In the case of real α_i and β_j, (1.40) is impossible. Moreover, it follows from (1.40), preserving the assumptions of Theorem 1.1, that instead of certain β_{i_1} and β_{i_2} we can choose numbers of the form

$$\beta'_{i_1} = 2\pi i/\alpha_1, \qquad \beta'_{i_2} = 2\pi i/\alpha_2.$$

We repeat the entire argument for the new set of numbers α_i, β'_j, where, for simplicity, $i_1 = 1$ and $i_2 = 2$. Then for some $p \in \mathfrak{P}$, as a consequence of 1.7 and

1.9, the degree of the field $\mathbf{K}(e^{\alpha_1\beta_2/p},\ldots,e^{\alpha_1\beta_4/p})$ over $\mathbf{K}$ is equal to p^3 and the degree of $\mathbf{K}(e^{\alpha_2\beta_1/p}, e^{\alpha_2\beta_3/p},\ldots)$ over $\mathbf{K}$ is equal to p^3. Therefore, we can apply the method of proof of 1.10, considering first $z_p = (y_1/p)\alpha_1 + y_2\alpha_2$ and reducing m_2, m_3, m_4 (mod p), then considering $w_p = x_1\alpha_1 + (x_2/p)\alpha_2$ and reducing m_1, m_3, m_4 (mod p). We need only to use the integral representation in the proof of 1.10 three times and replace $S_i = [S_0/4^i]$ in (1.10) by $S_i = [S_0/8^i]$. Some corollaries of 1.2 are given in the next section.

2. New cases of algebraically independent numbers. We give several corollaries of Theorem 1.2, which was proved in §1. Despite its unwieldy formulation involving the presence of 13 numbers, this theorem leads to some essentially new examples of algebraically independent numbers included in sets of numbers connected with the exponential function.

THEOREM 2.1. *Suppose β is a quadratic irrationality and φ, ψ are arbitrary (nonzero) complex numbers such that $\varphi/\psi \notin \mathbf{Q}(\beta)$. Then among the numbers φ, e^{φ}, $e^{\varphi\beta}$, e^{ψ}, $e^{\psi\beta}$ there are two that are algebraically independent.*

PROOF. Put $\alpha_1 = 1$, $\alpha_2 = \beta$; $\beta_1 = \varphi$, $\beta_2 = \varphi\beta$, $\beta_3 = \psi$, $\beta_4 = \psi\beta$. Then 2.1 follows from 1.2.

COROLLARY 2.2. *Suppose β is a quadratic irrationality and* $\ln a$, $\ln b$ *are linearly independent logarithms over* $\mathbf{Q}$ *of the algebraic numbers a, b. Then among the numbers* $\ln a$, a^{β}, b^{β} *there are two that are algebraically independent.*

COROLLARY 2.3. *Suppose* $\ln a$ *is the logarithm of the algebraic number a and β is a quadratic irrationality. Then either* $\ln a$, a^{β} *are algebraically independent or for any algebraic number b such that* $\ln a/\ln b$ *is irrational the numbers a^{β} and b^{β} are algebraically independent.*

COROLLARY 2.4. *Suppose* $\ln a$ *is the logarithm of the algebraic number a, and β is a quadratic irrationality. If* $\ln\ln a/\ln a \notin \mathbf{Q}(\beta)$, *then among the numbers* $\ln a$, $(\ln a)^{\beta}$, a^{β} *there are two that are algebraically independent.*

PROOF. In the notation of 2.1, put $\varphi = \ln a$, $\psi = \ln\ln a$. The condition $\ln\ln a/\ln a \notin \mathbf{Q}(\beta)$ seems superfluous; let us consider some instances in which it is eliminated. Without loss of generality, β can be represented in the form $\beta = \sqrt{\mathfrak{D}}$ where $\mathfrak{D}$ is a square-free rational integer.

PROPOSITION 2.5. *For any algebraic number a there exists at most one $\mathfrak{D}$ such that among* $\ln a$, $(\ln a)^{\sqrt{\mathfrak{D}}}$, $a^{\sqrt{\mathfrak{D}}}$ *there are two numbers that are algebraically independent.*

PROOF. Assume that for a given algebraic number a there exist two $\mathfrak{D}$, $\mathfrak{D}'$, where $\mathfrak{D} \neq \mathfrak{D}'$, such that the transcendence degrees of the fields

$$\mathbf{Q}(\ln a, (\ln a)^{\sqrt{\mathfrak{D}}}, a^{\sqrt{\mathfrak{D}}})$$

and $\mathbf{Q}(\ln a, (\ln a)^{\sqrt{\mathfrak{D}'}}, a^{\sqrt{\mathfrak{D}'}})$ are at most 1. Then, by Corollary 2.4, we have $\ln\ln a = (x + y\sqrt{\mathfrak{D}})\ln a$, $\ln\ln a = (z + w\sqrt{\mathfrak{D}'})\ln a$, and therefore, since $[\mathbf{Q}(\sqrt{\mathfrak{D}}, \sqrt{\mathfrak{D}'} : \mathbf{Q}] = 4$, it follows that $\ln a$ is an algebraic number, which is impossible.

PROPOSITION 2.6. *Suppose a is a positive real algebraic number, $a \neq 1$, $\ln a$ is the principal value of the logarithm, and $\mathfrak{D} < 0$. Then among the numbers $\ln a$, $(\ln a)^{\sqrt{\mathfrak{D}}}$, $a^{\sqrt{\mathfrak{D}}}$ there are two that are algebraically independent.*

PROOF. Otherwise, as 2.4 shows, $\ln a = a^x a^{y\sqrt{\mathfrak{D}}}$ with $\mathfrak{D} < 0$, $\sqrt{\mathfrak{D}} = i\sqrt{-\mathfrak{D}}$. Since $\ln a$ is the principal value of the logarithm, i.e. a real number, we obtain $a^{2y\sqrt{\mathfrak{D}}} \pm 1$, which is impossible by Gel'fond's theorem.

PROPOSITION 2.7. *If $\mathfrak{D} > 0$, then among the numbers π, $\pi^{\sqrt{\mathfrak{D}}}$, $e^{\pi i\sqrt{\mathfrak{D}}}$ there are two that are algebraically independent.*

PROOF. From 2.4, in the case $\operatorname{tr\,deg}\mathbf{Q}(\pi, \pi^{\sqrt{\mathfrak{D}}}, e^{\pi i\sqrt{\mathfrak{D}}}) \leqslant 1$, we obtain $\pi i = (-1)^x e^{\pi i y\sqrt{\mathfrak{D}}}$ with $x, y \in \mathbf{Q}$. Then $e^{\pi i y\sqrt{\mathfrak{D}}}$ is an algebraic number, which contradicts Gel'fond's theorem of 1929 **[9]**.

From the lemmas of Stark's type in §1 we obtain

LEMMA 2.8. *Suppose θ is transcendental, $(m, n) = 1$, and $\mathbf{K}$ is an algebraic number field. Then the degree of $\theta^{m/n}$ over $\mathbf{K}(\theta)$ is equal to n.*

THEOREM 2.9. *Suppose $a \neq 1$ is an algebraic number. If*

$$\ln\ln a/\ln a \neq x + y\sqrt{\mathfrak{D}},$$

for any $x, y \in \mathbf{Q}$ with $H(x), H(y) \leqslant H$, then the three numbers $\ln a$, $(\ln a)^{\sqrt{\mathfrak{D}}}$, $a^{\sqrt{\mathfrak{D}}}$ cannot be expressed algebraically in terms of one of them by polynomials of degree $< H$.

Theorem 2.9 follows from 2.4 and 2.8.

Let us turn to other consequences of 1.2. One pertains to the sets of numbers $\{e, e^e, e^{e^2}\}$ and $\{e, a^e, a^{e^2}\}$, where $a \neq 0, 1$ is algebraic. With respect to the problem of the algebraic independence of two numbers in such sets, the only results were obtained by A. O. Gel'fond **[8]**:

$$\operatorname{tr\,deg}\mathbf{Q}(e, e^e, e^{e^2}, e^{e^3}) \geqslant 2 \quad \text{and} \quad \operatorname{tr\,deg}\mathbf{Q}(e, a^e, a^{e^2}, a^{e^3}, a^{e^4}) \geqslant 2.$$

COROLLARY 2.10. *Suppose a is an algebraic number, $a \neq 0, 1$. Then among the five numbers e, e^e, e^{e^2}, a^e, a^{e^2} there are two that are algebraically independent.*

PROOF. Put $\alpha_1 = 1$, $\alpha_2 = e$, $\beta_1 = 1$, $\beta_2 = e$, $\beta_3 = \ln a$, $\beta_4 = e\ln a$. If $\ln a \notin \mathbf{Q}(e)$, then 2.10 follows from 1.2. Assume that $\ln a(x + ye) = z + we$, where $x, y, z, w \in \mathbf{Z}$. If $y = 0$, then $\ln a = r + se$, where $r, s \in \mathbf{Q}$. Then we also consider two numbers $\alpha_1' = 1$, $\alpha_2' = e$, $\alpha_3' = e^2$, $\beta_1' = 1$, $\beta_2' = r + se = \ln a$. We apply Gel'fond's theorem **[8]** to $\overline{\mathbf{Q}}(\alpha_i', \beta_j', e^{\alpha_i'\beta_j'}) = \overline{\mathbf{Q}}(e, e^e, e^{e^2}, a^e, a^{e^2})$. Corollary 2.10 is automatically obtained. Suppose $y \neq 0$. Then $\ln a = (z + we)/(x + ye)$, and here $\ln a \notin \mathbf{Q}$. Then consider the numbers $\alpha_1'' = 1$, $\alpha_2'' = e$, $\beta_1'' = 1$, $\beta_2'' = e$, $\beta_3'' = (z + we)/(x + ye) = \ln a$. It is easy to see that β_1'', β_2'', β_3'' are linearly independent. Therefore, we can apply the theorem of **[8]** to the set of numbers $\{\alpha_i'', \beta_j'', e^{\alpha_i''\beta_j''}\}$ and we obtain $\operatorname{tr deg} \mathbf{Q}(e, e^e, e^{e^2}, a^e) \geqslant 2$, which is enough to prove 2.10.

As an example of the general Theorem 1.1 we have

COROLLARY 2.11. *Suppose $a \neq 0, 1$ is an algebraic number. Then among the numbers e, e^e, a^e, a^{e^2}, a^{e^3} there are two that are algebraically independent.*

3. A new general theorem on two algebraically independent numbers. Using Baker's method and the arguments of §1, we will prove new general theorems on two algebraically independent numbers, not improvable from the point of view of Baker's method as we now know it. We use the considerations from §1 in the proofs.

THEOREM 3.1. *Suppose $\alpha_1, \ldots, \alpha_N$ and $\beta_1, \ldots, \beta_M$ are sets of complex numbers, linearly independent over $\mathbf{Q}$. Assume that $\mathbf{K}$ is an arbitrary subfield of $\mathbf{C}$ of transcendence degree at most 1. Let r_1 denote the number of elements in $\{\beta_1, \ldots, \beta_M\}$ that are linearly independent over $\mathbf{K}$, and $r_2 + 1$ the number of elements in $\{1, \beta_1, \ldots, \beta_M\}$ that are linearly independent over $\mathbf{K}$.*

If $MN \geqslant M + N + r_1 N$, then the numbers $e^{\alpha_i\beta_j}$ $(1 \leqslant i \leqslant N, 1 \leqslant j \leqslant M)$ cannot all belong to $\mathbf{K}$.

If $MN \geqslant M + N + r_2 N$, then the numbers α_i, $e^{\alpha_i\beta_j}$ $(1 \leqslant i \leqslant N, 1 \leqslant j \leqslant M)$ cannot all belong to $\mathbf{K}$.

The case $r_2 = 1$ was considered in §1; the case $r_1 = 1$ follows from **[8]**. Therefore, we consider $r_1 = 2$.

THEOREM 3.2. *Suppose α_1, α_2 and $\beta_1, \ldots, \beta_6$ are sets of numbers, linearly independent over $\mathbf{Q}$. Then for any subfield $\mathbf{K} \subset \mathbf{C}$ of transcendence degree $\leqslant 1$, $\dim_{\mathbf{K}}\{\beta_1, \ldots, \beta_6\} \leqslant 2$ implies that $e^{\alpha_i\beta_j} \notin \mathbf{K}$ for some $i = 1, 2$ and $j = 1, \ldots, 6$.*

PROOF OF THEOREM 3.2. Suppose κ_1, κ_2 are certain complex numbers and $\beta_j = a_j\kappa_1 + b_j\kappa_2$: $j = 1, \ldots, 6$, with $a_j, b_j \in \mathbf{K}$, where $\mathbf{K} = \mathbf{Q}(\theta, \omega)$ is a field of transcendence degree 1 and ω is an algebraic integer of degree $\nu \geqslant 1$ over $\mathbf{Q}(\theta)$.

Consider first the auxiliary function

$$F(z_1, z_2) = \sum_{\lambda_1=0}^{L_0-1} \cdots \sum_{\lambda_6=0}^{L_0-1} C_{\lambda_1, \ldots, \lambda_6} e^{\kappa_1(a_1\lambda_1 + \cdots + a_6\lambda_6)z_1} \times e^{\kappa_2(b_1\lambda_1 + \cdots + b_6\lambda_6)z_2}$$

and its partial derivatives $F_{\sigma_1, \sigma_2}(z_1, z_2)$.

Assume that $e^{\alpha_i\beta_j} \in \mathbf{K}$ $(i = 1, 2$ and $j = 1, \ldots, 6)$. Then there exist S_j^1, S_j^2, $S_{i,j}^3$ in $\mathbf{Z}[x, y]$ and T_j^1, T_j^2, $T_{i,j}^3$ in $\mathbf{Z}[x]$ $(i = 1, 2$ and $j = 1, \ldots, 6)$ such that

$$(3.1) \qquad a_j = \frac{S_j^1(\theta, \omega)}{T_j^1(\theta)}, \qquad b_j = \frac{S_j^2(\theta, \omega)}{T_j^2(\theta)}, \qquad e^{\alpha_i\beta_j} = \frac{S_{i,j}^3(\theta, \omega)}{T_{i,j}^3(\theta)}.$$

We denote by $c_i > 0$ certain constants depending on α_i and β_j, and by X, Y, Z, T certain sufficiently large natural numbers. We put

$$(3.2)\qquad L_0 = [X\ln^{-1/2}X], \qquad X_0 = X, \qquad S_0 = [c_1 X^2 \ln^{-3/2} X].$$

We write down a system of equations with unknowns $C_{\lambda_1,\ldots,\lambda_6}$ in which the coefficients are numbers in the ring $\mathbf{Z}[\theta, \omega]$:

$$(3.3)\qquad \kappa_1^{-\sigma_1}\kappa_2^{-\sigma_2}T^{[c_2X^2\ln^{-1/2}X]}F_{\sigma_1,\sigma_2}(x_1\alpha_1 + x_2\alpha_2) = 0,$$

where $T_0 = \Pi_{i,j}T_{i,j}^3(\theta)T_j^1(\theta)T_j^2(\theta)$ and σ_1, σ_2, x_1, x_2 are natural numbers such that $0 \leqslant \sigma_1 + \sigma_2 \leqslant S_0$, $0 \leqslant x_1, x_2 \leqslant X_0$.

Applying Siegel's Lemma or 1.4, we conclude that there exist numbers $C_{\lambda_1,\ldots,\lambda_6}$ such that

$$(3.4)\qquad C_{\lambda_1,\ldots,\lambda_6} = \sum_{l=0}^{L_2} C_{\lambda_1,\ldots,\lambda_{6,l}}\theta^l$$

for any $\lambda_1,\ldots,\lambda_6$ assuming the values $0,\ldots,L_0 - 1$, and the equations (3.3) are satisfied for $F_{\sigma_1,\sigma_2}(z)$,

$$(3.5)\qquad L_2 = [X^2\ln^{-1/2}X] \quad \text{and} \quad \max|C_{\lambda_1,\ldots,\lambda_{6,l}}| < \exp(c_2X^2\ln^{-1/2}X).$$

As in §1, we use the function $F(z_1, z_2)$ to prove the following

LEMMA 3.3. *There exists $c_3 > 0$ such that for any $T > 0$ there is a polynomial $P(x) \in \mathbf{Z}[x]$ with $t(P) = \max\{\ln H(P), d(P)\} \leqslant T$ and*

$$|P(\theta)| < \exp(-c_3T^2\ln^{1/2}T).$$

PROOF OF 3.3. The proof is analogous to that of 1.1. Consider a sufficiently large constant $c_4 > 0$ such that the set $\mathfrak{P}$ of primes p less than c_4 has cardinality $\geqslant \frac{1}{2}c_4/\ln c_4$. Let p denote any prime in $\mathfrak{P}$, $p > 11$. Assume that the degree of the field $\mathbf{K}(e^{\alpha_1\beta_1/p},\ldots,e^{\alpha_1\beta_6/p})$ over $\mathbf{K}$ is equal to p^6. The case where $[\mathbf{K}(e^{\alpha_1\beta_1/p},\ldots,e^{\alpha_1\beta_6/p})\colon \mathbf{K}] < p^6$ and $[\mathbf{K}(e^{\alpha_2\beta_1/p},\ldots,e^{\alpha_2\beta_6/p})\colon \mathbf{K}] < p^6$ is handled exactly as in §1.

For each rational integer $i \geqslant 0$ we put

$$(3.6)\qquad L_i = [L_0/p^i], \qquad X_i = [X_0p^i], \qquad S_i = [S_0/4^i].$$

We choose a bound $\mathfrak{T}$ for i so that $L_{\mathfrak{T}} = 0$, i.e. $\mathfrak{T} = [\ln L_0/\ln p] + 1$. We also introduce the following notation: for a fixed sequence $\bar{l} = (l_1,\ldots,l_6)$ of numbers $l_j = 0,\ldots,p^i - 1$ we put

$$(3.7)\qquad C^{\bar{l},i}_{m_1,\ldots,m_6} = C_{m_1p^i+l_1,\ldots,m_6p^i+l_6},$$

if $m_1p^i + l_1 < L_0,\ldots,m_6p^i + l < L_0$. If $m_jp^i + l_j \geqslant L_0$, we put

$$(3.8)\qquad C^{\bar{l},i}_{m_1,\ldots,m_6} = 0.$$

In (3.7) and (3.8) we consider only $m_j \leqslant L_i$. We define a function $F^{\bar{l},i}(z_1, z_2)$:

$$F^{\bar{l},i}(z_1, z_2) = \sum_{m_1=0}^{L_i} \cdots \sum_{m_6=0}^{L_i} C^{\bar{l},i}_{m_1,\ldots,m_6} \times e^{\kappa_1(a_1m_1+\cdots+a_6m_6)z_1} e^{\kappa_2(b_1m_1+\cdots+b_6m_6)z_2}. \tag{3.9}$$

LEMMA 3.4. *For each* $i \geqslant 0$, $i \leqslant \mathfrak{T}$, *and any sequence* $\bar{l} = (l_1,\ldots,l_6)$ *of numbers in* $\{0, 1,\ldots,p^i - 1\}$, *we have for the function* $F^{\bar{l},i}$ *in* (3.9) *and its partial derivatives* $F^i_{\sigma_1,\sigma_2}(z_1, z_2) \equiv F^{\bar{l},i}_{\sigma_1,\sigma_2}(z_1, z_2)$ *the equations*

$$F^{\bar{l},i}_{\sigma_1,\sigma_2}(x_1\alpha_1 + x_2\alpha_2) = 0 \tag{3.10}$$

whenever $\sigma_1 + \sigma_2 \leqslant S_i$ *and* $x_1, x_2 \leqslant X_i$. *In* (3.10), $F_{\sigma_1,\sigma_2}(z)$ *stands for* $F_{\sigma_1,\sigma_2}(z, z)$.

Let Ω_i denote the set of all quadruples of natural numbers σ_{j_1}, x_{j_2} such that $\sigma_1 + \sigma_2 \leqslant S_i$ and $x_1, x_2 \leqslant X_i$.

PROOF OF 3.4. For $i = 0$, (3.10) holds by virtue of (3.3). Assume that (3.10) holds for all $i' < i$, $\bar{l} = (l_1,\ldots,l_6)$ with $l_j \leqslant p^{i'-1}$, and $(\sigma_1, \sigma_2, x_1, x_2) \in \Omega_{i'}$. We will show that (3.10) holds for i.

Fix $\bar{l} = (l_1,\ldots,l_6)$; if $l_j = \lambda_j^{i-1} + \cdots + \lambda_j^0 p^{i-1}$ with $\lambda_j^0,\ldots,\lambda_j^{i-1} \in \{0,\ldots,p - 1\}$, we put $\bar{l}' = (l'_1,\ldots,l'_6)$, where $l'_j = \lambda_j^{i-1} + \cdots + \lambda_j^1 p^{i-2}$, i.e.

$$l_j = l'_j + p^{i-1}\lambda_j^0: \qquad j = 1,\ldots,6. \tag{3.11}$$

Consider the function $\mathbf{F}(z_1, z_2) = F^{\bar{l}',i-1}(z_1, z_2)$. According to (3.10), for $i - 1$, we have $\mathbf{F}_{\sigma_1,\sigma_2}(x_1\alpha_1 + x_2\alpha_2) = 0$ for $(\sigma_1, \sigma_2, x_1, x_2) \in \Omega_{i-1}$. Therefore,

$$\mathbf{F}^{(s)}_{\sigma_1,\sigma_2}(x_1\alpha_1 + x_2\alpha_2) = 0 \tag{3.12}$$

whenever $\sigma_1 + \sigma_2 \leqslant \frac{1}{2}S_{i-1}$, $s \leqslant \frac{1}{2}S_{i-1}$, and x_1, $x_2 \leqslant X_i$. Thus, when $\sigma_1 + \sigma_2 \leqslant \frac{1}{2}S_{i-1}$ we have the integral representation

$$\left|\mathbf{F}^{(s)}_{\sigma_1,\sigma_2}(z)\right| = \left|\frac{1}{2\pi i}\int_\Gamma \prod_{x_1=0}^{X_{i-1}} \prod_{x_2=0}^{X_{i-1}} \left\{\frac{z - x_1\alpha_1 - x_2\alpha_2}{\zeta - x_1\alpha_1 - x_2\alpha_2}\right\}^{[S_{i-1}/2]} \frac{\mathbf{F}_{\sigma_1,\sigma_2}(\zeta)\,d\zeta}{z - \zeta}\right|, \tag{3.13}$$

where Γ is the circle $|\zeta| = X^{9/4}$ and $|z| < |\zeta|$. Recall that $i \leqslant \mathfrak{T}$, i.e. $X_{i-1} \leqslant XL_0 \leqslant X^2$. Since we also have $L_{i-1}X_{i-1} \leqslant L_0X$ and $S_{i-1}X^2_{i-1} \geqslant S_0X_0^2(\frac{1}{2}p)^i$, according to (3.6), the estimate

$$\left|\mathbf{F}_{\sigma_1,\sigma_2}(z)\right| < \exp\left(-c_5X^2_{i-1}S_{i-1}\ln X\right) \tag{3.14}$$

for $|z| \leqslant X^2$ follows from (3.13). Consider $z = (x_1/p)\alpha_1 + x_2\alpha_2$ for natural numbers x_1, $x_2 \leqslant X_{i-1}$ and consider the number

$$\varphi = \kappa_1^{-\sigma_1}\kappa_2^{-\sigma_2}\mathbf{F}_{\sigma_1,\sigma_2}\left((x_1/p)\alpha_1 + x_2\alpha_2\right) \tag{3.15}$$

in the field $\mathbf{K}(e^{\alpha_1\beta_1/p},\ldots,e^{\alpha_1\beta_6/p})$. If we multiply φ by $T_0^{[c_6X^2\ln^{-1/2}X]}$ and consider the norm of φ in the field $\mathbf{K}_p/\mathbf{Q}(\theta)$, where $\mathbf{K}_p = \mathbf{Q}(\theta, \omega, e^{\alpha_1\beta_1/p},\ldots,e^{\alpha_1\beta_6/p})$, we arrive at a polynomial $\mathbf{P}(x) \in \mathbf{Z}[x]$ such that

$$t(\mathbf{P}) \leqslant c_7X^2\ln^{-1/2}Xp^6. \tag{3.16}$$

Using the estimate (3.14) for $|\varphi|$, we obtain an estimate for its norm:

$$(3.17)\qquad \begin{aligned}|\mathbf{P}(\theta)| &< \exp\left(-c_8 X_{i-1}^2 S_{i-1} \ln X + c_9 X^2 \ln^{-1/2} X p^6\right)\\ &\leqslant \exp\left(-c_{10} X_{i-1}^2 S_{i-1} \ln X\right);\end{aligned}$$

here $X_{i-1}^2 S_{i-1} \geqslant X_0^2 S_0 \geqslant c_1 X^4 \ln^{-3/2} X$. If $\mathbf{P}(x) \not\equiv 0$, then (3.16) and (3.17) imply Lemma 3.3 with $T = [c_7 X^2 \ln^{-1/2} X p^6]$. In this regard, $c_3 \geqslant c_{10}/c_7^2 p^6$. If $\mathbf{P}(x) \equiv 0$, then $\varphi = 0$ by properties of the norm. Therefore, it follows from (3.15) that

$$(3.18)\qquad \mathbf{F}_{\sigma_1,\sigma_2}((x_1/p)\alpha_1 + x_2\alpha_2) = 0$$

for all $\sigma_1 + \sigma_2 \leqslant \frac{1}{2} S_{i-1}$ and $x_1,\ x_2 \leqslant X_{i-1}$. We use these equalities for the inductive step. We first write out (3.18) explicitly, taking into account (3.9):

$$(3.19)\qquad \sum_{m_1=0}^{L_{i-1}} \cdots \sum_{m_6=0}^{L_{i-1}} \bar{C}^{l',i-1}_{m_1,\dots,m_6} (a_1 m_1 + \cdots + a_6 m_6)^{\sigma_1} \times (b_1 m_1 + \cdots + b_6 m_6)^{\sigma_2} \prod_{j=1}^{6} (e^{\alpha_1 \beta_j})^{x_1 m_j/p} \cdot \prod_{j=1}^{6} (e^{\alpha_2 \beta_j})^{x_2 m_j} = 0.$$

We consider that $a_j,\ b_j,\ e^{\alpha_1\beta_j},\ e^{\alpha_2\beta_j} \in \mathbf{K}$ and recall the above assumption that $[\mathbf{K}(e^{\alpha_1\beta_1/p},\dots,e^{\alpha_1\beta_6/p})\colon \mathbf{K}] = p^6$. Then for natural numbers $\sigma_1,\ \sigma_2,\ x_1,\ x_2$ with $(x_1,\ p) = 1$ equality (3.19) is equivalent to the system of equalities

$$(3.20)\qquad \begin{aligned}&\sum_{m_1=0,m_1\equiv\lambda_1}^{L_{i-1}} \cdots \sum_{m_6=0,m_6\equiv\lambda_6}^{L_{i-1}} \bar{C}^{l',i-1}_{m_1,\dots,m_6}\\ &\times (a_1 m_1 + \cdots + a_6 m_6)^{\sigma_1} (b_1 m_1 + \cdots + b_6 m_6)^{\sigma_2}\\ &\times \prod_{j=1}^{6} (e^{\alpha_1\beta_j})^{x_1 m_j/p} \prod_{j=1}^{6} (e^{\alpha_2\beta_j})^{x_2 m_j} = 0,\end{aligned}$$

for all $\lambda_1,\dots,\lambda_6$ in $\{0,\dots,p-1\}$, where the congruences $\equiv$ are to be understood as (mod p). Put $\lambda_1 = \lambda_1^0,\dots,\lambda_6 = \lambda_6^0$ (see (3.11)). Each $m_j = 0,\dots,L_{i-1}$, $m_j \equiv \lambda_j^0$ (mod p), can be written in the form $m_j = m_j' p + \lambda_j^0$. Since $m_j \leqslant L_{i-1}$, it follows that $m_j' \leqslant L_i$. Divide (3.20) by

$$\prod_{j=1}^{6} (e^{\alpha_1\beta_j})^{x_1\lambda_j^0/p} \cdot \prod_{j=1}^{6} (e^{\alpha_2\beta_j})^{x_2\lambda_j^0}.$$

Recall that $(\gamma + e_0)^\sigma = \sum_{s=0}^{\sigma} C_\sigma^s e^{\sigma-s} \gamma^s$ and the fact that (3.20) holds for all $\sigma_1 + \sigma_2 \leqslant S_{i-1}/2$. Adding the equalities (3.20), we obtain

$$(3.21)\qquad \begin{aligned}&\sum_{m_1=0,m_1=m_1'p+\lambda_1^0}^{L_{i-1}} \cdots \sum_{m_6=0,m_6=m_6'p+\lambda_6^0}^{L_{i-1}} \bar{C}^{l',i-1}_{m_1,\dots,m_6}\\ &\times (a_1 m_1' + \cdots + a_6 m_6')^{\sigma_1} (b_1 m_1' + \cdots + b_6 m_6')^{\sigma_2}\\ &\times \prod_{j=1}^{6} (e^{\alpha_1\beta_j})^{x_1 m_j'} \prod_{j=1}^{6} (e^{\alpha_2\beta_j})^{x_2 p m_j'} = 0.\end{aligned}$$

In view of definition (3.7) and (3.11), we obtain for the $m'_j = 0,\dots,L_i$ such that $m'_j p^i + l_j < L_0$: $j = 1,\dots,6$, the equations

$$(3.22)\quad C^{\bar{l},i}_{m'_1,\dots,m'_6} = C_{m'_1p^i+l_1,\dots,m'_6p^i+l_6} = C_{m_1p^{i-1}+l'_1,\dots,m_6p^{i-1}+l'_6} = C^{\bar{l}',i-1}_{m_1,\dots,m_6},$$

in which $m_j = m'_j p + \lambda^0_j$ and $\lambda^0_j p^{i-1} + l'_j = l_j$: $j = 1,\dots,6$. From (3.21) and (3.22) according to (3.9) we obtain the important condition

$$(3.23)\qquad F^{\bar{l},i}_{\sigma_1,\sigma_2}(x_1\alpha_1 + x_2 p\alpha_2) = 0$$

for all natural numbers $\sigma_1 + \sigma_2 \leqslant \frac{1}{2}S_{i-1}$ and $x_1,\ x_2 \leqslant X_{i-1}$, $(x_1, p) = 1$.

We apply (3.23) to the function $f_i(z) \equiv F^{\bar{l}_0,i}_{\sigma_1^0,\sigma_2^0}(z)$ for fixed σ_1^0, σ_2^0 such that $\sigma_1^0 + \sigma_2^0 \leqslant S_i \leqslant [\frac{1}{4}S_{i-1}]$. We obtain, for any $s \leqslant S_i$,

$$(3.24)\qquad f_i^{(s)}(x_1\alpha_1 + px_2\alpha_2) = 0$$

for all natural numbers x_1, x_2 such that $x_1,\ x_2 \leqslant X_{i-1}$, $(x_1, p) = 1$. It is easy to see that the number of such pairs (x_1, x_2) is equal to or greater than $(1 - 1/p)X^2_{i-1}$, which in turn is equal to or greater than $\frac{1}{2}X^2_{i-1}$. Using (3.24), we have

$$(3.25)\quad |f_i(z)| = \left|\frac{1}{2\pi i}\int_\Gamma \prod_{x_1=0,(x_1,p)=1}^{X_{i-1}} \prod_{x_2=0}^{X_{i-1}} \left\{\frac{z - x_1\alpha_1 - x_2\alpha_2 p}{\zeta - x_1\alpha_1 - x_2\alpha_2 p}\right\}^{S_i} \frac{f_i(\zeta)\,d\zeta}{z - \zeta}\right|,$$

where Γ is the circle $|\zeta| = X^{9/4}$ and $|z| < |\zeta|$. Estimating $\max|f_i(\zeta)|$ on Γ, we obtain from (3.2) that

$$(3.26)\qquad \max_{|\zeta| = X^{9/4}} |f_i(\zeta)| \leqslant \exp(c_{11}X^{13/4}).$$

If we take into account that the number of factors of the form

$$(z - x_1\alpha_1 - x_2\alpha_2 p)/(\zeta - x_1\alpha_1 - x_2\alpha_2 p)$$

in the right-hand side of (3.25) is at least $\frac{1}{2}X^2_{i-1}S_i$ and that

$$(3.27)\qquad X^2_{i-1}S_i \geqslant c_{12}X^4 \ln^{-3/2} X (p^2/4)^{i-1},$$

we can deduce from (3.25)–(3.27) that

$$(3.28)\qquad |f_i(z)| < \exp(-c_{13}X^2_{i-1}S_i \ln X)$$

for $|z| \leqslant X^2$. Since $i \leqslant \mathfrak{T}$ and $X_i \leqslant Xp^i \leqslant 2XL_0 \leqslant 2X^2\ln^{-1/2}X$, it follows from (3.28) that

$$(3.29)\qquad \left|F^{\bar{l},i}_{\sigma_1,\sigma_2}(x_1\alpha_1 + x_2\alpha_2)\right| < \exp(-c_{14}X^2_{i-1}S_i \ln X)$$

for all x_1, x_2 such that $|x_1|$, $|x_2| \leqslant X_i$ and for all $\sigma_1 + \sigma_2 \leqslant S_i$. Since $X^2_{i-1}S_i \geqslant \frac{1}{2}X^4\ln^{-3/2}X$, we have

$$(3.30)\qquad \left|F^{\bar{l},i}_{\sigma_1,\sigma_2}(x_1\alpha_1 + x_2\alpha_2)\right| < \exp(-c_{15}X^4\ln^{-1/2}X)$$

for all $(\sigma_1, \sigma_2, x_1, x_2) \in \Omega_i$.

From (3.30) we obtain either 3.3 or (3.10). Indeed, assume that

(3.31) $$F^{\bar{l},i}_{\sigma_1,\sigma_2}(x_1\alpha_1 + x_2\alpha_2) \neq 0$$

for certain $(\sigma_1, \sigma_2, x_1, x_2) \in \Omega_i$. Instead of $F^{\bar{l},i}_{\sigma_1,\sigma_2}(x_1\alpha_1 + x_2\alpha_2)$ consider the number

$$\psi = T_0^{X_iL_i}\kappa_1^{-\sigma_1}\kappa_2^{-\sigma_2}F^{\bar{l},i}_{\sigma_1,\sigma_2}(x_1\alpha_1 + x_2\alpha_2) \neq 0$$

of the field **K**. Consider also the norm $p = \operatorname{Nm}\psi$ in $\mathbf{K}/\mathbf{Q}(\theta)$. Then $p \in \mathbf{Z}[\theta]$. From (3.2) and the definitions of $F^{\bar{l},i}_{\sigma_1,\sigma_2}(z)$ and $(\sigma_1, \sigma_2, x_1, x_2)$ we obtain an estimate of the type of ψ as an element of $\mathbf{Z}[\theta, \omega] = \mathfrak{L}$,

$$t_{\mathfrak{L}}(\psi) \leqslant c_{16}X^2\ln^{-1/2}X + c_{17}X_iL_i + c_{18}S_i\ln X \leqslant c_{19}X^2\ln^{-1/2}X.$$

Therefore, $p = P(\theta)$, where $P(x) \in \mathbf{Z}[x]$ and

(3.32) $$t(P) \leqslant c_{20}X^2\ln^{-1/2}X.$$

Moreover, $P(x) \not\equiv 0$, since $\psi \neq 0$ and $p \neq 0$. We also obtain from (3.30) the estimate

(3.33) $$|P(\theta)| < \exp\left(-c_{21}X^4\ln^{-1/2}X\right).$$

Consequently, (3.32) and (3.33) show that Lemma 3.3 holds with $T = [c_{20}X^2\ln^{-1/2}X]$ and we may assume that (3.31) does not hold. This means that for all $(\sigma_1, \sigma_2, x_1, x_2) \in \Omega_i$ we have $F^{\bar{l},i}_{\sigma_1,\sigma_2}(x_1\alpha_1 + x_2\alpha_2) = 0$ and (3.10) is proved. Thus, Lemma 3.4 has been established for all $i \leqslant \mathfrak{T}$.

We apply (3.10) with $i = \mathfrak{T}$. Thus, we can apply (3.10) with $\sigma_1 = \sigma_2 = 0$. For all sequences $\bar{l} = (l_1,\ldots,l_6)$ of numbers in $\{0, 1,\ldots,p^{\mathfrak{T}} - 1\}$ we have

(3.34) $$C^{\bar{l}}_{0,\ldots,0} = 0.$$

Since $p^{\mathfrak{T}} > L_0$, it follows from the definition of $C^{\bar{l}}_{m_1,\ldots,m_6}$ that

(3.35) $$C_{\lambda_1,\ldots,\lambda_6} = 0$$

for all $\lambda_1,\ldots,\lambda_6$ assuming the values $0, 1,\ldots,L_0$, which contradicts the choice of the C_λ. From the given proof of 3.4 we see that 3.3 holds for $T = [c_{21}X^2\ln^{-1/2}X]$ and certain absolute constants c_{21} and c_4. Since $X > c_{22}$ is arbitrary, Lemma 3.3 is proved for all sufficiently large T. By Lemma VII of Chapter III of **[8]**, 3.3 implies that θ is algebraic and 3.2 is proved.

To prove 3.1 we use the above scheme, changing the auxiliary function. Consider the case $MN \geqslant M + N + rN$, where r is the dimension over **K** of the vector space generated by $\{\beta_1,\ldots,\beta_M\}$ and where $e^{\alpha_i\beta_j} \in \mathbf{K}$: $i = 1,\ldots,N$; $j = 1,\ldots,M$. Then there exist $\kappa_1,\ldots,\kappa_r$ such that $\beta_i = a_{1,i}\kappa_1 + \cdots + a_{r,i}\kappa_r$: $i = 1,\ldots,M$, with $a_{j,i} \in \mathbf{K}$. Consider the auxiliary function

$$F(z_1,\ldots,z_2) = \sum_{\lambda_1=0}^{L-1}\cdots\sum_{\lambda_M=0}^{L-1} C_{\lambda_1,\ldots,\lambda_M}e^{\kappa_1(\lambda_1 a_{1,1}+\cdots+\lambda_M a_{1,M})z_1} \times \cdots$$
$$\times e^{\kappa_r(\lambda_1 a_{r,1}+\cdots+\lambda_M a_{r,M})z_2}.$$

If $\mathbf{K} = \mathbf{Q}(\theta, \omega)$, we determine numbers $C_{\lambda_1,\ldots,\lambda_M}$ in $\mathbf{Z}[\theta, \omega]$ such that

$$F_{\sigma_1,\ldots,\sigma_2}(z,\ldots,z) = 0$$

for any natural numbers $\sigma_1,\ldots,\sigma_r$; $\sigma_1 + \cdots + \sigma_r \leqslant S_0$ and $z = x_1\alpha_1 + \cdots + x_N\alpha_N$ with natural numbers $x_1,\ldots,x_N \leqslant X_0$. Here, for any sufficiently large natural number L,

$$S_0 = \left[L^{(M+N)/(N+r)}\ln^{-N/(N+r)}L\right]; \qquad X_0 = \left[c_1L^{(M-r)/(N+r)}\ln^{r/(N+r)}L\right].$$

Now, as above, we establish by means of a variant of Stark's method the existence of a nontrivial polynomial $P(x) \in \mathbf{Z}[x]$ such that

$$|P(\theta)| < \exp(-c'L^{(MN+M+N-rN)/(N+r)}\ln^{(N+1)r/(N+r)}L)$$

with

$$t(P) \leqslant c''L^{(N+M)/(N+r)}\ln^{r(N+r)}L.$$

Consequently, we can apply Gel'fond's lemma concerning a test for the transcendence of a number to deduce that θ is algebraic.

Now consider the case $MN \geqslant M + N + rN$, where $r + 1$ is the dimension over $\mathbf{K}$ of the vector space generated by $\{1, \beta_1,\ldots,\beta_M\}$ and where $\mathbf{K}$ is a field of transcendence degree 1 such that α_i, $e^{\alpha_i\beta_j} \in \mathbf{K}$: $i = 1,\ldots,N$; $j = 1,\ldots,M$. We consider the following function:

$$F_1(z) = \sum_{\lambda_0=0}^{L_0-1}\sum_{\lambda_1=0}^{L-1}\cdots\sum_{\lambda_M=0}^{L-1} C_{\lambda_0,\lambda_1,\ldots,\lambda_M}z^{\lambda_0}e^{(\lambda_1\beta_1+\cdots+\lambda_M\beta_M)z}.$$

We put $\beta_i = a_{0,i} + \kappa_1a_{1,i} + \cdots + \kappa_ra_{r,i}$: $i = 1,\ldots,M$, where $1, \kappa_1,\ldots,\kappa_r$ is a basis of the vector space over $\mathbf{K}$ generated by $\{1, \beta_1,\ldots,\beta_M\}$, and we put

$$F(z_0, z_1,\ldots,z_r) = \sum_{\lambda_0=0}^{L_0-1}\sum_{\lambda_1=0}^{L-1}\cdots\sum_{\lambda_M=0}^{L-1} C_{\lambda_0,\lambda_1,\ldots,\lambda_M}z_0^{\lambda_0}e^{(\lambda_1a_{0,1}+\cdots+\lambda_Ma_{0,M})z_0}$$
$$\times e^{\kappa_1(\lambda_1a_{1,1}+\cdots+\lambda_Ma_{1,M})z_1}\cdots e^{\kappa_r(\lambda_1a_{r,1}+\cdots+\lambda_Ma_{r,M})z_r}.$$

We can choose $C_{\vec\lambda}$ in $\mathbf{Z}[\theta]$ so that

$$F_{\sigma_0,\ldots,\sigma_r}(z,\ldots,z) = 0$$

for any natural numbers $\sigma_0,\ldots,\sigma_r$ such that $\sigma_0 + \cdots + \sigma_r \leqslant S_0$ and $z = x_1\alpha_1 + \cdots + x_N\alpha_N$ for natural numbers $x_1,\ldots,x_N \leqslant X_0$. Here, for any sufficiently large natural number L

$$L_0 = \left[L^{(M+N)/(N+r)}\ln^{-N/(N+r)}L\right], \qquad S_0 = \left[c_2L^{(M+N)/(N+r)}\ln^{-N/(N+r)}L\right],$$
$$X_0 = \left[c_3L^{(M-r)/(N+r)}\ln^{r/(N+r)}L\right].$$

Then we can establish the existence of $R(x) \in \mathbf{Z}[x]$, $R(x) \not\equiv 0$, such that

$$|R(\theta)| < \exp(-c^{iii}L^{(MN+M+N-rN)/(N+r)}\ln^{(N+1)r/(N+r)}L);$$
$$t(R) \leqslant c^{iv}L^{(M+N)/(N+r)}\ln^{r/(N+r)}L.$$

Finally, note that by combining the methods of Chapter 1, §§1 and 3 we can generalize 3.1 to a field **K** of transcendence degree n. For this purpose it suffices to replace the inequalities in 3.1 by

$$MN \geqslant (2^n - 1)(M + N) + rN.$$

References

1. A. Baker, *Linear forms in the logarithms of algebraic numbers*. I, II, III, IV, Mathematika **13** (1966), 204–216; **14** (1967), 102–107, 220–228; **15** (1968), 204–216.

2. ______, *Transcendental number theory*, Cambridge Univ. Press, Cambridge, 1979.

3. A. Baker and H. M. Stark, *On a fundamental inequality in number theory*, Ann. of Math. (2) **94** (1971), 190–199.

4. H. M. Stark, *Further advances in the theory of linear forms in logarithms*, Diophantine Approximation and its Applications, Academic Press, New York, 1973, pp. 255–293.

5. S. Lang, *Algebra*, Addison-Wesley, Reading, Mass., 1965.

6. G. V. Chudnovsky, *Baker's method in the theory of transcendental numbers*, Uspehi Mat. Nauk **31** (1976), 281–282. (Russian)

7. ______, *The Gelfond-Baker method in problems of Diophantine approximation*, Colloq. Math. Soc. János Bolyai **13** (1976), 19–30.

8. A. O. Gel'fond, *Transcendental and algebraic numbers*, Translated from the first Russian ed. by L. F. Baron, Dover, New York, 1960.

9. ______, *Sur les nombres transcendants*, C. R. Acad. Sci. Paris **189** (1929), 1224–1226.

10. M. Waldschmidt, *Utilisation de la methode de Baker dans des problems d'independance*, C. R. Acad. Sci. Paris **275** (1972), 1215–1217.

11. ______, *Nombres transcendants*, Lecture Notes in Math., vol. 402, Springer-Verlag, Berlin and New York, 1974.

Translated by G.A. KANDALL

CHAPTER 3

ON THE WAY TO SCHANUEL'S CONJECTURE

ALGEBRAIC CURVES NEAR A POINT. I. GENERAL THEORY OF COLORED SEQUENCES

As is evident from the title, we are interested in some specific point and curves passing near it. It is not difficult to guess that the existence of some singular family passing near a point is related to the singular arithmetic nature of the point. We consider a point (in space) whose coordinates are a transcendence basis for a family of numbers associated with Schanuel's conjecture [**2**]:

$$\alpha_1,\dots,\alpha_n,\quad e^{\alpha_1},\dots,e^{\alpha_n}.$$

This interest in algebraic curves and their families is related to a phenomenon which always strikes the author as unexpected. This is the existence, known for more than 20 years, of a dense sequence of algebraic curves passing near this "key" point with transcendental coordinates. This phenomenon, of course, has always been used in investigations. For example, in the papers of Lang [**10**], Gel′fond [**5**], Brownawell [**3**], and the author [**6**] a family of algebraic curves played a significant role. Later, in the author's investigations [**7**, **8**] related to Schanuel's conjecture, various families of curves and hypersurfaces near a transcendence basis became, generally speaking, the main object of study.

In this chapter we will consider in detail the two-dimensional situation and interesting sets of curves, without concern for their origins. This is the beginning of a sequence of papers by the author concerning estimates of transcendence degrees of fields arising from Schanuel's conjecture. The final estimates are close to those desired and are significantly better than those of [**7**, **8**].

0. Preliminary material. All notation, definitions, and terminology are standard [**10**, **14**]. We mention only one definition. For a polynomial $P(x_1,\dots,x_k)\in \mathbf{C}[x_1,\dots,x_k]$ we denote by $H(P)$ the maximum of the moduli of the coefficients

0076-5376/84 $1.00 + $.25 per page

of $P(x_1,\ldots,x_k)$ and by $d(P)$ (the degree of $P(x_1,\ldots,x_k)$) the number $d_{x_1}(P) + \cdots + d_{x_k}(P)$. We put $t(P) = \max\{\ln H(P), d(P)\}$ and call $t(P)$ the type of $P(x_1,\ldots,x_k)$.

We mention at the outset that all arguments in this paper apply to the case of simultaneously increasing heights and degrees of polynomials. We will therefore make use of an old idea **[5, 10]** and simply look at everything from the point of view of the growth of the type of polynomials. It is much more important to do this than to assume boundedness of the degree, as is sometimes done. Furthermore, consideration of the type is also important for applications to algebraic independence **[5]**. Note that it is very easy to transfer our arguments to the case of simultaneously employed heights and degrees. Therefore, we will focus our attention only on $t(P)$ and will say no more about this question.

To make it clear with which sequences of polynomials it will be necessary to deal we give a canonical example.

EXAMPLE 0.1. Suppose $\vec{\theta} = (\theta_1,\ldots,\theta_n) \in \mathbf{C}^n$, where $\theta_1,\ldots,\theta_n$ are algebraically independent over $\mathbf{Q}$. Then for any natural number H there exists a polynomial $P_{H,\vec{\theta}} = P_H$, $P_H(x_1,\ldots,x_n) \in \mathbf{Z}[x_1,\ldots,x_n]$, such that

$$P_H(x_1,\ldots,x_n) \not\equiv 0, \tag{1}$$

$$t(P_H) \leqslant H, \tag{2}$$

$$|P_H(\theta_1,\ldots,\theta_n)| < \exp(-c_0 H^{n+1}), \tag{3}$$

where $c_0 = c_0(n) > 0$ is a constant depending only on n.

The family of hypersurfaces $\{P_{H,\vec{\theta}}(x_1,\ldots,x_n) = 0\colon H \geqslant 1\}$ satisfying conditions (1)–(3) is called a Dirichlet family near the point $\vec{\theta}$.

This name stems from the fact that the existence of polynomials satisfying (1)–(3) is guaranteed by the Dirichlet "pigeon-hole principle". The proof can be found in **[19, 18]**. Incidentally, there are several interesting (inequivalent) proofs of 0.1. The assertion of 0.1 is proved in **[19]** by means of Minkowski's theorem.

The situation is already clear. We are interested in sequences of polynomials $P_H(x_1,\ldots,x_n) \in \mathbf{Z}[x_1,\ldots,x_n]$ satisfying conditions (1) and (2) of 0.1 and, instead of (3), a stronger condition, for example,

$$|P_H(\theta_1,\ldots,\theta_n)| < \exp(-H^{n+1}\varphi(H)), \tag{3$'$}$$

where $\varphi(H)$ is a nonnegative, monotone increasing function such that $\lim_{H\to\infty}\varphi(H) = \infty$.

Here arises the main question, crucial for an investigation of Schanuel's conjecture.

QUESTION 0.2. For which points $\vec{\theta} = (\theta_1,\ldots,\theta_n) \in \mathbf{C}^n$ does there exist a family of polynomials $\{P_H(x_1,\ldots,x_n) \in \mathbf{Z}[x_1,\ldots,x_n]\colon H \geqslant 1\}$ satisfying (1), (2), and (3$'$)?

Since conditions (1)–(3$'$) impose serious arithmetic-analytic restrictions on $\vec{\theta}$, Lang **[10]** naturally conjectured that $\theta_1,\ldots,\theta_n$ satisfying the conditions of 0.2 must be algebraically dependent over $\mathbf{Q}$.

He had good reason for this conjecture. This was the transcendence criterion for one number (i.e. $n = 1$) which he proved in [**10**, **11**], but which was established earlier by Gel'fond [**5**], although not in complete generality.

CRITERION 0.3. Suppose $\theta \in \mathbf{C}$. Suppose also that F is a nonnegative, monotone increasing function of a natural argument H such that $\lim_{H\to\infty} F(H) = \infty$ and there exists a constant a_0 such that $F(H+1) \leqslant a_0 F(H)$ for all $H \geqslant a_1$. If for every $H \geqslant a_2$ there exists a polynomial $P_H(x) \in \mathbf{Z}[x]$, $P_H(x) \not\equiv 0$, $t(P_H) \leqslant F(H)$, such that

$$|P_H(\theta)| < \exp\left(-c_1 F(H)^2\right),$$

for some constant $c_1 = c_1(a_0, a_1, a_2) > 0$, then the number θ is algebraic.

In particular, if $F(H) \equiv H$: $H \geqslant 1$, we obtain that for $n = 1$ every $\theta \in \mathbf{C}$ satisfying the conditions of Question 0.2 must be algebraic.

Therefore, the appearance of Lang's conjecture seemed natural. No one took any particular notice of it until E. Bombieri (see [**11**]) observed that it was simply false. Moreover, it was false for any function $\varphi(H)$ in (3′).

The situation here is as follows. In studying singular linear forms, J. Cassels [**1**] proved this assertion:

LEMMA 0.4. *For any nonnegative monotone function $\varphi(t)$: $t \in \mathbf{R}^+$, there exist two algebraically independent numbers θ_1, θ_2 such that the inequality*

$$|x\theta_1 + y\theta_2 + z| < \exp(-\varphi(H))$$

is solvable in rational integers x, y, z: $X = \max(|x|, |y|, |z|) \neq 0$, $X \leqslant H$, for all $H \geqslant c_2$. Here $c_2 = c_2(\varphi) \geqslant 0$ is some constant.

Consequently, for $n \geqslant 2$ we cannot expect a precise answer to Question 0.2. Nevertheless, the points $\vec{\theta}$ satisfying 0.2 have a specific arithmetic nature. We have to consider so-called fields of "finite transcendence type".

DEFINITION 0.5. Suppose $\mathcal{K} = \mathbf{Q}(\vartheta_1, \ldots, \vartheta_d, \omega)$ is a subfield of $\mathbf{C}$ of transcendence degree d over $\mathbf{Q}$, and $\vartheta_1, \ldots, \vartheta_d$ is its transcendence basis. The field $\mathcal{K}$ has transcendence type $\leqslant \tau$ $(< \infty)$, if there exists a constant $c(\mathcal{K}) > 0$ such that for every polynomial $P(x_1, \ldots, x_d) \in \mathbf{Z}[x_1, \ldots, x_d]$, $P(x_1, \ldots, x_d) \not\equiv 0$, we have

$$|P(\vartheta_1, \ldots, \vartheta_d)| > \exp\left(-c(\mathcal{K}) t(P)^\tau\right). \tag{0.1}$$

REMARK 0.6. The field $\mathcal{K}$ has transcendence type $\leqslant (\tau, \tau')$ if (0.1) is replaced by

$$|P(\vartheta_1, \ldots, \vartheta_d)| > \exp\left(-c(\mathcal{K}) \cdot t(P)^\tau (\ln t(P))^{\tau'}\right). \tag{0.2}$$

These definitions were given in [**3**, **10**, **14**].

As Example 0.1 shows, a field $\mathcal{K} \subset \mathbf{C}$ of transcendence degree d cannot have transcendence type $\leqslant d + 1$. Finiteness of transcendence type is an arithmetic restriction and does not always hold. However, for almost all points $\vec{\theta} \in \mathbf{C}^n$ the field $\mathcal{K}_{\vec{\theta}} = \mathbf{Q}(\theta_1, \ldots, \theta_n)$ has finite transcendence type; for $n = 1$ this type is even equal to 2. Therefore, it is extremely valuable (if only for applications) that to the coordinates θ_i of a $\vec{\theta} = (\theta_1, \ldots, \theta_n)$ satisfying 0.2 correspond fields $\mathbf{Q}(\theta_i)$ of transcendence type > 2.

More precisely, we have the following result of Brownawell [**3**]. We give this result in the form in which it will be used in the present paper.

THEOREM 0.7. *Suppose $\theta_1, \theta_2 \in \mathbf{C}$ and θ_1, θ_2 are algebraically independent over $\mathbf{Q}$. If for every $H \geqslant 1$ there exists a polynomial $P_H(x, y) \in \mathbf{Z}[x, y]$, $P_H(x, y) \not\equiv 0$, $t(P_H) \leqslant H$, such that*

$$|P_H(\theta_1, \theta_2)| < \exp(-c_3 H^\mu) \tag{0.3}$$

for some constant $c_3 = c_3(\theta_1, \theta_2) > 0$, then the field $\mathbf{Q}(\theta_1)$ (and the field $\mathbf{Q}(\theta_2)$) has transcendence type $\geqslant \frac{1}{2}\mu$. If (0.3) is replaced by

$$|P_H(\theta_1, \theta_2)| < \exp(-c_3 H^\mu \varphi(H)) \tag{0.4}$$

where $\varphi(t)$: $t \in \mathbf{R}^+$, is an unbounded nonnegative function, then $\mathbf{Q}(\theta_1)$ (and $\mathbf{Q}(\theta_2)$) has transcendence type $> \frac{1}{2}\mu$.

In the latter case this means that for any $c_4 > 0$ the inequality

$$|P(\theta_1)| < \exp\left(-c_4 t(P)^{\mu/2}\right)$$

is satisfied by an infinite number of distinct polynomials $P(x) \in \mathbf{Z}[x]$, $P(x) \not\equiv 0$.

As will be clear from what follows, this is a weak assertion, in comparison with the results employed. The precise nature of the estimate will be discussed in this and subsequent papers of the author on the same theme.

To strengthen this result in subsequent papers using the methods set forth below we will draw upon various algebraic, analytic, and combinatorial techniques. Clearly, the most we can expect is that the type of $\mathbf{Q}(\theta_1)$ is at least μ. But this expectation is too optimistic, as is shown by the example of a Dirichlet family corresponding to a point (θ_1, θ_2) in "general position" in $\mathbf{C}^2$. Then $\mu = 3$, although $\mathbf{Q}(\theta_1)$ and $\mathbf{Q}(\theta_2)$ "almost always" have type $\tau = 2$. In any event, the type of $\mathbf{Q}(\theta_1)$ will be closer to μ than to $\frac{1}{2}\mu$.

At the end of the first section of the chapter we will explain in detail why such estimates are necessary.

1. A dense family of curves. Coloring. In this section we analyze from general viewpoints a family of curves passing near a point $\vec{\theta} \in \mathbf{C}^2$ and satisfying the conditions of Question 0.2. The curves $P_H(x, y) = 0$ corresponding to the various $H \geqslant 1$ are "colored" with two colors, red and blue. The coloring is of a conventional nature and is introduced for the sake of illustration. It will also be useful for the investigation of families of curves related to Schanuel's conjecture.

Of all the auxiliary facts we will need, at present, only one.

LEMMA 1.1. *Suppose $P_1(x_1, \ldots, x_n), \ldots, P_m(x_1, \ldots, x_n)$, $P(x_1, \ldots, x_n) \in \mathbf{C}[x_1, \ldots, x_n]$. If $P(x_1, \ldots, x_n) = P_1(x_1, \ldots, x_n) \times \cdots \times P_m(x_1, \ldots, x_n)$, then for the height $H(P)$ of the polynomial $P(x_1, \ldots, x_n)$ we have the lower estimate*

$$H(P) \geqslant e^{-d(P)} H(P_1) \cdots H(P_m).$$

The proof of this lemma, its history, and some refinements can be found in [5]. We give a useful corollary of 1.1.

COROLLARY 1.2. *Suppose* $P_1(x_1,\ldots,x_n)$, $P_2(x_1,\ldots,x_n) \in \mathbf{Z}[x_1,\ldots,x_n]$. *If the polynomial* $(P_1(x_1,\ldots,x_n))^s$, $s \geq 1$, *divides* $P_2(x_1,\ldots,x_n)$, *then for the type* $t(P_1)$ *of* $P_1(x_1,\ldots,x_n)$ *we obtain the estimate*

$$t(P_1)s \leq 2t(P_2).$$

We can now turn to an investigation of families of polynomials. We will first prove a lemma in the spirit of [3, 6], but noteworthy in that there is no elimination of variables. The method of proof of this lemma appears in one paper after another with minor alterations.

LEMMA 1.3. *Suppose* $P(x_1,\ldots,x_n) \in \mathbf{Z}[x_1,\ldots,x_n]$ *and* $\vec{\theta} = (\theta_1,\ldots,\theta_n) \in \mathbf{C}^n$. *If*

$$|P(\theta_1,\ldots,\theta_n)| \leq \varepsilon < 1 \tag{1.1}$$

then either there exists a divisor $Q(x_1,\ldots,x_n) \in \mathbf{Z}[x_1,\ldots,x_n]$ *of* $P(x_1,\ldots,x_n)$ *which is a power of a* $\mathbf{Q}$*-irreducible polynomial and for which*

$$|Q(\theta_1,\ldots,\theta_n)| \leq \varepsilon^{1/3}, \tag{1.2}$$

or else there exist two coprime divisors $P_1(x_1,\ldots,x_n)$, $P_2(x_1,\ldots,x_n) \in \mathbf{Z}[x_1,\ldots,x_n]$ *of* $P(x_1,\ldots,x_n)$ *for which*

$$|P_1(\theta_1,\ldots,\theta_n)| \leq \varepsilon^{1/3}, \qquad |P_2(\theta_1,\ldots,\theta_n)| \leq \varepsilon^{1/3}. \tag{1.3}$$

PROOF. Let $R_1^{h_1},\ldots,R_k^{h_k}$ denote all distinct divisors of $P(x_1,\ldots,x_n)$ which are powers of $\mathbf{Q}$-irreducible polynomials in $\mathbf{Z}[x_1,\ldots,x_n]$. By Gauss's lemma,

$$P(x_1,\ldots,x_n) = A(R_1(x_1,\ldots,x_n))^{h_1}\cdots(R_k(x_1,\ldots,x_n))^{h_k}, \tag{1.4}$$

where $A \in \mathbf{Z}$, $A \neq 0$. To eliminate trivial cases we assume that $\theta_1,\ldots,\theta_n$ are algebraically independent or, more precisely, that

$$P(\theta_1,\ldots,\theta_n) \neq 0. \tag{1.5}$$

If (1.5) did not hold, we would at once obtain $R_i(\theta_1,\ldots,\theta_n) = 0$ for some i, $1 \leq i \leq k$, and (1.2) holds with $Q(x_1,\ldots,x_n) = R_i(x_1,\ldots,x_n)$. Therefore, we may assume that (1.5) holds.

Furthermore, we number the polynomials R_i: $1 \leq i \leq k$, so that

$$|R_1(\theta_1,\ldots,\theta_n)|^{h_1} \leq \cdots \leq |R_k(\theta_1,\ldots,\theta_n)|^{h_k}. \tag{1.6}$$

Since $\varepsilon < 1$, it follows from (1.1) and (1.4) that for some l, $1 \leq l \leq k$, we have

$$\begin{aligned}|R_1(\theta_1,\ldots,\theta_n)|^{h_1} \leq \cdots \leq |R_l(\theta_1,\ldots,\theta_n)|^{h_l} < |A|^{-1}\\ \leq |R_{l+1}(\theta_1,\ldots,\theta_n)|^{h_{l+1}} \leq \cdots \leq |R_k(\theta_1,\ldots,\theta_n)|^{h_k}.\end{aligned} \tag{1.7}$$

Clearly, there exists i, $1 \leq i \leq l$, such that

$$\begin{aligned}|A|\cdot|R_1(\theta_1,\ldots,\theta_n)|^{h_1}\cdots|R_i(\theta_1,\ldots,\theta_n)|^{h_i}\\ < |R_{i+1}(\theta_1,\ldots,\theta_n)|^{h_{i+1}}\cdots|R_k(\theta_1,\ldots,\theta_n)|^{h_k}\end{aligned} \tag{1.8}$$

and also

$$(1.9)\qquad |A|\cdot|R_1(\theta_1,\dots,\theta_n)|^{h_1}\cdots|R_{i-1}(\theta_1,\dots,\theta_n)|^{h_{i-1}}$$
$$\geqslant|R_i(\theta_1,\dots,\theta_n)|^{h_i}\cdots|R_k(\theta_1,\dots,\theta_n)|^{h_k}.$$

We first use (1.8). Let

$$P_1'(x_1,\dots,x_n)=A\cdot(R_1(x_1,\dots,x_n))^{h_1}\cdots(R_i(x_1,\dots,x_n))^{h_i}$$

and

$$P_2'(x_1,\dots,x_n)=(R_{i+1}(x_1,\dots,x_n))^{h_{i+1}}\cdots(R_k(x_1,\dots,x_n))^{h_k}.$$

Then $P_1'(x_1,\dots,x_n)$, $P_2'(x_1,\dots,x_n)$ are coprime divisors of $P(x_1,\dots,x_n)$. Assuming that (1.3) does not hold for these divisors, it follows from (1.8) that

$$(1.10)\qquad |R_{i+1}(\theta_1,\dots,\theta_n)|^{h_{i+1}}\cdots|R_k(\theta_1,\dots,\theta_n)|^{h_k}>\varepsilon^{1/3}.$$

We now use (1.9). Let

$$P_1''(x_1,\dots,x_n)=A\cdot(R_1(x_1,\dots,x_n))^{h_1}\cdots(R_{i-1}(x_1,\dots,x_n))^{h_{i-1}}$$

and

$$P_2''(x_1,\dots,x_n)=(R_i(x_1,\dots,x_n))^{h_i}\cdots(R_k(x_1,\dots,x_n))^{h_k}.$$

Once again, $P_1''(x_1,\dots,x_n)$, $P_2''(x_1,\dots,x_n)$ are coprime divisors of $P(x_1,\dots,x_n)$. Therefore, if (1.3) does not hold for these divisors, it follows from (1.9) that

$$(1.11)\qquad |A|\cdot|R_1(\theta_1,\dots,\theta_n)|^{h_1}\cdots|R_{i-1}(\theta_1,\dots,\theta_n)|^{h_{i-1}}>\varepsilon^{1/3}.$$

Multiplying (1.10) and (1.11), we obtain

$$(1.12)\quad |A|\cdot|R_1(\theta_1,\dots,\theta_n)|^{h_1}\cdots|R_{i-1}(\theta_1,\dots,\theta_n)|^{h_{i-1}}$$
$$\times|R_{i+1}(\theta_1,\dots,\theta_n)|^{h_{i+1}}\cdots|R_k(\theta_1,\dots,\theta_n)|^{h_k}>\varepsilon^{2/3}.$$

In view of (1.4), this means that

$$(1.13)\qquad |P(x_1,\dots,x_n)|>\varepsilon^{2/3}\cdot|R_i(\theta_1,\dots,\theta_n)|^{h_i}.$$

Using (1.1), we can rewrite the last inequality

$$(1.14)\qquad \varepsilon^{1/3}>|R_i(\theta_1,\dots,\theta_n)|^{h_i}.$$

In view of (1.4), inequality (1.14) proves (1.2) with $Q(x_1,\dots,x_n)=(R_i(x_1,\dots,x_n))^{h_i}$. Lemma 1.3 is proved.

Lemma 1.3 can be restated as follows:

PROPOSITION 1.4. *Assume there exists a polynomial* $P(x_1,\dots,x_n)\in\mathbf{Z}[x_1,\dots,x_n]$ *such that for a given* $(\theta_1,\dots,\theta_n)\in\mathbf{C}^n$ *we have*

$$|P(\theta_1,\dots,\theta_n)|\leqslant\varepsilon<1,$$

and suppose $t(P)\leqslant T$. *Then either there exists a polynomial* $Q(x_1,\dots,x_n)\in\mathbf{Z}[x_1,\dots,x_n]$ *which is a power of a* **Q**-*irreducible polynomial* $R(x_1,\dots,x_n)\in\mathbf{Z}[x_1,\dots,x_n]$, *i.e.* $Q(x_1,\dots,x_n)=(R(x_1,\dots,x_n))^s$, *where* $t(R)s\leqslant 2T$ *and* $t(Q)\leqslant 2T$, *and for which*

$$|Q(\theta_1,\dots,\theta_n)|\leqslant\varepsilon^{1/3}$$

or else there exist two coprime polynomials $P_1(x_1,\ldots,x_n)$, $P_2(x_1,\ldots,x_n) \in \mathbf{Z}[x_1,\ldots,x_n]$ *for which*

$$|P_1(\theta_1,\ldots,\theta_n)| \leq \varepsilon^{1/3}, \qquad |P_2(\theta_1,\ldots,\theta_n)| \leq \varepsilon^{1/3}$$

and $t(P_1)$, $t(P_2) \leq 2T$.

In this form, Proposition 1.4 follows from 1.3 and 1.2.

These preliminary facts enable us to begin our investigation.

We consider for simplicity the plane case, which is the most interesting one. We fix a point $\vec{\theta} = (\theta_1, \theta_2) \in \mathbf{C}^2$ such that θ_1, θ_2 are algebraically independent over $\mathbf{Q}$. Assume there exists a set of algebraic curves defined by the equations $P_H(x, y) = 0$, passing near $\vec{\theta}$, and satisfying the following conditions:

(C.1) For every natural number H, $H \geq c_5 \geq 1$ (c_5 being some constant) there exists a polynomial $P_H(x, y) \in \mathbf{Z}[x, y]$ such that

(1) $P_H(x, y) \not\equiv 0$;

(2) we have an estimate for the type of $P_H(x, y)$,

$$t(P_H) \leq c_6 H, \tag{1.15}$$

where c_6 is some constant;

(3) for some nonnegative, unbounded, monotone increasing function $\varphi(t)$: $t \in \mathbf{R}^+$, we have the estimate

$$|P_h(\theta_1, \theta_2)| < \exp(-H^{\mu}\varphi(H)) \tag{1.16}$$

for all $H > c_7 \geq 1$.

Of course, in the conditions (C.1) we are mainly interested in the inequalities (1.16). The constant μ occurring in the exponent of the right-hand side of (1.16) is of special interest. It is suitable for assertions like Theorem 0.7, but in the present paper its form is inessential. Incidentally, to eliminate the trivial case of the Dirichlet family 0.1 we may assume that $\mu \geq 3$. How can we bring the family of curves $\{P_H(x, y) = 0\colon H \in \mathbf{N}\}$ to a more convenient form? It would be ideal if all of the curves were irreducible and continued to satisfy the conditions (C.1). It is very easy to satisfy conditions (1) and (2) in (C.1); the situation regarding condition (3) is much more complicated. Since $P_H(x, y)$ may have a large number of irreducible components, $P_H = R_1 \cdots R_k$, it follows that the values of the R_i at the point $\vec{\theta}$ can be significantly larger than the value of P_H at $\vec{\theta}$. Accordingly, we introduce a partition of the numbers H (i.e. of the corresponding polynomials P_H) into "red" and "blue".

For the moment, we provisionally color "blue" those H for which there exists $P_H(x, y) \in \mathbf{Z}[x, y]$ satisfying conditions (1)–(3) of (C.1), perhaps with other constants.

We naturally color all remaining H "red". The problem, of course, is to transform the "red" polynomials into polynomials more interesting than simply elements of $\mathbf{Z}[x, y]$.

The main result of the subsection 1 is that for each "red" H we can choose a polynomial $P_H(x, y) \in \mathbf{Z}[x, y]$ satisfying conditions (1)–(3) of (C.1) such that $P_H(x, y)$ is coprime to $P_{H-1}(x, y)$ and to $P_{H+1}(x, y)$.

The corresponding precise assertion will be proved below.

To prove all of the necessary details in the construction of such "red" polynomials we need a simple auxiliary result. We give a general result which will also be used in subsequent papers.

LEMMA 1.5. *Suppose that* $\{P_1(x_1,\dots,x_n),\dots,P_m(x_1,\dots,x_n)\}$ *is a family of polynomials in* $\mathbf{C}[x_1,\dots,x_n]$ *having no nonconstant common divisor. Let* $R(x_1,\dots,x_n)$, $Q(x_1,\dots,x_n)$ *be any two polynomials in* $\mathbf{C}[x_1,\dots,x_n]$, *each of degree at most* d. *Then there exist rational integers* $A_1,\dots,A_m$, *not all zero, such that*

$$|A_i| \leq ((m-1)d+1)^{m-1}: \qquad 1 \leq i \leq m, \tag{1.17}$$

and the polynomial

$$A_1P_1(x_1,\dots,x_n) + \cdots + A_mP_m(x_1,\dots,x_n)$$

is coprime to $R(x_1,\dots,x_n)$ *and to* $Q(x_1,\dots,x_n)$.

PROOF. By two polynomials in $\mathbf{C}[x_1,\dots,x_n]$ being coprime we mean, of course, that they have no nonconstant common divisor.

Let $h_1(x_1,\dots,x_n),\dots,h_s(x_1,\dots,x_n)$ and $q_1(x_1,\dots,x_n),\dots,q_t(x_1,\dots,x_n)$ denote all of the distinct $\mathbf{C}$-irreducible divisors of $R(x_1,\dots,x_n)$ and $Q(x_1,\dots,x_n)$, respectively. Then, by definition of degree, $s \leq d(R)$ and $t \leq d(Q)$, i.e.

$$s \leq d, \qquad t \leq d. \tag{1.18}$$

For any rational integer a, $a \neq 0$, we denote by $P^a(x_1,\dots,x_n)$ the polynomial

$$P^a(x_1,\dots,x_n) = P_1(x_1,\dots,x_n) + aP_2(x_1,\dots,x_n) + \cdots + a^{m-1}P_m(x_1,\dots,x_n). \tag{1.19}$$

We will show that for some a, $a \neq 0$, $|a| \leq (m-1)d+1$, the polynomial $P^a(x_1,\dots,x_n)$ is coprime to $R(x_1,\dots,x_n)$ and to $Q(x_1,\dots,x_n)$. Assume this is not true for any a, $a \neq 0$, $|a| \leq (m-1)d+1$. Then, by definition of the h_i and q_j, for any a, $a \neq 0$, $|a| \leq (m-1)d+1$, the polynomial $P^a(x_1,\dots,x_n)$ is divisible in $\mathbf{C}[x_1,\dots,x_n]$ by one of the polynomials $h_i(x_1,\dots,x_n)$, or $q_j(x_1,\dots,x_n)$: $1 \leq i \leq s$, $1 \leq j \leq t$. We define a mapping F of the set $\{-(m-1)d-1,\dots,-1,1,\dots,(m-1)d+1\}$ into $\{1,\dots,s,s+1,\dots,s+t\}$ as follows: for $|a| \leq (m-1)d+1$, $a \neq 0$, we put

$$F(a) = \begin{cases} i, & \text{if } P^a \text{ is divisible by } h_i; \\ s+j, & \text{if } P^a \text{ is divisible by } q_j. \end{cases} \tag{1.20}$$

Since (1.18) holds, we have $s+t \leq 2d$, while the cardinality of the set $\{-(m-1)d-1,\dots,-1,1,\dots,(m-1)d+1\}$ is equal to $2(m-1)d+2$. Consequently, some element of $\{1,\dots,s+t\}$ has more than $m-1$ preimages. This means that for m distinct numbers a_i, $a_i \neq 0$, $|a_i| \leq (m-1)d+1$: $i = 1,\dots,m$, we have $F(a_1) = \cdots = F(a_m)$.

In view of (1.20), it follows that all of the polynomials $P^{a_i}(x_1,\ldots,x_n)$: $i = 1,\ldots,m$, are divisible by the same **C**-irreducible polynomial $r(x_1,\ldots,x_n)$ (equal to some $h_{i_0}(x_1,\ldots,x_n)$ or $q_{j_0}(x_1,\ldots,x_n)$).

Taking into account definition (1.19), we can write this information as a series of equalities:

(1.21)

$$P_1(x_1,\ldots,x_n)+\cdots+a_1^{m-1}P_m(x_1,\ldots,x_n)=r_1(x_1,\ldots,x_n)r(x_1,\ldots,x_n),$$
$$\vdots \qquad \vdots \qquad \vdots$$
$$P_1(x_1,\ldots,x_n)+\cdots+a_m^{m-1}P_m(x_1,\ldots,x_n)=r_m(x_1,\ldots,x_n)r(x_1,\ldots,x_n),$$

where $r_1(x_1,\ldots,x_n),\ldots,r_m(x_1,\ldots,x_n) \in \mathbf{C}[x_1,\ldots,x_n]$.

We view (1.21) as a system of equations in $\mathbf{C}[x_1,\ldots,x_n]$ relative to the unknowns $P_1(x_1,\ldots,x_n),\ldots,P_m(x_1,\ldots,x_n)$.

The determinant of this system is a Vandermonde determinant:

$$\begin{vmatrix} 1 & a_1 & a_1^2 & \cdots & a_1^{m-1} \\ 1 & a_2 & a_2^2 & \cdots & a_2^{m-1} \\ \vdots & \vdots & \vdots & \cdots & \vdots \\ 1 & a_m & a_m^2 & \cdots & a_m^{m-1} \end{vmatrix} \neq 0,$$

since all of the a_i are distinct. Therefore, solving (1.21) by Cramer's rule, we obtain a series of equalities

$$(1.22) \qquad \begin{aligned} P_1(x_1,\ldots,x_n) &= S_1(x_1,\ldots,x_n)\cdot r(x_1,\ldots,x_n), \\ &\vdots \\ P_m(x_1,\ldots,x_n) &= S_m(x_1,\ldots,x_n)\cdot r(x_1,\ldots,x_n), \end{aligned}$$

where $S_i(x_1,\ldots,x_n) \in \mathbf{C}[x_1,\ldots,x_n]$: $i = 1,\ldots,m$. But this means that the family of polynomials $\{P_1(x_1,\ldots,x_n),\ldots,P_m(x_1,\ldots,x_n)\}$ has common divisor $r(x_1,\ldots,x_n)$. This contradicts our hypothesis.

Therefore, for some rational integer a, $a \neq 0$, $|a| \leqslant (m-1)d+1$, the polynomial $P^a(x_1,\ldots,x_n)$ in (1.19) is coprime to $R(x_1,\ldots,x_n)$ and to $Q(x_1,\ldots,x_n)$. Since

$$\max\{1,|a|,\ldots,|a|^{m-1}\} \leqslant ((m-1)d+1)^{m-1},$$

the lemma is proved with $A_i = a^{i-1}$: $i = 1,\ldots,m$.

REMARK 1.6. The estimate $|A_i| \leqslant ((m-1)d+1)^{m-1}$ is quite excessive; it is satisfactory only in the case $m = 2$ used below. In fact, the bound for $\max(|A_i|: 1 \leqslant i \leqslant m)$ can be chosen in the form $O(d^{1/m})$; the corresponding result is easily obtained.

We give a corollary of 1.5 in the case $m = 2$ which interests us.

COROLLARY 1.7. *Suppose* $P_1(x_1,\dots,x_n)$, $P_2(x_1,\dots,x_n)$ *are two coprime polynomials in* $\mathbf{C}[x_1,\dots,x_n]$. *If* $R(x_1,\dots,x_n)$, $Q(x_1,\dots,x_n)$ *are arbitrary polynomials, there exist rational integers* a, b *such that* $|a|+|b|\neq 0$, $\max(|a|,|b|)\leqslant \max(d(R), d(Q))$, *and the polynomial* $aP_1(x_1,\dots,x_n)+bP_2(x_1,\dots,x_n)$ *is coprime to* $R(x_1,\dots,x_n)$ *and to* $Q(x_1,\dots,x_n)$.

We now make a preliminary coloring of the numbers H, $H\geqslant c_5$.

DEFINITION 1.8. A natural number H, $H\geqslant c_5$, is called blue ($H\in B$) if the polynomial $P_H(x,y)$ in (C.1) has a divisor $P'_H(x,y)$ which is a power of a $\mathbf{Q}$-irreducible polynomial and which satisfies, instead of (1.16), the inequality

$$|P'_H(x,y)|<\exp(-\tfrac{1}{3}H^{\mu}\varphi(H)). \tag{1.23}$$

Otherwise, H is called red ($H\in R$). We obtain a partition of $N\setminus\{1,\dots,[c_5]\} = R\cup B$ into two disjoint subsets.

In accordance with this definition, we will prove in the next section a theorem on the representation of a dense sequence (C.1) as a colored sequence. As we have already mentioned, the purpose of such a result will be explained in §3.

2. Existence theorem for colored sequences. In this section we prove the main result of this section of the chapter.

THEOREM 2.1. *Suppose there exists a family of polynomials* $\{P_H(x,y)\colon H\geqslant c_5\}$ *satisfying conditions* (1)–(3) *of* (C.1). *Then there exists a family of polynomials* $\{R_H(x,y)\colon H\geqslant c_8\}$ *in* $\mathbf{Z}[x,y]$ *satisfying the following conditions*:

(C.2) *For every natural number* $H\geqslant c_8$, $R_H(x,y)$ *is a polynomial in* $\mathbf{Z}[x,y]$ *and*

(1) $R_H(x,y)\not\equiv 0$;

(2) *we have an estimate for the type of* $R_H(x,y)$,

$$t(R_H)\leqslant 4c_6H=c_9H; \tag{2.1}$$

(3) *for the nonnegative unbounded function* $\varphi(t)\colon t\in\mathbf{R}^+$, *in* (C.1) *we have the estimate*

$$|R_H(\theta_1,\theta_2)|<\exp(-\tfrac{1}{4}H^{\mu}\varphi(H)) \tag{2.2}$$

for all $H\geqslant c_{10}$;

(4) *for any* $H\geqslant c_8+1$, *either* $R_H(x,y)$ *is a power of a* $\mathbf{Q}$*-irreducible polynomial or else* $R_H(x,y)$ *is coprime to* $R_{H-1}(x,y)$ *and to* $R_{H+1}(x,y)$. *In the first case, H is called "blue", and in the second case "red".*

In particular, a family of polynomials $\{R_H(x,y)\colon H\geqslant c_8\}$ in $\mathbf{Z}[x,y]$ satisfying (1)–(3) of (C.2) also satisfies (1)–(3) of (C.1), if we replace c_5 by c_8, c_6 by c_9, c_7 by c_{10}, and $\varphi(t)$ by $\frac{1}{4}\varphi(t)$.

PROOF OF THEOREM 2.1. Assume there exists a family of polynomials $\{P_H(x,y)\colon H\geqslant c_5\}$ satisfying (1)–(3) of (C.1).

Consider Definition 1.8. We will show that the "blue" and "red" natural numbers in the sense of 1.8 are, respectively, "blue" and "red" in the sense of condition (4) of (C.2).

We begin with the definition of the constants c_8, c_9, c_{10}. Since $\varphi(t)$: $t \in \mathbf{R}^+$, is a nonnegative, unbounded, monotone increasing function, there exists $H_0 \geqslant c_5$ such that for $H \geqslant H_0$ we have $\varphi(H) \geqslant 48(c_6 + 1)$ and $H \cdot c_6 \geqslant 48(c_6 + 1)$. Then we put $c_8 = c_{10} = H_0$.

We can now already define the polynomials $R_H(x, y) \in \mathbf{Z}[x, y]$ for $H \geqslant c_8$. The definition of $R_H(x, y)$ is the simplest when H is "blue" in the sense of 1.8. Suppose H is "blue", i.e. $H \in B$. According to 1.8, there exists a polynomial $Q(x, y) \in \mathbf{Z}[x, y]$ which is a divisor of $P_H(x, y)$ and is a power of a $\mathbf{Q}$-irreducible polynomial and which satisfies the estimate

(2.3) $$|Q(\theta_1, \theta_2)| < \exp(-\tfrac{1}{3}H^\mu\varphi(H)).$$

In this case we put $Q(x, y) = R_H(x, y)$, i.e. we define $R_H(x, y)$ by means of the equality

(2.4) $$R_H(x, y) \equiv Q(x, y), \quad \text{if } H \in B.$$

Clearly, condition (1) is satisfied; condition (4) is also satisfied, since $Q(x, y)$ is a power of a $\mathbf{Q}$-irreducible polynomial; condition (3) is satisfied because of (2.3). It remains to verify (2). According to condition (2) of (C.1), we have $t(P_H) \leqslant c_6 H$, and $Q(x, y)$ is a divisor of $P_H(x, y)$. Thus, Corollary 1.2 yields the estimate $t(R_H) = t(Q) \leqslant 2c_6 H < 4c_6 H = c_9 H$. Therefore, the polynomial $R_H(x, y)$ defined by (2.4) satisfies conditions (1)–(4) of (C.2). This takes care of the case where H is "blue" ($H \in B$) in the sense of 1.8.

We now turn to the case where H is "red" ($H \in R$) in the sense of 1.8. This means that $H \notin B$, i.e. for no divisor of $P_H(x, y)$ which is a power of an irreducible polynomial in $\mathbf{Z}[x, y]$ does (2.3) hold. We apply Lemma 1.3 in the case $n = 2$, $P(x, y) \equiv P_H(x, y)$, and $\varepsilon = \exp(-H^\mu\varphi(H))$. We have $\varepsilon < 1$, since $H \geqslant 1$ and $\varphi(H) > 0$. In view of Lemma 1.3 and the fact that $H \notin B$, there exist two coprime divisors of $P_H(x, y)$, namely polynomials $P_1^H(x, y)$ and $P_2^H(x, y)$ in $\mathbf{Z}[x, y]$, such that

(2.5) $$|P_1^H(\theta_1, \theta_2)| \leqslant \exp(-\tfrac{1}{3}H^\mu\varphi(H)); \quad |P_2^H(\theta_1, \theta_2)| \leqslant \exp(-\tfrac{1}{3}H^\mu\varphi(H)).$$

The rest of the construction of the polynomials $\{R_H(x, y): H \geqslant c_8\}$ is effected by induction on the natural numbers $H \geqslant c_8$.

First, for all "blue" H ($H \in B$) we define, as in (2.4), $R_H(x, y) \in \mathbf{Z}[x, y]$ so that all conditions in (C.2) are satisfied.

Assume that for all "red" H', i.e. $H' \notin B$, such that $c_8 \leqslant H' < H$ we have constructed polynomials $R_{H'}(x, y) \in \mathbf{Z}[x, y]$ satisfying conditions (1)–(4) in (C.2).

We will show that for $H \geqslant c_8$ there exists a polynomial $R_H(x, y) \in \mathbf{Z}[x, y]$ also satisfying conditions (1)–(4) in (C.2).

Suppose first that $H \in B$. Then a polynomial $R_H(x, y)$ satisfying (C.2) has been defined above. We therefore consider the case where H is "red" ($H \in R$, $H \notin B$) and $H \geqslant c_8$. We use further the proof of Corollary 1.7. The construction of the polynomial $R_H(x, y)$ must take into account the already existing "blue" polynomials. We therefore distinguish two cases.

First case. The number $H+1$ is "red", i.e. the polynomial $R_{H+1}(x, y)$ has not yet been constructed.

Consider two coprime polynomials $P_1^H(x, y)$, $P_2^H(x, y) \in \mathbf{Z}[x, y]$ which are divisors of $P_H(x, y)$ and satisfy (2.5). It follows at once from Lemma 1.1 or Corollary 1.2 that we have the following estimate for the types of $P_1^H(x, y)$ and $P_2^H(x, y)$:

$$\max(t(P_1^H), t(P_2^H)) \leqslant t(P_1^H) + t(P_2^H) \leqslant 2t(P_H) \leqslant 2c_6 H. \tag{2.6}$$

Applying Corollary 1.7, we obtain two rational integers a, b, not both zero, such that the polynomial

$$aP_1^H(x, y) + bP_2^H(x, y) \tag{2.7}$$

is coprime to $R_{H-1}(x, y)$ and

$$\max(|a|, |b|) \leqslant d + 1, \tag{2.8}$$

where $d = d(R_{H-1})$. Note that in the case $H = [c_8 + 1]$ we can replace $R_{H-1}(x, y)$ by a unit (constant) polynomial. It is now clear that as $R_H(x, y)$ we can choose the polynomial (2.7):

$$R_H(x, y) \equiv aP_1^H(x, y) + bP_2^H(x, y). \tag{2.9}$$

Since $|a| + |b| \neq 0$ and $P_1^H(x, y)$, $P_2^H(x, y)$ are coprime, we have $R_H(x, y) \not\equiv 0$, i.e. condition (1) in (C.2) is satisfied. It is very easy to estimate the type of $R_H(x, y)$:

$$\begin{aligned} t(R_H) &\leqslant t(P_1^H) + t(P_2^H) + 2\ln(\max(|a|, |b|)) \\ &\leqslant t(P_1^H) + t(P_2^H) + 2\ln(d(R_{H-1}) + 1) \\ &\leqslant t(P_1^H) + t(P_2^H) + 2\ln(4c_6(H-1) + 1). \end{aligned}$$

Since $Hc_6 \geqslant 48$ for $H \geqslant H_0$ (see (1.15) and condition (2) in (C.1)), it follows that $\ln(4c_6(H-1)+1) \leqslant c_6 H$. Therefore,

$$t(P_1^H) + t(P_2^H) + 2c_6 H \geqslant t(R_H). \tag{2.10}$$

If we take into account (2.6), inequality (2.10) yields the estimate required by condition (2) in (C.2):

$$t(R_H) \leqslant 4c_6 H = c_9 H. \tag{2.11}$$

Condition (3) also holds. Indeed,

$$\begin{aligned} |R_H(\theta_1, \theta_2)| &\leqslant \max(|a|, |b|) \cdot \max(|P_1^H(\theta_1, \theta_2)|, |P_2^H(\theta_1, \theta_2)|) \\ &\leqslant (d(R_{H-1}) + 1) \cdot \max(|P_1^H(\theta_1, \theta_2)|, |P_2^H(\theta_1, \theta_2)|). \end{aligned}$$

Using (2.5), we obtain

$$|R_H(\theta_1, \theta_2)| \leqslant \exp\left(\ln(d(R_{H-1}) + 1) - \tfrac{1}{3}H^\mu \varphi(H)\right). \tag{2.12}$$

But for $H \geqslant c_{10}$ we have, by definition of c_{10}, $\varphi(H) \geqslant 48(1 + c_6)$, and since $\mu \geqslant 3$, then for $H \geqslant c_{10}$ we have

$$\tfrac{1}{12}H^\mu \varphi(H) \geqslant \tfrac{48}{12}H^\mu c_6 > 4c_6 H > \ln(4c_6 H) \geqslant \ln(d(R_{H-1}) + 1).$$

From this and (2.12) we at once obtain

$$(2.13) \qquad |R_H(\theta_1, \theta_2)| < \exp(-\tfrac{1}{4}H^\mu \varphi(H)).$$

Thus, condition (3) in (C.2) holds. Finally, condition (4) in (C.2) is satisfied because $R_H(x, y)$ and $R_{H-1}(x, y)$ are coprime and $H + 1$ is not "blue". Since $H + 1$ is "red", $R_{H+1}(x, y)$ is coprime to $R_H(x, y)$ by construction. This takes care of the first case.

Second case. Assume that $H + 1$ is "blue", i.e. a polynomial $R_{H+1}(x, y) \in \mathbf{Z}[x, y]$ satisfying (1)–(4) in (C.2) already exists. The proof in this case does not differ from that of the preceding one. We again apply Corollary 1.7 to a pair of coprime polynomials $P_1^H(x, y), P_2^H(x, y) \in \mathbf{Z}[x, y]$ satisfying (2.5) and (2.6).

Putting $d = \max(d(R_{H-1}), d(R_{H+1}))$, we obtain two rational integers a, b, not both zero, such that the polynomial

$$(2.14) \qquad aP_1^H(x, y) + bP_2^H(x, y)$$

is coprime to $R_{H-1}(x, y)$ and to $R_{H+1}(x, y)$, with

$$(2.15) \qquad \max(|a|, |b|) \leqslant d + 1 = \max(d(R_{H-1}), d(R_{H+1})) + 1.$$

We put $R_H(x, y)$ equal to the polynomial (2.14):

$$(2.16) \qquad R_H(x, y) \equiv aP_1^H(x, y) + bP_2^H(x, y).$$

Then $R_H(x, y) \not\equiv 0$, since $P_1^H(x, y)$, $P_2^H(x, y)$ are coprime. We can estimate the type of $R_H(x, y)$ as follows:

$$\begin{aligned} t(R_H) &\leqslant t(P_1^H) + t(P_2^H) + 2\max(\ln(d(R_{H-1}) + 1), \ln(d(R_{H+1}) + 1)) \\ &\leqslant t(P_1^H) + t(P_2^H) + 2\max(\ln(4c_6(H-1) + 1), \ln(2c_6(H+1) + 1)). \end{aligned}$$

However, as we already know,

$$\ln(4c_6(H-1) + 1) \leqslant c_6 H \quad \text{and} \quad \ln(2c_6(H+1) + 1) \leqslant c_6 H$$

for any $H > 1$, inasmuch as $\ln(3n + 1) \leqslant n$ for all $n \geqslant 2$. Therefore,

$$(2.17) \qquad t(R_H) \leqslant t(P_1^H) + t(P_2^H) + 2c_6 H.$$

It follows from (2.6) and (2.17) that

$$(2.18) \qquad t(R_H) \leqslant 4c_6 H,$$

i.e. condition (2) in (C.2) is proved. In a similar way we can establish condition (3) in (C.2):

$$|R_H(\theta_1, \theta_2)| \leqslant \max(|a|, |b|) \cdot \max(|P_1^H(\theta_1, \theta_2)|, |P_2^H(\theta_1, \theta_2)|).$$

Applying (2.5) and (2.15), we obtain

(2.19)

$$|R_H(\theta_1, \theta_2)| \leqslant \exp(\max(\ln(d(R_{H-1}) + 1), \ln(d(R_{H+1}) + 1)) - \tfrac{1}{3}H^\mu \varphi(H)).$$

Since $\varphi(H) \geqslant 48c_6$ for $H \geqslant c_{10}$, it follows that for $H \geqslant c_{10}$ we have

$$\begin{aligned}\tfrac{1}{12}H^{\mu}\varphi(H) &\geqslant 4c_6 H^{\mu} \geqslant 4c_6 H \\ &> \ln(4c_6(H-1)+1) + \ln(2c_6(H+1)+1) \\ &\geqslant \ln(d(R_{H-1})+1) + \ln(d(R_{H+1})+1).\end{aligned}$$

Therefore, (2.19) means that for $H \geqslant c_{10}$ we have

$$|R_H(\theta_1, \theta_2)| < \exp(-\tfrac{1}{4}H^{\mu}\varphi(H)) \tag{2.20}$$

and condition (3) in (C.2) holds. Finally, by construction, $R_H(x, y)$ is coprime to $R_{H-1}(x, y)$ and to $R_{H+1}(x, y)$. Thus, all conditions in (C.2) are satisfied.

So for all "red" $H \geqslant c_8$ we have constructed polynomials $R_H(x, y) \in \mathbf{Z}[x, y]$ satisfying conditions (1)–(4) in (C.2). Theorem 2.1 is proved.

For certain technical reasons it is convenient to express Theorem 2.1 in a somewhat different form, in which the separation of polynomials into "blue" and "red", as in condition (4) of (C.2), is eliminated. The formulation of the following theorem is a little reminiscent of the works of Davenport and Schmidt [**20, 21**], since the problems considered here and in [**20, 21**] are similar in spirit.

Intuitively, the idea of the assertion given below can be explained as follows. Suppose there exists a sequence of polynomials $\{R_H(x, y)\colon H \geqslant c_8\}$ satisfying (C.2). We seek to extract from the natural numbers $H \geqslant c_8 \geqslant 1$ an infinite sequence X_n: $n = 1, 2, \ldots$, such that

(a) the polynomials $R_{X_n}(x, y)$ and $R_{X_{n+1}}(x, y)$ are coprime for any $n = 1, 2, 3, \ldots$,

(b) for any $n = 1, 2, \ldots$, if $X_n + 1 < X_{n+1}$, then $R_{X_n}(x, y)$ is a power of an irreducible polynomial in $\mathbf{Z}[x, y]$, and for $X_n < H < X_{n+1}$ every $R_H(x, y)$ is a power of the same polynomial.

This goal is realized by the following

THEOREM 2.2. *Suppose there exists a family of polynomials $\{P_H(x, y)\colon H \geqslant c_5\}$ in $\mathbf{Z}[x, y]$ satisfying conditions* (1)–(3) *in* (C.1). *Then there exists a family of polynomials $\{R_H(x, y)\colon H \geqslant c_8\}$ in $\mathbf{Z}[x, y]$ satisfying the following conditions*:

(C.3) *For every natural number $H \geqslant H_0 = [c_8 + 1]$ the conditions* (1)–(4) *in* (C.2) *are satisfied. Moreover, there exists an infinite increasing sequence X_n: $n < \infty$, of natural numbers such that*

$$H_0 = X_0 < X_1 < X_2 < \cdots < X_n < \cdots : \qquad n < \infty, \tag{2.21}$$

and also

(i) *for any $n = 0, 1, 2, \ldots$ the polynomials $R_{X_n}(x, y)$, $R_{X_{n+1}}(x, y)$ are coprime,*

(ii) *if $X_n + 1 < X_{n+1}$, then $R_{X_n}(x, y)$ is a power of a $\mathbf{Q}$-irreducible polynomial $Q_{X_n}(x, y) \in \mathbf{Z}[x, y]$,*

(iii) *if $X_n < H < X_{n+1}$, then $R_H(x, y)$ is also a power of the polynomial $Q_{X_n}(x, y)$ in* (ii),

(iv) *for any $n = 0, 1, 2, \ldots$ we have inequalities connecting the type $t(R_{X_n})$ of the polynomial $R_{X_n}(x, y)$ with $|R_{X_n}(\theta_1, \theta_2)|$,*

$$t(R_{X_n}) \leqslant c_9 X_n, \tag{2.22}$$

(2.23) $$|R_{X_n}(\theta_1, \theta_2)| < \exp(-\tfrac{1}{4}X_n^{\mu}\varphi(X_n)),$$

(v) *for any* $n = 0, 1, 2, \ldots$ *there exists a natural number* $s_n \geqslant 1$ *such that the following inequalities connecting* X_n *with* X_{n+1} *hold,*

(2.24) $$s_n \cdot t(R_{X_n}) \leqslant c_{11}X_{n+1},$$

(2.25) $$|R_{X_n}(\theta_1, \theta_2)| < \exp(-\tfrac{1}{4}(X_{n+1} - 1)^{\mu}\varphi(X_{n+1} - 1)/s_n),$$

where c_{11} *is a constant,* $c_{11} = 18c_9$.

REMARK 2.3. Intuitively, s_n can be thought of as that power to which it is necessary to raise $R_{X_n}(x, y)$ in order to obtain $R_{X_{n+1}-1}(x, y)$ when $X_n \leqslant X_{n+1} - 1$ (see (iii)).

PROOF OF THEOREM 2.2. Suppose there exists a family $\{P_H(x, y)\colon H \geqslant c_5\}$ satisfying conditions (1)–(3) in (C.1). Then, by Theorem 2.1, there exists a family of polynomials $\{R_H(x, y)\colon H \geqslant c_8\}$ satisfying conditions (1)–(4) in (C.2). We retain the constants c_8, c_9, c_{10} introduced in 2.1, whose meanings were explained at the beginning of the proof of Theorem 2.1.

A sequence X_n: $n = 0, 1, 2, \ldots,$ of natural numbers satisfying conditions (i)–(v) in (C.3) will be constructed by induction. First, we put

(2.26) $$X_0 = H_0 = [c_8 + 1].$$

Then conditions (i)–(iii) and (v) are trivial, and (iv) is a reformulation of conditions (2) and (3) in (C.2).

Let us now assume that we have constructed a finite sequence

(2.27) $$H_0 = X_0 < X_1 < \cdots < X_m$$

of natural numbers, $m \geqslant 0$, such that conditions (i)–(v), including inequalities (2.22)–(2.25), are satisfied for all $n \leqslant m$. We will now construct $X_{m+1} > X_m$ to extend the sequence (2.27). We will first show that there exists a natural number $Y > X_m$ such that the following conditions are satisfied:

(2.28) the polynomial $R_Y(x, y)$ is coprime to $R_{X_m}(x, y)$.

Indeed, suppose that no $Y > X_m$ satisfies (2.28). Apply condition (4) of (C.2). We obtain for $Y = X_m + 1$ that $R_{X_m+1}(x, y)$ is a power of a **Q**-irreducible polynomial $T_{X_m+1}(x, y) \in \mathbf{Z}[x, y]$ and that $R_{X_m}(x, y)$ is also a power of a **Q**-irreducible polynomial $T_{X_m}(x, y) \in \mathbf{Z}[x, y]$. Moreover, since $R_{X_m}(x, y)$, $R_{X_m+1}(x, y)$ are not coprime, $T_{X_m}(x, y) \equiv T_{X_m+1}(x, y)$. Applying condition (4) in (C.2) with $Y = X_m + 2$ and so on, we obtain that all of the polynomials

$$R_{X_m}(x, y), R_{X_m+1}(x, y), \ldots, R_H(x, y)$$

with $H > X_m$ are powers of the same irreducible polynomial $T_{X_m}(x, y)$. Again, using the negation of (2.28) with $Y = H + 1$ and condition (4) in (C.2), we see that $R_{H+1}(x, y)$ is a power of $T_{X_m}(x, y)$. Finally, the polynomials

$$R_{X_m}(x, y), R_{X_m+1}(x, y), \ldots, R_H(x, y), \ldots$$

for all $H > X_m$ are powers of the polynomial $T_{X_m}(x, y) \equiv T_0(x, y) \in \mathbf{Z}[x, y]$. Then for $H > X_m$ we have

$$R_H(x, y) \equiv (T_0(x, y))^{m_H} \tag{2.29}$$

where $m_H \geqslant 1$. Using Corollary 1.2 and condition (2) in (C.2), we deduce from (2.29) that for $H > X_m$ we have

$$m_H t(T_0) \leqslant 2c_9 H. \tag{2.30}$$

From (2.29) and condition (3) in (C.2) it also follows that for all $H > X_m$ we have

$$|T_0(\theta_1, \theta_2)| < \exp(-\tfrac{1}{4}H^\mu \varphi(H)/m_H). \tag{2.31}$$

Combining (2.30) and (2.31), we obtain for $H > X_m$ the inequality

$$|T_0(\theta_1, \theta_2)| < \exp(-\tfrac{1}{4}H^\mu \varphi(H) t(T_0)/2c_9 H). \tag{2.32}$$

But $\mu > 3$ and $\lim_{H\to\infty} \varphi(H) = \infty$. Consequently, if we pass to the limit as $H \to \infty$ in (2.32), we obtain $T_0(\theta_1, \theta_2) = 0$, and since $T_0(x, y) \not\equiv 0$ in view of (2.29) and condition (1) in (C.2), it follows that θ_1, θ_2 are algebraically independent over $\mathbf{Q}$, despite the assumption of (C.1).

Thus, our assumption that (2.28) is false for all $Y > X_m$ leads to a contradiction, so there exists some $Y > X_m$ satisfying (2.28). Denote by X_{m+1} the smallest $Y = X_{m+1} > X_m$ satisfying (2.28).

Such an $X_{m+1} < \infty$, $X_m < X_{m+1}$, exists. We will show that it satisfies our requirements.

First, by definition, $X_{m+1} = Y$ satisfies (2.28). This proves property (i).

Let us prove (ii) and (iii). Suppose $X_{m+1} > X_m + 1$. Since X_{m+1} is the smallest $Y > X_m$ satisfying (2.28), it follows that $Y = X_m + 1$ does not satisfy (2.28). Consequently, by property (4) in (C.2), the polynomial $R_{X_m}(x, y)$ is a power of a $\mathbf{Q}$-irreducible polynomial $Q_{X_m}(x, y) \in \mathbf{Z}[x, y]$. Thus, (ii) has been established. Suppose, however, that (iii) does not hold. Let H denote the smallest natural number, $X_m < H < X_{m+1}$, such that $R_H(x, y)$ is not a power of $Q_{X_m}(x, y)$. Then $R_{H-1}(x, y)$ is a power of $Q_{X_m}(x, y)$. Since $Y = H < X_{m+1}$ does not satisfy (2.28), $R_H(x, y)$ is not coprime to $R_{X_m}(x, y)$, i.e. to $Q_{X_m}(x, y)$. Therefore, $R_H(x, y)$, $R_{H-1}(x, y)$ have a common divisor. According to condition (4) in (C.2), this means that $R_H(x, y)$ is a powser of an irreducible polynomial $Q'(x, y)$. But $R_H(x, y)$ and a power of $Q_{X_m}(x, y)$ have a common divisor, and therefore $Q'(x, y) \equiv Q_{X_m}(x, y)$. Property (iii) is proved.

Property (iv), as we have already mentioned, is a reformulation of properties (2) and (3) in (C.2), and inequalities (2.22) and (2.23) are equivalent to (2.1) and (2.2), respectively.

It remains to prove (v). First note that if $X_{m+1} = X_m + 1$, then (2.24) and (2.25) with $s_m = 1$ follow from (2.22) and (2.23) for any $c_{11} \geqslant c_9$. We therefore consider the case $X_{m+1} > X_m + 1$ (see (ii) and (iii)). According to (ii),

$$R_{X_m}(x, y) = (Q_{X_m}(x, y))^q \tag{2.33}$$

for some natural number $q \geqslant 1$. Also, applying (iii) with $H = X_{m+1} - 1$, we obtain

$$R_H(x, y) = \left(Q_{X_m}(x, y)\right)^p \tag{2.34}$$

for $H = X_{m+1} - 1 > X_m$, where p is a natural number, $p \geqslant 1$. Now let $s_m \geqslant 1$ denote the smallest natural number larger than p/q. In other words,

$$p/q \leqslant s_m, \qquad s_m - p/q < 1. \tag{2.35}$$

We will show that (2.24) and (2.25) are satisfied with $n = m$ for such a choice of s_m. Indeed, from (2.33) and Lemma 1.1 we obtain

$$\tfrac{1}{3}qt(Q_{X_m}) \leqslant t(R_{X_m}) \leqslant 3qt(Q_{X_m}). \tag{2.36}$$

Similarly, from (2.34) and also 1.1 and 1.2 we obtain

$$\tfrac{1}{3}pt(Q_{X_m}) \leqslant t(R_H) \tag{2.37}$$

with $H = X_{m+1} - 1$.

It follows from (2.36) that

$$s_m t(R_{X_m}) \leqslant 3 s_m qt(Q_{X_m}). \tag{2.38}$$

In view of (2.35) and (2.38),

$$s_m t(R_{X_m}) \leqslant 3(p/q + 1)qt(Q_{X_m})$$

if $p \geqslant q$, or

$$s_m t(R_{X_m}) \leqslant 3qt(Q_{X_m})$$

if $p < q$.

In any event, therefore,

$$s_m t(R_{X_m}) \leqslant \max\{6p, 3q\}t(Q_{X_m}). \tag{2.39}$$

From (2.36) and (iv) we obtain

$$3qt(Q_{X_m}) \leqslant 9t(R_{X_m}) \leqslant 9c_9 X_m < 9c_9 X_{m+1}; \tag{2.40}$$

from (2.37) and condition (2) in (C.2) we obtain

$$6pt(Q_{X_m}) \leqslant 18t(R_H) \leqslant 18c_9(X_{m+1} - 1). \tag{2.41}$$

Therefore, putting $c_{11} = 18c_9 \geqslant c_9, c_6$ (since $c_9 = 4c_6 > c_6$), we obtain by combining (2.39)–(2.41) that

$$s_m t(R_{X_m}) \leqslant c_{11} X_{m+1}. \tag{2.42}$$

But (2.42) is equivalent to (2.24). Let us prove (2.25).

In view of (2.33) and (2.34), we obtain

$$|R_{X_m}(\theta_1, \theta_2)| = |R_H(x, y)|^{q/p} \tag{2.43}$$

with $H = X_{m+1} - 1$. It follows from (2.43) and condition (3) of (C.2) that

$$|R_{X_m}(\theta_1, \theta_2)| < \exp\left(-q/4p \cdot (X_{m+1} - 1)^\mu \varphi(X_{m+1} - 1)\right). \tag{2.44}$$

But according to (2.35), $s_m \geqslant p/q$, i.e. $q/p \geqslant 1/s_m$, i.e. $-q/4p \leqslant -1/4s_m$. Therefore, (2.44) yields

$$(2.45) \qquad |R_{X_m}(\theta_1, \theta_2)| < \exp\left(-\frac{1}{4s_m}(X_{m+1}-1)^{\mu}\varphi(X_{m+1}-1)\right).$$

Thus, (v) is proved. Consequently, the X_{m+1} constructed above satisfies conditions (i)–(v), and the sequence of numbers (2.27), satisfying (C.3), can be extended indefinitely. Theorem 2.2 is proved.

It is in this form that we will use the sequences of polynomials which we have called colored.

3. Some words of explanation. In this section we will explain why we need a result like Theorem 2.1 or 2.2, which describes properties of dense sequences satisfying the conditions (C). In the first place, dense sequences satisfying (C.1) arise in studies related to Schanuel's conjecture in the style of [**3**, **5**–**7**].

For example, for the Schanuel set $\{\beta_j, e^{\alpha_i\beta_j}\}$, where $\alpha_1,\ldots,\alpha_N$ and $\beta_1,\ldots,\beta_M$ are two collections of linearly independent numbers, there arise sequences of type (C.1) for polynomials in n variables, where $n = \deg \mathbf{Q}(\beta_j, e^{\alpha_i\beta_j})$. As a parameter in this case there appears the well-known (see [**3**, **7**]) invariant connecting M and N:

$$\mu = \kappa_2 = \frac{M(N+1)}{M+N}.$$

A sequence with $\mu = \kappa_2$ satisfying (C.1) in conjunction with Theorem 0.7 was used to construct the first examples of numbers of the form α^β among which three were algebraically independent. For results of this type see Brownawell [**3**, **14**].

These results were then strengthened by the author [**6**], where instead of (C.1) he used much more precise conditions [**12**]. These conditions also lead to colored sequences of polynomials and will be studied later in the general case.

Theorem 0.7 turns out to be useful in studying algebraic independence of four numbers. Using a method of the author [**6**], Waldschmidt [**16**] gave an example of a set of numbers of the form α^β containing four that are algebraically independent. Even though this example is worse than the results of [**7**, **8**], it is of interest because of the relative simplicity of the reasoning involved.

Therefore, it is natural to expect that theorems like 0.7 in conjunction with stronger algebraic considerations would lead, in particular, to new results on the algebraic independence of numbers of the form α^β. These applications to the measures of algebraic independence of values of elliptic and Abelian functions appear in [**27**, **28**].

There is another reason for studying colored sequences. In addition to Schanuel's conjecture there are other problems pertaining to the algebraic independence of values of meromorphic functions. For example, there is an analogue of Schanuel's conjecture for values of elliptic functions and Abelian integrals. In this case sequences of type (C.1) exist after the analytic part of the corresponding proofs.

Therefore, investigations of sequences satisfying (C.1)–(C.3) and then sequences satisfying more complicated conditions give additional information for the study of elliptic functions. It is indeed the case, as the studies in later papers [**22**–**26**] show.

Thus we see that an investigation of the sequences we have provisionally called "colored" is important for various problems in the theory of transcendental numbers.

REFERENCES

1. J. W. S. Cassels, *An introduction to Diophantine approximation*, Cambridge Univ. Press, Cambridge, 1957.

2. S. Lang, *Algebra*, Addison-Wesley, Reading, Mass., 1965.

3. W. D. Brownawell, *Gelfond's method for algebraic independence*, Trans. Amer. Math. Soc. **210** (1975), 1–26.

4. R. Tijdeman, *On the algebraic independence of certain numbers*, Indag. Math. **33** (1971), 146–162.

5. A. D. Gelfond, *Transcendental and algebraic numbers*, Dover, New York, 1960.

6. G. V. Chudnovsky, *Algebraic independence of several values of the exponential function*, Mat. Notes **15** (1974), 391–398.

7. ______, *Some analytic methods in the theory of transcendental numbers*, Inst. of Math. Ukrain. SSR Acad. Sci., Preprint IM-74-8, Kiev, 1974, pp. 1–48; and see this volume, Chapter 1.

8. ______, *Analytical methods in Diophantine approximations*, Inst. of Math. Ukrain. SSR Acad. Sci., Preprint IM-74-9, Kiev, 1974, pp. 1–52; and see this volume, Chapter 1.

9. S. O. Gelfond, *On the algebraic independence of transcendental numbers of certain classes*, Uspehi Mat. Nauk **4** (1949), 14–48 = Amer. Math. Soc. Transl. (1) **2** (1962), 125–169.

10. S. Lang, *Introduction to transcendental numbers*, Addison-Wesley, Reading, Mass., 1966.

11. ______, *Transcendental numbers and Diophantine approximations*, Bull. Amer. Math. Soc. **77** (1971), 635–677.

12. W. D. Brownawell, *Sequences of Diophantine approximations*, J. Number Theory **6** (1974), 11–21.

13. ______, *The algebraic independence of certain values of the exponential function*, Norske Vid. Selsk. Skr. (1972), no. 23; J. Number Theory **6** (1974), 22–31.

14. M. Waldschmidt, *Numbers transcendents*, Lecture Notes in Math., vol. 402, Springer, New York, 1974.

15. ______, *Indépendance algébrique par la méthode de G. V. Chuduovsky*, Séminaire Délange-Pisot-Poitou **16** (1975), no. 8.

16. ______, *Les travaux de G. V. Chudnovsky sur les nombres transcendants*, Séminaire Bourbaki, Lecture Notes in Math., vol. 567, 1977, pp. 274–292.

17. ______, *Suite colorées*, Séminaire Délange-Pisot-Poitou, 1975/76, No. G21, pp. G21-1—G21-11.

18. H. Davenport and W. M. Schmidt, *Dirichlet's theorem on Diophantine approximation*. II, Acta Arith. **16** (1969/70), 413–424.

19. W. M. Schmidt, *Approximation to algebraic numbers*, L'Enseignement Math. **17** (1971), 187–253.

20. H. Davenport and W. M. Schmidt, *Approximation to real numbers by algebraic integers*, Acta. Arith. **15** (1969), 393–416.

21. ______, *A theorem on linear forms*, Acta Arith. **14** (1967/68), 209–223.

22. G. V. Chudonovsky, *Algebraic independence of values of exponential and elliptic functions*, Proc. Internat. Congr. Math. (Helsinki, 1978), vol. 1, Acad. Sci. Fenn., Helsinki, 1980, pp. 339–350.

23. ______, *Algebraic grounds for the proof of algebraic independence. How to obtain measures of algebraic independence using elementary methods. Part I. Elementary algebra*, Comm. Pure Appl. Math. **34** (1981), 1–28.

24. ______, *Measures of irrationality, transcendence and algebraic independence*, Journées Arithmétiques 1980, (J. V. Armitage (ed.)), Cambridge Univ. Press, Cambridge, 1982, pp. 11–82.

25. ______, see this volume, Introductory paper and Chapter 4.

26. ______, *Independence algébrique dans la methode de Gelfond-Schneider*, C. R. Acad. Sci. Paris **291A** (1980), A365–A368.

27. ______, *Algebraic independence of the values of the elliptic function at algebraic points. Elliptic analogue of the Lindemann-Weierstrass theorem*, Invent. Math. **61** (1980), 267–290.

28. ______, see this volume, Chapters 7 and 8.

ALGEBRAIC CURVES NEAR A POINT. II. FIELDS OF FINITE TRANSCENDENCE TYPE AND COLORED SEQUENCES. RESULTANTS

This part is a direct continuation of the preceding one, to which we will briefly refer as [**1**]. The definitions in [**1**] will often be used without further explanation.

In the course of the present investigation the general scheme of "colored" sequences of [**1**] will be used to study fields arising by adjoining to $\mathbf{Q}$ the coordinates θ_i of a point $\vec{\theta} \in \mathbf{C}^n$. As before [**1**], we consider the plane case ($n = 2$), and instead of treating the degree and the height separately we consider at the outset the type of a polynomial. However, the methods employed in this paper are general and carry over simply to the spatial case ($n > 3$). The requirements of this transfer, which will be realized in other papers of this series, are also served by the general lemmas of [**1**] and the lemmas of this present paper.

An essential feature of this paper is the use of elimination of an unknown by means of the Sylvester resultant. This method was intentionally chosen as central to this paper, since it is well known and has long been used in investigations on this theme. Therefore, establishing powerful new results, though obtained by an old imprecise method, demonstrates the possibilities inherent in a detailed study of properties of colored sequences.

In subsequent papers we will considerably improve the concrete estimates derived below. However, the present paper is the key to understanding the substance of all arguments on the same theme. We therefore deliberately aim now for inferior estimates.

0. Discussion. In §3 of [**1**] we gave reasons for studying properties of "colored" sequences. We mentioned that the existence of a colored sequence near a point $\vec{\theta}$ implies nontrivial lower bounds for the transcendence type τ of the fields generated by the coordinates θ_i. One estimate (the only one up to now) of this form for τ in terms of the exponent μ of the colored sequence was given in Theorem 0.7 of [**1**]. In the present paper we will give a significant improvement of this result. Let us first recall the basic

DEFINITION 0.1. Suppose $\mathcal{K} = \mathbf{Q}(\vartheta_1,\dots,\vartheta_d, \omega)$ is a subfield of $\mathbf{C}$ of transcendence degree d over $\mathbf{Q}$, and $\vartheta_1,\dots,\vartheta_d$ is a transcendence basis. We say that the field $\mathcal{K}$ has transcendence type $\leqslant \tau$ ($< \infty$) if there exists a constant $C(\mathcal{K}) > 0$ such that for every polynomial $P(x_1,\dots,x_d) \in \mathbf{Z}[x_1,\dots,x_d]$, $P(x_1,\dots,x_d) \not\equiv 0$, we have

$$|P(\vartheta_1,\dots,\vartheta_d)| > \exp(-C(\mathcal{K})t(P)^{\tau}). \tag{0.1}$$

If in the right-hand side of (0.1) we replace $t(P)^{\tau}$ by $t(P)^{\tau}(\ln t(P))^{\tau'}$, then we say that $\mathcal{K}$ has type $\leqslant (\tau, \tau')$.

Our whole investigation is carried out within the scope of this definition. Since we are considering the plane case, our investigation pertains to the case $d = 1$, i.e. to fields of transcendence degree 1. For such fields with $\tau < \infty$, it follows that $\tau \geqslant 2$ and "almost all" (but not all) such fields have $\tau = 2$.

Fields of finite transcendence type arise in the investigation of colored sequences. Let us therefore recall the fundamental definition of a colored sequence given in [**1**]. We will often refer to this definition in what follows.

We fix, once and for all, a point $\vec{\theta} \in \mathbf{C}^2$, $\vec{\theta} = (\theta_1, \theta_2)$, such that θ_1 and θ_2 are algebraically independent over $\mathbf{Q}$. We also fix a constant $\mu > 0$.

DEFINITION 0.2. We say that condition (C) with the fixed μ is satisfied (near $\vec{\theta}$) if there exist constants $c_1, c_2, c_3 > 0$ such that:

(C.1) There exists a family $\{P_H(x, y)\colon H \geqslant c_1\}$ of polynomials in $\mathbf{Z}[x, y]$, indexed by all natural numbers $\geqslant c_1$, such that

(1) $P_H(x, y) \not\equiv 0$ for all natural numbers $H \geqslant c_1$;

(2) for all $H \geqslant c_1$ we have

$$t(P_H) = \max\{\ln H(P_H), d(P_H)\} \leqslant c_2 H; \tag{0.2}$$

(3) for a fixed nonnegative, unbounded, monotone increasing function $\varphi(t)$: $t \in \mathbf{R}^+$ and any natural number $H \geqslant c_3$ we have an estimate of $P_H(x, y)$ at $\vec{\theta}$:

$$|P_H(\theta_1, \theta_2)| < \exp(-H^{\mu}\varphi(H)). \tag{0.3}$$

We immediately make an important remark. Since a Dirichlet family for $\vec{\theta}$ (see Example 0.1 of [**1**]) shows the fulfillment of (C) for any point $\vec{\theta}$ with any $\mu < 3$, it is natural, in order to eliminate trivial cases, to assume at the outset that $\mu \geqslant 3$. We do precisely this.

ASSUMPTION 0.3. In the definition of condition (C) in (0.2), $\mu \geqslant 3$.

However, in the present paper the hypothesis $\mu \geqslant 4$ would seem more natural.

The first result, with which we can compare all of ours, is the above-mentioned theorem of Brownawell [**2**] (Theorem 0.7 of [**1**]). We can state it as follows.

THEOREM 0.4 [**2**]. *Suppose that condition* (C) *with a given* μ *is satisfied for* $\vec{\theta} = (\theta_1, \theta_2)$. *Then neither the field* $\mathbf{Q}(\theta_1)$ *nor the field* $\mathbf{Q}(\theta_2)$ *has transcendence type* $\leqslant \frac{1}{2}\mu$.

Let us briefly discuss the extent to which this result can be improved. As is evident from (C), the best we can expect is an estimate from below of the form $\tau > \mu$ (if $P_H(x, y)$ lacks each of the variables infinitely often). But the case $\mu < 3$, which holds (according to 0.1 of [**1**]) for all $\vec{\theta} \in \mathbf{C}^2$, and the impossibility of $\tau > 2$ for all transcendental θ convince us of the unreality of this conjecture. It is clear that an assertion of the form $\tau > \mu - 1 + \varepsilon$ for any $\varepsilon > 0$ also cannot be proved. On the other hand, in this paper it is proved that $\tau \geqslant \mu - 2$, which is significantly better than 0.4 in all nontrivial cases (with $\mu \geqslant 4$).

At the end of this paper we will briefly discuss possible generalizations and applications of our results.

1. Auxiliary assertions. In this section we mention some auxiliary assertions from [**2, 3, 1**] which will be needed in our investigation.

First, we have a lemma on an estimate of the type of a divisor of a polynomial (Lemma 1.1 of [**1**]). Variants of this lemma and their proofs can be found in [**13**].

LEMMA 1.1. *Suppose* $P(x_1,\dots,x_n)$, $P_1(x_1,\dots,x_n),\dots,P_m(x_1,\dots,x_n)$ *are polynomials in* $\mathbf{C}[x_1,\dots,x_n]$. *Let* $d(P)$ *denote the degree of* $P(x_1,\dots,x_n)$, *i.e.* $d(P) = d_{x_1}(P)+\cdots+d_{x_n}(P)$. *If*

$$(1.1)\qquad P(x_1,\dots,x_n) = P_1(x_1,\dots,x_n)\cdots P_m(x_1,\dots,x_n),$$

then for the heights $H(P)$, $H(P_1),\dots,H(P_m)$ *we have*

$$(1.2)\qquad H(P)\cdot e^{d(P)} \geqslant H(P_1)\cdots H(P_m).$$

In particular, it follows from Lemma 1.1 that if (1.1) holds, we obtain for the types $t(P)$, $t(P_1),\dots,t(P_m)$ the estimate (instead of (1.2)):

$$(1.3)\qquad t(P_1)+\cdots+t(P_m) \leqslant 3t(P).$$

Therefore, Lemma 1.1 will be used in the form (1.3), rather than in the form (1.2).

Since we are going to use resultants, we need some facts about them. For our purposes, the form or description of the resultant of two polynomials is unimportant. We need only the following lemma.

LEMMA 1.2. *Suppose that* $P(x, y)$, $Q(x, y) \in \mathbf{Z}[x, y]$ *are two nonzero polynomials having no nonconstant common divisor.*[1] *Then there exists a polynomial* $R(x) \in \mathbf{Z}[x]$, *the resultant* $R(P, Q)_y$ *of* $P(x, y)$ *and* $Q(x, y)$ *with respect to* y, *such that*

(1) $R(x) \not\equiv 0$,

(2) *the type* $t(R)$ *of* $R(x)$ *can be estimated in terms of the degrees* $d(P)$, $d(Q)$ *and the types* $t(P)$, $t(Q)$ *of* $P(x, y)$, $Q(x, y)$,

$$(1.4)\qquad t(R) \leqslant c_4(d(P)t(Q) + d(Q)t(P)).$$

In particular,

$$(1.5)\qquad t(R) \leqslant 2c_4 t(P)t(Q).$$

In (1.4) *and* (1.5), $c_4 > 0$ *is some absolute constant*;

(3) *for any complex numbers* ϑ_1, ϑ_2 *we have*

$$(1.6)\qquad \max\{|P(\vartheta_1,\vartheta_2)|, |Q(\vartheta_1,\vartheta_2)|\} \geqslant \exp(-c_5(d(P)t(Q) + d(Q)t(P)))\cdot|R(\vartheta_1)|,$$

where $c_5 > 0$ *is some absolute constant.*

In particular, we have

$$(1.7)\qquad \max\{|P(\vartheta_1,\vartheta_2)|, |Q(\vartheta_1,\vartheta_2)|\} \geqslant \exp(-2c_5 t(P)t(Q))\cdot|R(\vartheta_1)|.$$

---

[1]As in [**1**], we call such polynomials coprime.

A construction of the polynomial $R(x)$ in terms of the coefficients of $P(x, y)$ and $Q(x, y)$ as polynomials in y can be found in many places **[2]**. Property (1) is well known. Estimates (1.4)–(1.5) and (1.6)–(1.7) are given in **[2]**, with a simple method of proof. Finally, the author's paper **[4]** contains Lemma 1.2 and even much more general assertions. In **[4]** there is no construction of $R(x)$, but this is not needed in our investigation.

In studying colored sequences we rely on the results of **[1]**. Of all the results proved in **[1]**, the one that is absolutely necessary is Theorem 2.2 of **[1]**, which is the main one for all construction.

THEOREM 1.3 (THEOREM 2.2 OF **[1]**). *Suppose that condition* (C) *is satisfied for a given* $\vec{\theta} \in \mathbf{C}^2$ *and a given* μ $(\geqslant 3)$. *Then there exist constants* $c_6, c_7, c_8 > 0$, *effectively defined in terms of* c_1, c_2, c_3, μ, *and* $\varphi(t)$, *such that*:

(C.4) *there exists an infinite increasing sequence* X_n: $n = 0, 1, 2, \ldots,$ *of natural numbers*

$$[c_6 + 1] = X_0 < X_1 < \cdots < X_n < \cdots : \qquad n < \infty, \tag{1.8}$$

and a sequence of corresponding polynomials $R_{X_n}(x, y) \equiv R_n(x, y)$,

$$R_0(x, y), R_1(x, y), \ldots, R_n(x, y), \ldots : \qquad n < \infty, \tag{1.9}$$

from $\mathbf{Z}[x, y]$ *such that*:

(1) *for any* $n = 0, 1, 2, \ldots$ *we have* $R_n(x, y) \not\equiv 0$;

(2) *for any* $n = 0, 1, 2, \ldots$ *the polynomials* $R_n(x, y)$, $R_{n+1}(x, y)$ *are coprime*;

(3) *if* $X_n + 1 < X_{n+1}$: $n = 0, 1, 2, \ldots,$ *then* $R_n(x, y)$ *is a power of a* $\mathbf{Q}$*-irreducible polynomial* $Q_n(x, y)$ *in* $\mathbf{Z}[x, y]$;

(4) *for any* $n = 0, 1, 2, \ldots$ *we have*:

$$t(R_n) \leqslant c_7 X_n, \tag{1.10}$$

$$|R_n(\theta_1, \theta_2)| < \exp\left(-\tfrac{1}{4} X_n^{\mu} \varphi(X_n)\right). \tag{1.11}$$

(5) *for any* $n = 0, 1, 2, \ldots$ *there exists a natural number* $s_n \geqslant 1$ *such that the following inequalities connecting* X_n *and* X_{n+1} *are satisfied*:

$$s_n t(R_n) \leqslant c_8 X_{n+1}, \tag{1.12}$$

$$|R_n(\theta_1, \theta_2)| < \exp\left(-\tfrac{1}{4}(X_{n+1} - 1)^{\mu} \varphi(X_{n+1} - 1)/s_n\right). \tag{1.13}$$

This formulation of Theorem 2.2 of **[1]** differs only slightly from the original. Of all the conditions in (C.2) satisfied by $\{R_H(x, y)\}$ there remains here only one, namely condition (1) in (C.2), which is (1) in (C.4). Instead of $R_{X_n}(x, y)$ we write simply $R_n(x, y)$. Condition (i) in (C.3) is (2) in (C.4); (ii) in (C.3) is (3) in (C.4); (iv) and (v) in (C.3) are (4) and (5) in (C.4); (iii) in (C.3) is just omitted.

We can now simply renounce the Definition 0.2 of a colored sequence and use instead the sequence of polynomials $\{R_n(x, y): n = 0, 1, 2, \ldots\}$ whose existence is claimed in Theorem 1.3.

We can now begin the exposition of the main result of this paper.

In concluding this preliminary section we give two more results (on polynomials in one variable), which will be used only as auxiliary assertions in §2. One lemma is rather well known.

LEMMA 1.4. *Suppose $P(x) \in \mathbf{Z}[x]$ and θ is an arbitrary complex number, and suppose that*

$$|P(\theta)| \leqslant \varepsilon < H(P)^{-4d(P)}, \tag{1.14}$$

where $H(P)$ is the height and $d(P)$ the degree of $P(x)$. Then there exists a divisor $Q(x)$ of $P(x)$ which is a power of a $\mathbf{Q}$-irreducible polynomial $P_0(x) \in \mathbf{Z}[x]$,

$$Q(x) = (P_0(x))^s, \tag{1.15}$$

such that

$$|Q(\theta)| \leqslant \varepsilon H(P)^{4d(P)} < 1. \tag{1.16}$$

For the polynomial $P_0(x) \in \mathbf{Z}[x]$ we obtain the estimates

$$t(P_0) \leqslant 3t(P)/s; \tag{1.17}$$

$$|P_0(\theta)| \leqslant \varepsilon^{1/s} H(P)^{4d(P)/s} < 1. \tag{1.18}$$

Clearly, (1.17) and (1.18) follow directly from (1.15), (1.16), and Lemma 1.1 in the form (1.3). Proofs of Lemma 1.4 can be found in [**3, 5, 6, 7**].

The second lemma essentially involves the resultant for polynomials in one variable. This lemma can be obtained by simply assuming that the $P(x, y)$ and $Q(x, y)$ in Lemma 1.2 do not depend on x. Nevertheless, we state it separately for the sake of completeness.

LEMMA 1.5. *Suppose that $P(x)$, $Q(x) \in \mathbf{Z}[x]$ are nonzero polynomials having no nonconstant common divisor. Then for any complex number θ we have*

$$\begin{aligned} \max\{|P(\theta)|, |Q(\theta)|\} &\geqslant \exp(-c_9(d(P)t(Q) + d(Q)t(P))) \\ &\geqslant \exp(-2c_9 t(P)t(Q)), \end{aligned} \tag{1.19}$$

where $c_9 > 0$ is an absolute constant.

As we have said, 1.5 is obtained from 1.2 if $P(x, y)$ and $Q(x, y)$ do not depend on x. In this case, (1.19) follows from (1.6)–(1.7), i.e. $R(x) \equiv R_0 \in \mathbf{Z}$ and $|R_0| \geqslant 1$. A direct proof of Lemma 1.5 is given in [**3, 7**].

2. Transcendence types of fields connected with colored sequences. The auxiliary assertions established in §1 provide us with everything needed to prove the main result of this paper, namely Theorem 2.1, which significantly generalizes Theorem 0.4.

THEOREM 2.1. *Suppose that condition* (C) *is satisfied for a point $\vec{\theta} = (\theta_1, \theta_2) \in \mathbf{C}^2$ with a given μ. Then neither the field $\mathbf{Q}(\theta_1)$ nor the field $\mathbf{Q}(\theta_2)$ has transcendence type $\leqslant \mu - 2$.*

REMARK 2.1′. This theorem is more precise than 0.4 in all nontrivial cases: where $\mu \geqslant 4$. If $\mu < 4$, both 0.4 and 2.1 yield nothing: fields of transcendence type < 2 are algebraic.

PROOF OF THEOREM 2.1. Suppose that condition (C) in 0.2 is satisfied for a point $\vec{\theta} = (\theta_1, \theta_2)$ and a given μ. In view of 0.3 and Remark 2.1′, we may assume that $\mu \geqslant 4$. However, this restriction is not necessary.

Applying Theorem 1.3, we obtain an infinite increasing sequence of natural numbers

$$X_0 < X_1 < X_2 < \cdots < X_n < \cdots : \quad n < \infty, \tag{2.1}$$

and a sequence of polynomials $R_0(x, y), \ldots, R_n(x, y), \ldots$: $n < \infty$, satisfying (C.4).

The idea behind the rest of the argument is to consider the resultants of the polynomials $R_n(x, y)$ and $R_{n+1}(x, y)$ with respect to y. Since $R_n(x, y)$, $R_{n+1}(x, y)$ are coprime according to condition (2) in (C.4), the resultant $R_{n,n+1}(x) \not\equiv 0$ exists and, by Lemma 1.2, we have good estimates for $|R_{n,n+1}(\theta_i)|$. However, the larger that X_{n+1} is in comparison to X_n the better these estimates will be. Therefore, it is necessary to consider all resultants $R_{n,n+1}(x)$ simultaneously.

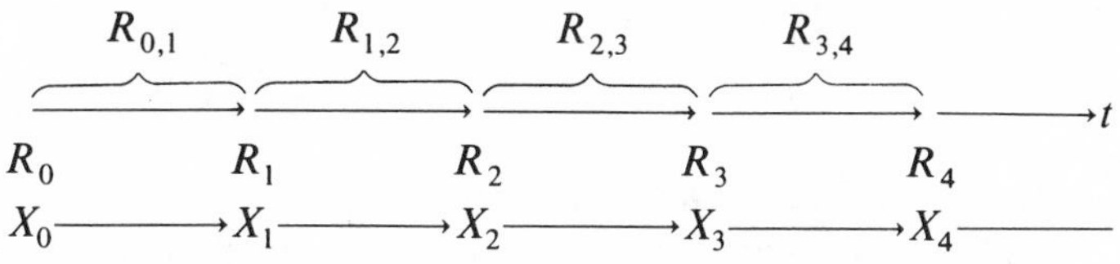

FIGURE 1

Figure 1 depicts a symbolic table. On the t-axis we have marked the types $t(R_n)$ of the polynomials $R_n(x, y)$ corresponding to the points $X_n \in N$. An arrow from R_n to R_{n+1}, and from X_n to $X_{n+1} - 1$, means that according to (C.3) and (C.4) (see conditions (3) and (5) in (C.4) and Remark 2.3 of [**1**]) all polynomials between X_n and $X_{n+1} - 1$ are powers of the same irreducible polynomial, the power of which is $R_n(x, y)$ when $X_n + 1 < X_{n+1}$ (see condition (3) in (C.4)). The exponent $s_n \geqslant 1$ in condition (5) of (C.4) is the power to which we must raise $R_n(x, y) \equiv R_{X_n}(x, y)$ in order to obtain the polynomial $R_{X_{n+1}-1}(x, y)$ in (C.3). Thus, Figure 1 restores the purely "colored" sequence in (C.3).

We will now realize the idea described above. According to (1) and (2) in (C.4), $R_n(x, y)$ and $R_{n+1}(x, y)$ are nonzero coprime polynomials. Therefore, by Lemma 1.2, for any $n = 0, 1, 2, \ldots$ there exists a polynomial $R_{n,n+1}(x)$, the resultant of $R_n(x, y)$ and $R_{n+1}(x, y)$ with respect to y, satisfying the conditions of Lemma 1.2. This means that $R_{n,n+1}(x) \in \mathbf{Z}[x]$, $R_{n,n+1}(x) \not\equiv 0$, and we have the estimates

$$t(R_{n,n+1}) \leqslant 2c_4 t(R_n) t(R_{n+1}), \tag{2.2}$$

(2.3)

$$|R_{n,n+1}(\theta_1)| \leqslant \exp(2c_5 t(R_n) t(R_{n+1})) \max\{|R_n(\theta_1, \theta_2)|, |R_{n+1}(\theta_1, \theta_2)|\}.$$

Using (2.2), (2.3), and (C.4), we will now prove that the field $\mathbf{Q}(\theta_1)$ does not have transcendence type $\leqslant \mu - 2$. Since the problem is symmetric, we obtain, interchanging θ_1 and θ_2, that the transcendence type of $\mathbf{Q}(\theta_2)$ is greater than $\mu - 2$.

We have

$$(2.4) \qquad |R_{n,n+1}(\theta_1)| \leqslant \exp\left(2c_5 t(R_n) t(R_{n+1}) - \tfrac{1}{4} X_n^\mu \varphi(X_n)\right).$$

We transform (2.2) and (2.3) using inequalities (1.10) and (1.11) in (C.4). We obtain

$$(2.5) \qquad t(R_{n,n+1}) \leqslant c_{10} X_n X_{n+1},$$

where $c_{10} = 2c_4 c_7^2 > 0$. Moreover, since $\varphi(t)$ is monotone increasing, it follows from (1.11) and (2.3) that

$$(2.6) \qquad |R_{n,n+1}(\theta_1)| \leqslant \exp\left(c_{11} X_n X_{n+1} - \tfrac{1}{4} X_n^\mu \varphi(X_n)\right),$$

where $c_{11} = 2c_5 c_7 > 0$.

We return to the proof of the fact that the field $\mathbf{Q}(\theta_1)$ has transcendence type $> \mu - 2$. Assume, to the contrary, that $\mathbf{Q}(\theta_1)$ has transcendence type $\leqslant \mu - 2$. According to Definition 0.1, this means that there exists a constant $C_\theta > 0$ such that for all polynomials $P(x) \in \mathbf{Z}[x]$, $P(x) \not\equiv 0$, we have

$$(2.7) \qquad |P(\theta_1)| > \exp\left(-C_\theta t(P)^{\mu-2}\right).$$

We will show that (2.7) leads to a contradiction.

To do this we will deduce from (2.7) (the assumption that $\mathbf{Q}(\theta_1)$ has type $\tau \leqslant \mu - 2$), a bound for $\overline{\lim}_{n\to\infty} \ln(X_{n+1})/\ln X_n$. In Lemma 2.2 it is shown that if $\tau \leqslant \mu - 2$, this limit is at most $\mu - 3$. After proving 2.2, we will show that this limit is greater than $\mu - 3$. We should mention that the existence of an upper bound for $\overline{\lim}_{n\to\infty} \ln(X_{n+1})/\ln X_n$ can even be obtained from the estimate $\tau < \mu$ for any $\mu > 4$. This will be done in subsequent papers of the author; see [**8**].

LEMMA 2.2. *There exists $n_0 \geqslant 0$ such that for all $n \geqslant n_0$ we have*

$$(2.8) \qquad X_{n+1} \leqslant X_n^{\mu-3}.$$

PROOF OF LEMMA 2.2. Assume, to the contrary, that there exists an infinite sequence of natural numbers $n_0 < n_1 < \cdots < n_k < \cdots : k < \infty$, such that

$$(2.9) \qquad X_{n_k+1} > X_{n_k}^{\mu-3}: \qquad k < \infty.$$

We turn to property (5) in (C.4) instead of (2.6). For a natural number $s_n \geqslant 1$ we have

$$(2.10) \qquad s_n t(R_n) \leqslant c_8 X_{n+1},$$

$$(2.11) \qquad |R_n(\theta_1, \theta_2)| < \exp\left(-\tfrac{1}{4}(X_{n+1} - 1)^\mu \varphi(X_{n+1} - 1)/s_n\right).$$

Consequently, it follows from (2.3) that

(2.12)

$$|R_{n,n+1}(\theta_1)| \leqslant \exp(c_{12}X_{n+1}^2/s_n)\max\{\exp(-\tfrac{1}{4}(X_{n+1}-1)^\mu\varphi(X_{n+1}-1)/s_n), \times\exp(-\tfrac{1}{4}X_{n+1}^\mu\varphi(X_{n+1}))\}.$$

Since $\varphi(t)$: $t \in \mathbf{R}^+$, is monotone increasing and $s_n \geqslant 1$, it follows from (2.12) that

$$|R_{n,n+1}(\theta_1)| \leqslant \exp(c_{12}X_{n+1}^2/s_n - \tfrac{1}{4}(X_{n+1}-1)^\mu\varphi(X_{n+1}-1)/s_n). \tag{2.13}$$

Since $\mu \geqslant 4$ and $\varphi(t)$: $t \in \mathbf{R}^+$ increases without bound as $t \to \infty$, then there exists $n_1 < \infty$ and $c_{14} > 0$ such that (2.13) yields, for $n \geqslant n_1$,

$$|R_{n,n+1}(\theta_1)| \leqslant \exp(-c_{14}X_{n+1}^\mu\varphi(X_{n+1}-1)/s_n). \tag{2.14}$$

In (2.14) the constants $n_1 < \infty$ and $c_{14} > 0$ satisfy the conditions

$$\varphi(X_{n_1+1}-1) \geqslant 1 + c_{12}, \qquad X_{n_1+1} > 2\mu, \qquad c_{14} = \tfrac{1}{8}\cdot 2^\mu.$$

Note that neither here nor elsewhere in this paper are the parameters chosen in the best way; their choice is subject only to specific requirements.

We apply (2.14) to the sequence X_{n_k} in (2.9). We choose $k_0 < \infty$ so that $n_{k_0} \geqslant n_1$. Then (2.14) can be written in the form

$$|R_{n_k,n_k+1}(\theta_1)| \leqslant \exp(-c_{14}X_{n_k+1}^\mu\varphi(X_{n_k+1}-1)s_{n_k}^{-1}) \tag{2.15}$$

for $k \geqslant k_0$. We now estimate the type $t(R_{n,n+1})$ directly from (2.2) and (1.10):

$$t(R_{n,n+1}) \leqslant 2c_4t(R_n)t(R_{n+1}) \leqslant 2c_4c_7t(R_n)\cdot X_{n+1}. \tag{2.16}$$

By construction, $R_{n,n+1}(x) \not\equiv 0$ for any n. Therefore, we can apply inequality (2.7) to $R_{n,n+1}(x)$. We obtain, in view of (2.16),

$$|R_{n,n+1}(\theta_1)| \geqslant \exp(-C_\theta(c_{15}t(R_n)X_{n+1})^{\mu-2}) \tag{2.17}$$

with $c_{15} = 2c_4c_7$. Comparing (2.15) and (2.17) with $n = n_k$ for any $k \geqslant k_0$, we obtain

$$c_{16}t(R_{n_k})^{\mu-2}X_{n_k+1}^{\mu-2} \geqslant c_{14}X_{n_k+1}^\mu\varphi(X_{n_k+1}-1)\cdot s_{n_k}^{-1}. \tag{2.18}$$

Inequality (2.18) can be rewritten in the form

$$c_{17}s_{n_k}t(R_{n_k})^{\mu-2} \geqslant X_{n_k+1}^2\varphi(X_{n_k+1}-1). \tag{2.19}$$

for $k \geqslant k_0$. According to (2.10),

$$s_{n_k}t(R_{n_k}) \leqslant c_8X_{n_k+1}.$$

Therefore, (2.19) takes the form

$$c_8c_{17}t(R_{n_k})^{\mu-3} \geqslant X_{n_k+1}\varphi(X_{n_k+1}-1) \tag{2.20}$$

for $k \geqslant k_0$. In view of (1.10) for $t(R_{n_k})$:

$$t(R_{n_k}) \leqslant c_7X_{n_k}.$$

It then follows from (2.20) that

$$c_{18} X_{n_k}^{\mu-3} \geqslant X_{n_k+1} \varphi\left(X_{n_k+1} - 1\right) \tag{2.21}$$

with $c_{18} = c_8 c_{17} c_7$. Choose $k_1 \geqslant k_0$ so that $\varphi(X_{n_{k_1}+1} - 1) > c_{18}$. Then for $k \geqslant k_1$ we have $\varphi(X_{n_k+1} - 1) > c_{18}$. Consequently, for $k \geqslant k_1$ ($\geqslant k_0$) inequality (2.21) yields

$$X_{n_k}^{\mu-3} > X_{n_k+1}. \tag{2.22}$$

This last inequality contradicts (2.9). Therefore, (2.9) is impossible and (2.8) holds. Lemma 2.2 is proved.

We will show that for all polynomials $R_{n,n+1}(x)$ the estimates (2.2) and (2.3) together with Lemma 1.4 lead to an extraction of a nontrivial irreducible divisor of $R_{n,n+1}(x)$ with a good upper bound at the point θ_1.

To do this, for any $n \geqslant n_1$ consider inequality (2.13), which is a direct consequence of (2.3), (2.10), and (2.11). We rewrite this inequality for $n \geqslant n_1$ as

$$|R_{n,n+1}(\theta_1)| \leqslant \exp\left(-c_{14} X_{n+1}^{\mu} \varphi(X_{n+1} - 1) \cdot s_n^{-1}\right). \tag{2.23}$$

According to (2.10), we have

$$c_8^{-1} t(R_n) \leqslant s_n^{-1} X_{n+1}. \tag{2.24}$$

Applying (2.23) and (2.24), we obtain

$$|R_{n,n+1}(\theta_1)| \leqslant \exp\left(-c_{19} X_{n+1}^{\mu-1} t(R_n) \varphi(X_{n+1} - 1)\right) \tag{2.25}$$

for all $n \geqslant n_1$ with $c_{19} = c_{14} c_8^{-1}$. Now along with (2.25) we consider an estimate of $t(R_{n,n+1})$. Namely, according to (2.2), (1.10), and (1.11),

$$t(R_{n,n+1}) \leqslant 2c_4 t(R_n) t(R_{n+1}) \leqslant 2c_4 c_7 X_{n+1} t(R_n).$$

Consequently, with $c_{20} = 2c_4 c_7$ we obtain

$$t(R_{n,n+1}) \leqslant c_{20} t(R_n) X_{n+1}. \tag{2.26}$$

We will find an $n_2 \geqslant n_1$ such that, for $n \geqslant n_2$,

$$4\left(t(R_{n,n+1})\right)^2 < \frac{c_{19}}{2} X_{n+1}^{\mu-1} t(R_n) \varphi(X_{n+1} - 1). \tag{2.27}$$

In view of (2.26), it is necessary that

$$4c_{20}^2 t(R_n)^2 X_{n+1}^2 < \frac{c_{19}}{2} X_{n+1}^{\mu-1} t(R_n) \varphi(X_{n+1} - 1).$$

This last inequality is equivalent to the following one:

$$4c_{20}^2 t(R_n) < \frac{c_{19}}{2} X_{n+1}^{\mu-3} \varphi(X_{n+1} - 1). \tag{2.28}$$

To prove (2.28), according to (1.10), it suffices to show that

$$4c_{20}^2 c_7 < \frac{c_{19}}{2} \varphi(X_{n+1} - 1). \tag{2.29}$$

Indeed, if (2.29) holds, then $\mu \geqslant 4$, i.e. $X_{n+1}^{\mu-4} \geqslant 1$, and

$$4c_{20}^2 c_7 < \frac{c_{19}}{2} X_{n+1}^{\mu-4} \varphi(X_{n+1} - 1).$$

Since $X_n < X_{n+1}$, it follows that

$$4c_{20}^2(c_7 X_n) < \frac{c_{19}}{2} \cdot X_{n+1}^{\mu-3}\varphi(X_{n+1} - 1).$$

Then (1.10) yields (2.28).

We now choose n_2 so that $n_2 > n_1$ and

(2.30) $$4c_{20}^2 c_7 \cdot (c_{19}/2)^{-1} < \varphi\left(X_{n_2+1} - 1\right).$$

Since $\varphi(t)$: $t \in \mathbf{R}^+$ is a monotone function, (2.29) follows from (2.30) for all $n \geqslant n_2$. Therefore, (2.28) and (2.27) hold for all $n \geqslant n_2$.

Inequality (2.27) enables us to apply Lemma 1.4 to $R_{n,n+1}(x)$. In the notation of Lemma 1.4 we have $P(x) = R_{n,n+1}(x)$ and, in view of (2.25),

(2.31) $$\varepsilon = \exp\left(-c_{19} X_{n+1}^{\mu-1} t(R_n)\varphi(X_{n+1} - 1)\right).$$

Condition (2.27) guarantees (1.14) for $n \geqslant n_2$:

$$|R_{n,n+1}(\theta_1)| \leqslant \varepsilon < \exp\left(-4(t(R_{n,n+1}))^2\right).$$

Now by Lemma 1.4, for any $n \geqslant n_2$ there exists a $\mathbf{Q}$-irreducible polynomial $Q_n(x) \in \mathbf{Z}[x]$ such that $(Q_n(x))^{f_n}$ is a divisor of $R_{n,n+1}(x)$ and (1.17) and (1.18) hold with $P_0(x) = Q_n(x)$. Therefore, for $n \geqslant n_2$ we have

(2.32) $$t(Q_n) \leqslant 3t(R_{n,n+1})/f_n,$$

(2.33) $$|Q_n(\theta_1)| \leqslant \varepsilon_n^{1/f_n} \cdot \exp\left(4(t(R_{n,n+1}))^2/f_n\right).$$

In view of (2.26) and (2.31), we can rewrite (2.32) and (2.33) in the form

(2.34) $$t(Q_n) \leqslant c_{21} t(R_n) \cdot X_{n+1}/f_n,$$

(2.35) $$|Q_n(\theta_1)| \leqslant \exp\left(-c_{19} X_{n+1}^{\mu-1} t(R_n)\varphi(X_{n+1} - 1)\cdot f_n^{-1} + 4(t(R_{n,n+1}))^2 \cdot f_n^{-1}\right)$$

where $c_{21} = 3c_{20}$. According to (2.27), for $n \geqslant n_2$ inequality (2.35) can be rewritten as

(2.36) $$|Q_n(\theta_1)| \leqslant \exp\left(-c_{22} X_{n+1}^{\mu-1} t(R_n)\varphi(X_{n+1} - 1) \cdot f_n^{-1}\right)$$

where $c_{22} = \frac{1}{2}c_{19}$ for $n \geqslant n_2$. Taking into account (2.34), we obtain

$$-t(R_n) \cdot X_{n+1} f_n^{-1} \leqslant -c_{21}^{-1} t(Q_n).$$

Therefore, it follows from (2.36) that, for $n \geqslant n_2$,

(2.37) $$|Q_n(\theta_1)| \leqslant \exp\left(-c_{23} X_{n+1}^{\mu-2} t(Q_n)\varphi(X_{n+1} - 1)\right)$$

where $c_{23} = c_{22}c_{21}^{-1}$.

For a given $n \geqslant n_2$ we denote by $m(n)$, $n < m(n) < \infty$, the smallest m such that $Q_n \neq \pm Q_m$. Then

(2.38) $$Q_n = \pm Q_m: \quad n \leqslant m < m(n), \quad Q_n \neq \pm Q_{m(n)}.$$

Such an $m(n) < \infty$ exists for any n. If it did not, then, in view of (2.37), we would have for all m, $n \leqslant m < \infty$,

$$|Q_n(\theta_1)| \leqslant \exp\left(-c_{23} X_{m+1}^{\mu-2} t(Q_n)\varphi(X_{m+1} - 1)\right).$$

As $m \to \infty$ this gives $Q_n(\theta_1) = 0$, which contradicts the transcendence of θ_1. Therefore, there exists $m(n) < \infty$ satisfying (2.38). Using (2.37) and (2.38), we obtain

(2.39) $$|Q_n(\theta_1)| \leqslant \exp\left(-c_{23} X_{m(n)}^{\mu-2} t(Q_n) \varphi(X_{m(n)} - 1)\right)$$

and

(2.40) $$|Q_{m(n)}(\theta_1)| \leqslant \exp\left(-c_{23} X_{m(n)+1}^{\mu-2} t(Q_{m(n)}) \varphi(X_{m(n)+1} - 1)\right).$$

In view of (2.38), the polynomials $Q_n(x)$, $Q_{m(n)}(x)$, which are **Q**-irreducible, are coprime, i.e. they have no nonconstant common divisor. We can therefore apply Lemma 1.5. According to Lemma 1.5, we obtain

(2.41) $$\max\{|Q_n(\theta_1)|, |Q_{m(n)}(\theta_1)|\} \geqslant \exp\left(-2c_9 t(Q_n) \cdot t(Q_{m(n)})\right).$$

We compare (2.39)–(2.40) with (2.41). If $t(Q_{m(n)}) \leqslant t(Q_n)$, then (2.39) and (2.40) yield

(2.42) $$\max\{|Q_n(\theta_1)|, |Q_{m(n)}(\theta_1)|\} \leqslant \exp\left(-c_{23} X_{m(n)}^{\mu-2} t(Q_{m(n)}) \varphi(X_{m(n)} - 1)\right).$$

for $t(Q_{m(n)}) \leqslant t(Q_n)$. Then from (2.41) we obtain

(2.43) $$c_{23} X_{m(n)}^{\mu-2} \varphi(X_{m(n)} - 1) \leqslant 2c_9 t(Q_n).$$

But by definition, $m(n) > n$. Thus, (2.43) implies

(2.44) $$c_{24} X_{n+1}^{\mu-2} \varphi(X_{n+1} - 1) < t(Q_n)$$

where $c_{24} = c_{23}(2c_9)^{-1}$. According to (2.34), for any $n \geqslant n_2$ we have

(2.45) $$t(Q_n) \leqslant c_{21} t(R_n) X_{n+1} \leqslant c_{21} c_7 X_n X_{n+1} \leqslant c_{21} c_7 X_{n+1}^2.$$

But inequalities (2.44) and (2.45) are incompatible for $n \geqslant n_3$, where $n_3 \geqslant n_2$ is such that

$$\varphi(X_{n_3+1} - 1) > c_{21} c_7 (c_{24})^{-1}.$$

Consequently, the case $t(Q_{m(n)}) \leqslant t(Q_n)$ is impossible for all $n \geqslant n_3$.

Therefore, we consider the case $n \geqslant n_3$ and $t(Q_n) \leqslant t(Q_{m(n)})$. In this case, from (2.39) and (2.40) we obtain, instead of (2.42),

(2.46) $$\max\{|Q_n(\theta_1)|, |Q_{m(n)}(\theta_1)|\} \leqslant \exp\left(-c_{23} X_{m(n)}^{\mu-2} t(Q_n) \varphi(X_{m(n)} - 1)\right)$$

for all $n \geqslant n_3$. Comparing (2.42) and (2.46), we obtain

(2.47) $$c_{23} X_{m(n)}^{\mu-2} \varphi(X_{m(n)} - 1) \leqslant 2c_9 t(Q_{m(n)}).$$

Now (2.34) together with (2.47) yield

(2.48) $$c_{23} X_{m(n)}^{\mu-2} \varphi(X_{m(n)} - 1) \leqslant 2c_9 c_{21} c_7 X_{m(n)} X_{m(n)+1}.$$

for $m(n) > n \geqslant n_3 \geqslant n_1$. If we put $c_{25} = 2c_9 c_{21} c_7 c_{23}^{-1}$, we obtain from (2.48) that

(2.49) $$X_{m(n)}^{\mu-3} \varphi(X_{m(n)} - 1) \leqslant c_{25} X_{m(n)+1}.$$

Inequality (2.49) holds for any $n \geqslant n_3$.

We can now compare (2.49) and Lemma 2.2, which is a consequence of (2.7). Here we use the assumption that the field $\mathbf{Q}(\theta_1)$ has transcendence type $\leqslant \mu - 2$. *We emphasize that all of the calculations* (2.1)–(2.6), (2.10)–(2.14), *and* (2.23)–(2.49) *in no way depend on this assumption and follow from the conditions in* $(C.4)$.[2]

We can find an $n_4 \geqslant n_3$ such that

$$\varphi(X_{n_4} - 1) > c_{25}. \tag{2.50}$$

Since $\varphi(t)$ is monotone increasing and $m(n) > n$, it follows from (2.50) that

$$\varphi(X_{m(n)} - 1) > c_{25}, \tag{2.51}$$

for all $n \geqslant n_4$. Then from (2.49) and (2.51) we obtain, for $n \geqslant n_4 \geqslant n_3$,

$$X_{m(n)}^{\mu-3} < X_{m(n)+1}. \tag{2.52}$$

This contradicts (2.8) for any $n \geqslant n_4$, n_0, since $m(n) > n$. Thus, Lemma 2.2, and with it assumption (2.7), is false.

Therefore, the field $\mathbf{Q}(\theta_1)$ cannot have transcendence type $\leqslant \mu - 2$. The same is true (by symmetry) for $\mathbf{Q}(\theta_2)$. Theorem 2.1 is completely proved.

3. Subsequent papers in this series. The result proved in this present paper is best from the point of view of resultants. We can even introduce an assumption of the existence of numbers $\theta_1, \theta_2 \in \mathbf{C}$, a sequence X_i: $i < \infty$, and polynomials $P_i(x, y) \in \mathbf{Z}[x, y]$ such that

$$t(P_i) \leqslant X_i; \qquad |P_i(\theta_1, \theta_2)| < \exp(-X_{i+1}^{\mu-1} X_i); \tag{3.1}$$

$$\lim_{i \to \infty} \ln X_{i+1} / \ln X_i = \chi < \infty. \tag{3.2}$$

Conditions (3.1) and (3.2) guarantee that the system of polynomials

$$\left\{P_H(x, y) = (P_i(x, y))^{[H/X_i]}: X_i < H \leqslant X_{i+1};\ 1 \leqslant i < \infty\right\}$$

is colored. For this family, resultants give an estimate for the type τ of $\mathbf{Q}(\theta_1)$: $\tau \geqslant \mu - 2$. This would be true if $\chi = \mu - 3$. However, the case $\chi = \mu - 3$ is impossible. Nevertheless, the estimate $\tau \geqslant \mu - 2$ can be improved.

These assertions will be proved in the next papers.

We quote the following result of **[8]**:

THEOREM 3.1. *Under the assumption* (C) *of Definition* 0.2 *for a given* μ *and* $\vec{\theta} = (\theta_1, \theta_2)$, *neither of the fields* $\mathbf{Q}(\theta_1)$ *or* $\mathbf{Q}(\theta_2)$ *has the transcendence type* $\tau \leqslant \mu - 2 + 1/(\mu - 1)$. *Moreover, if for a given* μ *and* $\vec{\theta}$ *the condition* (C.1) *of Definition* 0.2 *is satisfied and* $\mu \geqslant (\tau + 3 + (\tau^2 + 2\tau - 3)^{1/2})/2$, *then for an arbitrary* $C > 0$, *there exist infinitely many algebraic numbers* (ξ_1, ξ_2) *such that*

$$\max\{|\theta_1 - \xi_1|, |\theta_2 - \xi_2|\} < \exp\{-Ct(\vec{\xi})^{\tau}\},$$

[2] It is necessary to make this remark because the quantity $\mu - 2$ in Theorem 2.1 can be improved on the basis of (2.23)–(2.49).

where

$$t(\bar{\xi}) = [\mathbf{Q}(\xi_1, \xi_2) : \mathbf{Q}] \cdot \left\{ \frac{\log t(\xi_1)}{d(\xi_1)} + \frac{\log t(\xi_2)}{d(\xi_2)} \right\}$$

and $t(\xi_1)$, $t(\xi_2)$ *are types of* ξ_1, ξ_2.

These and other similar strong results are proved using more detailed studies of singular points of intersections of algebraic curves. For complete proofs using these methods see **[9]** (an elementary in troduction to this approach) and **[10]** and especially **[11]**.

REFERENCES

1. G. V. Chuduovsky, *On the way to the Schanuel conjecture*. Part I of this chapter.

2. W. D. Brownawell, *Gelfond's method for algebraic independence*, Trans. Amer. Math. Soc. **210** (1975), 1–26.

3. A. O. Gelfond, *Transcendental and algebraic numbers*, Dover, New York, 1960.

4. G. V. Chudnovsky, *Analytical methods in Diophantine approximations*, Inst. of Math. Ukrain. SSR Acad. Sci. Preprint IM-74-9, Kiev, 1974, pp. 1–52; see this volume, Chapter 1.

5. M. Waldschmidt, *Nombres transcendents*, Lecture Notes in Math., vol. 402, Springer, New York, 1974.

6. S. Lang, *Introduction to transcendental numbers*, Addison-Wesley, Reading, Mass., 1966.

7. A. O. Gelfond, *On the algebraic independence of algebraic powers of algebraic numbers*, Uspehi Math. Nauk **5** (1949), 14–48 = Amer. Math. Soc. Transl. (1) **2** (1962), 125–169.

8. G. V. Chudnovsky, *Independance algébrique dans la methode de Gelfond-Schneider*, C. R. Acad. Sci. Paris **291A** (1980), A365–A368.

9. ______, *Algebraic grounds for the proof of algebraic independence. How to obtain measures of algebraic independence using elementary methods. Part I. Elementary algebra*, Comm. Pure Appl. Math. **34** (1981), 1–28.

10. ______, *Measures of irrationality, transcendence and algebraic independence*, Journées Arithmétiques 1980, (J. V. Armitage (ed.)), Cambridge Univ. Press, Cambridge, 1982, pp. 11–82.

11. ______, *Criteria of algebraic independence of several numbers*, see this volume, Chapter 4.

CHAPTER 4

CRITERIA OF ALGEBRAIC INDEPENDENCE OF SEVERAL NUMBERS

1. While for a single number one has a very powerful Gelfond criterion [**1**] of transcendence, for more than one number there are only a few results that can be considered satisfactory. We present a brief survey of them, and formulate and prove their improvements. Let us stress one new feature of the proposed criteria. We formulate them in such a way that they immediately imply results on the measure of transcendence or the measure of algebraic independence. Such an approach explains why the Gelfond criterion must be changed to a different kind of statement.

Let us start with the well-known Gelfond criterion, that we present in the case of splitting of size and degree.

GELFOND LEMMA 1.1 (BROWNAWELL [**4**] AND WALDSCHMIDT [**6**]). *Let $\theta \in \mathbf{C}$ and $a > 1$. Let δ_N and σ_N be monotonically increasing sequences of positive numbers such that $\sigma_N \to \infty$ and*

$$\delta_{N+1} \leqslant a\delta_N, \qquad \sigma_{N+1} < a\sigma_N.$$

If for every $N \geqslant N_0$ there is a nonzero polynomial $P_N(z) \in \mathbf{Z}[z]$ such that

$$\deg P_N < \delta_N, \qquad t(P_N) = \deg P_N + \log H(P_N) < \sigma_N$$

and

$$|P_N(\theta)| < \exp(-6a\delta_N\sigma_N),$$

then θ is algebraic and $P_N(\theta) = 0$: $N \geqslant N_1$.

Usually this criterion is applied for $\delta_N = \sigma_N = O(N^\lambda)$ (with some interesting exceptions).

When one tries to generalize this criterion for two numbers, one finds the famous Cassels' [**11**] counterexample of a pair of numbers $(\vartheta_{1,f}, \vartheta_{2,f})$ associated with any monotone functions $f(h) \to \infty$ such that inequalities

$$|x\vartheta_{1,f} + y\vartheta_{2,f} + z| < \exp(-f(h)), \qquad \max(|x|, |y|, |z|) \leqslant h$$

have solutions in integers x, y, z for $h \geqslant h_0$.

However, the construction of the numbers $(\vartheta_{1,f}, \vartheta_{2,f})$ by Cassels shows that both $\vartheta_{1,f}$ and $\vartheta_{2,f}$ are "too well" approximated by rational numbers.

In order to take this feature into account, one tries to use the fact that the subsequence of a given sequence $(\theta_1, \ldots, \theta_n)$ has small "transcendence type" in the sense of Lang.

DEFINITION 1.2. Let $(\theta_1, \ldots, \theta_n) \in \mathbf{C}^n$ and $\tau \geqslant n + 1$. The set $(\theta_1, \ldots, \theta_n)$ is said to have transcendence type at most τ ($\leqslant \tau$) if there exists a constant $C > 0$ such that for every nonzero $P(x_1, \ldots, x_n) \in \mathbf{Z}[x_1, \ldots, x_n]$ we have

$$\log|P(\theta_1, \ldots, \theta_n)| > C \cdot t(P)^\tau.$$

We can afford to interrupt the exposition and attract attention to two problems.

PROBLEM 1.3. Let $\varepsilon > 0$ and $n \geqslant 2$. Is is true that almost all numbers $(\theta_1, \ldots, \theta_n) \in \mathbf{C}^n$ (or $\mathbf{R}^n$) with respect to a Lebesgue measure, have transcendence type $\leqslant n + 1 + \varepsilon$?

While $n = 2$ is within the possibilities of the existing methods, $n > 2$ seems a hard question.

PROBLEM 1.4. To find for $n \geqslant 1$ an example of a set $(\theta_1, \ldots, \theta_n) \in \mathbf{R}^n$ with type of transcendence $n + 1$.

One can suspect that for $n = 1$ this is π; for $n = 2$ this is $(\pi, \Gamma(1/3))$ or $(\pi, \Gamma(1/4))$: their type of transcendence is known to be $\leqslant n + 1 + \varepsilon$ for any $\varepsilon > 0$.

For sequences, whose subsequences have bounded types of transcendence we do have natural generalizations of the Gelfond criterion.

PROPOSITION 1.5 (BROWNAWELL). *Let $(\theta_1, \ldots, \theta_n)$ have transcendence type $\leqslant \tau$. Let $\theta \in \mathbf{C}$, $a > 1$, δ_N and σ_N be monotonically increasing sequences of positive numbers such that $\sigma_N \to \infty$ and*

$$\delta_{N+1} < a\delta_N, \qquad \sigma_{N+1} < a\sigma_N.$$

There is a $C_1 > 0$ such that, if for every $N \geqslant N_0$ there is a nonzero polynomial $P_N(x_0, x_1, \ldots, x_n) \in \mathbf{Z}[x_0, x_1, \ldots, x_n]$ with

$$\deg_{x_0}(P_N) < \delta_N, \qquad t(P_N) < \sigma_N$$

and

$$\log|P_N(\theta, \theta_1, \ldots, \theta_n)| < -C_1(\delta_N\sigma_N)^\tau$$

then θ is algebraic over $\mathbf{Q}(\theta_1, \ldots, \theta_n)$ and $P_N(\theta, \theta_1, \ldots, \theta_n) = 0$ for all $N \geqslant N_1$.

Certainly, Proposition 1.5 is not very sharp and one can wonder whether the upper bound $-C_1(\delta_N\sigma_N)^\tau$ for $\log|P_N(\theta, \theta_1, \ldots, \theta_n)|$ can be replaced by $-C_2\delta_N\sigma_N^\tau$.

There is a possibility of a more careful analysis of the situation in Proposition 1.5, which was considered by the author under the name of "colored sequences". We present the early result in this direction following the reformulation of D. Brownawell [**12**].

PROPOSITION 1.6. *Let* $(\theta_1,\dots,\theta_n)$ *have transcendence type* $\leqslant \tau_1$ *and* $(\theta_2,\dots,\theta_n)$ *have transcendence type* $\leqslant \tau_2$ *and let* $\theta \in \mathbf{C}$. *There exists a constant* $C_3 > 0$ *such that, if there exists a sequence of nonzero polynomials* $P_N(x_0, x_1,\dots,x_n) \in \mathbf{Z}[x_0, x_1,\dots,x_n]$ *for* $N \geqslant N_0$ *with* $t(P_N) \leqslant N$ *and*

$$\log|P_N(\theta,\theta_1,\dots,\theta_n)| < -C_3 N^{\max\{4\tau_2,\tau_1+3\tau_2-1\}},$$

then θ *is algebraic over* $\mathbf{Q}(\theta_1,\dots,\theta_n)$.

Usually this statement is considered for $n = 1$, when $\tau_2 = 1$ and the upper bound for $\log|P_N(\theta,\theta_1)|$ is $-C_3 N^{\max(4,\tau_1+2)}$ (in any case $\tau_1 \geqslant 2$). It is very easy to improve considerably this last result, which will be done later.

However, one wants unconditional generalizations of the Gelfond Lemma [**1**], without any references to the diophantine properties of subsequences. For this there are two options: (i) to impose some algebraic conditions on the polynomials $P_N(\bar{x})$ in the Gelfond Lemma; (ii) to add more analytic restrictions on $|P_N(\bar{\theta})|$. Possibility (i) is the most interesting one and opens a big future when it will be combined with the abstract definition of the auxiliary function as an abstract Padé approximation. Nevertheless, possibility (ii) is also interesting, though less algebraic in its formulation.

The generalization of the Gelfond criterion using (ii) was proposed by the author in 1975. In its initial form it was formulated as follows [**2**]:

PROPOSITION 1.7. *Let* $n \geqslant 1$, $(\theta_1,\dots,\theta_n) \in \mathbf{C}^n$, $a > 1$ *and let* σ_N *be a monotonically increasing function with* $\sigma_N \to \infty$ *as* $N \to \infty$ *and* $\sigma_{N+1} < a\sigma_N$.

Let us assume that for every $N \geqslant N_0$ *there is a nonzero polynomial* $P_N(x_1,\dots,x_n) \in \mathbf{Z}[x_1,\dots,x_n]$ *such that*

$$t(P_N) \leqslant \sigma_N, \qquad -C_4\sigma_N^{2^n} < \log|P_N(\theta_1,\dots,\theta_n)| < -C_5\sigma_N^{2^n}$$

for $N \geqslant N_0$. *Then* $\theta_1,\dots,\theta_n$ *are algebraically dependent* (*over* $\mathbf{Q}$).

The exponent 2^n is the best only for $n = 1$. Naturally, the lower and upper bounds for $\log|P_N(\theta_1,\dots,\theta_n)|$ can be improved and the statement of this criterion of algebraic independence can be reformulated in a better way.

However, this criterion still looks slightly artificial, because in practice one cannot find a sequence $(\theta_1,\dots,\theta_n)$ such that $\log|P_N(\theta_1,\dots,\theta_n)|$ is bounded below and above by a function of $t(P_N)$ of order $O(t(P_N)^{n+1})$ or $o(t(P_N)^{n+1})$ for $P_N(x_1,\dots,x_n) \in \mathbf{Z}[x_1,\dots,x_n]$.

Moreover I do not even know whether such a sequence $(\theta_1,\dots,\theta_n) \in \mathbf{C}^n$ exists and, if it does, what is the measure of such a set of sequences?

Essentially, all previously formulated criteria of algebraic independence are presented. In this paper we will present new ones. Most of them copy the style of existing criteria but with some new features. For example, the most important development we are trying to pursue is an attempt, once and for all, to get the results not in the form of algebraic independence, but rather in the form of the measure of algebraic independence. We take this approach because we want to present a unified approach and measure of algebraic independence.

As one will see, the changes in the criteria are rather minor.

In order to get an idea of what kind of criteria of algebraic independence we can propose, I can suggest to you the following reformulation of Brownawell's criterion.

LEMMA 1.8. *Let $\theta = (\theta_1, \theta_2)$ be algebraically independent numbers, $a > 1$ and let σ_N be a monotonically increasing function $\sigma_N \to \infty$ such that $\sigma_{N+1} < a\sigma_N$.*

We suppose that for every $N \geqslant N_0$ there exists a polynomial $P_N(x, y) \in \mathbf{Z}[x, y]$, $P_N \not\equiv 0$, such that

$$t(P_N) \leqslant \sigma_N \quad \text{and} \quad \log|P_N(\theta_1, \theta_2)| < -\sigma_N^{\mu}.$$

Then $\bar{\theta}$ has transcendence type $\geqslant \mu/2$ and, moreover, there are infinitely many algebraic numbers ξ_1, ξ_2 such that

$$[\mathbf{Q}(\xi_1, \xi_2):\mathbf{Q}] \leqslant L, \qquad [\mathbf{Q}(\xi_1, \xi_2):\mathbf{Q}] \cdot \left(\frac{t(\xi_1)}{d(\xi_1)} + \frac{t(\xi_2)}{d(\xi_2)}\right) \leqslant L$$

and

$$|\theta_1 - \xi_1| + |\theta_2 - \xi_2| < \exp(-C \cdot L^{\mu/2}).$$

In particular, one gets Brownawell's statement. However one immediately sees that the simultaneous approximation is of nontrivial type. Roughly speaking, in the "generic case" $[\mathbf{Q}(\xi_1, \xi_2): \mathbf{Q}]$ is the product $d(\xi_1) \cdot d(\xi_2)$. If this *would* be true, then

$$d(\xi_1)d(\xi_2) \leqslant L, \qquad d(\xi_2)t(\xi_1) + d(\xi_1)t(\xi_2) \leqslant L.$$

This implies for one of θ_1, θ_2, say θ_1, the satisfaction of infinitely many inequalities in algebraic numbers ξ_1,

$$|\theta_1 - \xi_1| < \exp\left(-C\left(d(\xi_1)t(\xi_1)\right)^{\mu/2}\right),$$

almost as if θ_1 is of the transcendence type $\geqslant \mu$! (Not $\mu/2$.) A simple counterexample will show that this is impossible (say $\mu = 3$ e.g.), which means that $[\mathbf{Q}(\xi_1, \xi_2): \mathbf{Q}] \ll d(\xi_1)d(\xi_2)$. The real truth is of course in the statement like

$$[\mathbf{Q}(\xi_1, \xi_2):\mathbf{Q}] = d(\xi_1) = d(\xi_2),$$

i.e. both ξ_1, ξ_2 are generators of $\mathbf{Q}(\xi_1, \xi_2)$. One knows e.g. that in this case the Liouville theorem for $|P(\xi_1, \xi_2)|$ looks much better than in a general case.

In any case, we can speak about simultaneous approximation (and promote this direction).

2. We propose first of all the following form of the criterion of the measure of transcendence of a single number.

LEMMA 2.1. *Let $\theta \in \mathbf{C}$ and let $\mathfrak{M}$ be a linearly ordered countable set and functions $\delta_{\mathfrak{N}}$, $\sigma_{\mathfrak{N}}$, $F_{\mathfrak{N}}$, $G_{\mathfrak{N}}$, $K_{\mathfrak{N}}$ be monotonically increasing as functions of $\mathfrak{N} \in \mathfrak{M}$ such that $\sigma_{\mathfrak{N}} \to \infty$, $F_{\mathfrak{N}} \to \infty$, $G_{\mathfrak{N}} \to \infty$, $K_{\mathfrak{N}} \to \infty$ as $\mathfrak{N} \to \infty$ together with the conditions $\delta_{\mathfrak{N}} \leqslant \sigma_{\mathfrak{N}}$ and $\sigma_\mu < a \cdot \sigma_{\mathfrak{N}}$ for a successor μ of $\mathfrak{N}$ in $\mathfrak{M}$. For any $\mathfrak{N} \geqslant \mathfrak{N}_0$ either*

(i) *there exists a polynomial $P_{\mathfrak{N}}(x) \in \mathbf{Z}[x]$ such that*

$$d(P_{\mathfrak{N}}) \leqslant \delta_{\mathfrak{N}}, \qquad t(P_{\mathfrak{N}}) \leqslant \sigma_{\mathfrak{N}}$$

and

$$-G_{\mathfrak{N}} < \log|P_{\mathfrak{N}}(\theta)| < -F_{\mathfrak{N}},$$

or

(ii) *there is a system of polynomials $C_l(x) \in \mathbf{Z}[x]$: $l \in \mathfrak{L}_{\mathfrak{N}}$ without a common factor such that*

$$d(C_l) \leqslant \delta_{\mathfrak{N}}, \qquad t(C_l) \leqslant \sigma_{\mathfrak{N}}, \qquad \log|C_l(\theta)| < -K_{\mathfrak{N}}:$$

$l \in \mathfrak{L}_{\mathfrak{N}}$. Let $\lim \sigma_{\mathfrak{N}}/F_{\mathfrak{N}} = 0$ as $\mathfrak{N} \to \infty$ in $\mathfrak{M}$. Then for any algebraic number ζ of degree $\leqslant d(\zeta)$ and height $\leqslant H(\zeta)$ we get for any $\mathfrak{N} \in \mathfrak{M}$, $\mathfrak{N} \geqslant \mathfrak{N}_0$

$$\log|\theta - \zeta| > -\max\{G_{\mathfrak{N}} + c_1\sigma_{\mathfrak{N}},\ F_{\mathfrak{N}} + c_1\sigma_{\mathfrak{N}},\ K_{\mathfrak{N}} + c_1\sigma_{\mathfrak{N}}\}, \tag{2.2}$$

provided that

$$\min\{K_{\mathfrak{N}}, F_{\mathfrak{N}}\} \geqslant \sigma_{\mathfrak{N}} d(\zeta) + \delta_{\mathfrak{N}} \log H(\zeta) + 1.$$

In particular, let $F_{\mathfrak{N}}$, $G_{\mathfrak{N}}$, $K_{\mathfrak{N}}$ be of the same order of magnitude

$$\lim G_{\mathfrak{N}}/F_{\mathfrak{N}} = \gamma, \qquad \lim K_{\mathfrak{N}}/F_{\mathfrak{N}} = \chi$$

and $\gamma, \chi \neq 0, \infty$. We choose $\mathfrak{N}_1$ from the condition

$$c_2 F_{\mathfrak{N}_1} \geqslant \sigma_{\mathfrak{N}_1} d(\zeta) + \delta_{\mathfrak{N}_1} \log H(\zeta) \geqslant c_2^{-1} F_{\mathfrak{N}_1},$$

for some $c_2 > 0$. Then we have

$$|\theta - \zeta| > \exp(-c_3 F_{\mathfrak{N}_1}) \tag{2.3}$$

for $c_3 > 0$ depending on γ, χ and c_2.

REMARK 2.4. This statement covers all the cases: (a) when the transcendence type is estimated and $\mathfrak{M} = \mathbf{N}$, $\delta_N = \sigma_N$; (b) when type and degree are estimated simultaneously, $\mathfrak{M} \subseteq \{(x, y) \in \mathbf{N}^2: x \leqslant y\}$ with a lexicographic order; or (c) when the degree is bounded, $\mathfrak{M} = \mathbf{N} \times \{1, \ldots, d\}$. E.g. let us consider case (a). We get

COROLLARY 2.5. *Let $\mathfrak{M} = \mathbf{N}$, $\delta_N = \sigma_N$ and $f(t)$ be a function inverse to $F(t)/\sigma(t)$, where $F(N) = F_N$, $\sigma(N) = \sigma_N$, etc. Then under the assumption of Lemma 2.1,*

$$|\theta - \zeta| > \exp\{-c_4 F(f(t(\zeta)))\} \tag{2.6}$$

where

$$t(\zeta) = d(\zeta) + \log H(\zeta) \geqslant t_0.$$

Lemma 2.1 contains, in particular, the sharpened form of Gelfond's criterion. Indeed, when $F_{\mathfrak{N}}$ grows faster than $\delta_{\mathfrak{N}}\sigma_{\mathfrak{N}}$ and $K_{\mathfrak{N}}$ grows faster than $\delta_{\mathfrak{N}}\sigma_{\mathfrak{N}}$, then the bound (2.2) contradicts Dirichlet's bound.

The bound (2.2) or (2.3), (2.6) can be, naturally, reformulated for $|P(\theta)|$ with $P(x) \in \mathbf{Z}[x]$ instead of $|\theta - \zeta|$.

PROPOSITION 2.7. *Let us assume that the conditions of* 2.1 (*or* 2.5) *are satisfied. Then for* $P(x) \in \mathbf{Z}[x]$, $P(x) \not\equiv 0$, *the same bounds* (2.2), (2.3) (*or* (2.6)) *are satisfied for* $|\theta - \zeta|$ *replaced by* $|P(\theta)|$, *if one replaces* $d(\zeta)$ *by* $d(P)$, $H(\zeta)$ *by* $H(P)$ (*and* $t(\zeta)$ *by* $t(P)$). *In particular, in the situation of Corollary* 2.5 *we get*

$$|P(\theta)| > \exp(-c_5 F(f(t(P)))) \quad \textit{for some } c_5 > 0.$$

PROOF OF LEMMA 2.1. Let us take a sufficiently large $C > 0$ and $\mathfrak{N}_0 \in \mathfrak{M}$ such that

$$\min\{F_{\mathfrak{N}}, G_{\mathfrak{N}}, K_{\mathfrak{N}}\} > C\sigma_{\mathfrak{N}}$$

if $\mathfrak{N} \geqslant \mathfrak{N}_0$. Let ζ be an algebraic number and $P(\zeta) = 0$ with $d(\zeta) \leqslant d(P)$, $H(\zeta) \leqslant H(P)$ for $P(x) \in \mathbf{Z}[x]$, $P(x) \not\equiv 0$. One can take $P(x)$ as a minimal polynomial of ζ, but in view of Proposition 2.7, there is no need to do this.

Let us take $\mathfrak{N} \geqslant \mathfrak{N}_0$. If for a given $\mathfrak{N}$ the alternative (ii) takes place, there is an $l_0 \in \mathfrak{L}_{\mathfrak{N}}$ such that $C_{l_0}(\zeta) \neq 0$, because at least one of the $C_l(x)$ is relatively prime with the minimal polynomial of ζ. If $P(x)$ is the power of an irreducible polynomial, then again there is a $C_{l_0}(x)$ relatively prime with $P(x)$. However, if $P(x)$ is a reducible polynomial, some modifications of the arguments are necessary. Namely, one finds integer coefficients n_l: $l \in \mathfrak{L}_{\mathfrak{N}}$, such that

$$\sum_{l \in \mathfrak{L}_{\mathfrak{N}}} |n_l| \ll d(P)$$

and $\sum_{l \in \mathfrak{L}_{\mathfrak{N}}} n_l C_l(x)$ is relatively prime with $P(x)$. This leads only to slight changes in the estimates—in constants.

Henceforth, in the case (ii) we can bound below $|C_{l_0}(\zeta)|$ using the Liouville theorem. An alternative approach is to bound $\operatorname{res}(C_{l_0}, P)$ according to a famous formula **[1]**,

$$\operatorname{res}(C_{l_0}, P) \leqslant \{d(C_{l_0})H(P)|P(\zeta)| + d(P)H(C_{l_0})|C_{l_0}(\zeta)|\} \times H(C_{l_0})^{d(P)-1} \cdot H(P)^{d(C_{l_0})-1}.$$

When we are bounding $|C_{l_0}(\zeta)|$ we get

$$|C_{l_0}(\zeta)| \geqslant \exp(-d(\zeta)t(C_{l_0}) - d(C_{l_0}) \log H(\zeta)),$$

or

$$|C_{l_0}(\zeta)| \geqslant \exp(-d(\zeta)\sigma_{\mathfrak{N}} - \log H(\zeta)\delta_{\mathfrak{N}}).$$

At the same time, if

$$K_{\mathfrak{N}} \geqslant d(\zeta)\sigma_{\mathfrak{N}} + \log H(\zeta)\delta_{\mathfrak{N}} + 1,$$

then we have

$$C_{\mathfrak{N}}(\theta) = C_{\mathfrak{N}}(\zeta) + C'_{\mathfrak{N}}(\theta - \zeta)$$

with $\log |C'_{\mathfrak{N}}| \leqslant \sigma_{\mathfrak{N}} + c \cdot \delta_{\mathfrak{N}}$, and we have

$$|\theta - \zeta| \geqslant |C'_{\mathfrak{N}}|^{-1} \big| |C_{\mathfrak{N}}(\theta)| - |C_{\mathfrak{N}}(\zeta)| \big| \geqslant \exp\{-K_{\mathfrak{N}} - \sigma_{\mathfrak{N}} - C\delta_{\mathfrak{N}}\}$$

by the inequality on $K_{\mathfrak{N}}$.

Now, let us consider the alternative (i). In this case there are two possibilities that can happen, $P_{\mathfrak{N}}(\zeta) = 0$ or $P_{\mathfrak{N}}(\zeta) \neq 0$.

If $P_{\mathfrak{N}}(\zeta) \neq 0$, then, as above, one gets

$$|P_{\mathfrak{N}}(\zeta)| \geqslant \exp(-\sigma_{\mathfrak{N}} d(\zeta) - \delta_{\mathfrak{N}} \log H(\zeta)).$$

We have again

$$|\theta - \zeta| \geqslant |P'_{\mathfrak{N}}|^{-1} \cdot \big| |P_{\mathfrak{N}}(\theta)| - |P_{\mathfrak{N}}(\zeta)| \big|,$$

where $\log |P'_{\mathfrak{N}}| \leqslant \sigma_{\mathfrak{N}} + C \cdot \delta_{\mathfrak{N}}$.

If

$$F_{\mathfrak{N}} \geqslant d(\zeta)\sigma_{\mathfrak{N}} + \log H(\zeta) \cdot \delta_{\mathfrak{N}} + 1,$$

then

$$|\theta - \zeta| > \exp\{-F_{\mathfrak{N}} - \sigma_{\mathfrak{N}} - C\delta_{\mathfrak{N}}\}.$$

Let, at last, $P_{\mathfrak{N}}(\zeta) = 0$. Then

$$|P_{\mathfrak{N}}(\theta)| \leqslant \exp\{\sigma_{\mathfrak{N}} + C\delta_{\mathfrak{N}}\} \cdot |\theta - \zeta|.$$

The inequalities in (i) show

$$|\theta - \zeta| > \exp\{-G_{\mathfrak{N}} - \delta_{\mathfrak{N}} \cdot C - \sigma_{\mathfrak{N}}\}.$$

The inequality (2.2) is proved.

3. We formulate a version of the Liouville theorem in the case of an arbitrary set $S \subset \mathbf{C}^n$ of algebraic numbers of the dimension zero, i.e. being a set of common zeros of a zero-dimensional ideal in $\mathbf{Z}[x_1,\ldots,x_n]$. Namely, we consider the following situation. We have an ideal J in $\mathbf{Z}[x_1,\ldots,x_n]$ which is zero dimensional in the sense, e.g. that the set

$$S(J) = \{\vec{x}_0 \in \mathbf{C}^n \colon P(\vec{x}_0) = 0 \text{ for every } P \in J\}$$

is a finite set. We are working now in the affine situation since projective considerations do not add anything. Every element of $S(J)$ has a prescribed multiplicity, defined e.g. in **[16]**.

Let $P_1,\ldots,P_k$ be certain generators of J, whose degrees and types we know as

$$d(P_i) \leqslant D_i, \qquad t(P_i) \leqslant T_i\colon \qquad i = 1,\ldots,k.$$

Naturally, $k \geqslant n$. By an intersection theory (say, Bezout's theorem) we have

$$|S(J)| \leqslant D_1 \cdots D_k.$$

This bound is far from optimal whenever $k > n$. In this case we can even use the following bound

$$|S(J)| \leqslant \Big(\max_{i=1,\ldots,k} D_i\Big)^n.$$

Instead of treating different cases, we assume already that $P_1(x),\ldots,P_n(x)$ have only finitely many common zeros. Then we consider an ideal $I = (P_1,\ldots,P_n)$ and $S(I)$ instead of $S(J)$. Let $\vec{x}_0 \in S(I)$ and $m(\vec{x}_0)$ be a multiplicity of $\vec{x}_0$ in $S(I)$. Then we have

$$\sum_{\vec{x}_0\in S(I)} m(\vec{x}_0) \leqslant D_1 \cdots D_n.$$

We can apply to I the theory of u-resultants in the form of Kronecker (cf. van der Waerden [**18**] or Hodge-Pedoe [**15**]). One gets the following main statement.

LEMMA 3.1. *In the above notation, the coordinates of common zeros $\vec{x}_0 \in S(J)$ are bounded above in terms of $D_1,\ldots,D_n$ and $T_1,\ldots,T_n$. Namely, let*

$$\vec{x}_0 = (x_{10},\ldots,x_{n0}) \quad \textit{for } \vec{x}_0 \in S(I).$$

Then for every $i = 1,\ldots,n$ there is a rational integer A_i, $A_i \neq 0$, such that for any distinct elements $\vec{x}^1,\ldots,\vec{x}^l$ of $S(I)$ and $n_i \leqslant m(\vec{x}^i)$: $i = 1,\ldots,l$, the number

$$A_i \cdot \prod_{j=1}^{l} (\vec{x}^j)_i^{n_j}$$

is an algebraic integer. Moreover, for any $i = 1,\ldots,n$ one has

$$|A_i| \cdot \prod_{\vec{x}\in S(I)} \max\{1,(\vec{x})_i\}^{m(\vec{x})} \leqslant \exp\left(C_1 \cdot \sum_{r=1}^{n} T_r \cdot \prod_{\substack{s=1\\ s\neq r}}^{n} D_s\right),$$

for a constant $C_1 > 0$ depending only on n.

The analogue of the Liouville theorem applied to the elements of the set $S(I)$ has the following form.

LEMMA 3.2. *Let, as before, $I = (P_1,\ldots,P_n)$ where $d(P_i) \leqslant D_i$, $t(P_i) \leqslant T_i$: $i = 1,\ldots,n$, and the set $S(I)$ is a finite one.*

Let $R(x_1,\ldots,x_n) \in \mathbf{Z}[x_1,\ldots,x_n]$, $R \not\equiv 0$. Let us assume that for several distinct $\vec{x}^1,\ldots,\vec{x}^l$ from $S(I)$ of multiplicities $m_1,\ldots,m_l$, respectively, we have

$$R(\vec{x}^j) \neq 0: \qquad j = 1,\ldots,l.$$

Then we have the following lower bound:

$$\prod_{j=1}^{l} |R(\vec{x}^j)|^{m_j} \geqslant \exp\left\{-C_2\left\{\sum_{i=1}^{n} t(P_i)d(R)\cdot\prod_{s\neq i} d(P_i) + t(R)d(P_1)\cdots d(P_n)\right\}\right\}, \tag{3.3}$$

where $C_2 > 0$ depends only on n.

PROOF OF LEMMA 3.2. We consider the following auxiliary object, taking into account the notation of Lemma 3.1:

$$\mathfrak{M} = \prod_{i=1}^{n} A_i^{d(R)} \prod_{\substack{\vec{x}\in S(I)\\ R(\vec{x})\neq 0}} R(\vec{x})^{m(\vec{x})}. \tag{3.4}$$

In (3.4) the product is over only those elements $\vec{x}$ of $S(I)$ for which $R(\vec{x}) \neq 0$. This is a usual "seminorm". Then the definition of the set $S(I)$ (invariant under the algebraic conjugation) and the choice of A_i: $i = 1,\dots,n$, in Lemma 3.1 we get $\mathfrak{M} \in \mathbf{Z}$. From the form of $\mathfrak{M}$ it follows that $\mathfrak{M} \neq 0$, so that

$$|\mathfrak{M}| \geq 1. \tag{3.5}$$

We represent $\mathfrak{M}$ as a product of two factors, $\mathfrak{M} = \mathfrak{A} \cdot \mathfrak{B}$, where $\mathfrak{A}$ is the product in the left-hand side of (3.3). The product $\mathfrak{B}$ is bounded from above by Lemma 3.1:

(3.6)

$$|\mathfrak{B}| \leq \left(\prod_{i=1}^{n} A_i \prod_{\substack{\vec{x}\in S(I)\\ \vec{x}\notin\{\vec{x}^j:\, j=1,\dots,L\}\\ R(\vec{x})\neq 0}} \max\left(1, |(\vec{x})_i|\right)^{m(\vec{x})} \right)^{d(R)} \exp\left(2t(R)\prod_{i=1}^{n} d(P_i)\right)$$

$$\leq \exp\left(C_4 \left\{ t(R) \prod_{i=1}^{n} d(P_i) + d(R) \sum_{i=1}^{n} t(P_i) \prod_{s\neq i} d(P_s) \right\} \right).$$

Combining (3.5) and (3.6) one gets (3.3).

In order to prove our results in a straightforward way, we make some agreements on the notations.

If we start, in the general case with the ideal $I = (P_1,\dots,P_n)$ in $\mathbf{Z}[x_1,\dots,x_n]$ of dimension zero, then the set

$$S(I) = \{\vec{x}_0 \in \mathbf{C}^n:\, P_i(\vec{x}_0) = 0:\, i = 1,\dots,n\}$$

is a set of vectors in $\mathbf{C}^n$ with algebraic coordinates. The set $S(I)$ is naturally divided into components S_α closed under the conjugation

$$S(I) = \bigcup_{\alpha\in A} S_\alpha.$$

The partition into sets S_α can be made in such a way that all elements of S_α have the same multiplicity m_α of its occurring in $S(I)$. We call S_α an irreducible component of $S(I)$ and we have

$$\sum_{\alpha\in A} m_\alpha |S_\alpha| \leq d(P_1)\cdots d(P_n).$$

We can define a type and degree of the component S_α. The degree is naturally $|S_\alpha|$ itself and the type is defined using the sizes of the coordinates of elements of $S(I)$.

For this we remained at Lemma 3.1, where we had nonzero rational integers A_i: $i = 1,\ldots,n$, such that

$$A_i \cdot \prod_{\vec{x}\in S'} (\vec{x})_i^{n(\vec{x})}$$

is an algebraic integer for any $S' \subseteq S(I)$ and $n(\vec{x}) \leqslant m(\vec{x})$: $\vec{x} \in S'$, and we have a bound

$$|A_i| \cdot \prod_{\vec{x}\in S(I)} \max\left\{1, |(\vec{x})_i|^{m(\vec{x})}\right\} \leqslant \exp\left\{C_1 \cdot \sum_{j=1}^{n} t(P_j) \cdot \prod_{s\neq j} d(P_s)\right\}:$$

$$i = 1,\ldots,n.$$

Naturally, the quantity

$$\log\left\{\prod_{i=1}^{n} |A_i| \cdot \prod_{\vec{x}\in S(I)} \max\left\{1, |(\vec{x})_i|\right\}^{m(\vec{x})}\right\}$$

can be called *the size* of $S(I)$. We can define in a similar way the size of the component S_α as

$$\log\left\{\prod_{i=1}^{n} |a_i^\alpha| \cdot \prod_{\vec{x}\in S_\alpha} \max\left\{1, |(\vec{x})_i|\right\}\right\},$$

where a_i^α are the smallest nonzero rational integers such that

$$a_i^\alpha \cdot \prod_{\vec{x}\in S_i} (\vec{x})_i$$

are algebraic integers for any $S_1 \subseteq S_\alpha$: $i = 1,\ldots,n$.

By the *type* of the set S_α we understand the sum of the degree $|S_\alpha|$ and its size. We denote the type of S_α by $t(S_\alpha)$. Similarly one can define a type of $s(I)$ as a sum of its degree and size. The type of $S(I)$ is also denoted by $t(S(I))$. We note that the degree of $S(I)$ is not $|S(I)|$ but rather $\Sigma_{\vec{x}\in S(I)} m(\vec{x})$, when elements of $S(I)$ are counted with multiplicities.

By the definition of types we have

$$\sum_{\alpha\in A} t(S_\alpha) \cdot m_\alpha \leqslant t(S(I)) \leqslant \exp\left\{C_2 \sum_{j=1}^{n} t(P_j) \prod_{s\neq j} d(P_s)\right\}.$$

In fact, in order to derive this inequality, in the part concerning the relationship between a_i^α: $\alpha \in A$ and A_i, one uses the Gauss lemma and properties of u-resultants [**17**].

The case $n = 2$ is very easy to understand using resultants [**3**]. In this case we have two relatively prime polynomials $P(x, y)$, $Q(x, y) \in \mathbf{Z}[x, y]$ and the set S of their common zeros

$$S = \left\{(x, y) \in \mathbf{C}^2 \colon P(x, y) = Q(x, y) = 0\right\}.$$

Their coordinates can be determined using the resultants of $P(x, y)$, $Q(x, y)$. We make a change of the coordinates to a "normal" form and get new polynomials $P(x', y')$, $Q(x', y')$. In "normal" coordinates distinct elements of S have both their coordinates distinct.

The resultants $R_1(x')$ and $R_2(y')$ of $P'(x', y')$, $Q'(x', y')$ (in "normal" coordinates) with respect to y' and x', respectively, can be written as

$$R_1(x') = a_1 \prod_{i=1}^{k} (x' - \zeta'_{1,i})^{m_i}, \qquad R_2(y') = a_2 \prod_{i=1}^{k} (y' - \zeta'_{2,i})^{m_i},$$

where $(\zeta'_{1,i}, \zeta'_{2,i})$ is an element of S of the multiplicity m_i. In particular, one can represent $R_1(x')$, $R_2(y')$ in terms of the powers of irreducible polynomials,

$$R_1(x') = \prod_{\alpha\in A} P^1_\alpha(x')^{m_\alpha}, \qquad R_2(y') = \prod_{\alpha\in A} P^2_\alpha(y')^{m_\alpha},$$

where $P^1_\alpha(x') = a_1^\alpha \Pi_j(x' - \zeta^\alpha_{1,j})$ and $P^2_\alpha(y') = a_2^\alpha \Pi_j(y' - \zeta^\alpha_{2,j})$ and $(\zeta^\alpha_{1,j}, \zeta^\alpha_{2,j})$ are elements of S. We naturally define

$$S_\alpha = \{(\zeta^\alpha_{1j}, \zeta^\alpha_{2j})\},$$

so that $\bigcup_{\alpha\in\beta} S_\alpha = S$. We can now look on the type of S and S_α: $\alpha \in A$, from the point of view of resultants.

E.g. one can define the size of S as the sum of the sizes of R_1 and R_2. This will be equivalent to the previous definition of the size.

Consequently we can say

$$t(S) \leqslant t(R_1) + t(R_2),$$

while for $\alpha \in A$,

$$t(S_\alpha) \leqslant t(P^1_\alpha) + t(P^2_\alpha).$$

This definition of the type, expressed not in coordinate form is, as a matter of fact, more useful in the higher-dimensional case, when we are working with mixed ideals, nonnormal intersections, etc.

We want to remark that our decomposition of $S(I)$ into the union of (disjoint) irreducible components is, certainly, not unique. Indeed, a given irreducible component can split if the corresponding polynomial is decomposed over $\mathbf{Z}$. However, it is easier to work with our definition, though, using a simple geometry and ideal theory, one can define a canonical decomposition of $S(I)$ into the (maximal) union of irreducible components over $\mathbf{Z}$.

4. We will now present a complete proof of the criterion of the algebraic independence of two numbers that generalizes our previous statements.

THEOREM 4.1. *Let $(\theta_1, \theta_2) \in \mathbf{C}^2$, $a > 1$ and σ_N, σ'_N be monotonically increasing functions such that $\sigma'_N > \sigma_N > N$;*

$$\sigma_{N+1} < a\sigma_N.$$

For any $N \geqslant N_0$ let there exist either

(i) a polynomial $P_N(x, y) \in \mathbf{Z}[x, y]$ such that

$$t(P_N) \leqslant N, \qquad -\sigma'_N < \log|P_N(\theta_1, \theta_2)| < -\sigma_N$$

or

(ii) *a system* $C_l(x, y) \in \mathbf{Z}[x, y]$: $l \in \mathfrak{L}_N$, *of polynomials without a common factor such that*

$$t(C_l) \leqslant N, \qquad \log|C_l(\theta_1, \theta_2)| < -\sigma_N: \qquad l \in \mathfrak{L}_N.$$

Now if σ_N *is growing faster than* N^3: $\lim_{N\to\infty} \sigma_N/N^3 = \infty$ *and* $\lim \sigma'_N/\sigma_N < \infty$, *then numbers* θ_1, θ_2 *are algebraically dependent* (*over* $\mathbf{Q}$).

PROOF. Let us assume that θ_1, θ_2 are algebraically independent. We can also assume that $\sigma'_N = C_3 \cdot \sigma_N$. This last assumption can be lifted.

First of all we must start with a pair of relatively prime polynomials of type (i) or (ii).

For simplicity in this proof we call for any given $N \geqslant N_0$, the case (i) "blue", and if case (ii) is satisfied, the corresponding situation (or the number N) is called "red".

First of all, we use a very simple argument to get two relatively prime polynomials.

LEMMA 4.2. *Let us assume that there is a polynomial* $P(x_1,\ldots,x_n) \in \mathbf{Z}[x_1,\ldots,x_n]$ *such that for a given* $(\theta_1,\ldots,\theta_n) \in \mathbf{C}^n$ *we have*

$$|P(\theta_1,\ldots,\theta_n)| \leqslant \varepsilon < 1$$

and $t(P) \leqslant T$. *Then either there exists a polynomial* $Q(x_1,\ldots,x_n) \in \mathbf{Z}[x_1,\ldots,x_n]$ *which is a power of an irreducible polynomial and*

$$|Q(\theta_1,\ldots,\theta_n)| \leqslant \varepsilon^{1/3} \tag{4.3}$$

and $t(Q) \leqslant 2T$, *or there are two relatively prime polynomials* $P_1(x_1,\ldots,x_n)$, $P_2(x_1,\ldots,x_n) \in \mathbf{Z}[x_1,\ldots,x_n]$ *such that*

$$\max\{|P_1(\theta_1,\ldots,\theta_n)|, |P_2(\theta_1,\ldots,\theta_n)|\} \leqslant \varepsilon^{1/3} \tag{4.4}$$

and $t(P_1) \leqslant 2T, t(P_2) \leqslant 2T$.

The proof is straightforward, follows Gelfond's arguments and has already been presented many times.

Let us call the alternative (4.3) in Lemma 4.2 "green" and that in (4.4) "orange". If given N is "red", then, independently of what ("green" or "orange") occurs, we have two relatively prime polynomials.

CLAIM 4.5. *For a "red"* $N \geqslant N_0$ *there are always two polynomials* $P^r_N(x, y)$, $Q^r_N(x, y) \in \mathbf{Z}[x, y]$ *that are relatively prime and satisfy*

$$\begin{gathered} \max\{|P^r_N(\theta_1, \theta_2)|, |Q^r_N(\theta_1, \theta_2)|\} \leqslant \exp(-\sigma_N/3), \\ \max\{t(P^r_N), t(Q^r_N)\} \leqslant 2N. \end{gathered} \tag{4.6}$$

Indeed, we use Lemma 4.2 for $n = 2$ and any of the nonzero polynomials $C_{l_0}(x, y) = P(x, y)$: $l_0 \in \mathfrak{L}_N$. In the "orange" case (4.4) we get (4.6) at once. In the "green" case (4.3) we take a polynomial $Q(x, y) \stackrel{\text{def}}{=} Q^r_N(x, y)$ and find a polynomial $C_{l_1}(x, y)$: $l_1 \in \mathfrak{L}_N$, relatively prime with $Q_N^r(x, y)$. Such a polynomial always exists, since $Q_N^r(x, y)$ is a power of an irreducible polynomial. We put $P_N^r(x, y) = C_{l_1}(x, y)$, so that (4.6) is satisfied.

As a consequence of Claim 4.5 we obtain that if there is any $N \geqslant N_1$ (for sufficiently large $N_1 > N_0$) which is "red", then for this N we have two relatively prime polynomials with the properties (4.6). Hence, we can consider the situation when all $N \geqslant N_1$, for some N_1, are "blue".

CLAIM 4.7. *Let all $N \geqslant N_1$ be "blue". Then for any $N_2 > N_1$ there is $N \geqslant N_2$ such that one has two relatively prime polynomials $P_N^b(x, y) \in \mathbf{Z}[x, y]$, $Q_N^b(x, y) \in \mathbf{Z}[x, y]$ satisfying*

$$(4.8)\qquad \begin{aligned} \max\{|P_N^b(\theta_1, \theta_2)|, |Q_N^b(\theta_1, \theta_2)|\} &\leqslant \exp(-\sigma_N/3), \\ \max\{t(P_N^b), t(Q_N^b)\} &\leqslant 3N. \end{aligned}$$

Indeed, if for some $N \geqslant N_2$ we have an "orange" case (4.4), then (4.8) is true. Let, however, all $N \geqslant N_2$ be "green" (4.3). Then we have a polynomial $Q'_N(x, y) \in \mathbf{Z}[x, y]$, being a power of an irreducible one, such that

$$(4.9)\qquad \log|Q'_N(\theta_1, \theta_2)| < -\sigma_N/3, \qquad t(Q'_N) \leqslant 2N;$$

moreover, $Q'_N = (P'_N)^{s_N}$ for an irreducible P'_N and $s_N \geqslant 0$. The two polynomials P'_N and P'_{N+1} must be different for some $N > N_2$, since otherwise for $P'_N = P'_{N+1} = P'_{N+2} = \cdots$, we get from (4.9) that

$$\log|P'_N(\theta_1, \theta_2)| < -\sigma_M/3s_M, \qquad s_M \leqslant 2M,$$

and $M \to \infty$ gives us $P'_N(\theta_1, \theta_2) = 0$. If, however, $P'_N(x, y)$ and $P'_{N+1}(x, y)$ are different (relatively prime), then two polynomials $Q'_N(x, y)$ and $Q'_{N+1}(x, y)$ satisfy (4.8).

As a result of Claims 4.5 and 4.7 we obtain for any $N_3 \geqslant N_0$ the existence of $N > N_3$ such that there are two relatively prime polynomials $\overline{P}_N(x, y)$, $\overline{Q}_N(x, y) \in \mathbf{Z}[x, y]$ satisfying

$$(4.10)\qquad \begin{aligned} \max\{|\overline{P}_N(\theta_1, \theta_2)|, |\overline{Q}_N(\theta_1, \theta_2)|\} &< \exp(-\sigma_N/3) \\ \max\{t(\overline{P}_N), t(\overline{Q}_N)\} &\leqslant 3N. \end{aligned}$$

We take a sufficiently large N such that $\sigma_N/3$ is sufficiently large with respect to $(3N)^3$; symbolically $N^3 = o(\sigma_N)$ or $\sigma_N > C \cdot 3^4 N^3$ for a large constant $C > 0$ and take two polynomials $\overline{P}_N(x, y)$, $\overline{Q}_N(x, y)$ satisfying (4.10).

Our main object becomes a set $\overline{S}_N = S(\overline{P}_N, \overline{Q}_N)$ of common zeros of $\overline{P}_N(x, y)$ and $\overline{Q}_N(x, y)$. The set $\overline{S}_N$ has a degree (elements counted with multiplicities) at most $3^2 \cdot N^2$, and type (estimated through the resultants) at most $8 \cdot 3^2 \cdot N^2$: $t(\overline{S}_N) \leqslant 72N^2$. We are looking at irreducible components $S_{N,\alpha}$: $\alpha \in A_N$, of $\overline{S}_N$,

$$\overline{S}_N = \bigcup_{\alpha \in A_N}{}' S_{N,\alpha}$$

($\bigcup'$ indicates that elements of the union are disjoint sets).

We have

$$\sum_{\alpha \in A_N} m_\alpha d(S_{N,\alpha}) \leqslant d(\bar{S}_N) \leqslant 9N^2, \tag{4.11}$$

where $d(S_{N,\alpha}) = |S_{N,\alpha}|$ is a degree of $S_{N,\alpha}$ and m_α is a multiplicity of the component, and similarly for types

$$\sum_{\alpha \in A_N} m_\alpha t(S_{N,\alpha}) \leqslant t(\bar{S}_N) \leqslant 72N^2. \tag{4.12}$$

We are looking now on those elements of $\bar{S}_N$ and $S_{N,\alpha}$ that are close to $\bar{\theta} = (\theta_1, \theta_2)$. All evaluations are made in l^1-norm in $\mathbf{C}^2$.

For a "green" $\bar{\xi} \in \bar{S}_N$ we want to bound above $\|\bar{\theta} - \bar{\xi}\|$. We use for this the following convenient notations:

$$\|\bar{\theta} - \bar{\xi}\| \leqslant \exp(-E(\bar{\xi})). \tag{4.13}$$

($E(\bar{\xi})$ is defined since $\bar{\theta} \neq \bar{\xi}$) where, for simplicity, we always assume

$$E(\bar{\xi}) \leqslant \sigma_N/3 \quad \text{for any } \bar{\xi} \in \bar{S}_N. \tag{4.14}$$

In this notation we can express the main auxiliary result that is formulated in a rather general form:

PROPOSITION 4.15. *Let $\bar{\theta} = (\theta_1, \theta_2) \in \mathbf{C}^2$ (with the l^1-norm) and $P(x, y)$, $Q(x, y)$ be two relatively prime polynomials from $\mathbf{Z}[x, y]$.*

Let $S = S(P, Q)$ be the set of the zeros of an ideal (P, Q) and

$$S = {\bigcup_{\alpha \in A}}' S_\alpha$$

its representation through irreducible components. If m_α is a multiplicity of S_α, then one has

$$\sum_{\alpha \in A} m_\alpha t(S_\alpha) \leqslant t(S) \leqslant 4(d(P)t(Q) + d(Q)t(P)) \leqslant 8t(P)t(Q).$$

Let us assume now that

$$\max\{|P(\bar{\theta})|, |Q(\bar{\theta})|\} \leqslant \exp(-E) \tag{4.16}$$

for $E > 0$. One can define the distance of $\bar{\xi} \in S$ to $\bar{\theta}$ as

$$\|\bar{\theta} - \bar{\xi}\| \leqslant \exp(-E(\bar{\xi})).$$

Let us put

$$\mathcal{E}(S, \bar{\theta}) = \sum_{\bar{\xi} \in S, E(\bar{\xi}) > 0} E(\bar{\xi}) \tag{4.17}$$

and

$$\mathcal{E}(S_\alpha, \bar{\theta}) = \sum_{\bar{\xi} \in S_\alpha, E(\bar{\xi}) > 0} E(\bar{\xi}) \tag{4.18}$$

as the definition of logarithmic distance from $\bar{\theta}$ to S or S_α.

In order to express relations between E, $\mathfrak{S}(S, \bar{\theta})$ *and* $\mathfrak{S}(S_\alpha, \bar{\theta})$ *we put*

$$(4.19)\quad T = \gamma_0(d(P)t(Q) + d(Q)t(P) + d(P)d(Q)\log(d(P)d(Q) + 2))$$

for an absolute constant $\gamma_0 > 0$ $(\gamma_0 \leqslant 2)$ *such that* $T(S) \leqslant T$.

Similarly for every $\alpha \in A$ *there is a bound* T_α *of* $t(S_\alpha)$ *of the form*

$$(4.20)\qquad t(S_\alpha) \leqslant T_\alpha \leqslant t(S_\alpha) + d(S_\alpha)\log(d(P)d(Q) + 2).$$

In terms of T_α *we can formulate results on* $\mathfrak{S}(S, \bar{\theta})$, $\mathfrak{S}(S_\alpha, \bar{\theta})$. *If we assume* $E \geqslant 4T$, *then*

$$(4.21)\qquad \mathfrak{S}(S, \bar{\theta}) = \sum_{\alpha \in A} m_\alpha(S_\alpha, \bar{\theta}) \geqslant E - 2T.$$

We denote the element of S_α *nearest to* $\bar{\theta}$ *by* $\bar{\xi}_\alpha$, $E(\bar{\xi}_\alpha) = \min_{\zeta \in S_\alpha} E(\zeta)$. *Then we have*

$$(4.22)\qquad E(\bar{\xi}_\alpha) \geqslant \min\{c_1\mathfrak{S}(S_\alpha, \bar{\theta}), c_1\mathfrak{S}(S_\alpha, \bar{\theta})^2/d(S_\alpha)T_\alpha\}$$

and for the other $\bar{\zeta} \in S_\alpha$ *close to* $\bar{\theta}$ *we have*

$$(4.23)\qquad \sum_{\bar{\zeta} \in S_\alpha, E(\bar{\xi}) \geqslant B_\alpha} E(\bar{\zeta}) \geqslant \mathfrak{S}(S_\alpha, \bar{\theta})/4$$

where

$$(4.24)\qquad B_\alpha = c_2\mathfrak{S}(S_\alpha, \bar{\theta})^{3/2} \cdot \left(d(S_\alpha)T_\alpha^{1/2}\right)^{-1}$$

for $\alpha \in A$.

As an application of these bounds we have the following result, where $d(S_\alpha)$ is replaced by its upper bound T_α. We remark that in addition to (4.20),

$$(4.25)\qquad \sum_{\alpha \in A} m_\alpha T_\alpha \leqslant T.$$

In particular, there exists $\alpha_0 \in A$ such that

$$(4.26)\qquad \mathfrak{S}(S_{\alpha_0}, \bar{\theta})/T_{\alpha_0} \geqslant \mathfrak{S}(S, \bar{\theta})/T \geqslant E/T - 2.$$

Under conditions (4.26) one has as a corollary of (4.22)–(4.24),

$$(4.27)\qquad \begin{gathered} E(\bar{\xi}_{\alpha_0}) \geqslant \min\{c_1\mathfrak{S}(S_{\alpha_0}, \bar{\theta}), c_1\mathfrak{S}(S_{\alpha_0}, \bar{\theta})^2/T_{\alpha_0}^2\}, \\ \sum\left\{E(\bar{\xi}) \colon \bar{\xi} \in S_{\alpha_0}, E(\bar{\xi}) \geqslant c_2\left(\mathfrak{S}(S_{\alpha_0}, \bar{\theta})/T_{\alpha_0}\right)^{3/2}\right\} \geqslant \mathfrak{S}(S_{\alpha_0}, \bar{\theta})/4. \end{gathered}$$

PROOF (SEE **[3]** AND APPENDIX BELOW). First of all we must change the system of the coordinates to a "normal" one **[3]** with respect to a system of polynomials $P(x, y)$, $Q(x, y)$. We use the Appendix for this. According to it or Lemma 3.3 **[3]** there is a *nonsingular* transformation

$$(\pi)\qquad \begin{cases} x = x'a + y'c, \\ y = x'b + y'd \end{cases}$$

for rational integers a, b, c, d such that

$$\max(|a|,|b|,|c|,|d|) \leqslant M \leqslant \gamma_1 d(P)d(Q), \tag{4.28}$$

and which is normal with respect to $P(x, y)$, $Q(x, y)$. It means, from our point of view, that for $\bar{\theta}$ written in new coordinates (x', y') as $\bar{\theta}' = (\theta_1', \theta_2')$ and for *any* common zero $\bar{\zeta}$ of $P(x, y) = 0$, $Q(x, y) = 0$, i.e. element of S, in new coordinates $\bar{\zeta}' = (\zeta_1', \zeta_2')$ we have

$$\|\bar{\theta}' - \bar{\zeta}'\|_{l_1} \leqslant 4M^2 \min\{|\theta_1' - \zeta_1'|, |\theta_2' - \zeta_2'|\}. \tag{4.29}$$

The property (4.29) is, certainly, central from all "normality" properties.

In new "normal" variables we consider the resultant $R(x')$ of $P'(x', y')$ ($\equiv P(x, y)$) and $Q'(x', y')$ ($\equiv Q(x, y)$) taken with respect to y'. The polynomial $R(x')$ is a polynomial of degree $\leqslant d(P)d(Q)$, but the type of $R(x')$ might be slightly higher than that of $R(x)$. This explains why we change the type $t(S)$ to be a slightly higher quantity T, defined in (4.19).

We prefer, however, to work with T and T_α because the proof is straightforward.

Now irreducible components $S_\alpha' = \pi^{-1}(S_\alpha)$ (we write now in transformed coordinates) are connected with irreducible components of $R(x')$. Namely, we have

$$R(x') = \prod_{\alpha \in A} P_\alpha(x')^{m_\alpha}$$

where $P_\alpha(x')$ is an irreducible polynomial from $\mathbf{Z}[x']$. Here zeros of $P_\alpha(x')$ are exactly an x'-projection of the set S_α': $\alpha \in A$.

Hence we can work with $P_\alpha(x')$ rather than with $S_\alpha' = \pi^{-1}(S_\alpha)$. According to (4.29) it is enough to bound above $|\zeta_1' - \theta_1'|$ in order to bound $\|\bar{\zeta}' - \bar{\theta}'\|$ with $\bar{\zeta}' = (\zeta_1', \zeta_2') \in S'$. Here and everywhere in the proof of Proposition 4.15, $\bar{\zeta}' = \pi^{-1}(\bar{\zeta})$, $\bar{\theta}' = \pi^{-1}(\bar{\theta})$.

First of all we can evaluate $\log|R(\theta_1')|$ in terms of

$$E \leqslant -\min\{|\log|P(\bar{\theta}\ )||\log|Q(\bar{\theta}\ )||\}.$$

For this we use the property of the resultants in the classical form [**1**].

Namely, we can use the formula [**17**]

$$|\mathrm{res}_x(p, q)| \leqslant \{d(p)H^+(q)|q(x_0)| + d(q)H^+(p)|p(x_0)|\}$$
$$\times H^+(q)^{d(p)-1} \cdot H^+(p)^{d(q)-1}$$

for arbitrary polynomials $p(x)$, $q(x) \in \mathbf{C}[x]$ and $H^+(p) = \max\{1, H(p)\}$. We put $p(x') = P'(\theta_1', x)$, $q(x') = Q'(\theta_1', x')$ and $x_0 = \theta_2'$.

We obtain this way the inequality

$$\log|R'(\theta_1')| \leqslant -E + T.$$

In particular, writing $R'(x)$ in the form

$$R'(x') = a \prod_{\alpha \in A} (x' - \zeta_1'^{\alpha})^{m_\alpha} = a \prod_{\bar{\zeta}' \in S'} (x' - (\bar{\zeta}')_1)$$

for $a \in \mathbf{Z}$, and using (4.29) one immediately obtains (4.21).

In order to get other statements (4.22)–(4.24) we simply use the irreducible polynomial $P_\alpha(x')$, whose zeros are an x'-projection of S' and apply to the polynomial $P_\alpha(x')$ the statement of Proposition A.2.2 of the Appendix.

At last, we derive the statements (4.6)–(4.27). Indeed, if (4.26) is false for every $\alpha_0 \in A$, we get

$$\sum_{\alpha \in A} \mathcal{E}(S_\alpha, \bar{\theta})m_\alpha \cdot T < \sum_{\alpha \in A} T_\alpha \cdot m_\alpha \cdot \mathcal{E}(S, \bar{\theta}),$$

which contradicts (4.25). Results (4.27) are a consequence of (4.22)–(4.24).

REMARK. In (4.27), the quantity $E(\bar{\xi}_{\alpha_0}) \leqslant -\log\|\bar{\theta} - \bar{\xi}_{\alpha_0}\|$ is bounded below by $c_1 \cdot \min\{\mathcal{E}(S_{\alpha_0}, \bar{\theta}), (\mathcal{E}(S_{\alpha_0}, \bar{\theta})/T_{\alpha_0})^2\}$. From these two terms, "usually" (in practice), $(\mathcal{E}(S_{\alpha_0}, \bar{\theta})/T_{\alpha_0})^2$ is a smaller one. Indeed, if $\mathcal{E}(S_{\alpha_0}, \bar{\theta})$ is large, then (4.27) shows simply that "most" of the measure $\mathcal{E}(S_{\alpha_0}, \bar{\theta})$ is concentrated in a single term of it, $E(\bar{\xi}_{\alpha_0})$: $E(\bar{\xi}_{\alpha_0}) \geqslant c_1\mathcal{E}(S_{\alpha_0}, \bar{\theta})$.

Let us continue the proof of Theorem 4.1. We apply Proposition 4.15 to irreducible components $S_{N,\alpha}$. We introduce the quantities T_α, T_N,

$$T_N = \gamma_2 N^2 \log N, \tag{4.30}$$

and similarly $T_\alpha = T_{N,\alpha}$ such that

$$t(S_{N,\alpha}) \leqslant T_{N,\alpha} = \gamma_2 t(S_{N,\alpha}) \log N \tag{4.31}$$

and

$$\sum_{\alpha \in A_N} m_\alpha T_{N,\alpha} \leqslant T_N. \tag{4.32}$$

If we apply Proposition 4.15, we obtain an $\alpha_0 \in A_N$ such that the following conditions are satisfied:

CLAIM 4.33. *There is an* $\alpha_0 \in A_N$ *and* $\bar{\xi}_0 \in S_{N,\alpha_0}$ *with*

$$E(\bar{\xi}_0) = \max\{E(\bar{\xi}): \bar{\xi} \in S_{N,\alpha_0}\}$$

such that for

$$\mathcal{E}_{N,0} \stackrel{\text{def}}{=} \sum \{E(\bar{\xi}): \bar{\xi} \in S_{N,\alpha_0}, E(\bar{\xi}) > 0\} \tag{4.34}$$

we have

$$\mathcal{E}_{N,0}/T_{N,\alpha_0} \geqslant \sigma_N/3T_N - 2, \tag{4.35}$$

and

$$\begin{gathered} E(\bar{\xi}_0) \geqslant \min\left\{c_1\mathcal{E}_{N,0}, c_1\left(\mathcal{E}_{N,0}/T_{N,\alpha_0}\right)^2\right\}, \\ \sum\left\{E(\bar{\xi}): \bar{\xi} \in S_{N,\alpha_0}, E(\bar{\xi}) \geqslant c_2\left(\mathcal{E}_{N,0}/T_{N,\alpha_0}\right)^{3/2}\right\} \geqslant \mathcal{E}_{N,0}/4. \end{gathered} \tag{4.36}$$

REMARK 4.37. It is possible to assume that the equality in (4.35) is satisfied, $\mathcal{E}_{N,0} = T_{N,\alpha_0}(\sigma_N/3T_N - 2)$ and take this as a definition of $\mathcal{E}_{N,0}$ only in terms of T_{N,α_0} (cf. (4.31), (4.32)), σ_N and T_N.

Our main data is set down and we can proceed with the actual proof. The idea of the proof is very simple. We take $\bar{\xi} \in S_{N,\alpha_0}$ and substitute it into polynomials $P_M(\bar{x})$ in the "blue" case or $C_l(\bar{x})$: $l \in \mathfrak{L}_M$, in the "red" case for $M < N$, hoping to get a contradiction using the properties of S_{N,α_0} and the Liouville theorem.

This can be understood better in the language of ideal theory. We take an ideal I_0 corresponding to a component S_{N,α_0}, $I_0 \supset I = (P_N, Q_N)$ and we want to show that the ideal (I_0, P_M) for the blue M or $(I_0, C_l: l \in \mathfrak{L}_M)$ for the red M gives us $\mathbf{Z}[x, y]$. Then the contradiction *may* follow from the evaluation of these ideals at the point $\bar{\theta}$. One understands that this is a rather general program, useful in the n-dimensional situation. However, it should be handled with greatest care when applied to bounds of different types, sizes and degrees, because none of the polynomials is annihilating at $\bar{x} = \bar{\theta}$.

First of all, one very useful general remark is

CLAIM 4.38. *For a polynomial $R(x, y) \in \mathbf{Z}[x, y]$, if $R(\bar{\xi}) = 0$ for some $\bar{\xi} \in S_{N,\alpha_0}$, then $R(\bar{\xi}_1) = 0$ for all $\bar{\xi}_1 \in S_{N,\alpha_0}$.*

This follows from the minimal irreducibility of the component S_{N,α_0}. For the future part of the proof we write

$$S_N^0 = S_{N,\alpha_0} \quad \text{and} \quad T_N^0 = T_{N,\alpha_0}.$$

We consider M_0 defined in a way that the following condition is satisfied (here $\gamma_3 \leqslant 2\max\{\log|\theta_1|, \log|\theta_2|, 1\} + 2$):

$$E(\bar{\xi}_0) > \sigma'_M + \gamma_3 M. \tag{4.39}$$

CLAIM 4.40. *If the condition* (4.39) *is satisfied, M is "blue" and $P_M(x, y)$ is the corresponding polynomial from* (i), *then*

$$P_M(\bar{\xi}_0) \neq 0.$$

Indeed, let us assume $P_M(\xi_0) = 0$. Then we have

$$\log|P_M(\theta)| \leqslant \log\|\bar{\theta} - \xi_0\| + \gamma_3 M.$$

According to the definition of σ'_M in (i) and (4.39), this contradicts the definition of $E(\bar{\xi}_0)$.

Let us now consider the consequences of $P_M(\bar{\xi}_0) \neq 0$ for M satisfying assumptions different from (4.39).

CLAIM 4.41. *Let $M \geqslant N_0$ and for some $K \geqslant 2$ let the following conditions be satisfied*:

$$c_2(\mathcal{E}_{N,0}/T_N^0)^{3/2} \geqslant \frac{2K}{K-1}\gamma_3 M; \qquad K\sigma_M/3 > E(\bar{\xi}_0). \tag{4.42}$$

If $P(x, y) \in \mathbf{Z}[x, y]$ is any polynomial such that $t(P) \leqslant 2M$ and

$$\log|P(\theta_1, \theta_2)| < -\sigma_M/3,$$

then from $P(\bar{\xi}_0) \neq 0$ *it follows that*

$$c_3 Mt\left(S_N^0\right) > \mathcal{E}_{N,0}/K. \tag{4.43}$$

PROOF OF CLAIM 4.41. Since $P(\bar{\xi}_0) \neq 0$, by Claim 4.38, $P(\bar{\xi}) \neq 0$ for any $\bar{\xi} \in S_N^0$. We denote $c_2(\mathcal{E}_{N,0}/T_N^0)^{3/2}$ by B_0. We apply now the Liouville theorem to the quantity

$$\alpha = \prod_{\bar{\xi} \in S_N^0, E(\bar{\xi}) \geqslant B_0} P(\bar{\xi}).$$

According to the Liouville theorem (§3) we obtain

$$|\alpha| \geqslant \exp\left(-c_4 Mt\left(S_{N,0}\right)\right) \tag{4.44}$$

for a type $t(S_{N,0})$ of the set $S_{N,0}$. Let us obtain an upper bound for $|\alpha|$. For this we notice that

$$|P(\bar{\xi})| \leqslant |P(\bar{\theta})| + \|\bar{\theta} - \bar{\xi}\| \cdot \exp\{\gamma_3 2M\}. \tag{4.45}$$

We have $\|\bar{\theta} - \bar{\xi}\| \cdot \exp(-E(\bar{\xi}))$. For $E(\bar{\xi}) \geqslant B_0$, we have, according to (4.42),

$$\|\bar{\theta} - \bar{\xi}\| \exp\{2\gamma_3 M\} \leqslant \exp\left(-E(\bar{\xi})/K\right).$$

Since we have $|P(\bar{\theta})| < \exp(-\sigma_M/3)$ and take into account (4.42) we obtain from (4.45)

$$|P(\bar{\xi})| \leqslant 2\exp\left(-E(\bar{\xi})/K\right): \qquad \bar{\xi} \in S_N^0, \quad E(\bar{\xi}) \geqslant B_0.$$

In particular, by the definition of α we obtain

$$|\alpha| \leqslant 2^{d(S_N^0)} \cdot \exp\left\{-\sum_{\bar{\xi} \in S_N^0, E(\bar{\xi}) \geqslant B_0} E(\bar{\xi})/K\right\}.$$

Using (4.36) we get

$$|\alpha| \leqslant 2^{d(S_N^0)} \cdot \exp\{-\mathcal{E}_{N,0}/4\}. \tag{4.46}$$

Comparing (4.44) and (4.46) we deduce (4.43). Claim 4.41 is proved.

In order to simplify the statement of Claim 4.41 we make the following simple

CLAIM 4.47. *Let us have*

$$M < \frac{1}{2c_3} \cdot \frac{\sigma_N}{KN^2} \quad \textit{for } N \geqslant N_4.$$

Then condition (4.43) is false, while the first of the conditions (4.42) is satisfied. Indeed, we use (4.35) and we get by (4.30)

$$\mathcal{E}_{N,0}/T_N^0 \geqslant \sigma_N/\gamma_2 N^2 \log N - 2; \tag{4.48}$$

this shows that for large N, when $N^3 = O(\sigma_N)$, the first of the conditions (4.42) is satisfied. From (4.48) it follows also that

$$\mathcal{E}_{N,0}/t\left(S_N^0\right) \geqslant \sigma_N/N^2 - 2\gamma_2 \log N$$

according to a definition (4.31). The choice of M immediately implies that for a large M, (4.43) is false.

COROLLARY 4.49. *Let $M < \sigma_N/2c_3KN^2$, but $K\sigma_M/3 > E(\bar{\xi}_0)$ for $K \geqslant 2$. Then for every polynomial $P(x, y) \in \mathbf{Z}[x, y]$ such that $t(P) \leqslant 2M$ and*

$$\log|P(\bar{\theta})| < -\sigma_M/3,$$

we have $P(\bar{\xi}_0) = 0$.

In particular we have

COROLLARY 4.50. *Let $M < \sigma_N/2c_3KN^2$ and*

$$K\sigma_M/3 > E(\bar{\xi}_0) > \sigma'_M + \gamma_3 M, \tag{4.51}$$

for some $K \geqslant 2$; then M is "red" but not "blue".

Indeed, we combine Claim 4.40 and Corollary 4.49 to get for the "blue" M, both $P_M(\bar{\xi}_0) = 0$ and $P_M(\bar{\xi}_0) \neq 0$.

We now notice that for some constant $K > 2$ there is always a sufficiently large M satisfying (4.51). Indeed, let N be sufficiently large so that

$$\sigma'_M \leqslant \gamma_5\sigma_M, \quad \text{if } M \geqslant N_6,$$

while for a large constant c_6 ($> \gamma_5$, etc.) we have $\sigma_M > c_6M^3$ if $M \geqslant N_6$. Then (4.51) can be changed to

$$K/3\sigma_M > E(\bar{\xi}_0) > (\gamma_5 + \gamma_3)\sigma_M. \tag{4.52}$$

Taking into account the definition of σ_M one can always satisfy these inequalities for K depending only on γ_5, γ_3 and a. One more restriction on M reads

$$M < \sigma_N/2c_3KN^2. \tag{4.53}$$

However, the definition of $E(\bar{\xi}_0)$ in (4.14) requires that $E(\bar{\xi}) \leqslant \sigma_N/3$. This ensures that for a sufficiently large N the condition (4.53) is satisfied for M, satisfying (4.52). Indeed, for a large N we can substitute $M < c_6/2c_3 \cdot N$ for (4.53), where c_6 is a large constant.

Let us denote M, satisfying (4.52) (and consequently, (4.53)) by $M = M_0$. Then Corollary 4.50 asserts that M_0 is "red".

CLAIM 4.54. *The set S_N^0 has the type $\leqslant 8\gamma_0M_0^2$.*

PROOF OF CLAIM 4.54. According to Claim 4.5 there are two relatively prime polynomials $P_{M_0}^r(x, y)$, $Q_{M_0}^r(x, y) \in \mathbf{Z}[x, y]$ such that

$$\max\{t(P_{M_0}^r), t(Q_{M_0}^r)\} \leqslant 2M_0,$$

$$\max\{|P_{M_0}^r(\theta)|, |Q_{M_0}^r(\bar{\theta})|\} < \exp(-\sigma_{M_0}/3).$$

Corollary 4.49 shows that $P_{M_0}^r(\bar{\xi}_0) = 0$, $Q_{M_0}^r(\bar{\xi}_0) = 0$, since all the conditions of 4.49 are satisfied for M_0 in view of (4.52)–(4.53). Now S_N^0 is a subset of a set of common zeros of $Q_{M_0}^r(x, y)$, $P_{M_0}^r(x, y)$. Because of the bounds of types of $Q_{M_0}^r$, $P_{M_0}^r$ we obtain by a standard argument that $t(S_N^0) \leqslant 8\gamma_0M_0^2$.

We are entering the final part of the proof. Here is the only place where we use the transcendence of $\bar{\theta}$! First of all one should remark that the function of N,

$t(S_N^0)$ is unbounded as $N \to \infty$. Indeed from (4.36) one obtains for a large N that $E(\bar{\xi}_0) \geqslant N$. This means

$$\|\bar{\theta} - \bar{\xi}_0\| \leqslant \exp(-N)$$

and since $t(\bar{\xi}_0) \leqslant t(S_N^0)$ and $\bar{\theta}$ is transcendental (from $\mathbf{C}^2$), $t(\bar{\xi}_0) \to \infty$ and thus $t(S_N^0) \to \infty$.

In order to finish the proof one defines a number M_1 in such a way as

$$M_1 = \left[\sqrt{t(S_N^0)/8\gamma_0}\right] - 1. \tag{4.55}$$

Definition (4.55) means that

$$8\gamma_0 M_1^2 < t(S_N^0) < 8\gamma_0(M_1 + 2)^2.$$

Claim 4.54 shows that

$$M_1 < M_0. \tag{4.56}$$

CLAIM 4.57. *Let* $R(x, y) \in \mathbf{Z}[x, y]$, $t(R) \leqslant 2M_1$ *and* $\log|R(\bar{\theta})| < -\sigma_{M_1}/3$. *Then we have* $R(\bar{\xi}_0) = 0$.

PROOF. Let us assume that $R(\bar{\xi}_0) \neq 0$. Then we can use the Liouville theorem in order to estimate $|R(\bar{\xi}_0)|$ below. We get, using (4.55) from the Liouville theorem that

$$|R(\bar{\xi}_0)| \geqslant \exp(-\gamma_6 M_1 t(S_{N,0})) \geqslant \exp(-\gamma_7 M_1^3) \tag{4.58}$$

for some absolute constant $\gamma_7 > 0$. On the other hand we can estimate $|R(\bar{\xi}_0)|$ above using the bound for $\|\bar{\theta} - \bar{\xi}_0\|$,

$$\begin{aligned} |R(\bar{\xi}_0)| &\leqslant |R(\bar{\theta})| + \|\bar{\theta} - \bar{\xi}_0\| \exp(\gamma_3 M_1) \\ &\leqslant \exp(-\sigma_{M_1}/3) + \exp(-E(\bar{\xi}_0) + \gamma_3 M_1). \end{aligned}$$

Now by the choice of (4.56) we have, using the definition (4.52) of M_0: $E(\bar{\xi}_0) > (\gamma_5 + \gamma_3)\sigma_{M_0} \geqslant \gamma_7\sigma_{M_1} + \gamma_3 M_1$, because M_0 is sufficiently large. This implies

$$|R(\bar{\xi}_0)| < \exp(-\gamma_8\sigma_{M_1}). \tag{4.59}$$

Now we note that, by the remark of the unboundedness of $t(S_N^0)$ as $N \to \infty$ and the definition (4.55) we have $M_1 \geqslant N_6$ provided $N \geqslant N_7$. Then $\sigma_{M_1} > C_6 M_1^3$ and for $C_6 > \gamma_7 \cdot \gamma_8^{-1}$ the inequalities (4.58) and (4.59) contradict each other. Claim 4.57 is proved.

Now it is enough to prove that M_1 is not colored; it is neither "blue" nor "red".

CLAIM 4.60. *The number* M_1 *is not "blue"*.

PROOF OF CLAIM 4.60. Let us assume that M_1 is blue and let $P_{M_1}(x, y)$ be the corresponding polynomial from (i). Then, by Claim 4.57, $P_{M_1}(\bar{\xi}_0) = 0$. However, by (4.56) and the choice of M_0 in (4.52), the condition (4.39),

$$E(\bar{\xi}_0) > \sigma'_{M_0} + \gamma_3 M_0 \geqslant \sigma'_{M_1} + \gamma_3 M_1,$$

is satisfied. Then by Claim 4.40, we have $P_{M_1}(\bar{\xi}_0) \neq 0$. Hence, M_1 is not "blue".

CLAIM 4.61. *The number M_1 from* (4.55) *is not "red"*.

PROOF OF CLAIM 4.61. Let us assume that M_1 is "red". We again recall that $t(S_N^0) \to \infty$ as $N \to \infty$; so for $N \geqslant N_7$, $M_1 \geqslant N_0$. We apply Claim 4.5 and get two relatively prime polynomials $P_{M_1}^r(x, y) \in \mathbf{Z}[x, y]$, $Q_{M_1}^r(x, y) \in \mathbf{Z}[x, y]$ such that

$$\max\{t(P_{M_1}^r), t(Q_{M_1}^r)\} \leqslant 2M_1,$$
$$\max\{|P_{M_1}^r(\bar{\theta})|, |Q_{M_1}^r(\bar{\theta})|\} < \exp(-\sigma_{M_1}/3).$$

According to Claim 4.37 we have $P_{M_1}^r(\bar{\xi}_0) = 0$, $Q_{M_1}^r(\bar{\xi}_0) = 0$. Hence, by Claim 4.38 the whole set S_N^0 contains in the set of common zeros of $P_{M_1}^r(x, y)$, $Q_{M_1}^r(x, y)$. In other words we can bound the type $t(S_N^0)$ in terms of $P_{M_1}^r$, $Q_{M_1}^r$. This way we get a bound

$$t(S_N^0) \leqslant 8\gamma_0 M_1^2.$$

However this contradicts the choice (4.55) of M_1. Claim 4.61 is proved.

The number M_1 is uncolored. However $t(S_N^0) \to \infty$ as $N \to \infty$, so that $M_1 \geqslant N_0$ if $N \geqslant N_7$. This means, according to the statement of Theorem 4.1, that M_1 is, indeed, colored. Since it is not, θ_1 and θ_2 are not algebraically independent. Theorem 4.1 is proved.

REMARK 4.62. There is no need to demand, in general, $\lim \sigma'_N/\sigma_N < \infty$ if $\lim \sigma_N/N^3 = \infty$. One of the possible improvements in this direction is the following: if

$$\lim_{N \to \infty} \sigma_N/N^3 = \infty,$$

then we demand only

$$\sigma'_N \ll \sigma_{\sqrt{\sigma_N/N}}.$$

We notice that the only place in the proof (ad absurdum) where we used any additional information about $\bar{\theta} \in \mathbf{C}^2$ was the moment, where we required $\|\bar{\theta} - \bar{\xi}\| > 0$ for any $\bar{\xi} \in \overline{\mathbf{Q}}^2$. Hence one can change the end of the statement of Theorem 4.1.

THEOREM 4.1′. *Under the conditions of Theorem* 4.1 *both numbers θ_1 and θ_2 are algebraic.*

REMARK. In principle one can deduce this statement from Theorem 4.1 directly, without looking into the proof, by using in addition the Gelfond criterion.

5. Let us present now one of the possible generalizations of the criterion of the algebraic dependence (we should say, rather than algebraic independence) for more than two numbers. In this case the result is not that strong, though.

PROPOSITION 5.1. *Let* $\bar{\theta} = (\theta_1, \ldots, \theta_n) \in \mathbf{C}^n$, $a > 1$ *and* σ_N, σ'_N *be monotonically increasing functions such that* $\sigma'_N > \sigma_N$ *and* $\sigma_{N+1} < a\sigma_N$. *For any* $N \geqslant N_0$ *let there be a polynomial* $P_N(x_1, \ldots, x_n) \in \mathbf{Z}[x_1, \ldots, x_n]$ *such that*

$$t(P_N) \leqslant N, \qquad -\sigma'_N < \log |P_N(\theta_1, \ldots, \theta_n)| < -\sigma_N.$$

We assume that as $N \to \infty$, $\lim \sigma'_N/\sigma_N < \infty$ *and* $\lim_{N\to\infty} \sigma_N/N^{n+1} = \infty$. *Then the numbers* $\theta_1, \ldots, \theta_n$ *are algebraically dependent.*

Moreover, each of θ_i: $i = 1, \ldots, n$, *is algebraic.*

REMARK 5.2. The statement of Proposition 5.1 can be improved in the case $\sigma'_N \gg \sigma_N \gg N^{n+1}$. We need in general statements like this:

$$\lim_{N\to\infty} \sigma_N/N^{n+1} = \infty$$

and for $\sigma'_M \gg\ll \sigma_N$ we have

$$\frac{\sigma_N}{N^n} \gg \frac{\sigma'_M}{\sigma_M} \cdot M,$$

i.e. if $\sigma_M/M \gg\ll N^n$, then we can take σ'_M as $\sigma'_M \ll \sigma_N$ or we can take σ'_M as

$$\sigma'_M \ll \sigma_{(\sigma_M/M)^{1/n}};$$

this is the statement closest to best possible.

It is most desirable to formulate in the n-dimensional situations statements like Theorem 5.1.

The most natural assumption can be as follows. We take $a > 1$ as before and formulate

ASSUMPTION 5.3. *Let* $\bar{\theta} = (\theta_1, \ldots, \theta_n) \in \mathbf{C}^n$ *and* σ_N, σ'_N *be monotonically increasing functions such that* $\sigma'_N > \sigma_N < N$, $\sigma_N \to \infty$ *as* $N \to \infty$ *and* $\sigma_{N+1} < a\sigma_N$.

We assume that for every $N \geqslant N_0$, *either*

(i) *there exists a polynomial* $P_N(x_1, \ldots, x_n) \in \mathbf{Z}[x_1, \ldots, x_n]$ *such that*

$$t(P_N) \leqslant N, \qquad -\sigma'_N < \log |P_N(\theta_1, \ldots, \theta_n)| < -\sigma_N,$$

or

(ii) *there exists a system* $C_l(x_1, \ldots, x_n) \in \mathbf{Z}[x_1, \ldots, x_n]$: $l \in \mathfrak{L}_N$, *of polynomials without common factor such that*

$$t(C_l) \leqslant N, \qquad \log |C_l(\theta_1, \ldots, \theta_n)| < -\sigma_N$$

for $l \in \mathfrak{L}_N$.

Let us assume that $\lim_{N\to\infty} \sigma_N/N^{n+1} = \infty$ *and, say,* $\lim_{N\to\infty} \sigma'_N/\sigma_N < \infty$.

We can assume either that $\lim \sigma_N/N^{n+1} = \infty$ or $\sigma'_N \ll \sigma_{(\sigma_N/N)^{1/n}}$.

We assume that under Assumption 5.3 all numbers $\theta_1, \ldots, \theta_n$ are algebraic (or, say, only algebraically dependent).

6. Looking for the proper form of the criterion of the algebraic independence, one naturally looks for the criterion applicable to the usual problems of the Transcendence Number Theory. We mean first of all the criterion useful for the

Gelfond-Schneider or Thue-Siegel methods. One finds at once that there is something peculiar in the Gelfond-Schneider method that is reasonable to adopt in the criterion.

The classical Gelfond-Schneider method can be presented briefly in the following way, see [**12**]. One has a system of functions $f_{\vec{\lambda}}(\vec{z})$: $\vec{\lambda} \in \mathfrak{L}$ (for $\vec{z} \in \mathbf{C}^d$), such that for a given set $S \subset \mathbf{C}^d$ one has

$$\left.\frac{\partial^{|\vec{k}|}}{\partial z_1^{k_1} \cdots \partial z_d^{k_d}} f_{\vec{\lambda}}(\vec{z})\right|_{\vec{z}=\vec{w}} \in \mathbf{K}$$

for all $\vec{w} \in S$ and $\vec{k} = (k_1,\ldots,k_d) \in \mathbf{N}^d$, $|\vec{k}| = k_1 + \cdots + k_d \leqslant N_{\vec{w}} - 1$. Here **K** is a fixed field of the finite transcendence degree n over **Q**:

$$\mathbf{K} = \mathbf{Q}(\theta_1,\ldots,\theta_n, \vartheta)$$

and ϑ is algebraic over $\mathbf{Q}(\theta_1,\ldots,\theta_n)$.

Usually one takes $f_{\vec{\lambda}}(\vec{z})$ in the form

$$f_{\vec{\lambda}}(\vec{z}) = f_1(\vec{z})^{\lambda_1} \cdots f_m(\vec{z})^{\lambda_m}$$

and $\vec{\lambda} = (\lambda_1,\ldots,\lambda_m) \in \mathbf{N}^m$ with $0 \leqslant \lambda_i \leqslant L_i - 1$: $i = 1,\ldots,m$, for a fixed set of functions $f_1(\vec{z}),\ldots,f_m(\vec{z})$. These functions satisfy differential (partial differential) equations if $N_{\vec{w}} > 1$: $\vec{w} \in S$, or the law of addition (when $N_{\vec{w}} = 1$). We must note that for a clever form of the Gelfond-Schneider method associated with Abelian or degenerate Abelian (group) varieties, the form of $f_{\vec{\lambda}}(\vec{z})$ as powers of fixed functions is not necessary. The same is true in the Thue-Siegel method, when $f_{\vec{\lambda}}(\vec{z})$ are of the form

$$f_{\vec{\lambda}}(\vec{z}) = z_1^{L_1} \cdots z_d^{L_d} f_j(\vec{z})$$

for $0 \leqslant L_i \leqslant M_i - 1$: $i = 1,\ldots,m$, and functions $f_j(\vec{z})$ satisfy linear differential equations.

The main object auxiliary function $F(\vec{z})$, is defined as

$$F(\vec{z}) = \sum_{\vec{\lambda} \in \mathfrak{L}} C_{\vec{\lambda}} \cdot f_{\vec{\lambda}}(\vec{z})$$

with undetermined coefficients $C_{\vec{\lambda}}$ from the ring $\mathbf{Z}[\theta_1,\ldots,\theta_n]$ (one may take $C_{\vec{\lambda}}$ from the ring of integers of **K**, but this usually only creates problems, cf. A. Gelfond [**1**], without adding anything new).

These undetermined coefficients $C_{\vec{\lambda}}$, written as polynomials $C_{\vec{\lambda}}(\theta_1,\ldots,\theta_n) = \sum_{\vec{\mu}} C_{\vec{\lambda},\vec{\mu}} \theta_1^{\mu_1} \cdots \theta_n^{\mu_n}$ with undetermined integer coefficients $C_{\vec{\lambda},\vec{\mu}}$, are determined from the system of linear equations

$$\left.\frac{\partial^{\vec{k}}}{\partial z_1^{k_1} \cdots \partial z_d^{k_d}} F(\vec{z})\right|_{\vec{z}=\vec{w}} = 0$$

for all $\vec{k} \in \mathbf{N}^d$, $|\vec{k}| \leqslant N_{\vec{w}} - 1$ and $\vec{w} \in S$.

This system of linear equations can be written explicitly. If

$$\partial_{\vec{z}}^{\vec{k}} f_{\vec{\lambda}}(\vec{z})\big|_{\vec{z}=\vec{w}} = \frac{R_{\vec{k},\vec{w},\vec{\lambda}}(\theta_1,\ldots,\theta_n, \vartheta)}{P_{\vec{k},\vec{w}}(\theta_1,\ldots,\theta_n, \vartheta)}$$

for polynomials $R_{\vec{k},\vec{w},\vec{\lambda}}(\theta_1,\ldots,\theta_n,\vartheta)$, $P_{\vec{k},\vec{w}}(\theta_1,\ldots,\theta_n,\vartheta)$ with integer coefficients, then the system of equations on $F(\vec{z})$ can be represented in the explicit form

$$\sum_{\vec{\lambda}\in\mathfrak{L}} C_{\vec{\lambda}}(\theta_1,\ldots,\theta_n)R_{\vec{k},\vec{w},\vec{\lambda}}(\theta_1,\ldots,\theta_n,\vartheta)=0$$

for $\vec{k}\in\mathbf{N}^d$, $|\vec{k}|\leqslant N_{\vec{w}}-1$, $\vec{w}\in S$. For the solution of this system of equations one uses the Thue-Siegel lemma. If one knows the bound for the degree and height of polynomials $R_{\vec{k},\vec{w},\vec{\lambda}}(\theta_1,\ldots,\theta_n,\vartheta)$, then we can bound both the degree and the height of polynomials $C_{\vec{\lambda}}(\theta_1,\ldots,\theta_n,\vartheta)$.

We decided to include the corresponding computations, since it will be more convenient in the future to refer to a single general scheme than to start over each time.

We assume that ϑ has degree ν over $\mathbf{Q}(\theta_1,\ldots,\theta_n)$ and that $R_{\vec{k},\vec{w},\vec{\lambda}}(\theta_1,\ldots,\theta_n,\vartheta)$ can be written in the form

$$R_{\vec{k},\vec{w},\vec{\lambda}}(\theta_1,\ldots,\theta_n,\vartheta)=\sum_{j_1=0}^{D_1}\cdots\sum_{j_n=0}^{D_n}\sum_{j_{n+1}=0}^{\nu-1}R_{\vec{k},\vec{w},\vec{\lambda},\vec{j}}\cdot\theta_1^{j_1}\cdots\theta_n^{j_n}\vartheta^{n+1},$$

where $R_{\vec{k},\vec{w},\vec{\lambda},\vec{j}}$ are rational integers (or elements of the ring of integers $\mathcal{O}$ of a fixed algebraic number field $\mathbf{L}$, $[\mathbf{L}:\mathbf{Q}]=\nu_1$). If we write polynomials $C_{\vec{\lambda}}(\theta_1,\ldots,\theta_n)$ through their coefficients,

$$C_{\vec{\lambda}}(\theta_1,\ldots,\theta_n)=\sum_{\mu_1=0}^{X_1-1}\cdots\sum_{\mu_n=0}^{X_n-1}C_{\vec{\lambda},\vec{\mu}}\theta_1^{\mu_1}\cdots\theta_n^{\mu_n},$$

then our system of equations takes the form of the system of linear equations on $C_{\vec{\lambda},\vec{\mu}}$,

$$\sum_{\vec{\lambda}\in\mathfrak{L}}\sum_{j_1+\mu_1=i_1}\cdots\sum_{j_n+\mu_n=i_n}C_{\vec{\lambda},\vec{\mu}}\cdot R_{\vec{k},\vec{w},\vec{\lambda},\vec{j}}=0$$

for all $\vec{k}\in\mathbf{N}^d$, $|\vec{k}|\leqslant N_{\vec{w}}-1$: $\vec{w}\in S$, $j_{n+1}=0,1,\ldots,\nu-1$ and $i_1=0,1,\ldots,D_1+X_1-1;\ldots;i_n=0,1,\ldots,D_n+X_n-1$.

The number of unknowns $C_{\vec{\lambda},\vec{\mu}}$ is

$$\mathfrak{N}_u=|\mathfrak{L}|\cdot X_1\cdots X_n,$$

while the number of equations these unknowns are supposed to satisfy is

$$\mathfrak{N}_e=\sum_{\vec{w}\in S}\binom{N_{\vec{w}}+d-1}{d}\nu(D_1+X_1)\cdots(D_n+X_n).$$

Then, the solution $C_{\vec{\lambda},\vec{\mu}}$ of these equations exists provided $\mathfrak{N}_u>\mathfrak{N}_e$. If all $R_{\vec{k},\vec{w},\vec{\lambda},\vec{j}}$ are rational integers, then the nontrivial solution (not all $C_{\vec{\lambda},\vec{\mu}}$ are zeros) exists in rational integers if $\mathfrak{N}_u>\mathfrak{N}_e$. If all $R_{\vec{k},\vec{w},\vec{\lambda},\vec{j}}$ are integers from the field $\mathbf{L}$, $[\mathbf{L}:\mathbf{Q}]=\nu_1$, then the solution $C_{\vec{\lambda},\vec{\mu}}$ in rational integers (not all of which are zero) exists for $\mathfrak{N}_u>\nu_1\mathfrak{N}_e$.

If one knows the bound for the heights of integers $R_{\vec{k},\vec{w},\vec{\lambda},\vec{j}}$ then, under stronger assumptions on $\mathfrak{N}_u/\mathfrak{N}_e$, one gets a reasonable bound for sizes of solution $C_{\vec{\lambda},\vec{\mu}}$.

Siegel's lemma which we use has the form of

LEMMA 6.1 (THUE-SIEGEL). *Let* $a_{i,j}$: $i = 0,\dots,\mathfrak{N}_u - 1$; $j = 0,\dots,\mathfrak{N}_e - 1$, *be elements of an algebraic number field* $\mathbf{L}$, $[\mathbf{L}:\mathbf{Q}] = \nu_1$, *that are integers over* $\mathbf{Z}$ *and are polynomials in* $\theta_1,\dots,\theta_n$, *and, as polynomials in* n *variables (note that* $(\theta_1,\dots,\theta_n)$ *are algebraically independent), they may have a common factor,*

$$C_{\vec{\lambda}}(\theta_1,\dots,\theta_n) = C^{(0)}(\theta_1,\dots,\theta_n)\cdot C_{\vec{\lambda}}(\theta_1,\dots,\theta_n),$$

where $C^{(0)}(\theta_1,\dots,\theta_n) \in \mathbf{Z}[\theta_1,\dots,\theta_n]$ *and polynomials* $C_{\vec{\lambda}}(\theta_1,\dots,\theta_n) \in \mathbf{Z}[\theta_1,\dots,\theta_n]$ *are already without a common factor. One notices immediately that a new function*

$$F'(\vec{z}) = \left(C^{(0)}(\theta_1,\dots,\theta_n)\right)^{-1}F(\vec{z}),$$

which differs only by a constant from $F(\vec{z})$, *satisfies the same equations as* $F(\vec{z})$. *It has a form*

$$F'(\vec{z}) = \sum_{\vec{\lambda}\in\mathfrak{L}} C'_{\vec{\lambda}}(\theta_1,\dots,\theta_n) f_{\vec{\lambda}}(\vec{z}),$$

with polynomial coefficients $C'_{\vec{\lambda}}(\theta_1,\dots,\theta_n)$: $\vec{\lambda}\in\mathfrak{L}$, *having no common factor. The bound for degree and height of* $C'_{\vec{\lambda}}(\theta_1,\dots,\theta_n)$ *is similar to that of* $C_{\vec{\lambda}}(\theta_1,\dots,\theta_n)$,

$$\deg_{\theta_i} C'_{\vec{\lambda}} \leqslant \deg_{\theta_i} C_{\vec{\lambda}} \leqslant D_i - 1: \qquad i = 1,\dots,n,$$

and

$$\log H(C'_{\vec{\lambda}}) \leqslant \log H(C_{\vec{\lambda}}) + n \deg C_{\vec{\lambda}}: \qquad \vec{\lambda}\in\mathfrak{L}.$$

In other words, one can always choose the coefficients $C_{\vec{\lambda}}(\theta_1,\dots,\theta_n)$ of the auxiliary function $F(\vec{z})$ to be polynomials from $\mathbf{Z}[\theta_1,\dots,\theta_n]$ without a common factor.

Having constructed $F(\vec{z})$ one uses the typical machinery of the Gelfond-Schneider method. First of all, we extrapolate $F(\vec{z})$, i.e. we use the knowledge of $\Sigma_{\vec{w}\in S} N_{\vec{w}}$ zeros of $F(\vec{z})$ to get an upper bound for

$$\max_{j=0,\dots,\mathfrak{N}_e-1} \sum_{i=0}^{\mathfrak{N}_u-1} \overline{|a_{i,j}|} \leqslant A.$$

If $\mathfrak{N}_u > \nu_1\mathfrak{N}_e$, then the system of equations

$$\sum_{i=0}^{\mathfrak{N}_u-1} a_{i,j}\cdot x_i = 0: \qquad j = 0,\dots,\mathfrak{N}_e - 1,$$

has a nonzero solution $(x_0,\dots,x_{\mathfrak{N}_u-1})$ in rational integers such that

$$\max_{i=0,\dots,\mathfrak{N}_u-1} |x_i| \leqslant \left(\sqrt{2}A\right)^{\nu_1\mathfrak{N}_e/(\mathfrak{N}_u-\nu_1\mathfrak{N}_e)}.$$

We apply the Siegel lemma to our system of equations. Let

$$\max_{|\vec{k}|\leqslant N_{\vec{w}}-1,\vec{w}\in S} \sum_{\vec{\lambda}\in\mathfrak{L}} \sum_{\vec{j},0\leqslant j_1\leqslant D_1,\dots,0\leqslant j_n\leqslant D_n,0\leqslant j_{n+1}\leqslant\nu} \overline{|R_{\vec{k},\vec{w},\vec{\lambda},\vec{j}}|} \leqslant A_0,$$

where all $R_{\vec{k},\vec{w},\vec{\lambda},\vec{j}}$ are elements of the ring $\mathfrak{O}$ of integers of $\mathbf{L}$, $[\mathbf{L}:\mathbf{Q}] = \nu_1$.

We put, without the loss of generality (but with a certain loss in values of constants),

$$X_i = D_i: \qquad i = 1,\ldots,n.$$

Let $\tilde{\mathfrak{N}}_u = |\mathfrak{L}|$ and

$$\tilde{\mathfrak{N}}_e = \sum_{\vec{w}\in S} \binom{N_{\vec{w}}+d-1}{d}.$$

If $\mathfrak{N}_u > \nu\nu_1 2^n \tilde{\mathfrak{N}}_e$, there exists a nontrivial solution of our system of equations, satisfying the following bound on its sizes:

$$\max_{\lambda\in\mathfrak{L},0\leq\mu_1\leq D_1-1,\ldots,0\leq\mu_n\leq D_n-1} \log|C_{\vec{\lambda},\vec{\mu}}| \leq \frac{\nu\nu_1 2^n \tilde{\mathfrak{N}}_e}{\tilde{\mathfrak{N}}_u - \nu\nu_1 2^n \tilde{\mathfrak{N}}_e}\left(\log A_0 + \frac{1}{2}\log 2\right).$$

If the function $F(\vec{z})$ is defined, one first performs a simple but useful operation. The coefficients $C_{\vec{\lambda}}(\theta_1,\ldots,\theta_n) = \sum_{\vec{\mu}} C_{\vec{\lambda},\vec{\mu}}\theta_1^{\mu_1}\cdots\theta_n^{\mu_n}$, are polynomials in $\theta_1,\ldots,\theta_n$, and, as polynomials in n variables (note that $(\theta_1,\ldots,\theta_n)$ are algebraically independent), they may have a common factor

$$C_{\vec{\lambda}}(\theta_1,\ldots,\theta_n) = C^{(0)}(\theta_1,\ldots,\theta_n)\cdot C'_{\vec{\lambda}}(\theta_1,\ldots,\theta_n),$$

where $C^{(0)}(\theta_1,\ldots,\theta_n) \in \mathbf{Z}[\theta_1,\ldots,\theta_n]$ and polynomials $C'_{\vec{\lambda}}(\theta_1,\ldots,\theta_n) \in \mathbf{Z}[\theta_1,\ldots,\theta_n]$ are already without a common factor. One notices immediately that a new function

$$F'(\vec{z}) = \left(C^{(0)}(\theta_1,\ldots,\theta_n)\right)^{-1} F(\vec{z}),$$

which differs only by a constant from $F(z)$, satisfies the same equations as $F(\vec{z})$. It has a form

$$F'(\vec{z}) = \sum_{\vec{\lambda}\in\mathfrak{L}} C'_{\vec{\lambda}}(\theta_1,\ldots,\theta_n) f_{\vec{\lambda}}(\vec{z}),$$

with polynomial coefficients $C'_{\vec{\lambda}}(\theta_1,\ldots,\theta_n)$: $\vec{\lambda}\in\mathfrak{L}$, having no common factor. The bound for degree and height of $C'_{\vec{\lambda}}(\theta_1,\ldots,\theta_n)$ are similar to that of $C_{\vec{\lambda}}(\theta_1,\ldots,\theta_n)$,

$$\deg_{\theta_i} C'_{\vec{\lambda}} \leq \deg_{\theta_i} C_{\vec{\lambda}} \leq D_i - 1:$$

$i = 1,\ldots,n$, and

$$\log H(C'_{\vec{\lambda}}) \leq \log H(C_{\vec{\lambda}}) + n\deg C_{\vec{\lambda}}:$$

$\vec{\lambda}\in\mathfrak{L}$.

In other words, one can always choose the coefficients $C_{\vec{\lambda}}(\theta_1,\ldots,\theta_n)$ of the auxiliary function $F(\vec{z})$ to be polynomials from $\mathbf{Z}[\theta_1,\ldots,\theta_n]$ without a common factor.

Having constructed $F(\vec{z})$ one uses the typical machinery of the Gelfond-Schneider method. First of all, we extrapolate $F(\vec{z})$, i.e. we use the knowledge of $\sum_{\vec{w}\in S} N_{\vec{w}}$ zeros of $F(\vec{z})$ to get an upper bound for

$$\left|\frac{\partial^{k'}}{\partial z_1^{k'_1}\cdots\partial z_d^{k'_d}} F(\vec{z})\,|_{\vec{z}=\vec{w}_1}\right|$$

for $\vec{k}' \in \mathbf{N}^d$, $|\vec{k}'| \leq N'_{\vec{w}_1} - 1$: $\vec{w}_1 \in S_1$ and $S \subseteq S_1$ where

$$\sum_{\vec{w}_1 \in S_1} N'_{\vec{w}_1} > \sum_{\vec{w} \in S} N_{\vec{w}}$$

(though $N'_{\vec{w}_1}$ is not necessarily $\geqslant N_{\vec{w}_1}$ for $\vec{w}_1 \in S$, cf. Baker's method **[9]**, where $N'_{\vec{w}_1} < N_{\vec{w}_1}$ but $S \subset S_1$). Such extrapolation is usually achieved using the appropriate version of the Schwarz lemma.

After extrapolation comes the end of the proof in the form of zero and/or the Small Value Lemma. While during the extrapolation we showed the bound of the form, say

$$\max_{\substack{|\vec{k}'| \leq N'_{\vec{w}_1} - 1 \\ \vec{w}_1 \in S_1}} \log \left| \frac{\partial^{\vec{k}'}}{\partial z_1^{k'_1} \cdots \partial z_d^{k'_d}} F(\vec{z}) \Big|_{\vec{z} = \vec{w}_1} \right| < -\psi_1,$$

we show now that either

$$\text{(i)} \qquad -\psi_2 < \max_{\substack{|\vec{k}'| \leq N'_{\vec{w}_1} - 1 \\ \vec{w}_1 \in S_1}} \log \left| \frac{\partial^{\vec{k}'}}{\partial z_1^{k'_1} \cdots \partial z_d^{k'_d}} F(\vec{z}) \Big|_{\vec{z} = \vec{w}_1} \right|,$$

or

$$\text{(ii)} \qquad \max_{\vec{\lambda} \in \mathfrak{L}} \log |C_{\vec{\lambda}}(\theta_1, \ldots, \theta_n)| < -\psi_3.$$

Here usually ψ_1, ψ_2, ψ_3 are of the same order of magnitude. Since, in these cases we are considering

$$\frac{\partial^{\vec{k}'}}{\partial z_1^{k'_1} \cdots \partial z_d^{k'_d}} F(\vec{z}) \Bigg|_{\vec{z} = \vec{w}_1} \in \mathbf{K} \quad \text{for } \vec{w}_1 \in S,$$

alternative (i) is the same as in Proposition 1.7. This actually explains why criterion 1.7 arose. However, there is an alternative (ii) which is more like that in the Gelfond lemma **[1]**, *but* with one exception: the system of polynomials $C_{\vec{\lambda}}(\theta_1, \ldots, \theta_n)$, being all small, do *not* have a common factor. This is the first example of an algebraic restriction that can be added to any criterion.

We have formulated several criterion with this duality: either there is a polynomial bounded below and above at a given point (case (i)), or there is a system of polynomials without a common factor, bounded only above (case (ii)). This alternative is typical for *all* results obtained using the Gelfond-Schneider method and, hence, it will be rather a universal Criterion of Transcendence, the measure of transcendence, algebraic independence and the measure of algebraic independence.

There is one technical point that should be clarified. In case (i), we are dealing with a polynomial from $\mathbf{Z}[\theta_1, \ldots, \theta_n, \vartheta]$, not from $\mathbf{Z}[\theta_1, \ldots, \theta_n]$.

There is a very simple trick, which shows how to avoid this obstacle and always works only with polynomials in algebraically independent numbers instead of elements of a field of a finite transcendence degree. This is a lemma from our early paper (1974), see [**2**, **10b**].

LEMMA 6.2. *Let* ϑ *be algebraic over* $\mathbf{Q}(\theta_1,\dots,\theta_n)$ *and be of the degree* ν. *Let* $R(\theta_1,\dots,\theta_n,\vartheta)$ *be an element of the field* $\mathbf{K}=\mathbf{Q}(\theta_1,\dots,\theta_n,\vartheta)$ *of type* $\leqslant T$ *(with respect to a basis* $(\theta_1,\dots,\theta_n,\vartheta)$ *of* $\mathbf{K}$). *Let*

$$-E_1 \leqslant \log|R(\theta_1,\dots,\theta_n,\vartheta)| \leqslant -E_2$$

and the degree of $R(\theta_1,\dots,\theta_n,\vartheta)$ *is at most* D. *Then for some* $C>0$ *there is a polynomial* $P(\theta_1,\dots,\theta_n)\in\mathbf{Z}[\theta_1,\dots,\theta_n]$ *such that*

$$t(P)\leqslant T\cdot C, \qquad d(P)\leqslant D\cdot C$$

and

$$T\cdot C - E_1/n \leqslant \log|P(\theta_1,\dots,\theta_n)| \leqslant T\cdot C - E_2.$$

We now present applications of our criteria of algebraic independence to the proof of the algebraic independence of values of functions studied using the Gelfond-Schneider method. These are values of exponential, elliptic and Abelian functions. Our first application is to values of exponential functions. We preserve the notatinos of Sec. 3 of Chapter 1 [**2**] and the Introductory paper, Sec. 3. Following these notations, let $\alpha_1,\dots,\alpha_N$ and $\beta_1,\dots,\beta_M$ be two sequences of linearly independent (over $\mathbf{Q}$) complex numbers and let

$$S_1=\{e^{\alpha_i\beta_j}\},\ S_2=\{\beta_j,e^{\alpha_i\beta_j}\},\ S_3=\{\alpha_i,\beta_j,e^{\alpha_i\beta_j}\}\ (1\leqslant i\leqslant N,\ 1\leqslant j\leqslant M).$$

Then the standard application of the Gelfond-Schneider method, as above, leads to alternatives (i) or (ii). These alternatives and the corresponding estimates are summarized in Proposition 3.7 of Chapter 1 [**2**]. Applying Theorem 4.1 we obtain

COROLLARY 6.3. *Let*

$$\mathrm{Card}(S_i)/(M+N)\geqslant 3 \quad \textit{for } i=1,2$$

and

$$\mathrm{Card}(S_i)/(M+N)>3 \quad \textit{for } i=3.$$

Then there are at least 3 *algebraically independent numbers in* S_i: $i=1,2,3$.

This gives another elementary proof of Theorem 2 of Sec. 3 of the Introductory paper, this volume.

In the elliptic function case, applying the Gelfond-Schneider method, as in [**10(a)**], to the elliptic version of the Lindemann-Weierstrass theorem, we obtain another elementary proof of Theorem 11, §2, and Theorem 11, §4 of the Introductory paper. Similarly for Abelian functions the Gelfond-Schneider method, as presented in Chapter 7, and Criterion 4.1 give another elementary proof of Theorems 12, 13 (for three a.i. numbers), §2, and Theorem 13, §4 of the Introductory paper.

Appendix. Intersections of two curves and small values of polynomials.

A1. In the proof of the algebraic independence of three or more numbers we need, first of all, the possibility of effectivization of the Hörmander-Łojasiewicz theorem: we need to change the assertion of the smallness of some polynomials at a given point to the assertion that a given point is close to the zeroes of polynomials.

There exists a nonelementary approach that gives us *one* close zero. However we found an elementary method giving us a whole bunch of zeroes, rather close to a given point.

We will show how this method works for a plane situation. Now we will not require the coefficients of the polynomials to be rational integers, thus leaving the possibility of generalization of our results to a field of an arbitrary degree of transcendence.

We start as before with

$$(1.1) \qquad P(x, y), Q(x, y) \in \mathbf{C}[x, y]$$

—two nonzero polynomials without a common factor.

Further let

$$(1.2) \qquad \bar{\xi}_i = (\xi_{1i}, \xi_{2i}): \qquad i = 1,\dots,N,$$

be N points in $\mathbf{C}^2$ common to $P(x, y) = 0$, $Q(x, y)$.

According to the Bezout theorem, $N \leqslant d(P)d(Q)$ and some of the points $\bar{\xi}_i$ may have multiplicity $m_i \geqslant 1$ (we count $\bar{\xi}_i$ in this case m_i times in the sequence $\bar{\xi}_1,\dots,\bar{\xi}_N$).

Now let

$$(1.3) \qquad \bar{\zeta}_j = (\zeta_{1j}, \zeta_{2j}): \qquad j = 1,\dots,k,$$

be geometrically distinct points of the intersection of $P(x, y) = 0$, $Q(x, y) = 0$, where a point $\bar{\zeta}_j$ is counted with the multiplicity m_j,

$$\sum_{j=1}^{k} m_j = N \leqslant d(P) \cdot d(Q).$$

We now consider the situation typical to all problems of algebraic independence,

$$(1.4) \qquad |P(\theta_1, \theta_2)| < \varepsilon, \qquad |Q(\theta_1, \theta_2)| < \varepsilon$$

with some fixed point $\bar{\theta} = (\theta_1, \theta_2)$ and ε sufficiently small with respect to the types and degrees of P and Q, say,

$$\varepsilon < \exp(-O[d(P)d(Q)\{t(P) + t(Q)\}]).$$

Now we want, of course, to change (1.4) to something like

$$|\bar{\theta} - \bar{\zeta}_j| < c \cdot \varepsilon^{1/m_j} \cdots.$$

Using elementary methods, i.e. nothing but resultants, we can get something very weak but not too weak for applications.

However in the process of doing this we found new, typically multidimensional, obstacles. They are (cf. [3])

(1) in general, when x-coordinates, say, of different $\bar{\zeta}_j$ are equal, we cannot relate the multiplicity m_j of $\bar{\zeta}_j$ to the order of zero of $\mathrm{res}_y(P, Q)$ at ζ_{ij}, and

(2) in general three quantities

$$\min_j |\bar{\theta} - \bar{\zeta}_j|, \quad \min_j |\theta_1 - \zeta_{ij}| \quad \text{and} \quad \min_j |\theta_2 - \zeta_{2j}|$$

are not connected at all.

In order to get rid of all these problems we can use a rather elementary method which can be called a "rotation of two coordinates"; we can make a change of basis (not necessarily an orthogonal one).

We consider vectors $\bar{e} = (a, b)$ in $\mathbf{R}^2$ ($\subset \mathbf{C}^2$) such that a and b are integers (rational integers), $(a, b) = 1$ and b (or $|b|$) is a prime. For a given integer $k \geqslant 1$ we put

(1.5) $\mathfrak{J}_k = \{\bar{e} = (a, b) \colon a, b \in \mathbf{Z}, (a, b) = 1, |b| \text{ is a prime}; |a|, |b| \leqslant k\}.$

By $(\cdot, \cdot)$ we denote the usual scalar product in $\mathbf{C}^2$.

Let us suppose that we have k vectors $\bar{x}_i$: $i = 1, \ldots, k$, in $\mathbf{C}^2$ (here $\bar{x}_i$ are $\bar{\xi}_i$ or $\bar{\theta} - \bar{\xi}_i$ depending on the situation).

Also for any vector $\bar{e} \in \mathfrak{J}_k$ we consider a vector $\bar{e}'$ naturally orthogonal to $\bar{e}$; if $\bar{e} = (a, b)$, then $\bar{e}' = (b, -a)$. So

(1.6) $$(\bar{e}, \bar{e}') = 0, \qquad \|\bar{e}\| = \|\bar{e}'\|.$$

LEMMA 1.1. *If $\bar{e}_1, \bar{e}_2 \in \mathfrak{J}_k$ and $\bar{e}_1 \neq \bar{e}_2$, then $\bar{e}_1, \bar{e}_2$ are linearly independent.*

PROOF. If $\bar{e}_1 = (a_1, b_1)$; $\bar{e}_2 = (a_2, b_2)$ for primes b_1, b_2 and $a_1 b_2 - b_1 a_2 = 0$, then $b_2 \mid a_2$ and $b_1 \mid a_2$ for $\bar{e}_1 \neq \bar{e}_2$, and this is impossible.

For any vector $\bar{x}$ in $\mathbf{C}^2$ we put

(1.7) $$(\bar{x}, \bar{e}') \overset{\text{def}}{=} \alpha(\bar{x}, \bar{e}) \quad \text{if } \bar{e} \in \mathfrak{J}_k.$$

What does this mean? If $\bar{e}_1$, $\bar{e}_2$ are two distinct elements of $\mathfrak{J}_k$, then we can write $\bar{x}$ in a basis $\{\bar{e}_1, \bar{e}_2\}$ according to Lemma 1.1, $\bar{x} = A\bar{e}_1 + B\bar{e}_2$. Then $(\bar{x}, \bar{e}_1') = B(\bar{e}_2, \bar{e}_1')$ or

$$B = (\bar{x}, \bar{e}_1') / (\bar{e}_2, \bar{e}_1').$$

In the notation of (1.7) we have

(1.8) $$\bar{x} = \frac{-\alpha(\bar{x}, \bar{e}_2) \cdot \bar{e}_1 + \alpha(\bar{x}, \bar{e}_1) \cdot \bar{e}_2}{(\bar{e}_2, \bar{e}_1')}.$$

Now let us suppose that we have three vectors $\bar{e}_1, \bar{e}_2, \bar{e}_3 \in \mathfrak{J}_k$ and that they are all different. Then

(1.9) $$\bar{e}_3' = \alpha \bar{e}_1' + \beta \bar{e}_2',$$

where $\alpha, \beta \in \mathbf{Q}$. From the definition of $\mathfrak{J}_k$ (1.5) we derive

(1.10) $$H(\alpha), H(\beta) \leqslant 2k^2.$$

Then we have

$$\alpha(\bar{x}, \bar{e}_3) = \alpha \cdot \alpha(\bar{x}, \bar{e}_1) + \beta \cdot \alpha(\bar{x}, \bar{e}_2). \tag{1.11}$$

Trivially, from (1.10)–(1.11) we get

$$|\alpha(\bar{x}, \bar{e}_3)| \leqslant 4k^2 \max\{|\alpha(\bar{x}, \bar{e}_1)|, |\alpha(\bar{x}, \bar{e}_2)|\} \quad \text{if } \bar{e}_1 \neq \bar{e}_2. \tag{1.12}$$

Now we are ready to make the change of basis that will give us all the results we need. This change of basis will have the form

$$\{(1,0); (0,1)\} \rightleftarrows \{\bar{e}_1, \bar{e}_2\}$$

for some vectors $\bar{e}_1, \bar{e}_2 \in \mathcal{J}_k$ that are different.

We will formulate this lemma in the most general form possible.

LEMMA 1.2 ("ROTATION OF THE BASIC VECTORS"). *Let $\bar{x}_i$: $i = 1,\ldots,k$, be k geometrically distinct vectors from $\mathbf{C}^2$, where we assign to $\bar{x}_i$ the multiplicity m: $i = 1,\ldots,k$. Let $\bar{\theta} = (\theta_1, \theta_2) \in \mathbf{C}^2$ be fixed.*

There exists a change of coordinates

$$(\tilde{\pi}) \text{ or } (1.13) \qquad x = x'a + y'c, \qquad y = x'b + y'd,$$

$(\tilde{\pi})$ is nonsingular with b, d-primes, $c \neq d$ and integers a, b such that $(a, b) = 1$, $(c, d) = 1$ and

$$\max(|a|, |b|, |c|, |d|) \leqslant M.$$

Here the number M is chosen in such a way that

$$2M\pi(M) \geqslant k^2(3 + k), \tag{1.14}$$

e.g.

$$M \leqslant c_1 N^2 \leqslant c_1 d(P) d(Q).$$

For the sake of simplicity we also put

$$\bar{y}_i = \bar{x}_i - \theta: \qquad i = 1,\ldots,k.$$

We now have

(i) *if $i \neq j$ for $i, j = 1,\ldots,k$, then both coordinates of $\bar{x}'_i$, $\bar{x}'_j$ are different: for $\bar{x}'_i = (x'_{1i}, x'_{2i})$ we have $x'_{1i} \neq x'_{ij}$; $x'_{2i} \neq x'_{2j}$ whenever $i \neq j$;*

(ii) *there is one $i_0 \leqslant k$ such that for $\bar{\theta}' = (\theta'_1, \theta'_2)$ we have*

$$|\theta'_1 - x'_{1i_0}| = \min_{i=1,\ldots,k} |\theta'_1 - x'_{1i}|, \qquad |\theta'_2 - x'_{2i_0}| = \min_{i=1,\ldots,k} |\theta'_2 - x'_{2i}|;$$

(iii) *there exists a $j_0 \leqslant k$ such that*

$$|\theta'_1 - x'_{1j_0}|^{m_{j_0}} = \min_{i=1,\ldots,k} |\theta'_1 - x'_{1i}|^{m_i},$$

$$|\theta'_2 - x'_{2j_0}|^{m_{j_0}} = \min_{i=1,\ldots,k} |\theta'_2 - x'_{2i}|^{m_i},$$

(iv) (*the property of equality of the coordinates*): *for any $i = 1,\ldots,k$ we have*

$$\|\bar{\theta}'_1 - \bar{x}'_i\| = \sqrt{|\theta'_1 - x'_{1i}|^2 + |\theta'_2 - x'_{2i}|^2} \tag{1.15}$$

$$\leqslant \sqrt{1 + 16M^4} \min\{|\theta'_1 - x'_{1i}|, |\theta'_2 - x'_{2i}|\}.$$

Of course, property (iv) is the most important.

PROOF. For two $\{i, j\}$; $i, j \leqslant k$, $i \neq j$, we put $\mathcal{J} = \mathcal{J}_M$ as in (1.5) and

$$\Delta_{\{i,j\}} = \{\bar{e} \in \mathcal{J}: \alpha(\bar{x}_i, \bar{e}) = \alpha(\bar{x}_j, \bar{e})\}. \tag{1.16}$$

According to (1.8) the set $\Delta_{\{i,j\}}$ cannot consist of more than one element, since, otherwise $\bar{x}_i = \bar{x}_j$ which contradicts $i \neq j$. We put $\mathcal{J}_0 = \mathcal{J} \setminus \cup_{\{i,j\}} \Delta_{\{i,j\}}$ and

$$|\mathcal{J}_0| \geqslant |\mathcal{J}| \frac{k(k-1)}{2}. \tag{1.17}$$

Now for $\bar{e} \in \mathcal{J}_0$ we define $i(\bar{e}) \leqslant k$ as

$$|\alpha(\bar{y}_{i(\bar{e})}, \bar{e})| = \min_{i=1,\ldots,k} |\alpha(\bar{y}_i, \bar{e})|$$

and $j(\bar{e}) \leqslant k$ as

$$|\alpha(\bar{y}_{j(\bar{e})}, \bar{e})|^{m_{j(\bar{e})}} = \min_{i=1,\ldots,k} |\alpha(\bar{y}_i, \bar{e})|^{m_i}.$$

Then $i: \mathcal{J}_0 \to \{1,\ldots,k\}$ and $j: \mathcal{J}_0 \to \{1,\ldots,k\}$, so we can find

$$\mathcal{J}_1 \subseteq \mathcal{J}_0, \qquad |\mathcal{J}_1| \geqslant |\mathcal{J}_0|/k^2, \tag{1.18}$$

such that $i(\mathcal{J}_1) = i_0$, $j(\mathcal{J}_1) = j_0$ for two $i_0, j_0 \leqslant k$. Now let us take any two elements $\bar{e}_1, \bar{e}_2 \in \mathcal{J}_1$, $\bar{e}_1 \neq \bar{e}_2$ (assuming that $|\mathcal{J}_1| \geqslant 2$ of course). Then due to (1.16),

$$\alpha(\bar{x}_i, \bar{e}_j) \neq \alpha(\bar{x}_{i'}, \bar{e}_j) \tag{1.19}$$

for any $i, i' = 1,\ldots,k$; $i \neq i'$, $j = 1, 2$, and due to (1.18);

$$\alpha(\bar{y}_{i_0}, \bar{e}_j) = \min_i |\alpha(\bar{y}_i, \bar{e}_j)|, \qquad |\alpha(\bar{y}_{j_0}, e_j)|^{m_{j_0}} = \min_i |\alpha(\bar{y}_i, e_j)|^{m_i}$$

for $j = 1, 2$. Now in the new basis $\{\bar{e}_1, \bar{e}_2\}$, where $\bar{e}_1 = (a, b)$, $\bar{e}_2 = (c, d)$,

$$x = ax' + cy', \qquad y = bx' + dy',$$

we have by (1.8)

$$x'_{1i} \neq x'_{1j}, \quad x'_{2i} \neq x'_{2j} \quad \text{for } i \neq j$$

and (i)–(iii) are satisfied.

We remind the reader again that

$$\bar{y}_i = \bar{x}_i - \bar{\theta} = (x_{1i} - \theta_1, x_{2i} - \theta_2)$$

and M is chosen in such a way that

$$|\mathcal{J}_1| \geqslant |\mathcal{J}|/k^2 - \frac{k-1}{2k} \geqslant 2.$$

(In fact, for (i)–(iii) we need only $2M\pi(M) \geqslant 3k^2$.)

For any $i = 1,\ldots,k$ there exists at most one $e \in \mathcal{J}_1$, $\bar{e} = F(i)$ such that for all $\bar{e}_1, \bar{e}_2 \in \mathcal{J}_1$ different from $\bar{e} = F(i)$,

$$|\alpha(\bar{y}_i, \bar{e}_1)| \leqslant 4M^2 |\alpha(\bar{y}_i, \bar{e}_2)|. \tag{1.20}$$

[Indeed, we put

$$|\alpha(\bar{y}_i, \overline{F(i)})| = \min_{\bar{e}\in\mathcal{J}_1} |\alpha(\bar{y}_i, \bar{e})|$$

for any $i = 1,\ldots,k$. Then starting from (1.12) for any $\bar{e}_2 \neq \bar{e} = F(i)$, $\bar{e}_2 \in \mathcal{J}_1$, we have

$$\begin{aligned}|\alpha(\bar{y}_i, \bar{e}_1)| &\leq 4K^2 \max\{|\alpha(\bar{y}_i, \bar{e}_2)|, |\alpha(\bar{y}_i, F(i))|\}\\ &= 4M^2|\alpha(\bar{y}_i, \bar{e}_2)| \quad \text{Q.E.D.]}\end{aligned}$$

Let

$$\mathcal{J}_2 = \mathcal{J}_1 \setminus F^{-1}\{1,\ldots,k\},$$

i.e.

$$|\mathcal{J}_2| \geq |\mathcal{J}_1| - k \geq \frac{|\mathcal{J}|}{k^2} - \frac{k-1}{2k} - k \geq 2$$

for our choice of $|\mathcal{J}| \geq 2M\pi(M) \geq (3+k)k^2$.

Now let us take two elements $\bar{e}_1, \bar{e}_2 \in \mathcal{J}_2$, $\bar{e}_1 \neq \bar{e}_2$, and consider the change of coordinates $(x, y) \to (x', y')$ where

(1.21) $$\bar{x} = x'\bar{e}_1 + y'\bar{e}_2$$

for $\bar{x} = (x, y)$. Then according to our choice of $\bar{e}_1, \bar{e}_2 \in \mathcal{J}_2 \subset \mathcal{J}_1 \subset \mathcal{J}_0$, we have

(1.22) $$x'_{1i} \neq x'_{1j}, \quad x'_{2i} \neq x'_{2j} \quad \text{for } i \neq j,$$

$i, j = 1,\ldots,k$, as

$$x' = -\alpha(\bar{x}, \bar{e}_2)/(\bar{e}_2, \bar{e}'_1), \qquad y' = \alpha(\bar{x}, \bar{e}_1)/(\bar{e}_2, \bar{e}'_1)$$

in formula (1.10). Again for the same reason, as $\bar{e}_1, \bar{e}_2 \in \mathcal{J}_1$ because of (1.18),

(1.23) $$|y'_{\lambda i_0}| = \min_{i=1,\ldots,k} |y'_{\lambda i}|, \qquad \lambda = 1, 2,$$

(1.24) $$|y'_{\lambda j_0}|^{m_{j_0}} = \min_{i=1,\ldots,k} |y'_{\lambda i}|^{m_i}, \qquad \lambda = 1, 2,$$

and (i)–(iii) are proved as $\bar{y}'_i = \bar{x}'_i - \bar{\theta}'$: $i = 1,\ldots,k$.

At last as $\bar{e}_1, \bar{e}_2 \in \mathcal{J}_2$ we have

(1.25) $$|y'_{1i}| \leq 4M^2|y'_{2i}| \quad \text{and} \quad |y'_{2i}| \leq 4M^2|y'_{1i}|$$

for all $i = 1,\ldots,k$. In particular

(1.26) $$\|\bar{y}'_i\| = \|\bar{x}'_i - \bar{\theta}'\| \leq \sqrt{1 + 16M^4} \times \min\{|x'_{1i} - \theta_1|, |x'_{2i} - \theta_2|\}.$$

The proof is finished.

A2. We can now apply this lemma to our situation.

Let $P(x, y)$, $Q(x, y)$ be as in (1.1), and apply Lemma 1.2 to the points

$$\bar{x}_i = \bar{\zeta}_i\colon \ i = 1,\ldots,k; \qquad \bar{y}_i = \bar{\zeta}_i - \bar{\theta}\colon \ i = 1,\ldots,k,$$

where $\bar{\zeta}_i$: $i = 1,\ldots,k$, are geometrically distinct points of the intersections of $P(x, y) = 0$, $Q(x, y) = 0$ and $k \leq d(P) \cdot d(Q)$.

We make the change of the coordinates

$$(\tilde{\pi}') \qquad \begin{cases} x = ax' + cy', \\ y = bx' + dy' \end{cases}$$

by Lemma 1.2. We obtain polynomials $P'(x', y') = P(x, y)$, $Q'(x', y') = Q(x, y)$ from $\mathbf{C}[x', y']$ without a common factor as before.

Now the system of the coordinates is normal, i.e. for the intersection points $(\zeta'_{1i}, \zeta'_{2i})$ of $P'(x', y') = 0$, $Q'(x', y') = 0$, if $i \neq j$ then $\zeta'_{1i} \neq \zeta'_{1j}$, $\zeta'_{2i} \neq \zeta'_{2j}$.

This means that for the resultant $R(x') = \mathrm{res}_{y'}(P', Q')$ we have the following decomposition:

$$R(x') = a \prod_{i=1}^{k} (x' - \zeta'_{1i})^{m_i}.$$

So we can now turn to an arbitrary polynomial

$$R(x) = a \prod_{i=1}^{k} (x - \eta_i)^{m_i} \tag{2.1}$$

of degree $d = \sum_{i=1}^{k} m_i$, where $\eta_1, \ldots, \eta_k$ are distinct zeroes.

We also define the semiresultant of $R(x)$, $\Delta = \Delta(R)$:

$$\Delta(R) = a^{2d} \prod_{\substack{i,j=1 \\ i \neq j}}^{k} (\eta_i - \eta_j)^{m_i m_j} \tag{2.2}$$

which is a polynomial from the coefficients of $R(x)$ with easily defined height and degree, and

$$\Delta(R) \neq 0. \tag{2.3}$$

We consider a fixed $\theta \in \mathbf{C}$ such that

$$|R(\theta)| < |a|. \tag{2.4}$$

We put

$$|R(\theta)| \cdot |a|^{-1} \leqslant \exp(-E), \qquad E > 0, \tag{2.5}$$

and we arrange zeroes in an order such that

$$|\theta - \eta_i| < 1 \quad \text{for } i = 1, \ldots, r,$$

we put

$$|\theta - \eta_i| = \exp(-e_i): \qquad i = 1, \ldots, r, \tag{2.6}$$

and we assume

$$m_1 e_1 \geqslant \cdots \geqslant m_r e_r > 0. \tag{2.7}$$

From (2.5) we immediately obtain

$$\sum_{i=1}^{r} m_i e_i \geqslant E. \tag{2.8}$$

Now keeping in mind the bound

$$|a| \prod_{i=1}^{k} \max\{1, |\eta_i|^{m_i}\} \leqslant (d(R) + 1)H(R),$$

from the definition of the semiresultant $\Delta(R)$ we obtain

$$-\frac{\log|\Delta|}{2} + \sum_{i=1}^{r} m_i e_i \left\{ \sum_{j=1}^{i-1} \min\{m_i, m_j\} \right\} \leqslant T_1 = 8d(R)t(R). \tag{2.9}$$

COROLLARY 2.1. *If in the above notation*

$$E \geqslant c_1\left(8d(R)t(R) + \frac{\log|\Delta(R)|}{2}\right) = c_1\left(T_1 + \frac{\log|\Delta|}{2}\right)$$

for some $c_1 > 0$, then $m_1 e_1 \geqslant c_2 E$ for $c_2 = c_2(c_1) > 0$.

However in the cases we will consider $\Delta(R) \in \mathbf{Z}$, $\log|\Delta| > 0$ and always $E < c_3 T_1$ for a sufficiently small $c_3 > 0$.

In order to show what precise results we have, we restrict ourselves to the following situation:

(A) $R(x) \in \mathbf{Z}[x]$, so that $\log|\Delta(R)| > 0$,

(B) for simplicity we assume that everything is evaluated only in terms of the type of $R(x)$, i.e. we replace $d(R)$ by $t(R)$.

Let

$$R(x) = P_1(x) \cdots P_m(x), \tag{2.10}$$

where $P_j(x) \in \mathbf{Z}[x]$: $j = 1, \ldots, m$, are irreducible polynomials over $\mathbf{Q}$.

Now for

$$|P_j(\theta)| = \exp(-E_j): \quad j = 1, \ldots, m, \tag{2.11}$$

we take those $P_1(x), \ldots, P_n(x)$: $n \leqslant m$, where $E_i \geqslant 0$: $i = 1, \ldots, n$. Then

$$\sum_{j=1}^{n} E_j \geqslant E \quad \text{and} \quad \sum_{j=1}^{n} t(P_j) \leqslant c_4 t(P) \tag{2.12}$$

for an absolute constant $c_4 > 0$.

We always assume as before, that the number $t(P)^2$ is sufficiently large with respect to E.

From (2.12) it follows that there exists a $j = 1, \ldots, n$, say $j = 1$, such that

$$E_j \geqslant c_4 t(P_j) \cdot \frac{E}{t(P)}: \quad j = 1 \leqslant n. \tag{2.13}$$

Now by applying (2.9) to $P_1(x) = R(x) \in \mathbf{Z}[x]$ we obtain

$$\begin{gathered} \sum_{j=1}^{r} e_j^{(1)} \geqslant E_1, \\ \sum_{j=1}^{r} (j-1) e_j^{(1)} \leqslant 8d(P_1)t(P_1) \leqslant 8t(P_1)^2. \end{gathered} \tag{2.14}$$

Our principal result in this direction is

PROPOSITION 2.2. *For an irreducible polynomial* $P_1(x)$ *we have* (*of course* $P_1(x) \in \mathbf{Z}[x]$)

$$|P_1(\theta)| \leqslant \exp(-E_1)$$

and for its roots η_i: $i = 1,\ldots,r$,

$$|\theta - \eta_i| = \exp(-e_i): \qquad i = 1,\ldots,r.$$

We have

$$(2.14) \qquad \sum_{i=1}^{r} e_i \geqslant E_1, \qquad \sum_{i=1}^{r} (i-1)e_i \leqslant 8d(P_1)t(P_1).$$

Now if $E_1 \geqslant c_5 d(P_1)t(P_1)$, *then*

$$(2.15) \qquad e_1 \geqslant c_6 E_1 \quad \textit{for some } c_6 = c_6(c_5) > 0.$$

If, however, $E_1 < c_5 d(P_1)t(P_1)$ *for a sufficiently small* $c_5 > 0$, *then we have*

$$(2.16) \qquad E_1 > c_6 E_1^2 / d(P_1)t(P_1): \qquad c_7 > 0,$$

and moreover,

$$(2.17) \qquad \sum_{\substack{j=1 \\ e_j \geqslant B}}^{r} e_j \geqslant E_1/4 \quad \textit{for } B = c_8 E_1^{3/2}\left(d(P_1)t(P_1)^{1/2}\right)^{-1}.$$

REMARK. Here, in order to exclude trivial cases, we assume $E_1 \geqslant c_9 t(P_1)$ (cf. with (2.13)).

PROOF. We have (2.16) trivially from (2.14) (see Lemma 2.3 [**3**]). In order to prove (2.17) we put

$$I_C = \{j = 1,\ldots,r: e_j < C\}.$$

Let $A = E_1/8d(P_1)$. If

$$(2.18) \qquad \sum_{\substack{j=1 \\ j \notin I_A}}^{r} e_j \geqslant \frac{3}{4} E_1$$

is not true, then

$$\sum_{\substack{j=1 \\ j \in I_A}}^{r} e_j \geqslant \frac{1}{4} E_1.$$

Then

$$|I_A| \cdot A \geqslant \sum_{\substack{j=1 \\ j \in I_A}}^{r} e_j \geqslant \frac{1}{4} E_1.$$

But $|I_A| \leqslant r \leqslant d(P_1)$, i.e. $A \geqslant E_1/4d(P_1)$, which is impossible. We define B as in (2.17),

$$B = c_8 E_1^{3/2}\left(d(P_1)t(P_1)^{1/2}\right)^{-1}$$

for some small constant $c_8 > 0$. Then $B \geqslant A$ according to the previous remark.

Now we want to show (2.17). If (2.17) is not true, then by (2.18),

$$\sum_{\substack{j=1 \\ j \in J_B \setminus J_A}}^{r} e_j \geqslant E_1/2. \tag{2.19}$$

But according to (2.14) we have

$$|J_B \setminus J_A|^2 \cdot A \leqslant c_{10} d(P_1)t(P_1).$$

On the other hand, by (2.19),

$$|J_B \setminus J_A| \cdot B \geqslant E_1/2.$$

Thus

$$B \geqslant \frac{E_1^{3/2}}{16 d(P_1)t(P_1)},$$

which contradicts the choice of B with $c_8 < 1/16$. Q.E.D.

Taking Lemma 1.2 and Proposition 2.2 into account, we can now formulate the most general statement which is sufficient for all of our applications.

THEOREM 2.3. *Let $P(x, y) \in \mathbf{Z}[x, y]$, $Q(x, y) \in \mathbf{Z}[x, y]$ be two nontrivial polynomials without any nonconstant common factor, $\bar{\zeta}_j = (\zeta_{1j}, \zeta_{2j})$: $j = 1,\ldots,k$, are geometrically distinct points of the intersection of $P(x, y) = 0$, $Q(x, y) = 0$, where $\bar{\zeta}_j$ has the multiplicity m_j: $j = 1,\ldots,k$: $\sum_{j=1}^{k} m_j \leqslant d(P)d(Q)$.*

Let $\bar{\theta} = (\theta_1, \theta_2)$ be a fixed point in $\mathbf{C}^2$ such that

$$|P(\theta_1, \theta_2)| \leqslant \varepsilon, \qquad |Q(\theta_1, \theta_2)| < \varepsilon$$

for $\varepsilon = \exp(-E_1)$, where

$$E_1 \geqslant c_{11} d(P)d(Q)\{t(P) + t(Q)\}.$$

Let us put

$$\|\bar{\theta} - \bar{\zeta}_j\| = \exp(-e_j): \qquad j = 1,\ldots,k.$$

We put

$$T = d(Q)t(P) + d(P)t(Q) + d(P)d(Q)\log(d(P)d(Q) + 2).$$

Then there exists a $j_1 \leqslant k$ such that

$$e_{j_1} m_{j_1} \geqslant \min\left\{c_{12} E_1,\, c_{12} \frac{E_1^2}{d(P)d(Q)T}\right\}.$$

Moreover we have

$$\sum_{\substack{j=1 \\ e_j > B}}^{k} e_j m_j \geqslant \frac{E_1}{4} \quad \text{for } B = c_{13} E_1^{3/2} \left(d(P) d(Q) T^{1/2} \right)^{-1}$$

in the case $E_1 \leqslant c_{14} d(P) d(Q) T$.

We obtain this result by changing $P(x, y)$, $Q(x, y)$ to $P'(x', y')$, $Q'(x', y')$; that's why we get an extra term $d(P)d(Q)\log(d(P)d(Q) + 2)$ in T.

For applications of these results we should also mention that if

$$(\tilde{\pi}) \qquad \begin{cases} x = ax' + cy', \\ y = bx' + dy', \end{cases} \qquad \max(|a|, |b|, |c|, |d|) \leqslant M,$$

and (ξ_1', ξ_2') are algebraic numbers, then for (ξ_1, ξ_2) we have

$$\frac{t(\xi_1)}{d(\xi_1)} + \frac{t(\xi_2)}{d(\xi_2)} \leqslant c_{15} \left(\frac{t(\xi_1')}{d(\xi_1')} + \frac{t(\xi_2')}{d(\xi_2')} + c_{16} \log M \right).$$

(Here (ξ_1', ξ_2') is the intersection point of $P'(x', y') = 0$, $Q'(x', y') = 0$, i.e. $Q(\xi_1') = Q(\xi_2')$.)

References

1. A. O. Gelfond, *Transcendental and algebraic numbers*, Moscow, 1952.
2. G. V. Chudnovsky, *Analytical methods in diophantine approximations*, Preprint IM-74-9, Inst. Math., Kiev, 1974 (and see this volume, Chapter 1).
3. ______, *Algebraic grounds for the proof of algebraic independence*. I, Comm. Pure Appl. Math. **34** (1981), No. 1, 1–28.
4. W. D. Brownawell, *Sequences of diophantine approximations*, J. Number Theory **6** (1974), 11–21.
5. S. Lang, *Introduction to transcendental numbers*, Addison-Wesley, Reading, Mass., 1966.
6. M. Waldschmidt, *Nombres transcendants*, Lecture Notes in Math., vol. 437, Springer, New York, 1975.
7. ______, *Les travaux de G. V. Chudnovsky sur les nombres transcendants*, Sem. Bourbaki (1975–1976), No. 488, Lecture Notes in Math., vol. 567, Springer, New York, 1977, pp. 274–292.
8. T. Schneider, *Einfuhrung in die transzendenten zahlen*, Springer, Berlin, 1957.
9. A. Baker, *Transcendental number theory*, Cambridge Univ. Press, Cambridge, 1975.
10. (a) G. V. Chudnovsky, *Algebraic independence of the values of elliptic function at algebraic points. Elliptic analogue of the Lindemann-Weierstrass theorem*, Invent. Math. **61** (1980), 267–290.

    (b) *Algebraic independence of some values of the exponential functions*, Mathem. Not. Sov. Acad. **15** (1974), no. 4, 391–398.
11. J. W. S. Cassels, *An introduction to diophantine approximation*, Cambridge Univ. Press, Cambridge, 1957.
12. G. V. Chudnovsky, Proc. Conf. Number Theory (Carbondale, 1979), Lecture Notes in Math., vol. 751, Springer, New York, 1979, pp. 49–69.
13. ______, *Independence algebrique dans le methode de Gelfond-Schneider*, C. R. Acad. Sci. Paris, Ser A **291** (1980), A419–A422.
14. P. Cijsouw, *Transcendence measures*, North-Holland, Amsterdam, 1972.
15. W. Hodge and D. Pedoe, *Methods of algebraic geometry*, Cambridge Univ. Press, Cambridge, 1952.
16. I. R. Shafarevich, *Basic algebraic geometry*, Springer, New York, 1976.
17. *Transcendence theory*, Proc. Cambridge Conf. (1976), Academic Press, New York, 1977.
18. B. L. Van der Waerden, *Modern algebra*, 2 vols., Ungar, New York, 1951.

CHAPTER 5

METHODS OF THE THEORY OF TRANSCENDENTAL NUMBERS, DIOPHANTINE APPROXIMATIONS AND SOLUTIONS OF DIOPHANTINE EQUATIONS

Introduction. In this chapter methods of the theory of transcendental numbers (Siegel's method of 1929 [**1**], Gel'fond's method of 1934 [**18**], and Baker's method of 1966 [**5**]) are used to study Diophantine problems, in particular Diophantine approximations and solutions of Diophantine equations. In this chapter we will be mainly concerned with the study of measures of linear independence, irrationality (and approximations to algebraic numbers) for values of exponential, binomial (and logarithmic) functions. Two methods are used in our studies. One is Siegel's method [**1**] of the construction of families of linear forms approximating functions satisfying linear differential equations. Siegel's method generalizes upon Hermite's earlier explicit construction of rational approximations to exponential functions. In §§2–4 we transform and refine Siegel's method, to obtain the best possible results on measures of Diophantine approximations, and linear independence (and transcendence) for arbitrary numbers that are polynomials in exponents of rational numbers. Our new technique is generalized to other classes of analytic functions, including E-functions. We refer the reader to §2 for a discussion of Siegel's method and our new results.

Another method of the theory of transcendental numbers, that we use, is the Gel'fond-Schneider-Baker method. This method is based on the introduction of auxiliary functions and does not require Siegel's "minimality" property. In §1 we use different versions of this method to study measures of Diophantine approximations of $x^{j/n}$ for $x \in \mathbf{Q}$ in terms of $H(x)$ and $|x-1|$. Such results are of primary importance for the bounds of integer solutions (and the bounds of the number of solutions) of Diophantine equations, in connection with effective and noneffective versions of the Thue-Siegel theorems.

1. We study here approximations of algebraic numbers $\alpha^{j/n}$ by algebraic numbers of bounded degrees using Gel'fond's method of 1934 [**19**]. We apply these results for studying the number of integer points on algebraic curves.

0076-5376/84 $1.00 + $.25 per page

For the proof of our results on linear forms in two logarithms of algebraic numbers, or, equivalently, on approximations of the numbers $\alpha^{j/n}$ by algebraic numbers, we use several techniques. While the algebraic and analytic method of the proofs is Gel'fond's method [**19**], in the End of the Proof, we use Stark's method [**20**] of p-divisibility and a different, simpler analytic method of the author (close to Siegel's [**1**]).

We should mention at the outset that the most important thing in this section is not the theorems being proved, but rather the method of proof itself, which is a modification of the method developed by A. O. Gel'fond (see [**19**, **5**]).

It turns out that this method enables us to reprove the Thue-Siegel theorem [**21**] on approximations of algebraic numbers. This new proof, like the original, is noneffective and is based on the use of two approximations. However Gel'fond's method gives an effective expression for *all* constants, occuring in the Thue-Siegel theorem, for algebraic roots $\alpha^{j/n}$. The results we obtain differ from those obtained by the Thue-Siegel method, especially for large n.

For Thue's theorem we use

EXAMPLE 1.1. Suppose for a natural number a we have

$$\left|a^{1/n} - \frac{x_1}{y_1}\right| < H\left(\frac{x_1}{y_1}\right)^{-\kappa_1} \quad \text{and} \quad \left|a^{1/n} - \frac{x_2}{y_2}\right| < H\left(\frac{x_2}{y_2}\right)^{-\kappa_2} \tag{1.1}$$

for nontrivial distinct pairs of integers x_i, y_i, where κ_1, $\kappa_2 > 2$ to avoid trivial cases. Then

$$\left|\left(a\frac{y_1^n}{x_1^n}\right)^{1/n} - \frac{x_2 y_1}{y_2 x_1}\right| < H\left(\frac{x_2}{y_2}\right)^{-\kappa_2} \cdot \left(|a|^{1/n} + 1\right).$$

Thus, (1.1) can be replaced by

$$\left|b^{1/n} - \frac{u}{v}\right| < H\left(\frac{u}{v}\right)^{-O(\kappa_2)} \tag{1.2}$$

where $u, v \in \mathbf{Z}$, $b \in \mathbf{Q}$, and

$$|1 - b| < H(b)^{-O(\kappa_1/n)}. \tag{1.3}$$

THEOREM 1.2. *Suppose k, l are multiplicatively independent rational numbers > 1, j and n are natural numbers, $n \geqslant j$,*

$$|1 - k| \leqslant k_*^{-1} \quad \text{and} \quad H(l) > H(k),$$

where $H(\cdot)$ is the height. Then there exists an absolute constant $c_0 > 0$ such that

$$|k^{j/n} - l| > \exp\left(-c_0 \frac{\ln H(l) \ln H(k) \ln^2 n}{\ln(k_* + 1)}\right).$$

PROOF. We use Gel'fond's method [**18**, **19**] with improvements due to Baker [**5**] and Fel'dman. Consider the following auxiliary functions of a complex variable z:

$$f(z)=\sum_{\lambda_1=0}^{L_1}\sum_{\lambda_2=0}^{L_2} C_{\lambda_1,\lambda_2}k^{(\lambda_1+j\lambda_2/n)z},$$

$$\mathfrak{F}_\sigma(z)=\sum_{\lambda_1=0}^{L_1}\sum_{\lambda_2=0}^{L_2} C_{\lambda_1,\lambda_2}(n\lambda_1+j\lambda_2)^\sigma k^{\lambda_1 z}l^{\lambda_2 z}, \tag{1.4}$$

$$\Phi_\sigma(z)=\sum_{\lambda_1=0}^{L_1}\sum_{\lambda_2=0}^{L_2} C_{\lambda_1,\lambda_2}(n\lambda_1+j\lambda_2+1)\cdots(n\lambda_1+j\lambda_2+\sigma)k^{\lambda_1 z}l^{\lambda_2 z}.$$

Here the C_{λ_1,λ_2} are rational integers, chosen in accordance with Siegel's well-known lemma [**1**, **5**] so that

$$\Phi_\sigma(x)=0 \quad \text{for } \sigma=0,1,\ldots,S_0;\ x=0,1,\ldots,X_0. \tag{1.5}$$

We choose the parameters L_1, L_2, S_0, X_0 so that the usual scheme of Gel'fond's analytic method is applicable.

Suppose

$$|k^{j/n}-l|<\exp(-T), \tag{1.6}$$

where T denotes $c_0 \ln H(l)\ln H(k)\ln^2 n/\ln(k_*+1)$ for a sufficiently large constant $c_0>0$.

We choose L_1, L_2, X_0, S_0 in the form

$$L_1=\left[\frac{c_1\ln H(l)\ln n}{\ln(k_*+1)}\right],\qquad L_2=\left[\frac{c_1\ln H(k)\ln n}{\ln(k_*+1)}\right],$$
$$X_0=[c_2\ln n],\qquad S_0=\left[\frac{c_1c_2\ln H(l)\ln H(k)\ln n}{\ln^2(k_*+1)}\right], \tag{1.7}$$

for sufficiently large constants c_1, c_2, where $c_2=\frac{1}{4}c_1^{1/2}$. For such a choice of L_i, X_0, S_0 and c_1, c_2, we have, for c_0 sufficiently large in comparison with c_1 and c_2, the usual inequalities needed to apply Gel'fond's method:

$$\tfrac{1}{16}L_1L_2\geqslant S_0X_0,\quad L_1\ln H(k)X_0\ll T,\quad L_2\ln H(l)X_0\ll T;$$
$$S_0\ln(k_*+1)\ll T,\quad S_0\ln n\ll T,\quad S_0\ln\left(\frac{nL_1}{S_0}\right)\ll T, \tag{1.8}$$

where $x\ll y$ means that $c\cdot x\leqslant y$ for an arbitrarily large constant $c>0$.

According to [**1**, **5**] rational integers C_{λ_1,λ_2} (not all 0) satisfying (1.5) can be chosen so that

$$\max_{\lambda_1,\lambda_2}|C_{\lambda_1,\lambda_2}|\leqslant\exp\left(20c_1^{3/2}\frac{\ln H(l)\ln H(k)\ln^2 n}{\ln(k_*+1)}\right). \tag{1.9}$$

It follows from (1.4) and (1.5) that

$$\mathfrak{F}_\sigma(x) = 0: \qquad \sigma = 0, 1, \ldots, S_0;\ x = 0, 1, \ldots, X_0. \tag{1.10}$$

Comparing $\mathfrak{F}_\sigma(x)$ and $f^{(\sigma)}(x)$ and taking into account (1.6) and (1.7)–(1.10), we obtain

$$|f^{(\sigma)}(x)| < \exp(-T/2)\sigma! \tag{1.11}$$

for $0 \leqslant \sigma \leqslant S_0$, $0 \leqslant x \leqslant X_0$. Applying (1.11) and Schwarz's Lemma [**5**, **19**, **20**], we obtain from (1.7) and (1.8) that

$$|f^{(\sigma)}(x)| \leqslant \exp\left(-\frac{c_1^2 \ln H(k) \ln H(l) \ln^2 n}{2\ln(k_* + 1)}\right)(nL_1 + jL_2)^\sigma$$

for $\sigma = 0, 1, \ldots, [\frac{1}{2}S_0]$ and $|x| \leqslant c_3 \ln n$ for an arbitrarily large constant c_3 and $c_0 > 2c_1^2 c_3$. In particular, for the same bounds on σ and x we have

$$|\mathfrak{F}_\sigma(x)| < \exp\left(-c_1^2 \frac{\ln H(k) \ln H(l) \ln^2 n}{4\ln(k_* + 1)}\right)(nL_1 + jL_2)^\sigma. \tag{1.12}$$

Moreover,

$$\Phi_\sigma(x) = \sum_{s=0}^{\sigma} \tilde{C}_{s,\sigma} \mathfrak{F}_s(x) \tag{1.13}$$

for integers $\tilde{C}_{s,\sigma} \leqslant (\sigma/s)\sigma!(s!)^{-1}$. In view of (1.7), (1.8), and (1.12), we finally obtain

$$|\Phi_\sigma(x)/\sigma!| \leqslant \exp\left(-\frac{c_1^2 \ln H(l) \ln H(k) \ln^2 n}{8\ln(k_* + 1)}\right)$$

for $\sigma \leqslant \frac{1}{2}S_0$ and $x \leqslant c_3 \ln n$. But for rational integral $x \geqslant 0$, according to (1.4), $\Phi_\sigma(x)/\sigma!$ is a polynomial $P(k, l)$ in k and l of k-degree $\leqslant L_1 x$, l-degree $\leqslant L_2 x$, and (in view of (1.9)) height of at most

$$\exp\left((40c_1^{3/2} + c_3)\frac{\ln H(l) \ln H(k) \ln^2 n}{\ln(k_* + 1)}\right).$$

Since k and l are rational, it follows that for c_1^2 large in comparison with c_3 we have

$$\Phi_\sigma(x) = 0: \qquad x = 0, \ldots, [c_3 \ln n] = X_1;\ \sigma = 0, \ldots, [S_0/2] = S_1. \tag{1.14}$$

Choosing c_3 so that $c_3 c_2 > 2c_1^2$ we obtain a system in which the number of equations is greater than the number of variables. We will show below that a simple analytic Lemma 1.4 implies that (1.14) is impossible for $c_3 c_2 > 2c_1^2$. This simple method is employed below in Theorem 1.5. Here we use instead Baker-Stark's method, see [**20**] and [**5**], which is used in more general situations, for example in Chapter 2 of this volume.

Let p be any fixed prime > 5. Assume first that $[\mathbf{Q}(k^{1/p}, l^{1/p})\colon \mathbf{Q}] = p^2$. This condition will be removed in the final part of the proof of Theorem 1.2.

For each rational integer $i \geqslant 0$, we put

(1.15) $$L_{1i} = [L_1/p^i],\quad L_{2i} = [L_2/p^i],\quad S_{1i} = [S_1/4^i],\quad X_{1i} = X_1 p^i,$$

where $S_1 = [\frac{1}{2}S_0]$, $X_1 = [c_8 \ln n]$. We choose for i a bound $\mathfrak{T}$ so that $p^{\mathfrak{T}} > L_1 \geqslant L_2$, i.e.,

(1.16) $$\mathfrak{T} = \left[\frac{\ln L_1}{\ln p}\right] + 1.$$

For each set $\vec{\lambda} = (\lambda_{00}, \lambda_{01}, \ldots, \lambda_{0i-1}; \lambda_{10}, \lambda_{11}, \ldots, \lambda_{1i-1})$ of $2i$ numbers from $\{0, 1, \ldots, p-1\}$ we put $\vec{\lambda}_0 = \lambda_{00} + \lambda_{01}p + \cdots + \lambda_{0i-1}p^{i-1}$ and $\vec{\lambda}_1 = \lambda_{10} + \lambda_{11}p + \cdots + \lambda_{1i-1}p^{i-1}$. Then

(1.17) $$f^{\vec{\lambda},i}(z) \equiv \sum_{m_1=0}^{L_{1i}} \sum_{m_2=0}^{L_{2i}} C^{\vec{\lambda},i}_{m_1,m_2} k^{(m_1+jm_2/n)z},$$

where the coefficients $C^{\vec{\lambda},i}_{m_1,m_2}$ are defined as

(1.18) $$C^{\vec{\lambda},i}_{m_1,m_2} = C_{m_1p^i+\vec{\lambda}_0,\, m_2p^i+\vec{\lambda}_1}$$

for $m_1p^i + \vec{\lambda}_0 \leqslant L_1$, $m_2p^i + \vec{\lambda}_1 \leqslant L_2$; otherwise, $C^{\vec{\lambda},i}_{m_1,m_2} = 0$.

We also define, by analogy with (1.4),

$$\mathfrak{F}^{\vec{\lambda},i}_{\sigma}(z) = \sum_{m_1=0}^{L_{1i}} \sum_{m_2=0}^{L_{2i}} C^{\vec{\lambda},i}_{m_1,m_2}(m_1n + m_2j)^{\sigma} k^{m_1z} l^{m_2z}.$$

We establish by induction on i, $0 \leqslant i \leqslant \mathfrak{T}$, the following

LEMMA 1.3. *For each i, $0 \leqslant i \leqslant \mathfrak{T}$, we have*

(1.19) $$\mathfrak{F}^{\vec{\lambda},i}_{\sigma}(x) = 0$$

for any sequence $\vec{\lambda}$ of length $2i$ and natural numbers $\sigma \leqslant S_{1i}$ and $x \leqslant X_{1i}$.

PROOF. For $i = 0$, Lemma 1.3 follows from (1.14). Assume the lemma has been proved for all $i' \leqslant i$. We will prove the lemma for $i + 1$. Along with the sequence $\vec{\lambda} = (\lambda_{00}, \lambda_{01}, \ldots, \lambda_{0i}; \lambda_{10}, \ldots, \lambda_{1i})$ of length $2(i+1)$ consider the sequence $\vec{\lambda}' = (\lambda_{00}, \ldots, \lambda_{0i-1}; \lambda_{10}, \ldots, \lambda_{1i-1})$ and the corresponding function $f^i(z) = f^{\vec{\lambda}',i}(z)$. We apply the assumption (1.19); in view of (1.6), it follows immediately (since $L_{1i}X_{1i} \leqslant L_1X_1$; $L_{2i}X_{1i} \leqslant L_2X_2$; $S_{1i} \leqslant S_1$) that

(1.20) $$|(f^i(x))^{(\sigma)}| \leqslant \exp(-T/2)(nL_{1i} + jL_{2i})^{\sigma},$$

i.e., $|(f^i(x))^{(\sigma)}| \leqslant \exp(-T/4)\sigma!$ for all integers $x = 0, \ldots, X_{1i}$ and $\sigma \leqslant S_{1i}$. We will again use Schwarz's lemma [5, 10], taking into account that $|k - 1| \leqslant k_*^{-1}$.

Note that for $R = 4X_{1i+1}(k_* + 1)$ we have

(1.21)

$$\max_{|z|=R} |(f^i(z))^{(\sigma)}|$$
$$\leqslant \exp\left(40c_1^{3/2}\frac{\ln H(l)\ln H(k)\ln^2 n}{\ln(k_* + 1)} + L_{1i}X_{1i+1}\ln H(k) + L_{2i}X_{1i+1}\ln H(l)\right)$$
$$\times (nL_{1i} + jL_{2i})^\sigma$$
$$\leqslant \exp\left((40c_1^{3/2} + 2p)\frac{\ln H(l)\ln H(k)\ln^2 n}{\ln(k_* + 1)}\right)(nL_{1i} + jL_{2i})^\sigma.$$

Now choose c_1 sufficiently large in comparison with p. We consider that

$$X_{1i}S_{1i}\ln(k_* + 1) \geqslant 1/2c_1^2\frac{\ln H(l)\ln H(k)\ln^2 n}{\ln(k_* + 1)}(p/4)^i. \tag{1.22}$$

It follows from (1.20)–(1.22) when $p > 5$ that

$$|(f^i(z))^{(\sigma')}| \leqslant \exp\left(-\frac{c_1^2\ln H(l)\ln H(k)\ln^2 n}{8\ln(k_* + 1)}\right)(nL_{1i} + jL_{2i})^{\sigma'} \tag{1.23}$$

for all z, $|z| \leqslant X_{1i+1}$, and $\sigma' = 0, 1, \ldots, [\frac{1}{2}S_{1i}]$. In particular, for the same z and σ' we obtain

$$|\mathcal{F}^i_{\sigma'}(z)| \leqslant \exp\left(-\frac{c_1^2\ln H(l)\ln H(k)\ln^2 n}{16\ln(k_* + 1)}\right)(nL_{1i} + jL_{2i})^{\sigma'}. \tag{1.24}$$

As above, we introduce

(1.25)

$$\Phi^i_{\sigma'}(z) = \sum_{m_1=0}^{L_{1i}}\sum_{m_2=0}^{L_{2i}} C^{\vec{\lambda}',i}_{m_1,m_2}(nm_1 + jm_2 + 1)\cdots(nm_1 + jm_2 + \sigma')k^{m_1z}l^{m_2z}.$$

Using (1.13) and (1.24), we obtain

$$|\Phi^i_{\sigma'}(z)/\sigma'!| \leqslant \exp\left(-\frac{c_1^2\ln H(l)\ln H(k)\ln^2 n}{32\ln(k_* + 1)}\right), \tag{1.26}$$

since

$$\left(\frac{nL_{1i} + jL_{2i}}{\sigma'}\right)^{\sigma'} \leqslant \exp\left(\frac{2c_1^{3/2}\ln H(l)\ln H(k)\ln^2 n}{\ln^2(k_* + 1)}\right).$$

In (1.26) consider z to be of the form $z = x/p$ for a natural number x, $x \leqslant X_{1i+1}$. It follows directly from (1.25) that $\Phi^i_{\sigma'}(x/p)(\sigma'!)^{-1}$ has the form

$R(k^{1/p}, l^{1/p})$, where $R(x, y) \in \mathbf{Z}[x, y]$; $R(x, y)$ has x- and y-degrees

$$d_x(R) \leqslant L_{1i}X_{1i+1}; \qquad dy(R) \leqslant L_{2i}X_{1i+1}$$

and height

$$H(R) \leqslant \exp\left(\frac{40c_1^{3/2}\ln H(l)\ln H(k)\ln^2 n}{\ln^2(k_* + 1)}\right).$$

Then either $\Phi^i_\sigma(x/p) = 0$ or

$$(1.27) \qquad \left|\Phi^i_{\sigma'}\left(\frac{x}{p}\right)(\sigma'!)^{-1}\right| \geqslant \exp\left(-100p^2c_1^{3/2}\frac{\ln H(l)\ln H(k)\ln^2 n}{\ln(k_* + 1)}\right).$$

For c_1 sufficiently large in comparison with p, it follows from (1.26) and (1.27) that $\Phi^i_{\sigma'}(x/p) = 0$ for all $\sigma' = 0,\ldots,[\frac{1}{2}S_{1i}]$ and $x = 0, 1,\ldots, X_{1i+1}$. For the same σ' and x, therefore,

$$(1.28) \qquad \mathfrak{F}^i_{\sigma'}(x/p) = 0.$$

Suppose $(x, p) = 1$. In view of our assumption that $\mathbf{Q}(k^{1/p}, l^{1/p})$ has degree p^2, it follows from (1.28) that for any λ_1, λ_2 in $[0, p)$ we have

$$(1.29) \qquad \sum_{m_1=0, m_1\equiv\lambda_1}^{L_{1i}} \sum_{m_2=0, m_2\equiv\lambda_2}^{L_{2i}} C^{\vec{\lambda}',i}_{m_1,m_2}(m_1n + m_2j)^{\sigma'}k^{m_1x/p}l^{m_2x/p} = 0,$$

for all $\sigma' \leqslant \frac{1}{2}S_{1i}$, $x \leqslant X_{1i+1}$, $(x, p) = 1$, if the congruences $\equiv$ are understood to mean mod p. Each $m_j \equiv \lambda_j \pmod{p}$ can be represented in the form $m_j = m'_jp + \lambda_j$, where $m'_j \leqslant L_{1i+1}$. Dividing (1.29) by $k^{\lambda_1x/p} \cdot l^{\lambda_2x/p}$ and considering a linear combination of the equalities (1.29), we obtain

$$\sum_{m_1=0, m_1=m'_1p+\lambda_1}^{L_{1i}} \sum_{m_2=0, m_2=m'_2p+\lambda_2}^{L_{2i}} C^{\vec{\lambda}',i}_{m_1,m_2}(m'_1pn + m'_2pj)^{\sigma}k^{m'_1x}l^{m'_2x} = 0,$$

for all $\sigma \leqslant \frac{1}{2}S_{1i}$, $x \leqslant X_{1i+1}$, $(x, p) = 1$. Put $\lambda_1 = \lambda_{0i}$ and $\lambda_2 = \lambda_{1i}$, where $\vec{\lambda} = (\lambda_{00},\ldots,\lambda_{0i}; \lambda_{10},\ldots,\lambda_{1i})$. According to the definition (1.18), we obtain

$$\sum_{m'_1=0}^{L_{1i+1}} \sum_{m'_2=0}^{L_{2i+1}} C^{\vec{\lambda},i+1}_{m'_1,m'_2}(m'_1n + m'_2j)^{\sigma}k^{m'_1x}l^{m'_2x} = 0,$$

i.e. $\mathfrak{F}^{i+1}_\sigma(x) = 0$.

Thus, the function $f^{i+1}(z)$ satisfies the inequalities

$$(1.30) \qquad |(f^{i+1}(z))^{\sigma'}| < \exp(-T/2)(nL_{1i+1} + jL_{1i+1})^{\sigma'}$$

for all $\sigma' \leqslant \frac{1}{2}S_{12}$ and natural numbers $z \leqslant X_{1i+1}$, $(z, p) = 1$. The number of such $z \leqslant X_{1i+1}$, $(z, p) = 1$, is at least $\frac{1}{2}X_{1i+1}$. Taking into account (1.30), we apply, as above, Schwarz's lemma [**5**, **10**]. We obtain

$$|\Phi^{i+1}_{\sigma''}(x)(\sigma''!)^{-1}| < \exp\left(-c_1^2 \frac{\ln H(l) \ln H(k) \ln^2 n}{32 \ln(k_* + 1)}\right) \tag{1.31}$$

for $\sigma'' \leqslant S_{1i+1} \leqslant \frac{1}{4} S_{1i}$ and a natural number $x \leqslant X_{1i+1}$.

Since

$$\max\{L_{1i+1} X_{1i+1} \ln H(k), L_{2i+1} X_{1i+1} \ln H(l)\} \leqslant c_1^{3/2} \frac{\ln H(l) \ln H(k) \ln^2 n}{\ln(k_* + 1)},$$

it follows from (1.25) and (1.31) that

$$\Phi^{i+1}_{\sigma''}(x) = 0 \tag{1.32}$$

for $\sigma'' \leqslant S_{1i+1}$ and $x \leqslant X_{1i+1}$. Lemma 1.3 is proved.

We apply 1.4 with $i = \mathfrak{T}$. We have $L_{1\mathfrak{T}} = L_{2\mathfrak{T}} = 0$ by (1.16), i.e. for $\sigma = 0$,

$$C^{\vec{\lambda}, \mathfrak{T}}_{0,0} = 0. \tag{1.33}$$

Since any $\lambda_i \leqslant L_i$ can be represented in the form $\lambda_i = \lambda_{i0} + \lambda_{i1} p + \cdots + \lambda_{i\mathfrak{T}-1} p^{\mathfrak{T}-1}$, it follows from (1.33) and (1.18) that $C_{\lambda_1, \lambda_2} = 0$ for all $\lambda_1 = 0, 1, \ldots, L_1$ and $\lambda_2 = 0, 1, \ldots, L_2$. This contradicts the choice of the C_{λ_1, λ_2}.

It remains to consider the case avoided in Lemma 1.3, i.e. $[\mathbf{Q}(k^{1/p}, l^{1/p}) : \mathbf{Q}] < p^2$ for a given prime $p > 5$. We will show that this case can be reduced to a previous one. Suppose $\mathbf{Q}(k^{1/p}, l^{1/p})$ has degree $< p$ over $\mathbf{Q}(k^{1/p})$ for any $p > 5$. We use Lemma [**20**]: if $\mathbf{Q}(\alpha^{1/p}, \beta^{1/p})$ does not have degree p over $\mathbf{Q}(\alpha^{1/p})$, then $\beta = \alpha^i \gamma^p$ for $\gamma \in \mathbf{Q}$ and a natural number $i < p$.

We use the so-called "Thue lemma" (which is a simple consequence of Dirichlet's principle): if $(a, p) = 1$, there exist natural numbers $x, y \leqslant p^{1/2}$ such that $ax + y$ or $ax - y$ is divisible by p. We apply this lemma to the case $a = i$. Raising the equality $l = k^i \gamma^p$ to a suitable power, we obtain an equality of the form $l^x k^y = \gamma_1^p$, where $\gamma_1 \in \mathbf{Q}$ and x, y are rational integers such that $|x|$, $|y| \leqslant p^{1/2}$. For the height $H(\gamma_1)$ we obtain

$$H(\gamma_1) \leqslant H(l)^{x/p} \cdot H(k)^{y/p} \leqslant H(l)^{2/p^{1/2}}. \tag{1.34}$$

Moreover, $l = \gamma_1^{p/x} k^{-y/x}$, i.e. inequality (1.6) for $|k^{j/n} - l|$ can be replaced by an inequality for $|k^{j/n} - k^{-y/x} \gamma_1^{p/x}|$. Indeed, it follows from (1.6) that for an arbitrarily large constant $c_0' > 0$ we have

$$|k^{(jx+yn)/np} - \gamma_1| < \exp\left(-c_0' \frac{\ln H(k) \ln H(l) \ln^2 n}{\ln(k_* + 1)}\right), \tag{1.35}$$

and, obviously, $np \geqslant 2p^{1/2} n \geqslant jx + yn$, since $p > 5$.

We will show that for $l_1 = \gamma_1$, $n_1 = np$, and $j_1 = jx + yn$ we have (1.6) with the natural replacement of $H(l)$ by $H(l_1)$ and $\ln n$ by $\ln n_1$. Since k remains unchanged, it suffices to show that

$$2 \ln H(\gamma_1) \ln^2(np) < \ln H(l) \ln^2 n. \tag{1.36}$$

Inequality (1.36) clearly follows from (1.34), otherwise we obtain $n < c_{10}p$, which says that n is bounded and in this case Theorem 1.2 is obvious. Thus, for $l_1 = \gamma_1$ we have (1.35), which is analogous to (1.6), and, in view of (1.34) and $p > 5$, we have $H(\gamma_1) < H(l)$. Continuing this procedure, we obtain a sequence of rational numbers $l, l_1, \ldots,$ with decreasing heights satisfying the inequality for $|k^{j_s/n_s} - l_s|$. Then for the bounded l_i an upper bound of the form

$$\exp\left(-O\left(\frac{\ln H(k)\ln^2 n}{\ln(k_* + 1)}\right)\right)$$

is impossible for $n > O(1)$.

The theorem is completely proved.

We now give an entirely different proof of Theorem 1.2 based on new arguments and a very simple analytic Lemma 1.4 similar to Normality Lemma 1 of §2 below. Multidimensional versions of Lemma 1.4 can be used similarly in Baker's method [**5**].

Instead of the p-adic methods of Stark [**20**], we can use methods similar to Gel′fond's method of 1934.

LEMMA 1.4. *Let us denote*

$$F_\sigma(\alpha, \beta) = \sum_{\lambda_1=0}^{L_1} \sum_{\lambda_2=0}^{L_2} C_{\lambda_1,\lambda_2}(\lambda_1 + \theta\lambda_2 + 1) \cdots (\lambda_1 + \theta\lambda_2 + \sigma)\alpha^{\lambda_1} \cdot \beta^{\lambda_2}$$

with $C_{\lambda_1,\lambda_2} \in \mathbf{C}^1$ *and nonzero* $\alpha, \beta \in \mathbf{C}^1$, α *not a root of unity.*

Let $\lambda_1 + \lambda_2\theta \neq \lambda_1' + \lambda_2'\theta$ *for* $0 \leqslant \lambda_1, \lambda_1' \leqslant L_1$, $0 \leqslant \lambda_2, \lambda_2' \leqslant L_2$ *and* $(\lambda_1, \lambda_2) \neq (\lambda_1', \lambda_2')$. *Assume that* $X \geqslant 1$, $S \geqslant 1$ *and* $F_\sigma(\alpha^i, \beta^i) = 0$ *for* $i = 0, \ldots, X-1$ *and* $\sigma = 0, 1, \ldots, S-1$.

If $XS > L_1(L_2 + 1) + XL_2$, *then all* C_{λ_1,λ_2} *are zero*: $\lambda_1 = 0, \ldots, L_1$; $\lambda_2 = 0, \ldots, L_2$.

PROOF. Let us denote for $\lambda_2 = 0, \ldots, L_2$, $P_{\lambda_2}(x) \overset{\text{def}}{=} \sum_{\lambda_1=0}^{L_1} C_{\lambda_1,\lambda_2}x^{\lambda_1}$. Then we get

$$F_\sigma(\alpha, \beta) = \frac{d^\sigma}{dx^\sigma}\left(\sum_{\lambda_2=0}^{L_2} P_{\lambda_2}(x)(\beta/\alpha)^{\theta\lambda_2}x^{\theta\lambda_2+\sigma}\right)\Bigg|_{x=\alpha}$$

for any $\sigma = 0, 1, \ldots$. If we now denote for $i = 0, \ldots, X-1$,

$$F_i(x) = \sum_{\lambda_2=0}^{L_2} P_{\lambda_2}(x)x^{\theta\lambda_2} \cdot (\beta/\alpha)^{\theta i\lambda_2},$$

then, according to the assumptions we obtain:

$$F_i^{(\sigma)}(\alpha^i) = 0: \qquad \sigma = 0, 1, \ldots, S-1$$

for any $i = 0, \ldots, X-1$. Let us assume that not all C_{λ_1,λ_2} ($\lambda_1 = 0, \ldots, L_1$; $\lambda_2 = 0, \ldots, L_2$) are zero; then we take those polynomials $P_{\lambda_j}(x)$: $j = 0, \ldots, J$

among $\{P_{\lambda_2}(x)\colon \lambda_2 = 0,\ldots,L_2\}$ that are not zero; $0 \leqslant J \leqslant L_2$. According to the assumptions, $\lambda_1 + \lambda_2\theta \neq \lambda_1' + \lambda_2'\theta$ whenever $0 \leqslant \lambda_1, \lambda_1' \leqslant L_1$, $0 \leqslant \lambda_2, \lambda_2' \leqslant L_2$ and $(\lambda_1, \lambda_2) \neq (\lambda_1', \lambda_2')$. Hence, functions $P_{\lambda_j}(x)x^{\theta\lambda_j}$: $0 \leqslant J \leqslant L_2$ are linearly independent (over **C**). We define $W(x) = \det((d/dx)^i(P_{\lambda_j}(x)x^{\theta\lambda_j}))_{i,j=0,\ldots,J}$, so that $W(x) \not\equiv 0$. Then we have $W(x) = D(x) \cdot x_1^{\delta}$, $\delta = \sum_{j=0}^{J}\{\theta\lambda_j - j\}$, where $D(x)$ is a polynomial in x of degree of at most $L_1 \cdot (J+1)$. Now the functions $F_i(x)$ are linear combinations of $P_{\lambda_j}(x)x^{\theta\lambda_j}$:

$$F_i(x) = \sum_{j=0}^{J} P_{\lambda_j}(x)x^{\theta\lambda_j} \cdot (\beta/\alpha)^{\theta i \lambda_j}: \qquad i = 0,\ldots,X-1.$$

Making a linear transformation of a determinant $W(x)$ and taking into account that numbers α^i are nonzero, we obtain

$$\operatorname*{ord}_{x=\alpha^i}(W(x)) \geqslant \operatorname*{ord}_{x=\alpha^i}(F_i(x)) - J$$

for $i = 0,\ldots,X-1$ and hence $\operatorname{ord}_{x=\alpha^i}(D(x)) \geqslant S - J$: $i = 0,\ldots,X-1$. Consequently,

$$L_1(J+1) \geqslant \sum_{i=0}^{X-1} \operatorname*{ord}_{x=\alpha^i}(D(x)) \geqslant X(S-J),$$

or $XS \leqslant XJ + L_1(J+1) \leqslant XL_2 + L_1(L_2+1)$. The Lemma is proved.

We proved results much more general than this one in the context of Padé approximations in **[17]**, see especially Theorem 8.3.1. The author wants to point out that R. Tijdeman obtained results of the same form for binomial and exponential functions (1979).

We can now apply Lemma 1.4 in the proof of Theorem 1.2 directly to (1.14) to arrive at a contradiction with c_0 being sufficiently large. The advantage of Lemma 1.4 is the possibility of obtaining a "relatively small value" of constant c_0. We will present an example of such estimates below.

Based on our Lemma 1.4 we now give a particularly short proof of the bound of Theorem 1.2 with an explicit form of constant c_0. In particular, our result gives a simple effective measure of diophantine approximations to an algebraic number $k^{j/n}$. Aiming at such measures of irrationality, we consider only the case of $H(1)$ sufficiently large with respect to $H(k)$ (and n). We obtain

THEOREM 1.5. *Let k and 1 be rational numbers and j, n be relatively prime positive integers, $n \geqslant j$ and let $|1-k| \leqslant k_*^{-1}$ for $k_* > 1$. Then we have*

$$|k^{j/n} - l| > \exp\left(-c_0 \cdot \frac{\ln H(l)\ln H(k)}{\ln(1+k_*)} \cdot \left(1 + \frac{\ln n}{\ln(1+k_*)}\right)^2\right)$$

with $c_0 \leqslant e^{10}$.

PROOF. We consider the following auxiliary functions:

$$f_\sigma(z) = \sum_{\lambda_1=0}^{L_1} \sum_{\lambda_2=0}^{L_2} C_{\lambda_1,\lambda_2} \binom{\lambda_1 + \lambda_2 \cdot j/n}{\sigma} k^{(\lambda_1 + j\lambda_2/n)z},$$

$$(1.37) \qquad \Phi_\sigma(z) = \sum_{\lambda_1=0}^{L_1} \sum_{\lambda_2=0}^{L_2} C_{\lambda_1,\lambda_2} \binom{\lambda_1 + \lambda_2 \cdot j/n}{\sigma} k^{\lambda_1 z} \cdot l^{\lambda_2 z},$$

$$\mathcal{F}_\sigma(z) = \sum_{\lambda_1=0}^{L_1} \sum_{\lambda_2=0}^{L_2} C_{\lambda_1,\lambda_2} (\lambda_1 + j\lambda_2/n)^\sigma k^{\lambda_1 z} \cdot l^{\lambda_2 z}.$$

We use Siegel's lemma [**1, 15**] to find rational integers so that the following system of linear equations on C_{λ_1,λ_2} is satisfied:

$$(1.38) \qquad \Phi_\sigma(x) = 0: \qquad \sigma = 0, \ldots, S_0;\ x = 0, \ldots, X_0.$$

We assume that k is close to 1: $|1 - k| \leqslant k_*^{-1}$, $k_* > 1$ and that

$$|k^{j/n} - l| \leqslant \exp(-c_0 \cdot T),$$

$$(1.39) \qquad T = \frac{\ln H(k) \cdot \ln H(l)}{\ln(1 + k_*)} \cdot \Lambda^2, \qquad \Lambda = 1 + \frac{\ln n}{\ln(1 + k_*)}.$$

Parameters L_1, L_2, S_0, X_0 are chosen in a way similar to (1.7):

$$(1.40) \qquad L_1 = \left[\frac{c_1 \ln H(l)}{\ln(1 + k_*)} \cdot \Lambda\right], \qquad L_2 = \left[\frac{c_1 \ln H(k)}{\ln(1 + k_*)} \cdot \Lambda\right],$$

$$X_0 = [c_2 \Lambda], \qquad S_0 = \left[c_2' \cdot \frac{\ln H(l) \cdot \ln H(k)}{\ln^2(1 + k_*)} \Lambda\right].$$

We note now that for $\mu_n = \prod_{p|n} p^{1/(p-1)}$ the number $(n\mu_n)^\sigma \binom{\lambda_1 + \lambda_2 j/n}{\sigma}$ is always a rational integer, provided that λ_1, λ_2, σ are (nonnegative) rational integers. We put $N = n\mu_n$. In the estimates below we assume that $H(1)$ is sufficiently large with respect to $H(k)$. (In fact, it is enough to assume only that $\ln H(l) \geqslant \lambda \ln H(k)$ for some $\lambda < 2$ and $H(1) \geqslant c_0'$ for a sufficiently large constant c_0').

From (1.37) and (1.40) it follows that the sizes of the coefficients in system (1.38) defining C_{λ_1,λ_2} are bounded from above by

$$U = \exp\left\{\left(c_2' \frac{\ln N}{\ln(1 + k_*)} + 2c_1 c_2\right)\right\} \cdot T.$$

Hence Siegel's lemma [**1, 5, 15**] implies that C_{λ_1,λ_2} can be chosen as rational integers, not all identically zero such that system (1.38) is satisfied and such that

$$(1.41) \qquad \max\left(|C_{\lambda_1,\lambda_2}| : 0 \leqslant \lambda_1 \leqslant L_1;\ 0 \leqslant \lambda_2 \leqslant L_2\right) \leqslant |C|,$$

$$|C| \leqslant U^{c_2 c_2'/(c_1^2 - c_2 c_2')} \cdot e^{o(L_1)}.$$

Functions $\mathcal{F}_\sigma(x)$ are simple linear combinations of $\Phi_s(x)$: $s = 0,\ldots,\sigma$. This immediately implies:

(1.42) $$\mathcal{F}_\sigma(x) = 0: \qquad \sigma = 0,\ldots,S_0;\ x = 0,\ldots,X_0.$$

We have for $s \geqslant 0$,

(1.43)
$$\left(\frac{d}{dx}\right)^s f_\sigma(x) = \sum_{\lambda_1=0}^{L_1}\sum_{\lambda_2=0}^{L_2} C_{\lambda_1,\lambda_2}\binom{\lambda_1+\lambda_2\cdot j/n}{\sigma}(\lambda_1+\lambda_2 j/n)^s\cdot k^{(\lambda_1+j\lambda_2/n)x}.$$

From (1.42) it follows that for $\sigma + s \leqslant S_0$ and an integer x, $0 \leqslant x \leqslant X_0$, the number obtained by replacing $k^{j/n}$ by l in (1.43), is zero. Hence we obtain

$$\left|\frac{1}{s!}\left(\frac{d}{dx}\right)^s f_\sigma(x)\right| \leqslant |C|\cdot|k|^{L_1X_0}\cdot N^{S_0}\cdot(2l)^{L_1}\cdot|k^{j/n}-l|: \qquad x = 0,1,\ldots,X_0$$

with integers σ, s: $\sigma + s \leqslant S_0$.

We now apply Schwarz's lemma and interpolation formulas to the function $f_\sigma(x)$ (see Gel'fond **[19]**, Baker **[5]** and Cijsouw **[10]**). Rough estimates (Lemma 2.1, Chapter 1, this volume) give:

(1.44)
$$|f_\sigma(z)|_R \leqslant 2\cdot\left(\frac{2}{A}\right)^{(X_0+1)\cdot(S_0-\sigma)}\cdot|f_\sigma(z)|_{AR}$$
$$+\left(\frac{6R}{X_0+1}\right)^{(X_0+1)(S_0-\sigma)}\cdot\max\left\{\left|\frac{1}{s!}f_\sigma^{(s)}(x)\right|: s \leqslant S_0-\sigma;\ x \leqslant X_0\right\}.$$

Here we take $R = c_5\Lambda$ and we put $A = 2k_*^\delta$ for $0 < \delta \leqslant 1$. We obtain

(1.45) $$|f_\sigma(z)|_{AR} \leqslant L_1L_2\cdot 2^\sigma\cdot|C|\cdot|k|^{(L_1+L_2j/n)AR}$$
$$\leqslant |C|\cdot\exp\left\{\left(2c_1c_5k_*^{\delta-1}\ln^{-1}H(k)-\delta c_2(c_2'-c_3)\right)\cdot T\right\},$$

where $\sigma \leqslant c_3S_0/c_2'$.

We also remark that (1.37) implies

(1.46) $$|f_\sigma(x)-F_\sigma(x)| \leqslant L_1L_2\cdot|C|\cdot 2^{L_1}\cdot|k|^{L_1x}\cdot|k^{j/n}-l|,$$

cf. above. Since $N^\sigma\cdot\binom{\lambda_1+\lambda_2 j/n}{\sigma}$ is always a rational integer, then for denominators $\operatorname{den}(k)$ and $\operatorname{den}(l)$ of k and l, respectively,

$$\Phi_\sigma(x)N^\sigma\cdot\left\{\operatorname{den}(k)^{L_1}\cdot\operatorname{den}(l)^{L_2}\right\}^x$$

is a rational integer. Hence, for any integer x, $x \leqslant c_5\Lambda$ either $\Phi_\sigma(x) = 0$ or

$$|\Phi_\sigma(x)| \geqslant N^{-\sigma}\cdot e^{-2c_1c_5T}.$$

We now immediately use our Lemma 1.5 in the End of the Proof. It follows from Lemma 1.5 that if we have equations

$$\Phi_\sigma(x) = 0: \qquad \sigma = 0,1,\ldots,\sigma_x$$

for $x = 0, 1, \ldots, X_1 = [c_5\Lambda]$ and if $\Sigma_{x=0}^{X_1} \sigma_x > L_1L_2 + L_2X_1$, then all numbers C_{λ_1,λ_2} are zero.

Since not all C_{λ_1,λ_2} are zeroes, for any $c_3 \leqslant c_2$ and such that $c_3c_5 > c_1^2$ there always exists an integer x_0, $x_0 \leqslant c_5\Lambda$ such that

$$|\Phi_{\sigma_0}(x_0)| \geqslant \exp\{-(2c_1c_5 + c_3)\cdot T\}. \tag{1.47}$$

Combining inequalities (1.44)–(1.47) we obtain a simple system of algebraic inequalities that give a lower bound for c_0. It is easy to see that one can now take for c_0 number $\geqslant e^{10}$.

Baker's method of bounds of linear forms in logarithms of n algebraic numbers yields explicit expressions for constants for an arbitrary n (cf. references in **[5]** for papers of various authors on this subject). The method of proof of Theorem 1.5 is completely different from those used for $n > 2$ (cf. e.g. Stark's method used above). A simple scheme of the proof of Theorem 1.5 can be considerably modified to obtain a better value of constant c_0 (and constants in Theorem 1.7); and a better dependence on n.

Modifications in the proofs of Theorems 1.5 and 1.7 include a different choice of systems of linear equations (1.5) or (1.42) for auxiliary functions of the form

$$f_\sigma(x) = \sum_{\lambda_1=0}^{L_1}\sum_{\lambda_2=0}^{L_2} C_{\lambda_1,\lambda_2}\cdot\binom{\lambda_1+\lambda_2\cdot j/n}{\sigma}\cdot k^{(\lambda_1+\lambda_2 j/n)x},$$

and

$$f_{\sigma,i}(x) = \sum_{\lambda_1=0}^{L_1}\sum_{\lambda_2=0}^{L_2} C_{\lambda_1,\lambda_2}\binom{\lambda_1+\lambda_2 j/n}{\sigma}\lambda_2^i\cdot k^{(\lambda_1+\lambda_2 j/n)x}.$$

We apply results of the form of Lemma 1.4 or similar results of the form of Normality Lemma 1 of §2.

Our considerations give an upper bound for the constants c_0, c_0' in Theorems 1.5, 1.6, 1.7: $c_0 \leqslant e^4$, $c_0' < 4$.

Our result provides an effective determination of constants in the context of the (noneffective) Thue-Siegel theorem.

The first application of Theorems 1.2 and 1.5 pertains to the case of the Thue-Siegel theorem **[21, 19]**.

COROLLARY 1.6. *Suppose a is a rational number and n is a natural number. If x_i and y_i are distinct pairs of integers such that*

$$\left|a^{1/n} - \frac{x_1}{y_1}\right| < c_6|y_1|^{-\kappa_1}, \qquad \left|a^{1/n} - \frac{x_2}{y_2}\right| < c_6|y_2|^{-\kappa_2},$$

then $\kappa_1\kappa_2 \leqslant c_0'n$. Here $c_6 > 0$ depends on a and n, and $c_0' > 0$ is an absolute constant.

PROOF. It suffices to apply Example 1.1 and use Theorem 1.5.

We have a generalization of Theorem 1.5 to the case of algebraic numbers.

THEOREM 1.7. *Suppose α, β are algebraic numbers of a field* $\mathbf{K}$ *of degree d which have degrees d_1, d_2 and heights $H(\alpha)$, $H(\beta)$. If $n \geqslant j$ are natural numbers and* $|1-\alpha| \leqslant \alpha_*^{-1}$, *then*

$$|\alpha^{j/n} - \beta| \geqslant \exp\left(-c_0 \frac{\ln H(\alpha) \ln H(\beta) \ln^2 n}{\ln(\alpha_* + 1)} \cdot \frac{d^4}{d_1 d_2}\right).$$

PROOF. Let $\mathbf{K} = \mathbf{Q}(\theta)$; the coefficients C_{λ_1,λ_2} of the auxiliary functions can be expressed in the form

$$C_{\lambda_1,\lambda_2} = \sum_{i=0}^{d-1} C_{\lambda_1,\lambda_2,i}\theta^i.$$

We use the same auxiliary functions as in (1.4) or (1.37):

$$\mathcal{F}_\sigma(x) = \sum_{\lambda_1=0}^{L_1} \sum_{\lambda_2=0}^{L_2} C_{\lambda_1,\lambda_2}(\lambda_1 + j\lambda_2/n)^\sigma k^{\lambda_1 z} l^{\lambda_2 z};$$

$$\Phi_\sigma(z) = \sum_{\lambda_1=0}^{L_1} \sum_{\lambda_2=0}^{L_2} C_{\lambda_1,\lambda_2}\binom{\lambda_1 + \lambda_2 j/n}{\sigma} k^{\lambda_1 z} l^{\lambda_2 z}.$$

It is easy to see that in this case the parameters can be chosen as follows:

$$L_1 = \left[\frac{\tilde{c}_1 \ln H(\beta) \ln n}{\ln(\alpha_* + 1)} \cdot \frac{d^2}{d_2}\right], \qquad L_2 = \left[\frac{\tilde{c}_2 \ln H(\alpha) \ln n}{\ln(\alpha_* + 1)} \cdot \frac{d^2}{d_1}\right],$$

$$X_0 = [\tilde{c}_3 d \ln n], \qquad S_0 = \left[\frac{\tilde{c}_4 \ln H(\alpha) \ln H(\beta) \ln n}{\ln^2(\alpha_* + 1)} \cdot \frac{d^3}{d_1 d_2}\right].$$

Then the usual method of Gel′fond, Schwarz's lemma, and Siegel's lemma lead to the End of the Proof, where it is necessary to show that all C_{λ_1,λ_2} are equal to zero. For this, as above, we apply Lemma 1.4.

Because of the applications of Lemma 1.4 one can choose as an upper bound for c_0 the same one as in Theorem 1.5, provided that $H(\alpha)$ is sufficiently large with respect to $H(\beta)$ and that α^* is sufficiently large (with respect to d).

The results proved above can also be used to establish the existence of at most two solutions of a series of Diophantine equations (cf. [**13**]). We illustrate this by means of the equation

$$ax^n - by^n = c, \tag{1.48}$$

where a, b, c are integers. This equation was first investigated in detail by Siegel [**13**] by means of his technique of hypergeometric functions. His theorem asserts:

If $(ab)^{n/2-1} \geqslant 4c^{2n-2}(n\mu_n)^n$, where $\mu_n = \prod_{p|n} p^{1/(p-1)}$, the equation (1.48) has at most one solution in natural numbers. Theorems 1.2 and 1.5 permit a significant improvement of Siegel's estimate. In particular, from 1.5 we obtain

PROPOSITION 1.8. *For $n \geqslant c_0$, equation* (1.48) *has at most two solutions in natural numbers. Moreover, for $ab > c_6 \cdot cn$,* (1.48) *has at most one solution.*

Of interest are those cases, when (1.48) does not have integer rational solutions at all, except trivial ones. For example, let us consider the inequality

$$(1.49) \qquad |ax^n - (a+c)y^n| \leqslant |c|.$$

This inequality has the solution $x = y = 1$.

PROPOSITION 1.9. *For* $|a| \geqslant (|c|/\ln(|c|+1))^{1+\Lambda_0/n}$ *for an absolute constant* $\Lambda_0 > 0$ *the inequality* (1.49) *does not have non-trivial solutions.*

PROOF. We obtain the estimate

$$\left|\left(\frac{x}{y}\right)^n - \left(1 + \frac{c}{a}\right)\right| \leqslant \left|\frac{c}{ay^n}\right|.$$

Then Theorems 1.2 and 1.5 provide us with the desired estimate.

Proposition 1.9 cannot be improved in the sense that a simple example provides the satisfiability of (1.49) with

$$|a| \geqslant \left(|c|/\ln(|c|+1)\right)^{1+1/n},$$

i.e. $\Lambda_0 > 1$. Different results on the Diophantine approximations to algebraic roots $\alpha^{j/n}$ for rational (algebraic) α, and solutions of Diophantine equations $ax^n - by^n = c$ are presented in **[14]**. These results use the Thue-Siegel method, and are closer to the methods used in §§2–4. The methods of **[14]** and the methods of "minimal" approximating forms (Padé approximations) can be used for the approximation of algebraic roots and bounds of linear forms in logarithms (cf. **[26]**).

As we showed, Gel'fond's method gives another proof of the Thue-Siegel theorem for roots $\alpha^{j/n}$ of algebraic numbers. Similarly our considerations, and the standard reductions **[5, 1]** of solutions of an arbitrary Thue equation $f(x, y) = A$ to the form $\alpha x^n - \beta y^n = \gamma$ with algebraic α, β, γ, x, y give the possibility of bounding the number of **K**-integral solutions of Thue equations. The existence of a large number of **K**-integral solutions of $f(x, y) = A$ then implies the existence of a large number of algebraic approximations to $(\alpha/\beta)^{1/n}$. In this way we can get

THEOREM 1.10. *Suppose* **K** *is an algebraic number field of degree* m *and* $f(x, y)$ *is an irreducible form of degree* $n \geqslant 3$ *with* **K***-integral coefficients. Then the number of* **K***-integral solutions of the equation* $f(x, y) = A$ *does not exceed* $c(m, n, \Psi(A))$, *where* $\Psi(A)$ *is the number of prime divisors in the decomposition of* (A).

For the proof of Theorem 1.10, suppose $f(x, y)$ is a **K**-irreducible binary form of degree $n \geqslant 3$ with coefficients in $\mathbf{I}_{\mathbf{K}}$. As usual, $\mathbf{I}_{\mathbf{K}}$ is the ring of integers of the algebraic number field **K** of degree m.

Suppose $A \in \mathbf{I}_{\mathbf{K}}$ and $\#(f, A, \mathbf{K})$ is the number of **K**-integral solutions of the equation

$$(1.50) \qquad f(x, y) = A.$$

The following remark is important. Suppose Δ is the discriminant of $f(x, y)$, $\Delta \neq 0$, and $\Delta \in \mathbf{I}_{\mathbf{K}}$. A simple argument shows that for any $\gamma > 0$ there exist $\leqslant c(n, m, \gamma)$ solutions of (1.50) with $|\overline{x}|, |\overline{y}| \leqslant |\overline{\Delta}|^{\gamma}$.* Similarly, for any $\gamma > 0$ there exist $\leqslant c(n, m, \Psi(A), \gamma)$ solutions of (1.50) such that $|\overline{x}|, |\overline{y}| < |\overline{A}|^{\gamma}$. Therefore, assuming that $\#(f, A, \mathbf{K})$ is sufficiently large in comparison with n and m, we obtain (cf. [13]):

(1.51) For any constant $\gamma = \gamma(m, n) > 0$ there exist $O(\#(f, A, \mathbf{K}))$ solutions in integers x, y of $\mathbf{I}_{\mathbf{K}}$ such that $|\overline{x}|, |\overline{y}| > (|\overline{A}| \cdot |\overline{\Delta}|)^{\gamma}$.

Suppose α is a root of the equation $f(x, 1) = 0$. Then, on introducing the new field $\mathbf{K}_1 = \mathbf{K}(\alpha)$, (1.50) can be rewritten in the form

$$\operatorname{Norm}_{\mathbf{K}_1/\mathbf{K}}(x - \alpha y) = A. \tag{1.52}$$

Let $\psi_1(A)$ denote the number of pairwise nonassociated elements of $\mathbf{K}_1$ for which $\operatorname{Norm}_{\mathbf{K}_1/\mathbf{K}}$ is equal to A. Let $\varepsilon_1, \ldots, \varepsilon_w$ be a fundamental system of units of the field $\mathbf{K}_1$. By a standard procedure [5], the equation (1.52) can be rewritten in the form

$$x - \alpha y = \gamma \varepsilon_1^{h_1} \cdots \varepsilon_w^{h_w}, \tag{1.53}$$

where γ is one of the $\psi_1(A)$ numbers of norm A, $|\overline{\gamma}| \leqslant |\overline{A}|^{1/n} |R(\mathbf{K}_1)|$, $R(\mathbf{K}_1)$ being the regulator of $\mathbf{K}_1$, and $h_1, \ldots, h_w$ are rational integers. Put $\mu_i = x - \alpha_i y$, where $\alpha_1 = \alpha, \ldots, \alpha_n$ are the conjugates of α over $\mathbf{K}$. We have $\mu_1 \cdots \mu_n = A$. Condition (1.51) shows that there exist indices l and j such that $|\mu_l^{(j)}| = \min |\mu_k^{(i)}|$ and

$$|\mu_l^{(j)}| < c_3(H(x))^{-\kappa}. \tag{1.54}$$

Here $\kappa \leqslant mn - 1$ and $c_3 > 0$ depends only on m and n, provided that (x, y) is one of the solutions indicated in (1.51). In the sequel we will restrict ourselves to such solutions (x, y).

From (1.52) and (1.53) we obtain the fundamental identity

$$(\alpha_k^{(j)} - \alpha_l^{(j)})\mu_i^{(j)} - (\alpha_i^{(j)} - \alpha_l^{(j)})\mu_k^{(j)} = (\alpha_k^{(j)} - \alpha_i^{(j)})\mu_l^{(j)},$$

where i, k, l are distinct (they can be so chosen since $n \geqslant 3$). Therefore,

$$\beta_1^{h_1} \cdots \beta_w^{h_w} - \beta_{w+1} = \omega, \tag{1.55}$$
$$\beta_1 = \varepsilon_{1,i}^{(j)}/\varepsilon_{1,k}^{(j)}, \ldots, \beta_w = \varepsilon_{w,i}^{(j)}/\varepsilon_{w,k}^{(j)},$$

where

$$\beta_{w+1} = \frac{(\alpha_i^{(j)} - \alpha_l^{(j)})\gamma_k^{(j)}}{(\alpha_k^{(j)} - \alpha_l^{(j)})\gamma_i^{(j)}}, \qquad \omega = \frac{(\alpha_k^{(j)} - \alpha_i^{(j)})\mu_l^{(j)}\gamma_k^{(j)}}{(\alpha_k^{(j)} - \alpha_l^{(j)})\mu_k^{(j)}\gamma_i^{(j)}}.$$

For fixed $f(x, y)$ and A, there are at most $(nm)^3\psi_1(A)^3$ numbers of the form β_{w+1}. Fix some natural number N and consider remainders h_i', $0 \leqslant h_i' < N$, such

*For an algebraic number α and its conjugates $\alpha^{(1)} = \alpha, \ldots, \alpha^{(n)}$, $|\overline{\alpha}| \overset{\text{def}}{=} \max_{i=1,\ldots,n} |\alpha^{(i)}|$.

that $h_i \equiv h_i'$ (mod N). Let β_0 stand for the number $\beta_1^{h_1} \cdots \beta_w^{h_w}$. Then (1.55) can be written in the form

$$\beta_0(\beta_1^{x_1} \cdots \beta_w^{x_w})^N - \beta_{w+1} = \omega. \tag{1.56}$$

Now assume that $\#(f, A, \mathbf{K}) > c_4(n, m, N)$. Then equation (1.56) has $O(\#(f, A, \mathbf{K}))$ solutions with constant β_0. Assume in addition that $\#(f, A, \mathbf{K}) > c_5(\psi_1(A), n, m)$ and regard β_{w+1} as constant. Denote $\beta_1^{x_1} \cdots \beta_w^{x_w}$ by δ. It follows from (1.54) that

$$\left| \delta^N - \frac{\beta_{w+1}}{\beta_0} \right| < H(x)^{-c_6} \tag{1.57}$$

for some $c_6 > 0$, and $\ln H(\delta) \ll \ln H(x)/N$. Since $\#(f, A, \mathbf{K}) > c_7(\psi_1(A), n, m)$, we obtain many solutions of (1.57). We then apply Theorem 1.7 with $N > c_8(m, n)$.

The proof of 1.10 is completed by estimating $\psi_1(A)$ in terms of $\tau(A)$, the number of prime divisors of A: $\psi_1(A) \leqslant \tau(A)^{nm}$, where nm is the degree of $\mathbf{K}_1$.

REMARK. If to assume in 1.10 that $|\mathrm{Norm}\,\Delta| \geqslant c(n, m, A)$, then the estimate of $\#(f, A, \mathbf{K})$ will depend only on m and n.

It is possible to prove p-adic variants of Theorems 1.5 and 1.7. Using these theorems, it is possible to establish a p-adic variant of Theorem 1.10. This result can be applied to the determination of elliptic curves with prescribed points of bad reductions.

2. To studies of the diophantine approximations of values of exponential functions we apply the general principles of construction of auxiliary approximating forms that were explained in Siegel's fundamental paper [**1**]. For completeness we formulate the basic principles of Siegel's approach, which consist of the replacement of the arithmetic problem by a problem of interpolation and approximation of functions in the complex plane. In order to study the measure of diophantine approximations (or the linear independence) of numbers $\omega_0, \ldots, \omega_r$, we, following Siegel [**1**], assume the existence of $r + 1$ forms $L_k = h_{k,0}\omega_0 + \cdots + h_{k,r}\omega_r$: $k = 0, \ldots, r$ that are linearly independent (i.e. $\det(h_{k,l}) \neq 0$), whose integer rational coefficients satisfy $|h_{k,l}| \leqslant H$, and such that

$$\max(|L_0|, \ldots, |L_r|) \leqslant \mu$$

for a sufficiently small μ. Let us now take any linear form in $\omega_0, \ldots, \omega_r$: $L = h_0\omega_0 + \cdots + h_r\omega_r$ with $\max(|h_0|, \ldots, |h_r|) = h$, that we want to estimate. Since $r + 1$ forms $L_0, \ldots, L_r$ are linearly independent, then from these $r + 1$ forms we can choose r forms, which, together with the form L, constitute a system of linearly independent forms. Let these be the forms $L_1, L_2, \ldots, L_r$. Let $(\lambda_{k,l})$ be a matrix inverse to the matrix of the coefficients of $L, L_1, \ldots, L_r$. Then it is easy to obtain the inequalities for the coefficients $\lambda_{k,l}$:

$$|\lambda_{k,0}| < r!H^r: \qquad k = 0, 1, \ldots, r;$$

$$|\lambda_{k,l}| < r!H^{r-1}h: \qquad k = 0, \ldots, r;\ l = 1, \ldots, r.$$

Then from the equation

$$\omega_k = \lambda_{k,0}L + \lambda_{k,1}L_1 + \cdots + \lambda_{k,r}L_r: \qquad k = 0, 1, \ldots, r,$$

immediately follows the inequality for L:

$$|L| \geqslant |\omega_k| / (r!H^r) - r_\mu h/H.$$

Now if $\mu \cdot H^{r-1} \to 0$ as $H \to \infty$, then from the last inequality we obtain an effective positive lower bound for $|L|$ in terms of h. In particular, it gives us a sufficient condition of the linear independence of $\omega_0, \ldots, \omega_r$ over $\mathbf{Q}$.

The construction of the systems of linear forms $L_0, \ldots, L_r$ is a key to the method presented above. This construction can be achieved if ω_k are values of functions $\omega_k(x)$ given by formal power series in x and satisfying differential equations. Now let $\omega_0(x), \omega_1(x), \ldots, \omega_r(x)$ be formal power series, not identically zero (and, without a loss of generality, linearly independent over $\mathbf{C}(x)$). Then Siegel proposes the consideration of "minimal" forms in $\omega_0(x), \ldots, \omega_r(x)$ with polynomial coefficients:

$$L_k(x) = h_{k,0}\omega_0(x) + \cdots + h_{k,r}\omega_r(x): \qquad k = 0, \ldots, r,$$

where the coefficients $h_{k,l} = h_{k,l}(x)$ are polynomials in x of degrees of at most H, and μ is the smallest exponent of x which enters the expansion of any of $L_0(x), \ldots, L_r(x)$ in powers of x. The "minimality" of $L_0(x), \ldots, L_r(x)$ is the condition that $\mu - rH \to +\infty$ as $H \to \infty$ (similar to the number-theoretic case above). The existence of such "minimal" forms is obvious from the pigeon-hole principle, and such linear forms are called, in modern day terminology, remainder functions in the Hermite-Padé approximation problem to functions $\omega_0(x), \ldots, \omega_r(x)$ **[2]**. As Siegel **[1]** pointed out, the main difficulty is the proof of $\det(h_{k,l}(x)) \not\equiv 0$. Siegel himself showed how easily this difficulty can be avoided, if $\omega_k(x)$ satisfy a system of linear differential equations over $\mathbf{C}(x)$:

$$(2.1) \qquad \frac{d\omega_k}{dx} = a_{k,0}\omega_0(x) + \cdots + a_{k,r}\omega_r(x); \qquad k = 0, 1, \ldots, r.$$

Hence, if we have a single "minimal" form

$$L(x) = h_0(x)\omega_0(x) + \cdots + h_r(x)\omega_r(x),$$

then, after its differentiation in x, and its multiplication by the common denominator of $a_{k,l}(x)$, we arrive at a form of "minimal" type too. Repeating this operation of differentiation and multiplication r times, we obtain $r + 1$ "minimal" forms. It is easy to see that the determinant of these $r + 1$ "minimal" forms is identically zero if and only if the determinant $\Delta(x)$ of forms $L(x), L'(x), \ldots, L^{(r)}(x)$ is also identically zero. Siegel **[1]** gave a very simple criterion for $\Delta(x) \not\equiv 0$, *independent of the "minimality" of the form* $L(x)$. Here is Siegel's criterion:

NORMALITY LEMMA 1. *Let the general solution of the system of linear differential equations* (2.1) *for the functions* ω_k *be*

$$\omega_k = C_0\omega_{k,0} + C_1\omega_{k,1} + \cdots + C_r\omega_{k,r}: \qquad k = 0, 1, \ldots, r,$$

where $C_0, C_1, \ldots, C_r$ *are arbitrary constants. Then the determinant of a system of* $r+1$ *linear forms (with polynomial coefficients)*

$$L(x) = h_0(x)\omega_0(x) + \cdots + h_r\omega_r(x), \qquad \frac{dL}{dx}, \ldots, \frac{d^rL}{dx^r}$$

is identically zero if and only if there is a linear homogeneous relation with constant coefficients between $r+1$ *functions*

$$h_0\omega_{0,l} + \cdots + h_r\omega_{r,l}: \qquad l = 0, 1, \ldots, r.$$

The absence of such relations was studied by Siegel [**1**, **3**] in the case of exponential and Bessel functions. The Galois and monodromy methods prove $\Delta(x) \not\equiv 0$ in most of the interesting cases [**4**].

It turns out, however, that for the "minimal" form $L(x)$, with $\mu = \text{ord}_{x=0}(L(x)) \gg rH$, the determinant $\Delta(x)$ is always not identically zero, if H is sufficiently large: $H \geqslant H_0$ (though H_0 is ineffective). Proofs of this result (Schidlovsky's lemma) can be found in Baker [**5**] and Mahler [**6**] with a particularly elegant proof given in Lemma 2, Chapter 11 [**5**].

It is crucial in Siegel's method to pass from the system of functional "minimal" forms $L(x)$, $L'(x), \ldots, L^{(r)}(x)$, with the determinant $\Delta(x) \not\equiv 0$, to the system of number forms $L(\xi)$, $L_1(\xi), \ldots, L_r(\xi)$ with the determinant $\Delta(\xi) \neq 0$ and integer coefficients. Then, under appropriate arithmetic conditions, we obtain estimates of a measure of the diophantine approximations of numbers $\omega_0(\xi)$, $\omega_1(\xi), \ldots, \omega_r(\xi)$ where ξ is a rational (or algebraic) number.

Let, as above, $\mu = \text{ord}_{x=0}(L(x))$ for a given "minimal" form $L(x)$. Successfully differentiating $L(x)$ and multiplying by the common denominator $d(x)$ of rational functions $a_{k,l}$ from (2.1), we obtain a sequence of "minimal" forms:

$$L_s(x) = d(x)\frac{d}{dx}L_{s-1}(x): \qquad s = 1, \ldots. \tag{2.2}$$

Then the order of $L_s(x)$ at $x = 0$ is at least $\mu - s$: $s = 0, 1, \ldots$. Let ν denote the maximum of the degrees of polynomials $d(x)$, $d(x)a_{k,l}(x)$: $k, l = 0, \ldots, r$, and let H be the maximum of the degrees of polynomial coefficients $h_0(x), \ldots, h_r(x)$ of $L(x)$. Then the determinant $D(x)$ of forms $L, L_1, \ldots, L_r$ has a degree in x of at most $(r+1)H + r(r+1)\nu/2$. Expressing $\omega_k(x)$ in terms of forms $L, L_1, \ldots, L_r$ we get

$$D(x)\omega_k(x) = A_{k0}(x) \cdot L + \cdots + A_{kr}(x) \cdot L_r: \qquad k = 0, 1, \ldots, r, \tag{2.3}$$

for polynomials $A_{k,l}(x)$.

Since the right side of (2.3) has at least $x = 0$ a zero of order $\mu - r$, then $D(x)\omega_k(x)$ is divisible by $x^{\mu - r}$. Without a loss of generality we can always assume that $\sum_{k=0}^r |\omega_k(0)| > 0$. Thus $D(x)$ is divisible by $x^{\mu - r}$, i.e. for any $\xi \neq 0$, the polynomial $D(x)$ (which is not identically zero) has a zero at $x = \xi$ of an order of at most

$$s \leqslant (r+1)H + \frac{r(r+1)}{2}\nu + r - \mu.$$

For "minimal" forms L with $\mu = (r+1)H + o(H)$ as $H \to \infty$, we obtain $s = o(H)$. We have $D^{(s)}(\xi) \neq 0$, $D^{(p)}(\xi) = 0$: $p = 0, 1, \ldots, s-1$. We now differentiate the identity (2.3) s times in x and use differential equations (2.1). In the notations of (2.2) we obtain the following identities in $\omega_0(\xi), \ldots, \omega_r(\xi)$:

$$(2.4) \qquad d(\xi)^s \cdot D^{(s)}(\xi) = \sum_{l=0}^{s+r} B_{k,l}(\xi) \cdot L_l(\xi): \qquad k = 0, 1, \ldots, r.$$

Assuming that $x = \xi$ is not a singularity of a system of equations (2.1), i.e. $d(\xi) \neq 0$, we obtain from (2.4) that among $r + s + 1$ forms $L(\xi), L_1(\xi), \ldots, L_{r+s}(\xi)$ there exist $r + 1$ linearly independent forms in variables $\omega_0(\xi), \omega_1(\xi), \ldots, \omega_r(\xi)$—whose determinant is not zero.

In this way, from the construction of a "minimal" functional form $L(x)$, we arrive at $r + 1$ approximating forms in numbers $\omega_0(\xi), \ldots, \omega_r(\xi)$ for rational (algebraic) $\xi \neq 0$. In order to obtain good bounds of the measure of diophantine approximations of $\omega_0(\xi), \ldots, \omega_r(\xi)$ it is necessary to control the growth of the absolute values of coefficients of polynomials in $L_s(x)$ after s differentiations of $L(x)$. Also the choice of a "minimal" form $L(x)$ is achieved using Siegel's Lemma [**1**] (see also Chapter 7, §7 of this volume). Hence, additional assumptions about the common denominators of the coefficients of Taylor expansions of $\omega_0(x), \ldots, \omega_r(x)$ must be imposed in order to apply Siegel's method. Such assumptions are satisfied if $\omega_i(x)$ are combinations of exponential functions, leading to the Lindemann-Weierstrass theorem and its qualitative form. Siegel also showed how his methods are applied to E-functions and G-functions, including various hypergeometric functions with rational parameters. We refer the reader to [**3**, **5**, **6**] for the developments of these methods for E-functions. An important analytic assumption in Siegel's method is the existence of the system (2.1) of first order linear differential equations on ω_k. Though the assumption can be relaxed, one sees from definition (2.2) that any linear form $L_s(\xi)$ obtained in the way described above, will be a linear combination of $\omega_k(\xi), \omega_k'(\xi), \ldots, \omega_k^{(s)}(\xi)$. That is why all results on the measures of the linear independence of numbers $\omega_0(\xi), \ldots, \omega_r(\xi)$ according to Siegel's method give the exponent of linear independence $\geqslant \Omega$, where Ω is the number of linearly independent functions over $\mathbf{C}(x)$ among the functions $\omega_k^{(s)}(x)$: $k = 0, 1, \ldots, r$; $s = 0, 1, \ldots$ (Liouville's type of estimate). In particular, in the problem of the measures of irrationality and transcendence for numbers of the form $\Sigma_{k=0}^{r} A_k(\beta) e^{\alpha_k \beta}$ with polynomials $A_k(z) \in \mathbf{Q}[z]$ and (rational) algebraic α_k, β, these measures depend heavily on the heights of $A_k(z)$ and on α_k and β, see Moreno [**7**]. Another important problem arising here is the case of ξ which is an algebraic irrational number. In this case, as for exponents e^{ξ}, the measure of transcendence (irrationality), depends on the degree of ξ (again with the Liouville-type of exponent).

It turns out, however, that Siegel's method can be considerably modified in order to obtain the best possible results independently of the orders of differential equations satisfied by $\omega_k(x)$, and improving the measures of diophantine approximations of $\omega_k(\xi)$ for algebraic ξ. To do this we use the background of

Siegel's approach outlined above, but for the construction of the auxiliary approximating "minimal" forms we rely on the technique of "partially G-invariant" auxiliary functions described in detail in Chapter 9 of this volume (cf. also [**8, 9**]). Using these methods of "G-invariance", we combine the operation of the differentiation of $L(x)$ (as above) with the action of the group ring $\mathbf{Z}[G]$, as in Chapter 9 [**16**]. Detailed constructions and proofs of important results are given below.

Our new methods allow us to improve considerably the measures of linear independence, irrationality and transcendence of numbers that are values of functions satisfying linear differential equations with constant coefficients (polynomials in exponential functions). In particular, new measures of transcendence for exponents e^{ξ} with algebraic ξ are obtained, improving the existing ones that were obtained by Siegel's method [**1**] and by Gel'fond-Schneider's method, see [**10**] and Chapter 1, of this volume. Our results show that the values of any exponential functions (rational functions in exponents) at rational points: $\theta_0,\dots,\theta_r$ admit the best measure of linear independence: $|h_0\theta_0 + \cdots + h_r\theta_r| > h^{-r-\varepsilon}$ for any $\varepsilon > 0$ and $h = \max(|h_0|,\dots,|h_r|) \geqslant c_0(\theta_0,\dots,\theta_r,\varepsilon)$. Similar results take place in the context of the author's elliptic generalization of the Lindemann-Weierstrass theorem [**11**] (see the Introductory paper and Chapter 7 of this volume). Namely, let $\wp(x)$ be a Weierstrass elliptic function with algebraic invariants g_2, g_3 and periods ω_1, ω_2 with imaginary quadratic ω_1/ω_2 (CM-case). Let $\alpha_1,\dots,\alpha_n$ be algebraic numbers linearly independent over $\mathbf{Q}(\omega_1/\omega_2)$, and let θ_k be a rational function in $\wp(\alpha_1)$, $\wp'(\alpha_i)$: $i = 1,\dots,n$ with coefficients from $\mathbf{Q}[g_2, g_3]$: $k = 0, 1,\dots,r$. Then we obtain a measure of the linear independence of $\theta_0, \theta_1,\dots,\theta_r$ close to the best possible ones given by Dirichlet's bound: for arbitrary rational integers $h_0,\dots,h_r$ we have $|h_0\theta_0 + \cdots + h_r\theta_r| > h^{-c_1 r}$ where $c_1 > 0$ depends only on the degree of the field $\mathbf{Q}(g_2, g_3, \alpha_1,\dots,\alpha_n)$, provided that

$$h = \max(|h_0|,\dots,|h_r|) \geqslant c_2(\theta_0,\dots,\theta_r,\alpha_1,\dots,\alpha_n) \quad \text{and} \quad (h_0,\dots,h_r) = 1.$$

This provides an important addition to the author's earlier result

$$|P(\wp(\alpha_1),\dots,\wp(\alpha_n))| > H(P)^{-c_3 d(P)^n} \quad \text{for } P(x_1,\dots,x_n) \in \mathbf{Z}[x_1,\dots,x_n]$$

[**11, 12**] (cf. the Introductory paper and Chapter 7 of this volume).

Our improvements in Siegel's method also give us results close to the best possible for arbitrary values of Siegel's E-functions at rational points in §4.

3. We show how, using Siegel's method [**1**] and the method of G-invariant systems of functions of Chapter 9 [**16**] we can prove the best possible results on the measure of irrationality, linear independence and transcendence of polynomials at exponents of algebraic numbers.

We start with n sequences of distinct algebraic numbers $\{\beta_{i,1},\dots,\beta_{i,k_i}\}$: $i = 1,\dots,n$, such that any sequence $\{\beta_{i,1},\dots,\beta_{i,k_i}\}$ contains all the conjugates of each of its elements for $i = 1,\dots,n$. Then we consider numbers

$$\theta_i = \sum_{j=1}^{k_i} C_{i,j} e^{\beta_{i,j}}: \qquad i = 1,\dots,n, \tag{3.1}$$

where $C_{i,j}$ are rational numbers, and such that $C_{i,j_1} = C_{i,j_2}$ whenever β_{i,j_1} is algebraically conjugate to β_{i,j_2}: $j_1, j_2 = 1,\dots,k_i$; $i = 1,\dots,n$. As we prove below, the measure of linear independence of numbers $\theta_1,\dots,\theta_n$ is close to the best possible. This means that for any $\varepsilon > 0$ and arbitrary rational integers $h_1,\dots,h_n$, we have $|h_1\theta_1 + \cdots + h_n\theta_n| > h^{-n+1-\varepsilon}$ provided that $h = \max(|h_1|,\dots,|h_n|) \geqslant c_4(\theta_1,\dots,\theta_n, \varepsilon)$.

The key element in the proof is the method of G-invariant systems of functions and the action of the Galois group of the algebraic number field generated by $\beta_{i,j}$: $j = 1,\dots,k_i$: $i = 1,\dots,n$. We refer the reader to Chapter 9 [**16**], especially §§2 and 4, where the notations of group rings are introduced and applied to the construction of "conjugate" auxiliary functions. Here we use these constructions to build auxiliary approximating "minimal" forms in the context of Siegel's method [**1**]. As in [**16**], let $\mathbf{K}$ be a Galois algebraic number field containing all $\beta_{i,j}$: $j = 1,\dots,k_i$; $i = 1,\dots,n$ with the Galois group $G = \mathrm{Gal}(\mathbf{K}/\mathbf{Q})$, and let $[\mathbf{K}:\mathbf{Q}] = |G| = d$. We use the action of a group ring $\mathbf{Z}[G]$, the elements α of which have the form $\alpha = \sum_{g\in G} n_g \cdot g$ with $n_g \in \mathbf{Z}$: $g \in G$, with the natural operations as described in [**16**].

We introduce lattices in $\mathbf{K}$ generated by the action of $\mathbf{Z}[G]$ on algebraic numbers $\beta_{i,j}$: $j = 1,\dots,k_i$; $i = 1,\dots,n$. Let us denote by $\mathfrak{G}$, as in [**16**], a lattice in $\mathbf{K}$ consisting of $\mathfrak{A} = \mathfrak{s}_1 + \cdots + \mathfrak{s}_n$, whose ith component $\mathfrak{s}_i$ has the form $\sum_{j=1}^{k_i} a_{i,j}\beta_{i,j}$ for rational integers $a_{i,j}$: $j = 1,\dots,k_i$; $i = 1,\dots,n$. I.e.

$$\mathfrak{G} = \left\{\mathfrak{A} = \mathfrak{s}_1 + \cdots + \mathfrak{s}_n \in \mathbf{K}: \mathfrak{s}_i = \sum_{j=1}^{k_i} a_{i,j}\cdot\beta_{i,j} \quad \text{for } a_{i,j} \in \mathbf{Z}: j = 1,\dots,k_i;\ i = 1,\dots,n\right\}. \tag{3.2}$$

Since $\{\beta_{i,1},\dots,\beta_{i,k_i}\}$ is invariant under the action of G (is invariant under the algebraic conjugation), $i = 1,\dots,n$; the lattice $\mathfrak{G}$ is G-invariant. Similar to [**16**] we have a natural homomorphism of the $\sum_{i=1}^n k_i$-dimensional lattice $\mathbf{Z}^{k_1}\times\cdots\times\mathbf{Z}^{k_n}$ on $\mathfrak{G}$. This homomorphism λ: $\mathbf{Z}^{k_1}\times\cdots\times\mathbf{Z}^{k_n} \to \mathfrak{G}$ is defined as follows. For $\vec{v} = (v_1,\dots,v_n)$, and $v_i = (a_{i,1},\dots,a_{i,k_i}) \in \mathbf{Z}^{k_i}$: $i = 1,\dots,n$ we define $\lambda(\vec{v}) = \mathfrak{A}$, $\mathfrak{A} = \mathfrak{s}_1 + \cdots + \mathfrak{s}_n$ with $\mathfrak{s}_i = \sum_{j=1}^{k_i} a_{i,j}\beta_{i,j}$.

Let us now display the action of G on the lattice $\mathfrak{G}$. For $g \in G$ and any algebraic number $\beta \in \mathbf{K}$ we denote by $\beta^{(g)} = g(\beta)$ an algebraic number, conjugate to β under the action of G. Then the action of g on $(\beta_{i,1},\dots,\beta_{i,k_i})$ induces a permutation $\pi_{i,g}$ of $(1,\dots,k_i)$ in the following way: $g(\beta_{i,j}) = \beta_{i,\pi_{i,g}(j)}$: $j = 1,\dots,k_i$; $i = 1,\dots,n$. Any permutation π_i of $(1,\dots,k_i)$ acts naturally on $\mathbf{Z}^{k_i}$: $\pi_i(n_1,\dots,n_{k_i}) = (n_{\pi_i(1)},\dots,n_{\pi_i(k_i)})$: $i = 1,\dots,n$. Hence we can easily describe the action of g on $\mathfrak{A} \in \mathfrak{G}$. If $\mathfrak{A} = \mathfrak{s}_1 + \cdots + \mathfrak{s}_n$, $\mathfrak{s}_i = \sum_{j=1}^{k_i} a_{i,j}\beta_{i,j}$: $i = 1,\dots,n$, then $\mathfrak{A}^{(g)} = \mathfrak{s}_1^{(g)} + \cdots + \mathfrak{s}_n^{(g)}$, where

$$\mathfrak{s}_i^{(g)} \overset{\text{def}}{=} \sum_{j=1}^{k_i} a_{i,j}\beta_{i,j}^{(g)} = \sum_{j=1}^{k_i} a_{i,j}\beta_{i,\pi_{i,g}(j)}: \qquad i = 1,\dots,n.$$

In $\mathbf{Z}^{k_1}\times\cdots\times\mathbf{Z}^{k_n}$ we get $\lambda(\vec{v})=\mathfrak{A}$, $\lambda(\vec{v}^{(g)})=\mathfrak{A}^{(g)}$, where

$$\vec{v}^{(g)}=(\pi_1(v_1),\ldots,\pi_n(v_n))\quad\text{for } \vec{v}=(v_1,\ldots,v_n).$$

For a rational integer $N\geqslant 1$ we denote by $D(N)$ the set of $\vec{v}=(v_1,\ldots,v_n)\in\mathbf{Z}^{k_1}\times\cdots\times\mathbf{Z}^{k_n}$ such that $v_i=(v_{i,1},\ldots,v_{i,k_i})$, all numbers $v_{i,j}$ are nonnegative: $j=1,\ldots,k_i$ and $\sum_{j=1}^{k_i}v_{i,j}=N$: $i=1,\ldots,n$.

Then we denote by $\mathcal{G}(N)$ the image of a simplex $D(N)$ in $\mathcal{G}$ under the map λ: $\mathbf{Z}^{k_1}\times\cdots\times\mathbf{Z}^{k_n}\to\mathcal{G}$. This means that
(3.3)

$$\mathcal{G}(N)=\Big\{\mathfrak{A}=\mathfrak{s}_1+\cdots+\mathfrak{s}_n\in\mathcal{G}:\mathfrak{s}_i=\sum_{j=1}^{k_i}a_{i,j}\beta_{i,j},$$
$$\text{where } a_{i,j}\in\mathbf{Z},\ a_{i,j}\geqslant 0\text{ and }\sum_{j=1}^{k_i}a_{i,j}=N: j=1,\ldots,k_i;\ i=1,\ldots,n\Big\}.$$

We also define for any $i=1,\ldots,n$,

$$\mathcal{G}_i(N)=\Big\{\mathfrak{A}=\mathfrak{s}_1+\cdots+\mathfrak{s}_n\in\mathcal{G};\ \mathfrak{s}_{i_1}=\sum_{j=1}^{k_{i_1}}a_{i_1,j}\beta_{i_1,j},\ \text{where } a_{i_1,j}\in\mathbf{Z},\ a_{i_1,j}\geqslant 0$$
$$\text{and }\sum_{j=1}^{k_{i_1}}a_{i_1,j}=N-\delta_{i,i_1}: j=1,\ldots,k_{i_1};\ i_1=1,\ldots,n\Big\}.$$

Below, for a given $\mathfrak{A}=\mathfrak{s}_1+\cdots+\mathfrak{s}_n\in\mathcal{G}$, we denote by $\mathfrak{A}-e_{i,j}$ an element of the lattice $\mathcal{G}$ of the form: $\mathfrak{A}-e_{i,j}\overset{\text{def}}{=}\mathfrak{s}_1'+\cdots+\mathfrak{s}_n'$, where $\mathfrak{s}_{i_1}'=\mathfrak{s}_i-\delta_{ii_1}\beta_{i,j}$: $j=1,\ldots,k_i$: $i,i_1=1,\ldots,n$.

We remark that for $\mathfrak{A}\in\mathcal{G}(N)$, we do not necessarily have $\mathfrak{A}-e_{i,j}\in\mathcal{G}_i(N)$, because $\mathfrak{s}_i-e_{i,j}$ might not be represented as $\sum_{j_1=1}^{k_i}a_{i,j},\ \beta_{i,j_1}$ with nonnegative rational integers a_{i,j_1}.

Using the notations above we construct a system of G-conjugate "minimal" forms to systems of exponential functions of the form $e^{-\mathfrak{A}x}$ for $\mathfrak{A}=\sum_{i=1}^{n}\sum_{j=1}^{k_i}a_{i,j}\beta_{i,j}$ with nonnegative integers $a_{i,j}$: $j=1,\ldots,k_i$; $i=1,\ldots,n$.

The "minimal" approximating forms that we consider have the following structure. They are enumerated by vectors $\mathfrak{A}=\mathfrak{s}_1+\cdots+\mathfrak{s}_n\in\mathcal{G}(N)$ for a sufficiently large $N\geqslant 1$, and are linear combinations of $e^{-(\mathfrak{A}-\beta_{i,j})x}$ with polynomial coefficients:

$$L_{\mathfrak{A}}(x)=\sum_{i=1}^{n}\sum_{j=1}^{k_i}P_{\mathfrak{A}-e_{i,j},i}e^{-(\mathfrak{A}-\beta_{i,j})x},$$

where $\mathfrak{A}=\mathfrak{s}_1+\cdots+\mathfrak{s}_n\in\mathcal{G}(N)$ and $P_{\mathfrak{A}-e_{i,j},i}=P_{\mathfrak{A}-e_{i,j},i}(x)$ are polynomials with coefficients from $\mathbf{K}$, that satisfy the G-invariance statements. These approximating forms are constructed using Siegel's Lemma (cf. [**15**], §6) similar to

[1]. Our G-invariance conditions on $P_{\mathfrak{A},i}(x)$ are determined, similar to [16], by the action of G on $\mathcal{G}$. Namely, for $\mathfrak{A} \in \mathcal{G}_i(N)$ and polynomials $P_{\mathfrak{A},i}(x) \in \mathbf{K}[x]$, we require that for $g \in G$, the conjugate $(P_{\mathfrak{A},i}(x))^{(g)}$ to the polynomial $P_{\mathfrak{A},i}(x)$ under the action of g be of the form $(P_{\mathfrak{A},i}(x))^{(g)} = P_{\mathfrak{A}^{(g)},i}(x)$. In order to express this condition of G-invariance in a simpler form, we fix an algebraic integer ω of $\mathbf{K}$ such that $\omega^{(g)}$: $g \in G$ is a basis of the field $\mathbf{K}/\mathbf{Q}$.

We fix a sufficiently large integer $N \geqslant 1$ and an arbitrary ε, $1 > \varepsilon > 0$. As in Siegel's method [1], we consider an integer parameter H, bounding the degrees of polynomial coefficients $P_{\mathfrak{A},i}(x)$. We construct polynomials $P_{\mathfrak{A},i}(x) \in \mathbf{K}[x]$, $\mathfrak{A} \in \mathcal{G}_i(N)$; $i = 1,\dots,n$, and the corresponding approximating forms $L_{\mathfrak{A}}(x)$ in the next theorem.

THEOREM 3.1. *There exist polynomials $P_{\mathfrak{A},i}(x) \in \mathbf{K}[x]$: $\mathfrak{A} \in \mathcal{G}_i(N)$; $i = 1,\dots,n$ of degrees H, not all identically zero such that the following conditions are satisfied*:

(1) $(P_{\mathfrak{A},i}(x))^{(g)} = P_{\mathfrak{A}^{(g)},i}(x)$ *for* $\mathfrak{A} \in \mathcal{G}_i(N)$; *and we put for* $\mathfrak{A} \in \mathcal{G}\backslash\mathcal{G}_i(N)$, $P_{\mathfrak{A},i}(x) \equiv 0$: $i = 1,\dots,n$.

(2) *For any* $\mathfrak{A} = \mathfrak{s}_1 + \cdots + \mathfrak{s}_n \in \mathcal{G}(N)$, *the linear form*

$$L_{\mathfrak{A}}(x) \overset{\text{def}}{=} \sum_{i=1}^{n} \sum_{j=1}^{k_i} P_{\mathfrak{A}-e_{i,j},i}(x) \cdot e^{-(\mathfrak{A}-\beta_{i,j})x} \cdot C_{i,j} \tag{3.4}$$

has a zero at $x = 0$ *of an order of at least* $[(\mu - \varepsilon) \cdot H]$, *with*

$$\mu = \sum_{i=1}^{n} \operatorname{Card}(\mathcal{G}_i(N))/\operatorname{Card}(\mathcal{G}(N)). \tag{3.5}$$

(3) *We have for* $\mathfrak{A} \in \mathcal{G}_i(N)$, $i = 1,\dots,n$:

$$P_{\mathfrak{A},i}(x) = \sum_{k=0}^{H} \frac{H!}{k!} p_{\mathfrak{A},i,k} x^k, \tag{3.6}$$

where $p_{\mathfrak{A},i,k}$ are algebraic integers from $\mathbf{K}$ whose sizes are bounded as follows:

$$\max\left\{\left|\overline{p_{\mathfrak{A},i,k}}\right| : k = 0,\dots,H;\ \mathfrak{A} \in \mathcal{G}_i(N);\ i = 1,\dots,n\right\} \leqslant \exp\{c_5 H/\varepsilon\}$$

for $c_5 > 0$ depending only on N, $\mathbf{K}$ and $\beta_{i,j}$.

PROOF OF THEOREM 3.1. The coefficients of expansion of $L_{\mathfrak{A}}(x)$ at $x = 0$ can be represented in terms of coefficients $p_{\mathfrak{A}-e_{i,j},i,k}$ of $p_{\mathfrak{A}-e_{i,j}}(x)$ as follows. Let

$$L_{\mathfrak{A}}(x) = \sum_{m=0}^{\infty} \frac{H!}{m!} c_{\mathfrak{A},m} x^m.$$

Then

$$c_{\mathfrak{A},m} \overset{\text{def}}{=} \sum_{i=1}^{n} \sum_{j=1}^{k_i} \sum_{k=0}^{\min(m,H)} \left\{\binom{m}{k}\left(-\mathfrak{A} + \beta_{i,j}\right)^{m-k} C_{i,j} \cdot p_{\mathfrak{A}-e_{i,j},i,k}\right\}. \tag{3.7}$$

The conditions (2) on polynomials $P_{\mathfrak{A}',i}(x)$: $\mathfrak{A}' \in \mathcal{G}_i(N)$; $i = 1,\dots,n$ are equivalent to the following system of equations:

$$c_{\mathfrak{A},m} = 0: \qquad \mathfrak{A} \in \mathcal{G}(N);\ m = 0,\dots,[(\mu - \varepsilon)H] - 1 \tag{3.8}$$

in the notations of (1.7). To represent the G-invariance conditions (1) in an explicit form, we represent the numbers $p_{\mathfrak{A}',i,k}$ from $\mathbf{K}$ as follows:

$$p_{\mathfrak{A}',i,k} = \sum_{g \in G} p_{\mathfrak{A}',i,k,g}\omega^{(g)} \tag{3.9}$$

for rational integers $p_{\mathfrak{A}',i,k,g}$: $\mathfrak{A}' \in \mathcal{G}_i(N)$; $i = 1,\dots,n$; $k = 0,1,\dots,H$ and $g \in G$. Here, as above, $\omega^{(g)}$: $g \in G$ is a basis of $\mathbf{K}/\mathbf{Q}$ for an algebraic integer $\omega \in \mathbf{K}$. The conditions (1) mean that $(p_{\mathfrak{A}',i,k})^{(g)} = p_{(\mathfrak{A}')^{(g)},i,k}$. Hence we can put

$$p_{\mathfrak{A}',i,k,g} \overset{\text{def}}{=} p_{(\mathfrak{A}')^{(h)},i,k,e} \tag{3.10}$$

for $h = g^{-1} \in G$, a unit element $e \in G$, and for all $\mathfrak{A}' \in \mathcal{G}_i(N)$: $i = 1,\dots,n$; $k = 0,1,\dots,H$ and $g \in G$.

In the notations of (3.9)–(3.10) and (3.7), the system of equations (3.8): $c_{\mathfrak{A},m} = 0$, is a system of equations on unknown rational integers $p_{\mathfrak{A}',i,k,e}$ with algebraic number coefficients. In order to reduce this system of equations to a system of equations with rational number coefficients, we need the G-invariance properties of $c_{\mathfrak{A},m}$, that follow from (1).

Let $g \in G$, then g acts as a permutation π_i on $(1,\dots,k_i)$: $\beta_{i,j}^{(g)} = \beta_{i,\pi_i(j)}$; $j = 1,\dots,k_i$; $i = 1,\dots,n$. We have, according to (1),

$$(p_{\mathfrak{A}-e_{i,j},i,k})^{(g)} = p_{\mathfrak{A}^{(g)}-e_{i,\pi(j)},i,k}.$$

Thus we have from (3.7):

$$(c_{\mathfrak{A},m})^{(g)} = c_{\mathfrak{A}^{(g)},m}: \qquad m = 0,1,\dots, \tag{3.11}$$

$\mathfrak{A} \in \mathcal{G}(N)$, $g \in G$. The Siegel Lemma that we use has the following form.

SIEGEL'S LEMMA (cf. [**1**, **5**, **15**]). *Let M, K denote integers with $K > M > 0$ and let $u_{i,j}$ $(1 \leq i \leq M,\ 1 \leq j \leq K)$ denote real numbers with absolute values of at most U (≥ 1). Then there exist integers $x_1,\dots,x_N$ not all zero, with absolute values of at most $2(KU)^{M/(K-M)}$ such that*

$$\left|\sum_{j=1}^{N} u_{i,j}x_j\right| < 1: \qquad i = 1,\dots,M.$$

REMARK. If all numbers $u_{i,j}$ are rational integers, then, obviously, $\sum_{j=1}^N u_{i,j}x_j = 0$: $i = 1,\dots,M$. We apply Siegel's lemma, however, in the situation, where $u_{i,j}$ are algebraic numbers.

Let b be the common denominator of all algebraic numbers $\beta_{i,j}$: $j = 1,\dots,k_i$; $i = 1,\dots,n$. Then according to (3.7) each of the expressions

$$b^H \cdot c_{\mathfrak{A},m}: \qquad \mathfrak{A} \in \mathcal{G}(N),\ m = 0,\dots,[(\mu - \varepsilon)H] - 1 \tag{3.12}$$

is a linear form in rational integer unknowns

$$p_{\mathfrak{A}',i,k,e}: \qquad \mathfrak{A}' \in \mathcal{G}_i(N);\ i = 1,\dots,n;\ k = 0,1,\dots,H, \tag{3.13}$$

with algebraic integer coefficients (from **K**) of sizes bounded by c_6^H for some $c_6 > 0$ depending only on N, **K** and $\beta_{i,j}$, $C_{i,j}$ ($j = 1,\ldots,k_i$; $i = 1,\ldots,n$). There are at most $\mathrm{Card}(\mathcal{G}(N)) \cdot (\mu - \varepsilon) \cdot H$ linear forms (1.12) with

$$\sum_{i=1}^{n} \mathrm{Card}(\mathcal{G}_i(N)) \cdot (H+1)$$

unknowns (3.13).

From Siegel's lemma above, it follows that there are rational integers $p_{\mathfrak{A}',i,k,e}$ (3.13), not all zero, such that all inequalities

$$(3.14) \qquad |b^H c_{\mathfrak{A},m}| < 1: \qquad \mathfrak{A} \in \mathcal{G}(N), m = 0,\ldots,[(\mu-\varepsilon)H] - 1$$

are satisfied. We now apply G-invariance relations (3.11), and conclude that expressions (3.12) form a set of algebraic integers, closed under algebraic conjugations. This implies that for $\mathfrak{A} \in \mathcal{G}(N)$, $m = 0,\ldots,[(\mu-\varepsilon)H] - 1$,

$$c_{\mathfrak{A},m} = 0.$$

Indeed, any number algebraically conjugate to $c_{\mathfrak{A},m}$ has the form $(c_{\mathfrak{A},m})^{(g)} = c_{\mathfrak{A}^{(g)},m}$, and $\mathfrak{A}^{(g)} \in \mathcal{G}(N)$: $g \in G$. All numbers $b^H \cdot c_{\mathfrak{A}^{(g)},m}$ are algebraic integers and, thus $\prod_{g\in G}(b^H \cdot c_{\mathfrak{A}^{(g)},m})$ is a rational integer. But according to (3.14), $|\prod_{g\in G}(b^H \cdot c_{\mathfrak{A}^{(g)},m})| < 1$. Hence, $c_{\mathfrak{A},m} = 0$, and the system of equations (3.8) is satisfied. Theorem 3.1 is proved.

The system of approximating forms $L_{\mathfrak{A}}(x)$ is used, following Siegel's method [**1**], described above, to obtain measures of the linear independence of numbers $\theta_1,\ldots,\theta_n$ in (3.1). We need upper bounds on the sizes of coefficients of derivatives of linear forms $L_{\mathfrak{A}}(x)$.

For $m = 0, 1, \ldots$ we define

$$L_{\mathfrak{A}}^{(m)}(x) = \left(\frac{d}{dx}\right)^m L_{\mathfrak{A}}(x): \qquad \mathfrak{A} \in \mathcal{G}(N).$$

Then from (3.4) it follows that

$$(3.15) \qquad L_{\mathfrak{A}}^{(m)}(x) \stackrel{\mathrm{def}}{=} \sum_{i=1}^{n} \sum_{j=1}^{k_i} P_{\mathfrak{A}-e_{i,j},i}^{\langle m\rangle}(x) \cdot e^{-(\mathfrak{A}-\beta_{i,j})x} \cdot C_{i,j}.$$

The polynomials $P_{\mathfrak{A}',i}^{\langle m\rangle}(x)$ for $\mathfrak{A}' \in \mathcal{G}_i(N)$: $i = 1,\ldots,n$ are defined, according to (3.4) as follows. If $\mathfrak{A}' = \mathfrak{z}_1' + \cdots + \mathfrak{z}_n' \in \mathcal{G}_i(N)$: $i = 1,\ldots,n$, then we put

$$(3.16) \qquad P_{\mathfrak{A}',i}^{\langle m\rangle}(x) = \left(\frac{d}{dx} - \mathfrak{A}'\right)^m P_{\mathfrak{A}',i}(x): \qquad m = 0, 1, \ldots.$$

We need a simple upper bound on the sizes of polynomials $P_{\mathfrak{A}',i}^{\langle m\rangle}(x)$: $\mathfrak{A}' \in \mathcal{G}_i(N)$: $i = 1,\ldots,n$ and an upper bound on $L_{\mathfrak{A}}^{(m)}(x)$: $m = 0, 1, \ldots$.

LEMMA 3.2. *In the notations above, $P_{\mathfrak{A}',i}^{\langle m\rangle}(x)$ are polynomials of degree of at most H in x with algebraic integer coefficients from* **K**. *They satisfy the following G-invariance properties: for $g \in G$, $(P_{\mathfrak{A}',i}^{\langle m\rangle}(x))^{(g)} = P_{\mathfrak{A}'^{(g)},i}^{\langle m\rangle}(x)$: $\mathfrak{A}' \in \mathcal{G}_i(N)$, $i = 1,\ldots,n$. The sizes of the coefficients of polynomials $P_{\mathfrak{A}',i}^{\langle m\rangle}(x)$ are bounded by $H^H \cdot (H+m)^m \cdot \exp\{c_7 H/\varepsilon + c_8 m\}$ for $c_7 > 0$, $c_8 > 0$ depending on* **K**, *N and*

$\beta_{i,j}$, $C_{i,j}$: $\mathfrak{A}' \in \mathcal{G}_i(N)$; $i = 1,\ldots,n$ *and* $m = 0, 1,\ldots$. *For* $m = 0, 1,\ldots$ *and a given* $x \neq 0$ *we also have*

$$(3.17) \qquad |L_{\mathfrak{A}}^{(m)}(x)| \leq m^m \cdot c_9^m \cdot c_{10}^{H/\varepsilon} \cdot H^{(1-\mu+\varepsilon)H},$$

where $c_9 > 0$, $c_{10} > 0$ *depends on* **K**, N *and* $\beta_{i,j}$, $C_{i,j}$ *and on* x.

PROOF OF LEMMA 3.2. The upper bound for the degrees of $P_{\mathfrak{A}',i}^{\langle m\rangle}(x)$ follows from Theorem 3.1 and (3.16). The G-invariance properties follow from (1) of Theorem 3.1 and (3.16); the upper bounds for the sizes of $P_{\mathfrak{A}',i}^{\langle m\rangle}(x)$ are corollaries of (3) of Theorem 3.1 and the application of definition (3.16). To estimate from above the absolute value of $L_{\mathfrak{A}}^{(m)}(x)$ at $x = 0$ we use property (2) of Theorem 3.1. We have the following expansion of $L_{\mathfrak{A}}(x)$, in the notations of the proof of Theorem 3.1:

$$L_{\mathfrak{A}}(x) = \sum_{m \geq [(\mu-\varepsilon)H]}^{\infty} \frac{H!}{m!} c_{\mathfrak{A},m} x^m.$$

From expression (3.7) and the upper bounds of the sizes of polynomials $P_{\mathfrak{A},i}(x)$ we deduce that

$$|c_{\mathfrak{A},m}| \leq c_{11}^m c_{12}^{H/\varepsilon}: \qquad m = [(\mu - \varepsilon)H],\ldots$$

for $c_{11} > 0$, $c_{12} > 0$ depending only on **K**, N and $C_{i,j}$, $\beta_{i,j}$. This immediately implies (3.17) with c_9, $c_{10} > 0$ depending on N, **K**, $\beta_{i,j}$, $C_{i,j}$ and $x \neq 0$. Lemma 3.2 is proved.

The system of Card($\mathcal{G}(N)$) approximating linear forms $L_{\mathfrak{A}}(x)$ will now be used to construct a special "minimal" form $L(x)$ approximating a system of n functions $f_i(x) = \sum_{j=1}^{k_i} C_{i,j} e^{\beta_{i,j}x}$: $i = 1,\ldots,n$ in the way described in §2. We then use Siegel's method [**1**] to construct, using differentiation and G-conjugation, n linearly independent "minimal" forms approximating functions $f_i(x) = \sum_{j=1}^{k_i} C_{i,j} e^{\beta_{i,j}x}$: $i = 1,\ldots,n$.

Now let $C_{i,j}$ be, as in (3.1), a system of rational numbers such that $C_{i,j_1} = C_{i,j_2}$ whenever β_{i,j_1} is algebraically conjugate to β_{i,j_2}: $j_1, j_2 = 1,\ldots,k_i$; $i = 1,\ldots,n$.

REMARK. It is enough to assume, in fact, that $C_{i,j}$ are algebraic numbers from the field **K**, and for an arbitrary $g \in G$, $(C_{i,j})^{(g)} = C_{i,j'}$ when $(\beta_{i,j})^{(g)} = \beta_{i,j'}$: $j, j' = 1,\ldots,k_i$; $i = 1,\ldots,n$.

Let us now denote for $i = 1,\ldots,n$ and $m = 0, 1, 2,\ldots$

$$(3.18) \qquad P_i^{\langle m\rangle}(x) = \sum_{\mathfrak{A}' \in \mathcal{G}_i(N)} P_{\mathfrak{A}',i}^{\langle m\rangle}(x).$$

According to Theorem 3.1, $P_i^{\langle m\rangle}(x)$ is a polynomial with coefficients from **K**; we show, using G-invariance that $P_i^{\langle m\rangle}(x)$ has, in fact, rational number coefficients: $i = 1,\ldots,n$; $m = 0, 1, 2,\ldots$.

LEMMA 3.3. *For* $g \in G$, $\mathfrak{A}' \in \mathcal{G}(N)$ *we have* $(P_{\mathfrak{A}',i}^{\langle m\rangle}(x))^{(g)} = P_{\mathfrak{A}'^{(g)},i}^{\langle m\rangle}(x)$ *with* $i = 1,\ldots,n$; $m = 0, 1,\ldots$.

PROOF OF LEMMA 3.3. We have, according to property (2) of Theorem 3.1,

$$\left(P_{\mathfrak{A}',i}^{\langle m\rangle}(x)\right)^{(g)} = \left(\frac{d}{dx} - \mathfrak{A}'^{(g)}\right)^m P_{\mathfrak{A}'^{(g)},i}(x) = P_{\mathfrak{A}'^{(g)},i}^{\langle m\rangle}(x),$$

whenever $g \in G$, $\mathfrak{A}' \in \mathcal{G}_i(N)$; $i = 1,\ldots,n$ and $m = 0, 1,\ldots.$

LEMMA 3.4. *For $i = 1,\ldots,n$ and $m = 0, 1,\ldots$ the polynomial $P_i^{\langle m\rangle}(x)$ defined in* (3.18) *has rational number coefficients.*

PROOF OF LEMMA 3.4. Since $\mathcal{G}_i(N)$ is G-invariant, Lemma 3.4 follows from Lemma 3.3 and (3.18).

Now we can construct, by consecutive differentiations, a sequence of forms approximating functions $f_1(x),\ldots,f_n(x)$ with the "minimality" property. Namely, for $m = 0, 1,\ldots$ we define

$$L^{\langle m\rangle}(x) \stackrel{\text{def}}{=} \sum_{\mathfrak{A}\in\mathcal{G}(N)} L_{\mathfrak{A}}^{(m)}(x)e^{\mathfrak{A}x}, \quad \text{where } L_{\mathfrak{A}}^{(m)}(x) = \frac{d^m}{dx^m}L_{\mathfrak{A}}(x). \tag{3.19}$$

Then we obtain the representation of $L^{\langle m\rangle}(x)$ as a linear form in $f_1(x),\ldots,f_n(x)$ with polynomial coefficients, similar to that of (3.4):

LEMMA 3.5. *In the notations of* (3.18), (3.19) *we have*

$$L^{\langle m\rangle}(x) = \sum_{i=1}^{n} P_i^{\langle m\rangle}(x)f_i(x): \qquad m = 0, 1,\ldots. \tag{3.20}$$

PROOF OF LEMMA 3.5. We have

$$\left(\frac{d}{dx}\right)^m L_{\mathfrak{A}}(x) = \sum_{i=1}^{n}\sum_{j=1}^{k_i}\left(\frac{d}{dx}\right)^m \cdot \left(P_{\mathfrak{A}-e_{i,j},i}(x)e^{-(\mathfrak{A}-\beta_{i,j})x}C_{i,j}\right)$$

$$= \sum_{i=1}^{n}\sum_{j=1}^{k_i} P_{\mathfrak{A}-e_{i,j},i}^{\langle m\rangle}(x)e^{-(\mathfrak{A}-\beta_{i,j})x}C_{i,j}.$$

Hence, we have

$$L^{\langle m\rangle}(x) = \sum_{\mathfrak{A}\in\mathcal{G}(N)} L_{\mathfrak{A}}^{(m)}(x)e^{\mathfrak{A}x} = \sum_{i=1}^{n}\sum_{j=1}^{k_i} C_{i,j}e^{\beta_{i,j}x}\sum_{\mathfrak{A}\in\mathcal{G}(N)} P_{\mathfrak{A}-e_{i,j},i}^{\langle m\rangle}(x).$$

We then use property (2) of Theorem 3.1 that always $P_{\mathfrak{A}',i}(x) \equiv 0$ whenever $\mathfrak{A}' \notin \mathcal{G}_i(N)$: $i = 1,\ldots,n$. Hence, for any $j = 1,\ldots,k_i$ and $i = 1,\ldots,n$ we have

$$\sum_{\mathfrak{A}\in\mathcal{G}(N)} P_{\mathfrak{A}-e_{i,j},i}^{\langle m\rangle}(x) = \sum_{\mathfrak{A}'\in\mathcal{G}(N)} P_{\mathfrak{A}',i}^{\langle m\rangle}(x).$$

This shows that

$$\sum_{\mathfrak{A}\in\mathcal{G}(N)} L_{\mathfrak{A}}^{\langle m\rangle}(x)e^{\mathfrak{A}x} = \sum_{i=1}^{n}\sum_{j=1}^{k_i} C_{i,j}e^{\beta_{i,j}x}\sum_{\mathfrak{A}'\in\mathcal{G}_i(N)} P_{\mathfrak{A}',i}^{\langle m\rangle}(x)$$

$$= \sum_{i=1}^{n}\sum_{j=1}^{k_i} C_{i,j}e^{\beta_{i,j}x}P_i^{\langle m\rangle}(x) = \sum_{i=1}^{n} P_i^{\langle m\rangle}(x)\left\{\sum_{j=1}^{k_i} C_{i,j}e^{\beta_{i,j}x}\right\},$$

according to definition (3.18). Lemma 3.5 is proved.

We now apply Siegel's method [**1**] in order to find n linearly independent "minimal" forms approximating $f_1(x),\dots,f_n(x)$ among $L^{\langle m\rangle}(x)$: $m = 0, 1, 2\dots$. For this we assume from now on that functions $f_1(x),\dots,f_n(x)$ are linearly independent over $\mathbf{C}(x)$.

The condition of linear independence of functions $f_1(x),\dots,f_n(x)$ can be expressed using the following notations. Let $B = \bigcup_{i=1}^n\{\beta_{i,1},\dots,\beta_{i,k_i}\}$ and $C_{i,\beta} = C_{i,j}$ for $\beta = \beta_{i,j} \in B$: $j = 1,\dots,k_i$; $i = 1,\dots,n$. The functions $f_1(x),\dots,f_n(x)$ are linearly independent over $\mathbf{C}(x)$ if and only if the rank of the matrix $(C_{i,\beta}$: $\beta \in B$; $i = 1,\dots,n)$ is n.

Our main auxiliary result is the following.

THEOREM 3.6. *Let polynomials $P_{\mathfrak{A},i}(x)$: $\mathfrak{A} \in \mathcal{G}_i(N)$; $i = 1,\dots,n$ and linear forms $L_{\mathfrak{A}}(x)$: $\mathfrak{A} \in \mathcal{G}(N)$, satisfy all the conditions of Theorem* 3.1. *Let, as in* (3.20),

$$L^{\langle m\rangle}(x) = \sum_{\mathfrak{A}\in\mathcal{G}(N)}\left(\frac{d}{dx}\right)^m L_{\mathfrak{A}}(x)e^{\mathfrak{A}x}: \qquad m = 0, 1,\dots,$$

so that $L^{\langle m\rangle}(x) = \sum_{i=1}^n P_i^{\langle m\rangle}(x)f_i(x)$—is a linear form in $f_1(x),\dots,f_n(x)$ with polynomial coefficients. Then for N sufficiently large, $N \geqslant N_0(\beta_{i,j}, C_{i,j})$ and $H \geqslant H_0(\beta_{i,j}, C_{i,j}, N, \varepsilon)$, there are n linearly independent forms among

$$L^{\langle m\rangle}(x_0): \qquad m = 0, 1,\dots, c_{13}\varepsilon H$$

where $c_{13} > 0$ depends only on N. This means that the rank of the matrix $(P_i^{\langle m\rangle}(x_0)$: $i = 1,\dots,n)$, $m = 0, 1,\dots, c_{13}\varepsilon H$ is n.

The upper bound on m can be, in general, improved, especially when algebraic numbers $\beta_{i,j}$ are linearly independent (see below).

For the proof of Theorem 3.6 we follow Siegel's original method [**1**] (rather than its subsequent improvements) and the method of the proof of the Normality lemma 1 of §2 given in [**1**]. For modern versions of Siegel's proof see [**17**], especially Theorem 8.3.1.

For the applications of Siegel's method we remark that, for arbitrary polynomials $P_\alpha(x) \in \mathbf{C}[x]$: $\alpha \in A \subset \mathbf{C}$, the system of functions $P_\alpha(x)e^{\alpha x}$: $\alpha \in A$ is linearly independent over $\mathbf{C}$ if and only if $P_\alpha(x) \not\equiv 0$ for $\alpha \in A$.

For the proof of Theorem 3.6 we consider a scalar linear differential equation with rational function coefficients $L[\varphi] = 0$, whose space V of solutions is generated by $P_{\mathfrak{A},i}(x)e^{-\mathfrak{A}x}$ for $\mathfrak{A} \in \mathcal{G}_i(N)$: $i = 1,\dots,n$, as a vector space over $\mathbf{C}$. The dimension of the vector space V over $\mathbf{C}$ is d_N, and, according to Theorem 1.1, $d_N \leqslant \sum_{i=1}^n \mathrm{Card}(\mathcal{G}_i(N))$. Let us denote by $P_{\mathfrak{A}_\alpha}(x)e^{-\mathfrak{A}_\alpha x}$: $\alpha \in A_N$, $\mathrm{Card}(A_N) = d_N$, the basis of V over $\mathbf{C}$—the maximal set of linearly independent functions among $\{P_{\mathfrak{A},i}(x)e^{-\mathfrak{A}x}$: $\mathfrak{A} \in \mathcal{G}_i(N)$; $i = 1,\dots,n\}$.

In Lemmas 3.7, 3.8 below we assume that H is sufficiently large with respect to N and ε^{-1}, $\varepsilon \cdot \mathrm{Card}(\mathcal{G}(N))^2 < 1$.

LEMMA 3.7. *Let, as above, $P_{\mathfrak{A}_\alpha}(x)e^{-\mathfrak{A}_\alpha x}$: $\alpha \in A_N$, $\mathrm{Card}(A_N) = d_N$, be the basis of V over $\mathbf{C}$. Let e_N denote the maximal number of functions $L_{\mathfrak{A}}(x)$: $\mathfrak{A} \in \mathcal{G}(N)$ from* (3.4), *linearly independent over $\mathbf{C}$. Then $|e_N(\mu - \varepsilon) - d_N|/\varepsilon$ is bounded as*

$\varepsilon \to 0$. *I.e. for some* $c_{14} > 0$, $e_N(\mu - \varepsilon) - c_{14}\varepsilon \leqslant d_N \leqslant e_N(\mu - \varepsilon) + c_{14}\varepsilon$ *provided that* $N \geqslant N_1$.

In the case of linearly independent numbers $\beta_{i,j}$ a stronger assertion is valid.

LEMMA 3.8. *Let algebraic numbers* $\beta_{i,j}$: $j = 1,\ldots,k_i$; $i = 1,\ldots,n$ *be linearly independent over* **Q**, *and let* $\varepsilon \cdot \operatorname{Card}(\mathcal{G}(N))^2 < 1$. *Then for a sufficiently large* H, *we have*: $d_N = \sum_{i=1}^{n} \operatorname{Card}(\mathcal{G}_i(N))$ *and all functions* $P_{\mathfrak{A},i}(x)e^{-\mathfrak{A}x}$: $\mathfrak{A} \in \mathcal{G}_i(N)$; $i = 1,\ldots,n$ *are linearly independent over* **C**.

For the proof of Lemma 3.7 and 3.8 we need the following auxiliary statement following from Normality lemma 1.

LEMMA 3.9. *Let us assume that we have d linearly independent* (*over* **C**) *functions of the form* $P_\alpha(x)e^{\omega_\alpha x}$: $\alpha \in A$ *for complex* ω_α *and polynomials* $P_\alpha(x) \in \mathbf{C}[x]$ *of degrees of at most* H. *Let there be s linearly independent* (*over* **C**) *functions* $l_1(x),\ldots,l_s(x)$, *that are linear combinations of* $P_\alpha(x)e^{\omega_\alpha x}$: $\alpha \in A$. *Then* $M \overset{\text{def}}{=} \sum_{i=1}^{s} \operatorname{ord}_{x=0}(l_i(x)) \leqslant d\{H + (d-1)/2\}$. *For arbitrary* $x_0 = 0$, *the rank of the matrix*

$$\left(\left(\frac{d}{dx}\right)^m \cdot \left(P_\alpha(x)e^{\omega_\alpha x}\right)\Big|_{x=x_0} : \alpha \in A\right): \quad m = 0,1,\ldots,d\{H + (d-1)/2\} - M$$

is d. Namely, for

$$P_{\alpha,m}(x) \overset{\text{def}}{=} \left(\frac{d}{dx} + \omega_\alpha\right)^m P_\alpha(x): \qquad \alpha \in A;\ m = 0,1,\ldots,$$

the rank of the matrix

$$\left(P_{\alpha,m}(x_0): \alpha \in A;\ m = 0,1,\ldots,d \cdot \{H + (d-1)/2\} - M\right)$$

is d.

PROOF OF LEMMA 3.9. Let us denote for $m \geqslant 0$, $P_{\alpha,m}(x) = (d/dx + \omega_\alpha)^m P_\alpha(x)$: $\alpha \in A$. Then we have

$$\det\left(\left(\frac{d}{dx}\right)^m \left(P_\alpha(x)e^{\omega_\alpha x}\right): \alpha \in A;\ m = 0,1,\ldots,d-1\right)$$
$$= \exp\left\{\sum_{\alpha \in A} - \omega_\alpha x\right\}\det\left(P_{\alpha,m}(x): \alpha \in A;\ m = 0,1,\ldots,d-1\right).$$

We denote $D(x) = \det(P_{\alpha,m}(x): \alpha \in A;\ m = 0,1,\ldots,d-1)$. Since $P_\alpha(x)e^{\omega_\alpha x}$ are linearly independent over **C**, $D(x) \not\equiv 0$ (cf. Siegel's lemma 1 of §2). Polynomials $P_\alpha(x)$ have degrees of at most H in x. This implies that $D(x)$ is a polynomial of degree of at most $d \cdot H$. Since $l_1(x),\ldots,l_s(x)$ are linearly independent combinations of $P_\alpha(x)e^{\omega_\alpha x}$: $\alpha \in A$, we make a linear transformation of the determinant $D(x)$ and obtain:

$$\operatorname*{ord}_{x=0}(D(x)) \geqslant \sum_{i=1}^{s}\left\{\operatorname*{ord}_{x=0}(l_i(x)) - (d-i)\right\} \leqslant M - d(d-1)/2.$$

Hence, $M \leqslant d\{H + (d-1)/2\}$.

Now, as in §2, let $D(x) = x^{\gamma} \cdot D_1(x)$, where

$$\deg(D_1(x)) \leqslant \mu_0 \overset{\text{def}}{=} d\{H + (d-1)/2\} - M.$$

Hence $x_0 \neq 0$ is a zero of $D(x)$ of order $k \leqslant \mu_0$. Then the kth derivative $(d/dx)^k D(x)$ of $D(x)$ is a linear combination of determinants

$$\det\left(P_{\alpha,m_i}(x) \colon \alpha \in A, i = 1,\ldots,d\right);$$

for $0 \leqslant m_1 < \cdots < m_d \leqslant d + k - 1$. Since $(d/dx)^k D(x)\,|_{x=x_0} \neq 0$, d rows in

$$\left(P_{\alpha,m}(x) \colon \alpha \in A;\, m = 0, 1,\ldots,d + k - 1\right)$$

are linearly independent (over $\mathbf{C}$). Lemma 3.9 is proved.

PROOF OF LEMMA 3.8. Let e_N be the number of linearly independent functions among $L_{\mathfrak{A}}(x)$: $\mathfrak{A} \in \mathcal{G}(N)$. Since numbers $\beta_{i,j}$: $j = 1,\ldots,k_i$; $i = 1,\ldots,n$ are linearly independent over $\mathbf{Q}$, the map λ: $\mathbf{Z}^{k_1} \times \cdots \times \mathbf{Z}^{k_n} \to \mathcal{G}$ is one-to-one. In particular, according to the remark above, the number d_N of linearly independent functions among $P_{\mathfrak{A},i}(x)e^{-\mathfrak{A}x}$: $\mathfrak{A} \in \mathcal{G}_i(N)$; $i = 1,\ldots,n$ is equal to the number of nonzero polynomials $P_{\mathfrak{A},i}(x)$: $\mathfrak{A} \in \mathcal{G}_i(N)$; $i = 1,\ldots,n$. On the other hand, according to Lemma 3.9 and property (2) of Theorem 3.1 we have

$$e_N \cdot (\mu - \varepsilon)H \leqslant d_N \cdot \{H + (d_N - 1)/2\},$$

$$\mu = \sum_{i=1}^{n} \operatorname{Card}(\mathcal{G}_i(N))/\operatorname{Card}(\mathcal{G}(N))$$

or

$$\mu = \left\{\sum_{i=1}^{n} \binom{N + k_i - 2}{k_i - 1} \times \prod_{j=1, j\neq i}^{n} \binom{N + k_j - 1}{k_j - 1}\right\} \Big/ \prod_{i=1}^{n} \binom{N + k_i - 1}{k_i - 1}.$$

Because of the linear independence of $\beta_{i,j}$, the only linear relations between functions $L_{\mathfrak{A}}(x)$: $\mathfrak{A} \in \mathcal{G}(N)$ are of the form $P_{\mathfrak{A}',i}(x) = 0$ for $\mathfrak{A}' \in \mathcal{G}_i(N)$; $i = 1,\ldots,n$. Then for a sufficiently large H, $H \geqslant H_1(N, \varepsilon)$ we obtain

$$\mu - \varepsilon \leqslant d_N/e_N \cdot \{1 + (d_N - 1)/(2H)\}.$$

The linear independence of $\beta_{i,j}$, Lemma 3.9 and the assumption $\varepsilon \cdot \operatorname{Card}(\mathcal{G}(N))^2 < 1$ then implies $d_N = \sum_{i=1}^{n} \operatorname{Card}(\mathcal{G}_i(N))$ and $e_N = \operatorname{Card}(\mathcal{G}(N))$. Lemma 3.8 is proved.

The proof of Lemma 3.7 is based on Lemma 3.9 and on studies of the combinatorics of the lattice $\mathcal{G}$; cf. [26] for similar results. We note in this connection that for a sufficiently large N, $\operatorname{Card}(\mathcal{G}(N))$ is an integer valued polynomial in N, and $\operatorname{Card}(\mathcal{G}(N))/\operatorname{Card}(\mathcal{G}_i(N)) \to 1$ as $N \to \infty$: $i = 1,\ldots,n$.

PROOF OF THEOREM 3.6. We use the notations of Lemmas 3.7–3.8. Applying Lemma 3.9 and condition (2) of Theorem 3.1 we obtain

$$M \overset{\text{def}}{=} \sum_{j=1}^{e_N} \operatorname{ord}_{x=0}\left(L_{\mathfrak{A}_j}(x)\right) \geqslant e_N \cdot (\mu - \varepsilon) \cdot H,$$

where $L_{\mathfrak{A}_j}(x)$: $j = 1,\dots,e_N$ is the maximal set of linearly independent functions $L_{\mathfrak{A}}(x)$: $\mathfrak{A} \in \mathcal{G}(N)$. According to the statement of Lemma 3.9, $M \leqslant d_N \cdot \{H + (d_N - 1)/2\}$. We now use Lemma 3.7 and $e_N(\mu - \varepsilon) \geqslant d_N - c_{14}\varepsilon$ as $N \to \infty$. Hence, according to Lemma 3.9 again, the rank of the matrix $M_{x_0} = (P_{\mathfrak{A}_\alpha}^{\langle m \rangle}(x_0)$: $\alpha \in A_N$; $m = 0, 1,\dots,c_{13}\varepsilon H)$ is d_N for $c_{14} > 0$ depending only on N, $N \geqslant N_2$.

Let us now deduce the existence of n linearly independent forms among $L^{\langle m \rangle}(x)$: $m = 0, 1,\dots,c_{14}\varepsilon H$, from rank $(M_{x_0}) = d_N$.

We consider the following functions from V:

$$(3.21) \qquad P_i(x; x_0) = \sum_{\mathfrak{A}' \in \mathcal{G}_i(N)} P_{\mathfrak{A}',i}(x) \cdot e^{-\mathfrak{A}'(x - x_0)}: \qquad i = 1,\dots,n.$$

Since exponents $e^{-\mathfrak{A}'x}$: $\mathfrak{A}' \in \mathcal{G}_i(N)$ are linearly independent over $\mathbf{C}(x)$, the function $P_i(x;\ x_0)$ can be identically zero in x if and only if all polynomials $P_{\mathfrak{A}',i}(x)$ are identically zero: $\mathfrak{A}' \in \mathcal{G}_i(N)$: $i = 1,\dots,n$. Hence, for any given x_0, not all functions $P_i(x;\ x_0)$: $i = 1,\dots,n$ are identically zero. It might happen, however, that all functions $P_i(x;\ x)$ (at $x = x_0$) are zeroes: $i = 1,\dots,n$. Nevertheless, since not all $P_i(x;\ x_0)$ are zeroes: $i = 1,\dots,n$, there exists a minimal nonnegative integer $k \geqslant 0$, such that at least one of the functions

$$(3.22) \qquad \left(\frac{d}{dx}\right)^k P_i(x; x_0)\Big|_{x_0 = x}: \qquad i = 1,\dots,n$$

is nonzero (as a function of x). We now denote, as in (3.19), (3.20):

$$(3.23) \qquad L(x; x_0) \stackrel{\text{def}}{=} \sum_{\mathfrak{A} \in \mathcal{G}(N)} L_{\mathfrak{A}}(x) \cdot e^{\mathfrak{A}x_0}.$$

We have, as in the proof of Lemma 3.5,

$$\begin{aligned} L(x; x_0) &= \sum_{\mathfrak{A} \in \mathcal{G}(N)} L_{\mathfrak{A}}(x) e^{\mathfrak{A}x_0} \\ &= \sum_{\mathfrak{A} \in \mathcal{G}(N)} \sum_{i=1}^{n} \sum_{j=1}^{k_i} P_{\mathfrak{A} - e_{i,j},i}(x) \cdot e^{-(\mathfrak{A} - \beta_{i,j})x} \cdot e^{\mathfrak{A}x_0} \cdot C_{i,j} \\ &= \sum_{i=1}^{n} \sum_{j=1}^{k_i} C_{i,j} \cdot e^{\beta_{i,j}x} \cdot \sum_{\mathfrak{A} \in \mathcal{G}(N)} P_{\mathfrak{A} - e_{i,j},i}(x) \cdot e^{-\mathfrak{A}(x - x_0)}. \end{aligned}$$

We have for $j = 1,\dots,k_i$; $i = 1,\dots,n$,

$$\sum_{\mathfrak{A} \in \mathcal{G}(N)} P_{\mathfrak{A} - e_{i,j},i}(x) \cdot e^{-\mathfrak{A}(x - x_0)} = e^{\beta_{i,j}(x - x_0)} \sum_{\mathfrak{A}' \in \mathcal{G}_i(N)} P_{\mathfrak{A}',i}(x) \cdot e^{-\mathfrak{A}'(x - x_0)}.$$

Hence,

$$L(x; x_0) = \sum_{i=1}^{n} \sum_{j=1}^{k_i} C_{i,j} \cdot e^{\beta_{i,j}x_0} \sum_{\mathfrak{A}' \in \mathcal{G}_i(N)} P_{\mathfrak{A}',i}(x) \cdot e^{-\mathfrak{A}'(x - x_0)}.$$

Thus, according to (3.21):

$$(3.24) \qquad L(x; x_0) = \sum_{i=1}^{n} P_i(x; x_0) \cdot f_i(x_0),$$

with $f_i(x_0) = \sum_{j=1}^{k_i} C_{i,j} e^{\beta_{i,j} x_0}$: $i = 1,\dots,n$. According to the definition of k in (3.22), we have $(d/dx_0)^l P_i(x; x_0)|_{x_0=x} = 0$ for all $l = 0,\dots,k-1$ and $i = 1,\dots,n$. Hence applying $(d/dx_0)^k$ to (3.24), we obtain

$$(3.25)\qquad \left(\frac{d}{dx_0}\right)^k L(x; x_0)\bigg|_{x_0=x} = \sum_{i=1}^{n} f_i(x_0)\cdot\left(\frac{d}{dx_0}\right)^k \cdot P_i(x; x_0)\bigg|_{x_0=x}.$$

Combining (3.25) with expressions (3.23) and (3.21), we arrive at

COROLLARY 3.10. *We obtain n polynomials*

$$Q_i(x) = \sum_{\mathfrak{A}' \in \mathcal{G}_i(N)} \mathfrak{A}'^k \cdot P_{\mathfrak{A}',i}(x): \qquad i = 1,\dots,n$$

not all identically zero and such that

$$\sum_{i=1}^{n} Q_i(x)\cdot f_i(x) = \sum_{\mathfrak{A} \in \mathcal{G}(N)} \mathfrak{A}^k \cdot L_{\mathfrak{A}}(x)\cdot e^{\mathfrak{A}x}.$$

PROOF OF COROLLARY 3.10. According to (3.21),

$$Q_i(x) = (d/dx_0)^k \cdot P_i(x; x_0)|_{x_0=x}$$

and, according to (3.23), $(d/dx_0)^k \cdot L(x; x_0)|_{x_0=x} = \sum_{\mathfrak{A}\in\mathcal{G}(N)} \mathfrak{A}^k \cdot L_{\mathfrak{A}}(x)\cdot e^{\mathfrak{A}x}$. Hence, Corollary 3.10 follows from (3.25).

Let us now prove that polynomials $Q_1(x),\dots,Q_n(x)$ from Corollary 3.10 are linearly independent over **C**.

LEMMA 3.11. *Under the assumptions of Theorem* 3.1 *and in the notations of Corollary* 3.10, *polynomials* $Q_1(x),\dots,Q_n(x)$ *are linearly independent over* **C**.

PROOF OF LEMMA 3.11. Let, on the contrary, the polynomials $Q_1(x),\dots,Q_n(x)$ be linearly dependent over **C**. Let us choose a maximal set of polynomials among $Q_1(x),\dots,Q_n(x)$ that are linearly independent over **C**. We can assume, without a loss of generality, that these are $Q_1(x),\dots,Q_l(x)$ for $l < n$. Then the function

$$R(x) = \sum_{i=1}^{n} Q_i(x)\cdot f_i(x)$$

from Corollary 3.10 can be represented in the form: $R(x) = \sum_{i=1}^{l} Q_i(x) h_i(x)$, where functions $h_1(x),\dots,h_l(x)$ are linear combinations of functions $f_1(x),\dots,f_n(x)$ with constant coefficients. Hence $h_1(x),\dots,h_l(x)$ are linear combinations of $e^{\beta_{i,j}x}$: $j = 1,\dots,k_i$; $i = 1,\dots,n$ with constant coefficients. For an integer $K \geqslant 1$ we take a sublattice of $\mathcal{G}$:

$$\mathcal{G}_K = \left\{\mathfrak{A} = \sum_{i=1}^{n}\sum_{j=1}^{k_i} a_{i,j}\beta_{i,j} \in \mathcal{G}: a_{i,j} \in \mathbf{Z},\ a_{i,j} \geqslant 0 \right.$$
$$\left. j = 1,\dots,k_i;\ i = 1,\dots,n \text{ and } \sum_{i=1}^{n}\sum_{j=1}^{k_i} a_{i,j} = K \right\}.$$

Then the functions $Q_i(x)e^{\mathfrak{A}x}$: $\mathfrak{A} \in \mathcal{G}_K$, $i = 1,\dots,l$ are all linearly independent over $\mathbf{C}$ because $Q_1(x),\dots,Q_l(x)$ are linearly independent. We now apply Lemma 3.9 to $d \overset{\text{def}}{=} l \cdot \mathrm{Card}(\mathcal{G}_K)$ linearly independent (over $\mathbf{C}$) functions $Q_i(x)e^{\mathfrak{A}x}$: $\mathfrak{A} \in \mathcal{G}_K$; $i = 1,\dots,l$.

According to the definition of $\mathcal{G}_K$, for $\mathfrak{A}_1 \in \mathcal{G}_{K-1}$ and $j = 1,\dots,k_i$; $i = 1,\dots,n$, $e^{\beta_{i,j}x} \cdot e^{\mathfrak{A}_1 x} = e^{\mathfrak{A}x}$ for some $\mathfrak{A} \in \mathcal{G}_K$. Hence, for any function $h_i(x)$: $i = 1,\dots,l$ and $\mathfrak{A}_1 \in \mathcal{G}_{K-1}$, $h_i(x)e^{\mathfrak{A}_1 x}$ is a linear combination (with constant coefficients) of $e^{\mathfrak{A}x}$: $\mathfrak{A} \in \mathcal{G}_K$. This implies that for any $\mathfrak{A}_1 \in \mathcal{G}_{K-1}$, the function

$$l_{\mathfrak{A}_1}(x) = R(x) \cdot e^{\mathfrak{A}_1 x}$$

is a linear combination (with constant coefficients) of functions $Q_i(x)e^{\mathfrak{A}x}$: $\mathfrak{A} \in \mathcal{G}_K$; $i = 1,\dots,n$. There are $s = \mathrm{Card}(\mathcal{G}_{K-1})$ linearly independent (over $\mathbf{C}$) functions $l_{\mathfrak{A}_1}(x)$: $\mathfrak{A}_1 \in \mathcal{G}_{K-1}$. According to (2) of Theorem 3.1 and the representation of $R(x)$ from Corollary 3.10, $\mathrm{ord}_{x=0}(R(x)) \geqslant (\mu - \varepsilon)H$. Hence, $\mathrm{ord}_{x=0}(l_{\mathfrak{A}_1}(x)) \geqslant (\mu - \varepsilon)H$ for any $\mathfrak{A}_1 \in \mathcal{G}_{K-1}$. Applying Lemma 3.9, we conclude that

$$\mathrm{Card}(\mathcal{G}_{K-1}) \cdot (\mu - \varepsilon) \cdot H \leqslant l \cdot \mathrm{Card}(\mathcal{G}_K) \cdot \{H + (l \cdot \mathrm{Card}(\mathcal{G}_K) - 1)/2\}.$$

We note now that $\mathrm{Card}(\mathcal{G}_K) \sim \lambda_0 \cdot K^\nu + O(K^{\nu-1})$ for $K \geqslant K_0$ and K_0 depending only on $\beta_{i,j}$ ($j = 1,\dots,k_i$; $i = 1,\dots,n$). Hence $\mu - \varepsilon \leqslant l \cdot (K/(K-1))^\nu + l^2\lambda_0 \cdot K^{2\nu}/(K-1)^\nu \cdot H^{-1} + o(H^{-1})$. For a sufficiently large K and $H \geqslant H_0(K)$, we obtain $\mu - \varepsilon \leqslant l$. On the other hand, $\mu = \Sigma_{i=1}^n \mathrm{Card}(\mathcal{G}_i(N))/\mathrm{Card}(\mathcal{G}(N))$, so that $\mu - \varepsilon > n - 1$ for $N \geqslant N_1(\varepsilon)$. This implies for $H \geqslant H_1(N, \varepsilon)$ that $n - 1 < l$. Consequently, polynomials $Q_1(x),\dots,Q_n(x)$ are linearly independent over $\mathbf{C}$ and Lemma 3.11 is proved.

To end the proof of Theorem 3.6, we express $P_i(x; x_0)$: $i = 1,\dots,n$ from (3.22) in terms of the basis $P_{\mathfrak{A}_\alpha}(x) \cdot e^{-\mathfrak{A}_\alpha x}$: $\alpha \in A_N$ of V. Since exponents $e^{-\mathfrak{A}'x}$: $\mathfrak{A}' \in \mathcal{G}_i(N)$, are all linearly independent over $\mathbf{C}(x)$, we obtain the following representation of $P_i(x; x_0)$:

$$P_i(x; x_0) = \sum_{\alpha \in A_N} B_{\alpha,i} \cdot P_{\mathfrak{A}_\alpha}(x) \cdot e^{-\mathfrak{A}_\alpha(x - x_0)}$$

for constants $B_{\alpha,i}$ (independent of x_0): $\alpha \in A_N$; $i = 1,\dots,n$. The rank of the matrix ($B_{\alpha,i}$: $\alpha \in A_N$; $i = 1,\dots,n$) is equal to n. Indeed, let there exist constants $u_1,\dots,u_n$, not all zero, such that $\Sigma_{i=1}^n u_i \cdot B_{\alpha,i} = 0$: $\alpha \in A_N$. Then

$$\sum_{i=1}^n u_i \cdot P_i(x; x_0) = \sum_{\alpha \in A_N} P_{\mathfrak{A}_\alpha}(x) \cdot e^{-\mathfrak{A}_\alpha(x - x_0)} \cdot \sum_{i=1}^n u_i B_{\alpha,i} \equiv 0.$$

Hence, functions $P_1(x; x_0),\dots,P_n(x; x_0)$ are linearly dependent (over $\mathbf{C}$), which contradicts the linear independence of $Q_i(x) = (d/dx_0)^k \cdot P_i(x; x_0)|_{x_0 = x}$: $i = 1,\dots,n$, established in Lemma 3.11.

Thus $\mathrm{rank}(B_{\alpha,i}$: $\alpha \in A_N$; $i = 1,\dots,n) = n$.

We now obtain $(d/dx)^m P_i(x;\ x_0) = \sum_{\alpha\in A_N} B_{\alpha,i} \cdot P^{\langle m\rangle}_{\mathfrak{A}_\alpha}(x) \cdot e^{-\mathfrak{A}_\alpha(x-x_0)}$, for $m = 0, 1, \ldots$. Next, by (3.21) and (3.24),

$$\left(\frac{d}{dx}\right)^m L(x;\ x_0) = \sum_{i=1}^n \left(\frac{d}{dx}\right)^m \cdot P_i(x;\ x_0) \cdot f_i(x_0),$$

$$\left(\frac{d}{dx}\right)^m P_i(x;\ x_0) = \sum_{\mathfrak{A}'\in\mathcal{G}_i(N)} P^{\langle m\rangle}_{\mathfrak{A}',i}(x) \cdot e^{-\mathfrak{A}'(x-x_0)}.$$

Thus, in the notations of Lemma 3.5 and (3.20), (3.19)

$$P_i^{\langle m\rangle}(x_0) = \sum_{\alpha\in A_N} B_{\alpha,i} \cdot P^{\langle m\rangle}_{\mathfrak{A}_\alpha}(x_0),$$

$$L^{\langle m\rangle}(x_0) = \sum_{i=1}^n P_i^{\langle m\rangle}(x_0) \cdot f_i(x_0),$$

$m = 0, 1, \ldots$.

The rank of the matrix $M_{x_0} = (P^{\langle m\rangle}_{\mathfrak{A}_\alpha}(x_0)\colon \alpha \in A_N;\ m = 0, 1, \ldots, c_{13}\,\varepsilon H)$ is d_N and the rank of $(B_{\alpha,i}\colon \alpha \in A_N;\ i = 1, \ldots, n)$ is n. Hence the rank of the matrix $(P_i^{\langle m\rangle}(x_0)\colon m = 0, 1, \ldots, c_{13}\varepsilon H;\ i = 1, \ldots, n)$ is n. Theorem 3.6 is proved.

Normality Theorem 3.6 implies the existence of n linearly independent forms approximating numbers $\theta_1, \ldots, \theta_n$ from (3.1).

Let us assume, as above, that $(\beta_{i,1}, \ldots, \beta_{i,k_i})$ are sequences of distinct algebraic numbers, closed under the algebraic conjugation, and let $(C_{i,1}, \ldots, C_{i,k_i})$ be rational numbers such that $C_{i,j} = C_{i,j'}$, whenever $\beta_{i,j}$ is algebraically conjugate to $\beta_{i,j'}$: $j, j' = 1, \ldots, k_i$: $i = 1, \ldots, n$. Also let $B = \bigcup_{i=1}^n \{\beta_{i,1}, \ldots, \beta_{i,k_i}\}$ and let the rank of the matrix $(C_{i,\beta})$ $(\beta \in B,\ i = 1, \ldots, n)$ be n.

THEOREM 3.12. *Let, in the notations above, $\theta_i = \sum_{j=1}^{k_i} C_{i,j} \cdot e^{\beta_{i,j}}$: $i = 1, \ldots, n$. Then for any $\delta > 0$ and any sufficiently large integer D, $D \geq D_2(\beta_{i,j}, C_{i,j}, \delta)$ there exist rational integers $P_{i,j}$: $i, j = 1, \ldots, n$ such that the following conditions are satisfied. The determinant of $P_{i,j}$ — $\det(P_{i,j})_{i,j=1}^m$ is nonzero; the absolute values of numbers $P_{i,j}$ are bounded by $D^{(1+\varepsilon)D}$ and*

$$\left|\sum_{j=1}^n P_{i,j}\theta_j\right| \leq D^{(1+\delta-n)D}: \qquad i, j = 1, \ldots, n.$$

PROOF OF THEOREM 3.12. Let, as above, N be a sufficiently large integer, and D be a sufficiently large integer with respect to N and δ^{-1}; $D \geq D_0(\beta_{i,j}, C_{i,j}, \delta)$. We apply Theorem 3.6 and obtain n linearly independent forms (at $x_0 = 1$):

$$L^{\langle m\rangle}(1) = \sum_{i=1}^n P_i^{\langle m\rangle}(1) \cdot \theta_i: \qquad m = 0, 1, \ldots, \delta D$$

with a matrix $(P_i^{\langle m\rangle}(1)\colon i = 1, \ldots, n;\ m = 0, 1, \ldots, \delta D)$ of rank n. We choose n distinct integers $0 \leq m_1 < \cdots < m_n \leq \delta D$ such that $\det(P_i^{\langle m_j\rangle}(1))_{i,j=1}^n \neq 0$.

According to Lemmas 3.3 and 3.5, polynomials $P_i^{\langle m_j\rangle}(x)$ have rational number coefficients, whose common denominator divides b^{m_j} for an integer $b \geq 1$. We define $P_{i,j} = b^{m_i} P_j^{\langle m_i \rangle}$: $i, j = 1,\ldots,n$. Then $P_{i,j}$ are rational integers with $\det(P_{i,j})_{i,j=1}^n \neq 0$. From Lemma 3.2 it follows that for a sufficiently large D, $D \geqslant D_1(\beta_{i,j}, C_{i,j}, \delta)$, $\max\{|P_{i,j}|: i, j = 1,\ldots,n\} < D^{(1+\delta)D}$. For a sufficiently large N we have, according to (3.5), $\mu \to n$ as $N \to \infty$. Thus for D sufficiently large, $D \geqslant D_2(\beta_{i,j}, C_{i,j}, \delta)$, (3.17) implies

$$|L_{\mathfrak{A}}^{\langle m \rangle}(1)| \leqslant D^{(-n+1+\delta/2)D}: \qquad m = 0, 1,\ldots,\delta D$$

for any $\mathfrak{A} \in \mathcal{G}(N)$. Now $L^{\langle m \rangle}(1) = \Sigma_{\mathfrak{A} \in \mathcal{G}(N)} L_{\mathfrak{A}}^{(m)}(1) \cdot e^{\mathfrak{A}}$, so that we obtain for $D \geqslant D_2(\beta_{i,j}, C_{i,j}, \delta)$:

$$\left| \sum_{j=1}^{n} P_{i,j} \cdot \theta_j \right| \leqslant D^{(-n+1+\delta)D},$$

Theorem 3.12 is proved.

Now Siegel's method of §2 allows us to prove our main result on the approximation of numbers $\theta_1,\ldots,\theta_n$.

THEOREM 3.13. *In the notations above, let* $\theta_i = \Sigma_{j=1}^{k_i} C_{i,j} \cdot e^{\beta_{i,j}}$: $i = 1,\ldots,n$. *Then for any* $\delta > 0$ *and arbitrary rational integers* $h_1,\ldots,h_n$ *with* $h = \max(|h_1|,\ldots,|h_n|)$ *we have*

$$|h_1\theta_1 + \cdots + h_n\theta_n| > h^{-n+1-\delta} \quad \text{when } h \geqslant h_0(\beta_{i,j}, C_{i,j}, \delta).$$

PROOF OF THEOREM 3.13. Let $h_1,\ldots,h_n$ be rational integers not all zero such that $\max(|h_i| : i = 1,\ldots,n) = h$ and such that

$$h_1\theta_1 + \cdots + h_n\theta_n = l: \qquad 0 < |l| < 1.$$

We apply Theorem 3.12 and obtain for any $\delta > 0$ and for the integer $D \geqslant D_2(\beta_{i,j}, C_{i,j}, \delta)$, n linearly independent forms in $\theta_1,\ldots,\theta_n$ with rational integer coefficients of the form

$$L_i = \sum_{j=1}^{n} P_{i,j}\theta_j: \qquad i = 1,\ldots,n$$

such that $\max(|P_{i,j}|: i, j = 1,\ldots,n) \leqslant D^{(1+\delta)D}$ and $\max(|L_i|: i = 1,\ldots,n) \leqslant D^{(1+\delta-n)D}$. The linear independence of forms $L_1,\ldots,L_n$ means that $\det(P_{i,j})_{i,j=1}^n \neq 0$. Hence we can always find $n-1$ forms, say, $L_1,\ldots,L_{n-1}$, that are linearly independent together with the form l. Let V be a matrix formed from coefficients of forms $L_1,\ldots,L_{n-1}, l$; let Δ_1 be a determinant of V, and $\Delta_{i,j}$ be the algebraic complement of the (i, j)th element of V. Then

$$\theta_i \Delta_1 = \sum_{j=1}^{n-1} L_j \Delta_{j,i} + l\Delta_{n,i}: \qquad i = 1,\ldots,n.$$

From the definition of V it follows that $|\Delta_{n,i}| \leqslant (n-1)! D^{(n-1)(1+\delta)D}$,

$$\max(|\Delta_{j,i}| : j = 1,\ldots,n-1) \leqslant (n-1)! h \cdot D^{(n-2)(1+\delta)D}: \qquad i = 1,\ldots,n.$$

Let $i = 1, 2, \ldots, n$ be chosen in a way such that $\theta_i \neq 0$, which is possible since $l \neq 0$. Then

$$|\theta_i \Delta_1| \leq n!\left(hD^{(-1+\delta(n-1))D} + |l| D^{(n-1+\delta(n-1))D}\right).$$

Since $\Delta_1 \neq 0$ and all elements of V are rational integers, $|\Delta_1| \geq 1$. We choose now an integer D to be the smallest integer $\geq D_2$ such that $D^{nD}|l| \geq h$.

For this definition of D we get $|\theta_i| \leq 2 \cdot n! \cdot |l| \cdot D^{(n-1)\cdot(1+\delta)D}$. This implies a lower bound on $|l|$:

$$|l| \geq c_{14} \cdot h^{-(n-1)(1+\delta)/(1+\delta(n-1))}$$

and $c_{14} = c_{14}(\beta_{i,j}, C_{i,j}, \delta)$. For a sufficiently small δ this proves Theorem 3.13.

Our main result, Theorem 3.13 gives measures of diophantine approximations (transcendence) of various numbers connected with the exponential function. For example, we get

COROLLARY 3.14. *Let $\xi \neq 0$ be an algebraic number of degree d. Let $\varepsilon > 0$. Then for arbitrary rational integers p and q we obtain*

$$|e^{\xi} - p/q| > |q|^{-d(d+1)-\varepsilon}$$

provided that $|q| \geq q_0(\xi, \varepsilon)$.

PROOF OF COROLLARY 3.14. Let $\xi_1 = \xi, \ldots, \xi_d$ be a complete set of (distinct) algebraic numbers conjugate to ξ. We use Weierstrass' method (cf. [**5**, **6**]) to reduce the bound of the linear form $|qe^{\xi} - p|$ to the statement of the form of Theorem 3.13. We have $\prod_{j=1}^{d}(qe^{\xi_j} - p) = \sum_{\alpha \in A} h_\alpha \theta_\alpha$ where θ_α are sums of exponents $\exp\{k_1\xi_{\pi(1)} + \cdots + k_d\xi_{\pi(d)}\}$ over all permutations π of $(1, \ldots, d)$ for $0 \leq k_1 \leq \cdots \leq k_d \leq 1$. Hence $\mathrm{Card}(A) \leq d + 1$ and $\max_{\alpha \in A}|h_\alpha| \leq 2^d \max(|p|, |q|)^d$. From Theorem 3.13 we deduce $|\sum_{\alpha \in A} h(\alpha)| \geq |q|^{d(-d-\varepsilon)}$ or $|qe^{\xi} - p| > |q|^{-d^2-d\varepsilon-d+1}$. Corollary 3.14 is proved.

Corollary 3.14 is so far the strongest result on the measure of irrationality of e^{ξ} with algebraic irrational ξ. The best result known before [**25**] has the exponent $-4d^2 + 2d - \varepsilon$, instead of $-d(d+1) - \varepsilon$ in Corollary 3.14.

We have similar results on the measure of the transcendence of numbers e^{ξ}.

COROLLARY 3.15. *Let $\xi \neq 0$ be an algebraic number of degree d. Then for the measure of transcendence of e^{ξ} we have the following bounds. Let $\varepsilon > 0$. Then for an arbitrary polynomial $P(x) \in \mathbf{Z}[x]$ of degree of at most $d(P)$ and height of at most $h(P)$ we have*

$$|P(e^{\xi})| > h(P)^{-d\binom{d(P)+d}{d}+1-\varepsilon}.$$

For an arbitrary algebraic number ζ of degree of at most $d(\zeta)$ and of height of at most $h(\zeta)$ we have

$$|e^{\xi} - \zeta| > h(\zeta)^{-d\binom{d(\zeta)+d}{d}-\varepsilon}.$$

Here $h(P)$ is sufficiently large (with respect to $h(\xi)$, ε^{-1} and $d(P)$), and $h(\zeta)$ is sufficiently large with respect to $h(\xi)$, ε^{-1} and $d(\xi)$.

For an algebraic ξ of special form, better results can be deduced from Theorem 3.13 directly. Also results similar to Corollary 3.15 hold for arbitrary elements of fields generated by e^{ξ}. For a bounded $d(P)$ our results improve considerably the existing measures of the transcendence of e^{ξ}, see [**7**, **10**] and Chapter 1, this volume. As we mentioned in §2, results of the form of Corollaries 3.14–3.15 are valid for values of arbitrary elliptic functions with a complex multiplication at algebraic points that supplement our elliptic version of the Lindemann-Weierstrass theorem [**11**, **12**].

4. We show here how, using the methods introduced above for exponential functions, one can prove strong results on the measures of the linear independence of values of arbitrary E-functions of Siegel [**1**] at rational points. Our studies refine and extend Siegel's results from [**1**, **3**].

Following [**1**] we call $f(x) = \Sigma_{n=0}^{\infty} a_n x^n/n!$ an E-function, if $f(x)$ satisfy a linear differential equation over $\mathbf{Q}(x)$, $a_n \in \mathbf{Q}$: $n = 0, 1, \ldots$ and if, for any $\varepsilon > 0$, $|a_n| \leqslant n^{\varepsilon n}$ and $\operatorname{denom}\{a_0, \ldots, a_n\} \leqslant n^{\varepsilon n}$ for $n \geqslant n_0(\varepsilon)$. Our main result is a measure of the linear independence of $f_1(r), \ldots, f_n(r)$ for E-functions $f_i(x)$: $i = 1, \ldots, n$ and $r \in \mathbf{Q}$, $r \neq 0$, which is close to the best possible. The proof we present below relies, as do Siegel's results [**1**], on the assumption of the normality of functions $f_1(x), \ldots, f_n(x)$. As the method shows, this assumption can be removed, and the corresponding result is valid in its full generality.

Let $f_i = f_i(x)$: $i = 1, \ldots, n$ be a component of a nonzero solution of a matrix linear differential equation of the first order over $\mathbf{Q}(x)$:

$$\frac{d}{dx} f_i^{(j)} = \sum_{l=1}^{k_i} A_{j,l}^{(i)} \cdot f_i^{(l)}: \qquad j = 1, \ldots, k_i \tag{4.1}$$

for $A_{j,l}^{(i)} \in \mathbf{Q}(x)$: $j, l = 1, \ldots, k_i$ and $f_i^{(1)} \stackrel{\text{def}}{=} f_i$: $i = 1, \ldots, n$. Equivalently, $f_i = f_i(x)$ is a solution of a scalar linear differential equation of the order k_i over $\mathbf{Q}(x)$

$$\frac{d^{k_i}}{dx^{k_i}} f_i(x) + a_{k_i-1}^{(i)} \frac{d^{k_i-1}}{dx^{k_i-1}} f_i(x) + \cdots + a_0^{(i)} f_i(x) = 0, \tag{4.1$'$}$$

$a_l^{(i)} \in \mathbf{Q}(x)$: $l = 0, \ldots, k_i - 1$; $i = 1, \ldots, n$. Any equation (4.1′) can be reduced to the form (4.1) if one puts

$$f_i^{(j)} = \frac{d^{j-1}}{dx^{j-1}} f_i(x), \qquad A_{j,l}^{(i)} = \delta_{j+1,l} - \delta_{j,k_i} \cdot a_{l-1}^{(i)}$$

for $j, l = 1, \ldots, k_i$; $i = 1, \ldots, n$.

THEOREM 4.1. *Let $f_1(x), \ldots, f_n(x)$ be E-functions satisfying linear differential equations over $\mathbf{Q}(x)$. Then for any $\varepsilon > 0$ and a rational number r, $r \neq 0$, which is not a singularity of linear differential equations satisfied by $f_1(x), \ldots, f_n(x)$ there exists a constant $c_0 = c_0(\varepsilon, r, f_1, \ldots, f_n) > 0$ with the following property. For arbitrary rational integers $H_1, \ldots, H_n$ and $H = \max(|H_1|, \ldots, |H_n|)$ if $H_1 f_1(r) + \cdots + H_n f_n(r) \neq 0$, then*

$$|H_1 f_1(r) + \cdots + H_n f_n(r)| > H^{-n+1-\varepsilon}$$

provided that $H \geqslant c_0$.

PROOF OF THEOREM 4.1. Let $f_1 = f_1(x), \ldots, f_n = f_n(x)$ be arbitrary E-functions satisfying linear differential equations over $\mathbf{Q}(x)$, and linear independent over $\mathbf{C}(x)$.

Let f_i satisfy a scalar linear differential equation of the form (4.1′) over $\mathbf{Q}(x)$ of order k_i: $i = 1, \ldots, n$. In the notations above of (4.1), let

$$f_i^{(j)}(x) = \left(\frac{d}{dx}\right)^{j-1} \cdot f_i(x): \qquad j = 1, \ldots, k_i;\ i = 1, \ldots, n.$$

We fix an ε, $0 < \varepsilon < \frac{1}{2}$ and a sufficiently large parameter N, $N \geqslant N_0(f_1, \ldots, f_n, \varepsilon)$.

As in §3, we follow the method of Chapter 9 [**16**], to construct a large family of conjugate auxiliary functions (approximating $f_1(x), \ldots, f_n(x)$) enumerated very similarly to §3, by elements of the lattice $\mathbf{Z}^{k_1} \times \cdots \times \mathbf{Z}^{k_n}$, i.e. by multi-indices $J = (\cdots; \alpha_{i,1}, \ldots, \alpha_{i,k_i}; \cdots)$ from $\mathbf{Z}^{k_1} \times \cdots \times \mathbf{Z}^{k_n}$ with nonnegative integers $\alpha_{i,j}$. To simplify notations we introduce generating functions in n groups of variables $\bar{c}_i = (c_{i,1}, \ldots, c_{i,k_i})$: $i = 1, \ldots, n$ and we put $\bar{c} = (\bar{c}_1, \ldots, \bar{c}_n)$. We consider sublattices of $\mathbf{Z}^{k_1} \times \cdots \times \mathbf{Z}^{k_n}$ very similar to §3. Now let $J = (\cdots; \alpha_{i,1}, \ldots, \alpha_{i,k_i}; \cdots) \in \mathbf{Z}^{k_1} \times \cdots \times \mathbf{Z}^{k_n}$ be a multi-index with nonnegative integers $\alpha_{i,j}$. We put $\|J\|_i = N$ if for any $j \neq i$, $\sum_{l=1}^{k_j} \alpha_{j,l} = N$ while $\sum_{l=1}^{k_i} \alpha_{i,l} = N - 1$; and we put $|J| = N$ if for all $i = 1, \ldots, n$, $\sum_{l=1}^{k_i} \alpha_{i,l} = N$.

We denote (for fixed $k_1, \ldots, k_n$):

$$M_N \stackrel{\text{def}}{=} \sum_{i=1}^{n} \binom{N + k_i - 2}{k_i - 1} \times \prod_{j=1, j \neq i}^{n} \binom{N + K_j - 1}{k_j - 1},$$

$$S_N \stackrel{\text{def}}{=} \prod_{i=1}^{n} \binom{N + k_i - 1}{k_i - 1}.$$

THEOREM 4.2. *For any $D \geqslant D_0(f_1, \ldots, f_n, \varepsilon, N)$ there exist polynomials $P_{i,J}(x)$: $J \in \mathbf{Z}^{k_1} \times \cdots \times \mathbf{Z}^{k_n}$; $\|J\|_i = N$: $i = 1, \ldots, n$, not all zero and with integer rational coefficients such that the following conditions are satisfied.*

(1) *Polynomials $P_{i,J}(x)$ have a degree in x of at most D, and height of at most $D^{D+\varepsilon D}$ and are of the form*

$$P_{i,J}(x) = \sum_{m=0}^{D} \frac{D!}{m!} p_{i,J,m} x^m, \tag{4.2}$$

$\|J\|_i = N$: $i = 1, \ldots, n$. Here $p_{i,J,m}$ are rational integers of absolute values of at most $D^{\varepsilon D}$ $(m = 0, \ldots, D)$. We also formally put $P_{i,J}(x) = 0$ if J has a negative component: $i = 1, \ldots, n$.

(2) *For any $I \in \mathbf{Z}^{k_1} \times \cdots \times \mathbf{Z}^{k_n}$, $|I| = N$, the (remainder) function*

$$R_I(x) \stackrel{\text{def}}{=} \sum_{i=1}^{n} \sum_{j=1}^{k_i} P_{i,I-e_{i,j}}(x) f_i^{(j)}(x) \tag{4.3}$$

has a zero at $x = 0$ of an order of at least M, where

$$f_i^{(j)}(x) = \left(\frac{d}{dx}\right)^{j-1} f_i(x) \quad \text{and} \quad M = \left[\frac{M_N - \varepsilon}{S_N} \cdot D\right].$$

PROOF. Let $f_i(x) = \sum_{m=0}^{\infty} a_{m,i} x^m/m!$ and $f_i^{(j)}(x) = \sum_{m=0}^{\infty} a_{m,i,j} x^m/m!$: $i = 1,\ldots,n$; $j = 0, 1,\ldots,k_i$. We obtain an expansion of $R_I(x)$ at $x = 0$ in terms of $a_{m,i,j}$ and $p_{i,J,m}$. Namely, for

$$R_I(x) = \sum_{m=0}^{\infty} \frac{D!}{m!} r_{I,m} x^m,$$

we obtain the following representation for $r_{I,m}$:

$$r_{I,m} \stackrel{\text{def}}{=} \sum_{i=1}^{n} \sum_{j=1, I(i,j)\geqslant 1}^{k_i} \sum_{l=0}^{\min(m,D)} \binom{m}{l} a_{m-l,i,j} \cdot p_{i,I-e_{i,j},l}: \tag{4.4}$$

$|I| = N$, and $I(i, j)$ denotes the (i, j)-th component of I. The only conditions that determine polynomials $P_{i,J}(x)$: $\|J\|_i = N$: $i = 1,\ldots,n$, are given by the following system of M_N linear equations on coefficients $p_{i,J,m}$:

$$r_{I,m} = 0: \qquad m = 0, 1,\ldots, M - 1;\ |I| = N \tag{4.5}$$

(in the notations of (4.4)). The total number of unknowns $p_{i,J,m}$ in (4.5) is $(D + 1)M_N$. Finally, according to the definition of E-functions, and (4.4), the coefficients at $p_{i,J,m}$ in the system (4.5) of linear equations, are rational numbers whose absolute values and whose common denominator are uniformly bounded by $D^{\delta D}$ for any $\delta > 0$ and $D \geqslant D_1(f_1,\ldots,f_n, \delta)$. Thus we can use Siegel's lemma [**1**, **15**] and find a nontrivial solution $p_{i,J,m}$ of the system of equations (4.5) in rational integers and such that $\max\{|p_{i,J,m}|: \|J\|_i = N: i = 1,\ldots,n;\ m = 0,\ldots,D\} \leqslant D^{\varepsilon D}$ provided that $D \geqslant D_0(f_1,\ldots,f_n, \varepsilon, N)$. Theorem 4.2 is proved.

We construct generating functions of $P_{i,J}(x)$ and $R_I(x)$ to obtain polynomials in auxiliary variables $c_{i,j}$: $j = 1,\ldots,k_i$; $i = 1,\ldots,n$:

$$P_i(x|\bar{c}) = \sum_{J, \|J\|_i = N} P_{i,J}(x) \cdot \bar{c}^J,\ R(x|\bar{c}) = \sum_{I, |I| = N} R_I(x) \cdot \bar{c}^I.$$

If $J(i, j)$ denotes the (i, j)-th component of J, we put $\bar{c}^J = \prod_{i=1}^n \prod_{j=1}^{k_i} c_{i,j}^{J(i,j)}$. For $i = 1,\ldots,n$ and $j = 1,\ldots,k_i$ we denote by $e_{i,j}$ the unit vector from $\mathbf{Z}^{k_1} \times \cdots \times \mathbf{Z}^{k_n}$ with 1 on the (k, j)-th place. Now let $d(x)$ be a common denominator of all rational functions $A_{j,l}^{(i)} = A_{j,l}^{(i)}(x)$: $j, l = 1,\ldots,k_i$; $i = 1,\ldots,n$ in (4.1).

Let $d = \max\{\deg(d(x)) - 1,\ \deg(d(x)A_{j,l}^{(i)}(x))$: $j, l = 1,\ldots,k_i$; $i = 1,\ldots,n\}$. We then define inductively

$$P_i^{\langle m\rangle}(x|\bar{c}) = \sum_{J, \|J\|_i = N} P_{i,J}^{\langle m\rangle}(x)\bar{c}^J,$$

where $P_i^{\langle 0\rangle}(x|\bar{c}) \stackrel{\text{def}}{=} P_i(x|\bar{c})$, and for $m = 0, 1,\ldots$

$$P_{i,J}^{\langle m+1\rangle}(x) = d(x) \cdot \left\{\frac{d}{dx} P_{i,J}^{\langle m\rangle}(x) - \sum_{j,l=1, J(i,j)\geqslant 1}^{k_i} A_{j,l}^{(i)}(x) \right. \\ \left. \times (J(i,l) + 1) P_{i,J+e_{i,l}-e_{i,j}}^{\langle m\rangle}(x)\right\}. \tag{4.6}$$

We obtain new approximating forms

$$R_I^{\langle m\rangle}(x) = \sum_{i=1}^{n} \sum_{j=1,I(i,j)\geq 1}^{k_i} P_{i,I-e_{i,j}}^{\langle m\rangle}(x) f_i^{(j)}(x)$$

with polynomials $P_{i,J}^{\langle m\rangle}(x)$, $\|J\|_i = N$: $i = 1,\ldots,n$ defined inductively for $m \geq 0$ in (4.6). We need a simple upper bound on the sizes of polynomials $P_{i,J}^{\langle m\rangle}(x)$ and upper bounds of $R_I^{\langle m\rangle}(x)$:

LEMMA 4.3. *In the notations above, polynomials $P_{i,J}^{\langle m\rangle}(x)$ have degrees of at most $D + md$ and heights of at most $D^{D+\varepsilon D} \cdot (D + dm)^{2m} \cdot c_1^m$, where $c_1 = c_1(f_1,\ldots,f_n, N) > 0$—depends only on the system of linear differential equations satisfied by $f_i^{(j)}$ and N: $j = 1,\ldots,k_i$; $\|J\|_i = N$, $i = 1,\ldots,n$, and $m = 0, 1,\ldots$. For $m = 0, 1,\ldots$ and given $x \neq 0$ we also have*

$$|R_I^{\langle m\rangle}(x)| \leq m^{3m} \cdot c_2^m \cdot c_3^m M^{(-1+2\varepsilon)M}$$

with $c_2 > 0$, $c_3 > 0$ depending only on the system of linear differential equations satisfied by $f_i^{(j)}(x)$, and N: $|I| = N$ and on x.

PROOF OF LEMMA 4.3. The upper bounds for degrees of $P_{i,J}^{\langle m\rangle}(x)$ and the upper bounds of heights of $P_{i,J}^{\langle m\rangle}(x)$ are direct consequences of the recurrence formula (4.6). To obtain the upper bounds of $R_I^{\langle m\rangle}(x)$ we use the recurrence definition (4.6), from which it follows that $R_I^{\langle m\rangle}(x)$ is a linear combination of functions $(d/dx)^{m'} R_{I'}(x)$ with coefficients bounded in absolute value by $m^{2m} \cdot c_2^m$. To estimate from above, $(d/dx)^{m'} R_{I'}(x)$ at $x \neq 0$, we use property (2) of Theorem 4.2. We have, in the notations of the proof of Theorem 4.2,

$$R_{I'}(x) = \sum_{m=M}^{\infty} \frac{D!}{m!} r_{I',m} x^m.$$

Using definition (4.4) and the upper bounds on heights of polynomials $P_{i,J}(x)$ in (1), Theorem 4.2 we immediately obtain

$$\left|\left(\frac{d}{dx}\right)^{m'} R_{I'}(x)\right| \leq D! \sum_{m=M}^{\infty} (m!)^{2\varepsilon} \frac{|x|^{m-m'}}{(m-m')!}$$

$$\leq m'! \sum_{m=M}^{\infty} (m!)^{-1+2\varepsilon} 2^m |x|^{m-m'} \leq m'! c_3^M (M!)^{-1+2\varepsilon}.$$

This proves Lemma 4.3.

In particular, we obtain a system of "minimal" approximating forms to $f_1(x),\ldots,f_n(x)$ at $x = 0$ if we put $I = I_0$, where $I_0 = (N, 0,\ldots,0;\ldots;N, 0,\ldots,0)$. We get for $m = 0, 1,\ldots,$

$$R_{I_0}^{\langle m\rangle}(x) = \sum_{i=1}^{n} P_{i,I_0-e_{i,1}}^{\langle m\rangle} f_i \stackrel{\text{def}}{=} \sum_{i=1}^{n} P_{i,I_i}^{\langle m\rangle}(x) f_i(x).$$

We need to have n linearly independent functions of the form $R_0^{\langle m\rangle}(x)$ (or linear forms approximating $f_1(x),\ldots,f_n(x)$ in the sense of Siegel [**1**]). These n linearly independent forms always exist as the following general result shows.

THEOREM 4.4. *Let $P_{i,J}^{\langle m\rangle}(x)$ for $\|J\|_i = N$, $i = 1,\dots,n$ and $m \geqslant 0$ be defined as above in* (4.6) *and let $I_i = I_0 - e_{i,1}$, for $I_0 = (N, 0,\dots,0;\dots)$ as above. Then for any $x_0 = 0$ such that $d(x_0) \neq 0$, and sufficiently large $D \geqslant D_0(f_1,\dots,f_n, \varepsilon, N)$, the rank of the following matrix is n:*

$$\left(P_{i,I_i}^{\langle m\rangle}(x_0): i = 1,\dots,n\right): \qquad m = 0,\dots, (n+\varepsilon)\cdot D - M.$$

REMARK. The upper bound for m, $m \leqslant (n+\varepsilon)D - M$ cannot in general be significantly improved (we note that $M \geqslant (n-\varepsilon)D$ for N sufficiently large with respect to ε^{-1}, $N \geqslant N_0(f_1,\dots,f_n, \varepsilon)$). However, as we see below under Siegel's normality condition, this bound improves considerably:

$$m \leqslant M_N \cdot \left\{D + \frac{d}{2}M_N\right\} - \left\{S_N \cdot M - \frac{1}{2}M_N^2\right\} + c_0.$$

This bound is already the best possible (for a sufficiently large D).

For $i = 1,\dots,n$ the equation (4.1) in matrix form is

$$\frac{d}{dx}\varphi_i = A^{(i)}\varphi_i, \tag{4.7}$$

where φ_i is a fundamental $k_i \times k_i$ matrix of solutions of (4.1) with the first column being $(f_i^{(j)}: j = 1,\dots,k_i)^t$. Linear differential equations adjoint to (4.7) have the form

$$\frac{d}{dy}F_i = -A^{(i)t}F_i \tag{4.8}$$

for $F_i(y) \overset{\text{def}}{=} (\varphi_i(y)^t)^{-1}$: $i = 1,\dots,n$.

As usual, for a $k \times k$ matrix B we denote by $P_N(B)$ the N-th induced matrix of B. For any $i = 1,\dots,n$ we denote by $A(i, N)$ the matrix

$$P_N(A_1) \otimes \cdots \otimes P_{N-1}(A_i) \otimes \cdots \otimes P_N(A_n),$$

and similarly we denote

$$F(i, N) \overset{\text{def}}{=} P_N(F_1) \otimes \cdots \otimes P_N(F_{i-1}) \otimes P_{N-1}(F_i) \otimes \cdots \otimes P_N(F_n).$$

Then $F(i, N)$ satisfies the natural differential equation

$$\frac{d}{dy}F(i, N) = -A(i, N)^t F(i, N): \qquad i = 1,\dots,n. \tag{4.9}$$

Our approach follows Siegel's original studies [**1**], and, following Siegel, we introduce normality conditions on (symmetric) powers of solutions of linear differential equations (4.8). Along the lines of [**1**] we call the system of functions $f_i(x)$: $i = 1,\dots,n$ normal, if the fundamental matrices $F_i(y)$: $i = 1,\dots,n$ of (4.8), and the corresponding fundamental matrices $F(i, N)$: $i = 1,\dots,n$ of (4.9) for $N \geqslant 1$, are linearly independent over $\mathbf{C}(y)$. The linear independence of $F_i = (F_{i,j,l})_{j,l=1}^{k_i}$: $i = 1,\dots,n$ over $\mathbf{C}(y)$ means that any linear relation $\sum_{i=1}^n \sum_{j,l=1}^{k_i} p_{i,j}(y) C_{i,l} F_{i,j,l}(y) \equiv 0$ with constants $C_{i,l}$ and polynomials $p_{i,j}(y)$ implies that all products $p_{i,j}C_{i,l}$ are zero.

We point to criterion of [4] (Example 12 of Section 6, Chapter VI) of normality of $f_i(x)$: $i = 1,\dots,n$ in terms of algebraic independence of elements of matrices $F_i(y)$: $i = 1,\dots,n$ only.

THEOREM 4.5. *Let, in the notations above, the system of functions $f_j(x)$: $i = 1,\dots,n$ be normal. Let $P_{i,J}^{\langle m\rangle}(x)$ for $\|J\|_i = N$: $i = 1,\dots,n$ and $m \geqslant 0$ be defined as above in* (4.6). *Then, if*

$$M > D\cdot(M_N - 1)/S_N - D/S_N^2 + \varepsilon D \text{ and } D \geqslant D_1(f_1,\dots,f_n, \varepsilon, N),$$

the determinant of the $M_N \times M_N$ matrix

$$M(x) = \left(P_{i,J}^{\langle m\rangle}(x): \|J\|_i = N;\, i = 1,\dots,n\right)_{m=0,\dots,M_N-1}$$

is not identically zero.

COROLLARY 4.6. *Let us preserve all the notations of Theorem* 4.5. *If*

$$M > D\cdot(M_N - 1)/S_n - D/S_N^2 + \varepsilon D$$

and $D \geqslant D_2(f_1,\dots,f_n, \varepsilon, N)$, and if $x_0 \neq 0$ is not a singularity of the system of equations (4.1) (*i.e. $d(x_0) \neq 0$*), *then for $M' = M_N\cdot\{D + dM_N/2\} - \{S_N\cdot M - \frac{1}{2}M_N^2\} + c_0$, the rank of the matrix*

$$\left(P_{i,J}^{\langle m\rangle}(x_0): \|J\|_i = N;\, i = 1,\dots,n\right)_{m=0,\dots,M'}$$

is exactly M_N.

It follows from Corollary 4.6 that the rank of the matrix

$$\left(P_{i,I_i}^{\langle m\rangle}(x_0): i = 1,\dots,n\right)_{m=0,\dots,M'}$$

is n, which is an improvement over Theorem 4.4 (see the Remark after Theorem 4.4).

PROOF OF THEOREM 4.5. Let $\bar{c}_i(y) = (c_{i,1},\dots,c_{i,k_i})$ be a solution of (4.8) with some initial conditions $\bar{s}_i$:

$$\bar{c}_i(y)^t = F_i(y)\cdot\bar{s}_i^t: \qquad i = 1,\dots,n.$$

We consider the following symmetric product of vectors

$$\bar{c}_1(y),\dots,\bar{c}_n(y): \bar{c}(i, N) \stackrel{\text{def}}{=} \bar{c}(y)^{*N}\otimes\cdots\otimes\bar{c}_i(y)^{*(N-1)}\otimes\cdots\otimes\bar{c}_n(y)^{*N},$$

where $\bar{c}(y)^{*N} = \bar{c}_i(y) * \cdots * \bar{c}_i(y)$ is N-th symmetric power of $\bar{c}_i(y)$: $i = 1,\dots,n$. The vector $\bar{c}(i, N)$ enumerates all monomials $\bar{c}^J$ in $P(i\,|\,\bar{c})$. We substitute $\bar{c}(i, N) = \bar{c}(i, N)(y)$ into the generating function $P_i(x\,|\,\bar{c})$ and put $x = y$:

$$P_i(y) \stackrel{\text{def}}{=} P_i(y\,|\,\bar{c}(i, N)): \qquad i = 1,\dots,n.$$

Then $P_i(y) = \Sigma_{J,\|J\|_i=N} P_{i,J}(y)\cdot\bar{c}(y)^J$ (is a scalar product of $\bar{c}(i, N)$ and of the vector $(P_{i,J}(y)$: $\|J\|_i = N)$). Since $\bar{c}(i, N)(y)$ satisfies a matrix first order system (4.9) any differentiation of $P_i(y)$ is again a linear combination of $\bar{c}(y)^J$ (with rational function coefficients). Moreover the comparison of (4.8), (4.9) with (4.6) immediately implies

$$\left(d(y)\frac{d}{dy}\right)^m P_i(y) = \sum_{\|J\|_i=N} P_{i,J}^{\langle m\rangle}(y)\cdot \bar{c}(y)^J, \tag{4.10}$$

$i = 1,\dots,n$ for $m = 0, 1,\dots.$

We use a property of normal systems of functions, cf. [**1**]. Let, as in [**1**], we have r systems of linear differential equations $dY_{k,t}/dy = \sum_{l=1}^{m_t} Q_{k,l,t}(y)Y_{l,t}$ ($k = 1,\dots,m_t$; $t = 1,\dots,r$) regular at $y = 0$, and $T = T(y)$ is a common denominator of rational functions $Q_{k,l,t}(y)$. For fixed polynomials $P_{k,t}(y)$ and an arbitrary linear combinations R of functions $P_{k,t}(y)Y_{k,t}(y)$,

$$R = \sum_{t=1}^{r}\sum_{k=1}^{m_t} C_{k,t}P_{k,t}(y)Y_{k,t}(y)$$

we define, using the differential equation for $Y_{k,t}(y)$:

$$R^{\langle m\rangle} = \left(T(y)\frac{d}{dy}\right)^m R,\ R^{\langle m\rangle} = \sum_{t=1}^{r}\sum_{k=1}^{m_t} C_{k,t}P_{k,t}^{\langle m\rangle}(y)Y_{k,t}(y).$$

Let

$$q = \max\{\deg(T(y)),\ \deg(T(y)Q_{k,l,t}(y))\}$$

and

$$\max\{\deg(P_{k,t}(y)) \colon k = 1,\dots,m_t;\ t = 1,\dots,r\} \leqslant D.$$

In these notations we have

LEMMA 4.7. *If there are s linearly independent functions of the form R, each of which has a zero at $y = 0$ of order at least u, then always*:

$$us - \frac{1}{2}\left\{\sum_{t=1}^{r} m_t\right\}^2 \leqslant p + \sum_{t=1}^{r} m_t\cdot\left\{D + \frac{q}{2}\left(\sum_{t=1}^{r} m_t\right)\right\}.$$

Now if

$$us - \frac{1}{2}\left\{\sum_{t=1}^{r} m_t - 1\right\}^2 > p + \left\{\sum_{t=1}^{r} m_t - 1\right\}\cdot\left\{D + \frac{q}{2}\cdot\left(\sum_{t=1}^{r} m_t - 1\right)\right\},$$

then the determinant $\Delta(y) = \det(P_{k,t}^{\langle m\rangle}(y)$: $k = 1,\dots,m_t$; $t = 1,\dots,r)$, $m = 0, 1,\dots,\sum_{t=1}^{r} m_t - 1$, *is not identically zero and has a zero at* $y = 0$ *of an order of at least*: $us - \frac{1}{2}\{\sum_{t=1}^{r} m_t\}^2 - p$.

Here p depends only on the maximal order of zeros of $Y_{k,t}$ at $y = 0$.

Lemma 4.7 is one of the versions of Normality lemma 1 of §2. For the proof cf. [**1, 5, 6**]. For exponential functions a similar assertion is Lemma 3.9.

We apply Lemma 4.7 to the following (normal) systems of functions. They are various symmetric products

$$\bar{c}(i, N) = \bar{c}_1(y)^{*N}\otimes\cdots\otimes\bar{c}_i(y)^{*(N-1)}\otimes\cdots:\qquad i = 1,\dots,n.$$

Normality of systems of functions $\bar{c}(i, N)$ and Lemma 4.7 imply that various functions R that are linear combinations of $P_{i,J}(y)\bar{c}(y)^J$ are linearly independent (by the assumption and Theorem 4.2). Then it follows from Lemma 4.7 that $\det M(x) \not\equiv 0$. Theorem 4.5 is proved.

PROOF OF COROLLARY 4.6. According to Theorem 4.5 $\det M(x) \not\equiv 0$. From the definition of $M(x)$ and Lemma 4.3 it follows that $\det M(x)$ is a polynomial in x of degree of at most $M_N D + dM_N^2/2$, since $\deg_x(P_{i,J}^{\langle m\rangle}(x)) \leqslant D + md$.

Moreover, from the proof of Theorem 4.5 it follows that $\det M(x)$ has a zero at $x = 0$ of order of at least $M \cdot S_N - \frac{1}{2}M_N^2 - p_0$, where p_0 depends only on $f_1, \ldots, f_n$. Thus $\det M(x) = x^a \cdot \Delta_0(x)$ is nonzero polynomial of degree of at most $M_0 \leqslant M_N \cdot D - S_N \cdot M + \frac{1}{2}M_N^2(1 + d) + p_0$. For a given $x_0 \neq 0$, let α be the order of zero of $\Delta_0(x)$ at $x = x_0$. Applying the differential operator $d(y)d/dy$ to

$$\left(d(y)\frac{d}{dy}\right)^m P_i(y) = \sum_{J, |J| = N} P_{i,J}^{\langle m\rangle}(y)\bar{c}(y)^J$$

a times we obtain, starting from $M(x_0)$, a matrix $(P_{i,J}^{\langle m\rangle}(x_0)$: $\|J\|_i = N$: $i = 1, \ldots, n)$: $m = 1, \ldots, M_N + \alpha - 1$ having rank M_N. Since $\alpha \leqslant M_0$, Corollary 4.6 is proved. For Theorem 4.4 we apply the representation (4.10) of $P_i(y)$ with $\bar{c}$ having the following initial conditions: $c_{i,j}(y)|_{y=x_0} = \delta_{j1} \cdot f_i(x_0)$.

We obtain now, following the method of §2, a system of linear forms approximating numbers $f_1(r), \ldots, f_n(r)$, $r \neq 0$.

THEOREM 4.8. *Let $r \neq 0$ be a rational number different from singularities of the system* (4.1). *Then for any $\delta > 0$ and any sufficiently large integer D, $D \geqslant D_3(f_1, \ldots, f_n, \delta)$ there exist rational integers $P_{i,j}$: $i, j = 1, \ldots, n$ such that the following conditions are satisfied. The determinant of $P_{i,j}$ — $\det(P_{i,j})_{i,j=1}^n$ is nonzero; the absolute values of numbers $P_{i,j}$ are bounded by $D^{(1+\delta)D}$ and*

$$\left|\sum_{j=1}^n P_{i,j} f_j(r)\right| \leqslant D^{(1+\delta-n)D}: \qquad i, j = 1, \ldots, n.$$

PROOF OF THEOREM 4.8. Let, as above, D be sufficiently large with respect to N and ε^{-1}. We use Lemma 4.3 for $I = I_0 = (N, 0, \ldots, 0; N, 0, \ldots, 0; \ldots; N, 0, \ldots, 0)$ and $I_i = I_0 - e_{i,1}$: $i = 1, \ldots, n$. According to Theorem 4.4, the matrix $(P_{i,I_i}^{\langle m\rangle}(r)$: $i = 1, \ldots, n)$: $m = 0, 1, \ldots, M_0$ has rank n, where $M_0 \leqslant (n + \varepsilon)D - M$ and a sufficiently large $D \geqslant D_4(f_1, \ldots, f_n, \varepsilon, N)$. Then there are n distinct integers $0 \leqslant m_1 < \cdots < m_n \leqslant M_0$ such that $\det(P_{i,I_i}^{\langle m_j\rangle}(r))_{i,j=1}^n \neq 0$. If $r = a/b$ for rational integers a and b, then, according to Lemma 4.3, $P_{i,I_i}^{\langle m_j\rangle}(a/b) \cdot b^{D+m_jd}$ is a rational integer and we define $P_{i,j} = P_{j,I_j}^{\langle m_i\rangle}(a/b) \cdot b^{D+m_id}$: $i, j = 1, \ldots, n$. Then $P_{i,j}$ are rational integers with $\det(P_{i,j})_{i,j=1}^n \neq 0$. From Lemma 4.3 and the upper bound $M_0 \leqslant (n + \varepsilon)D - M$ it follows that for D sufficiently large with respect to N, and N sufficiently large with respect to δ^{-1} we obtain

$$\max\{|P_{i,j}| : i, j = 1, \ldots, n\} < D^{D+\delta D}.$$

We have

$$R_{I_0}^{\langle m\rangle}(x) = \sum_{i=1}^{n} P_{i,I_i}^{\langle m\rangle}(x) f_i(x).$$

Substituting $x = r$ and using the upper bound of $R_{I_0}^{\langle m\rangle}(r)$ of Lemma 4.3 we obtain for D sufficiently large with respect to N and N sufficiently large with respect to δ^{-1},

$$\left|\sum_{j=1}^{n} P_{i,j} f_j(r)\right| \leq D^{-nD+(1+\delta)D}.$$

Theorem 4.8 is proved.

According to Siegel [**1**], the existence of n linearly independent approximating forms satisfying the conditions of Theorem 4.8 gives the lower bound for the measure of linear independence of numbers $f_1(r),\ldots,f_n(r)$.

Let us complete the proof of Theorem 4.1. Let $H_1,\ldots,H_n$ be rational integers not all zero such that $\max(|H_i|: i = 1,\ldots,n) = H$ and such that

$$H_1 f_1(r) + \cdots + H_n f_n(r) = l: \qquad 0 < |l| < 1.$$

We apply Theorem 4.8 and obtain for any $\delta > 0$ and for the integer $D \geq D_3(f_1,\ldots,f_n, \delta)$ n linearly independent forms in $f_1(r),\ldots,f_n(r)$ with rational integer coefficients of the form

$$L_i = \sum_{j=1}^{n} P_{i,j} f_j(r): \qquad i = 1,\ldots,n$$

such that $\max(|P_{i,j}|: i, j = 1,\ldots,n) \leq D^{(1+\delta)D}$ and $\max(|L_i|: i = 1,\ldots,n) \leq D^{(1+\delta-n)D}$. The linear independence of forms $L_1,\ldots,L_n$ means that $\det(P_{i,j})_{i,j=1}^{n} \neq 0$. Hence we can always find $n-1$ forms, say, $L_1,\ldots,L_{n-1}$, that are linearly independent together with the form l. Let V be a matrix formed from coefficients of forms $L_1,\ldots,L_{n-1}, l$; let Δ_1 be a determinant of V, and $\Delta_{i,j}$ be the algebraic complement of the (i, j)-th element of V. Then

$$f_i(r)\Delta_1 = \sum_{j=1}^{n-1} L_j \Delta_{j,i} + l\Delta_{n,i}: \qquad i = 1,\ldots,n.$$

From the definition of V it follows that

$$|\Delta_{n,i}| \leq (n-1)!D^{(n-1)(1+\delta)D},$$

$$\max\big(|\Delta_{j,i}| : j = 1,\ldots,n-1\big) \leq (n-1)!H \cdot D^{(n-2)(1+\delta)D}: \qquad i = 1,\ldots,n.$$

Let $i = 1, 2,\ldots,n$ be chosen in a way such that $f_i(r) \neq 0$, which is possible since $l \neq 0$. Then

$$|f_i(r)\Delta_1| \leq n!\left(H \cdot D^{(-1+\delta(n-1))D} + |l| D^{(n-1+\delta(n-1))D}\right).$$

Since $\Delta_1 \neq 0$ and all elements of V are rational integers, $|\Delta_1| \geq 1$. We choose now an integer D to be the smallest integer $\geq D_3$ such that $D^{nD}|l| \geq H$.

For this definition of D we get $|f_i(r)| \leqslant 2n! \cdot |l| \cdot D^{(n-1)\cdot(1+\delta)D}$. This implies a lower bound on $|l|$:

$$|l| \geqslant c_4 H^{-(n-1)(1+\delta)/(1+\delta(n-1))}$$

and $c_4 = c_4(f_1, \ldots, f_n, \delta, r)$. For a sufficiently small δ this proves Theorem 4.1.

REFERENCES

1. C. L. Siegel, *Über einige Anwendungen diophantischer Approximationen*, Abh. Preuss. Akad. Wiss. no. 1 (1929).

2. G. V. Chudnovsky, *Padé approximations to the generalized hypergeometric functions*. I, J. Math. Pure Appl., **58** (1979), 445–476.

3. C. L. Siegel, *Transcendental numbers*, Princeton, 1949.

4. E. L. Kolchin, *Algebraic groups and algebraic dependence*, Amer. J. Math. **90** (1968), 1151–1164.

5. A. Baker, *Transcendental number theory*, Cambridge Univ. Press, Cambridge, 1979.

6. K. Mahler, *Lectures on transcendental numbers*, Lecture Notes in Math., vol. 546, Springer, New York, 1976.

7. C. J. Moreno, *The values of exponential polynomials at algebraic points*. I, Trans. Amer. Math. Soc. **186** (1973), 17–31.

8. G. V. Chudnovsky, *A new method for investigation of arithmetic properties of analytic functions*, Ann. of Math. (2) **109** (1979), 353–377.

9. ______, *Values of meromorphic functions of order* 2, Séminaire Delange-Pisot-Poitou, No. 45, 1977-78, pp. 4501–4518.

10. P. L. Cijsouw, *Transcendence measures*, Amsterdam, 1972.

11. G. V. Chudnovsky, *Algebraic independence of the values of elliptic function at algebraic point. Elliptic analogue of Lindemann-Weierstrass theorem*, Invent. Math. **61** (1980), 267–280.

12. ______, *Indépendénce algébrique dans la méthode de Gelfond-Schneider*, C. R. Acad. Sci. Paris **291A** (1980), 485–487.

13. C. L. Siegel, *Die Gleichung* $ax^n - by^n = c$, Math. Ann. **114** (1937), 57–68.

14. G. V. Chudnovsky, *On the method of Thue-Siegel*, Ann. of Math. (2) **117** (1983), 325–382.

15. G. V. Chudnovsky, see Chapter 7, this volume.

16. ______, see Chapter 9, this volume.

17. ______, *Rational and Padé approximations to solutions of linear differential equations and the monodromy theory*, Lecture Notes Physics, vol. 126, Springer, New York, 1980, pp. 136–169.

18. A. O. Gel'fond, *Sur la septieme probleme de Hilbert*, Izvestia Acad. Sci. U.S.S.R. Math., **7** (1934), 623–630.

19. ______, *Transcendental and algebraic numbers*, Dover, New York, 1960.

20. H. M. Stark, *Further advances in the theory of linear forms in logarithms*, in Diophantine Approximations and its Applications, Academic Press, London, 1973, pp. 255–293.

21. C. L. Siegel, *Approximation algebraischer Zahlen*, Math. Zeit. **10** (1921), 173–213.

22. S. Lang, *Diophantine geometry*, Addison-Wesley, Reading, Mass., 1962.

23. ______, *Report on Diophantine approximations*, Bull. Soc. Math. France **93** (1965), 177–192.

24. V. A. Tartakovskiĭ, *A uniform estimate of the number of representations of a unit by a binary form of degree* $n \geqslant 3$, Soviet Math. Dokl. **11** (1970), 1026–1027.

25. L.-C. Kappe, *Zur Approximation von* e^α, Ann. Univ. Sci. Budapest Eötvös Sect. Math. **9** (1966), 3–14.

26. G. V. Chudnovsky, *Rational approximations to linear forms of exponentials and binomials*, Proc. Nat. Acad. Sci. U.S.A. **80** (1983), 3138–3141.

CHAPTER 6

SOME DIOPHANTINE PROBLEMS

0. It is know [**1**] that any recursively enumerable set S admits a representation in the form

$$S = \{x\colon (\exists y_1)(\forall y_2 \leqslant y_1)(\exists y_3) \cdots (\exists y_m)(P(x, n, y_1, \ldots, y_m) = 0)\},$$

where n is a fixed natural number depending on S and $P(x, n, y_1, \ldots, y_m)$ is a polynomial in $m + 2$ variables with integral coefficients.[1] In addition, according to [**2**], we can take $m = 6$.

Using this result, M. Davis, H. Putnam, and J. Robinson [**1**] and M. Davis [**4**] showed that any recursively enumerable set S can be represented in the form

$$S = \{x\colon (\exists x_1) \cdots (\exists x_n)(\exists u)(\exists v)((P(x, x_1, \ldots, x_n, u, v) = 0) \,\&\, \rho(u, v))\},$$

where $\rho(u, v)$ is an arbitrary predicate of exponential growth (a predicate of J. Robinson [**1**]).

From these and other results and from the existence of a nonrecursive, recursively enumerable set we obtain the algorithmic unsolvability of many problems in the theory of numbers, e.g. the problem of recognizing the existence of integral solutions of exponential Diophantine equations and the problem of recognizing when a polynomial with integral coefficients assumes powers of 2 as values.

It follows from the above-mentioned results that to prove the algorithmic unsolvability of the problem of recognizing the existence of integral solutions of Diophantine equations (i.e. to finally solve Hilbert's tenth problem) it suffices to show that at least one predicate of essentially exponential growth is Diophantine, i.e. existentially definable. It was in just this way that Davis and Putnam

[1]As early as 1931, Gödel showed that any recursively enumerable predicate admits an arithmetical representation.

established the unsolvability of a Diophantine problem in the polynomial ring $\mathbf{J}[\xi]$. They proved that the predicate $\rho(u, v) \leftrightarrow (v = 2_u \,\&\, u > 3)$ distinguishing solutions x of the Pell equation $x^2 - 3y^2 = 1$ is Diophantine in $\mathbf{J}[\xi]$.

In this present paper we show that the predicate $x = a_z$ distinguishing the zth solution x of the Pell equation $x^2 - (a^2 - 1)y^2 = 1$ is Diophantine. Since the predicate $x = a_z$ has exponential growth, any recursively enumerable set is Diophantine and, therefore, a Diophantine problem is algorithmically unsolvable.[2]

We remark that the question of the Diophantine nature of the predicate $x = a_z$ was raised in 1952 by J. Robinson [**3**].

The above result is deduced in §2 from properties considered in §§1 and 2 of certain recursion sequences A_n, B_n of the type appearing in the numerators and denominators of convergents of quadratic irrationalities. A brief account of the main result (Theorem 2.6) is given in [**9**].

In §3 we continue the investigation of the properties of such sequences (we study their periodicity mod A_n, B_n) and in §4 we obtain, on the bases of §§1, 3 and 4, a general result concerning an explicit Diophantine representation of recursion sequences u_n satisfying the equation $u_{n+2} = ku_{n+1} - u_n$.

In §4 we also give some consequences of our results (any recursively enumerable predicate is Diophantine, etc.).

On the basis of §2, we give in §5 an explicit Diophantine representation of the function $x = y^z$ and a Diophantine characterization of the Mersenne primes $2^n - 1$, and we indicate the possibility of a direct (not using §§3 and 4) solution of a problem in [**7**].

Model-theoretic aspects of the Diophantine nature of recursively enumerable sets are considered in §6. Here, in terms of ultrapowers, we characterize the Diophantine nature (or recursive enumerability) of a predicate and give applications to strong models of arithmetic.

In §7 we apply the results of elementary number theory in §§1–3 to specific Diophantine equations of the type $x^2 - dy^4 = 1$, $x^4 - dy^2 = 1$, and the like [**10**]. Here we also give a result on the solvability of the equation $1^n + \cdots + (m-1)^n = \lambda m^n$ for various values of λ. This answers a number of questions in [**6**].

1. The facts from elementary number theory which will be used below can be found in [**6**] (see also [**8**]).

In this section we study properties of the sequences of numerators and denominators of the convergents of a quadratic irrationality of a special form.

Here and below, the symbols a, b, c, d, m, n, k denote nonnegative integers (where stipulated, they may denote arbitrary integers).

Suppose α is irrational and

$$\alpha = d + \frac{1|}{|1} + \frac{1|}{|b} + \frac{1|}{|1} + \frac{1|}{|b} + \cdots, \tag{1}$$

---

[2] A similar result was obtained by Ju. V. Matijasevič [**15**].

i.e. α can be expanded into an infinite continued fraction with period 2. Symbolically, in accordance with [**6**], $\alpha = [d, \overline{1, b}]$.

Using simple properties of continued fractions [**6, 8**], we obtain

$$\alpha = d + \frac{1|}{|1} + \frac{1|}{|b - d + \alpha}.$$

Therefore,

$$\alpha = d + \frac{1}{1 + 1/(b - d + \alpha)} = \frac{db - d^2 + d\alpha + b + \alpha}{b - d + \alpha + 1};$$

$$\alpha^2 + (b - 2d)\alpha - db + d^2 - b = 0 \tag{2}$$

and $\alpha = (2d - b + \sqrt{b^2 + 4b})/2$, where $\alpha > 0$ since $[\alpha] = d$.

Let β denote the second root of (2).

Continued fractions of the form (1) with $d = b$ were studied in detail in [**6**].

EXAMPLE 1. Let $\alpha = \sqrt{a^2 - 1}$. Then, according to [**6**], we can expand α into a continued fraction

$$\alpha = a - 1 + \frac{1|}{|1} + \frac{1|}{|2a - 2} + \frac{1|}{|1} + \frac{1|}{|2a - 2} + \cdots, \tag{3}$$

i.e. an expansion of the form (1) with $d = a - 1$, $b = 2a - 2$.

Let us describe the sequences A_i, B_i (depending on α) with which we will work in the sequel.

Put

$$\begin{gathered} A_0 = 1, \qquad A_1 = d, \\ A_{2k+1} = bA_{2k} + A_{2k-1}, \qquad A_{2k+2} = A_{2k+1} + A_{2k}; \end{gathered} \tag{4}$$

$$B_0 = 0, \quad B_1 = 1, \quad B_{2k+1} = bB_{2k} + B_{2k-1}, \quad B_{2k+2} = B_{2k+1} + B_{2k}. \tag{5}$$

According to the rules of formation of the numerators and denominators of convergents [**6, 8**], it follows from (4) and (5) that A_{i+1} and B_{i+1} are, respectively, the numerator and denominator of the ith convergent of α.

We give the first few terms of the sequences A_n, B_n:

$$\begin{gathered} A_0 = 1, \quad A_1 = d, \quad A_2 = d + 1, \quad A_3 = bd + b + d, \\ A_4 = bd + b + 2d + 1; \\ B_0 = 0, \quad B_1 = 1, \quad B_2 = 1, \quad B_3 = b + 1, \quad B_4 = b + 2. \end{gathered}$$

In the sequel it will sometimes be necessary to consider separately the sequences A_{2n}, A_{2n+1} and B_{2n}, B_{2n+1}, because of their different properties (see [**6**]).

LEMMA 1.1.

$$\begin{aligned} A_{2n+2} &= (b + 2)A_{2n} - A_{2n-2}; \qquad & A_{2n+3} &= (b + 2)A_{2n+1} - A_{2n-1}; \\ B_{2n+2} &= (b + 2)B_{2n} - B_{2n-2}; \qquad & B_{2n+3} &= (b + 2)B_{2n+1} - B_{2n-1}. \end{aligned}$$

PROOF. According to (4), we have for the sequence A_{2n}:

$$A_{2n+2} = A_{2n+1} + A_{2n}.$$

Also, in view of (4),

$$A_{2n} - A_{2n-1} - A_{2n-2} = 0, \qquad A_{2n+1} - bA_{2n} - A_{2n-1} = 0.$$

It follows that

$$A_{2n+1} = (b+1)A_{2n} - A_{2n-2}.$$

Then

$$A_{2n+2} = (b+2)A_{2n} - A_{2n-2}.$$

The lemma can be proved for A_{2n+1}, B_{2n}, B_{2n+1} in a similar way.

It is clear from 1.1 that the sequences A_{2n+1}, A_{2n} and B_{2n+1}, B_{2n} satisfy the equation

$$u_{n+2} = ku_{n+1} - u_n; \tag{6}$$

where $k = b + 2$.

We isolate some easy results concerning sequences u_n satisfying (6), in order to simplify the exposition (see [**5**]).

PROPOSITION 1.2. *If the sequences u_n, v_n satisfy* (6) *and $u_0 = v_0$, $u_1 = v_1$, then $u_n = v_n$ for any n; if $u_n \equiv v_n \pmod{a}$ for $n = 0, 1$, then $u_n \equiv v_n \pmod{a}$ for any n.*

PROOF. If $u_i = v_i$ whenever $i < n$, $n \geqslant 2$, then $u_n = ku_{n-1} - u_{n-2} = v_n$. If $u_i \equiv v_i \pmod{a}$, then $u_n \equiv ku_{n-1} - u_{n-2} \equiv v_n \pmod{a}$.

In particular, if $a \mid u_0$ and $a \mid u_1$, then $a \mid u_n$ for any n.

PROPOSITION 1.3. *If the sequences u_n, v_n satisfy* (6) *and c, d are integers, then $cu_n + dv_n$ satisfies* (6).

PROOF. It is obvious that

$$cu_{n+2} + dv_{n+2} = k(cu_{n+1} + dv_{n+1}) - (cu_n + dv_n).$$

Note that if the sequences u_n, v_n satisfy (6) and $u_0v_1 \neq v_0u_1$, then any other sequence w_n satisfying (6) can be written in the form

$$w_n = xu_n + yv_n, \tag{7}$$

where x, y are rational integers determined by the conditions $w_i = xu_i + yv_i$, $i = 0, 1$.

Indeed, since $u_0v_1 \neq v_0u_1$, such x, y exist. Then the sequence $w_n' = xu_n + yv_n$ satisfies (6) by virtue of 1.3. But $w_i' = w_i$ for $i = 0, 1$, hence, by 1.2, $w_n' = w_n$, i.e. we have (7).

We apply this observation to the sequences A_{2n}, B_{2n} and A_{2n+2}, B_{2n+2}.

LEMMA 1.4.

$$A_{2n+2} = (d+1)A_{2n} + (bd+b-d^2)B_{2n},$$
$$B_{2n+2} = A_{2n} + (b-d+1)B_{2n}.$$

PROOF. It suffices to verify the formulas for $n = 0, 1$.
For $n = 0$ they are obvious, and for $n = 1$ we have

$$A_4 = bd + b + 2d + 1 = (d+1)^2 + (bd + b - d^2);$$
$$B_4 = b + 2 = d + 1 + (b - d + 1).$$

LEMMA 1.5.

$$A_{2n+1} = dA_{2n} + (bd+b-d^2)B_{2n};$$
$$B_{2n+1} = A_{2n} + (b-d)B_{2n}.$$

PROOF. In view of (4), (5), and Lemma 1.4,

$$A_{2n+1} = A_{2n+2} - A_{2n} = dA_{2n} + (bd+b-d^2)B_{2n};$$
$$B_{2n+1} = B_{2n+2} - B_{2n} = A_{2n} + (b-d)B_{2n}.$$

We give below some explicit expressions for A_n, B_n.

LEMMA 1.6. *The integers* A_{2n}, B_{2n} *are defined by the formula*

$$(A_2 - \alpha B_2)^n = A_{2n} - B_{2n}\alpha, \tag{8}$$

where α *is a root of* (2).

In view of the explicit expression for α, formula (8) can be rewritten as follows:

$$\left(\frac{b+2-\sqrt{b^2+4b}}{2}\right)^n = A_{2n} - B_{2n}\frac{2d-b}{2} - B_{2n}\frac{\sqrt{b^2+4b}}{2}. \tag{9}$$

PROOF OF LEMMA 1.6. If $n = 0$, we have $A_0 = 1$, $B_0 = 0$ and (8) holds.
When $n = 1$, the equality (8) is obvious.
Assume that (8) has been proved for n. Then for $n+1$ we have

$$\begin{aligned}(A_2 - B_2\alpha)^{n+1} &= (A_{2n} - B_{2n}\alpha)(A_2 - B_2\alpha)\\ &= A_{2n}A_2 - A_{2n}B_2\alpha - B_{2n}A_2\alpha + B_2B_{2n}\alpha^2.\end{aligned}$$

In view of Lemma 1.4,

$$\begin{aligned}(A_2 - B_2\alpha)^{n+1} &= A_{2n}(d+1) - A_{2n}\frac{2d-b+\sqrt{b^2+4b}}{2}\\ &\quad - B_{2n}(d+1)\frac{2d-b+\sqrt{b^2+4b}}{2}\\ &\quad + B_{2n}\frac{b^2+2d^2-2bd+2b+(2d-b)\sqrt{b^2+4b}}{2}\\ &= A_{2n}\left(d+1-\frac{2d-b}{2}\right)\\ &\quad + B_{2n}\left(bd+b-d^2-\frac{(b-d+1)(2d-b)}{2}\right)\\ &\quad - A_{2n}\frac{\sqrt{b^2+4b}}{2} - B_{2n}\frac{(b-d+1)\sqrt{b^2+4b}}{2}\\ &= A_{2n+2} - B_{2n+2}\frac{2d-b}{2} - B_{2n+2}\frac{\sqrt{b^2+4b}}{2}.\end{aligned}$$

Using Lemmas 1.5 and 1.6 we can prove

PROPOSITION 1.7. *For any natural number n,*

$$\left(\frac{b-\sqrt{b^2+4b}}{2}\right)^n = b^{[n/2]}(A_n - B_n\alpha). \tag{10}$$

PROOF. (a) Suppose $n = 2m$; then $[n/2] = m$. By Lemma 1.4,

$$\begin{aligned}\left(\frac{b-\sqrt{b^2+4b}}{2}\right)^{2m} b^{-m} &= \left(\frac{b^2+2b-b\sqrt{b^2+4b}}{2b}\right)^m\\ &= \left(\frac{b+2-\sqrt{b^2+4b}}{2}\right)^m = A_n - B_n\alpha.\end{aligned}$$

(b) Suppose $n = 2m+1$; then $[n/2] = m$. In view of (a) and Lemma 1.5,

$$\begin{aligned}\left(\frac{b-\sqrt{b^2+4b}}{2}\right)^{2m+1} b^{-m} &= (A_{2m} - B_{2m}\alpha)\left(\frac{b-\sqrt{b^2+4b}}{2}\right)\\ &= \frac{bA_{2m}}{2} - \frac{\alpha b B_{2m}}{2} - \frac{A_{2m}\sqrt{b^2+4b}}{2} + \frac{\alpha B_{2m}\sqrt{b^2+4b}}{2}\\ &= A_{2m}\frac{b}{2} + B_{2m}\frac{(d-b)\sqrt{b^2+4b}}{2} + B_{2m}\frac{b^2+2b-db}{2} - A_{2m}\frac{\sqrt{b^2+4b}}{2}\\ &= A_{2m+1} - B_{2m+1}\frac{2d-b}{2} - B_{2m+1}\frac{\sqrt{b^2+4b}}{2} = A_n + B_n\alpha.\end{aligned}$$

Proposition 1.7 can be used to establish many properties of the sequences A_n, B_n (see [**5**]).

PROPOSITION 1.8.

$$(A_i, B_i) = 1, \qquad (A_i, A_{i+1}) = 1, \qquad (B_i, B_{i+1}) = 1;$$
$$B_i \mid B_{ij} \quad \textit{for any } i,\ j.$$

PROOF. We use the following well-known identity for continued fractions [**6**, **8**]:

$$A_iB_{i+1} - B_iA_{i+1} = (-1)^i.$$

This implies that $(A_i, B_i) = 1$, $(A_i, A_{i+1}) = 1$, $(B_i, B_{i+1}) = 1$.

We will show that $B_i \mid B_{ij}$ with the aid of Proposition 1.7. From (10) we obtain

$$\begin{aligned} A_{ij} - B_{ij}\alpha &= A_{ij} - B_{ij}\frac{2d-b}{2} - B_{ij}\frac{\sqrt{b^2+4b}}{2} \\ &= b^{-[ij/2]}\left(b^{[i/2]}(A_i - \alpha B_i)\right)^j \\ &= b^{j[i/2]-[ij/2]}\left(A_i - B_i\frac{2d-b}{2} - B_i\frac{\sqrt{b^2+4b}}{2}\right)^j. \end{aligned}$$

Consequently,

$$\begin{aligned} (11)\quad B_{ij} = b^{j[i/2]-[ij/2]}\Bigg(& C_j^1 B_i\left(A_i - B_i\frac{2d-b}{2}\right)^{j-1} \\ & + C_j^3B_i^3\left(A_i - B_i\frac{2d-b}{2}\right)^{j-3}\frac{b^2+4b}{4} + \cdots \\ & + C_j^{2p+1}B_i^{2p+1}\left(A_i - B_i\frac{2d-b}{2}\right)^{j-1-2p}\left(\frac{b^2+4b}{4}\right)^p + \cdots\Bigg). \end{aligned}$$

It follows from (11) that $B_i \mid B_{ij}$.

In certain instances it is important to know whether $n^2 \mid B_i$ for certain n. We can answer this question with the aid of 1.8 and (11).

LEMMA 1.9. *Suppose $i, j \geqslant 0$; then $B_i^2 \mid B_{ij}$ if and only if $B_i \mid j$. Also, $B_i^3 \mid B_{ij}$ if and only if $B_i^2 \mid j$ and $2 \mid b$.*

PROOF. Note first that $(B_{2i+1}, b) = 1$, since $(B_1, b) = 1$ and $(B_3, b) = (b + 1, b) = 1$.

If $i = 2k+1$ is odd, then $j[i/2] - [ji/2] = kj - [j(2k+1)/2] = -[j/2]$. If i is even, then $j[i/2] - [ji/2] = 0$.

Thus, if i is odd, then

$$\begin{aligned} B_{ij} = b^{-[j/2]}\Bigg(& C_j^1B_i\left(A_i - B_i\frac{2d-b}{2}\right)^{j-1} + \cdots \\ & + C_j^{2p+1}B_i^{2p+1}\left(A_i - B_i\frac{2d-b}{2}\right)^{j-2p-1}\left(\frac{b^2+4b}{4}\right)^p + \cdots\Bigg). \end{aligned}$$

If i is even, then

$$B_{ij} = C_j^1 B_i \left(A_i - B_i \frac{2d-b}{2} \right)^{j-1} + \cdots$$

$$+ C_j^{2p+1} B_i^{2p+1} \left(A_i - B_i \frac{2d-b}{2} \right)^{j-2p-1} \left(\frac{b^2+4b}{4} \right)^p + \cdots.$$

Since for odd i we have $(B_i, b) = 1$ and since $(A_j, B_j) = 1$, it follows that $B_i^2 \mid B_{ij}$ if and only if $B_i \mid j$. Similarly, it can be seen that $B_i^3 \mid B_{ij}$ if and only if $B_i^2 \mid j$ and $2 \mid b$.

2. We apply the results of §1 in the case $\alpha = \sqrt{a^2-1}$ (i.e. $b = 2d = 2a - 2$; see Example 1). According to a well-known result [**6**, **3**], it follows from (4) and (5) that the sequences $A_{2n} = a_n$ and $B_{2n} = a'_n$ form the sequence (x_n, y_n) of solutions of the equation $x^2 - (a^2-1)y^2 = 1$.

Note that in the case under consideration, Lemmas 1.4 and 1.5 are given in [**6**] and Lemma 1.6 is given in [**6**, **3**].

LEMMA 2.1 [**3**]. *If* $b > 1$, *then*

$$a'_n \equiv b'_n \pmod{a-b} \quad \textit{and} \quad a'_n \equiv n \pmod{a-1}.$$

PROOF. In view of (8),

$$(12) \quad a'_n = C_n^1 a^{n-1} + C_n^3 a^{n-3}(a^2-1) + \cdots = \sum_{i=1}^{n}{}' C_n^i a^{n-i}(a^2-1)^{(i-1)/2},$$

where the symbol Σ' denotes summation over odd i. Formula (12) was given in Lemma 6 of [**3**]. It follows from (12) that

$$a'_n \equiv n \pmod{a-1}$$

and $a'_n \equiv b'_n \pmod{a-b}$, since $a^n \equiv b^n \pmod{a-b}$.

For what is to come (see 2.2), we extend the definition of $A'_n = B_n$ to the domain of negative integers n by putting

$$A'_{-2n} = a'_{-n} = -a'_n; \qquad A'_{-2n-1} = A'_{2n+1}.$$

Then for any integers n we have

$$A'_{2n+2} = 2aA'_{2n} - A'_{2n-2}; \qquad A'_{2n+3} = 2aA'_{2n+1} - A'_{2n-1}.$$

Note that $a - 1 \mid A_{2n+1}$ for any natural number n. Indeed, since $A_1 = a - 1$ and $A_3 = (a-1)(2a+1)$, we have $a - 1 \mid A_{2n+1}$ for $n = 0, 1$. By Proposition 1.2, $a - 1 \mid A_{2n+1}$.

The terms of the sequence $A_{2n+1}/(a-1)$ will be denoted by K_n (see §7 below).

According to (7), the sequence K_n can be expressed linearly in terms of A_{2n} and A'_{2n}. It suffices to express K_n in terms of A_{2n} and A'_{2n} for $n = 0, 1$. Obviously, $K_0 = A_1/(a-1) = 1$, $K_3 = A_3/(a-1) = 2a+1$. Therefore,

$$1 = x \cdot 1 + y \cdot 0 \quad \text{and} \quad 2a + 1 = x \cdot a + y \cdot 1.$$

It follows that

$$(13) \qquad K_n = A_{2n} + (a+1)A'_{2n}.$$

LEMMA 2.2. *For any natural numbers m, n we have*

(14) $$a'_{m+n} + a'_{m-n+1} = A'_{2n-1} \cdot A_{2m+1}/(a-1),$$

i.e. $a'_{m+n} + a'_{m-n+1} = A'_{2n+1} \cdot K_m$.

PROOF. It is obvious that the sequences $a'_{m+n} + a'_{m-n+1}$: $n = 0, 1, 2, \ldots$, and A'_{2n-1}: $n = 0, 1, 2, \ldots$, satisfy (6) with $k = 2a$.

Therefore, in view of Proposition 1.2, to prove (14) it suffices to show that

$$a'_{m+n} + a'_{m-n+1} = A'_{2n-1} \cdot K_m$$

for $n = 0, 1$. But for $n = 0, 1$ we have $A'_{2n-1} = 1$ and

$$a'_{m+n} + a'_{m-n+1} = A'_{2m} + A'_{2m+2} = A'_{2m} + aA'_{2m} + A_{2m} = K_m.$$

LEMMA 2.3. *For any natural numbers we have*

(15) $$a'_{(2m+1)i+l} - a'_l = \sum_{j=0}^{i-1} A'_{2(2m+1)j+2m+2l+1} \cdot K_m,$$

i.e.

$$a'_{(2m+1)i+l} \equiv a'_l \pmod{K_m};$$
$$(a-1)a'_{(2m+1)i+l} \equiv (a-1)a'_l \pmod{A_{2m+1}}.$$

PROOF. Put $n = m + 1 + l$ in (14). Then $a'_{2m+1+l} - a'_l = A'_{2m+2l+1}K_m$, i.e. (15) is true for $i = 1$. Suppose (15) holds for an $i \geqslant 1$. We will show that (15) holds for $i + 1$:

$$\begin{aligned} A'_{l+(2m+1)(i+1)} - a'_l &= a'_{l+2m+1+(2m+1)i} - a'_l \\ &= a'_{l+(2m+1)i} - a'_l + A'_{2l+2(2m+1)i+2m+1} \cdot K_m \\ &= \sum_{j=0}^{i} A'_{2(2m+1)j+2m+1+2l} \cdot K_m. \end{aligned}$$

LEMMA 2.4. *Suppose* $x = a'_r < a'_{m+1}$, $x \equiv a'_n \pmod{K_m}$, *and* $A'^2_{2i+1} \mid A'_{2m+1}$. *Then* $r \equiv n \pmod{A'_{2i+1}}$.

PROOF. By Lemma 1.9, $A'_{2i+1} \mid 2m+1$, since $A'^2_{2i+1} \mid A'_{(2i+1)(2m+1)}$. Choose j so that $n = (2m+1)j + l$, where $|l| \leqslant m$. Then, by Lemma 2.3,

$$x \equiv \operatorname{sgn} l \cdot a'_{|l|} \pmod{K_m}.$$

If $l \geqslant 0$, then obviously $x = a'_l$; if $l < 0$, then, since $K_m = a_m + (a+1)a'_m$, we have

$$x + a'_{|l|} \geqslant K_m \geqslant a \cdot a'_m + a_m + a'_{|l|}$$

and, since $x < a'_{m+1}$, the case $l < 0$ is impossible.

If $l \geqslant 0$, then $n \equiv l \pmod{2m+1}$ and $A'_{2i+1} \mid 2m+1$, i.e. $n \equiv l \pmod{A'_{2i+1}}$.

We now establish certain estimates for the sequences a_n, a'_n.

PROPOSITION 2.5 [**8**]. $a^n \leqslant a_n \leqslant (2a)^n$ *and* $a^{n-1} \leqslant a'_n \leqslant a_n$ *if* $n \geqslant 1$.

PROOF. That $a'_n \leqslant a_n$ is obvious. The remaining inequalities are valid for $n = 1, 2$.

If they are valid for n, then for $n + 1$ we have $a_{n+1} \leqslant 2aa_n \leqslant (2a)^{n+1}$ and $a_{n+1} = aa_n + (a^2 - 1)a'_n \geqslant a^{n+1}$, $a'_{n+1} \geqslant a^n$, since it is obvious that $a'_{n+1} = a_n + aa'_n$.

Using 2.1–2.5 we can prove

THEOREM 2.6. *The predicates* $y = a_x$ *and* $y = a'_x$ *are Diophantine predicates of exponential growth.*

PROOF. That $y = a_x$ and $y = a'_x$ have exponential growth follows from 2.5.

Obviously, it suffices to show that the predicate $y = a_x$ is Diophantine, since $y = a'_x$ can be expressed in Diophantine fashion in terms of $y = a_x$, namely, $y = a'_x \leftrightarrow (\exists z)(z^2 - (a^2 - 1)y^2 = 1 \,\&\, z = a_x)$.

We will show that $y = a_x$ $(x > 1)$ if and only if there exist z, w, t, p, b, u such that

(a) $y^2 - (a^2 - 1)w^2 = 1$, $x \leqslant y - (a - 1)w$,
(b) $t^2 - (a^2 - 1)p^2 = 1$, $z^2 - (b^2 - 1)u^2 = 1$,
(c) $(y - (a - 1)w)^2 \mid t - (a - 1)p$,
(d) $y - (a - 1)w \mid b - 1$, $(a + 1)p - t \mid b - a$,
(e) $\mathrm{Rem}(u; b - 1) = x$, $\mathrm{Rem}(u, b - a) = w$.

Assume there exist z, w, t, p, b, u such that (a)–(e) are satisfied. Then, according to 1.5 and 1.6,

$$y - (a - 1)w = A'_{2l+1}; \qquad t - (a - 1)p = A'_{2m+1}; \qquad (a + 1)p - t = K_m.$$

In view of (b), $u = b'_n$, hence, by Lemma 2.1, $u \equiv a'_n \pmod{b - a}$. Since it follows from (d) and (e) that $u \equiv w \pmod{b - a}$ and $a \equiv b \pmod{K_m}$, we have $w \equiv a'_n \pmod{K_m}$. Since, according to (c), $A'^2_{2l+1} \mid A'_{2m+1}$, $w = A'_{2l+2}$, and $A'_{2l+1} \mid 2m + 1$, it follows from Lemma 2.4 that $w = a'_r$, where $r \equiv n \pmod{A'_{2l+1}}$. In view of (d), (e), and Lemma 2.1, we have $n \equiv u \pmod{A'_{2l+1}}$, $u \equiv x \pmod{A'_{2l+1}}$, i.e. $x \equiv r \pmod{A'_{2l+1}}$. It follows from (a) that $A'_{2l+1} \geqslant x$ and $x = r$, since $r = l + 1 \leqslant A'_{2l+1}$.

Consequently, $w = a'_x$ and $y = a_x$.

Suppose $y = a_x$. If $w = A'_{2x}$, $h = A'_{2x-1}$, $t = A_{h(2x-1)+1}$, $p = A'_{h(2x-1)+1}$, then, by Lemmas 1.9 and 2.5, conditions (a)–(c) hold. Since $(a + 1)p - t = A_{h(2x-1)}/(a - 1)$, it follows that $(y - (a - 1)w, (a + 1)p - t) = 1$, inasmuch as $(A_{h(2x-1)}/(a - 1), A'_{2x-1}) = 1$; then, by the Chinese remainder theorem, there exists b such that (d) holds. Putting $u = b'_x$ and using Lemma 2.1, we obtain (e) and the theorem is proved.

The system of Diophantine equations (a)–(e) defining the predicate $\mathrm{Pl}(y, a, x) \leftrightarrow y = a_x$ is given in the paper of the author [**9**].

We remark that the technique of elementary number theory used in §§1, 2 to prove that $y = a_x$ is Diophantine *is rather simple and on a 17th century level* (cf. [**16**]) although somewhat tedious.

As we mentioned in §0, it follows from 2.6 that any recursively enumerable set is Diophantine. Consequences of 2.6 similar to this one are given below in §4.

3. We continue the investigation, begun in §1, of the properties of the general sequences A_n, B_n and we characterize A_n, B_n as the solutions of certain Diophantine equations.

THEOREM 3.1. *If $b \geqslant 2d$, then the pairs (A_{2n}, B_{2n}) constitute all solutions of the equation $x^2 - (2d - b)xy - (db - d^2 + b)y^2 = 1$. The pairs (A_{2n+1}, B_{2n+1}) constitute the solutions of the equation $x^2 - (2d - b)xy - (db - d^2 + b)y^2 = -b$.*

PROOF. Let α and β be the roots of (2). Then, by Lemma 1.6,

$$A_{2n} - \alpha B_{2n} = (A_2 - \alpha B_2)^n \quad \text{and} \quad A_{2n} - \beta B_{2n} = (A_2 - \beta B_2)^n.$$

Thus,

$$\begin{aligned}(A_{2n} - \alpha B_{2n})(A_{2n} - \beta B_{2n}) &= (A_2 - \alpha B_2)^n (A_2 - \beta B_2)^n \\ &= \left(A_2^2 - (2d - b)A_2 B_2 - (db - d^2 + b)B_2^2\right)^n \\ &= (d^2 + 1 + 2d - 2d^2 + bd - 2d + b - bd + d^2 - b)^n = 1.\end{aligned}$$

Similarly, by Proposition 1.9,

$$A_{2n+1} - \alpha B_{2n+1} = b^{-n}\left(\frac{b}{2} - \frac{\sqrt{b^2 + 4b}}{2}\right)^n = b^{-n}(A_1 - \alpha B_1)^{2n+1},$$

$$A_{2n+1} - \beta B_{2n+1} = b^{-n}(A_1 - \beta B_1)^{2n+1}.$$

Consequently,

$$\begin{aligned}(A_{2n+1} - \alpha B_{2n+1})(A_{2n+1} - \beta B_{2n+1}) &= b^{-2n}(A_1 - \alpha B_1)^{2n+1}(A_1 - \beta B_1)^{2n+1} \\ &= b^{-2n}\left(A_1^2 - (2d - b)A_1 B_1 - (db - d^2 + b)B_1^2\right)^{2n+1} \\ &= -b^{-2n} \cdot b^{2n+1} = -b.\end{aligned}$$

We will show that if (A_n, B_n) satisfies the equation

$$x^2 - (2d - b)xy - (db - d^2 + b)y^2 = 1,$$

then x/y is a convergent of α. This will complete the proof of Theorem 3.1.

If $x^2 - (2d - b)xy - (db - d^2 + b)y^2 = 1$, then $(x - \alpha y)(x - \beta y) = 1$. Therefore,

$$x - \alpha y = \frac{1}{x - \beta y} \quad \text{and} \quad \frac{x}{y} - \alpha = \frac{1}{y^2(x/y - \beta)}.$$

It follows that $x/y > \alpha$, since $\beta < 0$, and

$$\frac{x}{y} > \frac{2d - b}{2} + \frac{\sqrt{b^2 + 4b}}{2} > \frac{2d - b}{2} + 2 - \frac{\sqrt{b^2 + 4b}}{2},$$

since $b \geqslant 1$.

Thus, $x/y > \beta + 2$ and $x/y - \alpha > 0$.

Then $0 < x/y - \alpha < 1/2y^2$, and x/y is a convergent of α.

We extend the sequences A_n, B_n to negative integers.

To do this we define B_{-n} as follows:

$$B_{-n} = (-1)^{n+1} B_n. \tag{16}$$

Then for any integer n we have

$$B_{2n+3} = (b+2)B_{2n+1} - B_{2n-1}; \qquad B_{2n+2} = (b+2)B_{2n} - B_{2n-2}.$$

This result follows from (16) and Lemma 1.1.

We will investigate the periodicity of the sequence B_{2n} (mod B_m).

LEMMA 3.2. *For any natural numbers,*

$$B_{2n+2m} \equiv -B_{2m-2n} \pmod{B_{2m}}. \tag{17}$$

Therefore, $B_{2m+4Ni} \equiv B_{2m} \pmod{B_{2N}}$.

PROOF. Let us prove (17). For $n = 0$ it is obvious. For $n = 1$ we have

$$B_{2m+2} + B_{2m-2} = (b+2)B_{2m} - B_{2m-2} + B_{2m-2}.$$

The sequence $B_{2n+2m} + B_{2m-2n}$, $n = 0, 1, 2, \ldots$, satisfies (6) with $k = b + 2$. Therefore, by 1.2, (17) holds for any $n \geqslant 0$.

We will show that

$$B_{2m+4N} \equiv B_{2m} \pmod{B_{2N}}. \tag{18}$$

Indeed, according to (17),

$$B_{2n+2N} \equiv B_{2n-2N} \pmod{B_{2N}}.$$

Putting $2n - 2N = 2m$, we obtain (18).

It follows from (18) that

$$B_{2m+4Ni} \equiv B_{2m} \pmod{B_{2N}}.$$

In many cases where the sequence A_n (or B_n) is more explicitly given (the connection between d and b is indicated), we can prove other results similar to Lemma 3.2.

EXAMPLE 2. Consider the case $\alpha = (a^2 - 1)^{1/2}$. In view of 3.1, the sequences $A_{2n} = a_n$, $B_{2n} = a'_n$ constitute all solutions of the equation $x^2 - (a^2 - 1)y^2 = 1$.

For the sequences a_n, a'_n we have the following results:

$$\begin{gathered} a_{m+n} \equiv a_{m-n} \pmod{a'_m}, \qquad a'_{m+n} \equiv a'_{m-n} \pmod{a_m}, \\ a_{m+n} \equiv -a_{m-n} \pmod{a_m}. \end{gathered} \tag{19}$$

The formulas (19) are proved by induction on n. In the case $n = 0$ they are obvious. When $n = 1$, the third formula is obvious and the first two are proved as follows: by Lemma 1.4,

$$a_{m+1} = a \cdot a_m + (a^2 - 1)a'_m, \qquad a'_{m+1} = a_m + a \cdot a'_m.$$

Therefore, the formulas are true for $n = 1$. The rest of the proof follows from 1.2.

From these formulas we obtain the following ones:

$$a_{2Ni+m} \equiv (-1)^i a_m \pmod{a_N}, \qquad a_{2Ni+m} \equiv a_m \pmod{a'_N}, \tag{20}$$
$$a'_{2Ni+m} \equiv (-1)^i a'_m \pmod{a_N}.$$

It suffices to verify these formulas for $i = 1$. The proof in the case $i = 1$ is easily obtained from the formulas (19) proved above.

The formulas (20) can be applied to $\alpha = (a^2 + 1)^{1/2}$, for example.

Note that formulas of type (20) are given in [**12**] for some other sequences and are used there to study Diophantine equations.

We give further results on the periodicity of the sequence $A_{2n} \pmod{m}$.

LEMMA 3.3. *For any natural numbers m, n we have*

$$B_{2m+2n} + B_{2m-2n+2} = B_{2n-1}(B_{2m+2} + B_{2m}).$$

PROOF. For $n = 0, 1$ the formula is obvious. If the formula holds for all $i < n$, then for n we obtain

$$\begin{aligned} B_{2m+2n} + B_{2m-2n+2} &= (ab + 2)(B_{2m+2n-2} + B_{2m-2n+4}) \\ &\quad - (B_{2m+2n-4} + B_{2m-2n+2}) \\ &= ((ab + 2)B_{2n-3} - B_{2n-5})(B_{2m+2} + B_{2m}) \\ &= B_{2n-1}(B_{2m+2} + B_{2m}) \end{aligned}$$

and Lemma 3.3 is proved.

LEMMA 3.4.

$$B_{2m+2n} \equiv B_{2m-2n+2} \pmod{B_{2m+2} - B_{2m}}.$$

PROOF. For $n = 0, 1$ the formula is obvious. In view of (16), it follows in general from 1.2.

From Lemmas 3.3 and 3.4 we obtain

PROPOSITION 3.5.

$$B_{2m+2(2N+1)l} - B_{2m} = \sum_{j=0}^{l-1} B_{2(2N+1)j+2m+2N+1}(B_{2N+2} + B_{2N})$$

and

$$B_{2m+2(2N+1)l} \equiv (-1)^l B_{2m} \pmod{B_{2N+2} - B_{2N}}.$$

PROOF. Put $n = m + 1 + l$ in Lemma 3.3. Then we have

$$B_{(2m+1)2+2n} - B_{2n} = B'_{2m+2n+1}(B_{2m+2} + B_{2m})$$

from which we obtain the first part of Proposition 3.5 with $l = 1$. For $l > 1$ we have

$$B_{2m+2(2N+1)(l+1)} - B_{2m} = B_{2m+2(2N+1)+2(2N+1)l} - B_{2m}$$
$$= B_{2m+2(2N+1)l} - B_{2m} + B_{2m+2N+2(2N+1)l+1}\left(B_{2N+2} + B_{2N}\right)$$
$$= \sum_{j=0}^{l} B_{2(2N+1)j+2N+2m+1}\left(B_{2N+2} + B_{2N}\right).$$

It follows from Lemma 3.4 that

$$B_{2l+2(2N+1)} \equiv -B_{2l} \pmod{B_{2N+2} - B_{2N}},$$

where $l = n - m - 1$. Then

$$B_{2l+2(2N+1)(i+1)} \equiv -B_{2l+2(2N+1)i} \pmod{B_{2N+2} - B_{2l}},$$

hence

$$B_{2m+2(2N+1)l} \equiv (-1)^l B_{2m} \pmod{B_{2N+2} - B_{2N}}.$$

Using the results of §1 we can obtain an explicit expression for B_{2n} in terms of b and n. We will use this expression to examine the question of congruence $(\mathrm{mod}\, m)$ between the sequences B_{2n} for different α.

PROPOSITION 3.6.

$$B_{2n} = \sum_{l=1}^{n}{}' C_n^l \frac{(b^2 + 4b)^{(l-1)/2}}{2^{l-1}} \left(\frac{2+b}{2}\right)^{n-l}.$$

PROOF. This result is a consequence of the following formula, established in §1:

$$\left(\frac{2 + b - \sqrt{b^2 + 4b}}{2}\right)^n = A_n - \frac{2d - b}{2} B_n - B_{2n} \frac{\sqrt{b^2 + 4b}}{2}.$$

Suppose α is given in terms of d and b, and α' in terms of d' and b'. Then, by Proposition 3.6, for the sequences B_{2n} and B'_{2n} corresponding to α and α' we have

$$B_{2n} \equiv B'_{2n} \pmod{b' - b}.$$

Similarly, we obtain

LEMMA 3.7. *For α and α' satisfying $b, b' \geqslant 2$ we have $B_{2n} \equiv B'_{2n} \pmod{b' - b}$, $2^{n-1} B_{2n} \equiv n(2 + b)^{n-1} \pmod{b^2 + 4b}$.*
For example, $B_{2n} \equiv n \pmod{b}$.

Using 3.5–3.7 we can prove

LEMMA 3.8. *Suppose $x \equiv B_{2n} \pmod{B_{2m+2} + B_{2m}}$ and $x < B_{2m+2}$. Then $x = B_{2l}$, where $l \equiv n \pmod{2m + 1}$.*

PROOF. If $n = (2m + 1)j + l$, where $0 \leqslant l \leqslant 2m$, then, assuming $l \leqslant m$, we have, by Proposition 3.5, $x \equiv B_{2l} \pmod{B_{2m} + B_{2m+2}}$, from which it follows that $x = B_{2l}$.

If $m + 1 \leqslant l \leqslant 2m$, then, by 3.3 and 3.5,

$$x \equiv -B_{2(2m+1)-2l} \pmod{B_{2m+2} + B_{2m}},$$

which is impossible in view of the condition $x < B_{2m+2}$.

By means of §§1 and 2 we can also obtain results concerning the number $j(p)$, the smallest natural number n such that $B_{2n} \equiv 0 \pmod p$, where p is a prime. For example, in the case $\alpha = (a^2 - 1)^{1/2}$ the existence of $j(p)$ is easily proved. We can obtain estimates of $j(p)$ of the type $j(p) \leqslant p + 1$ and the like.

4. We will establish the Diophantine nature of sequences satisfying the condition $u_{n+2} = ku_{n+1} - u_n$.

In view of what was said in §1, sequences satisfying this condition can be linearly expressed in terms of the sequence A_{2n} (B_{2n}) with $d = 1$, $b = k - 2$.

Therefore, we need only establish the Diophantine nature of B_{2n} (A_{2n}).

THEOREM 4.1. *The predicates $y = B_{2x}$ and $y = A_{2x}$ are Diophantine.*[3]

PROOF. Obviously, it suffices to show that the predicate $y = B_{2x}$ is Diophantine, since $y = A_{2x}$ can be expressed in Diophantine fashion in terms of $y = B_{2x}$:

$$y = A_{2x} \leftrightarrow (\exists z)\big(z = B_{2x} \,\&\, y^2 - (2-b)zy - (2b-1)z^2 = 1\big).$$

We will show that $s = B_{2x}$ if and only if there exist natural numbers y, w, z, u, t, p, e such that

(f) $y^2 - (2-b)yw - (2b-1)w^2 = 1$,

(g) $z^2 - (2-e)zu - (2e-1)u^2 = 1$,

(h) $t^2 - (2-b)tp - (2b-1)p^2 = 1$,

(i) $\mathrm{Rem}(u, e) = x$, $\mathrm{Rem}(u, e-b) = s$,

(j) $(y + (b-1)w)^2 \mid t + (b-1)p$,

(k) $y + (b-1)w \mid e$, $t + (b+1)p \mid e - b$,

(l) $x \leqslant s \leqslant w$.

The proof is analogous to that of 2.6.

In view of (f), (h), and Theorem 3.1, we have $y = A_{2l}$, $w = B_{2l}$, $t = A_{2m}$, $p = B_{2m}$, $u = B'_{2n}$, where B'_{2n} corresponds to $b' = e$.

In view of 1.4 and 1.5,

$$B_{2m+1} = t + (b-1)p, \qquad B_{2l+1} = y + (b-1)w,$$

$$B_{2m} + B_{2m+2} = t + (b+1)p.$$

According to (i), $u \equiv s \pmod{e-b}$; by Lemma 3.7, $u \equiv B_{2n} \pmod{e-b}$, i.e. $B_{2n} \equiv s \pmod{e-b}$.

Since $t + (b+1)p \mid e - b$, it follows that $s \equiv B_{2n} \pmod{B_{2m} + B_{2m+2}}$.

From (j) and Lemma 1.9 we obtain $B_{2l+1} \mid 2m + 1$, i.e. $B_{2l+1} \leqslant 2m + 1$, hence $w < B_{2m+2}$ and, according to (l) and 3.8, $s = B_{2r}$, where $r \equiv n \pmod{2m+1}$, $r \equiv n \pmod{B_{2l+1}}$. However, $B_{2l+1} \mid e$ and, according to 3.7, $u \equiv n \pmod e$.

[3] Theorem 4.1 in the case $b = 1$ was cited by Ju. V. Matijasevič in [**15**].

Therefore, $u \equiv r \pmod{B_{2l+1}}$; it follows from (i) that $x \equiv r \pmod{B_{2l+1}}$, and since $x \leqslant w < B_{2l+1}$, we have $x = r$ and $s = B_{2x}$.

If $s = B_{2x}$, then $s \geqslant x$. Put $y = A_{12x}$, $w = B_{12x} \geqslant s$, i.e. (f) and (l) are satisfied. Also, $t = A_{(12x+1)B_{12x+1}-1}$, $p = B_{(12x+1)B_{12x+1}-1}$, and, since B_{12x+1} is odd (for any b), (h) is satisfied. According to 1.9, $B^2_{12x+1} \mid B_{(12x+1)B_{12x+1}}$, i.e. (j) holds.

In view of 1.8, $(B_{(12x+1)B_{12x+1}-1}, B_{12x+1}) = 1$, hence

$$(t + (b+1)p, y + (b-1)w) = (p, y + (b-1)w) = 1.$$

By the Chinese remainder theorem [6], there exists $e \geqslant 2$ such that (k) holds. Put $u = B'_{2x}$, where B'_{2n} corresponds to $b' = e$. Then (i) holds by virtue of 3.7.

According to §1, (7), any sequence of natural numbers u_n satisfying the condition $u_{n+2} = ku_{n+1} - u_n$ can be expressed in terms of A_{2n}, B_{2n}: $u_n = xA_{2n} + yB_{2n}$, where x, y are determined from the conditions $u_0 = x \cdot 1 + y \cdot 0$, $u_1 = x \cdot 2 + y \cdot 1$ and where A_i, B_i correspond to $b = k - 2 \geqslant 1$, $d = 1$.

Therefore, the predicate $R(z, n, k, x, y)$ signifying that $z = u_n$, where u_n satisfies (6) and $u_0 = x$, $u_1 = y$, is Diophantine. It is easy to see that this predicate is of exponential growth, since it follows from [8], Chapter 1, §4, that $B_{n+1} \geqslant 2^{(n-1)/2}$ and it is obvious that $(b+2)^n \geqslant B_{2n}$.

An explicit Diophantine representation of the predicate $R(z, n, k, x, y)$ can be obtained from (f)–(l), since the finite system (f)–(l) is equivalent (see [1]) to a single Diophantine predicate.

Similarly, using the method of §§1–3, a Diophantine representation can be obtained for sequences satisfying the condition $u_{n+2} = ku_{n+1} + u_n$ and the like.

In view of [4], any recursively enumerable predicate can be expressed in Diophantine fashion in terms of any predicate of exponential growth, hence from 2.6 (and 4.1) we obtain

THEOREM 4.2. *Any recursively enumerable predicate (recursively enumerable set) is Diophantine.*

Consequently, the problem of recognizing the existence of integral solutions of an arbitrary Diophantine equation is unsolvable, i.e. the answer to Hilbert's 10th problem is negative.

We give some consequences of the results of §4. For example, for any n there exists a polynomial $P_n(x_1,\ldots,x_n; y_1,\ldots,y_{k_n}; m)$ such that for any recursively enumerable predicate $R(x_1,\ldots,x_n)$ and some m we have

$$R(x_1,\ldots,x_n) \leftrightarrow (\exists y_1) \cdots (\exists y_{k_n})\big(P_n(x_1,\ldots,x_n; y_1,\ldots,y_{k_n}; m) = 0\big).$$

By a result of Davis [1], there exists a polynomial $Y(x_1,\ldots,x_k, m)$ such that for any recursively enumerable set S and some m, the set S coincides with the set of all positive integral values of $Y(x_1,\ldots,x_k, m)$, and the degree of $Y(x_1,\ldots,x_k, m)$ does not exceed 5.

When we speak of exponential Diophantine equations we mean equations of the form

$$\text{(A)} \qquad F(x_1^{y_1},\ldots,x_n^{y_n}) = 0,$$

where $F(\dots)$ is a polynomial with rational integer coefficients and $x_1,\dots,x_n$, $y_1,\dots,y_n$ denote everywhere further variables on the set of natural numbers. When we speak of solvability of equations we mean (unless it is stated otherwise) solvability in natural numbers.

There are several well-known problems connected with exponential Diophantine equations which have not been completely solved. Among these (apart from the already notorious problem of Fermat) is a series of problems known collectively as Euler's problem. It can be stated as follows:

(E) When does the equation

$$x_1^n + \cdots + x_k^n = y^n$$

have a solution?

The most interesting case of problem (E) is $k \leqslant n-1$. For a long time not one (nontrivial) solution of the equation

$$x_1^n + \cdots + x_{n-1}^n = y^n \tag{E_1}$$

was known for $n \geqslant 3$. Only recently, with the aid of a computer, was a solution found in the case $n = 5$. The question in the general case remains open.

Many exponential Diophantine equations are studied in the book of Hardy and Wright [**16**]. Some of these are:

$$x_1^n \pm x_2^n \pm \cdots \pm x_k^n = y_n, \tag{1}$$

$$x_1^n y_1^n + \cdots + x_k^n y_k^n = x_{k+1}^n y_{k+1}^n, \tag{2}$$

$$x_1^n + x_2^n = x_3^n y^n. \tag{3}$$

Up to the present time not one nontrivial exponential Diophantine equation similar to those above has been solved, i.e. it has not been (effectively) determined for which $n_1,\dots,n_k$ the equation (A): $F(x_1^{n_1},\dots,x_k^{n_k}) = 0$ has a solution in positive integers $x_1,\dots,x_k$ and for which it does not.

This leads to the conjecture that there exists a natural exponential Diophantine equation which is algorithmically unsolvable. In his report to the Moscow mathematical congress of 1966, A. I. Mal'cev [**17**] formulated this conjecture as follows:

(M) Do there exist a natural number s and integer-valued polynomials $f_{1i}(n),\dots,f_{si}(n)$: $1 \leqslant i \leqslant t$, such that the set of those n for which the equation

$$\sum_{i=1}^{t} \pm x_1^{f_{1i}(n)} \cdots x_s^{f_{si}(n)} = 0, \tag{M_1}$$

is solvable is nonrecursive? If so, what are the smallest values of s, t, and the degrees of the $f_{ki}(n)$?

We show below that the answer to Mal'cev's problem (M) is positive, and we present bounds on s, t, $\deg(f_{ki}(n))$.

Let us present the solution of problem (M). We make a number of preliminary remarks. The symbols K and P will denote sets (necessarily recursively enumerable) of natural numbers. Also, we define 2-Exp P as follows:

$$\text{2-Exp } P = \{m \colon m = 2^n,\ n \in P\}.$$

Clearly, if P is a recursively enumerable set, then so is 2-Exp P. We give several auxiliary results which will be needed later. By an (M_1)-equation, as in the statement of problem (M), we mean an equation of the form

$$\sum_{i=1}^{t} \pm x_1^{f_{1i}(n)} \cdots x_s^{f_{si}(n)} = 0,$$

where s, t are natural numbers and the $f_{ki}(n)$ are polynomials with integral coefficients. A system of equations

$$(\alpha) \qquad \begin{cases} F_1\left(x_1^{f_{11}(n)},\ldots,x_{k_1}^{f_{k_1 1}(n)}\right) = 0, \\ \cdots\cdots\cdots\cdots\cdots\cdots \\ F_m\left(x_1^{f_{1m}(n)},\ldots,x_{k_m}^{f_{k_m m}(n)}\right) = 0, \end{cases}$$

where $F_1,\ldots,F_m$ are polynomials with integer coefficients (as are the $f_{ij}(n)$), is said to be equivalent to the equation

$$(\beta) \qquad F_0\left(x_1^{g_1(n)},\ldots,x_l^{g_l(n)}\right) = 0,$$

if the set of those n for which system (α) is solvable (in natural numbers) coincides with the set of those n for which equation (β) is solvable.

LEMMA 4.3. *Any equation of the form*

$$(1) \qquad \sum_{i=1}^{t} \pm a_i x_1^{f_{1i}(n)} \cdots x_s^{f_{si}(n)} = 0,$$

where the a_i are fixed natural numbers, is equivalent to a single (M_1)*-equation.*

PROOF. Lemma 4.3 is obvious, since we need only to replace a natural number $a \geqslant 1$ in (1) by $x_{s+1}^0 + \cdots + x_{s+a}^0$ and put $f_{s+j}(n) \equiv 0$ for $1 \leqslant j \leqslant a$.

LEMMA 4.4. *Any system of* (M_1)*-equations is equivalent to a single* (M_1)*-equation.*

PROOF. Suppose (α) is a system of (M_1)-equations. Consider the following equation:

$$(\gamma) \qquad F_1^2\left(x_1^{f_{11}(n)},\ldots,x_{k_1}^{f_{k_1 1}(n)}\right) + \cdots + F_m^2\left(x_1^{f_{1m}(n)},\ldots,x_{k_m}^{f_{k_m m}(n)}\right) = 0.$$

Obviously (see [**1**]), equation (γ) is equivalent to system (α). Since $F_1,\ldots,F_m$ are polynomials with integral coefficients, the whole equation (γ) reduces to the form (1):

$$\sum_{i=1}^{t_0} \pm a_i x_1^{g_{1i}(n)} \cdots x_s^{g_{si}(n)} = 0,$$

and then, by Lemma 4.3, (γ) (and therefore (α)) is equivalent to a single (M_1)-equation.

We will now prove the main theorem of this subsection.

THEOREM 4.5. *There exist natural numbers* s, t *and polynomials* $f_{ki}(n)$: $1 \leqslant k \leqslant t$, $1 \leqslant i \leqslant s$, *with integral coefficients such that for a given* (M_1)*-equation*

$$\sum_{i=1}^{s} \pm x_1^{f_{1i}(n)} \cdots x_t^{f_{ti}(n)} = 0 \tag{2}$$

the set of those n *for which* (2) *is solvable in natural numbers is nonrecursive.*

As we have already observed, Theorem 4.5 provides an affirmative answer to question (M) of A. I. Mal′cev put forward by him in **[17]**.

PROOF. Let P_r be the set of all prime numbers (> 1). Choose $P \subset P_r$ to be a recursively enumerable, but nonrecursive, set. Then, as was mentioned above, 2-Exp P is a recursively enumerable set. Note the following simple fact:

(3) If $m > 1$, then $x^m \in$ 2-Exp $P \rightleftharpoons x = 2 \,\&\, m \in P$.

Indeed, if $x^m = 2^n$, where $n \in P \subset P_r$, then, since $m > 1$ and $n > 1$, we have $m = n$ and $x = 2$. Also, if $x = 2$ and $m \in P$, then $x^m \in$ 2-Exp P. Since 2-Exp P is a recursively enumerable set, it follows from Theorem 4.1 that there exists a polynomial $F(x_1, \ldots, x_k, y)$ such that

$$\text{2-Exp}\, P = \{n\colon (\exists x_1) \cdots (\exists x_k)(F(x_1, \ldots, x_k, n) = 0)\}. \tag{4}$$

If we write $F(x_1, \ldots, x_k, y)$ in the form

$$\sum_{i=1}^{r} \pm a_i x_1^{h_{1i}} \cdots x_k^{h_{ki}} y^{m_i},$$

where r, a_i, h_{l_i}, m_i: $1 \leqslant i \leqslant r$, $1 \leqslant l \leqslant k$, are fixed natural numbers (or zero), then (4) becomes

(5) 2-Exp P is the set of those n for which the equation

$$\sum_{i=1}^{r} \pm a_i x_1^{h_{1i}} \cdots x_k^{h_{ki}} n^{m_i} = 0$$

is solvable in natural numbers.

Consider the following equation of type (1):

$$\sum_{i=1}^{r} \pm a_i x_1^{h_{1i}} \cdots x_k^{h_{ki}} x_{k+1}^{m_i n} = 0. \tag{6}$$

Let K be the set of those n for which (6) is solvable (in natural numbers). Then, in view of (5), $n \in K \rightleftharpoons (\exists x)\,(x^n \in$ 2-Exp $P)$. Therefore, we obtain from (3)

(7) If $n > 1$, then $n \in K \rightleftharpoons n \in P$.

Since P is nonrecursive, it follows from (7) that K is nonrecursive.

By Lemma 4.3, the equation (6) is equivalent to some (M_1)-equation

$$\sum_{i=1}^{t} \pm x_1^{l_{1i}} \cdots x_s^{l_{si}} x_{s+1}^{m_i n} = 0, \tag{8}$$

where the l_{ki}: $1 \leqslant k \leqslant s$, $1 \leqslant i \leqslant t$, are integers ($\geqslant 0$).

Since K is nonrecursive, the set of those n for which (8) is solvable is nonrecursive. Note that the simple form of the (M_1)-equation (8) equivalent to (6) follows at once from the proof of Lemma 4.3. Theorem 4.5 is proved.

5. Using the theorem on the Diophantine representation of the predicate $\mathrm{Pl}(x, y, z) \leftrightarrow x = y_z$ and the results of M. Davis, H. Putnam, and J. Robinson [**1**, **4**], we obtained in §4 a Diophantine representation of any recursively enumerable predicate.

However, using the method of [**3**] and an explicit Diophantine representation of the predicate $\mathrm{Pl}(x, y, z)$ we can obtain an explicit Diophantine representation of the predicate $x = y^z$. From this, using ideas of J. Robinson, we can obtain an explicit Diophantine representation of a polynomial representing all primes, for example A, etc.

We give here a simple explicit Diophantine representation of the function $x = y^z$ in terms of the predicate $\mathrm{Pl}(x, y, z)$ (an explicit Diophantine representation of which was given in Theorem 2.6, (a)–(e)).

We use the following properties of the sequence a_n, which were proved in [**3**].

LEMMA 5.1 [**3**]. *If $x > 1$ and $a > x^n$, then*

$$a_n x^n \leqslant (ax)_n < a_n(x^n + 1).$$

PROOF. From formula (8),

$$\left(a_n + a'_n\sqrt{a^2 - 1}\right) = \left(a + \sqrt{a^2 - 1}\right)^n,$$

we obtain

$$a_n = \sum_{l=0}^{n}{}'' C_n^l a^{n-l}(a^2 - 1)^{l/2},$$

where Σ'' denotes summation over even l.

Therefore,

$$a_n x^n = \sum_{l=0}^{n}{}'' C_n^l (ax)^{n-l}(a^2x^2 - x^2)^{l/2},$$

$$(ax)_n = \sum_{l=0}^{n} C_n^l (ax)^{n-l}(a^2x^2 - 1)^{l/2}.$$

It follows that $a_n x^n \leqslant (ax)_n$. Moreover,

$$\frac{(ax)_n}{a_n x^n} \leqslant \frac{(a^2x^2 - 1)^{n/2}}{(a^2x^2 - x^2)^{n/2}} \leqslant \left(1 - \frac{1}{a^2}\right)^{-n}$$

and

$$\left(1 - \frac{1}{a^2}\right)^n > 1 - \frac{n}{a^2} > 1 - \frac{1}{a} \geqslant 1 - \frac{1}{x^n + 1}.$$

Then

$$a_n x^n \leqslant (ax)_n < a_n(x^n + 1).$$

LEMMA 5.2 [**3**]. *If $x > 1$ and $a > x^n$, then*

$$(a_n \leqslant (ax)_m \leqslant a \cdot a_n) \leftrightarrow (m = n).$$

PROOF. By Lemma 5.1, for $a > x^n$ we have

$$a_n \leqslant (ax)_n \leqslant a \cdot a_n.$$

It suffices to show that

$$(ax)_{n-1} < a_n \quad \text{and} \quad a \cdot a_n < (ax)_{n+1}.$$

By Lemma 5.1 (with n replaced by $n - 1$),

$$(ax)_{n-1} < a \cdot a_{n-1} \leqslant a_n.$$

In addition, obviously, $(ax)_{n+1} > (ax)(ax)_n > aa_n$.

Using these results and those of §2, we can give an explicit Diophantine representation of the function $x = y^z$, where $x > 1$ and $y > 1$.

By Lemma 5.1,

$$(a > y^z) \to ((x = y^z) \leftrightarrow a_z x \leqslant (ay)_z < a_z(x + 1)).$$

By Lemma 5.2,

$$(a > y^z) \to \big((u = (ay)_z) \leftrightarrow (\exists v)\big(u^2 - (a^2y^2 - 1)v^2 = 1 \,\&\, a_z \leqslant u \leqslant a \cdot a_z\big)\big).$$

Therefore,

$$\begin{aligned}(x < a \,\&\, a > y^z) \to ((x = y^z)\\ \leftrightarrow (\exists u)(\exists v)\big(u^2 - (a^2y^2 - 1)v^2 = 1 \,\&\, a_z x \leqslant u < a_z(x + 1)\big)\big).\end{aligned}$$

Choose a so that $a \geqslant (y + z)^{y+z-1} > y^z$. If b is a solution of the equation $p^2 - [(y + z)^2 - 1][y + z - 1]^2b^2 = 1$, then $a = b(y + z - 1) = (y + z)_n$, and $y + z - 1 \mid (y + z)_n$. By Lemma 2.1, $y + z - 1 \mid n$, i.e. $a \geqslant (y + z)_{y+z-1} \geqslant (y + z)^{y+z-1}$. Consequently, for x, $y > 1$ we have

$$\begin{aligned}(x = y^z) \leftrightarrow (\exists u)(\exists v)(\exists p)(\exists b)\\ \big(u^2 - (a^2y^2 - 1)v^2 = 1 \,\&\, xa_z \leqslant u < a_z(x + 1)\\ \leqslant (y + z - 1)a_z b \,\&\, p^2 - \big[(y + z)^2 - 1\big]\big[y + z - 1\big]^2 b^2 = 1\big),\end{aligned} \tag{21}$$

where $a = b(y + z - 1)$.

Since the predicates $x < y$ and $y = a_x$ are Diophantine, so is the predicate $x = y^z$. Formula (21) yields, on the basis of (a)–(e), an explicit Diophantine representation of the predicate $x = y^z$.

We have already observed that from an explicit Diophantine representation of the function $x = y^z$ there follows (see [**3**]) a Diophantine representation of the set of all primes. This explicit representation is very cumbersome and there is no need to present it here. However, many interesting recursively enumerable sets of primes admit a relatively simple Diophantine representation. This is true of the Mersenne primes $\mathfrak{M}_n$, i.e. the primes of the form $2^n - 1$.

To see this we use the following well-known result:

THEOREM 5.3 (LEHMER [5]).[4] *The number* $\mathfrak{M}_n$ *(n odd prime) is prime if and only if* $\mathfrak{M}_n$ *is a divisor of the* $(n-1)$*th term of the sequence* $S_1, S_2, \ldots$, *where* $S_1 = 4$ *and* $S_i = S_{i-1}^2 - 2$, $i = 2, 3, \ldots$.

Note the following expression for the sequence S_n in terms of 2_n (i.e. in terms of the sequence a_n with $a = 2$): $S_n = 2 \cdot 2_{2^{n-1}}$. The proof is given below.

For $n = 1$ we have $2_{2^{n-1}} = 2_1 = 2 = \frac{1}{2}S_1$. Assume that $S_n = 2 \cdot 2_{2^{n-1}}$ for an $n \geqslant 1$. Then

$$S_n^2 - 2 = 2\left(2 \cdot 2_{2^{n-1}}^2 - 1\right).$$

According to formula (11) with $b = 2d = 2, j = 2$, we have

$$2_{2^n} = 2_{2^n}^2 + 3 \cdot 2_{2^n}'^2 = 2 \cdot 2_{2^n}'^2 - 1.$$

Therefore, $S_n^2 - 2 = 2 \cdot 2_{2^n} = S_{n+1}$ and we have established the desired representation.

Theorem 5.3 can therefore be written in the following form:

(*) The number $2^n - 1 = \mathfrak{M}_n$ is prime (for a prime n) if and only if $\mathfrak{M}_n \mid 2_{2^{n-1}}$ (or $\mathfrak{M}_n \mid 2_{(\mathfrak{M}_n+1)/2}$).

The assertion (*) yields a simple explicit Diophantine representation of the function $y = \mathfrak{M}_x$ (for prime x). In view of (*), we obtain

$$y = \mathfrak{M}_x \leftrightarrow y + 1 = 2^x \,\&\, (\exists z)\left(zy = 2_{(y+1)/2}\right),$$

for prime x.

Thus, a system of Diophantine equations defining $y = \mathfrak{M}_x$ (for prime x) consists of equations (a)–(e) with y replaced by zy, x by $(y+1)/2$, and a by 2, and equation (21) with x replaced by $y + 1$, y by 2, and z by x.

In view of the above results and the fact that there exists a universal polynomial representing all recursively enumerable sets (which we denote by $Y(x_1, \ldots, x_k, n)$), it is of particular interest to know the smallest possible number of variables in this polynomial. A rough upper bound for the number of variables k can be obtained by means of the methods expounded in [**1**], a result of R. Robinson [**2**] (see §0), and an explicit Diophantine representation of the function $x = y^z$.

The estimate $l \leqslant 10^3$, for example, obviously does not reflect the true state of affairs, since it can be lowered by an argument somewhat more subtle than that of [**1**].

Obviously, it would be most desirable to have an estimate of the type found in [**2**], namely, $l \leqslant 6$. However, the likelihood is that the present method is not capable of yielding such a precise estimate.

To conclude this section, consider the following question raised by J. Robinson [**7**] ($n \geqslant 3$): Is the function $y = n^x$ Diophantine in the set Pow_n of all powers of n?

---

[4] See also [**6**].

It follows from §4 that the answer to this question is positive. However, it is possible to give a direct answer to this question, using only the methods and techniques of J. Robinson [7], which are similar in many respects to those of §2 (and §4).

In complete analogy with [7] we can prove

PROPOSITION A. *Any recursively enumerable set is Diophantine in the set of all numbers of the form n^{2^l} (n even).*

Also, in the case $n = 6$, for example, we obtain

PROPOSITION B. $y = 6^{2^x}$ *for some* $x > 0$ *if and only if* $y \in \mathrm{Pow}_6$ *and there exist natural numbers u and v such that* $y = 7(u^2 + v^2) + 1$.

It follows directly from Propositions A and B that, for example, for any recursively enumerable set S there exists a polynomial P with integral coefficients such that $S = \{x\colon P(x, y_1,\dots,y_n) = 6^t \text{ for certain } y_1,\dots,y_n, t\}$.

6. We now consider model-theoretic aspects for our results on Diophantine predicates.

We will first establish a model-theoretic characterization of Diophantine predicates (this characterization is analogous to that of definable predicates).

We use standard terminology and notation of the theory of models. If $\mathfrak{A}_i = \langle A_i, R^i_\xi\rangle_{\xi\in Z}$ $(i \in I)$ is a model of type $\langle \zeta, n_\xi\rangle_{\xi\in Z}$ (i.e. $R^i_\xi \subseteq A_i^{n_\xi}$), then $\mathrm{prod}\,\lambda i\mathfrak{A}_i$ denotes the direct product of the $\mathfrak{A}_i$, and if $\mathfrak{D}$ is an ultrafilter on I, then $\mathfrak{D}\text{-prod}\,\lambda i\mathfrak{A}_i$ denotes the factor-set of $\mathrm{prod}\,\lambda i\mathfrak{A}_i$ with respect to the relation $f \equiv_{\mathfrak{D}} g \leftrightarrow \{i\colon f(i) = g(i)\} \in \mathfrak{D}$ with the naturally defined predicates R_ξ (see [**10**]).

If $\mathfrak{A}_i = \mathfrak{A}$ for all $i \in I$, then we write $\mathfrak{D}\text{-prod}\,\lambda i\mathfrak{A}_i = \mathfrak{A}^I/\mathfrak{D}$. The model $\mathfrak{D}\text{-prod}\,\lambda i\mathfrak{A}_i$ is called an ultraproduct of the models $\mathfrak{A}_i$, and $\mathfrak{A}^I/\mathfrak{D}$ is called an ultrapower of $\mathfrak{A}$.

In these terms we can characterize Diophantine (i.e. existentially definable) predicates in the model $\mathfrak{N} = \langle N; +, \times, 0, 1, \dots\rangle$.

THEOREM 6.1. *A predicate $P(x_1,\dots,x_n)$ defined in $\mathfrak{N}$ is Diophantine if and only if, for any I, J and ultrafilters $\mathfrak{D}$ on I, $\mathfrak{E}$ on J, the predicate $\tilde{P}(x_1,\dots,x_n)$ defined in $\mathfrak{N}^I/\mathfrak{D}$ by $(\mathfrak{N}, P)^I/\mathfrak{D} = (\mathfrak{N}^I/\mathfrak{D}, \tilde{P})$ satisfies the following condition*:

(Im) *For any imbedding F of $\mathfrak{N}^I/\mathfrak{D}$ into $\mathfrak{N}^J/\mathfrak{D}$, it follows from $\mathfrak{N}^I/\mathfrak{D} \models \tilde{P}(f_1/\mathfrak{D},\dots,f_n/\mathfrak{D})$ that $\mathfrak{N}^J/\mathfrak{E} \models \tilde{P}(F(f_1/\mathfrak{D}),\dots,F(f_n/\mathfrak{D}))$.*

PROOF. Necessity is obvious, since it is known that existential formulas are preserved under passage to extensions.

Let us prove sufficiency. Let $K_1 = \{(\mathfrak{N}^I/\mathfrak{D}; c_1,\dots,c_n)\colon \mathfrak{D}$ is an ultrafilter on I, $\mathfrak{N}^I/\mathfrak{D} \models \tilde{P}(c_1,\dots,c_n)\}$, $K_2 = \{(\mathfrak{N}^I/\mathfrak{D}; c_1,\dots,c_n)\colon \mathfrak{D}$ is an ultrafilter on I, $\mathfrak{N}^I/\mathfrak{D} \models \neg\tilde{P}(c_1,\dots,c_n)\}$, and $K = K_1 \cup K_2$. We will show that the classes K_1 and K_2 are axiomatizable within K and that K_2 is universally axiomatizable.

Then K_2 will be defined within K by one universal axiom, i.e. K_1 will be defined within K by one existential axiom and the relation $P(x_1,\dots,x_n)$ will be Diophantine.

Indeed, if the K_i are defined by systems Σ_i: $i = 0, 1$, then the system $\Sigma_1 \cup \Sigma_2$ has no models within K. Since K is ultraclosed, it can be shown by the usual method **[10]** that $\Sigma_1' \cup \Sigma_2'$ has no models within K, where $\Sigma_1' \subset \Sigma_1$ and $\Sigma_2' \subset \Sigma_2$ are finite systems of axioms. Then $\&\Sigma_1'$ and $\&\Sigma_2'$ define K_1 and K_2, and K_1 is defined in K by the existential axiom $\neg(\&\Sigma_2')$.

Since K_2 is ultraclosed, to prove its universality in K it suffices to show heredity of K_2 in K. This is obvious, since from F: $(\mathfrak{N}^I/\mathfrak{D};\, c_1,\ldots,c_n) \to (\mathfrak{N}^J/\mathfrak{E};\, d_1,\ldots,d_n)$, where $\mathfrak{N}^J/\mathfrak{E} \models \neg\tilde{P}(d_1,\ldots,d_n)$, we obtain, according to (Im), $\mathfrak{N}^I/\mathfrak{D} \models \neg\tilde{P}(c_1,\ldots,c_n)$.

Similarly, in view of (Im), K_1 contains all subsystems of K of its models. Therefore, K_1 is axiomatizable in K, since K_1 is ultraclosed and from $\mathfrak{M}_0 \equiv \mathfrak{M}_1$, where $\mathfrak{M}_1 \in K_1$, follows F: $\mathfrak{M}_0 \prec \mathfrak{M}_1^I/\mathfrak{D}$, where $\mathfrak{M}_1^I/\mathfrak{D} \in K_1$, i.e. $\mathfrak{M}_0 \in K_1$.

Consequently, K_1 is defined in K by one existential axiom.

Actually, we have a somewhat more precise result.

PROPOSITION 6.2. *A predicate $P(x_1,\ldots,x_n)$ on N is Diophantine if and only if, for some nonprincipal ultrafilter $\mathfrak{D}$ on ω and any imbedding F of $\mathfrak{N}^\omega/\mathfrak{D}$ into $\mathfrak{N}^\omega/\mathfrak{D}$, $\mathfrak{N}^\omega/\mathfrak{D} \models \tilde{P}(f_1/\mathfrak{D},\ldots,f_n/\mathfrak{D})$ implies $\mathfrak{N}^\omega/\mathfrak{D} \models \tilde{P}(F(f_1/\mathfrak{D}),\ldots,F(f_n/\mathfrak{D}))$.*

Proposition 6.2 is proved under the assumption of the continuum hypothesis $2^\omega = \omega_1$; however, it follows from corollaries of Proposition 6.2 (e.g., 6.3 and 6.4) that the assumption $2^\omega = \omega_1$ can be eliminated in the usual way if we are dealing with specific arithmetical predicates.[5]

The proof of 6.2 is like that of 6.1, but we use results on ultrapowers which are valid under the assumption $2^\omega = \omega_1$ **[10]**.

A. The system $\mathfrak{N}^\omega/\mathfrak{D}$ is saturated for any nonprincipal ultrafilter $\mathfrak{D}$ on ω.

This implies

B. For any nonprincipal ultrafilters $\mathfrak{D}$ and $\mathfrak{E}$ on ω,

$$\mathfrak{A}^\omega/\mathfrak{D} \cong \mathfrak{M}^\omega/\mathfrak{D} \quad \text{for } \mathfrak{A} \equiv \mathfrak{M},\ \mathrm{card}(\mathfrak{A}),\ \mathrm{card}(\mathfrak{M}) \leqslant 2^\omega$$

(the symbol $\cong$ denotes isomorphism, and $\equiv$ elementary equivalence). For example, $(\mathfrak{A}^\omega/\mathfrak{D})^\omega/\mathfrak{D} \cong \mathfrak{A}^\omega/\mathfrak{D}$ if $\mathrm{card}(\mathfrak{A}) \leqslant 2^\omega$.[6]

Consequently, the question of the Diophantine nature of specific predicates in $\mathfrak{N}$ reduces to a purely algebraic question.

This reduction enables us to obtain another, model-theoretic, proof of Theorem 2.6 on the Diophantine nature of the predicate $y = a_x$ (or $y = 2_x$).

COROLLARY 6.3. *If $\mathfrak{D}$ is a nonprincipal ultrafilter on ω and F is an imbedding of $\mathfrak{N}^\omega/\mathfrak{D}$ into $\mathfrak{N}^\omega/\mathfrak{D}$, then $\mathfrak{N}^\omega/\mathfrak{D} \models f/\mathfrak{D} = a_{g/\mathfrak{D}}$ implies $\mathfrak{N}^\omega/\mathfrak{D} \models F(f/\mathfrak{D}) = a_{F(g/\mathfrak{D})}$.*

---

[5] Of course, arithmetical predicate means elementarily definable predicate.

[6] B follows from A in view of a theorem of Łoś (see **[10]**) to the effect that $\mathfrak{A}^I/\mathfrak{D} \equiv \mathfrak{A}$ and the fact that elementarily equivalent saturated models of the same cardinality are isomorphic.

The proof of Corollary 6.3 uses properties of the sequence $a_{f/\mathfrak{D}}$ in $\mathfrak{N}^\omega/\mathfrak{D}$ similar to those considered in §§1 and 2, but the proof is much simpler than that of Theorem 2.6.

The technique of ultraproducts enables us to establish the Diophantine nature of many auxiliary predicates needed in the proof of 2.6, and certain tedious calculations of §§1–3 turn out to be unnecessary.

By means of 6.2 we can give a direct proof that the function $x = y^z$ is Diophantine. However, the proof of this fact using 6.2 does not yield an explicit Diophantine predicate defining $x = y^z$.

In view of 6.2 (or 4.2), from 6.3 we obtain

COROLLARY 6.4. *A predicate $P(x_1,\ldots,x_n)$ in $\mathfrak{N}$ is recursively enumerable if and only if, for a nonprincipal ultrafilter $\mathfrak{D}$ on ω and any imbedding $F\colon \mathfrak{N}^\omega/\mathfrak{D} \to \mathfrak{N}^\omega/\mathfrak{D}$, $\mathfrak{N}^\omega/\mathfrak{D} \models \tilde{P}(f_1/\mathfrak{D},\ldots,f_n/\mathfrak{D})$ implies $\mathfrak{N}^\omega/\mathfrak{D} \models \tilde{P}(F(f_1/\mathfrak{D}),\ldots,F(f_n/\mathfrak{D}))$.*

The result that any recursively enumerable set is Diophantine has applications in the study of properties of models of arithmetic.

We consider only strong models $\mathfrak{M}$ of arithmetic, i.e. those models for which $\mathfrak{M} \equiv \mathfrak{N}$ (hence $\mathfrak{M} \succ \mathfrak{N}$).

If $\mathfrak{M}_1$, $\mathfrak{M}_2$ are strong models of arithmetic and $\mathfrak{M}_1 \subseteq \mathfrak{M}_2$, then $\mathfrak{M}_2 = \langle M_2; +,\ldots\rangle$ is called a cofinal extension of $\mathfrak{M}_1 = \langle M_1; +,\ldots\rangle$ if for any $x \in M_2$ there exists $y \in M_1$ such that $y >_{M_2} x$.[7]

PROPOSITION 6.5 [**11**]. *For any nonstandard strong model $\mathfrak{M}$ (i.e. $\mathfrak{M} \not\cong \mathfrak{N}$) of arithmetic there exists a nontrivial cofinal extension $\mathfrak{M}_0$ (which is also a nonstandard strong model of arithmetic).*

Note the following application of Theorem 4.2.

THEOREM 6.6.[8] *Suppose $\mathfrak{M}_1$, $\mathfrak{M}_2$ are strong models of arithmetic and $\mathfrak{M}_2$ is a cofinal extension of $\mathfrak{M}_1$. Then $\mathfrak{M}_2$ is an elementary extension of $\mathfrak{M}_1$.*

PROOF. According to Theorem 4.2, any recursively enumerable predicate is Diophantine, i.e. equivalent in $\mathfrak{N}$ to an existential formula. Since existential formulas are preserved under passage to supersystems, it follows that for any universally bounded formula ψ there exists an existential formula equivalent to it in $\mathfrak{N}$.

Note that since $\mathfrak{M}_1$ is a strong model of arithmetic, it follows from the truth of

$$(\forall x_1 \leqslant a_1)(\exists y_1 \leqslant b_1) \cdots (\forall x_{k+1})(\exists y_{k+1}) P(\bar{x}, \bar{y})$$

in $\mathfrak{M}_1$, where $a_i, b_i \in M_1$, that for any $a_{k+1} \in M_1$ there exists $b_{k+1} \in M_1$ such that

$$(\forall x_1 \leqslant a_1)(\exists y_1 \leqslant b_1) \cdots (\forall x_{k+1} \leqslant a_{k+1})(\exists y_{k+1} \leqslant b_{k+1}) P(\bar{x}, \bar{y})$$

is true in $\mathfrak{M}_1$.

[7]The symbol $>_{M_2}$ denotes an arithmetically definable order relation in $\mathfrak{M}_2$.

[8]In connection with this theorem cf. [**11**].

Applying this observation to the axiom $\varphi = (\forall x_1) \cdots (\exists y_n)\psi(\bar{x}, \bar{y})$, we see that for any $a_1, \ldots, a_n \in M_1$ there exist $b_1, \ldots, b_n \in M_1$ such that

$$\varphi' = (\forall x_1 \leqslant a_1) \cdots (\exists y_n \leqslant b_n)\psi(\bar{x}, \bar{y})$$

is true in $\mathfrak{M}_1$.

But, in view of 4.2, the predicate

$$P(\alpha_1, \ldots; \beta_1, \ldots) \leftrightarrow (\forall x_1 \leqslant \alpha_1) \cdots (\exists y_n \leqslant \beta_n)\psi(\bar{x}, \bar{y})$$

is Diophantine.

Therefore, φ' is true in $\mathfrak{M}_2$.

Since $\mathfrak{M}_2$ is a cofinal extension of $\mathfrak{M}_1$, it follows that if for any a_{k+1} there exists b_{k+1} such that

$$(\forall x_1 \leqslant a_1) \cdots (\forall x_{k+1} \leqslant a_{k+1})(\exists y_{k+1} \leqslant b_{k+1})P(\bar{x}, \bar{y})$$

is true in $\mathfrak{M}_2$, then

$$(\forall x_1 \leqslant a_1) \cdots (\forall x_{k+1})(\exists y_{k+1})P(\bar{x}, \bar{y})$$

is true in $\mathfrak{M}_2$.

Therefore, if φ' is true in $\mathfrak{M}_2$, then φ is true in $\mathfrak{M}_2$.

7. We will study some specific Diophantine equations using the results of elementary number theory obtained in §§1–3 concerning the sequence of solutions of Pell's equation.

We first consider an equation of the type $x^2 - Dy^4 = 1$ or $x^4 - Dy^2 = 1$ (see **[12]**).

We then consider (in some examples) the case of a system of two equations of Pell type. In conclusion, we examine equations of the forms $1^n + \cdots + (m-1)^n = \lambda m^n$ and

$$9(u^2 + 7v^2)^2 - 7(w^2 + 7y^2)^2 = 2.$$

Consider the problem of the solvability of the equation $x^2 - Dy^4 = 1$. Ljunggren (see **[13]**) showed that the equation $x^2 - Dy^4 = 1$, where $D > 0$ and D is not a perfect square, has at most two solutions in positive integers. But, as Mordell **[13]** noted, explicit formulas for D were not given for the cases where $x^2 - Dy^4 = 1$ has no solutions or has one or two positive solutions. It has been shown **[13]** that for primes $p \equiv 1 \pmod 4$ the equation $x^2 - py^4 = 1$ has no solutions except when $p = 5$, in which case there exists a solution $(9, 2)$.

In the investigations **[12]** and **[13]** of the equation $x^2 - Dy^4 = 1$ there were restrictions on D of the type $D \nmid 4$, $D \equiv 1, 2, 5 \pmod 8$, or the equation $x^2 - Dy^2 = -1$ has a solution.

Below we study the equation $x^2 - Dy^4 = 1$ in the general case, which reduces to an equation of the form $Ax^4 + By^4 = C$.

Consider equations $x^2 - Dy^4 = 1$, where $x^2 - Dy^2 = -1$ has no solution in integers.

The simplest examples of such equations are $x^2-(a^2-1)y^2=1$. We investigate them by means of §§1 and 2.

Suppose $a'_m=x^2$ and $n=2m$ is even. It follows from $a'_{2m}=2a'_m a_m=x^2$ that $a'_m=y^2$, $a_m=2z^2$ or $a'_m=2y^2$, $a_m=z^2$. Suppose $a'_m=y^2$ and $a_m=2z^2$. Then m is odd and a is even.

Thus, $a'_m=y^2$, $(2a-2)a'_m=2A'_mA_m$ in view of (11), since m is odd. According to 1.8, $(A'_m, A_m)=1$ and $a-1\mid A_m$. Let $K_m=A_m/(a-1)$ (see §2). Then $a'_m=K_mA'_m$ and $(K_m, A'_m)=1$; since $a'_m=y^2$, we have $K_m=u^2$, $A'_m=v^2$. Since, according to 3.1,

$$A_m^2-(a^2-1)A_m'^2=-(2a-2),$$

it follows that

$$(a-1)^2u^4-(a^2-1)v^4=-(2a-2),$$

or $(a+1)v^4-(a-1)u^4=2$.

In view of [**14**], this equation has only one solution $(\pm 1,\pm 1)$. Consequently, $m=1$. Since $a_m=2z^2$, we obtain $a=2z^2$, or $a/2=z^2$.

Consider the case $a'_m=2y^2$, $a_m=z^2$. Then m is even, $m=2k$, and $a'_{2k}=2a_ka'_k=2y^2$, from which it follows that $a_k=w^2$, $a'_k=l^2$.

Suppose $k=2^si$, $(i,2)=1$, and $s\geqslant 2$. Then $a'_{2^{s-1}\cdot i}=2c^2$, $a_{2^{s-1}\cdot i}=d^2$.

Then (a_m, a'_m) and $(a_{2^{s-1}\cdot i}, a'_{2^{s-1}\cdot i})$ for $i\neq 0$ are distinct solutions of $x^4-4(a^2-1)y^4=1$, which is impossible.

Consequently, $s\leqslant 1$. If $s=1$, then $k=2i$, $(i,2)=1$, and $a'_i=p^2$, $a_i=2h^2$, which implies that $i=1$, $a/2=z^2$, $a_2=w^2=2a^2-1=8z^4-1$.

But $w^2=8z^4-1$ is impossible, so $s=0$ and k is odd. Therefore, $k=1$, $a=e^2$, $m=2$, and $a_2=z^2=2a^2-1=2e^4-1$, i.e. $a=169$.

Let us now turn to the case $a'_n=x^2$, n odd. Then $a'_n=K_nA'_n$ and $K_n=z^2$, $A'_n=w^2$.

Therefore,

$$(a-1)^2z^4-(a^2-1)w^4=-(2a-2),$$
$$(a-1)z^4-(a+1)w^4=-2,$$

and the unique solution of this equation is $z=w=1$, i.e. $n=1$.

Consequently, we obtain

THEOREM 7.2. *The equation $x^2-(a^2-1)y^4=1$ has solutions $y=0,1$. It has the solution $2a$ when $a=2z^2$ and the solution $8a^3-4a=6214^2$ when $a=169$.*

We now consider the general case of the equation $x^2-Dy^4=1$, where D is not a perfect square.

Let (a,b) be the smallest solution of $x^2-Dy^2=1$. We define sequences u_n, v_n as follows:

$$u_0=1,\qquad v_0=0,\qquad u_1=a-1,\qquad v_1=b;$$
$$u_{2n+1}=(2a-2)u_{2n}+u_{2n-1};\qquad v_{2n+1}=(2a-2)v_{2n}+v_{2n-1};$$
$$u_{2n+2}=u_{2n+1}+u_{2n};\qquad v_{2n+2}=v_{2n+1}+v_{2n}.$$

Below, b^* denotes the square-free part of b.

Following 1.2, it is easy to show that $u_n = A_n$, $v_n = A'_n b$. It follows that $v_{2(2n+1)} = v_{2n+1}u_{2n+1}/(a-1)$ and $v_{2(2n)} = 2v_{2n}u_{2n}$.

In view of 3.1, the pairs (u_{2n}, v_{2n}) constitute all solutions of $x^2 - Dy^2 = 1$, and the pairs $(u_{2n+1}/(a-1), v_{2n+1}/b)$ constitute all solutions of $(a-1)x^2 - (a+1)y^2 = -2$.

We will find out when $v_{2m} = x^2$.

Suppose first that m is odd, $m = 2i+1$. Then, since each $u_{2i+1}/(a-1)$ is odd, we have $(u_{2i+1}/(a-1), v_{2i+1}) = 1$ and $v_{2i+1} = z^2$, $u_{2i+1}/(a-1) = y^2$. Thus, $v_{2i+1}/b = z^2/b = b^*w^2$. Therefore,

$$(*) \qquad (a-1)y^4 - (a+1)b^{*2}w^4 = -2.$$

This equation can be satisfied by at most one pair (y, w) when $a > 1$, and $2 \mid b^*$, since $u_{2i+1}/(a-1) \equiv \pm 1 \pmod{a+1}$.

Now suppose m is even, $m = 2n$; then either $v_{2n} = 2y^2$, $u_{2n} = z^2$ or else $v_{2n} = y^2$, $u_{2n} = 2z^2$. If $v_{2n} = y^2$, $u_{2n} = 2z^2$, then n is odd and a is even. By what was shown above, we obtain

$$(a-1)t^4 - (a+1)b^{*2}w^4 = -2,$$

where $t^2bb^*w^2 = y^2 = v_{2n}$ and also $u_{2n} = 2z^2$.

Assume that $v_{2n} = 2y^2$, $u_{2n} = z^2$, and $n = 2^s i$, $(i, 2) = 1$.

Then $s < 2$. Indeed, if $s \geqslant 2$, then $s - 2 \geqslant 0$ and $v_{2^s i} = c_1^2$, $u_{2^s i} = d_1^2$; $v_{2^{s-1}\cdot i} = c_2^2 \cdot 2$, $u_{2^{s-1}\cdot i} = d_2^2$, and $2^{s-1}i = 2 \cdot 2^{s-2}i$.

Consequently, (u_{2n}, v_{2n}) and $(u_{2^{s-1}\cdot i}, v_{2^{s-1}\cdot i})$ are distinct solutions for $i \neq 0$ of the equation $x^4 - 4Dy^4 = 1$, which is impossible in view of [**14**]. For $s = 0$ we obtain $(*)$ with $b^*/2 = (b/2)^*$.

Thus, $s = 1$, since we obviously have $s \geqslant 0$. In this case, $v_{2i} = c^2$, $u_{2i} = d^2$, and i is odd, i.e. $u_i/(a-1) \cdot v_i = c^2$ and, as above, we obtain

$$(a-1)t^4 - (a+1)b^{*2}w^4 = -2,$$

where $t^2bb^*w^2 = c^2$ and also $u_n = d^2$, $u_{2n} = z^2$, $n = 2i$, $(i, 2) = 1$.

From the above we obtain

THEOREM 7.3. *The equation $x^2 - Dy^4 = 1$ has at most two solutions in natural numbers. These solutions are determined from solutions of the equation $(*)$ for b^* or $(b/2)^*$.*

If $(*)$ has no solutions in natural numbers, then $x^2 - Dy^4 = 1$ has no solutions.

A theorem similar to 7.3 holds for the equation $x^4 - Dy^2 = 1$.

We illustrate our results by the examples $D = 37, 101, 141, 189, 197$. As noted in [**12**], these are the only numbers $D \equiv 5 \pmod 8$, $D < 200$, for which the equations $x^2 - Dy^4 = \pm 1$ cannot be solved by Cohn's method.

The smallest positive solutions of $x^2 - Dy^2 = 1$ for $D = 189, 141$ are, respectively, (55, 4) and (95, 8). Therefore, it follows from Theorem 7.3 (or 7.2) that the only solution of $x^2 - Dy^4 = 1$ for $D = 189$ is (55, 4).

For $D = 141$ we have $a = 95$, $b = 8$, and $b^* = 2$. Consider the equation $b^{*2}(a+1)x^4 - (a-1)y^4 = 2$, i.e. $2(a+1)x^4 - (a-1)y^4/2 = 1$, where y^2 is an element of the sequence y_n defined by the conditions $y_0 = 1$, $y_1 = 2a+1$, $y_{n+2} = 2ay_{n+1} - y_n$.

Obviously, $y_n \equiv (-1)^n \pmod{a+1}$, i.e. $(a-1)/2 \equiv -1 \pmod{a+1}$, which contradicts $a = 95 > 1$.

It follows from 7.3 that $x^2 - Dy^4 = 1$ has no positive solutions for $D = 141$.

For $D = a^2 + 1$ with $a = 6, 10, 14$ (i.e. for $D = 37, 101, 197$) it is easy to see that the smallest solution of $x^2 - Dy^2 = 1$ is $x = 2a^2 + 1$, $y = 2a$, where for $a = 6, 10, 14$ we have $y^* = a/2$.

Then, considering the equation

$$y^{*2}(x+1)z^4 - (x-1)w^4 = 2$$

we see that the left-hand side is divisible by $a/2$, which is impossible for $a > 1$.

Consequently, such equations have no solutions x, $y > 0$.

By means of formulas of type (19), (20) concerning the periodicity of the sequence a_n of solutions of Pell equations (mod a_m) it can be explicitly established in specific cases when it is that $a_n = x^2$. To do this we must consider the Jacobi symbol $(a_n \mid a_m)$ and use the congruences (20).

More concretely, suppose we are given an equation $x^2 - Dy^2 = 1$. Consider the sequence (u_n, v_n) of solutions of this equation. If (a, b) is the smallest solution, then $u_n = a_n$, $v_n = ba'_n$.

It follows that $u_{3n} \neq x^2$ and $v_{3n} \neq x^2$ for $n > 0$. Indeed, according to (11), $u_{3n} = u_n(4u_n^2 - 3)$, $v_{3n} = v_n(4u_n^2 - 1)$. Then $u_{3n} = x^2$ implies $4y^4 - 3 = x^2$ or $12y^4 - 1 = x^2$, which is impossible for $x > 1$.

From $v_{3n} = x^2$ follows $v_n = y^2$ and $4u_n^2 - 1 = z^2$, i.e. $4y^2 - z^2 = 1$. This equation has no positive solutions.

Consideration of only the cases $u_n = x^2$, $v_n = y^2$ for $n = 3i \pm 1$ enables us to directly apply (20).

The solution of systems of Diophantine equations

$$A_1x^2 + B_1y^2 = C_1 \quad \text{and} \quad A_2x^2 + B_2y^2 = C_2$$

also reduces to the study of properties of sequences of solutions of Pell equations.

For example, the system $x^2 - Dy^2 = 1$, $Ax^2 + By^2 = C$ is equivalent to the equation $u_n^2 = (C - By^2)/A$, i.e. $u_{2n} = 2(C - By^2)/A - 1$, where (u_n, v_n) is the sequence of solutions of $x^2 - Dy^2 = 1$.

Thus, we need only investigate the condition $u_{2n} = 2C/A - 1 - (2B/A)y^2$, where A, B, C are certain integers.

Formulas of type (20) are also applicable to an investigation of the condition $u_{2n} = 2C/A - 1 - 2By^2/A$.

Certain questions of elementary number theory lead to a system of two Pell equations.

For example, consider the question of Lucas (see **[6]**), which was solved by Watson and Ljunggren: When is $1^2 + \cdots + n^2$ a square, i.e. when does the equation $n(n+1)(2n+1)=6x^2$ have a solution? Since $(n, n+1)=(n+1, 2n+1)=(n, 2n+1)=1$, this equaiton is equivalent to a set of two systems

$$y^2-12x^2=1, \qquad y^2-2z^2=-1, \qquad n=6x^2.$$

$$2z^2-x^2=1, \qquad 3y^2-2x^2=1, \qquad n=x^2.$$

Using the technique discussed above, we see that the solution of the first system is $x=2$, $y=7$, $z=5$ and the solution of the second is $x=1$, $y=1$, $z=1$, i.e. $n=1, 24$.

Indeed, according to (11), the first system can be written as follows:

$$u_m^2=2z^2-1 \quad \text{or} \quad u_{2m}=4z^2-3,$$

where (u_n, v_n) is the sequence of solutions of $y^2-12x^2=1$.

It can be proved in an elementary way that $u_{2m}=4z^2-3$ is possible only when $m=1$, i.e. $z=5$, $y=7$, $x=2$.

It requires a somewhat more subtle argument to prove that the system $2z^2-x^2=1$, $3y^2-2x^2=1$ has the unique solution $x=y=z=1$.

Using a similar technique of elementary number theory we study the Diophantine equation $1^n+\cdots+(m-1)^n=\lambda m^n$, where m, n, λ are natural numbers, $m, n>1$. The question of the solvability of this equation in the case $\lambda=1$ was stated in [**6**] as a problem of Erdős.

Consider the case where λ is odd. It is easy to show that the equation $1^n+\cdots+(m-1)^n=\lambda m^n$ has a solution in this case only when $m=4k+3=2^c i+3$, where $c \geqslant 2$, $(i,2)=1$, $n=2^d j$, $(j,2)=1$.

We have

$$1^n+2^n+\cdots+(2^c\cdot i)^n+(1-\lambda)\big((2^c\cdot i+1)^n+(2^c\cdot i+2)^n\big)$$
$$=\lambda\big((2^c\cdot i+3)^n-(2^c\cdot i+2)^n-(2^c\cdot i+1)^n\big).$$

The right-hand side of this equation is divisible by 2^{d+2}, but not by 2^{d+3}. If $c>d+3$, then the left-hand side is divisible by 2^{d+2}, but not by 2^{d+3} when $\lambda=(m-1)/2-(2t+1)2^{d+2}$. On the other hand, it can be shown that it is impossible that n be odd.

In a similar way we can establish

THEOREM 7.4. *The equation* $1^n+\cdots+(m-1)^n=\lambda m^n$ *can have a solution only when* $\lambda=(m-1)/2-(2t+1)2^{d+2}$, $m=4k+3$, $n=2^d j$, *and*

$$(2t+1)2^{d+3}(n+1)/(n-1)<m<(2t+1)2^{d+3}(n+1)/(n-1)+1,$$

where $(m-1,\lambda)=1$, $c>d+3$.

Analogously, the equation $1^n+\cdots+(m-1)^n=\lambda m^n$, where $m, n>1$ and $m-1$ is prime, is unsolvable.

In conclusion, we remark that by studying the special equation

$$9(u^2+7v^2)^2-7(w^2+7y^2)^2=2$$

it is possible, in view of results of M. Davis, to obtain another proof of Theorem 4.2.

Solvability of this equation is connected with the question of when $v_n = r^2 + Ds^2$, where (u_n, v_n) is the sequence of solutions of $x^2 - Dy^2 = 1$.

The main results of §§1–5 were obtained by the author in 1969 and the beginning of 1970 (see **[9]**) and were reported at the Steklov Institute of Mathematics (Moscow) and Kiev University.

REFERENCES

1. M. Davis, H. Putnam and J. Robinson, *The decision problem for exponential diophantine equations*, Ann. of Math. (2) **74** (1961) No. 3, 425–436.
2. R. Robinson, *Arithmetical representations of recursively enumerable sets*, J. Symbolic Logic **21** (1956), 162–186.
3. J. Robinson, *Existential definability in arithmetic*, Trans. Amer. Math. Soc. **72** (1952), 437–449.
4. M. Davis, *Extensions and corollaries of recent work on Hilbert's tenth problem*, Illinois J. Math. **7** (1963), 246–250.
5. D. Lehmer, *An extend theory of Lucas functions*, Ann. of Math. (2) **31** (1930), 419–448.
6. W. Sierpinsky, *Elementary theory of numbers*, Warszawa, 1964.
7. J. Robinson, *Unsolvable diophantine problems*, Proc. Amer. Math. Soc. **22** (1969), 534–538.
8. A. Ja. Khinchine, *Continued fractions*, Chicago Univ. Press, 1964.
9. G. V. Chudnovsky, *Diophantine predicates*, Uspehi Mat. Nauk **25** (1970), no. 4 (154), 171, 185–186. (Russian)
10. H. J. Keisler, *Ultraproducts and saturated models*, Indag. Math. **26** (1964), 178–186.
11. M. Rabin, *Diophantine equations and non-standard models*, Logic, Method., Philos. of Science, Stanford, 1962 (reprinted in Mathematical Logic and its Applications, Moscow, 1967).
12. J. Cohn, *Five diophantine equations*, Math. Scand. **21** (1967), 61–70.
13. J. Mordell, *The Diophantine equation* $y^2 = Dx^4 + I$, J. London Math. Soc. **39** (1964), 161–164.
14. T. Scolem, *Diophantishe Gleichungen*, Springer-Verlag, 1938.
15. Ju. V. Matijasevič, *The Diophantineness of enumerable sets*, Dokl. Akad. Nauk SSSR **191** (1970), 279–282. (Russian)
16. G. Hardy and E. Wright, *An introduction to the theory of numbers*, Oxford, 1945.
17. A. I. Mal'cev, in Proceedings of IMC (Moscow, 1966), Nauka, Moscow, 1968.

Translated by G. A. KANDALL

CHAPTER 7

TRANSCENDENCES ARISING FROM EXPONENTIAL AND ELLIPTIC FUNCTIONS

Introduction. We present below a series of articles on the transcendence and algebraic independence of values of exponential and elliptic functions. Most of these materials were prepared in 1978 as preprints and appear here with some modifications. This chapter of the volume opens with the introductory paper based on the invited lecture presented at the Colloquium of Paris Universities of November 9, 1977. The sections that follow contain proofs of the results formulated in the Introduction to the volume and supply additional details to these results.

1. Principles of transcendence proofs. We will give the review on transcendental numbers arising from exponential and elliptic functions. In fact, only the investigation of these numbers as well as numbers connected with Abelian functions are in the center of transcendence number theory. It is clear why this happens: these numbers often occur in algebra and algebraic geometry. We will discuss the transcendence proofs, results obtained, conjectures and applications.

First of all we will give an example of a transcendence result and from it see how all the proofs are constructed. Interesting results in transcendence theory are always connected with algebraic points w of analytic functions $f_1(z),\ldots,f_n(z)$. The first natural question: how many w are there such that all the numbers $f_1(w),\ldots,f_n(w)$ are algebraic? Generally speaking we cannot say anything: there are many examples which show that the number of w can be infinite, etc. However if we suppose that $f_i(z)$ satisfy algebraic differential equations and are meromorphic functions of finite order of growth, then a partial answer can be given. In this direction there is a Schneider-Lang theorem [**S1**, **L1**], which in fact, is only the consequence of the Gelfond-Schneider method of 1934.

THEOREM S-L. *Let $f_1(z), \ldots, f_n(z)$ be meromorphic functions, $f_1(z)$ and $f_2(z)$ algebraically independent of order of growth at most ρ_1 and ρ_2 respectively. Suppose that* **K** *is a number field and d/dz maps $\mathbf{K}[f_1, \ldots, f_n]$ into itself. Then for numbers $w_1, \ldots, w_m$ such that*

$$f_i(w_j) \in \mathbf{K}: \qquad i = 1, \ldots, n; j = 1, \ldots, m \ (\textit{and}\ f_i(z)\ \textit{are regular at}\ w_j)$$

we have a bound for the power m of the set $\{w_1, \ldots, w_m\}$

$$m \leqslant [\mathbf{K}:\mathbf{Q}](\rho_1 + \rho_2).$$

The method of proof is typical and remains the same for all other proofs. We take a sufficiently large parameter L (with respect to m, $H(w_i)$, $[\mathbf{K}:\mathbf{Q}]$, $\rho_1 + \rho_2, \ldots$) and construct, using the Dirichlet box principle, a polynomial $P(x, y) \in \mathbf{Z}[x, y]$, $P(x, y) \not\equiv 0$, such that

$$d_x(P) \leqslant \left[L^{\rho_2/(\rho_1+\rho_2)}(\log L)^{1/2}\right]; \qquad d_y(P) \leqslant \left[L^{\rho_1/(\rho_1+\rho_2)}(\log L)^{1/2}\right];$$

$$H(P) = \text{height of } P \leqslant \exp\left(L(\log L)^{1/2}\right);$$

and for the function $F(z) = P(f_1(z), f_2(z))$;

(1.1) $$F^{(k)}(w_j) = 0: \qquad j = 1, \ldots, m; \quad k = 0, 1, \ldots, L-1.$$

Then we find s_0 minimal so that some $F^{(s_0)}(w_{j_0}) \neq 0$ and estimate $F^{(s_0)}(w_{j_0})$ because by (1.1) we have many zeroes of $F(z)$. We obtain an upper bound

$$\left|F^{(s_0)}(w_{j_0})\right| \leqslant \exp(s_0 \log s_0(1 - m/(\rho_1 + \rho_2) + o(1)));$$

and $F^{(s_0)}(w_{j_0}) \in \mathbf{K}$ by the assumptions of the theorem, so we have a lower bound, $d = [\mathbf{K}:\mathbf{Q}]$:

$$\left|F^{(s_0)}(w_{j_0})\right| \geqslant \exp(-s_0 \log s_0(d - 1 + o(1)));$$

thus $m \leqslant d(\rho_1 + \rho_2)$.

In the modern transcendence and algebraic independence results the situation is much more complicated: (1) the number m is not fixed and is connected with L and heights of numbers $f_i(w_j)$. (2) **K** is replaced by a field of finite degree of transcendence over **Q**. (3) It is necessary to obtain an estimate for s_0 in terms of $L, m, \ldots$. (4) In general we consider also the auxiliary functions $P(f_1(z_1), \ldots, f_n(z_2); g_1(z_2), \ldots, g_n(z_2))$, take partial derivatives $d^{k_1+k_2}/dz_1^{k_1}dz_2^{k_2}$ and put $z_1 = z_2 = w_i \ldots$ (this is Baker's method **[B2]** in transcendence theory instead of Gelfond's).

But Theorem S-L shows us various examples of transcendental constants. For this let $\wp(z)$ be a Weierstrass elliptic function satisfying $\wp'^2 = 4\wp^3 - g_2\wp - g_3$ for algebraic invariants g_2, g_3; E—the corresponding elliptic curve defined over $\overline{\mathbf{Q}}$. By $2\omega_1, 2\omega_2$ we understand two linearly independent periods of $\wp(z)$, by $\zeta(z)$ the Weierstrass ζ-function, $\zeta'(z) = -\wp(z)$ and by $2\eta_1, 2\eta_2$ quasi-periods: $\eta_i = \zeta(\omega_i)$, $i = 1, 2$. If ω_1/ω_2 is imaginary quadratic, then we say that $\wp(z)$ has complex multiplication (c.m.); if in this situation $\mathbf{Q}(\beta) = \mathbf{Q}(\omega_1/\omega_2)$, then we say that $\wp(z)$ has c.m. by β (or in the field $\mathbf{Q}(\beta)$). By an algebraic point u of $\wp(z)$ we understand a u such that either u is a period of $\wp(z)$ or $\wp(u) \in \overline{\mathbf{Q}}$. For given $\wp(z)$ we will denote by ω, η a period and corresponding quasi-period.

Now we can simply explain why we consider only exponential and elliptic functions (in a multi-dimensional situation, Abelian functions): they possess a law of addition and so from a given algebraic point we can construct infinitely many of them.

Applying Theorem S-L to exponential and elliptic functions we obtain classical results.

PROPOSITION 1.1. *If* $\alpha \in \overline{\mathbf{Q}}$, $\alpha \neq 0$, *then* e^{α} *is transcendental; if* $\alpha \neq 0, 1$, *then* $\log \alpha$ *is transcendental;*

$$f_1(z) = 2, \qquad f_2(z) = e^z; \qquad w_i = i \cdot \alpha\colon \quad i = 1, 2, \ldots.$$

PROPOSITION 1.2 (GELFOND-SCHNEIDER). *If* $\alpha_1, \alpha_2 \in \overline{\mathbf{Q}}$ *and* $\alpha_i \neq 0$, *and* β *is an algebraic irrational number, then* α_i^{β} *is transcendental; if* $\log \alpha_1/\log \alpha_2$ *is irrational, it is transcendental.*

Take $f_1(z) = e^z$; $f_2(z) = e^{\beta z}$: $w_i = i \cdot \log \alpha$: $i = 1, 2, \ldots$.

The results analogous to Propositions 1.1 and 1.2 were proved in 1935 by Schneider for elliptic functions.

PROPOSITION 1.3 (SCHNEIDER). *For* $\wp(z)$, *the numbers* ω *and* η *are transcendental, as are* η/ω *and* π/ω. *In fact if* u *is an algebraic point of* $\wp(z)$ *and* $\alpha \in \overline{\mathbf{Q}}$, $\alpha \neq 0, 1$, $\beta \in \overline{\mathbf{Q}}$ *then both* $u/\log \alpha$ *and* $\zeta(u) + \beta u$ *are transcendental. If* u *is an algebraic point of* $\wp(z)$ *and* u_2 *of* $\wp_2(z)$, *where* $\wp(uz)$, $\wp_2(u_2 z)$ *are algebraically independent functions, then* u/u_2 *is transcendental.*

Take $f_1(z) = \wp(z)$; $f_2(z) = \zeta(z) + \beta z$, $f_2(z) = z$ or $f_2(z) = e^{\beta z}$ or

$$f_2(z) = \wp_2(u_2/uz).$$

In 1966 Baker proposed a generalization of Gelfond's method and has shown that for $\mathbf{Q}$-linearly independent logarithms $\log \alpha_1, \ldots, \log \alpha_n$ of algebraic numbers $\alpha_i \in \overline{\mathbf{Q}}$ and algebraic $\beta_0, \beta_1, \ldots, \beta_n$ such that at least one of $\beta_1, \ldots, \beta_n$ is irrational, $\beta_0 + \beta_1 \log \alpha_1 + \cdots + \beta_n \log \alpha_n$ and $e^{\beta_0} \alpha_1^{\beta_1} \cdots \alpha_n^{\beta_n}$ are transcendental. He also obtained a bound on the measure of transcendence of these numbers—a fact important for algebraic applications (see below).

For elliptic functions Coates (cf. **[C21]**) has shown that for algebraic α_i, $\alpha_1\omega_1 + \alpha_2\omega_2 + \alpha_3\eta_1 + \alpha_4\eta_2 + \alpha_5\pi i$ is either zero or transcendental. An analogue of Baker's result on linear independence over $\overline{\mathbf{Q}}$ of algebraic points of $\wp(z)$ with c.m. was proven by Masser **[M3]**. We will give at the end the applications of estimation of these linear forms from below.

We must definitely say that except for the results listed before, there were no precise results on transcendence of constants arising from exponential and elliptic functions. However, we will show also some new achievements using instead algebraic independence proofs.

There remain a lot of unsolved problems of a rather simple nature, e.g. for linear independence over $\overline{\mathbf{Q}}$ (N. Katz): whether linear independence over $\overline{\mathbf{Q}}$ of η, ω, π/ω implies that E has complex multiplication (the converse is true). This is very interesting because for such E, N. Katz managed to obtain an expression for values of an L-function of E, $L_E(s)$ at $s = 1$ analogous to Dammerrel's theorem —important for the Birch-Swinnerton-Dyer conjecture.

The situation with algebraic independence (a.i.) of numbers listed before in Propositions 1.1–1.3 is very poor. Nothing is known about their algebraic independence. We do not know anything on the a.i. of logarithms. The most information we have concerns numbers α^β for algebraic α and β.

However there still remains a very difficult problem:

P1. Let $\alpha \neq 0, 1$ be algebraic and β is quadratic. Are $\log \alpha$ and α^β a.i.?

P2. Let the logarithms $\log \alpha_1$, $\log \alpha_2$ of algebraic numbers be linearly independent and β be quadratic. Are α_1^β and α_2^β a.i.?

Many of the next results are connected with these two problems.

Now fortunately we know comparatively much about the algebraic independence of values of the exponent. The first result was the Lindemann-Weierstrass theorem proved about a century ago. For algebraic $\alpha_1, \ldots, \alpha_n$ linearly independent over $\mathbf{Q}$, $e^{\alpha_1}, \ldots, e^{\alpha_n}$ are a.i. The next step was made in 1949 by Gelfond who found how to complete the transcendence proof with algebraic considerations in order to find two a.i. numbers in the set of the form $\{\alpha_i, \beta_j, e^{\alpha_i \beta_j}; i = 1, \ldots, N, j = 1, \ldots, M\}$. At that time he found the second (after Lindemann's theorem) example of a.i. numbers [**G1**].

THEOREM 1.4. *If $\alpha \neq 0, 1$ is algebraic and β is a cubic irrationality, then α^β and α^{β^2} are a.i.*

Unfortunately all other results obtained by this method have the form: there are two a.i. numbers among 3, 4, 5 or more given numbers. His arguments were repeated and refined as time passed, and we quote all *interesting* corollaries of Gelfond's methods. For brevity we will denote the maximal number of a.i. numbers in a given set S by $\#S$.

COROLLARY 1.5. $\#\{2^{\sqrt{2}}, 2^{\sqrt{3}}, 2^{\sqrt{6}}\} \geqslant 2$.

COROLLARY 1.6. *For algebraic α_1, α_2 with irrational* $\log \alpha_1/\log \alpha_2$ *and quadratic β,*

$$\#\{\log \alpha_1/\log \alpha_2, \alpha_1^\beta, \alpha_2^\beta\} \geqslant 2.$$

COROLLARY 1.7. *For $\alpha \in \overline{\mathbf{Q}}$, $\alpha \neq 0, 1$ and algebraic β of degree $n \geqslant 3$*

$$\#\{\alpha^\beta, \ldots, \alpha^{\beta^n}\} \geqslant 2.$$

Using Gelfond's method Waldschmidt [**W1**] and Brownawell [**B13**] have obtained such results.

COROLLARY 1.8. *Either e^e or e^{e^2} is transcendental.*

COROLLARY 1.9. *Either e^{π^2} is transcendental, or e and π are a.i.*

COROLLARY 1.10. *For rational r, either e^{e^r} or $e^{e^{2-r}}$ is transcendental.*

In Gelfond's method we consider the auxiliary function

$$F(z) = P(z, e^{\alpha_1 z}, \ldots, e^{\alpha_n z}),$$

taking $z = n_1\beta_1 + \cdots + n_M\beta_M$.

Using Baker's method I have shown in this direction **[C4]**

COROLLARY 1.11. *If $\alpha_1, \alpha_2 \in \overline{\mathbf{Q}}$, $\log\alpha_1/\log\alpha_2$ is transcendental and β is quadratic, then*

$$\#\{\log\alpha_1, \alpha_1^\beta, \alpha_2^\beta\} \geqslant 2.$$

COROLLARY 1.12. $\#\{\log 2, (\log 2)^i, 2^i\} \geqslant 2$.

COROLLARY 1.13. $\#\{\pi, \pi^{\sqrt{2}}, e^{\pi i\sqrt{2}}\} \geqslant 2$.

These results, Corollaries 1.6, 1.11, 1.12, 1.13 are partial answers to problems P1 and P2.

Fortunately now there is progress in the algebraic independence of three and more numbers connected with the exponential function. The first result in this direction was the following.

THEOREM 1.14. *If α is algebraic $\alpha \neq 0, 1$ and β has degree d, then there are $[\log_2(d+1)]$ a.i. numbers among*

$$\alpha^\beta, \ldots, \alpha^{\beta^{d-1}},$$

e.g. for $d(\beta) = 7$, $\#\{\alpha^\beta, \ldots, \alpha^{\beta^6}\} \geqslant 3$.

Now there are wide generalizations of these results. For example,

THEOREM 1.15. *For $\alpha \in \overline{\mathbf{Q}}$, $\alpha \neq 0, 1$ and β of degree 5, there are three a.i. numbers among*

$$\alpha^\beta, \alpha^{\beta^2}, \alpha^{\beta^3}, \alpha^{\beta^4}.$$

Also for $d(\beta) = d \geqslant 7$,

$$\#\{\alpha^\beta, \ldots, \alpha^{\beta^{d-1}}\} \geqslant \left[\frac{d+1}{2}\right].$$

Unfortunately it is only one "half" of Schanuel's conjecture.

Despite the number of unsettled questions we must say that the situation in algebraic independence for the exponential function is very good!

So we can try to say something on elliptic functions and generalize the exponential results to the elliptic function, changing e^z to $\wp(z)$; $\log\alpha$ to u; πi to ω, etc. in order to investigate a.i. in the following large family of numbers,

$$\{\alpha_i, \beta_j, \wp(\alpha_i\beta_j), \zeta(\alpha_i\beta_j), e^{\varphi_k\beta_j}\}.$$

However for a long time (up to 1975) there was no progress at all. We had no bounds for numbers of zeroes of polynomials in elliptic function, etc.

Using estimations for measures of transcendence and considering polynomials $P(\wp_1, \wp_2)$ we now can present many results on a.i. for elliptic functions. A special role is played by functions with c.m.

Our results take the form (as usual) there are 2 a.i. numbers in a set of $3, 4, \ldots$ elements; but for c.m. the power of set reduces (twice); and in fact for c.m. we will show the precise analogues of exponential results.

Exceptional interest (from my point of view) attaches to three new general theorems of algebraic independence of constants arising from both exponential and elliptic functions. In fact they are the first examples of two a.i. constants having an algebraic nature. The first result in this direction was my result in 1975.

THEOREM 1.16. *For* $\wp(z)$, $\#\{\omega_1, \omega_2, \eta_1, \eta_2\} \geqslant 2$.

This result is best possible for c.m. and we have

COROLLARY 1.17. *For c.m.* π *and* ω *are a.i.*

For example π, $\Gamma(\frac{1}{3})$ and π, $\Gamma(\frac{1}{4})$ and π, $\Gamma(\frac{1}{6})$ are a.i. The same for $m \geqslant 3$; π and $\prod_{i=1}^{m-1}\Gamma(i/m)^{\chi(i)}$ are a.i. for $\chi(i) = (i/m)$, the Legendre symbol.

Several versions of proofs of Theorem 1.16 are presented below in §2.

Now I can show a much more general and precise result valid for an arbitrary function $\wp(z)$.

THEOREM 1.18. *If* u *is an algebraic point of* $\wp(z)$, *then* $\zeta(u) - \eta u/\omega$ *and* η/ω *are a.i.*

We use an auxiliary function of the form

$$F(z) = P(\zeta(z) - \eta z/\omega, \wp(z)).$$

From this (and the Legendre relation) there follows the interesting

COROLLARY 1.19. *For arbitrary* $\wp(z)$ *the numbers* η/ω *and* π/ω *are a.i.*

(This contains also Corollary 1.17.)

We present one separate proof of Corollary 1.19 in §3 below (Theorem 3.1) together with a simple method of obtaining bounds of the measure of algebraic independence of numbers π/ω and η/ω from Corollary 1.19 (Theorem 3.4). The proof of Theorem 1.18 is given in §5 (Theorem 5.1). For recent progress in the estimation of measures of algebraic independence of numbers $\zeta(u) - \eta u/\omega$, η/ω (from Theorem 1.18) and π/ω, η/ω (from Corollary 1.19) see the author's paper **[C14]** and the Introduction to this volume.

Also as another generalization of 1.17 we have a general result for c.m. functions.

THEOREM 1.20. *If* $\wp(z)$ *has c.m. and* u *is an algebraic point of* $\wp(z)$, *then* u *and* $\zeta(u)$ *are a.i.*

For the proof of Theorem 1.20 see §5 (Theorem 5.3 and Corollary 5.4).

There are also much more complicated results.

THEOREM 1.21. *If* $\varphi \neq 0$, *then*

$$\#\{\omega_1, \omega_2, \varphi, e^{\varphi\omega_1}, e^{\varphi\omega_2}\} \geqslant 2, \qquad \#\{\eta_1, \eta_2, \varphi, e^{\varphi\omega_1}, e^{\varphi\omega_2}\} \geqslant 2.$$

For c.m. the following result holds.

THEOREM 1.22. *If* $\wp(z)$ *has c.m. in the field* $\mathbf{Q}(\beta)$, *then the two numbers* π/ω *and* $e^{\pi i\beta}$ *are a.i.* (η *and* $e^{\pi i\beta}$ *are also a.i.*).

Theorems 1.21 and 1.22 were presented for the first time together with their proofs in the author's paper [**C3**], and were announced for the first time in 1975 [**C24**]. The proof of Theorem 1.21 is given below in §4.

Now we want to present a collection of algebraic independence results. Before we had precise results, where numbers proved a.i. are very natural. The exposition below will resemble to some extent a zoo or paleontological exposition with wild numbers.

PROPOSITION 1.23. *For* $\wp(z)$ *with c.m. and u an algebraic point of* $\wp(z)$,

$$\#\left\{\wp\left(u\sqrt{2}\right), \wp\left(u\sqrt{3}\right), \wp\left(u\sqrt{6}\right)\right\} \geqslant 2.$$

This is an analogue of an exponential result,

$$\#\left(\alpha^{\sqrt{2}}, \alpha^{\sqrt{3}}, \alpha^{\sqrt{6}}\right\} \geqslant 2.$$

Another result of this sort is

PROPOSITION 1.24. *Let* $\wp_i(z)$ *have nonzero period* ω_i *and c.m. be in the field* $\mathbf{Q}(\tau_i)$, $i = 1, 2$. *Then*

$$\#\left\{\frac{\omega_1}{\omega_2}, \wp_1(\omega_1\tau_2), \wp_2(\omega_2\tau_1)\right\} \geqslant 2.$$

This is an analogue of an exponential result,

$$\#\left\{\frac{\pi}{\log\alpha}, e^{\pi}, \alpha^{i}\right\} \geqslant 2.$$

Proposition 1.24 is proved by using an auxiliary function of the following form:

$$F(z) = P\left(\wp_1(\omega_1 z), \wp_1(\omega_2 z)\right).$$

Now we enter into the real jungles of transcendental numbers.

THEOREM 1.25. *Let* $\wp(z)$ *have c.m. in the field* $\mathbf{Q}(\tau)$. *Then for linearly independent logarithms* $\log\alpha_1, \log\alpha_2$ *of algebraic numbers we have*

$$\#\{\omega, \alpha_1^{\tau}, \alpha_2^{\tau}\} \geqslant 2, \qquad \#\left\{\frac{\pi}{\omega}, \alpha_1^{\tau}, \alpha_2^{\tau}\right\} \geqslant 2.$$

The proof uses auxiliary functions of the form

$$F(z_1, z_2) = P(\wp(z_1), z_1; e^{\varphi_1 z_2}, e^{\varphi_2 z_2}, e^{\varphi_3 z_2}, e^{\varphi_4 z_2}),$$
$$F(z_1, z_2) = P(\wp(z_1), \zeta(z_1); e^{\varphi_1 z_2}, e^{\varphi_2 z_2}, e^{\varphi_3 z_2}, e^{\varphi_4 z_2})$$

where

$$\varphi_1 = \log\alpha_1, \qquad \varphi_2 = \log\alpha_2, \qquad \varphi_3 = \tau\log\alpha_1, \qquad \varphi_4 = \tau\log\alpha_2.$$

Using another variant of Baker's method we manage to prove interesting results of the form

THEOREM 1.26. *For $\wp(z)$ with c.m. in $\mathbf{Q}(\beta)$,*

$$\#\{\pi/\omega, e^{\pi/\omega}, e^{\pi\beta/\omega}\} \geqslant 2, \qquad \#\{gp/\omega, \log(\pi/\omega), (\pi/\omega)^{\beta}\} \geqslant 2.$$

The auxiliary function is $F(z) = P(\wp(\omega z_1), \zeta(\omega z_1), z_2, e^{\varphi z_2})$.

A special collection of interesting results can be deduced from these theorems for $\wp(z)$ with c.m. in $\mathbf{Q}(\beta)$ and $\alpha \in \overline{\mathbf{Q}}$, $\alpha \neq 0, 1$, we have

$$\#\{\omega, \log\alpha, \alpha^{\beta}\} \geqslant 2, \qquad \#\{\pi/\omega, \log\alpha, \alpha^{\beta}\} \geqslant 2,$$

$$\#\left\{\pi/\omega, \frac{\log(\pi/\omega)}{\omega}, (\pi/\omega)^{\beta}\right\} \geqslant 2.$$

Generalizing Theorem 1.21 we prove for c.m.

THEOREM 1.27. *Let $\wp(z)$ have c.m. in $\mathbf{Q}(\beta)$ and u be an algebraic point of $\wp(z)$; then for algebraic α, $\alpha \neq 0, 1$, and rational r*

$$\#\{u, \log\alpha, \alpha^{\beta}\} \geqslant 2, \qquad \#\{u, e^{u^r}, e^{u^r\beta}\} \geqslant 2, \qquad \#\{u, \log u, u^{\beta}\} \geqslant 2.$$

The auxiliary function in this situation has the form

$$F(z) = P(\wp(z), z, e^{\varphi z}, e^{\varphi\beta z}).$$

For the case when u is a period, this theorem is Theorem 1.21.

Proofs of results 1.23–1.27 are given in §§4, 5.

We have an analogue of the exponential result on e^{π^2}:

THEOREM 1.28. *If $\wp(z)$ has c.m. and $\wp(\omega^2)$ is algebraic, then ω and $\wp(1)$ are a.i., ω and $\wp(\omega^3)$ are a.i.; thus $\wp(\omega^2)$ or $\wp(\omega^3)$ is transcendental.*

At the end of this review we want to show how the transcendence methods are applied to more interesting algebraic questions. In fact we can obtain an upper bound for integer points on an elliptic curve with complex multiplication much better than the previous ones. The best existing bound on the integer points on elliptic curves with c.m., in particular for the curve $y^2 = x^3 + k$ over $\mathbf{Q}$, belong to Stark [**S9**], who used Baker's method of the linear forms in logarithms of algebraic numbers [**B2**, **B3**, **S8**]. We use instead bounds of the linear forms in (elliptic logarithms of) algebraic points on elliptic curves with c.m. in conjunction with the Birch-Swinnerton-Dyer conjecture and bounds on generators of the Mordell-Weil group of an elliptic curve.

For example, we can obtain a lower bound for linear points from combination of two algebraic points on elliptic curves.

Let u_1, u_2 be algebraic points of $\wp(z)$ of respective heights H_1 and H_2; β_1, β_2 be algebraic numbers; and the field $\mathbf{Q}(\wp(u_i), \beta_j)$ have degree d_1. Then for any $\varepsilon > 0$ we have

$$|\beta_1 u_1 + \beta_2 u_2| > \exp\left(-c(d, \varepsilon) \cdot (\log B)^{2+\varepsilon} (\log H_1 \cdot \log H_2)^{1+\varepsilon}\right),$$

where B is the maximum of heights β_1, β_2.

This lower bound can be applied to generators of the Mordell-Weil group of an elliptic curve with c.m. defined over $\mathbf{Q}$. In §6 we present upper bounds on the generators of this Mordell-Weil group in terms of the L-function of an elliptic curve using both the Birch-Swinnerton-Dyer conjecture and the Reimann hypothesis for this L-function. The bounds of §6 are applied to the curves E: $y^2 = x^3 + k$ over $\mathbf{Q}$ with rank $g = 1$. Assuming both the Birch-Swinnerton-Dyer conjecture and the Riemann hypothesis for $L_E(S)$ we obtain a bound of integer points x, y on E:

$$|x|, |y| \leqslant \exp\left(c(\varepsilon)|k|^{1/6+\varepsilon}\right).$$

2. The theorem of two algebraically independent numbers among periods and quasi-periods.

In this section we present several versions of the proof of the algebraic independence results 1.16, 1.17 from §1, concerning periods and quasi-periods of an arbitrary elliptic curve.

THEOREM 2.1. *For arbitrary $\wp(x)$ (with algebraic invariants), among the numbers $\omega_1, \omega_2, \eta_1, \eta_2$ there are two algebraically independent, or*

$$\#\{\omega_1, \omega_2, \eta_1, \eta_2\} \geqslant 2.$$

This result was first presented at a conference in Pullman, Washington, June 1975 [**C18**] and then published in [**C3**]. In fact, an elegant and clear exposition can be found in the papers [**B5** or **W3**].

VERSIONS OF PROOFS OF THEOREM 2.1. (A) The old method [**C3**, **W3**] using the estimation of type of transcendence. We use the fact that according to the Legendre relation $\pi i \in \mathbf{Q}(\omega_1, \omega_2, \eta_1, \eta_2)$. In other words, if

$$\#\{\omega_1, \omega_2, \eta_1, \eta_2\} \leqslant 1,$$

then there exists θ algebraic over $\mathbf{Z}[\pi]$ such that

$$\mathbf{Q}(\pi, \theta) \supseteq \mathbf{Q}(i, g_2, g_3, \omega_1, \omega_2, \eta_1, \eta_2).$$

Now we take ε, $1 > \varepsilon > 0$, and a sufficiently large natural number N and put

$$L = [N^{1-\varepsilon/4}], \qquad X_0 = [N^{1-\varepsilon/2}]. \tag{2.1}$$

Then we take an auxiliary function

$$F(z) = \sum_{\lambda_0=0}^{L} \sum_{\lambda_1=0}^{L} \sum_{\lambda_2=0}^{L} p(\lambda_0, \lambda_1, \lambda_2) z^{\lambda_0} \wp(z)^{\lambda_1} \zeta(z)^{\lambda_2}, \tag{2.2}$$

where $p(\lambda_0, \lambda_1, \lambda_2) \in \mathbf{Z}[\pi]$: $0 \leqslant \lambda_i \leqslant L$; $i = 0, 1, 2$. By Siegels' lemma (see [**G1** or **W1**]) it is possible to find a nonzero polynomial

$$P(x_0, x_1, x_2) \in \mathbf{Z}[\pi][x_0, x_1, x_2],$$
$$\deg_{x_i}(P) \leqslant L: \qquad i = 0, 1, 2,$$

such that all coefficients of $P(x_0, x_1, x_2)$ have, as polynomials from π, degree at most N and type at most $c_1 N$ and such that for an auxiliary function

$$F(z) = P(z, \wp(z), \zeta(z)) \tag{2.3}$$

we have conditions

$$F^{(k)}\left(\frac{\omega_1}{2} + 2n_1\omega_1 + 2n_2\omega_2\right) = 0 \tag{2.4}$$

for all $k = 0, \ldots, N-1$ and $0 \leqslant n_1 \leqslant X_0 - 1$, $0 \leqslant n_2 \leqslant X_0 - 1$. The existence of such $P(x_0, x_1, x_2) \not\equiv 0$ follows from Siegels' lemma because the number of conditions (2.4) is at most

$$N \cdot X_0^2 \leqslant N^{3-\varepsilon}$$

and the number of coefficients is

$$(L+1)^3 \geqslant N^{3-3\varepsilon/4}.$$

Now we find the smallest number N_1 such that

$$F^{(k)}\left(\frac{\omega_1}{2} + 2n_1\omega_1 + 2n_2\omega_2\right) \neq 0 \tag{2.5}$$

for some $k = 0, \ldots, N_1$ and $0 \leqslant n_i \leqslant (N_1 - 1)^{1-\varepsilon/2} - 1$: $i = 1, 2$. From (2.4) it follows that $N_1 \geqslant N$ and $N_1 < \infty$ because of algebraic independence of z, $\wp(z)$, $\zeta(z)$. Then we take

$$z_0 = \frac{\omega_1}{2} + 2n_1^0\omega_1 + 2n_2^0\omega_2; \qquad 0 \leqslant n_i^0 \leqslant (N_1 + 1)^{1-\varepsilon/2} - 1: \quad i = 1, 2,$$

and $k_0 \leqslant N_1$ such that

$$\xi = F^{(k_0)}(x_0) \neq 0, \tag{2.6}$$

but

$$F^{(k)}(x_0) = 0: \qquad 0 \leqslant k \leqslant k_0 - 1.$$

We use Schwarz' lemma applied to the function

$$\sigma(z)^{3L} \cdot F(z)$$

in the disk $|z| \leqslant N_1^{1-7\varepsilon/16}$ and obtain an upper bound for $|\xi|$,

$$|\xi| \leqslant \exp\left(-c_2 N_1^{3-\varepsilon} \log N_1\right). \tag{2.7}$$

Now $\xi \in \mathbf{Q}(\pi, \mathfrak{P})$ and considering its norm in $\mathbf{Q}(\pi)$ we come to the polynomial $P(x) \in \mathbf{Z}[x]$, $P(x) \not\equiv 0$, of type

$$t(P) = \max\{\log H(P), d(P)\} \leqslant c_3 N_1 \log N_1 \tag{2.8}$$

such that from (2.7) we have

(2.9) $$|P(\pi)| \leqslant \exp(-c_4 N_1^{3-\varepsilon} \log N_1).$$

Now we apply the well-known estimation (see **[G1]**) of the type of transcendence of π, and because π has the type of transcendence $\leqslant 2 + \delta$ for any $\delta > 0$, $P(x) \equiv 0$, which is impossible. Thus

$$\#\{\omega_1, \omega_2, \eta_1, \eta_2\} \geqslant 2.$$

(B) Another variant of the same proof. From (2.8) and (2.9) it follows that we need the type of transcendence of π (or any other number from $\mathbf{Q}(\omega_1, \omega_2, \eta_1, \eta_2)$) to be $\leqslant 3 - \delta$ for some $\delta > 0$. However the arguments can be slightly changed in order to use only *finiteness* of type of transcendence. We take the same function $F(z)$ from (2.3), satisfying (2.4).

We want to show the main features of proofs in each concrete case. So under the title "Outline of the proof" or "Proof" we will give the construction of the auxiliary function in each concrete case, necessary estimation and final arguments (so called "End of the Proof") leading to algebraic independence. In some particular cases it will be possible to give simultaneously several variants of the proof.

(C) In fact the proof does not depend at all on the estimation of the type of transcendence of π or any other number. To see this we can use D. Masser's **[M3]** kind of arguments.

LEMMA 2.2. *The function $F(z)$ from (2.3) has in the disk $|z| \leqslant R$ not more than $c_5 L^3 + c_6 L R^2$ zeroes.*

Assuming that Lemma 2.2 is false we, as in **[M3]**, apply Schwartz' lemma to $\sigma(z)^{3L} F(z)$ and then consider $F(z)$ at points

$$z = 2n_1\omega_1 + 2n_2\omega_2 + \frac{r_1\omega_1 + r_2\omega_2}{L}, \qquad 0 \leqslant n_1, n_2 \leqslant 2L,$$
$$0 \leqslant r_1, r_2 \leqslant (2L)^{1/2}.$$

If we put

$$z = 2n_1\omega_1 + 2n_2\omega_2 + \alpha_{r_1, r_2},$$

then

$$F(z) = P\big(2n_1\omega_1 + 2n_2\omega_2, \wp(\alpha_{r_1,r_2}), 2n_1\eta_1 + 2n_2\eta_2 + \zeta(\alpha_{r_1,r_2})\big)$$

and we use the fact that vectors (ω_1, ω_2) and (η_1, η_2) are linearly independent. Then Lemma 2.3 and Theorems A.2–A.3 from **[M3]** give Lemma 2.2.

From Lemma 2.2 it follows immediately that the number N_1 in variant (A) of the proof is bounded:

(2.10) $$N_1^{3-\varepsilon} \leqslant c_5 N^{3-3\varepsilon/4} + c_7 N^{1-\varepsilon/4} N_1^{2-\varepsilon},$$

and $N \leqslant N_1$.

Thus substituting from (2.10) to (2.8) and (2.9) we obtain for any sufficiently large N a polynomial $P_N(x) \in \mathbf{Z}[x]$ such that

$$P_N(x) \not\equiv 0, \qquad t(P_N) \leqslant c_8 N \log N, \qquad |P_N(\pi)| \leqslant \exp(-c_9 N^{3-\varepsilon} \log N). \tag{2.11}$$

Then for $1 > \varepsilon > 0$ from the usual Gelfond's lemma [**G1**] it follows that π is algebraic, which is impossible. This proves that $\mathbf{Q}(i, g_2, g_3, \omega_1, \omega_2, \eta_1, \eta_2)$ has a transcendence degree of at least two and Theorem 2.1 is proved.

This result has some interesting corollaries (see above and [**C11**, **B5**]). First of all the most interesting case is the case of complex multiplication.

COROLLARY 2.3. *If $\wp(z)$ has complex multiplication, then for an arbitrary period ω of $\wp(z)$*

ω and π are algebraically independent.

From the classical formulae and Corollary 2.3 it follows e.g. solution of some questions from [**S1**, **S7**] that

$$\Gamma(1/3), \Gamma(1/4), \Gamma(1/6) \text{ are transcendental.} \tag{2.12}$$

In fact if we use the Selberg-Chowla formula [**S11**, **W8**] we obtain for any discriminant $m \geqslant 3$,

$$\prod_{i=1}^{m-1} \Gamma\left(\frac{i}{m}\right)^{\chi(i)} \text{ and } \pi \text{ are algebraically independent,} \tag{2.13}$$

where $\chi(i) = (i/m)$ is the Legendre symbol.

In particular (in addition to (2.12)) we obtain that $\Gamma(1/3)$ and π, $\Gamma(1/4)$ and π, $\Gamma(1/6)$ and π are algebraically independent.

We remember that it is still an open question whether $\Gamma(1/5)$, $\Gamma(1/7)$ etc. are transcendental or even irrational. This question is in close connection with a much more general question.

QUESTION 2.4. Let **A** be a d-dimensional simple Abelian variety defined over $\overline{\mathbf{Q}}$ and $\Omega \cup H$ be the set of all (coordinates of) periods and quasi-periods of **A**. What is the degree of transcendence of $\mathbf{Q}(\Omega \cup H)$ in terms of d?

I have no hope for a nontrivial answer to this question. Some examples of G. Shimura, K. Ribert and others (see [**S3**, **R5**]) show that even for CM-varieties **A**, $\#\{\Omega \cup H\}$ can be comparatively small with respect to d.

For CM Abelian varieties the conjectural answer to Question 2.4 is given in [**C11**]; see also the Introduction to this volume. The proof of the Conjecture from [**C11**] is still unknown. The only fact that we know is

PROPOSITION 2.5. *Let* **A** *be a d-dimensional simple Abelian variety defined over $\overline{\mathbf{Q}}$ and $\Omega \cup H$ be the set of all (coordinates of) periods and quasi-periods of* **A**. *Then $\#\{\Omega \cup H\} \geqslant 2$.*

The method of proof of Proposition 2.5 is almost the same as of part (A) of the proof of Theorem 2.1. We use the fact that $\pi \in \overline{\mathbf{Q}(\Omega \cup H)}$ and construct an auxiliary function of the form

$$F(\bar{z}) = P(A_1(\bar{z}),\ldots,A_d(\bar{z}), H_1(\bar{z}),\ldots,H_d(\bar{z}), z_1,\ldots,z_d)$$

for $\bar{z} \in \mathbf{C}^d$, Abelian functions $A_1(\bar{z}),\ldots,A_d(\bar{z})$ and quasi-Abelian functions $H_1(\bar{z}),\ldots,H_d(\bar{z})$. Polynomial $P(x_1,\ldots,x_d,\ldots,x_{3d})$ has coefficients in $\mathbf{Z}[\pi]$ and we take $\bar{z}$ to be elements of a lattice of periods.

The first results—Theorem 2.1—was a good stimulus for further investigations of algebraic independence of constants connected with elliptic and exponential functions. Now we can present lots of interesting and (or) different new results in this direction.

Historically Corollary 2.3 was the first example of two algebraically independent numbers connected with an elliptic function with complex multiplication. In a year after Theorem 2.1 was proved (1975), we had another result.

THEOREM 2.6. *For arbitrary Weierstrass elliptic function $\wp(z)$ (with algebraic invariants) and pair (ω, η) of period and quasi-period of $\wp(z)$; $\zeta(\omega) = \eta$ we have*

π/ω and η/ω are algebraically independent.

This result contains as a particular case Theorem 2.1, and is proved in §3.

3. Algebraic independence and estimations of measure of algebraic independence of two numbers connected with elliptic functions. We give proof of algebraic independence of two numbers among g_2, g_3, π/ω, η/ω for period ω and corresponding quasi-period η of a Weierstrass elliptic function $\wp(z)$ with invariants g_2, g_3. We also estimate the type of transcendence.

THEOREM 3.1. *For any $\wp(z)$ among g_2, g_3, π/ω, η/ω there are two a.i. numbers.*

PROOF. Let $\mathbf{K} = \mathbf{Q}(g_2, g_3, \pi/\omega, \eta/\omega)$; we take an auxiliary function of the form

$$F(z) = P(\zeta(z) - \eta z/\omega, \wp(z)),$$

where

$$(3.1) \qquad d_x(P) = L_1, \qquad d_y(P) = L_2, \qquad L_2 \geqslant L_1,$$

and $P(x, y) \in \mathbf{K}[x, y]$ is defined by conditions

$$F^{(k)}(\omega/2 + 2n\omega + 2m\omega') = 0$$

where ω' and ω are periods of $\wp(z)$, linearly independent over $\mathbf{Q}$, and

$$(3.2) \qquad \begin{aligned} k &= 0,\ldots, [\gamma_1 L_2]; \qquad \gamma_1 > 1, \\ n, m &= 0,\ldots, [\gamma_2 L_1]; \qquad \gamma_2 < 1, \qquad \gamma_1\gamma_2 < 1. \end{aligned}$$

We use Siegel's lemma and remark that both functions $\zeta(z) - \eta z/\omega$, $\wp(z)$ are periodic with period 2ω.

Thus we find $P(x, y) \in \mathbf{K}[x, y]$, $P(x, y) \not\equiv 0$ with the following bound of the type of coefficients of $P(x, y)$:

$$t_{\mathbf{K}}(P) \leqslant \gamma_3 L_2 \log L_2.$$

If g_2, g_3 are algebraic numbers, then the degrees of coefficients of $P(x, y)$ as elements from $\mathbf{K}$ are bounded in the following way:

$$d_{\mathbf{K}}(P) \leqslant \gamma_4 L_1.$$

If g_1, g_3 are not algebraic, then the bound is $d_{\mathbf{K}}(P) \leqslant \gamma_4 L_2$. This is clear, because $F^{(k)}(\omega_1/2 + 2n\omega + 2m\omega') = R(\pi i/\omega, \eta/\omega)$; where R is of degree $\leqslant \gamma_5 L_1$. We also remark that $(\zeta(z) - \eta z/\omega)^{\lambda_1} \wp(z)^{\lambda_2 (k)}$—the derivative of the monomial—has the form $R_{k,\lambda_1,\lambda_2}(\pi i m/\omega; \eta/\omega)$, where $d(R_{k,\lambda_1,\lambda_2}) \leqslant L_1$ and $t(R_{k,\lambda_1,\lambda_2}) \leqslant L_2 + k + k \log k$.

Now we have not less than $\gamma_1\gamma_2 L_2 L_1^2$ zeroes of $F(z)$ in the disk $|z| \leqslant \gamma_5 L_1$. We apply Schwarz's lemma to the entire function

$$G(z) = F(z) \cdot \sigma(z)^{L_1 + L_2},$$

and obtain

$$\left|F^{(k_1)}(\omega/2 + 2n\omega + 2m\omega')\right| < \exp\left(-\gamma_6 L_1^2 L_2\right)$$

for $|z| \leqslant \gamma_5 L_1$ and $k_1 \leqslant \gamma_7 L_1^{2-\varepsilon} L_2$ for $\gamma_7 = \gamma_7(\varepsilon) > 0$ and any $\varepsilon > 0$.

Now we have the following bound for the numbers of zeroes of $F(z)$ (obtained by arguments of [**M3**]):

LEMMA 3.2. *The number of zeroes of $F(z) = P(\zeta(z) - \eta z/\omega, \wp(z))$ for $L_2 \geqslant L_1$ in disk of radius $|z| \leqslant R$ is at most*

$$\sum_{|w| \leqslant R} \operatorname{ord}_w(F) \leqslant \gamma_8 L_1^2 L_2 + \gamma_9 L_2 R^2.$$

In order to prove this lemma we use the usual type of arguments: we suppose that the lemma is not true, then we obtain many zeroes of $F(z)$. Then one applies Schwarz's lemma and comes to small values of $F(z)$,

$$|F(z)| \leqslant \exp\left(-\gamma_8 L_1^2 L_2\right),$$

for

$$z = m\omega' + r\omega/L_2: \qquad m \leqslant L_1;\ r \leqslant L_2.$$

Then one gets

$$F(z) = P\left(m\eta' - \frac{\eta\omega'}{\omega} m + \zeta\left(\frac{r\omega}{\nu_2}\right) - \frac{\eta r}{\nu_2}, \wp\left(\frac{r\omega}{\nu_2}\right)\right)$$

$$= P\left(\frac{m\pi i}{2\omega} + \zeta\left(\frac{r\omega}{\nu_2}\right) - \frac{\eta r}{\nu_2}, \wp\left(\frac{r\omega}{\nu_2}\right)\right).$$

One fixes r, and then one applies lemmas [**M3**] to $P(m\pi i/2\omega, \ldots)$ and then to $P(\wp(r\omega/\nu_2)), \ldots$.

Having the bound of Lemma 3.2 for the number of zeroes we find

$$k_0 \leqslant \gamma_{10} L_2, \qquad n_1^0, n_2^0 \leqslant L_1$$

such that

$$\xi = F^{(k_0)}\left(\omega/2 + 2n_1^0\omega + 2n_2^0\omega'\right) \neq 0.$$

Then $\xi \in \mathbf{K}$ and let also, contrary to the theorem, deg tr $\mathbf{K} = 1$. We have

$$|\xi| < \exp\left(-\gamma_6 L_1^2 L_2\right),$$

as discussed previously. Taking any transcendental element $\theta \in \mathbf{K}$ and taking $\mathrm{Norm}_{\mathbf{K}/\mathbf{Q}(\theta)}(\xi)$ we come to

$$P(x) \in \mathbf{Z}[x] \quad \text{with } P(x) \not\equiv 0;$$

(3.3)
$$\begin{cases} |P(\theta)| < \exp\left(-\gamma_{11} L_1^2 L_2\right), \\ t(P) \leqslant \gamma_{12} L_2 \log L_2, \\ d_1(P) \leqslant \gamma_{13} L_1 \quad \text{if } g_2, g_3 \in \overline{\mathbf{Q}} \text{ or} \\ d_2(P) \leqslant \gamma_{13} L_2 \quad \text{if not.} \end{cases}$$

Taking $L_1 = L_2$ and applying Gelfond's lemma we come to contradictions with deg tr $\mathbf{K} = 1$.

Thus Theorem 3.1 is proved in complete generality.

Now we proceed to the simple estimations of measure of algebraic independence of numbers involved in Theorem 3.1. For better estimates it is reasonable to restrict ourselves to the classical case when g_2, g_3 are algebraic.

We estimate the measure of a.i. of π/ω, η/ω for algebraic g_2 and g_3. As before L_1, L_2 are two parameters $L_2 \geqslant L_1$ and

(3.1′)
$$F(z) = P(\zeta(z) - \eta z/\omega, \wp(z));$$

where $P(x, y) \in \mathbf{K}[x; y]$ and $d_x(P) = L_1$, $d_y(P) = L_2$, where coefficients of $P(x, y)$ are defined by Siegel's lemma by the conditions

$$F^{(k)}(\omega/2 + n\omega + m\omega') = 0$$

for period ω', linearly independent with ω and

$$k = 0, \ldots, [c_1 L_2], \quad c_1 \geqslant 1, \qquad 0 \leqslant n, m \leqslant c_2 L_1, \qquad c_2 \ll 1,$$
$$c_1 c_2 \ll 1.$$

Then for coefficients of $P(x, y)$

(3.4)
$$P(x, y) = \sum_{\lambda_1=0}^{L_1} \sum_{\lambda_2=0}^{L_2} p(\lambda_1, \lambda_2) x^{\lambda_1} y^{\lambda_2},$$

we have

$$p(\lambda_1, \lambda_2) \in \mathbf{Z}[g_2, g_3, \pi/\omega, \eta/\omega]$$

and

(3.5)
$$t(p(\lambda_1, \lambda_2)) \leqslant c_3 L_2 \log L_2, \qquad d(p(\lambda_1, \lambda_2)) \leqslant c_2 L_1.$$

We recall a lemma proved before:

LEMMA 3.2. *The number of zeroes of* $F(z) = P(\zeta(z) - \eta z/\omega, \wp(z))$ *for* $L_2 \geqslant L_1$ *in a disk of radius* $|z| \leqslant R$ *is at most*

$$\sum_{|w| \leqslant R} \operatorname{ord}_w(F) \leqslant c_5 L_1^2 L_2 + c_6 L_2 R^2.$$

Thus we have

PROPOSITION 3.3. *For any sufficiently large* L_1, L_2; $L_2 \geqslant L_1 \geqslant c_7$, *we have either* (a) *polynomial* $P_{L_1,L_2}(x, y) \in \mathbf{Z}[x, y]$, $P_{L_1,L_2} \not\equiv 0$ *such that*

$$(3.6) \qquad t(P_{L_1,L_2}) \leqslant c_8 L_2 \log L_2, \qquad d(P_{L_1,L_2}) \leqslant c_9 L_1,$$

$$-c_{11} L_1^2 L_2 < \log\left|P_{L_1,L_2}(\pi/\omega, \eta/\omega)\right| < -c_{10} L_1^2 L_2.$$

(*This is simply* $P_{L_1,L_2}(\pi/\omega, \eta/\omega) = F^{(k)}(\omega/2 + m\omega')$—*not very small values of* $F(z)$; *if coefficients are not small*.)

(b) *System of polynomials* $C_l(x, y) \in \mathbf{Z}[x, y]$: $l \in \mathfrak{L}$, *without common factor such that*

$$(3.7) \qquad t(C_l) \leqslant c_{13} L_2 \log L_2, \qquad d(C_l) \leqslant c_{14} L_1,$$

$$\log\left|C_l(\pi/\omega, \eta/\omega)\right| < -c_{15} L_1^2 L_2.$$

(*This is the case when all coefficients are small and* $C_l(\pi/\omega, \eta/\omega) = p_{\lambda_1,\lambda_2}$.)

Let us suppose that we have a polynomial with small value at $(\pi/\omega, \eta/\omega)$: $P(x, y) \in \mathbf{Z}[x, y]$, $P \not\equiv 0$, $t(P) \leqslant t$, and

$$\log H(P) \leqslant h, \qquad d(P) \leqslant d,$$

$$(3.8) \qquad |P(\pi/\omega, \eta/\omega)| < \varepsilon = \exp\left(-C\left[hd^5 \log^8 t + d^6 \log^9 t\right]\right)$$

for a sufficiently large constant C. Then we take

$$(3.9) \qquad L_1 = \left[d^2 \log^3 t\right], \qquad L_2 = \left[(hd + d^2 \log t)\log^2 t\right].$$

Without loss of generality we can assume that $P(x, y)$ is irreducible. Then we compare P with P_{L_1,L_2}. It is clear that they are relatively prime, because

$$(3.10) \quad -\log\left|P_{L_1,L_2}(\pi/\omega, \eta/\omega)\right| < c_{11} L_1^2 L_2 = c_{11}'(hd^5 \log^8 t + d^6 \log^9 t)$$

for fixed c_{11}', while

$$(3.11) \qquad -\log|P(\pi/\omega, \eta/\omega)| > C(hd^5 \log^8 t + d^6 \log^9 t)$$

for sufficiently large $C \gg c_{11}'$. Thus $P \nmid P_{L_1,L_2}$ and we take the resultant

$$r(\eta/\omega) = r(P, P_{L_1,L_2}) \in \mathbf{Z}[\eta/\omega],$$

$$(3.12) \qquad \begin{cases} |r(\eta/\omega)| < \exp\left(-c_{16}(hd^5 \log^8 t + d^6 \log^9 t)\right), \\ t(r) \leqslant c_{17}(hd^2 + d^3 \log t)\log^3 t, \\ d(r) \leqslant c_{18} d^3 \log^3 t. \end{cases}$$

In the same way in case (b) we take the resultant of P with some of the polynomials C_l.

Now we use E. Reyssart's estimates **[R1]**.

$$(3.13)\quad |r(\pi/\omega)| > \exp\left(-\gamma[\log H(r)]^2 - \gamma d(r)^2\log^2 d(r)\right),$$
$$|r(\eta/\omega)| > \exp\left[-\gamma'\{d(r)\log H(r)\cdot \log\log H(r) + d^2(r)\log^3 d(r)\}\right].$$

Comparing estimates (3.13) with those of (3.12) we come to a contradiction with (3.8) for some constant $C > 0$. In other words we obtain the estimate of measure of a.i. of two numbers π/ω and η/ω.

THEOREM 3.4. *Let $\wp(z)$ have algebraic invariants g_2, g_3 and (ω, η) be the pair of period, quasi-period. Let*

$$P(x, y) \in \mathbf{Z}[x, y], \qquad P \not\equiv 0,$$

and

$$\log H(P) \leqslant h, \qquad d(P) \leqslant d, \qquad t = d + h.$$

Then

$$(3.14)\qquad |P(\pi/\omega, \eta/\omega)| > \exp\{-c_1[hd^5\log^8 t + d^6\log^9 t]\}$$

for a constant $c_1 = c_1(g_2, g_3) > 0$.

In fact bounds (3.14) can be considerably improved if we will use more clever techniques than resultants in (3.12). These estimates are presented in the Introduction to this volume, with complete proofs given in Chapter 8 based on the methods of Chapter 4 of this volume.

4. Theorems of algebraic independence of values of elliptic functions. In this section we present several results on algebraic independence of numbers connected with elliptic and exponential functions. Some of these results are very interesting, some of them are not so interesting, but in any case we were trying to gather all possible results that we obtained in the direction of algebraic independence and transcendence for elliptic functions.

THEOREM 4.1. *For any complex $\varphi \neq 0$ and any Weierstrass elliptic function $\wp(z)$*

$$\text{(i)}\qquad \#\{g_2, g_3, \omega_1, \omega_2, \varphi, e^{\varphi\omega_1}, e^{\varphi\omega_2}\} \geqslant 2$$

and

$$\text{(ii)}\qquad \#\{g_2, g_3, \eta_1, \eta_2, \varphi, e^{\varphi\omega_1}, e^{\varphi\omega_2}\} \geqslant 2.$$

The proof is a stereotype and was presented for the first time in **[C3]**. We take fields $\mathbf{K}_1$ or $\mathbf{K}_2$, containing number, in (i) or (ii), respectively, supposing $\deg\operatorname{tr}\mathbf{K}_i = 1$, $\mathbf{K}_i = \mathbf{Q}(\theta, \vartheta)$, where ϑ is transcendental over $\mathbf{Q}(\theta)$. We choose an auxiliary function in the form

$$F_1(z) = P(z, \wp(z), e^{\varphi z}) \quad \text{or} \quad F_2(z) = P(\zeta(z), \wp(z), e^{\varphi z})$$

for $P(x, y, z) \in \mathbf{Z}[\theta][x, y, z]$ and

$$d_x(P) = [X^{3/2}\ln^{-1/2}X], \qquad d_y(P) = [X^{3/2}\ln^{-1/2}X],$$
$$d_z(P) = [X^{1/2}\ln^{1/2}X]$$

where the type of coefficients of P is

$$t_{\mathbf{K}_j}(P) \leqslant c_1 X^{3/2}\ln^{1/2}X: \qquad j = 1, 2,$$
$$(F(n_1\omega_1 + n_2\omega_2))^{(k)} = 0: \qquad k = 0, \ldots, \left[c_2 X^{3/2}\ln^{-1/2}X\right],$$
$$n_i\text{-integers } 0 \leqslant n_i \leqslant [c_3 X].$$

Then by transcendence criteria we find that the transcendence type of $\mathbf{K}_i$ is not less than $2 + 1/3$. However, we know already that this type is less than $2 + \varepsilon$ for every $\varepsilon > 0$. In other words $\deg \operatorname{tr} \mathbf{K}_i = 1$ is impossible.

Some other results are

THEOREM 4.2. *Let $\wp_1(z)$ and $\wp_2(z)$ be elliptic functions with invariants g_2^1, g_3^1; g_2^2, g_3^2, correspondingly and with the periods ω_1, ω_2, respectively. If $\wp_1(\omega_1 z)$ and $\wp_2(\omega_2 z)$ are algebraically independent, then for any complex numbers v_1, v_2, v_3 such that*

$$1, v_1, v_2, v_3 \textit{ are linearly independent over } \mathbf{Q},$$

two of the numbers

$$\left\{g_2^1, g_3^1; g_2^2, g_3^2, \omega_1/\omega_2, \wp_i(\omega_i v_j): i = 1, 2,\ j = 1, 2, 3\right\}$$

are algebraically independent.

We denote the set of numbers under consideration by S.

PROOF. We take $\mathbf{Q}(S) = \mathbf{K}$ and suppose that $\mathbf{K} = \mathbf{Q}(\theta, \vartheta)$; $[\mathbf{K} : \mathbf{Q}(\theta)] < \infty$. Then the auxiliary function has the form

$$F(z) = P(\wp_1(\omega_1 z), \wp_2(\omega_2 z)),$$

where $P(x, y) \in \mathbf{Z}[\theta][x, y]$ and

$$d_x(P) = d_y(P) = L,$$
$$\operatorname{type}(P) = \left[\nu_1 L^{(M+4)/(M+2)}(\log L)^{2/(M+2)}\right].$$

Here $P(x, y) \not\equiv 0$ is chosen according to a Siegel lemma in combination with Coates' lemma by conditions

$$F^{(k)}(n_1 v_1 + \cdots + n_M v_M) = 0$$

for $k = 0, \ldots, [\nu_2 L^{(M+4)/(M+2)}(\log L)^{-M/(M+2)}]$ and integers n_i,

$$|n_i| \leqslant \left[\nu_3 L^{1/(M+2)}(\log L)^{1/(M+2)}\right] = X_1,$$

where $v_1, \ldots, v_M$ are linearly independent with 1 and

$$n_1 v_1 + \cdots + n_M v_M \in L_{X_1}$$

in the notation of Ramachandra's lemma **[R1]**, see Lemma 9.3. Now we know the upper bound for the number of zeroes of $F(z)$ in the disc $|z| \leqslant R$; the number of zeroes $\mathfrak{N}(F;\, R)$ in a disk of radius R is at most $\nu_4 L^3 + \nu_5 L R^2$.

Thus using periodicity of $F(z)$ we find that the number of zeroes $\mathfrak{N}(F;\, \nu_8 L)$ of $F(z)$ in disk $|z| \leqslant \nu_6 L$ is not less than $\nu_7 L^3$, if we use Schwarz's lemma. Now again by Coates' lemma and by the upper bound for $\mathfrak{N}(F;\, \nu_8 L)$, we obtain

$$z_1 = n_1 v_1 + \cdots + n_M v_M \in L_{X_2},$$

$$X_2 = \left[\nu_9 L^{(M+4)/(M+2)} (\log L)^{2/(M+2)}\right],$$

such that

$$F^{(k_1)}(z_1) \neq 0 \quad \text{for } k_1 \leqslant \left[\nu_{10} L^{(M+4)/(M+2)} (\log L)^{-M/(M+2)}\right]$$

and

$$\left|F^{(k_1)}(z_1)\right| < \exp\left(-\nu_{11} L^3\right).$$

Thus we get the polynomial $R_L(x) \in \mathbf{Z}[x]$, $R_L(x) \not\equiv 0$, such that

$$|R_1(\theta)| < \exp\left(-\nu_{12} L^3\right), \qquad t(R_L) \leqslant \nu_{13} L^{(M+4)/(M+2)} (\log L)^{2/(M+2)}.$$

So for $M = 3$ by Gelfond's lemma we obtain our Theorem 4.2.

COROLLARY 4.3. *Let $\wp_i(z)$ be Weierstrass elliptic functions with algebraic invariants and complex multiplication by τ_i, and let ω_i be their periods*: $i = 1, 2$. *If $\mathbf{Q}(\tau_1)$ and $\mathbf{Q}(\tau_2)$ are different, then among the numbers*

$$\omega_1/\omega_2,\, \wp_1(\omega_1\tau_2),\, \wp_2(\omega_2\tau_1)$$

there are 2 *that are algebraically independent.*

This is an analog of the exponential result

$$\#\left\{\frac{\log \alpha_1}{\log \alpha_2},\, \alpha_1^\beta,\, \alpha_2^\beta\right\} \geqslant 2$$

for $\mathbf{Q}$-linearly independent logarithms $\log \alpha_1$ and $\log \alpha_2$ of algebraic numbers α_1, α_2 and for quadratic irrationality β.

We also have Baker's method that we can apply to

THEOREM 4.4. *For any complex $\varphi \neq 0$*

$$\#\left\{g_2,\, g_3,\, \eta_1,\, \eta_2,\, \omega_2/\omega_1,\, \varphi,\, e^{\varphi},\, \varphi^{\varphi\omega_2/\omega_1}\right\} \geqslant 2.$$

For the proof we use Baker's method and the auxiliary function is

$$F(z_1, z_2) = P\left(\wp(\omega_1 z_1),\, \zeta(\omega_1 z_1),\, z_2, e^{\varphi z_2}\right).$$

Among the corollaries we have

COROLLARY 4.5.

$$\#\left\{\pi/\omega,\, e^{\pi/\omega},\, e^{\pi\beta/\omega}\right\} \geqslant 2, \qquad \#\left\{\pi/\omega, \log(\pi/\omega), (\pi/\omega)^{\beta}\right\} \geqslant 2.$$

Another Baker-type result is the following:

THEOREM 4.6. *For $\varphi_1, \varphi_2, \varphi_3$ linearly independent over* **Q**

$$\#\{\omega_1, \omega_2, e^{\varphi_j \omega_i}, g_2, g_3 : j = 1, 2, 3,\ i = 1, 2\} \geqslant 2,$$
$$\#\{\eta_1, \eta_2, e^{\varphi_j \omega_i}, g_2, g_3 : j = 1, 2, 3,\ i = 1, 2\} \geqslant 2.$$

In the proofs of Theorems 4.4 and 4.6 the Baker method is used in complete analogy with the corresponding results of the author for values of the exponential function. See **[C4]** and Chapter 2 of this volume for the corresponding proofs.

Here the auxiliary functions are the following:

$$F_1(z_1, z_2) = P(\wp(z_1), z_1, e^{\varphi_1 z_2}, e^{\varphi_2 z_2}, e^{\varphi_3 z_2}),$$
$$F_2(z_1, z_2) = P(\wp(z_1), \zeta(z_1), e^{\varphi_1 z_2}, \ldots, e^{\varphi_3 z_2}).$$

As corollaries we have

COROLLARY 4.7.

$$\#\{\omega, \alpha_1^\beta, \alpha_2^\beta\} \geqslant 2, \qquad \#\{\pi/\omega, \alpha_1^\beta, \alpha_2^\beta\} \geqslant 2.$$

THEOREM 4.8. *For* **Q**-*linearly independent algebraic points u_1, u_2 of $\wp(z)$ and φ_1, φ_2 linearly independent,*

$$\#\{g_2, g_3, u_i, \varphi_j, e^{\varphi_j u_i} : j = 1, 2; i = 1, 2\} \geqslant 2.$$

PROOF. We take

$$F(z) = P(z, e^{\varphi_1 z}, e^{\varphi_2 z}, \wp(z)), \qquad d_x(P) \leqslant \left[X^2 (\log X)^{-1/3}\right],$$
$$d_y(P) = d_z(P) \leqslant \left[X (\log X)^{-1/3}\right], \quad d_w(P) \leqslant \left[(\log X)^{2/3}\right]$$

and the coefficients of the polynomial $P(x, y, z, w)$ are defined by equations

$$F^{(k)}(nu_1 + nu_2) = 0: \qquad k = 0, \ldots, S_0 = \left[\gamma X^2 (\log X)^{-1/3}\right],$$
$$|n|, |m| \leqslant [\gamma' X].$$

5. Algebraic independence of numbers connected with nonperiodic points of elliptic curves. In preceding sections we have shown some examples of two algebraically independent numbers (e.g. π/ω, η/ω, etc.) arising from periodic points on elliptic curves. Now we give two very important examples of pairs of algebraically independent numbers that arise from an arbitrary *algebraic* point on an elliptic curve.

Let u, linearly independent with ω, be an algebraic point on $\wp(z)$ ($\wp(u) \in \overline{\mathbf{Q}}$) where $\wp(z)$ is a Weierstrass elliptic function with algebraic invariants. Then we show (Theorem 5.1) that

(5.1) $$\zeta(u) - \eta u/\omega \text{ and } \eta/\omega$$

are algebraically independent numbers for any pair (ω, η) of period, quasi-period of $\wp(z)$. In particular for $u = \omega'/2$ we come by the Legendre identity to the known case of $(\pi/\omega, \eta/\omega)$. Even the corollary of (5.1) that

$$\zeta(u) - \eta u/\omega \text{ is transcendental} \tag{5.2}$$

was proved for the first time. Before Ramachandra [**R1**] and M. Waldschmidt (cf. [**W1**]) had shown that among 3 numbers $\zeta(u_i) - \eta u_i/\omega$; $i = 1, 2, 3$, for linearly independent algebraic points u_1, u_2, u_3 there is one transcendental.

It is very important to note that (5.2) can be proved as a corollary of the algebraic independence result (5.1).

Another important result (definitely of "nonperiodic" nature) for $\wp(z)$ with complex multiplication (c.m.) gives us algebraic independence of

$$\zeta(u) \text{ and } u. \tag{5.3}$$

We also mention as a corollary of a more general result that for $\wp(z)$ with c.m. by β ($\notin \mathbf{Q}$) and ω_β the period of $\wp(z)$

$$\pi/\omega_\beta \text{ and } e^{\pi i\beta} \text{ are a.i.} \tag{5.4}$$

Here and below $\wp(z)$ denotes the Weierstrass elliptic function with algebraic invariants g_2, g_3. Let (ω, η) be a pair of (nonzero) period and quasi-period of $\wp(z)$, $\zeta(z)$.

When necessary, complex multiplication will be indicated.

THEOREM 5.1. *Let u be an algebraic point on $\wp(z)$: $\wp(u) \in \overline{\mathbf{Q}}$, and u and ω be linearly independent over $\mathbf{Z}$. Then*

$$\zeta(u) - \eta u/\omega \textit{ and } \eta/\omega \textit{ are algebraically independent.}$$

PROOF. Let $\mathbf{E} = \mathbf{Q}(g_2, g_3, \eta/\omega, \zeta(u) - \eta u/\omega, \wp(u), \wp'(u))$ and let us suppose that $\deg\operatorname{tr}\mathbf{E} = 1$. We take an auxiliary function in the form

$$F(z) = P(\zeta(z) - \eta z/\omega, \wp(z)) \tag{5.5}$$

(as in §3). Now we take two independent parameters L, X (which are sufficiently large) and a polynomial $P(x, y)$ such that

$$P(x, y) \in \mathbf{E}[x, y] \tag{5.6}$$

and

$$d_x(P) = L_1, \qquad d_y(P) = L_2 \tag{5.7}$$

and

$$L_1 = [c_1 X^3], \qquad L_2 = [LX^3]. \tag{5.8}$$

By the generalized Siegel lemma we find a polynomial $P(x, y) \in \mathbf{E}[x, y]$ (5.6),

$$P(x, y) \not\equiv 0 \tag{5.9}$$

such that the auxiliary function $F(z)$ satisfies the following equations:

$$F^{(k)}(\omega/4 + n\omega + mu) = 0 \tag{5.10}$$

for positive integers n, m, k with

$$m = 0, \ldots, X; \qquad k = 0, \ldots, S; \qquad n = 0, \ldots, Y \tag{5.11}$$

for

$$S = [c_2 LX^5], \qquad Y = [c_3 X^3]. \tag{5.12}$$

We recall that the function $F(z)$ is periodic with period ω. In this situation type and degree of coefficients of the polynomial $P(x, y)$ can be estimated in the following way:

$$\deg_{\mathbf{E}}(P) \leqslant c_4 X^3; \tag{5.13}$$

$$t_{\mathbf{E}}(P) \leqslant c_4 LX^5 \log(LX). \tag{5.14}$$

Here, as usual, $\deg_{\mathbf{E}}(P)$ and $t_{\mathbf{E}}(P)$ denotes, correspondingly, maximum of the degree and type of the coefficients of $P(x, y)$ as elements from $\mathbf{E}$.

Now we can find first nonzero $F(z)$ among points of the form $\omega/4 + n\omega + mu$ by means of Lemma 3.2 from §3.

LEMMA 3.2. *The number of zeroes of $F(z)$ in a disk of radius $|z| \leqslant R$ is not more than*

$$\sum_{|w| \leqslant R} \operatorname{ord}_w(F) \leqslant c_5(LX^9 + LX^3R^2).$$

Thus we find some $k_0 \leqslant c_6 LX^5$, $m_0 \leqslant c_6 X$, $n_0 \leqslant c_6 X^3$ for some $c_6 > c_5 + 1$ such that

$$\xi_0 = F^{(k_0)}(\omega/4 + n_0\omega + m_0 u) \neq 0. \tag{5.15}$$

Now using (5.5), (5.10)–(5.12) we apply Schwarz's lemma to a function $F(z)$ in the region

$$|z| \ll X^3. \tag{5.16}$$

We obtain an upper bound for ξ_0 (5.15) from Schwarz's lemma

$$|\xi_0| < \exp(-c_7 LX^9). \tag{5.17}$$

Now $\xi_0 \in \mathbf{E}$ and using choice of k_0, m_0, n_0 and (5.6)–(5.8), (5.13) and (5.14) we obtain for this element of $\mathbf{E}$ the following estimates of type and degree (in $\mathbf{E}$).

$$t_{\mathbf{E}}(\xi_0) \leqslant c_8 LX^5 \log(LX), \tag{5.18}$$

$$d_{\mathbf{E}}(\xi_0) \leqslant c_9 X^3. \tag{5.19}$$

Here, of course, $\xi_0 \neq 0$. Now we take any transcendental $\theta \in \mathbf{E}$ and $[\mathbf{E}: \mathbf{Q}(\theta)] < \infty$, as $\deg \operatorname{tr} \mathbf{E} = 1$. Taking the norm of ξ_0 in $\mathbf{E}/\mathbf{Q}(\theta)$ and then specializing $L = X$ we come in (5.17)–(5.19) to the contradiction with generalized Gelfond criteria **[G1]** of transcendence of θ. Theorem 5.1 is proved.

REMARK. We have in the proof of this theorem two independent parameters because we need them to estimate the measure of algebraic independence independently on degree and height. These bounds for the measure of the algebraic independence of $\zeta(u) - \eta u/\omega$ and η/ω, close to the best possible, are proved below in Chapter 8.

As a corollary of Theorem 5.1 we obtain the following

COROLLARY 5.2. *For an algebraic u, $\wp(u) \in \overline{\mathbf{Q}}$, linearly independent with the number ω,*

$$\zeta(u) - \eta u/\omega \text{ is transcendental.}$$

Another main result of this chapter is the following.

THEOREM 5.3. *For two linearly independent over $\mathbf{Z}$ algebraic points u_1, u_2 of $\wp(z)$ ($\wp(u_i) \in \overline{\mathbf{Q}}$: $i = 1, 2$) two of the numbers*

$$u_1, u_2, \zeta(u_1), \zeta(u_2)$$

are algebraically independent.

PROOF. Let $\mathbf{F} = \mathbf{Q}(g_2, g_3, \wp(u_i), \wp'(u_i), u_i, \zeta(u_i)$: $i = 1, 2)$ and suppose that $\deg \operatorname{tr} \mathbf{F} = 1$. The choice of auxiliary function and all analytical parts of the proof remain the same as in the proof of

$$(5.20) \qquad \#\{\omega_1, \omega_2, \eta_1, \eta_2\} \geqslant 2.$$

Indeed, (5.20) is simply part of Theorem 5.3 for $u_i = \omega_i$: $i = 1, 2$.

We take fixed $\varepsilon > 0$, $1 > \varepsilon > 0$, and two sufficiently large parameters L_0, L_1 satisfying

$$(5.21) \qquad L_1 \gg L_0 \gg L_1^{\varepsilon}.$$

Then we take the auxiliary function $F(z)$ in the usual form,

$$(5.22) \qquad F(z) = P(z, \wp(z), \zeta(z)) = \sum_{\lambda_0=0}^{L_0} \sum_{\lambda_1=0}^{L_1} \sum_{\lambda_2=0}^{L_2} p(\lambda_0, \lambda_1, \lambda_2) z^{\lambda_0} \wp(z)^{\lambda_1} \zeta(z)^{\lambda_2},$$

where

$$(5.23) \qquad L_2 = L_0$$

and $P(x, y, z) \in \mathbf{F}[x, y, z]$. Again by general Siegel's lemma we find a nonzero polynomial $P(x, y, z) \in \mathbf{F}[x, y, z]$, $P(x, y, z) \not\equiv 0$, such that the following equations are satisfied:

$$(5.24) \qquad F^{(k)}(nu_1 + mu_2) = 0$$

for positive integers satisfying

$$(5.25) \qquad k = 0, 1, \ldots, S-1, \qquad n, m = 0, 1, \ldots, X-1,$$

for

$$(5.26) \qquad S = \left[c_{10} L_0 L_1 (\log L_1)^{-1/2}\right], \qquad X^2 = \left[c_{11} L_0 (\log L_1)^{1/2}\right].$$

The estimates for type of $p(\lambda_0, \lambda_1, \lambda_2)$ as elements from $\mathbf{F}$ are also very good (mainly because of using Coates lemma).

$$(5.27) \qquad \max_{\lambda_i} t_{\mathbf{F}}(p(\lambda_0, \lambda_1, \lambda_2)) \leqslant c_{12} L_0 L_1 (\log L_1)^{1/2},$$

(5.28) $$\max_{\lambda_i} d_{\mathbf{F}}(p(\lambda_0, \lambda_1, \lambda_2)) \leqslant c_{15} L_0.$$

Using the known bound of the number of zeroes of the function $F(z)$ we come by Schwarz's lemma to a number

$$\zeta = F^{(k_0)}(n_0 u_1 + m_0 u_2) \neq 0$$

such that

(5.29) $$|\zeta| < \exp(-c_{16} L_0^2 L_1 (\log L_1))$$

and

(5.30) $$\zeta \in \mathbf{F}, \qquad \zeta \neq 0,$$

where

(5.31) $$t_{\mathbf{F}}(\zeta) \leqslant c_{17} L_0 L_1 (\log L_1)^{1/2}, \qquad d_{\mathbf{F}}(\zeta) \leqslant c_{18} L_0.$$

Now by generalized criteria **[G1, B13]** of transcendence we find (taking e.g. $L_0 = O(L_1)$) that (5.29)–(5.31) is inconsistent with $\deg \operatorname{tr} \mathbf{F} = 1$.

Again two independent parameters L_0, L_1 were introduced in order to obtain a measure of algebraic independence with different behavior in height and degree (the dependence on height is not bad).

COROLLARY 5.4. *For $\wp(z)$ with complex multiplication and u is an algebraic point on $\wp(z)$, then*

u and $\zeta(u)$ are algebraically independent.

The two parameters L_0 and L_1 from the proof of Theorem 5.3 are used to obtain the measure of algebraic independence of numbers $\zeta(u)$ and u of Corollary 5.4. These measures of algebraic independence are formulated in the Introduction to this volume; for the proof see **[C14]**.

As the last example of algebraically independent numbers we present

THEOREM 5.5. *For $\wp(z)$ with complex multiplication by β two numbers*

$$\pi/\omega \text{ and } e^{\pi i \beta}$$

are algebraically independent.

This theorem follows immediately from the results of §4. In fact it is necessary simply to apply Theorem 4.1. In order to do this we remark that instead of a "normalized" quasi-periodic function $\zeta(z)$ we can consider an arbitrary quasi-periodic function

$$\zeta(z) + \alpha z \quad \text{for algebraic } \alpha \in \overline{\mathbf{Q}},$$

satisfying as $\zeta(z)$ an algebraic differential equation. Changing $\zeta(z)$ to $\zeta(z) + \alpha z$ and η_i to $\eta_i + \alpha\omega_i$ we come to a variant of Theorem 4.1, §4.

THEOREM 5.6. *For any algebraic* $\alpha_1, \alpha_2, |\alpha_1| + |\alpha_2| > 0$ *and any complex* $\varphi \neq 0$ *we have*

$$\#\{\alpha_1\omega_1 + \alpha_2\eta_1, \alpha_1\omega_2 + \alpha_2\eta_2, \varphi, e^{\varphi\omega_1}, e^{\varphi\omega_2}\} \geqslant 2.$$

For $\varphi = \pi i/\omega$ Theorem 5.6 implies 5.5.

6. *L*-functions of elliptic curves.

In this section we collect various auxiliary results on properties of L-functions of elliptic curves and Abelian varieties and their values, in connection with applications of the Birch-Swinnerton-Dyer conjecture to bounds of integer points on curves. We refer reader to **[M2]** for a discussion of L-functions of Abelian varieties in connection with the Birch-Swinnerton-Dyer conjecture.

For an Abelian variety $\mathbf{A}$, the Dirichlet's L-series $L_{\mathbf{A}}(s)$ or $L(\mathbf{A}; s)$ has the form

$$L_{\mathbf{A}}(s) = \sum_{n=1}^{\infty} a_n n^{-s},$$

when we have Euler product $L_{\mathbf{A}}(s) = \Pi_v P_v(Nv^{-s})^{-1}$, where $P_v(t) \in \mathbf{Z}[t]$, $P_v(0) = 1$.

For points v of good reduction, denoted by $v \notin T$, one has the representation

$$P_v(t) = \prod_{i=1}^{2d} (1 - \varphi_{i,v}t),$$

where $\varphi_{i,v}$ are characteristic roots of Frobenius endomorphism of reduction (mod v), while

$$v \in T \text{ implies } P_v(\zeta N(v)^{-1}) \neq 0$$

for any root of unity ζ.

If $d = \dim \mathbf{A} = 1$, then

$$P_v(t) = (1 - \pi_v t)(1 - \bar{\pi}_v t)$$

for good reduction; otherwise, $P_v(t) = 1 - t,\ 1 + t, 1$ if $\mathbf{A}$ (mod v) is a trivial torus, torsion trivial torus or additive group, respectively.

LEMMA 6.1. *If* $d = 1$ *and* $[\mathbf{K} : \mathbf{Q}] < \infty$ *for an Abelian field* $\mathbf{K}$ *and if* $\mathbf{A}$ *has good reduction in points of ramifications of* $\mathbf{K}$, *then* $\Pi_\chi L_\chi(\mathbf{A}, s)$ *is L-series for* $\mathbf{A} \otimes \mathbf{K}$.

Here, for some L we have $\mathbf{K} \subset \mathbf{Q}(e^{2\pi i/L})$; for $[\mathbf{K} : \mathbf{Q}]$ characters χ.

Now we define the Birch-Swinnerton-Dyer conjecture (see **[C23]** and **[M2]**) for an Abelian variety $\mathbf{A}$. Let A be defined over $\mathbf{K}$.

CONJECTURE. Near $s = 1$ we have the asymptotics

$$L(\mathbf{A}; s) \backsim (s-1)^{\operatorname{rank} \mathbf{A}(\mathbf{K})} \cdot \frac{[(\mathbf{A}(\mathbf{K}))] \cdot H}{[\tilde{\mathbf{A}}(\mathbf{K})][\tilde{\mathbf{A}}'(\mathbf{K})]} \cdot \frac{2^{dr_2}}{|D_{\mathbf{K}}|^{d/2}} \cdot \prod_{v \in T} m_v.$$

Here m_v: $v: \in T$, correspond to archimedean or nonarchimedean points of bad reduction, and $H = H[\mathbf{A}(\mathbf{K})]$.

Here for $d = 1$ and nonarchimedean points $v \in T$

$$m_v = [\pi_0(\mathbf{A} \bmod v)]$$

is the number of connected components of the closed bundle of a Neron model of $\mathbf{A}$ at v, rational over the field of class of residue.

For archimedean v,

$$m_v = [\pi_0(\mathbf{A}(\mathbf{R}))] \cdot \left|\det\left(\int_{\gamma_i} \omega_j\right)\right| : \qquad \mathbf{K}_v = \mathbf{R},$$

$$m_v = 2^{-d}\left|\det\left(\int_{\gamma_j} \omega_j, \overline{\int_{\gamma_j} \omega_j}\right)\right| : \qquad \mathbf{K}_v = \mathbf{C}.$$

Let now $d = 1$, W^+ be real, W^- be imaginary period. Then

$$m_\infty = \beta^{-1} | W^+ \cdot W^- |,$$

where $\beta = 2$ if $\mathbf{A}(\mathbf{R})$ is connected and $\beta = 1$ otherwise. Let us suppose also that

$$f_A(z) = \sum_{n=1}^{\infty} a_n e^{2\pi i n z}$$

is $\Gamma_0(N)$ a cusp (parabolic) form

$$\Gamma_0(N) = \left\{\begin{pmatrix} a & b \\ c & d \end{pmatrix} : ad - bc = 1, a, b, c, d \in \mathbf{Z} : c \equiv 0 \bmod N\right\}.$$

We can write the L-series as an Euler product in $d = 1$,

$$L_E(s) = \prod_p \left(1 + (N_p - p - 1)p^{-s} + p^{1-2s}\right)^{-1}$$

for

$$N_p = |E(\mathbf{F}_p)|, \qquad N_p - p - 1 = a_p.$$

For an elliptic curve E defined over $\mathbf{K}$ we consider a uniformization by Weierstrass elliptic functions $\wp(u)$, $\wp'(u)$, with invariants from $\mathbf{K}$, generated by a lattice with periods W^+, W^-.

Lemma 6.2. *Let $u_1, \ldots, u_g$ be generators of the Mordell-Weil group of E over* $\mathbf{K}$.

$$\prod_{i=1}^{g} h(u_i) \leqslant \textit{constant (depending on } g) \cdot H[E(\mathbf{K})].$$

Here by $h(u)$ we understand Weil height:

$$h(u) = \max\{H(\wp(u)); H(\wp'(u))\}.$$

Also for an Abelian variety $\mathbf{A}$ over $\mathbf{K}$ we consider $\hat{h}(x) = \hat{h}(x, x)$—the Tate height. Here

$$|\hat{h}(x) - h(x)| \leqslant \delta$$

for any $x \in \mathbf{A}(\mathbf{K})$ for any Abelian variety $\mathbf{A}$ over $\mathbf{K}$ and constant δ depending on $\mathbf{K}$ and $\mathbf{A}(\mathbf{K})$ only.

For the estimate of $|L^{(g)}(1)|$ we need a good interference formula for coefficients of L-series. E.g. this is the upper bound of

$$\left|\sum_{n=1}^{X} a_n\right|$$

for a large X in terms of X and the conductor of the L-series. Such formulae can be deduced under the assumption that L-series satisfy functional equations. This is the Voronoj-Hardy formula and the general theorem belonging to E. Landau. We give the formulation of the Landau Theorem.

LANDAU THEOREM. *Let $\varphi(s) = \sum_{n=1}^{\infty} a(n) n^{-s}$ be a Dirichlet series, absolutely convergent for* $\mathrm{Re}(s) > k$, *and let the sum be regular for all s, excluding a possible pole for $s = k$ with residue ρ and having functional equation*

$$\left(\frac{\lambda}{2\pi}\right)^{s} \Gamma(s)\varphi(s) = \gamma\left(\frac{\lambda}{2\pi}\right)^{k-s} \Gamma(k-s)\varphi(k-s).$$

Then

$$\frac{1}{q!}\sum_{n=0}^{x} a(n)(x-n)^{q} = \rho \cdot \frac{\Gamma(k)}{\Gamma(k+q-1)} x^{k+q}$$

$$+\gamma \frac{x^{(q+k)/2}}{(2\pi/\lambda)^{q}} \sum_{n=1}^{\infty} \frac{a(n)}{n^{(k+q)/2}} \cdot J_{k+q}\left[\frac{4\pi}{\lambda}(nx)^{1/2}\right].$$

This formula was already used to obtain an estimate for the "interference property" of coefficients of cusp forms over $\Gamma(1)$ (or $\Gamma(N)$) by Moreno.

For the case of elliptic curves E, satisfying the Weil conjecture (e.g. for elliptic curves with complex multiplication) we have

COROLLARY 6.3. *If E is defined over* $\mathbf{Q}$ (*or over an abelian field* $\mathbf{K}$) *and has conductor N* (*correspondingly the norm of conductor is N*) *in the sense of Weil conjecture, then*

$$\left|L_E^{(g)}(1)\right| \leq c \cdot N^{1/4}$$

for c depending on g and $\mathbf{K}$.

Unfortunately in the case of complex multiplication this gives us almost nothing, when the conductor N is square in this case.

Only assuming the Riemann hypothesis for $L_E(s)$, namely that all zeroes of $L_E(s)$ lie on $\mathrm{Re}(s) = 1$ we can obtain a better estimate.

COROLLARY 6.4. *Let $L_E(s)$ satisfy a functional equation connected with given N and the Riemann hypothesis. Then*

$$\left|L_E^{(g)}(1)\right| \leq c(g, \varepsilon) N^{\varepsilon}$$

for any $\varepsilon > 0$.

The functional equation for $F(z) = \Sigma a_n q^n$ (the Mellin transform of $L_E(s)$) is said to be connected with N if it has the form

$$F(z) = -cN^{-1}z^{-2}F\left(-\frac{1}{Nz}\right), \qquad c = (-1)^q.$$

These bounds now can be applied to the study of integer points on an elliptic curve E in terms of linear forms in algebraic points of E.

If u_0 is an integer point of $E(\mathbf{K})$ and $u_1,\dots,u_g$ are the generators of the Mordell-Weil group $E(\mathbf{K})$ and

$$u_0 = \sum_{i=1}^{g} n_i u_i,$$

where $M = \Sigma_{i=1}^{g} |n_i|$ any $B = \log M$, then for $B \geqslant B_0$, we have

$$\left|\sum_{i=1}^{g} n_i u_i\right| < \exp(-cM^2)$$

where B_0 and c depend on δ, $[\mathbf{K}:\mathbf{Q}]$ and g.

To obtain an upper bound on the height $h(u_0)$ of u_0 in terms of N we come to the question of the estimate (see §1 and **[C11]**)

$$\left|\sum_{i=1}^{g} n_i u_i\right| < \exp\left(-e^{B^{\varepsilon}}\left[\prod_{i=1}^{g} h(u_i)\right]^x\right)$$

for $x \geqslant 1$, $\varepsilon > 0$.

We conclude this section with explicit formulas for elliptic curves with c.m. where the Weierstrass elliptic function, uniformizing $E(\mathbf{K})$ is, in fact, defined already over the field of complex multiplications of E (rather than over $\mathbf{K}$).

All the difficulties come from period ω (instead of π in Dirichlet's case of formula for $h \cdot \operatorname{Reg}(\mathbf{K})$).

For the curve

$$y^2 = 4x^2 + k \quad \text{or} \quad y^2 = x^3 + k \qquad (E_k),$$

with the period $\omega_k = \omega_1/k^{1/6}$. Here

$$\wp(z;k) = k^{1/3}\wp(k^{1/6}z;1).$$

If u is algebraic for (E_k), over $\mathbf{K}$, then

$$k^{1/3}\wp(k^{1/6}u;1) \in \mathbf{K}, \qquad k^{1/2}\wp'(k^{1/6}u;1) \in \mathbf{K}.$$

Also for

$$y^2 = 4x^3 - 4Dx, \qquad y_2 = x^3 - Dx, \qquad (E_D)$$

the period is $\omega_D = \omega_1/D^{1/4}$.

For these curves we have the representation

$$L_{E_D}^{(g)}(1) = \frac{\omega_1}{D^{1/4}} \cdot \frac{Ш_D H_D \alpha}{\beta} g!, \qquad L_{E_k}^{(g)}(1) = \frac{\omega_1}{k^{1/6}} \cdot \frac{Ш_k H_k \alpha'}{\beta'} g!.$$

7. Auxiliary algebraic results. Here we present algebraic results that are used in the algebraic independence proof. It is the definition of type and height of elements of fields of finite degree of transcendence and Siegel's lemma.

These results are used for the proof of criteria of transcendence and criteria of algebraic independence, in the relation to values of the exponential and elliptic functions.

A. In the fields of finite degree of transcendence instead of a very simple definition of height we must work with a following version of the notion of type (or size) of elements.

First of all for the transcendental extension $\mathbf{K} = \mathbf{Q}(x_1, \ldots, x_n)$ with $x_1, \ldots, x_n$ algebraically independent over $\mathbf{Q}$ and $I_K = \mathbf{Z}[x_1, \ldots, x_n]$ and any element $P(x_1, \ldots, x_n) \in \mathbf{Z}[x_1, \ldots, x_n]$,

$$P(x_1, \ldots, x_n) = \sum_{i_1, \ldots, i_n} a_{i_1, \ldots, i_n} x_1^{i_1} \cdots x_n^{i_n},$$

we put

$$\deg P = \sum_{i=1}^{n} \deg_{x_i}(P),$$

$$H(P) = \max |a_{i_1, \ldots, i_n}|, \qquad t(P) = \max\{\deg(P); H(P)\}.$$

Similarly for $\alpha \in \mathbf{Q}(x_1, \ldots, x_n)$ and $\alpha = P/Q$, where $P(\bar{x})$ and $Q(\bar{x})$ have no nontrivial common factors we put

$$\deg \alpha = \max\{\deg P; \deg Q\},$$

$$H(\alpha) = \max\{H(P); H(Q)\}, \qquad t(\alpha) = \max\{\deg \alpha, H(\alpha)\}.$$

Also for polynomial $P(\bar{x}) \in \mathbf{C}[\bar{x}]$ we define a euclidean norm

$$\|P\|_{L_2} = \left(\sum_{i_1, \ldots, i_n} |a_{i_1, \ldots, i_n}|^2 \right)^{1/2} \text{-}L_2\text{-norm}$$

of coefficients.

For the nonpurely transcendental extension $\mathbf{K} = \mathbf{Q}(x_1, \ldots, x_m; y_1, \ldots, y_n)$, where $\{y_1, \ldots, y_n\}$ is a vector basis for $\mathbf{K}$ over the purely transcendental extension $\mathbf{Q}(x_1, \ldots, x_m)$ we also define the corresponding analogue of type.

If $\alpha \in \mathbf{K}$, then multiplication by α is a linear transformation on $\mathbf{K}$. For $\alpha \neq 0$ let

$$(P_{ij}/Q)_{i,j=1, \ldots, n}$$

be the matrix of this transformation over $\mathbf{Q}(x_1, \ldots, x_m)$ with respect to a fixed basis $(y_1, \ldots, y_n)$ for $\mathbf{K}$; such that Q and the greatest common divisor of all P_{ij} are relatively prime. Then we call Q the denominator of α and define

$$\deg \alpha = \max_{1 \leq i, j \leq n} \{\deg Q; \deg P_{ij}\}$$

$$H(\alpha) = \max_{1 \leq i, j \leq n} \{H(Q), H(P_{ij})\}, \qquad t(\alpha) = \max\{\deg \alpha, H(\alpha)\}.$$

Below we will change the definition of type a little in the situation when $y_1, \ldots, y_n$ will not be fixed.

Such definition of size is very useful and gives us the following.

LEMMA 7.1. *For arbitrary elements* $\alpha_1, \ldots, \alpha_k$ *of* $\mathbf{K}$ *we have*

$$t(\alpha_1 + \cdots + \alpha_k) \leqslant 3(t(\alpha_1) + \cdots + (\alpha_k)) + \log k,$$
$$t(\alpha_1 \cdots \alpha_k) \leqslant 3(t(\alpha_1) + \cdots + t(\alpha_k)) + k \log n.$$

The main auxiliary result in this direction is the well-known Koksma-Popken-Gelfond lemma **[G1]**, giving the relation between the height of polynomials and their factors.

GELFOND'S LEMMA 7.2 **[G1]**. *Let* $P_1, \ldots, P_m$ *be the polynomials of* $\mathbf{C}[x_1, \ldots, x_q]$ *and let* $d(P) = \deg(P_1 \cdots P_m)$. *Then*

$$\|P_1 \cdots P_m\|_{L_2} \geqslant 2^{-d(P)+q/2} \cdot \|P_1\|_{L_2} \cdots \|P_m\|_{L_2}.$$

Now

$$e^x > \left(\frac{x+1}{2}\right)^{1/2} \cdot 2^x: \qquad x \geqslant 1.$$

We have

$$H(P) \leqslant \|P\| \leqslant H(P) \cdot \prod_{i=1}^{q} \left(1 + \deg_{x_i}(P)\right)^{1/2}.$$

GELFOND'S LEMMA 7.3. $H(P_1 \cdots P_m) \geqslant e^{-d(P_1 \cdots P_m)} \cdot H(P_1) \cdots H(P_m)$.

The main idea of the proof of this formula is the following: Let $Q(z) = q_n z^n + \cdots + q_0$-polynomial with zeroes $\beta_1, \ldots, \beta_n$ and $1 < \rho < |\beta_i|$ for all $|\beta_i| > 1$: $i = m+1, \ldots, n$. We have Jensens' formula

$$\int \log|Q(\rho e^{2\pi i\theta})| d\theta = m \log \rho + \log|q_n \beta_{m+1} \cdots \beta_n|,$$

or

$$\log|q_n \beta_{i_1} \cdots \beta_{i_k}| \leqslant \int_0^1 \log|Q(\rho e^{2\pi i\theta})| d\theta.$$

Thus

$$\log|q_k| \leqslant \log\binom{n}{k} + \int_0^1 \log|Q(\rho e^{2\pi i\theta})| d\theta.$$

In other words,

$$\log H(P_i) \leqslant (N(P_i) - 1)\log 2 + \int_0^1 \log|P_i(\rho e^{2\pi i\theta})| d\theta,$$
$$|P(z)| \leqslant (N+1)H\rho^N, \qquad |z| = \rho > 1.$$

B. *General Siegel lemma.* Here we give general versions of the Siegel lemma based on **[G1, W1]**.

SIEGEL'S LEMMA 7.4. *Let* $\mathbf{K}$ *be the field of finite degree* q *of transcendence and* $(\theta_1,\ldots,\theta_q,\omega)$ *be a system of generators of* $\mathbf{K}$ *on* $\mathbf{Q}$, *i.e.* $(\theta_1,\ldots,\theta_q)$ *is the basis of transcendence of* $\mathbf{K}$ *and* ω *is algebraic over* $\mathbf{Q}(\theta_1,\ldots,\theta_q)$. *There exists a constant* $C>0$, *depending on* $(\theta_1,\ldots,\theta_q,\omega)$ *with the following properties.*

For any integers n, r; $n \geqslant 2r > 0$, *and elements*

$$a_{ij}:\qquad 1\leqslant i\leqslant n;\quad 1\leqslant j\leqslant n,$$

from the ring $\mathbf{I}_{\mathbf{K}} = \mathbf{Z}[\theta_1,\ldots,\theta_q,\omega]$ *and system of equations*

$$\sum_{i=1}^{n} a_{ij}x_i = 0:\qquad j=1,\ldots,r,$$

there exist elements $\xi_1,\ldots,\xi_n$ *from* $\mathbf{I}_{\mathbf{K}}$, *not all zeroes, such that*

$$\sum_{i=1}^{n} a_{ij}\xi_i = 0 \quad \text{for all } j=1,\ldots,r,$$

where we have the following bound for types of ξ_i:

$$\max_{1\leqslant i\leqslant n} t(\xi_i) \leqslant C\Big[\max_{i,j} t(a_{i,j}) + \log n\Big].$$

This lemma follows immediately from the following classical Siegel lemma on the solutions of equations in integers.

USUAL SIEGEL LEMMA 7.5. *Let* R *and* S *be positive integers,* $R<S$ *and* $a_{i,j}\in\mathbf{Z}$ *be arbitrary integers* $1\leqslant i\leqslant R$, $1\leqslant j\leqslant S$, *and*

$$\max_{1\leqslant i\leqslant R,1\leqslant j\leqslant S} |a_{ij}| \leqslant A.$$

Then the system of equations

$$\begin{aligned} a_{11}z_1 + \cdots + a_{1S}z_S &= 0,\\ \text{-- -- -- -- -- -- -- --}&\\ a_{R1}z_1 + \cdots + a_{RS}z_S &= 0, \end{aligned}$$

has a solution in integers z_i, *not all zero, with*

$$|z_i| \leqslant (SA)^{R/(S-R)}.$$

PROOF OF LEMMA 7.5. We have a linear map: $\mathfrak{L}\colon \mathbf{Z}^S \to \mathbf{Z}^R$. Now if $B=[(SA)^{R/(S-R)}]$, then $(B+1)^{S-R} > (SA)^R$ and

$$(B+1)^S > (B+1)^R\cdot(SA)^R > (BSA+1)^R$$

and there are $(B+1)^S$ elements from $\mathbf{Z}^S$ with coordinates in $[0,B]$. Their images under $\mathfrak{L}$ in $\mathbf{Z}^R$ have jth coordinate $(1\leqslant j\leqslant R)$ lying in $[-N_jAB;(S-N_j)AB]$, where N_j is the number of negative elements among a_{ji}. Thus the number of images is not more than $(SAB+1)^R < (B+1)^S$. So two elements from $[0,B]^S$ have the same image, $\bar z_1, \bar z_2 \in [0,B]^S$ and their difference $\bar z = \bar z_1 - \bar z_2$ is the solution of our system.

8. Algebraic independence for values of exponent. For exponential functions we know good bounds for the number of zeroes (Gelfond-Tijdeman) and we have some precise results for algebraic independence. Thus it is natural to give a short review on algebraic independence of constants connected with exponential functions.

The first and most striking result belongs to Gelfond **[G1]** 1949, and since that time, most of the results have been formulated in the following way:

"Among the numbers $\xi_1,\dots,\xi$ (in the set S) there are 2 a.i. numbers".

For shortness, as before, we denote for the set $S \subset \mathbf{C}$, the number of a.i. over $\mathbf{Q}$ elements from S as

$$\#S = \deg.\operatorname{tr}.\mathbf{Q}(S).$$

For exponential functions we consider always three types of sets of numbers which we investigate.

ASSUMPTION. Let $\alpha_1,\dots,\alpha_N$ and $\beta_1,\dots,\beta_M$ be two sequence of complex numbers, linearly independent over $\mathbf{Q}$.

Now we introduce three sets of numbers.

(i) $S_1 = \{e^{\alpha_i\beta_j}\colon 1 \leqslant i \leqslant N;\ 1 \leqslant j \leqslant M\}$,

(ii) $S_2 = \{\beta_j,\ e^{\alpha_i\beta_j}\colon 1 \leqslant i \leqslant N;\ 1 \leqslant j \leqslant M\}$,

(iii) $S_3 = \{\alpha_i,\ \beta_j,\ e^{\alpha_i\beta_j}\colon 1 \leqslant i \leqslant N;\ 1 \leqslant j \leqslant M\}$.

To each of these sets there corresponds a system of numbers.

DEFINITION. $\kappa_i = |S_i|/(M+N)$: $i = 1,2,3$, i.e.

$$\kappa_1 = \frac{MN}{M+N}; \qquad \kappa_2 = \frac{M(N+1)}{M+N}; \qquad \kappa_3 = \frac{MN+M+N}{M+N},$$

and also $\gamma_1 = 1$, $\gamma_2 = N/(M+N)$, $\gamma_3 = 0$.

Let $\mathfrak{D}_i$ be the degree of transcendence of $\mathbf{Q}(S_i)$, $\mathfrak{D}_i = \#S_i$. Then in general we have the following result [**C2a**, **C2b**]:

THEOREM 8.1. *For $\kappa_i > 1$, $\mathfrak{D}_i \geqslant 1$, i.e. in S_i there is at least one transcendental number. For $n \geqslant 1$ and $\kappa_i \geqslant 2n+1$, $\mathfrak{D}_i \geqslant n+1$: $i = 1,2$; if $\kappa_3 > 2n+1$, then $\mathfrak{D}_3 \geqslant n+1$.*

We present sketches of proofs of these results using an approach based on a simpler form of auxiliary functions, than those usually employed. That is why Theorem 8.1 is not as good as author's best result on the algebraic independence of values of exponent: Theorem 3 of subsection 3 of the Introduction, this volume:

THEOREM 8.1′. *For $n = 1$ and $\kappa_i \geqslant n$, $\mathfrak{D}_i = n+1$, $i = 1,2$; if $\kappa_3 > n$, then $\mathfrak{D}_3 \geqslant n+1$.*

The proof for n $= 2$ is contained in Chapter 4, this volume.

Let $\mathbf{Q}(S_i) = \mathbf{Q}(\theta_1,\dots,\theta_{\mathfrak{D}_i}, \vartheta)$, where $\theta_1,\dots,\theta_{\mathfrak{D}_i}$ are the basis of transcendence of $\mathbf{Q}(S_i)$ and ϑ is algebraic over $\mathbf{Q}(\theta_1,\dots,\theta_{\mathfrak{D}_i})$ and

$$A_i = \mathbf{Z}\left[\theta_1,\dots,\theta_{\mathfrak{D}_i}\right].$$

We take an integer parameter $X \geqslant 1$ and take three auxiliary functions of the form

$$(8.1) \qquad F_1^X(z) = P_1(e^{\beta_1 z}, \ldots, e^{\beta_M z}),$$
$$P_1(x_1, \ldots, x_M) \in \mathbf{Z}[x_1, \ldots, x_M], \qquad \deg(P_1) \leqslant [X^{N/(M+N)}],$$
$$(8.2) \qquad F_2^X(z) = P_2(e^{\beta_1 z}, \ldots, e^{\beta_M z}),$$
$$P_2(x_1, \ldots, x_M) \in \mathbf{Z}[x_1, \ldots, x_M], \qquad \deg(P_2) \leqslant X,$$
$$(8.3) \qquad F_3^X(z) = P_3(z, e^{\beta_1 z}, \ldots, e^{\beta_M z}),$$
$$P_3(x_0, x_1, \ldots, x_M) \in \mathbf{Z}[x_0, x_1, \ldots, x_M]$$
$$\deg_{x_0}(P_3) \leqslant [X^{(M+N)/N} \log^{-1} X], \qquad \deg_{x_i}(P_3) \leqslant X\colon i = 1, \ldots, M.$$

The coefficients of the polynomials $P_i(\bar{x})$ are chosen using a version of Siegel's lemma in a way that makes the functions $F_i^X(z)$ small in discs of sufficiently large radius. For this we can choose $F_i^X(z)$ to be Padé-type approximations to $e^{\beta_i z}$ at $z = 0$. Though, the coefficients in the expansion of $e^{\beta_i z}$ at $z = 0$ are complex numbers and the coefficients of $P_i(\bar{x})$ are integers, we can always choose $P_i(\bar{x})$ in such a way that the coefficients of the expansion of $F_i^X(z)$ at $z = 0$ are small:

LEMMA 8.2. *For every* $i = 1, 2, 3$ *there exists a nonidentically zero polynomial* $P_i(\bar{x}) \in \mathbf{Z}[\bar{x}]$ *such that the bounds* (8.1)–(8.3) *are satisfied and the following conditions are satisfies too*:

The height of the polynomials $P_i(\bar{x})$ *is bounded by* A, *and we have the following upper bound on the coefficients of* $F_i^X(z)$ *at* $z = 0$:

$$\left|\left(\frac{d^s}{dz^s}\right) F_i^X(z)\,|_{z=0}\right| \leqslant \exp\left\{-c_1 \frac{X^{MN/(M+N)}}{S} \log A\right\}$$

for $i = 1$, $A = \exp(X)$, $S = X^{(\kappa_1+1)/2} \log^{-1/2} X$;

$$\left|\left(\frac{d^s}{dz^s}\right) F_i^X(z)\,|_{z=0}\right| \leqslant \exp\left\{-c_2 \frac{X^M}{S} \log A\right\}$$

for $i = 2$,

$$A = \exp(X^{(M+N)/(N+1)} \log^{1/(N+1)} X),$$
$$S = X^{(MN+2M+N)/2(N+1)} \log^{-N/2(N+1)} X;$$

$$\left|\left(\frac{d^s}{dz^s}\right) F_i^X(z)\,|_{z=0}\right| \leqslant \exp\left\{-c_3 \frac{X^{(MN+M+N)/N} \log^{-1} X}{S} \log A\right\}$$

for $i = 3$, $A = \exp(X^{(M+N)/N})$, $S = X^{(MN+2M+2N)/2N} \log^{-1} X$; *and for all non-negative integers* $s = 0, 1, \ldots, S$

The polynomials $P_i(\bar{x})$ constructed in Lemma 8.2 give rise to auxiliary functions $F_i^X(z)$ that have small values, together with their derivatives at complex points of the form $z = n_1\alpha_1 + \cdots + n_N\alpha_N$ for integers n_i. To verify this we expand $F_i^X(z)$ in a Taylor expansion in the neighborhood of $z = 0$. We obtain from Lemma 8.2 the following bounds

$$(8.4)\qquad |F_1^X(n_1\alpha_1 + \cdots + n_N\alpha_N)| \leq \exp\left\{-c_4\frac{X^{MN/(M+N)}}{S}\log A\right\},$$

for integers n_i, $0 \leq n_i \leq [c_5X^{M/(N+M)}]$: $i = 1,\ldots,N$;

$$(8.5)\qquad |(F_2^X(n_1\alpha_1 + \cdots + n_N\alpha_N))^{(k)}| \leq \exp\left\{-c_6\frac{X^M}{S}\log A\right\},$$

for integers $k = 0,\ldots,[X^{(M+N)/(N+1)}\log^{-N/(N+1)}X]$ and integers n_i, $0 \leq n_i \leq [c_7X^{(M-1)/(N+1)}\log^{1/(N+1)}X]$: $i = 1,\ldots,N$; and

$$(8.6)\qquad |(F_3^X(n_1\alpha_1 + \cdots + n_N\alpha_N))^{(k)}| \leq \exp\left\{-c_8\frac{X^{(MN+M+N)/N}\log^{-1}X}{S}\log A\right\},$$

for integers $k = 0,\ldots,[X^{(M+N)/N}\log^{-1}X]$ and integers n_i, $0 \leq n_i \leq [c_9X^{M/N}]$: $i = 1,\ldots,N$.

We can use now Tijdeman's lemma **[T1]** on the bound of the number of zeroes of an auxiliary function $P(z, e^{\alpha_i z})$. We can formulate this result as follows.

TIJDEMAN'S LEMMA. *If $F(z) = P(z, e^{\beta_1 z},\ldots,e^{\beta_M z})$, where $P(z_0, z_1,\ldots,z_M) \in \mathbf{C}[\bar{z}]$, $P \not\equiv 0$, then there are constants $c_1(M) > 0$, $c_2(M) > 0$, such that the number of zeroes of $F(z)$ in disk $|z| \leq R$ (counted with their multiplicities) is bounded by the following quantity*:

$$c_1 d_{z_0}(P) \cdot \prod_{i=1}^{M} d_{z_i}(P) + c_2 \sum_{i=1}^{M} d_{z_i}(P) \cdot R.$$

This result does not depend on any assumptions on the measure of linear independence of numbers α_i or β_i, and so Theorem 8.1 is valid without these additional assumptions. Tijdeman's Lemma and the bounds (8.1)–(8.3) imply that for every $i = 1, 2, 3$ one of the numbers in the left hand side of (8.4), (8.5) or (8.6), respectively, is non-zero. In this way we arrive at a sequence of non-zero polynomials from numbers of the sets S_i having small values. In particular, we obtain the following (see Chapter 1, §3):

PROPOSITION 8.3. *If $\kappa_i > 1$, then for any sufficiently large $Z \geq Z_0$, there exists a polynomial*

$$R_{iZ}(x_1,\ldots,x_{\mathfrak{D}_i}) \in \mathbf{Z}[x_1,\ldots,x_{\mathfrak{D}_i}],$$

$R_{iZ} \not\equiv 0$, such that

$$|R_{iZ}(\theta_1,\ldots,\theta_{\mathfrak{D}_i})| < \exp(-c_7 Z^{\kappa_i}\log^{\gamma_i} Z)$$

and

$$\mathrm{type}(R_{iZ}) \leq c_8 Z.$$

Explicit expressions of polynomials $R_{iZ}(\theta_1,\dots,\theta_{\mathfrak{D}_i})$ from Proposition 8.3 can be obtained taking into account expressions (8.1)–(8.3). For this let us denote

$$P_1(x_1,\dots,x_M) = \sum_{l_1=0} \cdots \sum_{l_M=0} p_{1;l_1,\dots,l_M} x_1^{l_1}\cdots x_M^{l_M}, \qquad l_i \leqslant [X^{N/(M+N)}];$$

$$P_2(x_1,\dots,x_M) = \sum_{l_1=0}^{X} \cdots \sum_{l_M=0}^{X} p_{2;l_1,\dots,l_M} x_1^{l_1}\cdots x_M^{l_M},$$

$$P_3(x_0, x_1,\dots,x_M) = \sum_{l_0=0}^{[X^{(M+N)/N}\log^{-1}X]} \sum_{l_1=0}^{X} \cdots \sum_{l_M=0}^{X} p_{3;l_0,l_1,\dots,l_M} x_0^{l_0} x_1^{l_1}\cdots x_M^{l_M}.$$

Here, according to Lemma 8.2, all coefficients $p_{1;l_1,\dots,l_M}$, $p_{2;l_1,\dots,l_M}$, $p_{3;l_1,\dots,l_M}$ are rational integers. Hence for nonzero numbers a_1, a_2, a_3 of the form presented in the left hand side of (8.4), (8.5), (8.6) have explicit representations in terms of elements of sets S_1, S_2, S_3, respectively. In the case $i = 1$, introducing new variables $x_{i,j}$, $i = 1,\dots,N; j = 1,\dots,M$ we obtain

$$a_1 = S_1(e^{\alpha_i\beta_j}: i = 1,\dots,N; j = 1,\dots,M)$$

for

$$S_1(x_{i,j}: i = 1,\dots,N; j = 1,\dots,M) \overset{\text{def}}{=} \sum_{l_1=0} \cdots \sum_{l_M=0} p_{1;l_1,\dots,l_M} \prod_{i=1}^{N}\prod_{j=1}^{M} (x_{i,j})^{l_j n_i}$$

and polynomials $S_1(x_{i,j})$ are parametrized by an integer vector $\vec{n} = (n_1,\dots,n_N)$. Similarly, in the cases $i = 2, 3$ we introduce variables x_i, y_j, $x_{i,j}$, $i = 1,\dots,N$, $j = 1,\dots,M$ such that a_2 and a_3 have the form

$$a_2 = S_2(\beta_j, e^{\alpha_i\beta_j}), \qquad a_3 = S_3(\alpha_i, \beta_j, e^{\alpha_i\beta_j})$$

for polynomials $S_2(y_j, x_{i,j})$ and $S_3(x_i, y_j, x_{i,j})$ of variables $x_i, y_j, x_{i,j}$: $i = 1,\dots,N$; $j = 1,\dots,M$, having the following explicit form:

$$S_2(y_j, x_{i,j}) \overset{\text{def}}{=} \sum_{l_1=0}^{X} \cdots \sum_{l_M=0}^{X} p_{2;l_1,\dots,l_M}(l_1y_1 + \cdots + l_My_M)^k \prod_{i=1}^{N}\prod_{j=1}^{M} x_{i,j}^{l_jn_i};$$

$$S_3(x_i, y_j, x_{i,j}) = \sum_{l_0=0}^{[X^{(M+N)/N}\log^{-1}X]} \sum_{l_1=0}^{X} \cdots \sum_{l_M=0}^{X} p_{3;l_0,l_1,\dots,l_M}$$
$$\times \sum_{s=0}^{k} \binom{k}{s} l_0 \cdots (l_0 - k + s + 1)(x_1d_1 + \cdots + x_N\alpha_N)^{l_0-k+s}$$
$$\times (l_1y_1 + \cdots + l_My_M)^s \prod_{i=1}^{N}\prod_{j=1}^{M} x_{i,j}^{l_jn_i}.$$

Here polynomials $S_2(y_j, x_{i,j})$ and $S_3(x_i, y_j, x_{i,j})$ are parametrized by integers k and $(n_1,\dots,n_N)$ subject to constraints by inequalities (8.5), (8.6). Hence inequalities (8.4), (8.5), (8.6) for the quantities a_1, a_2, a_3, respectively, take the form of small values of polynomials with integer coefficients in variables x_i, y_j, $x_{i,j}$. The simple structure of these polynomials S_1, S_2, S_3 for a_1, a_2, a_3, respectively, allows us to show their (linear) independence and to compute the dimension of graded modules generated by them for varying (integer) parameters k, $(n_1,\dots,n_N)$. In doing this we avoid the use of Tijdeman's Small Value Lemma and remove any conditions on the measure of the linear independence of α_i and β_j. To show linear independence for each of the classes of polynomials S_1, S_2 or S_3 with varying k, $(n_1,\dots,n_N)$ we consider the polynomials $S_1(x_{i,j})$. The problem of linear independence of various polynomials of the form S_1 is equivalent to the following one. Now many linearly independent vectors are there among $(\Pi_{i=1}^{N}\Pi_{j=1}^{M}(x_{i,j})^{l_j n_i}$: $l_j = 0,\dots,[X^{N/(M+N)}]; j = 1,\dots,M)$ for $0 \leqslant n_i \leqslant Y$?

The rank of the matrix above is maximal and equal to $([X^{N/(M+N)}] + 1)^M$ provided that $Y^N \geqslant c_{10} X^{MN/(M+N)}$ for an effective constant $c_{10} > 0$ depending only on N and M.

This bound on Y is exactly the one for which the upper bound (8.4) is established. Hence we obtain $([X^{N/(M+N)}] + 1)^M$ linearly independent inequalities on values of polynomials of the form $S_1(x_{i,j})$ under the specialization $x_{i,j} = e^{\alpha_i\beta_j}$: $i = 1,\dots,N; j = 1,\dots,M$. Similarly, we can determine dimension of a graded subalgebra generated by various polynomials $S_1(x_{i,j})$ in

$$\mathbf{Z}\left[x_{i,j}\colon i = 1,\dots,N; j = 1,\dots,M\right],$$

if we take into account the algebraic relations between $e^{\alpha_i\beta_j}$ that exist according to the condition deg tr $\mathbf{Q}(S_1) = \mathfrak{D}_1 \leqslant n$ with

$$MN/(M+N) \geqslant 2n+1.$$

Under the specialization $x_{i,j} = e^{\alpha_i\beta_j}$ we arrive at the system of equations and inequalities

$$Q_\alpha(x_{i,j}) = 0\colon \quad \alpha \in A_1; \tag{8.7}$$

$$|S_1(x_{i,j})| \leqslant \exp\left\{-c_4 \cdot S^{-1} \cdot X^{MN/(M+N)} \log A\right\},$$

for a system of algebraic equations $Q_\alpha(x_{i,j}) = 0$: $\alpha \in A_1$ with $Q_\alpha(x_{i,j}) \in \mathbf{Z}[x_{i,j}]$ $(i = 1,\dots,N; j = 1,\dots,M)$ defining an algebraic variety of dimension $\mathfrak{D}_1$ and an ideal $I = (Q_\alpha\colon \alpha \in A_1)$ in $\mathbf{Z}[x_{i,j}]$; $|A_1| \geqslant MN - \mathfrak{D}_1$. The system of equations and inequalities (8.7) is the main instrument in the proof of Theorem 81.1 in the case of $i = 1$ (similarly for $i = 2, 3$). Study of polynomials $S_1(x_{i,j})$ implies that the dimension of an algebraic variety defined by equations only: $Q_\alpha(x_{i,j}) = 0$: $\alpha \in A_1$ and $S_1(x_{i,j}) = 0$ for all S_1 with $0 \leqslant u_i \leqslant Y$, $i = 1,\dots,N$ is negative (i.e. an algebraic variety has no points). Hence the addition of polynomials $S_1(x_{i,j})$ to I generates $\mathbf{Z}[x_{i,j}]$. This allows us to apply elimination theory (in the form of the u-resultants of Kronecker) to the system of equations $Q_\alpha(x_{i,j}) = 0$, $\alpha \in A_1$, $S_1(x_{i,j}) = 0$. An appropriate choice of parameters A, $S\,A = \exp(X)$, $S = X^{(\kappa_1+1)/2}\log^{-1/2} X$ in (8.4)–(8.6) and X in (8.1)–(8.6) provides us with the End of the proof of Theorem 8.1 via the estimates of the determinants defining the corresponding u-resultants.

9. Algebraic independence in the elliptic function case. Now we come to the case of elliptic functions. In this situation it is natural to expect some analogues of exponential results. However the first theorems of Ramachandra [**R1**] and Waldschmidt (1967–1973) [**W7**] show unexpected increasing of the cardinality of sets involved in algebraic independence proof.

However there exists one case, which really has very much in common with the case of exponent. We even can make the following "dictionary" for translation:

exponent e^z	$\wp(z)$ with complex multiplication in $\mathbf{Q}(\tau)$
e^z	$\wp(z)$;
$\log \alpha$ for algebraic $\alpha \neq 0, 1$	u algebraic point of $\wp(z)$;
πi	ω period of $\wp(z)$;
$\alpha_1, \ldots, \alpha_N$ linearly independent over $\mathbf{Q}$	$\alpha_1, \ldots, \alpha_N$ linearly independent over $\mathbf{Q}(\tau)$

Unfortunately using this dictionary we can *prove* only some results.

Let us consider in general the problem of algebraic independence of constants connected with functions $\wp(z)$, $\zeta(z)$ and (or) z, e^z.

All constants, which we are investigating arise from

(a) algebraic numbers $\alpha \in \overline{\mathbf{Q}}$;

(b) algebraic points of exponent, i.e. $\log \alpha$ for $\alpha \neq 0, 1$, $\alpha \in \overline{\mathbf{Q}}$;

(c) from algebraic points of $\wp(x)$, e.g. ω_1, ω_2;

(d) from values $\zeta(u)$ for u-algebraic point of $\wp(x)$, e.g. η_1, η_2;

(e) from polynomials from these numbers;

(f) from operators e^x or $\wp(x)$ applied to these numbers.

In general the Schneider-Gelfond method enables us to consider the following general situation.

Let $\wp_1(z), \ldots, \wp_N(z)$ be Weierstrass functions (with invariants g_{i1}, g_{i2}: $1 \leqslant i \leqslant N$), $\zeta_1(z), \ldots, \zeta_N(z)$ be corresponding ζ-functions; $v_1, \ldots, v_M$ and $\varphi_1, \ldots, \varphi_k$ be two sequences of complex numbers, linearly independent over $\mathbf{Z}$. Let also

$$\wp_1(u_1 z), \ldots, \wp_N(u_N z)$$

be algebraically independent over $\mathbf{C}$. We consider a matrix of numbers

$$\text{(S)}\qquad \begin{array}{rl} \wp_i(u_i v_j): & 1 \leqslant i \leqslant N,\ 1 \leqslant j \leqslant M, \\ a_i \zeta_i(u_i v_j) + \beta_i v_j: & 1 \leqslant i \leqslant N,\ 1 \leqslant j \leqslant M, \\ u_1, \ldots, u_N: & a_i, \beta_i;\quad i = 1, \ldots, N, \\ e^{\varphi_i v_j}: & 1 \leqslant i \leqslant k,\quad 1 \leqslant j \leqslant M, \\ v_1, \ldots, v_m, & \varphi_1, \ldots, \varphi_k. \end{array}$$

In this set we consider combinations of some rows.

Analogously to exponential case we will consider the following situation:

ASSUMPTION 9.1. Let $u_1,\ldots,u_N$ be linearly independent over $\mathbf{Q}$ complex numbers such that the functions $\wp_1(u_1 z),\ldots,\wp_N(u_N z)$ are algebraically independent over $\mathbf{C}$. Let $v_1,\ldots,v_M$ be linearly independent over $\mathbf{Q}$ complex numbers.

We will also sometimes use the estimation of the measure of linear independence of $\{u_1,\ldots,u_N\}$ and $\{v_1,\ldots,v_M\}$ in one of the following forms. For arbitrary sequence $\{\eta_1,\ldots,\eta_k\}$ of numbers linearly independent over $\mathbf{Q}$ we propose the following criteria on measure of linear independence:

(H_1) there exists a constant $\tau_0,\ \tau_1 > 0$ such that

$$\left|\sum_{i=1}^{k} n_i\eta_i\right| > \exp\left(-\tau_0\left(\log\sum_{i=1}^{k}|n_i|\right)^{\tau_0}\right)$$

for $\sum_{i=1}^{k}|n_i| > \tau_1$ and integers $n_1,\ldots,n_k$;

Now we will always assume Assumption 9.1 to be true and so the sequences $\{u_1,\ldots,u_N\}$ and $\{v_1,\ldots,v_M\}$ will always satisfy these conditions. It is possible to introduce main systems of numbers connected with functions $\wp_i(z)$.

DEFINITION 9.2. *Let $\wp_i(z)$: $i = 1,\ldots,N$, be Weierstrass elliptic functions with invariants g_{i2}, g_{i3}: $i = 1,\ldots,N$, and nonzero discriminants $\Delta_i \neq 0$: $i = 1,\ldots,N$. We consider the following sets of numbers associated with $\wp_i(z)$: $i = 1,\ldots,N$, $\{u_1,\ldots,u_N\}$ and $\{v_1,\ldots,v_M\}$:*

$$S_{\mathrm{I}} = \{g_{i2}, g_{i3}, \wp_i(u_i v_j) : i = 1,\ldots,N; j = 1,\ldots,M\};$$

$$S_{\mathrm{II}} = \{g_{i2}, g_{i3}, u_i, \wp_i(u_i v_j) : i = 1,\ldots,N; j = 1,\ldots,M\};$$

$$S_{\mathrm{III}} = \{g_{i2}, g_{i3}, u_i, v_i, \wp_i(u_i v_j) : i = 1,\ldots,N; j = 1,\ldots,M\}.$$

Also it is possible to consider the set completely analogical to S_{II},

$$S'_{\mathrm{II}} = \{g_{i2}, g_{i3}, v_j, \wp_i(u_i v_j) : i = 1,\ldots,N; j = 1,\ldots,M\}.$$

For all interesting applications, however, invariants g_{i2}, g_{i3} of $\wp_i(z)$ will be algebraic, so from the transcendental point of view it is possible to remove them from S_i (see Remark below).

REMARK. Despite all interesting corollaries that really lead to algebraic invariants g_{i2}, g_{i3} there are several interesting cases when they can be transcendental. The first such case is the following. Let $\wp(z)$ be a Weierstrass elliptic function with invariants g_2, g_3 normalized in such a way that for modular invariant j of $\wp(z)$, $j = j(\omega_2/\omega_1)$ for fundamental periods ω_1, ω_2 of $\wp(z)$, $g_2, g_3 \in \mathbf{Q}(j)$. Then it is possible to consider $\wp(z)$ such that ω_1/ω_2 is algebraic, but not imaginary quadratic. Then $j(\omega_1/\omega_2)$ is of course transcendental and it is possible to consider the numbers of the following nature:

$$j(\omega_1/\omega_2),\ \omega_1,\ \wp\left(\omega_1 \cdot (\omega_2/\omega_1)^2\right), \text{ etc.}$$

Are they algebraically independent?

We will start with a usual scheme on the Gelfond-Schneider method and we will construct the corresponding polynomials with integer coefficients, having small values at points from S_i.

For $i = \mathrm{I}, \mathrm{II}, \mathrm{III}$ or $i = \mathrm{II}'$, i.e. $S_i = S'_{\mathrm{II}}$, we put $\mathbf{Q}(S_i) = \mathbf{Q}(\theta_1, \ldots, \theta_{\mathfrak{D}_i}, \vartheta)$, where $\{\theta_1, \ldots, \theta_{\mathfrak{D}_i}\}$ is the basis of transcendence of the field $\mathbf{Q}(S_i)$ and ϑ is algebraic over $\mathbf{Q}(\theta_1, \ldots, \theta_{\mathfrak{D}_i})$. We put also for subscript $i = \mathrm{I}, \mathrm{II}, \mathrm{II}', \mathrm{III}$,

$$A_i = \mathbf{Z}\left[\theta_1, \ldots, \theta_{\mathfrak{D}_i}\right]$$

for the ring of integers of $\mathbf{Q}(S_i)$ and, clearly,

$$\mathfrak{D}_i = \deg \operatorname{tr} \mathbf{Q}(S_i).$$

Auxiliary functions which we will construct will have the form

$$F(z) = P\left(\wp_1(u_1 z), \ldots, \wp_N(u_N z)\right)$$

or

$$F(z) = P\left(z, \wp_1(u_1 z), \ldots, \wp_N(u_N z)\right),$$

where $P(\bar{z})$ are polynomials from $A_i[\bar{z}]$ for appropriate choice of A_i. We will consider all cases in a similar framework.

Let X be a sufficiently large integer, $X \geqslant X_0 > 0$.

Case I. $i = \mathrm{I}$. We take an auxiliary function of the form

$$\begin{aligned} F_{\mathrm{I}}^X(z) &= P_{\mathrm{I}}\left(\wp_1(u_1 z), \ldots, \wp_N(u_N z)\right) \\ &= \sum_{\lambda_1=0}^{L_1} \cdots \sum_{\lambda_N=0}^{L_1} p_I^X(\lambda_1, \ldots, \lambda_N) \wp_1(u_1 z)^{\lambda_1} \cdots \wp_N(u_N z)^{\lambda_N}, \end{aligned}$$

where the polynomial

$$P_{\mathrm{I}}^X(x_1, \ldots, x_N) = \sum_{\lambda_1=0}^{L_1} \cdots \sum_{\lambda_N=0}^{L_1} p_{\mathrm{I}}^X(\lambda_1, \ldots, \lambda_n) x_1^{\lambda_1} \cdots x_n^{\lambda_n}$$

has the coefficients from the ring A_{I},

$$p_{\mathrm{I}}^X(\lambda_1, \ldots, \lambda_N) = \sum_{l_1=0}^{L_2} \cdots \sum_{l_{\mathfrak{D}_{\mathrm{I}}}=0}^{L_2} p_{\mathrm{I}}(\lambda_1, \ldots, \lambda_N, l_1, \ldots, l_{\mathfrak{D}_{\mathrm{I}}}) \theta_1^{l_1} \cdots \theta_{\mathfrak{D}_{\mathrm{I}}}^{l_{\mathfrak{D}_i}},$$

where $p_{\mathrm{I}}^X(\bar{\lambda}, \bar{l}) \in \mathbf{Z}$ and

$$L_1 = \left[c_1 X^{M/N}\right] \quad \text{and} \quad L_2 = \left[c_2 X^{(M+2N)/N}\right].$$

Case II. $i = \mathrm{II}$. We consider an auxiliary function of the form

$$\begin{aligned} F_{\mathrm{II}}^X(z) &= P_{\mathrm{II}}\left(\wp_1(u_1 z), \ldots, \wp_N(u_N z)\right) \\ &= \sum_{\lambda_1=0}^{L_3} \cdots \sum_{\lambda_N=0}^{L_3} p_{\mathrm{II}}^X(\lambda_1, \ldots, \lambda_N) \wp_1(u, z)^{\lambda_1} \cdots \wp_N(u_N z)^{\lambda_N}, \end{aligned}$$

where the polynomial

$$P_{\mathrm{II}}(x_1,\dots,x_N) = \sum_{\lambda_1=0}^{L_3} \cdots \sum_{\lambda_N=0}^{L_3} p_{\mathrm{II}}^X(\lambda_1,\dots,\lambda_N) x_1^{\lambda_1} \cdots x_N^{\lambda_N}$$

has the coefficients from the ring A_{II},

$$p_{\mathrm{II}}^X(\lambda_1,\dots,\lambda_N) = \sum_{l_1=0}^{L_4} \cdots \sum_{l_{\mathfrak{D}_{\mathrm{II}}}=0}^{L_4} p_{\mathrm{II}}^X(\lambda_1,\dots,\lambda_N, l_1,\dots,l_{\mathfrak{D}_{\mathrm{II}}}) \cdot \theta_1^{l_1} \cdots \theta_{\mathfrak{D}_{\mathrm{II}}}^{l_{\mathfrak{D}_{\mathrm{II}}}}$$

for $p_{\mathrm{II}}^X(\bar{\lambda}, \bar{l}) \in \mathbf{Z}$ and

$$L_3 = X, \qquad L_4 = \left[c_3 X^{(M+2N)/(M+2)} (\log L)^{2/(M+2)}\right].$$

Case II′, $i = \mathrm{II}'$. Let the auxiliary function have the form

$$\begin{aligned} F_{\mathrm{II}}'^X(z) &= P_{\mathrm{II}'}\left(z, \wp_1(u_1 z),\dots,\wp_N(u_N z)\right) \\ &= \sum_{\lambda_0=0}^{L_5} \sum_{\lambda_1=0}^{L_5} \cdots \sum_{\lambda_N=0}^{L_6} P_{\mathrm{II}'}^X(\lambda_0, \lambda_1,\dots,\lambda_N) z^{\lambda_0} \wp_1(u_1 z)^{\lambda_1} \times \cdots \times \wp_N(u_N z)^{\lambda_N}, \end{aligned}$$

where the polynomial

$$P_{\mathrm{II}'}(x_0,\dots,x_N) = \sum_{\lambda_0=0}^{L_5} \sum_{\lambda_1=0}^{L_6} \cdots \sum_{\lambda_N=0}^{L_6} P_{\mathrm{II}'}^X(\lambda_0, \lambda_1,\dots,\lambda_N) x_0^{\lambda_0} \cdots x_N^{\lambda_N}$$

has the coefficients from the ring A_{II}',

$$P_{\mathrm{II}'}^X(\lambda_0,\dots,\lambda_N) = \sum_{l_1=0}^{L_7} \cdots \sum_{l_{\mathfrak{D}_{\mathrm{II}}'}=0}^{L_7} p_{\mathrm{II}}'(\bar{\lambda}, \bar{l}) \theta_1^{l_1} \cdots \theta_{\mathfrak{D}_{\mathrm{II}}}^{l_{\mathfrak{D}_{\mathrm{II}}'}},$$

for $p_{\mathrm{II}}'(\bar{\lambda}, \bar{l}) \in \mathbf{Z}$ and

$$\begin{aligned} L_5 &= \left[c_4 X^{(N+2M)/(M+1)} (\log X)^{-M/(M+1)}\right], \\ L_6 &= \left[c_5 X^{(N-2)/(M+1)} (\log X)^{1/(M+1)}\right], \\ L_7 &= \left[c_6 X^{(N+2M)/(M+1)} (\log X)^{1/(M+1)}\right]. \end{aligned}$$

Case III, $i = \mathrm{III}$. The auxiliary function is of the form

$$F_{\mathrm{III}}^X(z) = P_{\mathrm{III}}^X\left(z, \wp_1(u, z),\dots,\wp_N(u_N z)\right),$$

where the polynomial

$$P_{\mathrm{III}}^X(x_0,\dots,x_N) = \sum_{\lambda_0=0}^{L_8} \sum_{\lambda_1=0}^{L_9} \cdots \sum_{\lambda_N=0}^{L_9} p_{\mathrm{III}}^X(\lambda_0, \lambda_1,\dots,\lambda_N) x_0^{\lambda_0} x_1^{\lambda_1} \cdots x_N^{\lambda_N}$$

has the coefficients from the ring A_{III}',

$$P_{\mathrm{III}}^X(\lambda_0,\dots,\lambda_N) = \sum_{l_1=0}^{L_{10}} \cdots \sum_{l_{\mathfrak{D}_{\mathrm{III}}}=0}^{L_{10}} p_{\mathrm{III}}^X(\bar{\lambda}, \bar{l}) \theta_1^{l_1} \cdots \theta_{\mathfrak{D}_{\mathrm{III}}}^{l_{\mathfrak{D}_{\mathrm{II}}}}$$

for $p_{\mathrm{III}}^X(\bar{\lambda}, \bar{l}) \in \mathbf{Z}$ and

$$L_8 = \left[c_7 X^{(2N+M)/N} (\log X)^{-1}\right],$$
$$L_9 = \left[c_8 X^{M/N}\right], \qquad L_{10} = \left[c_9 X^{(2N+M)/N}\right].$$

Now it is possible to choose $P_i(\bar{x})$ by Siegel's lemma (see **[W1]**) considering values of $F_i(z)$ at points $z = n_1 v_1 + \cdots + n_M v_M$ for integers n_i. However in order to do it correctly it is necessary to choose integers n_i such that none of the function $\wp_i(u_i z)$ becomes singular at $z = n_1 v_1 + \cdots + n_M v_M$. To do this we use a lemma of Ramachandra **[R1]** which enables us to choose a number of such z:

LEMMA 9.3. *Let us suppose that v_j are not poles of $\wp_i(u_i z)$: $1 \leqslant i \leqslant N$, $1 \leqslant j \leqslant M$ (this is very easy to obtain, replacing, if necessary, v_j by $v_j/2^k$ for an appropriate integer $k \geqslant 0$). Then there exists $\gamma > 0$, depending on g_{i2}, g_{i3}, u_i, v_j: $1 \leqslant i \leqslant N$, $1 \leqslant j \leqslant M$, such that the set*

$$L_R = \left\{ \sum_{j=1}^{M} n_j v_j : n_j \text{ are integers}, -R \leqslant n_j \leqslant R: \right.$$
$$\left. 1 \leqslant j \leqslant M \text{ and } \min_{i=1,\ldots,N} \min_{\omega \in \Lambda_i} \left| \sum_{j=1}^{M} n_j v_j - \omega \right| > \gamma \right\}$$

has power $|L_R| \geqslant R^M$ for $R \geqslant R_0$.

Here Λ_i is the lattice of periods of the function $\wp_i(u_i z)$, i.e. for periods ω_{1i}, ω_{2i} of $\wp_i(z)$,

$$\Lambda_i = \{ n\omega_{1i}/u_i + m\omega_{2i}/u_i : n, m \text{ are integers} \}.$$

It is possible to use only points from L_R as zeroes of auxiliary functions $F_i(z)$.

First of all we have the classical

LEMMA 9.4. *For an elliptic function $\wp(z)$ with invariants g_2, g_3 there are constants $\gamma_2 > 0$, $\gamma_3 > 0$, depending on g_2, g_3 and K such that for a function*

$$\wp(n_1 z_1 + \cdots + n_K z_K)^{\lambda}$$

and differential operator $\partial^{\bar{k}} = (\partial/\partial z_1)^{k_1} \cdots (\partial/\partial z_K)^{k_K}$ for $\bar{k} \in \mathbf{N}^K$ we have

$$\partial^{\bar{k}}\left[\wp(n_1 z_1 + \cdots + n_K z_K)^{\lambda}\right] = \frac{A_{\bar{k},\bar{n},\lambda}(\wp(z_1), \wp'(z_1), \ldots, \wp(z_K), \wp'(z_K))}{\Pi_{i=1}^{k} B_{\bar{n}}(\wp(z_1), \wp'(z_1), \ldots, \wp(z_K), \wp'(z_K))^{|\bar{k}|+\lambda}}$$

for integers n_i: $i = 1, \ldots, K$, where

$$A_{\bar{k},\bar{n},\lambda}(x_1, x_2, \ldots, x_{2K-1}, x_{2K}),\ B_{\bar{n}}(x_1, x_2, \ldots, x_{2K-1}, x_{2K})$$

are polynomials in $g_2, g_3, x_1, x_2, \ldots, x_{2K-1}, x_{2K}$ with integer coefficients, of total degree

$$\deg(A_{\bar{k},\bar{n},\lambda}) \leqslant \gamma_2(|\bar{k}| + \lambda) \sum_{i=1}^{K} |n_i|^2, \qquad \deg(B_{\bar{n}}) \leqslant \gamma_2 \sum_{i=1}^{K} |n_i|^2$$

and of type

$$\text{type}(A_{\bar{k},\bar{n},\lambda}) \leqslant \gamma_3 |\bar{k}| \log\left(\gamma_3 + (|\bar{k}| + \lambda) \sum_{i=1}^{K} |n_i|^2\right) + \gamma_3(|\bar{k}| + \lambda) \sum_{i=1}^{K} |n_i|^2,$$

$$\text{type}(B_{\bar{n}}) \leqslant \gamma_3 \sum_{i=1}^{K} |n_i|^2.$$

Here $\bar{k} = (k_1, \ldots, k_K)$, $\bar{n} = (n_1, \ldots, n_K)$ *and* $|\bar{k}| = \sum_{i=1}^{K} k_i$.

The upper bounds on the types and degrees of derivatives of elliptic functions from Lemma 9.4 can be improved using arguments introduced for the first time in Lemma 4.7 [**C7**]. For these improvements we refer the reader to [**C7**] and §§10, 11 below, particularly to the proof of Lemma 10.4 and Lemmas 11.5, 11.6.

Without any information of the number of zeroes of auxiliary functions $F_i^X(z)$ we come to the following

PROPOSITION 9.5. *For any sufficiently large* $X \geqslant X_1$ *in each of the cases we have*

Case I. *If* $N \geqslant 2$, $M \geqslant 3$ *and* $N + M \geqslant 7$, *then for some* $Y_{\mathrm{I}} \geqslant 1$, *such that the number* $\mathfrak{N}(F_{\mathrm{I}}; Y_{\mathrm{I}} X)$ *of zeroes of* $F_{\mathrm{I}}^X(z)$ *in the disc* $|z| \leqslant XY_{\mathrm{I}}$ *is not less than* $(XY_{\mathrm{I}})^M$,

$$\mathfrak{N}(F_{\mathrm{I}}; XY_{\mathrm{I}}) \geqslant (XY_{\mathrm{I}})^M,$$

we have a polynomial $R_{\mathrm{I}X}(x_1, \ldots, x_{\mathfrak{D}_{\mathrm{I}}}) \in \mathbf{Z}[x_1, \ldots, x_{\mathfrak{D}_{\mathrm{I}}}]$, $R_{\mathrm{I}X}(\bar{x}) \not\equiv 0$ *such that*

$$\left|R_{\mathrm{I}X}(\theta_1, \ldots, \theta_{\mathfrak{D}_{\mathrm{I}}})\right| < \exp\left(-e_1 X^M Y_{\mathrm{I}}^M \log(XY_{\mathrm{I}})\right)$$

and

$$t(R_{\mathrm{I}X}) \leqslant e_2 X^{(M+2N)/N} \cdot Y_{\mathrm{I}}^2.$$

Case II. *If* $N \geqslant 2$, $M \geqslant 1$, *then for some* $Y_{\mathrm{II}} \geqslant 1$ *such that the number* $\mathfrak{N}(F_{\mathrm{II}}; X^{(N-1)/(M+2)}(\log X)^{1/(M+2)} Y_{\mathrm{II}}^{1/2})$ *of zeroes of* $F_{\mathrm{II}}(z)$ *in the disk* $|z| \leqslant X^{(N-1)/(M+2)}(\log X)^{1/(M+2)} Y_{\mathrm{II}}^{1/2}$ *is not less than*

$$X^N \cdot Y_{\mathrm{II}}^{(M+2)/2}\left\{1 + \frac{\log Y_{\mathrm{II}}}{\log X}\right\}^{M/2},$$

$$\mathfrak{N}\left(F_{\mathrm{II}}; X^{(N-1)/(M+2)}(\log X)^{1/(M+2)} Y_{\mathrm{II}}^{1/2}\right) \geqslant X^N \cdot Y_{\mathrm{II}}^{(M+2)/2}\left\{1 + \frac{\log Y_{\mathrm{II}}}{\log X}\right\}^{M/2},$$

and there exists a polynomial $R_{\mathrm{II}X}(x_1, \ldots, x_{\mathfrak{D}_{\mathrm{II}}}) \in \mathbf{Z}[x_1, \ldots, x_{\mathfrak{D}_{\mathrm{II}}}]$, $R_{\mathrm{II}X} \not\equiv 0$, *with*

$$\left|R_{\mathrm{II}X}(\theta_1, \ldots, \theta_{\mathfrak{D}_{\mathrm{II}}})\right| < \exp\left(-e_3 X^N Y_{\mathrm{II}}^{(M+2)/2}\left\{\log(XY_{\mathrm{II}}) + \left(\frac{\log Y_{\mathrm{II}}}{\log X}\right)^{M/2}\right\}\right),$$

$$t(R_{\mathrm{II}X}) \leqslant e_1 X^{(M+2N)/(M+2)}(\log X)^{2/(M+2)} Y_{\mathrm{II}}\left\{1 + \frac{\log Y_{\mathrm{II}}}{\log X}\right\}.$$

Case III. *There exists* $Y_{\mathrm{III}} \geqslant 1$ *such that the number*

$$\mathfrak{N}\left(F_{\mathrm{III}}; XY_{\mathrm{III}}(\log(XY_{\mathrm{III}})/\log X)^{1/2}\right)$$

of zeroes of $F_{\mathrm{III}}^X(z)$ *in the disk* $|z| \leqslant XY_{\mathrm{III}} \cdot (\log(XY_{\mathrm{III}})/\log X)^{1/2}$ *is not less than*

$$X^{(MN+2N+M)/N} Y_{\mathrm{III}}^{(M+2)/2} \frac{\log(XY_{\mathrm{III}})}{\log X},$$

$$\mathfrak{N}\left(F_{\mathrm{III}};\, XY_{\mathrm{III}} \cdot \left[\frac{\log(XY_{\mathrm{III}})}{\log X}\right]^{1/2}\right) \leqslant X^{(MN+2N+M)/N} Y_{\mathrm{III}}^{(M+2)/2} \frac{\log(XY_{\mathrm{III}})}{\log X}$$

and there exists a polynomial $R_{\mathrm{III}X}(x_1,\ldots,x_{\mathfrak{D}_{\mathrm{III}}}) \in \mathbf{Z}[x_1,\ldots,x_{\mathfrak{D}_{\mathrm{III}}}]$, $R_{\mathrm{III}X} \not\equiv 0$ *such that*

$$\left|R_{\mathrm{III}X}(\theta_1,\ldots,\theta_{\mathfrak{D}_{\mathrm{III}}})\right| < \exp\left(-e_5 X^{(MN+M+2N)/N} Y_{\mathrm{III}}^{(M+2)/2} \left[\frac{\log(XY_{\mathrm{III}})}{\log X}\right]^{(M+2)/2}\right),$$

$$t(R_{\mathrm{III}X}) \leqslant X^{(2N+M)/N} \cdot \frac{\log(XY_{\mathrm{III}})}{\log X} Y_{\mathrm{III}}.$$

Proposition 9.5 in its generality was presented for the first time in the author's lectures on transcendental numbers during number theory year at the University of Maryland, spring 1978. Combining Proposition 9.5 with upper bounds on the number of zeroes of polynomials in elliptic functions of **[M8]** and with algebraico-geometric methods of proofs of algebraic independence (cf.**[C7–C9, C12]**) we arrive at

COROLLARY 9.6. *For three sets* S_{I}, S_{II}, S_{III} *we have the following lower bounds on the degrees of transcendence* $\mathfrak{D}_i$ *of* $\mathbf{Q}(S_i)$, $i = \mathrm{I}, \mathrm{II}, \mathrm{III}$.

Case I. *We have* $\mathfrak{D}_{\mathrm{I}} \geqslant MN/(M+2N)$, *provided that* $N \geqslant 2$, $M \geqslant 3$ *and* $N+M \geqslant 7$. *Hence there are at least* $MN/(M+2N)$ *a.i. elements in* S_{I}.

Case II. *We have* $\mathfrak{D}_{\mathrm{II}} \geqslant N(M+2)/(M+2N)$, *provided that* $N \geqslant 2$, $M \geqslant 1$. *Hence there are at least* $N(M+2)/(M+2N)$ *a.i. elements in* S_{II}.

Case III. *We have* $\mathfrak{D}_{\mathrm{III}} > (MN+M+N)/(M+2N)$. *Hence there are more than* $(MN+M+N)/(M+2N)$ *a.i. elements in* S_{III}.

10. The generalization of the Lindemann-Weierstrass theorem. In the two subsections below we study algebraic independence of values of functions uniformizing one-parametric subgroups of algebraic groups. We consider the situation, generalizing the Lindemann-Weierstrass theorem to Abelian varieties or arbitrary group varieties. For this we first study systems of functions satisfying algebraic laws of addition and differential equations, and then consider the particular case of Abelian varieties with complex multiplication.

General results on systems of functions, satisfying algebraic differential equations and laws of addition.

DEFINITION 10.1. Lef $f_1(z),\ldots,f_m(z)$ be a system of functions. We say that this system $\{f_1,\ldots,f_m\}$ satisfies algebraic differential equations and laws of addition if

(a) there exists an algebraic number field $\mathbf{K}$, $[\mathbf{K}:\mathbf{Q}] < \infty$,

(b) there are functions $f_{m+1}(z),\ldots,f_{m+k}(z)$ such that

(c) the operation of differentiation maps the ring $\mathbf{K}[f_1,\ldots,f_{m+1},\ldots,f_{m+k}]$ into itself,

(d) for any $j = 1,\dots,m+k$ the function $f_j(z+w)$ belongs to the field $\mathbf{K}(f_1(z),\dots,f_{m+k}(z), f_1(w),\dots,f_{m+k}(w))$.
Of course, condition (d) is more important than (c).

We recall that for

$$\deg\operatorname{tr} \mathbf{C}(f_1,\dots,f_{m+k}) = 1$$

the only examples of such systems reduce either to polynomials, or to exponentials $e^{\lambda z}$ or to elliptic functions $\wp(z)$ with algebraic invariants. However, in general, many examples of systems of functions satisfying algebraic differential equations and laws of addition are given by Abelian functions. E.g. for an Abelian variety $\mathbf{A} = (A_1(\bar{z}),\dots,A_d(\bar{z}), B(\bar{z}))$ of dimension d defined over $\overline{\mathbf{Q}}$ with Abelian functions $A_i(\bar{z})$, $B(\bar{z})$ regular at $\bar{z} = \bar{0}$ we have the system

$$A_1(z,0,\dots,0),\dots,A_d(z,0,\dots,0) \tag{10.1}$$

satisfying algebraic differential equations and laws of addition. As another example of a system of meromorphic functions, satisfying algebraic differential equations and laws of addition we can mention solutions of stationary Korteweg-de Vries equations of nth order (n-band potentials) and their derivatives [**M9, N1**].

The statement that any system of functions satisfying algebraic laws of addition are Abelian functions or their degenerations is commonly referred to as a "Weierstrass Theorem." This statement was proved by Painlevé in the case of at most two algebraically independent functions, i.e. when $\deg\operatorname{tr} \mathbf{C}(f_1,\dots,f_{m+k}) \leqslant 2$. In general, one can say that any system of analytic functions satisfying algebraic laws of addition arises from a one parameter subgroup of an algebraic group. This characteristic, which is due to A. Weil, allows us to classify systems of functions satisfing algebraic differential equations and laws of addition as particular choices of variables n a one parameter subgroup of a projective model of (a) an Abelian variety, (b) a matrix linear group, or (c) an extension of an Abelian variety by a linear group. We turn to the problem of algebraic independence of values of functions satisfying algebraic differential equations and laws of addition. Our result is the first result of algebraic independence of values of some functions not satisfying linear differential equations and may be considered as the natural generalization of the Schneider-Lang theorem on the case of algebraic independence.

First of all, we need a definition concerning the number of functions involved in the differential equations and in the laws of addition.

DEFINITION 10.2. Let $f_1,\dots,f_m$ be the system of functions satisfying algebraic differential equations and laws of addition and let $f_1,\dots,f_m, f_{m+1},\dots,f_{m+k}$ be the system of functions satisfying conditions (a)–(d) from Definition 10.1. Then we call the system $f_1,\dots,f_{m+k}$ a complete system of functions, satisfying algebraic differential equations and laws of addition over $\mathbf{K}$. The dimension of $f_1,\dots,f_{m+k}$

is defined as the maximal number of functions among $f_1,\dots,f_{m+k}$ algebraically independent over $\mathbf{C}(z)$,

$$\dim\{f_1,\dots,f_{m+k}\} = \deg \mathrm{tr}_{\mathbf{C}(z)} \mathbf{C}(f_1,\dots,f_{m+k}).$$

E.g. $\dim\{f_1,\dots,f_{m+k}\} \leqslant \deg \mathrm{tr}\, \mathbf{C}(f_1,\dots,f_{m-k})$, but it may happen that

$$\dim\{f_1,\dots,f_{m+k}\} = \deg \mathrm{tr}\, \mathbf{C}(f_1,\dots,f_{m+k}) - 1.$$

(For example

$$\dim\{z, \zeta(z), \wp(z)\} = 2 = \deg \mathrm{tr}\, \mathbf{C}(z, \zeta(z), \wp(z)) - 1.)$$

Some time ago we proved the first result on algebraic independence of values of elliptic functions at algebraic points [C7]. In the proof we used only the existence of the algebraic differential equation and the existence of the law of addition. Thus it becomes possible to generalize the results for arbitrary systems of meromorphic functions, satisfying algebraic differential equations and laws of addition. If $f_1,\dots,f_l$ are a complete system of functions satisfying algebraic differential equations and laws of addition, then for any integer $m \neq 0$

$$f_i(mz) = \mathfrak{F}_{m,i}(f_1(z),\dots,f_l(z)) \tag{10.2}$$

for $i = 1,\dots,l$, where $\mathfrak{F}_{m,i}(x_1,\dots,x_l)$ is a rational function with coefficients from $\mathbf{K}$, where the degree of the rational function $\mathfrak{F}_{m,i}(x_1,\dots,x_l)$ and the type of $\mathfrak{F}_{m,i}$ as a rational function with coefficients from $\mathbf{K}$ satisfy

$$\deg(\mathfrak{F}_{m,i}) \leqslant C_1 m^\rho, \qquad \mathrm{type}(\mathfrak{F}_{m,i}) \leqslant C_2 m^\rho \tag{10.3}$$

for all $i = 1,\dots,l$. The existence of a constant $\rho < \infty$ such that (10.2) and (10.3) are satisfied for any integer $m \neq 0$ is simply the consequence of the laws of addition (d) in Definition 10.1. The constant ρ depends on the degrees of the laws of addition (d). In this way we arrive at the following

DEFINITION 10.3. Let $\{f_1,\dots,f_l\}$ be a complete system of functions satisfying algebraic differential equations and laws of addition over a number field $\mathbf{K}$. If for any integer $m \neq 0$ we have

$$f_i(mz) = \mathfrak{F}_{m,i}(f_1(z),\dots,f_l(z))$$

for $\mathfrak{F}_{m,i}(x_1,\dots,x_l) \in \mathbf{K}(x_1,\dots,x_l)$, where the type $\mathfrak{F}_{m,i} \leqslant c_3 m^\rho$ for all $i = 1,\dots,l$, then we say that $\{f_1,\dots,f_l\}$ has arithmetical order $\leqslant \rho$.

The characterization of systems of functions satisfying the laws of addition as one parameter subgroups of algebraic groups shows that an appropriate change of variables allows one to bound ρ always as $\rho \leqslant 2$. Moreover, we see that $\rho = 2$ always in the case of one parameter subgroups of Abelian varieties, $\rho = 1$ for linear groups and, finally, in the case of extensions of Abelian varieties by linear groups, $\rho = 2$, but the degrees of $\mathfrak{F}_{m,i}\ (x_i,\dots,x_1)$ in some of the variables x_i are $\leqslant c_3' m$.

In order to construct auxiliary functions $F(z) = P(z, f_1(z))$ with zeroes at points $z = m_1\alpha_1 + \cdots + m_k\alpha_k$ for algebraic $\alpha_1,\dots,\alpha_k$ we must know that $f_1(z)$ is regular at points $m_1\alpha_1 + \cdots + m_k\alpha_k$ (at least in many such points). To do this, we use a method of Ramachandra **[R1]** and Lemma 9.3. Now we are ready to formulate the main analytical result connected with the values of $f_1(z)$ at algebraic points.

MAIN LEMMA 10.4. *Let $\{f_1,\dots,f_l\}$ be a complete system of functions, satisfying algebraic differential equations and laws of addition over a number field* **K** *and having arithmetical order $\leqslant \rho$. Let dimension of $\{f_1,\dots,f_l\}$ be $d \geqslant 1$ and f_1, say, be a transcendental meromorphic function of finite order of growth. Let $\alpha_1,\dots,\alpha_k$ be algebraic numbers, linearly independent over* **Q**. *We denote by*

$$\theta_1,\dots,\theta_m \tag{10.4}$$

some fixed basis of transcendence of the field

$$\mathbf{Q}\big(f_i(\alpha_j): i = 1,\dots,l, j = 1,\dots,k\big). \tag{10.5}$$

By $c_i > 0$ we denote constants depending on $f_i(z)$, k, $\alpha_1,\dots,\alpha_k$, $\theta_1,\dots,\theta_m$, and on the form of the laws of addition (d). *Let us suppose that we have sufficiently large integers L_0, L_1, S, X satisfying*

$$\min\{L_0, L_1, S_1, X\} > c_4, \tag{10.6}$$

(10.7) L_1 *is constant*, L_0, S, X *are sufficiently large with respect to* L,

$$L_0 L_1 = c_5 S X^k \quad \text{for some sufficiently large } c_5 > 0, \tag{10.8}$$

$$\log L_0 \geqslant c_6 X^{2k}. \tag{10.9}$$

Then we have either

(i) *a polynomial $P(x_1,\dots,x_m) \in \mathbf{Z}[x_1,\dots,x_m]$, with*

$$-c_7 S X^k \log L_0 < |P(\theta_1,\dots,\theta_m)| < -c_8 S X^k \log L_0, \tag{10.10}$$

$$t(P) \leqslant c_9 S \log L_0, \tag{10.11}$$

$$d(P) \leqslant c_{10} L_1 X^{\rho} \tag{10.12}$$

or

(ii) *a system $P_\lambda(x_1,\dots,x_m) \in \mathbf{Z}[x_1,\dots,x_m]$: $\lambda \in \mathfrak{L}$, of polynomials having no common factor and*

$$|P_\lambda(\theta_1,\dots,\theta_m)| < \exp\big(-c_{11} S X^k \log L_0\big), \tag{10.13}$$

$$t(P_\lambda) \leqslant c_{12} S \log L_0, \tag{10.14}$$

$$d(P_\lambda) \leqslant c_{13} L_1 X^{\rho} \tag{10.15}$$

for all $\lambda \in \mathfrak{L}$.

The proof of the Main Lemma 10.4 is absolutely similar to the proof of Corollary 4.10 in **[C7]**. We consider an auxiliary function of the form

$$F(z) = P(z, f_1(z)), \tag{10.16}$$

where

$$\mathbf{I} = \mathbf{Z}[\theta_1,\dots,\theta_m] \tag{10.17}$$

and

$$P(x, y) \in \mathbf{I}[x, y], \qquad P(x, y) \not\equiv 0, \tag{10.18}$$

$$d_x(P) = L_0, \qquad d_y(P) = L_1 \tag{10.19}$$

and the polynomial $P(x, y)$ is defined according to Siegel's lemma according to the conditions

$$F^{(s)}(m_1\alpha_1 + \cdots + m_k\alpha_k) = 0 \tag{10.20}$$

for

$$s = 0, 1, \ldots, S-1 \text{ and for } c_{14}X^k \text{ vectors } (m_1, \ldots, m_k)$$

with integer coefficients m_i,

$$0 < m_i \leqslant X - 1 : i = 1, \ldots, k, \text{ such that } f_1(z) \text{ is regular at } z = m_1\alpha_1 + \cdots + m_k\alpha_k. \tag{10.21}$$

We use here Ramachandra's lemma 9.3, see **[R11]**. In order to obtain bounds for types and degrees of coefficients of $P(x, y)$ and for types of

$$F^{(s)}(m_1\alpha_1 + \cdots + m_k\alpha_k)$$

we use a variant of the "Generalized Coates' lemma" 4.6 proved in **[C7]**.

The auxiliary result, formulated and proved as Lemma 4.6 in **[C7]** in the elliptic function case, is a consequence of a fairly general observation on the existence of special differential operators associated with systems of functions satisfying algebraic laws of addition. In this way we improve considerably on the bounds of the degrees and types in Lemma 9.4 above. In the general situation of the complete system $\{f_1(z), \ldots, f_l(z)\}$ of functions satisfying algebraic laws of addition we replace

$$F^{(s)}(m_1\alpha_1 + \cdots + m_k\alpha_k) = \left(\frac{\partial}{\partial z}\right)^s P(z, f_1(z))\big|_{z = m_1\alpha_1 + \cdots + m_k\alpha_k}$$

by the following one:

$$\left(\frac{\partial}{\partial w}\right)^s P(z_0 + w, f_1(z_0 + w))\big|_{w=0} \quad \text{for } z_0 = m_1\alpha_1 + \cdots + m_k\alpha_k.$$

Then we use the laws of addition for $\{f_1(z), \ldots, f_l(z)\}$ to represent $f_1(z_0 + w)$ as

$$\frac{P_0(f_1(z_0), \ldots, f_l(z_0), f_1(w), \ldots, f_l(w))}{Q_0(f_1(z_0), \ldots, f_l(z_0), f_1(w), \ldots, f_l(w))}$$

for rational functions $P_0(x_1, \ldots, x_l, y_1, \ldots, y_l)$, $Q_0(x_1, \ldots, x_l, y_1, \ldots, y_l)$ in $2l$ variables x_i, y_i, $i = 1, \ldots, l$. We assume here that all functions $f_i(w)$, $i = 1, \ldots, l$ are regular at $w = 0$. This assumption, however, does not cause a loss of generality because the origin $w = 0$ of the lattice $\mathbf{Z} \cdot \alpha_1 \oplus \cdots \oplus \mathbf{Z} \cdot \alpha_k$ is an arbitrary point and one can always find an element of the lattice at which all functions $f_i(w)$ are regular, dividing, if necessary $\alpha_1, \ldots, \alpha_k$ by integers of bounded sizes.

With this assumption we can represent

$$\left(\frac{\partial}{\partial w}\right)^s P(z_0+w, f_1(z_0+w))|_{w=0}$$

in the form

$$\left(\frac{\partial}{\partial w}\right)^s \{Q_0(\bar{f}(z_0), \bar{f}(w))^{-L_1}$$

$$\cdot G(z_0+w; P_0(\bar{f}(z_0), \bar{f}(w)), Q_0(\bar{f}(z_0), f(w)))\}|_{w=0},$$

where $G(x_0; x_1, x_2) = P(x_0; x_1/x_2)x_2^{L_1}$—is homogeneous in x_1, x_2 of degree L_1. Consequently the system of equations (10.20):

$$F^{(s)}(m_1\alpha_1 + \cdots + m_k\alpha_k) = 0; \qquad z_0 = m_1\alpha_1 + \cdots + m_k\alpha_k$$

for all $s = 0, 1, \ldots, S-1$ is equivalent to the following one:

$$\left(\frac{\partial}{\partial w}\right)^s G(z_0+w; P_0(\bar{f}(z_0), \bar{f}(w)), Q_0(\bar{f}(z_0), \bar{f}(w)))|_{w=0} = 0$$

for all $s = 0, 1, \ldots, S-1$. Here

$$G = G(z_0+w, P_0(\bar{f}(z_0), \bar{f}(w)), Q_0(\bar{f}(z_0), \bar{f}(w)))$$

is a polynomial in z_0, w of degree of at most L_0, and is a polynomial in $f_i(z_0)$, $f_i(w)$: $i = 1, \ldots, l$ of degree of at most $c_0'L_1$ for a constant c_0' depending only on degrees of P_0 and Q_0, i.e. on the law of addition for the system of functions $\{f_1, \ldots, f_l\}$.

An auxiliary statement that we formulated and proved above is Lemma 4.6 **[C7]** in the elliptic function case. For the reader's convenience we present it below as Lemma 11.5 in the most general multidimensional case.

Our arguments guarantee that it is possible to take $P(x, y)$ satisfying (10.18) and (10.19) such that (10.20) and (10.21) is true. Assuming the conditions (10.6)–(10.9) above for L_0, L_1, S, X, the types and degrees of coefficients $P_{\lambda_0,\lambda}$, of $P(x, y)$ as elements from **I** satisfy

(10.22) $$\max t(P_{\lambda_0,\lambda}) \leqslant c_{15} S \log L_0,$$

(10.23) $$\max d(P_{\lambda_0,\lambda}) \leqslant c_{16} L_1 X^{\rho},$$

because of the definition of arithmetical order of $\{f_1, \ldots, f_l\}$. Now we can apply Schwarz' lemma. We find an entire function $h(z)$ such that

$$h(z) \quad \text{and} \quad f_1(z)h(z)$$

are entire functions of order of growth $\rho' < \infty$. It is possible to apply Schwarz' lemma to the function

$$G(z) = h(z)^{L_1} \cdot F(z).$$

Thus

(10.24) $$|F^{(s_1)}(z_1)| < \exp(-c_{17} S X^k \log L_0)$$

for $s_1 \leqslant c_{18}S$, $|z_1| \leqslant c_{19}X$. It is possible now to use a Small Value Lemma, e.g. of Brownawell-Masser type **[B17]**. Applying the Small Value Lemma we immediately obtain Main Lemma 10.4.

From Main Lemma 10.4 we can deduce almost immediately results on algebraic independence. For this we can use methods of **[C8]**. For example, the result on the algebraic independence of two numbers among values of $f_i(\alpha_j)$ ($i = 1,\ldots,l$; $j = 1,\ldots,k$) follows directly from Lemma 10.4 using Gelfond's algebraic independence criteria **[G1]**. This gives us Theorem 10.6 below. Next, the result on three algebraically independent numbers among values of $f_i(\alpha_j)$ ($i = 1,\ldots,l$; $j = 1,\ldots,k$) is a direct consequence of Lemma 10.4 and the author's criteria for algebraic independence (cf. **[C8]**), **[C9]**. For the formulation and proofs see Chapter 4. This gives us Theorem 10.7 below.

We can give some examples of the results. We formulate them in the following notations:

General Assumption 10.5. Let $\{f_1(z),\ldots,f_l(z)\}$ be a complete system of functions, satisfying algebraic differential equations and laws of addition. Let d be the dimension of $\{f_1,\ldots,f_l\}$ and $f_1,\ldots,f_d$ be any d functions from $\{f_1,\ldots,f_l\}$ algebraically independent over $\mathbf{C}(z)$, such that f_1 is meromorphic of finite order of growth. Let $\{f_1,\ldots,f_l\}$ have arithmetical order $\leqslant \rho$.

According to the discussion above, we can always assume that $\rho \leqslant 2$.

Let us present first the two simplest corollaries of Main Lemma 10.4 that do not require any new algebraico-geometric methods for their proof.

Theorem 10.6. *Let $\{f_1,\ldots,f_l\}$ satisfy General Assumption* 10.5 *and $\alpha_1,\ldots,\alpha_k$ be algebraic numbers linearly independent over $\mathbf{Q}$. If $k > \rho$ then there are at least two algebraically independent numbers among*

$$f_i(\alpha_j): \quad i = 1,\ldots,d; j = 1,\ldots,k. \tag{10.25}$$

Theorem 10.7. *Let $\{f_1,\ldots,f_l\}$ satisfy General Assumption* 10.5 *and $\alpha_1,\ldots,\alpha_k$ be algebraic numbers linearly independent over $\mathbf{Q}$. If $k > 2\rho$ then among the numbers*

$$f_i(\alpha_j): \quad i = 1,\ldots,d; j = 1,\ldots,k,$$

there are at least three algebraically independent numbers.

Theorem 10.6 follows from Main Lemma 10.4 by Gelfond's generalized criteria **[B13]**.

Theorem 10.7 follows from Main Lemma 10.4 using elementary methods **[C8]**, **[C19]**.

In the general case we have extensions of Theorems 10.6, 10.7 for an arbitrary number of algebraically independent elements among the values of $f_i(\alpha_j)$, $i = 1,\ldots,d$; $j = 1,\ldots,k$. We present one of these results in the case, when ρ is bounded by 2. This is the case, in particular, of one parameter subgroups of Abelian varieties and extensions of Abelian varieties by linear groups.

THEOREM 10.8. *Let* $\{f_1,\ldots,f_l\}$ *satisfy General Assumptions* 10.5 *and let* $\alpha_1,\ldots,\alpha_k$ *be algebraic numbers linearly independent over* $\mathbf{Q}$. *Let the arithmetic order of growth of* $\{f_1,\ldots,f_l\}$ *be* $\rho = 2$. *If* $k > 2m$, *then there are at least* $m+1$ *algebraically independent numbers among the values*

$$f_i(\alpha_j): \qquad i = 1,\ldots,d; j = 1,\ldots,k.$$

The main part of the proof of Theorem 10.8 is the Main Lemma 10.4. The algebraic part of the proof is based on the system of polynomials from Lemma 10.4. We apply intersection theory (cf. **[S12]**) to polynomials $P(x_1,\ldots,x_m)$ and $P_\lambda(x_1,\ldots,x_m)$ from Lemma 10.4 constructed from $F^{(s)}(m_1\alpha_1 + \cdots + m_k\alpha_k)$ in the proof above. The inequality $m \geqslant k/2$ (with $\rho = 2$), equivalent to Theorem 10.8 in the notations of Lemma 10.4, is a consequence of a relatively simple computations in intersection theory. These computations show that for $k \geqslant 2m$, polynomials $P(x_1,\ldots,x_m)$ or $P_\lambda(x_1,\ldots,x_m)$ $(\lambda \in \mathfrak{L})$ for various L_0 generate zero-dimensional ideal in $\mathbf{Z}[x_1,\ldots,x_m]$. For results following from this method see **[C12]** and the introductory paper in this volume.

For $\rho = 2$ we have many examples of complete systems, satisfying General Assumption 10.5 and arising from Abelian functions (see §11).

These systems give most interesting applications of Lemma 10.4 and Theorems 10.6, 10.7 and 10.8.

These results are presented in the next section.

11. Values of Abelian functions at algebraic points: algebraic independence results. Let $\mathbf{A}$ be an Abelian variety defined over $\overline{\mathbf{Q}}$. Then, there are Abelian functions corresponding to $\mathbf{A}$: $A_1(\bar{z}),\ldots,A_d(\bar{z})$, $B(\bar{z})$, $\bar{z} = (z_1,\ldots,z_d)$ and an algebraic number field $\mathbf{K}$, $[\mathbf{K}:\mathbf{Q}] < \infty$ such that $\partial/\partial z_i$ maps $\mathbf{K}[A_1,\ldots,A_d, B]$: $i = 1,\ldots,d$, into itself, B is algebraic over $\mathbf{K}[A_1,\ldots,A_d]$: the functions $A_1,\ldots,A_d$, B are meromorphic of order 2 in $\mathbf{C}^d$, and $A_1,\ldots,A_d$, B satisfy an algebraic law of addition. We write simply $\mathbf{A} = (A_1,\ldots,A_d, B)$, where $\dim \mathbf{A} = d$. Of course all the functions $A_1(\bar{z}),\ldots,A_d(\bar{z})$, $B(\bar{z})$ are $2d$-periodic with periods $\vec{\omega}_1,\ldots,\vec{\omega}_{2d}$ such that the period matrix Ω satisfies Riemann's relations, $\Omega = (\vec{\omega}_1,\ldots,\vec{\omega}_{2d})$.

Without loss of generality we may also assume that $A_1(\bar{z}),\ldots,A_d(\bar{z})$, $B(\bar{z})$ are regular at $\bar{z} = 0$.

We will show how to generalize to Abelian functions a result **[C7]** on algebraic independence of several values of an elliptic function at algebraic points, thus giving a multidimensional generalization of Lindemann's theorem. However even in the case of CM-varieties in G. Shimura's sense **[S2]** we are unable to obtain the best possible results. There are still many unclear questions.

Let $\vec{\alpha} = (\alpha_1,\ldots,\alpha_d)$ be an algebraic vector (from $\overline{\mathbf{Q}}^d$). Then by **[L6]** one of the numbers $A_i(\vec{\alpha})$: $i = 1,\ldots,d$, is transcendental. We will try to investigate the degree of transcendence of a general set

$$A_i(\vec{\alpha}_j): \qquad i = 1,\ldots,d, j = 1,\ldots,n, \tag{11.1}$$

for algebraic vectors $\vec{\alpha}_j \in \overline{\mathbf{Q}}^d$, linearly independent over $\mathbf{Q}$: $j = 1,\ldots,n$. Of course the number of algebraically independent numbers in (11.1) is connected also with the linear dependence of $\vec{\alpha}_1,\ldots,\vec{\alpha}_n$ over the field of complex multiplication of **A**. We can obtain nontrivial results only in two cases:

(1) when in $\vec{\alpha}_j$: $j = 1,\ldots,n$, some coordinates are zero;

(2) when there are many complex multiplications in **A**.

For the proof we use the methods of §10, [**C7**, **C8**, **C19**].

THEOREM 11.1. *Let* $\mathbf{A} = (A_1,\ldots,A_d, B)$ *be as before and* $\vec{\alpha}_j = (0,\ldots,0, \alpha_j, 0,\ldots,0)$: $j = 1,\ldots,n$, *be vectors from* $\overline{\mathbf{Q}}^d$ *such that all coordinates but the* k_0*th are zero (for fixed* $k_0 \leqslant d$*). Then among the numbers*

$$A_i(\vec{\alpha}_j): \quad i = 1,\ldots,d, j = 1,\ldots,n, \tag{11.2}$$

there are at least $[(n+1)/2]$ *algebraically independent numbers.*

The proof of this result is indeed one dimensional. We consider the system of functions in **C**:

$$f_i(z) = A_i(0,\ldots,z_{k_0},0,\ldots,0): \qquad i = 1,\ldots,d,$$
$$f_{d+1}(z) = B(0,\ldots,z_{k_0},0,\ldots,0).$$

Then the system $\{f_1,\ldots,f_{d+1}\}$ satisfies algebraic differential equations and laws of addition in the sense of §10 (this is simply the consequence of the law of addition in **A**). Each of the functions $f_i(z)$ is meromorphic of order $\leqslant 2$ and $\{f_1,\ldots,f_{d+1}\}$ have arithmetic order $\leqslant 2$. Let us denote by $\theta_1,\ldots,\theta_m$ the basis of transcendence of $\mathbf{Q}(A_i(\vec{\alpha}_j))$: $i = 1,\ldots,d$, $j = 1,\ldots,n$. Now considering an auxiliary function of the form

$$F(z) = P(z, f_i(z))$$

we obtain from §10 the following result:

LEMMA 11.2. *Let us suppose that* L_0, L_1, S, X *are numbers such that*

$$\min\{L_0, L_1, S, X\} \geqslant c_1, \qquad L_0 L_1 = c_2 S X^n,$$
$$\log L_0 \geqslant X^{2n}, \qquad L_1 = c_3$$

for some constants $c_1 > 0$, $c_2 > 0$, $c_3 > 0$ *depending on* $f_1,\ldots,f_{d+1}$, $\vec{\alpha}_1,\ldots,\vec{\alpha}_n$, $\theta_1,\ldots,\theta_m$.

Then either

(a) *there exists a polynomial* $P(x_1,\ldots,x_m) \in \mathbf{Z}[x_1,\ldots,x_m]$ *such that*

$$-c_4 S X^n \log L_0 < |P(\theta_1,\ldots,\theta_m)| < -c_5 S X^n \log L_0, \tag{11.3}$$

$$t(P) \leqslant c_6 S \log L_0, \qquad d(P) \leqslant c_7 L_1 X^2, \tag{11.4}$$

or

(b) *there is a system of polynomials* $P_\lambda(x_1,\ldots,x_m) \in \mathbf{Z}[x_1,\ldots,x_m]$ *without common factor such that* ($\lambda \in \mathfrak{L}$),

$$|P_\lambda(\theta_1,\ldots,\theta_m)| < \exp(-c_8 S X^n \log L_0): \quad \lambda \in \mathfrak{L}, \tag{11.5}$$

$$t(P_\lambda) \leqslant c_9 S \log L_0, \qquad d(P_\lambda) \leqslant c_{10} L_1 X^2: \quad \lambda \in \mathfrak{L}. \tag{11.6}$$

Lemma 11.2 together with methods of **[C8, C9]** gives us Theorem 2.1.

The corresponding proofs are similar to the proofs of Theorem 10.8. The cases $n \leqslant 3$ and $n \leqslant 5$ are, in fact, immediate corollaries of the criteria of algebraic independence, see **[C7, C8, C9]** for the corresponding proofs. In the general case we use intersection theory to algebraic varieties defined by equations

$$P(x_1,\ldots,x_m)=0 \quad \text{or} \quad P_\lambda(x_1,\ldots,x_m)=0 \qquad (\lambda\in\mathfrak{L})$$

with various L_0 for polynomials $P(x_1,\ldots,x_m)$ and $P_\lambda(x_1,\ldots,x_m)$ from Lemma 11.2. These algebraic varieties do not have common points for $2m \leqslant n-1$, cf. **[C12]**.

Very nice results are proved for CM-varieties. Let us suppose that $\mathbf{A}$ is of CM-type **[S2]**, $\mathbf{E}$ is the field (CM-field) of complex multiplication of $\mathbf{A}$, $[\mathbf{E}:\mathbf{Q}]=2d$ and $\mathbf{E}$ is properly imbedded into End ($\mathbf{A}$). In the notations of **[C22]** we call $\mathbf{A}=(A_1,\ldots,A_d, B)$ strongly normalized.

THEOREM 11.3. *Let* $\mathbf{A}=(A_1,\ldots,A_d, B)$ *have CM-type corresponding to the field* $\mathbf{E}$, $[\mathbf{E}:\mathbf{Q}]=2d$ *in the above sense.*

If $\vec{\alpha}_j=(0,\ldots,\alpha_j,0,\ldots,0)$: $j=1,\ldots,k$, *are vectors from* $\overline{\mathbf{Q}}^d$, *linearly independent over* $\mathbf{E}$ *such that all coordinates but the* k_0*th are zero* ($k_0 \leqslant d$), *then all the numbers*

$$A_i(\vec{\alpha}_j): \qquad i=1,\ldots,d, j=1,\ldots,k,$$

are algebraically independent.

Theorem 11.3 follows from Theorem 11.1 and the strong normalization of CM-variety.

For general algebraic vectors results are not so precise.

THEOREM 11.4. *Let* $\mathbf{A}=(A_1,\ldots,A_d, B)$ *be a CM-variety with complex multiplication by* $\mathbf{E}$, $[\mathbf{E}:\mathbf{Q}]=2d$. *If* $\vec{\alpha}_1,\ldots,\vec{\alpha}_k$ *are vectors from* $\overline{\mathbf{Q}}^d$, *linearly independent over the field of complex multiplication, then among the kd numbers*

$$A_i(\vec{\alpha}_j): \qquad i=1,\ldots,d, j=1,\ldots,k,$$

there are k algebraically independent numbers.

For the proof of Theorem 11.4 we use an auxiliary function of the multidimensional form (cf. **[C22, M6, B4]** for similar but different auxiliary functions). This is the only case, when the results of §10 (Theorems, 10.6–10.8) cannot be immediately applicable because we are not choosing any particular one parameter subgroup of an Abelian variety $\mathbf{A}$. The new auxiliary function has the form

$$f(\bar{z})=P(\bar{z}, A_1(\bar{z}),\ldots,A_d(\bar{z})),$$

where $P(\bar{z}, y_1,\ldots,y_d)$ is a polynomial in $2d$ variables $z_1,\ldots,z_d, y_1,\ldots,y_d$ of total degree of at most L_0 in $z_1,\ldots,z_d$ and of at most L_1 in $y_1,\ldots,y_d$. The polynomial $P(\bar{z}, y_1,\ldots,y_d)$ is chosen in such a way that the following conditions on zeroes of the auxiliary functions $f(\bar{z})$ at points of the lattice over $\mathbf{E}$ generated by $\vec{\alpha}_1,\ldots,\vec{\alpha}_k$ are satisfied. These conditions have the following form:

$$\vec{\mathfrak{D}}\circ f(\mu_1\vec{\alpha}_1+\cdots+\mu_k\vec{\alpha}_k)=0$$

for $\overline{\mathfrak{D}} = \prod_{i=1}^{d}(\partial/\partial z_i)^{s_i}$ and nonnegative integers $s_1,\dots,s_d$ such that $s_1 + \cdots + s_d \leqslant S$, and for integers $\mu_1,\dots,\mu_k$ of $\mathbf{E}$ having sizes $|\mu_i| \leqslant X$: $i = 1,\dots,k$. The lattice points $\bar{z} = \mu_1\vec{\alpha}_1 + \cdots + \mu_k\vec{\alpha}_k$ are chosen so that the Abelian functions $A_1,\dots,A_d$ are regular at z (this can always be achieved using a particular choice of an order in $\mathbf{E}$ to which all μ_i belong). The sizes of μ_i can be determined in the following way: if $\nu_1,\dots,\nu_{2d}$ is a basis of $\mathbf{E}$ and $\mu_i = \sum_{j=1}^{2d} m_{i,j}\nu_j$, then $\max|m_{i,j}| \ll X$.

The parameters L_0, L_1, S and X are chosen in a way similar to that of Lemma 10.4 or Lemma 11.2. Namely, we have the following set of conditions:

$$L_0^d \cdot L_1^d = c_{11} \cdot S^d \cdot X^{2kd}, \qquad \log L_0 \geqslant c_{12} X^{2kd},$$

for some constants $c_{11} > 0$, $c_{12} > 0$ depending on $\mathbf{A}$ and $\vec{\alpha}_1,\dots,\vec{\alpha}_k$. The parameter L_1 is, as before, a sufficiently large constant. These definitions of parameters are equivalent to the conditions (10.6)–(10.9) of Main Lemma 10.4 with k there replaced by $2k$.

An important auxiliary statement similar to Lemma 4.6 of **[C7]**, that holds in the multidimensional case as well, allows us to improve the bounds of types and heights of arbitrary numbers of the form

$$\overline{\mathfrak{D}} \circ f(\mu_1\vec{\alpha}_1 + \cdots + \mu_k\vec{\alpha}_k)$$

for arbitrary $\overline{\mathfrak{D}} = \prod_{i=1}^{d}(\partial/\partial z_i)^{s_i}$ and $\mu_j \in \mathbf{E}$: $j = 1,\dots,k$. In view of the particular importance of this auxiliary result we present its complete proof:

LEMMA 11.5. *Let $P(\bar{z}, \bar{y}) = P(z_1,\dots,z_d, y_1,\dots,y_d)$ be an arbitrary polynomial of total degree D_0 in $\bar{z} = (z_1,\dots,z_d)$ and of total degree D_1 in $\bar{y} = (y_1,\dots,y_d)$. Let $\bar{a} = \sum_{i=1}^{l} n_i \cdot \bar{a}_i$ for arbitrary vectors $\bar{a}_1,\dots,\bar{a}_l$ and integers $n_1,\dots,n_l$. Let $m \geqslant 0$ be the smallest nonnegative integer such that*

$$\prod_{i=1}^{d}\left(\frac{\partial}{\partial z_i}\right)^{s_i} \cdot P(\bar{z}, A_1(\bar{z}),\dots,A_d(\bar{z}))\big|_{\bar{z}=\bar{a}} = 0$$

for all $s_1,\dots,s_d$ with $s_1 + \cdots + s_d < m$. Then for arbitrary nonnegative integers $s_1,\dots,s_d$ such that $s_1 + \cdots + s_d = m$ the number

$$\prod_{i=1}^{d}\left(\frac{\partial}{\partial z_i}\right)^{s_i} \cdot P(\bar{z}, A_1(\bar{z}),\dots,A_d(\bar{z}))\big|_{\bar{z}=a}$$

has the form

$$\frac{P_{s_1,\dots,s_d,n_1,\dots,n_l}\big(\bar{a}_j, A_i(\bar{a}_j), B(\bar{a}_j): i = 1,\dots,d; j = 1,\dots,l\big)}{Q_{s_1,\dots,s_d,n_1,\dots,n_l}\big(\bar{a}_j, A_i(\bar{a}_j), B(\bar{a}_j): i = 1,\dots,d; j = 1,\dots,l\big)},$$

where $P_{s_1,\dots,s_d,n_1,\dots,n_l}$ and $Q_{s_1,\dots,s_d,n_1,\dots,n_l}$ are polynomials in $\bar{a}_j$, $A_i(\bar{a}_j)$, $B(\bar{a}_j)$ of total degree of at most D_0 in the coordinates of $\bar{a}_j$: $j = 1,\dots,l$ and of total degree of at most

$$c_{13} \cdot \sum_{i=1}^{l} n_i^2 \cdot D_1$$

in $A_1(\bar{a}_j)$, $B(\bar{a}_j)$: $i = 1,\dots,d$; $j = 1,\dots,l$. *The types of polynomials* $P_{s_1,\dots,s_d,n_1,\dots,n_l}$ *and* $Q_{s_1,\dots,s_d,n_1,\dots,n_l}$ *are estimated from above in terms of the type of* P:

$$\max\left\{t\left(P_{s_1,\dots,s_d,n_1,\dots,n_l}\right), t\left(Q_{s_1,\dots,s_d,n_1,\dots,n_l}\right)\right\}$$
$$\leqslant t(P) + c_{14}\cdot\left\{m\log(m+1) + \sum_{i=1}^{l} n_i^2 \cdot D_1\right\}.$$

According to our assumptions about **A**, *all functions* $A_j(\bar{w})$, $B(\bar{w})$: $j = 1,\dots,d$ *are regular at* $\bar{w} = 0$ *and have algebraic coefficients in their expansion at* $\bar{w} = 0$. *We apply this observation to* (11.8) *along with the addition and multiplication formulas, expressing* $A_j(\bar{a})$, $B(\bar{a})$ *rationally in terms of* $A_j(\bar{a}_s)$, $B(\bar{a}_s)$: $j = 1,\dots,d$; $s = 1,\dots,l$. *A comparison of* (11.8) *with the definition of* G *immediately gives bounds for degrees and heights of a rational function representation* $P_{s_1,\dots,s_d,n_1,\dots,n_l}/Q_{s_1,\dots,s_d,n_1,\dots,n_l}$ *for the right hand side of* (11.8) *with* $\bar{a} = \sum_{s=1}^{l} n_s\bar{a}_s$.

The proof of Lemma 11.5 *gives us a more invariant statement which we present below.*

PROOF. To prove Lemma 11.5 we use the law of addition on **A**:

$$A_i(\bar{z} + \bar{w}) = \frac{P_i\left(A_j(\bar{z}), B(\bar{z}); A_j(\bar{w}), B(\bar{w})\right)}{Q\left(A_j(\bar{z}), B(\bar{z}); A_j(\bar{w}), B(\bar{w})\right)}$$

i, $j = 1,\dots,d$ for polynomials $P_i(\bar{z}', z_{d+1}; \bar{w}', w_{d+1})$, $Q(\bar{z}', z_{d+1}; \bar{w}', w_{d+1})$: $i = 1,\dots,d$. For $\mathfrak{D}_{\bar{z}} = \prod_{i=1}^{d}(\partial/\partial z_i)^{s_i}$ we have

$$\mathfrak{D}_{\bar{z}} \circ P(\bar{z}, A_1(\bar{z}),\dots,A_d(\bar{z}))\,|_{\bar{z}=\bar{a}}$$
$$= \mathfrak{D}_{\bar{w}} \circ P(\bar{a} + \bar{w}, A_1(\bar{a} + \bar{w}),\dots,A_d(\bar{a} + \bar{w}))\,|_{\bar{w}=0}.$$

Hence we have

$$\mathfrak{D}_{\bar{z}} \circ \{P(\bar{z}, A_1(\bar{z}),\dots,A_d(\bar{z}))\}\,|_{\bar{z}=\bar{a}}$$
$$(11.7)\qquad = \mathfrak{D}_{\bar{w}} \circ \left\{Q\left(A_j(\bar{a}), B(\bar{a}); A_j(\bar{w}), B(\bar{w})\right)^{-D_1}\right.$$
$$\left.\times G\left(\bar{a} + \bar{w}; A_j(\bar{a}), B(\bar{a}); A_j(\bar{w}), B(\bar{w})\right)\right\}\,|_{\bar{w}=0},$$

where

$$G(\bar{a} + \bar{w}; A_j(\bar{a}), B(\bar{a}); A_j(\bar{w}), B(\bar{w})) = Q(A_j(\bar{a}), B(\bar{a}); A_j(\bar{w}), B(\bar{w}))^{D_1}$$
$$\times P\left(\bar{a} + \bar{w}, \frac{P_i\left(A_j(\bar{a}), B(\bar{a}); A_j(\bar{w}), B(\bar{w})\right)}{Q\left(A_j(\bar{a}), B(\bar{a}); A_j(\bar{w}), B(\bar{w})\right)}\right): \qquad i, j = 1,\dots,d.$$

If m satisfies the assumptions of Lemma 11.5 and $s_1,\dots,s_d$ are such that $s_1 + \cdots + s_d = m$, then from (11.7) it follows immediately that

$$(11.8)\qquad \mathfrak{D}_{\bar{z}} \circ \{P(\bar{z}, A_1(\bar{z}),\dots,A_d(\bar{z}))\} = Q\left(A_j(\bar{a}), B(\bar{a}); A_j(\bar{0}), B(\bar{0})\right)^{-D_1}$$
$$\times\mathfrak{D}_{\bar{w}} \circ \left\{G\left(\bar{a} + \bar{w}; A_j(\bar{a}), B(\bar{a}); A_j(\bar{w}), B(\bar{w})\right)\right\}\,|_{w=0}.$$

LEMMA 11.6. *Let $d_{s_1,\ldots,s_d} = \Pi_{i=1}^d (\partial/\partial z_i)^{s_i}$ be a monomial differential operator. Then for every sequence of nonnegative integers $s_1,\ldots,s_d$ there exists a differential operator $\mathfrak{D}_{s_1,\ldots,s_d}$ of order in $\partial/\partial z_i$ of at most s_i: $i = 1,\ldots,d$ with coefficients being rational functions in $A_j(\bar{z})$, $B(\bar{z})$: $j = 1,\ldots,d$ with algebraic number coefficients, and having the only (leading) term of order $s_1 + \cdots + s_d$ in $\partial/\partial z_1,\ldots,\partial/\partial z_d$ of the form*

$$Q_0\big(A_j(\bar{z}), B(\bar{z})\big)^{-D_1} \cdot d_{s_1,\ldots,s_d}$$

(for a fixed D_1). For an arbitrary polynomial $P(\bar{z}, \bar{y})$ satisfying the conditions of Lemma 11.5, *we have*

$$\mathfrak{D}_{s_1,\ldots,s_d} \cdot P(\bar{z}, A_1(\bar{z}),\ldots,A_d(\bar{z})) = \frac{P'(\bar{z}, A_1(\bar{z}),\ldots,A_d(\bar{z}), B(\bar{z}))}{Q'(\bar{z}, A_1(\bar{z}),\ldots,A_d(\bar{z}), B(\bar{z}))},$$

where $P' = P'(\bar{z}, \bar{y}, y_{d+1})$, $Q' = Q'(\bar{z}, \bar{y}, y_{d+1})$ are polynomials in $\bar{z}$, $\bar{y}$, y_{d+1} of total degree of at most D_0 in $\bar{z}$, of total degree of at most $c_{15}D_1$ in $y_1,\ldots,y_d$, y_{d+1}. Moreover $Q' = Q_0^{D_1}$ and the coefficients of P' are linear combinations with algebraic number coefficients from **K** *of sizes of at most $c_{16}(\|s\|\log\|s\| + D_1)$, $\|s\| = s_1 + \cdots + s_d$, in coefficients of P.*

There are nonsingular linear transformations from the system of linear differential operators $\mathfrak{D}_{s_1,\ldots,s_d}$ to $d_{s_1,\ldots,s_d}$. This means that in addition to the representation of $\mathfrak{D}_{s_1,\ldots,s_d}$, an operator $d_{s_1,\ldots,s_d}$ is a linear combination of operators $\mathfrak{D}_{s'_1,\ldots,s'_d}$ with $s'_i \leqslant s_i$: $i = 1,\ldots,d$ with coefficients rational in $A_j(\bar{z})$, $B(\bar{z})$: $j = 1,\ldots,d$ over **K**. The leading term of this representation of $d_{s_1,\ldots,s_d}$ is $Q_0^{D_1} \cdot \mathfrak{D}_{s_1,\ldots,s_d}$.

The auxiliary results of Lemmas 11.5–11.6 are the main algebraic part of the proofs of Theorems 11.1, 11.3, 11.4.

The results of the form of Lemma 11.2 and the End of the Proof of Theorem 11.4 are proved similarly using intersection theory for algebraic subvarieties of Abelian varieties defined by algebraic equations of the form $P(\theta_1,\ldots,\theta_m) = 0$ or $P_\lambda(\theta_1,\ldots,\theta_m) = 0$ $(\lambda \in \mathfrak{L})$ for $\theta_1,\ldots,\theta_m$ of the form $A_i(\vec{\alpha}_j)$ (i.e. from $\mathbf{A}(\mathbf{Q}^d)$). Varying the pasrameter L_0 we obtain zero dimensional subvarieties, provided that $m < kd$. This proves Theorem 11.4 **[C12]**.

REMARK 11.7. For $d = 1$ we obtain from Theorem 11.4 the results of **[C7]** for an elliptic function with complex multiplication.

REFERENCES

A1. M. Anderson, *Inhomogeneous linear forms in algebraic points of an elliptic function*, Transcendence Theory: Advances and Applications, Academic Press, London and New York, 1977, Chapter 7, pp. 121–143.

B1. A. Baker, *On the quasi-periods of the Weierstrass ζ-function*, Göttingen Nachr. **N16** (1969), 145–157.

B2. ______, *Transcendental number theory*, Cambridge Univ. Press, Cambridge, 1975.

B3. ______, *The theory of linear forms in logarithms*, Transcendence Theory: Advances and Applications, Academic Press, London and New York, 1977, Chapter 1.

B4. D. Bertrand, *Approximations diophantiennes p-adique et courbes élliptiques*, Thèse, Paris, 1977 (see also, Compositio Math. **37** (1978), 21–50).

B5. ______, *Transcendence de valeurs de la fonction gamma d'après G. V. Chudnovsky*, Séminaire D-P-P. **17** (1975/1976), G8.

B6. ______, *Sur les périodes de formes modulaires*, C. R. Acad. Sci. Paris Sér A **288** (1979), 531–534.

B7. ______, *Fonctions modulaires, courbes de Tate et indépendence algébrique*, Séminaire D-P-P, **19** (1977/1978), no. 36.

B8. D. Bertrand and Y. Flicker, *Linear forms on abelian varieties over local fields*, Acta Arith. **38** (1979).

B9. D. Bertrand and D. Masser, *Linear forms in elliptic integrals*, Invent. Math. **58** (1980), 283–288.

B10. D. Bertrand and M. Laurent, *Propriétés de transcendence de nombres liés aux fonctions thêta*, C. R. Acad. Sci. Paris. Sér. A **292** (1981), 747–749.

B11. P. E. Blanksby and H. L. Montgomery, *Algebraic integers near the unit circle*, Acta Arith. **18** (1971), 355–369.

B12. E. Bombieri and S. Lang, *Analytic subgroups of group varieties*, Invent. Math. **11** (1970), 1–14. E. Bombieri, *Algebraic values of meromorphic maps*, Invent. Math. **10** (1979), 267–287.

B13. D. Brownawell, *Gelfond's method for algebraic independence*, Trans. Amer. Math. Soc. **210** (1975), 1–26.

B14. ______, *Algebraic independence of cubic powers of certain Liouville numbers*, Compositio Math. **38** (1979), 355–368.

B15. ______, *Some remarks on semi-resultants*, Transcendence Theory: Advances and Applications, Academic Press, London and New York, 1977, Chapter 14.

B16. D. Brownawell and K. Kubota, *The algebraic independence of Weierstrass functions and some related numbers*, Acta Arith. **39** (1977), 111–149.

B17. D. W. Brownawell and D. W. Masser, *Multiplicity estimates for analytic functions*. I, J. Reine Angew. Math. **314** (1980), 200–126; II Duke Math. J. **47** (1980), 273–295.

C1. G. V. Chudnovsky, *A mutual transcendence measure for some classes of numbers*, Soviet Math. Dokl. **15** (1974), 1424–1428.

C2a. ______, *Analytical methods in diophantine approximations*, Preprint IM-74-9. Inst. Math., Kiev, 1974 (and see this volume, Chapter 1).

C2b. ______, *Some analytical methods in the theory of transcendental numbers*. Preprint IM-74-8, Inst. Math., Kiev, 1974 (and see this volume, Chapter 1).

C3. ______, *Algebraic independence of constants connected with exponential and elliptic functions*, Dokl. Akad. Nauk Ukrain. SSR Ser. A **1976**, 698–707, 767. (Russian, English summary)

C4. ______, *Baker's method in the theory of transcendental numbers*, Uspehi Math. Sci **31** (1976), 281–282 (and see this volume, Chapter 2).

C5. ______, *A new method for the investigation of arithmetical properties of analytic functions*, Ann. of Math. (2) **109** (1979), 353–376.

C6a. ______, *Arithmetical properties of values of analytical functions*, Conf. on Diophantine Approximations (Oberwolfach, 1977).

C6b. ______, *Values of meromorphic functions of order* 2, Séminaire D-P-P (1977/1978), Novembre 1977, 19e année, University Publishing, Paris, 1978, pp. 4501–4518.

C6c. ______, *Singular points on complex hypersurfaces and multidimensional Schwarz lemma*, Séminaire D-P-P (1977/78), January 1977, 19e année, University Publishing, Paris, pp. 1–40. Reprinted in Progress in Math., Birkhauser-Verlag, vol. 12, 1980, pp. 29–69.

C7. ______, *Algebraic independence of the values of elliptic function at algebraic point. Elliptic analogue of Lindemann-Weierstrass theorem*, Invent. Math. **61** (1980), 267–280 (and see this volume, Introduction).

C8. ______, *Algebraic grounds for the proof of algebraic independence*. I *Elementary algebra*, Comm. Pure Appl. Math. **34** (1981), 1–28.

C9. ______, *Criteria of algebraic independence of several numbers*, Lecture Notes in Math., vol. 925, Springer-Verlag, Berlin and New York, 1982, 323–368.

C10. G. V. Chudnovsky and D. V. Chudnovsky, *Remark on the nature of spectrum of Lame equation. Problem from Transcendence Theory*, Letters Nuovo Cimento **29** (1980), 545–550.

C11. G. V. Chudnovsky, *Algebraic independence of values of exponential and elliptic functions*, Proc. Internat. Congr. Math. (Helsinki, 1978), vol. 1, Acad. Sci. Fenn., Helsinki, 1980, pp. 339–350 (and see this volume, Introduction).

C12. ______, *Indépendence algébrique dans la méthode de Gelfond-Schneider*, C. R. Acad. Sci. Paris Sér. A **291** (1980), A365–A368.

C13. ______, *Sur la mesure de la transcendence de nombres dépendent d'équations différentielles de type fuchsien*, C. R. Acad. Sci. Paris. Sér. A **291** (1980), A485–A487.

C14. ______, *Measures of irrationality, transcendence and algebraic independence. Recent progress*, Lectures at Exeter Conf. Number Theory (Exeter, April 1980) in Journees Arithmetiques, 1980, ed. by J. V. Armitage, Cambridge Univ. Press, 1982, pp. 11–83.

C15a. ______, *On Schanuel's hypothesis. Three algebraically independent numbers*. I, II, Notices Amer. Math. Soc. **23** (1976), A-272; Preliminary report, A-422.

C15b. ______, *The bound for linear forms in elliptic logarithms for the complex multiplication case*, Notices Amer. Math. Soc. **26** (1979), A-39.

C16. ______, *Padé approximation and the Riemann monodromy problem*, Bifurcation Phenomena in Mathematical Physics and Related Topics (Cargesé Lectures, June 1979), Reidel, Boston, 1980, pp. 448–510.

C17. ______, *Formules d'Hermite pour les approximations de Padé de logarithmes et de fonctions binomes, et mesures d'irrationalité*, C. R. Acad. Sci. Paris Sér. A **288** (1979), A965–A967.

C18. ______, *Algebraic independence of constants connected with exponential and elliptic functions*, Meeting of AMS, Pullman, Washington, June 1975, Notices Amer. Math. Soc. **22** (1975), A-486.

C19. ______, This volume, Chapters 4, 8.

C20. P. L. Cijsouw, *Transcendence measures*, Thesis, 1972.

C21. J. Coates, *Linear relations between $2\pi i$ and the periods of two elliptic curves*, Diophantine Approximation and its Applications, Academic Press, London, 1973, pp. 77–99.

C22. J. Coates and S. Lang, *Diophantine approximation on Abelian varieties with complex multiplication*, Invent. Math. **34** (1976), 129–133.

C23. J. Coates, *The arithmetic of elliptic curves with complex multiplication*, Proc. Internat. Congr. Math. (Helsinki, 1978), vol. 1, Acad. Sci. Fenn., Helsinki, 1980, pp. 351–355.

D1. P. Deligne, *Cycles de Hodge absolus et périods des intégrales des variétés abéliennes*, Bull. Soc. Math. France Mém. No. 2 (1980), 23–33.

D2. E. Dobrowolski, *On a question of Lehmer*, Bull. Soc. Math. France Mém. No. 2 (1980), 35–39.

D3. J. Drach, *Sur l'integration par quadratures de l'equation $d^2y/dx^2 = [\varphi(x) + h]y$*, C. R. Acad. Sci. Paris **168** (1919), 47–50, 337–340.

F1. Y. Z. Flicker, *Linear forms on arithmetic Abelian varieties: ineffective bound*, Bull. Soc. Math. France Mém. No. 2 (1980), 41–47.

G1. A. O. Gelfond, *Transcendental and algebraic numbers*, Dover, New York, 1960.

G2. H. Groscot, *Points entiers sur les courbes élliptiques*, Thése de troisième cycle (Paris VI), 1979.

L0. S. Lang, *Transcendental points on group varieties*, Topology **1** (1962), 313–318.

L1. ______, *Introduction to transcendental numbers*, Addison-Wesley, Reading, Mass., 1966.

L2. ______, *Diophantine approximation on Abelian varieties with complex multiplication*, Adv. in Math. **17** (1975), 281–336.

L3. ______, *Elliptic curves, diophantine analysis*, Springer-Verlag, Berlin and New York, 1978.

L4. M. Laurent, *Transcendence de périodes d'intégrales élliptiques*, J. Reine Angew. Math. **316** (1980), 123–139.

L5. F. Lindemann, *Über die Zahl π*, Math. Ann. **20** (1882), 213–225.

M1. K. Mahler, *Zur approximation der Exponentialfunktion und des Logarithmus*, J. Reine Angew. Math. **166** (1932), 118–150.

M2. Yu. I. Manin, *Cyclotomic fields and modular curves*, Uspekhi Math. Nauk **25** (1971), 7–71.

M3. D. Masser, *Elliptic functions and transcendence*, Lecture Notes in Math., vol. 437, Springer-Verlag, Berlin and New York, 1975.

M4. ______, *Linear forms in algebraic points of Abelian functions*, Parts I, II, Math. Proc. Cambridge Philos. Soc. **77** (1975), 499–513; **79** (1976), 55–70. Part III, Proc. London Math. Soc. **33** (1976), 549–564.

M5. ______, *The transcendence of certain quasi-periods associated with Abelian functions in two variables*, Compositio Math. **35** (1977), 239–258.

M6. ______, *Diophantine approximations and lattices with complex multiplication*, Invent. Math. **45** (1978), 61–82.

M7. ______, *On quasi-periods of Abelian functions with complex multiplication*, Bull. Soc. Math. France Mém. No. 2 (1980), 55–68.

M8. D. Masser and G. Wüstholz, *Zero estimates on group varieties*. (to appear).

M9. H. P. McKean, *Algebraic curves of infinite genus arising in the theory of nonlinear waves*, Proc. Internat. Congr. Math. (Helsinki, 1978), vol. 2, Acad. Sci. Fenn., Helsinki, 1980, pp. 777–783.

N1. S. P. Novikov, *Linear operators and integrable Hamiltonian systems*, Proc. Internat. Congr. Math. (Helsinki, 1978), vol. 1, Acad. Sci. Fenn., Helsinki, 1980, p. 187.

R1. K. Ramachandra, *Contributions to the theory of transcendental numbers*, Acta Arith. **14** (1968), 65–88.

R2. E. Reyssat, *Approximation algébrique de nombres liés aux fonctions élliptiques et exponentielles*, Bull. Soc. Math. France, **108** (1980), 47–79.

R3. ______, C. R. Acad. Sci. Paris, Sér. A **290** (1980),439–441.

R4. K. A. Ribet, *Dividing rational points on Abelian varieties of CM-type*, Compositio Math. **33** (1976), 69–74.

R5. ______, *Division fields of Abelian varieties with complex multiplication*, Bull. Soc. Math. France Mém. No. 2 (1980), 75–94.

S1. Th. Schneider, *Introduction aux nombres transcendants*, Gauthier-Villars, Paris, 1959.

S2. G. Shimura and Y. Taniyama, *Complex multiplication of abelian varieties*, Publ. Math. Soc. Japan, N6, 1961.

S3. G. Shimura, *On some problems of algebraicity*, Proc. Internat. Congr. Math. (Helsinki, 1978), vol. 1, Acad. Sci. Fenn., Helsinki, 1980, pp. 373–379.

S4. ______, *Automorphic forms and the periods of Abelian varieties*, J. Math. Soc. Japan **31** (1979), 561–592.

S5. ______, *The arithmetic of certain zeta functions and automorphic forms on orthogonal groups*, Ann. of Math. (2) **110** (1980).

S6. C. L. Siegel, *Über einige Anwendungen diophantischer Approximationen*. Abh. Preuss. Akad. Wiss. Phys-Math. Kl. No. 1, 1929.

S7. ______, *Transcendental numbers*, Ann. of Math. Studies, Princeton Univ. Press, Princeton, N. J., 1949.

S8. H. M. Stark, *Further advances in the theory of linear forms in logarithms*, Diophantine Approximation and its Applications, Academic Press, London, 1973, pp. 255–293.

S9. ______, *Effective estimates of solutions of some Diophantine equations*, Acta. Arith. **24** (1973), 251–259.

S10. C. L. Stewart, *On a theorem of Kronecker and related question of Lehmer*, Séminaire de Théorie des Nombres (Bordeaux, 1977–78), No. 7, 11pp.

S11. A. Selberg and S. D. Chowla, *On Epstein's zeta-function*, J. Reine Angew. Math. **227** (1967), 86–110.

S12. I. R. Shafarevic, *Foundation of algebraic geometry*, Springer-Verlag, New York, 1975.

T1. R. Tijdeman, *On the number of zeroes of general exponential polynomials*, Indag. Math. **33** (1971), 1–7.

T2. ______, *Exponential diophantine equations*, Proc. Internat. Congr. Math. (Helsinki, 1978), vol. 1, Acad. Sci. Fenn., Helsinki, 1980, pp. 381–387.

T3. *Transcendence theory*: *Advances and applications*, ed. A. Baker, D. Masser, Academic Press, New York and London, 1977.

W1. M. Waldschmidt, *Nombres transcendants*, Lecture Notes in Math., vol. 402, Springer-Verlag, Berlin and New York, 1974.

W2. ______, *Propriétés arithmétiques de fonctions de plusieurs variables*. I, II, Séminaire P. Lelong (Analyse), 15e année (1974/1975), Lecture Notes in Math., vol. 518, Springer-Verlag, Berlin and New Yoirk, 1976; ibid, 16e année (1975/1976), vol. 524, 1977.

W3. ______, *Les travaux de G. V. Choodnovsky sur les nombres transcendants*, Séminaire Bourbaki (1975/1976), no. 488, Lecture Notes in Math., vol. 567, Springer-Verlag, Berlin and New York, 1977, pp. 274-292.

W4. ______, *On functions of several complex variables having algebraic Taylor coefficients*, Transcendence Theory: Advances and Applications, Academic Press, London and New York, 1977, Chapter 11.

W5. ______, *Nombres transcendants et groupes algébriques*, Astérisque No. 69–70 (1980).

W6. ______, *Transcendence measures for exponentials and logarithms*, J. Austral. Math. Soc. **25** (1978), 445–465.

W7. ______, *Propriétés arithmétiques des valeurs de fonctions meromorphes algebriquement indépendantes*, Acta Arith. **23** (1973), 19–88.

W8. A. Weil, *Elliptic functions according to Eisenstein and Kronecker*, Springer, New York, 1976.

CHAPTER 8

MEASURE OF THE ALGEBRAIC INDEPENDENCE OF PERIODS AND QUASI-PERIODS OF ELLIPTIC CURVES

0. The purpose of this chapter is to present a complete proof of the bound of the measure of the algebraic independence of two numbers, connected with periods and quasi-periods of an elliptic curve. The presented proof is a complete one with all possible details and we give a detailed exposition of auxiliary results since our considerations are very general. The method of the proof presented is applied to any algebraic independence proof where the Gelfond-Schneider method is used. We decided to present the proof for the case of numbers π/ω, η/ω, for a period ω and the quasi-period $\eta = 2\zeta(\omega/2)$ of an elliptic curve, where the type of transcendence is $\leqslant 3 + \varepsilon$ for any $\varepsilon > 0$. This result has many important and interesting corollaries. For a complete proof we present all the auxiliary algebraic lemmas. The method of proof is elementary, using only the intersection theory and is based on two papers [**1a**, **1b**]. One can consider the proof below as a good introduction to more sophisticated studies of measures of algebraic independence of n algebraically independent numbers in the Gelfond-Schneider method with $n \geqslant 3$.

The presented method allows us to obtain the best bounds for the measure of the algebraic independence of other pairs of numbers connected with elliptic functions listed in [**3** and **4**]. Among those pairs of numbers are

(i) $\zeta(u) - \eta u/\omega$ and η/ω for an algebraic point u of the function $\wp(x)$ with algebraic invariants,

(ii) u and $\zeta(u)$ for an algebraic point u of $\wp(x)$ with a complex multiplication and algebraic invariants,

(iii) π/ω and $e^{\pi i\beta}$ for a period $\omega \neq 0$ of an elliptic curve defined over $\overline{\mathbf{Q}}$ with the complex multiplication by β.

0076-5376/84 \$1.00 + \$.25 per page

An important feature of the present proof is the "normality" [**9**] of the estimates or H^{-C} form of the bound for the height H. Our main example is a particular case of (i) for two numbers π/ω and η/ω for a period $\omega \neq 0$ and $\eta = 2\zeta(\omega/2)$, where the type of transcendence [**15**] is $\leqslant 3 + \varepsilon$ for every $\varepsilon > 0$.

1. Let us consider Weierstrass's elliptic function $\wp(x)$ with algebraic invariants g_2, g_3 and corresponding Weierstrass ζ-functions $\zeta'(x) = -\wp(x)$. Let (ω, η) be a pair of period and quasi-period of $\wp(x)$, i.e. $\eta = 2\zeta(\omega/2)$ (here $\omega \in L \backslash 2L$).

The main example deals only with π/ω, η/ω.

THEOREM 1.1. *Let $R(x, y) \in \mathbf{Z}[x, y]$; $R \not\equiv 0$. Let $\wp(x)$ be an elliptic function with algebraic invariants, and ω be a period of $\wp(x)$, $2\zeta(\omega/2) = \eta$ for a Weierstrass ζ-function $\zeta(x)$. Then*

$$|R(\pi/\omega, \eta/\omega)| > \exp\left(-C_0(\log H(R) + d(R)\log d(R))\, d(R)^2 \log^2 d(R)\right)$$

if $d(R) \geqslant d_0$ for some constant $C_0 > 0$.

REMARK. In particular, π/ω and η/ω have the type of transcendence $\leqslant 3 + \varepsilon$ for every $\varepsilon > 0$ [**3**]. However, the bound of Theorem 1.1 is much stronger than in [**3**] since for $H(R) \geqslant d(R)^{d(R)}$ it has the form

$$|R(\pi/\omega, \eta/\omega)| > H(R)^{-C_0' d(R)^2 \log^2 (d(R)+1)}.$$

This bound differs only by the factor $\log^2(d(R) + 1)$ from the upper bound given by Dirichlet's box principle. The same method as presented below gives H^{-C} bounds for pairs of numbers (i) and (ii) of §0.

PROOF OF THEOREM 1.1. Let $\wp(x)$ be a Weierstrass elliptic function with algebraic invariants g_2, g_3, and let $\zeta(x)$ be the corresponding Weierstrass ζ-function, $\zeta'(x) = -\wp(x)$. Let the pair (ω, η) be a period and quasi-period of $\wp(x)$ so that $\eta = 2\zeta(\omega/2)$ (and $\omega/2$ is not a period of $\wp(x)$). Let ω' be another period of $\wp(x)$, linearly independent with ω.

In order to get a bound for the measure of algebraic independence of π/ω, η/ω we work, following [**4**], with the auxiliary function of the form

$$F(z) = P(\zeta(z) - \eta/\omega \cdot z,\ \wp(z))$$

for $P(x, y) \in \mathbf{Z}[\pi/\omega, \eta/\omega][x, y]$ and $d_x(P) \leqslant L_1$, $d_y(P) \leqslant L_2$ and $L_2 \geqslant L_1 > 1$.

The function $F(z)$ above has the form

$$F(z) = \sum_{\lambda_1=0}^{L_1} \sum_{\lambda_2=0}^{L_2} C_{\lambda_1,\lambda_2} \cdot \left(\zeta(z) - \frac{\eta}{\omega} z\right)^{\lambda_1} \wp(z)^{\lambda_2}.$$

The one has the following representation for the derivatives of the function $F(z)$:

$$F^{(k)}(z) = \sum_{\lambda_1=0}^{L_1} \sum_{\lambda_2=0}^{L_2} C_{\lambda_1,\lambda_2} \sum_{s=0}^{k} \binom{k}{s} \left\{\left(\zeta(z) - \frac{\eta}{\omega} z\right)^{\lambda_1}\right\}^{(s)} \cdot \left\{\wp(z)^{\lambda_2}\right\}^{(k-s)}.$$

Consequently, at points $z = \omega/4 + m\omega'$ we have

$$(1.2)\qquad F^{(k)}\left(\frac{\omega}{4} + m\omega'\right) = \sum_{\lambda_1=0}^{L_1} \sum_{\lambda_2=0}^{L_2} C_{\lambda_1,\lambda_2} \cdot \mathcal{F}_{\lambda_1,\lambda_2,k,m}\left(\frac{\pi}{\omega}, \frac{\eta}{\omega}\right).$$

Here

$$(1.3)\qquad \mathcal{F}_{\lambda_1,\lambda_2,k,m}\left(\frac{\pi}{\omega}, \frac{\eta}{\omega}\right) = \sum_{j_1=0}^{\lambda_1} \sum_{j_2=0}^{\lambda_1} f_{\lambda_1,\lambda_2,k,m,j_1,j_2}\left(\frac{\pi}{\omega}\right)^{j_1}\left(\frac{\eta}{\omega}\right)^{j_2}$$

is a polynomial in π/ω, η/ω of degree of at most λ_1 in π/ω and in η/ω, with algebraic number coefficients $f_{\lambda_1,\lambda_2,k,m,j_1,j_2}$ from the field

$$\mathbf{K} = \mathbf{Q}(i, g_2, g_3, \wp(\omega/4), \wp'(\omega/4))$$

having sizes at most

$$(1.4)\qquad \max_{\substack{\lambda_1 \leqslant L_1, \lambda_2 \leqslant L_2 \\ j_1, j_2 \leqslant \lambda_1}} \overline{|f_{\lambda_1,\lambda_2,k,m,j_1,j_2}|}$$
$$\leqslant \exp\{C_1(L_2 + k) + k\log(k+1) + L_1\log(m+1)\}$$

for $C_1 > 0$ depending on g_2, g_3 only. From now on we will consider $\mathcal{F}_{\lambda_1,\lambda_2,k,m}$ to be polynomials $\mathcal{F}_{\lambda_1,\lambda_2,k,m}(x, y) \in \mathbf{K}[x, y]$ of sizes bounded in (1.4).

The coefficients C_{λ_1,λ_2}, as the polynomials $C_{\lambda_1,\lambda_2}(x, y)$, are constructed using the Thue-Siegel lemma (see [**6**, **7**]). Here is one of the definitions of $C_{\lambda_1,\lambda_2}(x, y)$.

LEMMA 1.5. *Let $L_2 \geqslant L_1$; then there exists a system of polynomials*

$$\{C_{\lambda_1,\lambda_2}(x, y) : 0 \leqslant \lambda_1 \leqslant L_1, 0 \leqslant \lambda_2 \leqslant L_2\}$$

from $\mathbf{Z}[x, y]$, without a common factor, such that the following conditions are satisfied:

$$(1.6)\qquad d(C_{\lambda_1,\lambda_2}) \leqslant L_1; \qquad t(C_{\lambda_1,\lambda_2}) \leqslant c_2 L_2 \log L_2;$$

$\lambda_1 = 0, 1, \ldots, L_1$; $\lambda_2 = 0, 1, \ldots, L_2$ *and the system of linear equations in C_{λ_1,λ_2} is satisfied*:

$$(1.7)\qquad \sum_{\lambda_1=0}^{L_1} \sum_{\lambda_2=0}^{L_2} C_{\lambda_1,\lambda_2}(x, y)\mathcal{F}_{\lambda_1,\lambda_2,k,m}(x, y) = 0$$

for $k = 0, 1, \ldots, c_3 L_2$, $m = 0, 1, \ldots, c_4 L_1$.

Here $c_3 c_4 < (8 \cdot [\mathbf{K}: \mathbf{Q}])^{-1}$. In particular, for the function

$$(1.8)\qquad F(z) = \sum_{\lambda_1=0}^{L_1} \sum_{\lambda_2=0}^{L_2} C_{\lambda_1,\lambda_2}\left(\frac{\pi}{\omega}, \frac{\eta}{\omega}\right)\left(\zeta(z) - \frac{\eta}{\omega}z\right)^{\lambda_1} \cdot \wp(z)^{\lambda_2},$$

we have

$$(1.9)\qquad F^{(k)}(\omega/4 + m\omega') = 0$$

for $k = 0, 1, \ldots, [c_3 L_2]$, $m = 0, 1, \ldots, [c_4 L_1]$.

Now for the function $F(z)$ (1.8), satisfying (1.9), we can use the usual methods of transcendental number theory (see [**11**, 7]). However from the point of view of applications, we consider a more general situation. We are not going too far in reaching generalizations however, because for functions of the form

$$P\left(\zeta(z) - \frac{\eta}{\omega}\cdot z,\ \wp(z)\right),$$

one has a Small Value Lemma. If there is no Small Value Lemma, it is necessary to consider a more general situation, where we actually have a family of functions.

We stick to π/ω and η/ω, and have in this case the following analytic statement:

LEMMA 1.10. *Let* C_{λ_1,λ_2} $(0 \leqslant \lambda_1 \leqslant L_1,\ 0 \leqslant \lambda_2 \leqslant L_2)$ *be complex numbers such that*

$$\max_{\substack{0\leqslant\lambda_1\leqslant L_1\\ 0\leqslant\lambda_2\leqslant L_2}} |C_{\lambda_1,\lambda_2}| \leqslant \exp \bar{C}$$

and that the function

$$G(z) = \sum_{\lambda_1=0}^{L_1} \sum_{\lambda_2=0}^{L_2} C_{\lambda_1,\lambda_2}\left(\zeta(z) - \frac{\eta}{\omega}\cdot z\right)^{\lambda_1} \cdot \wp(z)^{\lambda_2},$$

satisfies

$$\frac{1}{k!}\left|G^{(k)}\left(\frac{\omega}{4} + m\omega'\right)\right| \leqslant \exp(-E)$$

for $k = 0, 1, \ldots, [c_5 L_2]$, $m = 0, 1, \ldots, [c_6 L_1]$. *Then we have*

$$\frac{1}{k'!}\left|G^{(k')}\left(\frac{\omega}{4} + m'\omega'\right)\right| < \exp\left(-E + \bar{C} + c_7 L_1^2 L_2\right) + \exp\left(-c_9 L_1^2 L_2\right)$$

for $k' = 0, 1, \ldots, [c_8 L_2]$, $m' = 0, 1, \ldots, [c_8 L_1]$, *and we have at the same time a Small Value Lemma*:

$$\max\left\{|C_{\lambda_1,\lambda_2}| : 0 \leqslant \lambda_1 \leqslant L_1, 0 \leqslant \lambda_2 \leqslant L_2\right\}$$
$$\leqslant \max\left\{\frac{1}{k'!}|G^{(k')}\left(\frac{\omega}{4} + m'\omega\right) : k' = 0, 1, \ldots, [c_{10} L_2];\ m' = 0, 1, \ldots, [c_{10} L_1]\right\}$$
$$\cdot \exp\left(c_{11} L_1^2 L_2\right),$$

for some constant $c_{10} > 0$, *with* $c_{11} > 0$ *depending on the choice of* c_{10}.

The proof of Lemma 1.10 is usual, and depends mainly on the Schwarz lemma applied to a function

$$G_1(z) = \sigma(z)^{L_1 + 2L_2} G(z),$$

which is an entire function of z. Since $G(z)$ is periodic with a period ω, the function $G_1(z)$ has small values $< \exp(-E)$ of multiplicity K at points

$$z = \omega/4 + m\omega' + n\omega,$$

$m = 0, 1, \ldots, M$, $n = 0, 1, \ldots, M'$. We apply then, the usual Hermité interpolation formula **[6, 8]**, to $G_1(z)$ in the region $|z| = 2M_1$.

There is one nontrivial point in the proof of Lemma 1.5. It concerns the improvement of bound (1.6) to a better one,

$$t(C_{\lambda_1,\lambda_2}) \leqslant c_2 L_2 \log L_1 \tag{1.11}$$

for $\lambda_1 = 0, \ldots, L_1$; $\lambda_2 = 0, \ldots, L_2$. This is the place where we use the new arguments of **[9]** concerning (G, C)-functions **[10, 11]**. Indeed, the bound

$$t(C_{\lambda_1,\lambda_2}) \leqslant c_2 L_2 \log L_1$$

is the main point in establishing the "normal" measures of algebraic independence of π/ω, η/ω as announced above in 1.1:

$$\log |R(\pi/\omega, \eta/\omega)| > C_0 \log H(R)\, d(R)^2 \log^2(d(R) + 2)$$

for $R(x, y) \in \mathbf{Z}[x, y]$, $R \not\equiv 0$.

In order to get the bound for $t(C_\lambda)$, we make a change of variables

$$w = \wp(z), \qquad f(w) = \zeta(z) - \eta z/\omega. \tag{1.12}$$

Then $f(w)$ satisfies a differential equation

$$f'(w) = (-w - \eta/\omega)\sqrt{4w^3 - g_2 w - g_3}\,. \tag{1.13}$$

The function $f(w)$ is a (G, C)-function **[11]**. This means that the following condition is satisfied: if $k \geqslant 1$, then

$$\left(\sqrt{4w^3 - g_2 w - g_3}\right)^k f^{(k)}(w) = \mathcal{A}_k(w)\sqrt{4w^3 - g_2 w - g_3} + \mathcal{B}_k(w), \tag{1.14}$$

where

$$\mathcal{A}_k(w) = \mathcal{A}_{0,k}(w) + \frac{\eta}{\omega}\mathcal{A}_{1,k}(w), \qquad \mathcal{B}_k(w) = \mathcal{B}_{0,k}(w) + \frac{\eta}{\omega}\mathcal{B}_{1,k}(w)$$

and $\mathcal{A}_{0,k}(w)$, $\mathcal{A}_{1,k}(w)$, $\mathcal{B}_{0,k}(w)$, $\mathcal{B}_{1,k}(w)$ are polynomials in w with coefficients from $\mathbf{Q}[g_2, g_3]$ of degrees (in g_2, g_3) at most $\gamma_1 k$.

Moreover, and this is most essential for applications, for $K \geqslant 1$ there exists an integer $D_K \in \mathbf{Z}$, such that

$$|D_K| < \exp(\gamma_2 K) \tag{1.15}$$

with γ_2 depending only on g_2, g_3, and for which the polynomials

$$D_K \cdot \frac{1}{k!}\mathcal{A}_{0,k}(w), \qquad D_K \cdot \frac{1}{k!}\mathcal{A}_{1,k}(w), \qquad D_K \cdot \frac{1}{k!}\mathcal{B}_{0,k}(w),$$

$$D_K \cdot \frac{1}{k!}\mathcal{B}_{1,k}(w): \qquad k = 1, 2, \ldots, K,$$

all have *integer* coefficients from the field $\mathbf{Q}(g_2, g_3)$.

We can now rewrite

$$F(z) = \sum_{\lambda_1=0}^{L_1} \sum_{\lambda_2=0}^{L_2} C_{\lambda_1,\lambda_2}\left(\zeta(z) - \frac{\eta}{\omega}z\right)^{\lambda_1} \wp(z)^{\lambda_2}$$

in terms of a w-variable $F(z) = F_1(w)$, where

$$F_1(w) = \sum_{\lambda_1=0}^{L_1} \sum_{\lambda_2=0}^{L_2} C_{\lambda_1,\lambda_2} f(w)^{\lambda_1} w^{\lambda_2}. \tag{1.16}$$

Then we have

$$F_1^{(k)}(w) = \sum_{\lambda_1=0}^{L_1} \sum_{\lambda_2=0}^{L_2} C_{\lambda_1,\lambda_2} \cdot \sum_{s=0}^{k} \binom{k}{s}$$
$$\times \lambda_2 \cdots (\lambda_2 - s + 1) w^{\lambda_2 - s} \{ f(w)^{\lambda_1} \}^{(k-s)}.$$

Hence

$$\frac{1}{k!} F_1^{(k)}(w) = \sum_{\lambda_1=0}^{L_1} \sum_{\lambda_2=0}^{L_2} C_{\lambda_1,\lambda_2} \sum_{s=0}^{k} \binom{\lambda_2}{s}$$
$$\times w^{\lambda_2 - s} \frac{1}{(k-s)!} \{ f(w)^{\lambda_1} \}^{(k-s)}. \tag{1.17}$$

Because of the (G, C)-function property of $f(w)$, there exists $\mathfrak{D}_{L_1,k} \in \mathbf{Z}$ such that

$$|\mathfrak{D}_{L_1,k}| < \exp(\gamma_3 k \log(L_1 + 1)),$$

and we have

$$\frac{\mathfrak{D}_{L_1,k}}{(k-s)!} \{ f(w)^{\lambda_1} \}^{(k-s)} = \sum_{j=0}^{\lambda_1} \sum_{i=0}^{1} f(w)^j (4w^3 - g_2 w - g_3)^{i/2} \cdot R_{j,i,\lambda_1,s,k}(w),$$

with

$$R_{j,i,\lambda_1,s,k}(w) = R_{j,i,\lambda_1,s,k,0}(w) + \frac{\eta}{\omega} R_{j,i,\lambda_1,s,k,1}(w),$$

where $R_{j,i,\lambda_1,s,k,\chi}(w)$ is a polynomial of w with algebraic integer coefficients from $\mathbf{K}$ of degrees of at most $\gamma_4 k$, and of sizes of at most $\gamma_5(k \log(L_1 + 1) + L_1)$.

Hence the equations

$$\frac{1}{k!} F_1^{(k)}(w_0) = 0: \qquad k = 0, 1, \ldots, K, \tag{1.18}$$

can be replaced by

$$\sum_{\lambda_1=0}^{L_1} \sum_{\lambda_2=0}^{L_2} C_{\lambda_1,\lambda_2} \sum_{s=0}^{k} \binom{\lambda_2}{s} w_0^{\lambda_2 - s}$$
$$\times \sum_{j=0}^{\lambda_1} \sum_{i=0}^{1} f(w_0)^j (4w_0^3 - g_2 w_0 - g_3)^{i/2} \cdot R_{j,i,\lambda_1,s,k}(w_0) = 0:$$
$$k = 0, 1, \ldots, K. \tag{1.19}$$

Now we take a particular solution $f(w)$ of the differential equation (1.13) which corresponds to the following initial conditions:

$$(1.20) \qquad w_0 = \wp\left(\frac{\omega}{4}\right), \qquad f(w_0) = m\left(\frac{\omega\eta' - \omega'\eta}{\omega}\right)$$

for $m = 0, 1, \ldots, M$, with $M = [c_4 L_1]$ as above. We note that the differential equation (1.13) has a nontrivial monodromy, so a particular $f_0(w)$ defined in (1.20) can be considered to be one of the branches of $f(w)$.

Hence, we can determine polynomials $C_{\lambda_1,\lambda_2}(\pi/\omega, \eta/\omega)$ from the system of equations (1.18):

$$\frac{1}{k!}F_1^{(k)}(w) = 0 \quad \text{for } k = 0, 1, \ldots, [c_3 L_2] \text{ and } w = \wp(\omega/4),$$

$$f(w) = m \cdot \frac{\omega\eta' - \omega'\eta}{\omega}: \qquad m = 0, 1, \ldots, [c_4 L_1].$$

According to the previous discussion, this system of equations can be written in the usual form (1.19) as

$$(1.21) \qquad \sum_{\lambda_1=0}^{L_1} \sum_{\lambda_2=0}^{L_2} C_{\lambda_1,\lambda_2}\left(\frac{\pi}{\omega}, \frac{\eta}{\omega}\right) \mathcal{K}_{\lambda_1,\lambda_2,k,m}\left(\frac{\pi}{\omega}, \frac{\eta}{\omega}\right) = 0:$$

$$k = 0, 1, \ldots, [c_3 L_2]; \quad m = 0, 1, \ldots, [c_4 L_1].$$

Here $\mathcal{K}_{\lambda_1,\lambda_2,k,m}(\pi/\omega, \eta/\omega)$ is a polynomial in π/ω and η/ω of degrees $\leqslant \lambda_1$ (in each of the variables) with algebraic integer coefficients from the field $\mathbf{K} = \mathbf{Q}(g_2, g_3, \wp(\omega/4), \wp'(\omega/4), i)$ of sizes of at most

$$\gamma_6(k \log(L_1 + 1) + L_2 + L_1 \log(m + 1)).$$

We construct $C_{\lambda_1,\lambda_2}(\pi/\omega, \eta/\omega) \in \mathbf{Z}[\pi/\omega, \eta/\omega]$ using the Thue-Siegel lemma. One sees, however, that these $C_{\lambda_1,\lambda_2}(\pi/\omega, \eta/\omega)$ give the solution of the initial problem for $F(z)$. Indeed, by the definition of our change of variables, we have

$$F_1^{(k)}(w_0) = 0: \qquad k = 0, 1, \ldots, [c_3 L_2],$$

for

$$(w_0, f(w_0)) = \left(\wp\left(\frac{\omega}{4}\right), m_0\left(\frac{\omega\eta' - \omega'\eta}{\omega}\right)\right).$$

Hence, making the change of variables locally at w_0, we get

$$(1.22) \qquad F^{(k)}(z_0) = 0: \qquad k = 0, 1, \ldots, [c_3 L_2],$$

for $z_0 = \omega/4 + m_0\omega'$. The polynomials $C_{\lambda_1,\lambda_2}(\pi/\omega, \eta/\omega)$ are now the polynomials from $\mathbf{Z}[\pi/\omega, \eta/\omega]$, without the common factor and having bounds for height and degree

$$d(C_{\lambda_1,\lambda_2}) \leqslant c_1 \cdot L_1, \qquad t(C_{\lambda_1,\lambda_2}) \leqslant c_2 \cdot L_2 \log L_1.$$

LEMMA 1.5′. *For $L_2 \geqslant L_1$ there exists a system of polynomials*

$$\{C_{\lambda_1,\lambda_2}(x, y): 0 \leqslant \lambda_1 \leqslant L_1, 0 \leqslant \lambda_2 \leqslant L_2\}$$

from $\mathbf{Z}[x, y]$, *without a common factor, which satisfies all the conditions of Lemma* 1.5, *but the second inequality in* (1.6) *is replaced by a sharp one,*

$$t(C_{\lambda_1,\lambda_2}) \leqslant c_2 L_2 \log L_1: \qquad 0 \leqslant \lambda_1 \leqslant L_1, 0 \leqslant \lambda_2 \leqslant L_2. \tag{1.23}$$

We apply Lemma 1.10 to the function $F(z)$ from Lemma 1.5′. In this case we obtain immediately the usual statement

LEMMA 1.24. *For every* $L_2 \geqslant L_1 > 1$ *we have one of the following two possibilities: either*

(i) *there exists a polynomial* $P_{L_1,L_2}(x, y) \in \mathbf{Z}[x, y]$ *such that*

$$d(P_{L_1,L_2}) \leqslant c_{12} L_1, \qquad t(P_{L_1,L_2}) \leqslant c_{12} L_2 \log L_1,$$
$$-c_{14} L_1^2 L_2 < \log|P_{L_1,L_2}(\pi/\omega, \eta/\omega)| < -c_{13} L_1^2 L_2,$$

or

(ii) *there exists a system of polynomials* $C_l(x, y) \in \mathbf{Z}[x, y]$: $l \in \mathfrak{L}_{L_1,L_2}$ *without a common (nonconstant) factor such that*

$$d(C_l) \leqslant c_{15} L_1, \qquad t(C_l) \leqslant c_{16} L_2 \log L_1;$$
$$\log|C_l(\pi/\omega, \eta/\omega)| < -c_{17} L_1^2 L_2$$

for all $l \in \mathfrak{L}_{L_1,L_2}$.

We use these polynomials to bound the measure of the algebraic independence of π/ω, η/ω. For example, if we want to bound only the type of the transcendence of π/ω, η/ω, we can use polynomials only for $L_1 = L_2$.

Let us assume that the polynomial $R(x, y) \in \mathbf{Z}[x, y]$ is irreducible over $\mathbf{Z}$ (other cases follow looking at the factorization) and

$$\log|R(\pi/\omega, \eta/\omega)| = -t\varphi(d), \tag{1.25}$$

where $t = \log H(R) + d(R) \log d(R)$, $d = d(R)$, where $d \geqslant d_0$ is sufficiently large and $\varphi(d) > C d^2 \log^2 d$ for a sufficiently large $C > 0$.

First of all, we need a zero-dimensional set, approximating $\bar{\theta} = (\pi/\omega, \eta/\omega)$:

LEMMA 1.26. *There exists a polynomial* $P(x, y) \in \mathbf{Z}[x, y]$ *relatively prime with* $R(x, y)$ *such that*

$$\begin{aligned} &d(P) \leqslant c_{18} d^{1/3} \cdot \varphi(d)^{1/3} \log^{1/3} d, \\ &t(P) \leqslant c_{19} t d^{-2/3} \cdot \varphi(d)^{1/3} \log^{1/3} d, \\ &\log|P(\pi/\omega, \eta/\omega)| < -c_{20} t \varphi(d). \end{aligned} \tag{1.27}$$

PROOF OF LEMMA 1.26. We take a pair (L_1^0, L_2^0) with $L_2^0 \geqslant L_1^0$ such that

$$\begin{aligned} &L_1^0 = \left[c_{21} \cdot (d\varphi(d) \log d)^{1/3}\right], \\ &L_2^0 = \left[c_{22} \cdot t d^{-2/3} \varphi(d)^{1/3} \log^{-2/3} d\right] \qquad (\geqslant L_1^0) \end{aligned} \tag{1.28}$$

and

$$c_{14} (L_1^0)^2 L_2^0 < \tfrac{1}{2} t \varphi(d). \tag{1.29}$$

If the first case (i) of Lemma 1.24 is true, then we have a polynomial $P_{L_1^0,L_2^0}(x, y) \in \mathbf{Z}[x, y]$ such that

$$d\left(P_{L_1^0,L_2^0}\right) \leqslant c_{14}L_1^0, \qquad t\left(P_{L_1^0,L_2^0}\right) \leqslant c_{14} \cdot L_2^0 \log L_1^0$$

and

$$-c_{14}\left(L_1^0\right)^2 L_2^0 < \log|P_{L_1^0,L_2^0}(\bar{\theta})| < -c_{13}\left(L_1^0\right)^2 L_2^0. \tag{1.30}$$

Then $P_{L_1^0,L_2^0}(x, y)$ is relatively prime with $R(x, y)$. Indeed, otherwise $R(x, y) \mid P_{L_1^0,L_2^0}(x, y)$. We see that in this case, the lower bound for $|P_{L_1^0,L_2^0}(\bar{\theta})|$ in (1.30) contradicts the bound for $|R(\bar{\theta})|$ and the definition of (L_1^0, L_2^0) in (1.29). Thus, in the case (i) of Lemma 1.24, we put $P(x, y) = P_{L_1^0,L_2^0}(x, y)$.

In the case (ii) of Lemma 1.24, corresponding to (L_1^0, L_2^0), we take a polynomial $C_{\bar{\lambda}_0}(x, y)$: $\bar{\lambda} \in \mathfrak{L}_{L_1^0,L_2^0}$, relatively prime with $R(x, y)$. Such a polynomial $C_{\bar{\lambda}_0}(x, y)$ exists since $R(x, y)$ is irreducible. Then according to the choice of (L_1^0, L_2^0) in (1.28), (1.29), the polynomial $P(x, y) \overset{\text{def}}{=} C_{\bar{\lambda}_0}(x, y)$ satisfies all the requirements of Lemma 1.26.

Using Lemma 1.26, we will later approximate $\bar{\theta}$ by elements of an irreducible component of the zero-set S of $P(x, y)$, $R(x, y)$,

$$S = \left\{(x_0, y_0) \in \mathbf{C}^2 \colon P(x_0, y_0) = Q(x_0, y_0) = 0\right\}. \tag{1.31}$$

The main auxiliary result we use here is the statement that $\bar{\theta}$ is "approximated" by some component S_0 of S. This component S_0 comes from the ideal decomposition

$$(P, R) = \wp_1 \cdots \wp_{\mathcal{K}}$$

into primary components, and S_0 corresponds to the component $\wp_{i_0}$ which is "smallest" at the point $\bar{\theta}\,(= (\pi/\omega, \eta/\omega))$.

However, it is much more convenient not to use the algebraic language of ideals, but to use the geometric language, where irreducible variety stands for the primary ideal, and there is a notion of distance. E.g. one can use the Puiseux expansion to imagine how close a component can lie to a point $\bar{\theta}$.

Our arguments are completely general and are applied to a polynomial in n variables.

For this purpose we devote the next few pages to the description of the type properties of zero-dimensional sets and to the generalization of the Liouville theorem.

2. Let us formulate a version of the Liouville theorem in the case of an arbitary set $S \subset \mathbf{C}^n$ of algebraic numbers of dimension zero, i.e. a set of common zeroes of a zero-dimensional ideal in $\mathbf{Z}[x_1,\ldots,x_n]$. Namely, we consider the following situation. We have an ideal $\mathcal{I}$ in $\mathbf{Z}[x_1,\ldots,x_n]$ which is zero dimensional in the sense that the set

$$\mathcal{S}(\mathcal{I}) = \{\vec{x}_0 \in \mathbf{C}^n \colon P(\vec{x}_0) = 0 \text{ for every } P \in \mathcal{I}\}$$

is a finite set. We are now working in the affine situation since projective considerations do not add anything. Every element of $\mathbb{S}(\mathfrak{J})$ has a prescribed multiplicity, defined e.g. in Shafarevich's book [**12**].

Let $P_1,\ldots,P_k$ be certain generators of $\mathfrak{J}$, whose degrees and types we know,

$$d(P_i) \leqslant D_i, \qquad t(P_i) \leqslant T_i: \qquad i = 1,\ldots,k.$$

Naturally, $k \geqslant n$. By an intersection theory (say, Bezout theorem) we have

$$|\mathbb{S}(\mathfrak{J})| \leqslant D_1 \cdots D_k.$$

This bound is far from optimal whenever $k > n$. In this case we can use even the following bound:

$$|\mathbb{S}(\mathfrak{J})| \leqslant \Big(\max_{i=1,\ldots,k} D_i\Big)^n.$$

Instead of treating different cases, we assume already that $P_1(x),\ldots,P_n(x)$ have only finitely many common zeroes. Then we consider an ideal $I = (P_1,\ldots,P_n)$ and $\mathbb{S}(I)$ instead of $\mathbb{S}(\mathfrak{J})$. Let $\vec{x}_0 \in \mathbb{S}(I)$ and $m(\vec{x}_0)$ be a multiplicity of $\vec{x}_0$ in $\mathbb{S}(I)$. Then we have

$$\sum_{\vec{x}_0 \in \mathbb{S}(I)} m(\vec{x}_0) \leqslant D_1 \cdots D_n.$$

We can apply to I the theory of u-resultants in the form of Kronecker (cf. [**14** or **13**]). One gets the following main statement.

LEMMA 2.1. *In the notation above, the coordinates of common zeroes $\vec{x}_0 \in \mathbb{S}(\mathfrak{J})$ are bounded above in terms of $D_1,\ldots,D_n$ and $T_1,\ldots,T_n$. Namely, let*

$$\vec{x}_0 = (\vec{x}_{10},\ldots,x_{n0}) \quad \text{for } \vec{x}_0 \in \mathbb{S}(I).$$

Then for every $i = 1,\ldots,n$ there exists a rational integer A_i, $A_i \neq 0$, such that for any distinct elements $\vec{x}^1,\ldots,\vec{x}^l$ of $\mathbb{S}(I)$ and $n_i \leqslant m(\vec{x}^i)$: $i = 1,\ldots,l$, the number

$$A_i \cdot \prod_{j=1}^{l} (\vec{x}^j)_i^{n_j}$$

is an algebraic integer. Moreover, for any $i = 1,\ldots,n$ we have

$$(2.2) \qquad |A_i| \cdot \prod_{\vec{x} \in \mathbb{S}(I)} \max\{1, (\vec{x})_i\}^{m(\vec{x})} \leqslant \exp\left(C_1 \cdot \sum_{r=1}^{n} T_r \cdot \prod_{s\neq r, s=1}^{n} D_s\right),$$

for a constant $C_1 > 0$ depending only on n.

The analogue of the Liouville theorem applied to the elements of the set $\mathbb{S}(I)$ has the following form:

LEMMA 2.3. *Let, as before, $I = (P_1,\ldots,P_n)$ where $d(P_i) \leqslant D_i$, $t(P_i) \leqslant T_i$: $i = 1,\ldots,n$, and the set $\mathbb{S}(I)$ is a finite one.*

Let $R(x_1,\ldots,x_n) \in \mathbf{Z}[x_1,\ldots,x_n]$, $R \not\equiv 0$. Let us assume that for several distinct $\vec{x}^1,\ldots,\vec{x}^l$ from $\mathbb{S}(I)$ of multiplicities $m_1,\ldots,m_l$, respectively, we have

$$R(\vec{x}^j) \neq 0: \qquad j = 1,\ldots,l.$$

Then we have the following lower bound:

$$(2.4)\qquad \prod_{j=1}^{l} |R(\vec{x}^j)|^{m_j}$$
$$\geqslant \exp\left\{-C_2\left\{\sum_{i=1}^{n} t(P_i)\,d(R)\cdot \prod_{s\neq i} d(P_i) + t(R)\,d(P_1)\cdots d(P_n)\right\}\right\},$$

where $C_2 > 0$ *depends only on* n.

PROOF OF LEMMA 2.3. We consider the following auxiliary object, taking into account the notations of Lemma 2.1:

$$(2.5)\qquad \mathfrak{R} = \prod_{i=1}^{n} A_i^{d(R)} \prod_{\substack{\vec{x}\in\mathfrak{S}(I)\\ R(\vec{x})\neq 0}} R(\vec{x})^{m(\vec{x})};$$

in (2.5) the product is over only those elements $\vec{x}$ of $\mathfrak{S}(I)$ for which $R(\vec{x}) \neq 0$. This is a usual "seminorm" [**4**]. Then by the definition of the set $\mathfrak{S}(I)$ (invariant under the algebraic conjugation) and the choice of A_i: $i = 1,\ldots,n$, in Lemma 2.1 we get $\mathfrak{R} \in \mathbf{Z}$. From the form of $\mathfrak{R}$ it follows that $\mathfrak{R} \neq 0$, so that

$$(2.6)\qquad |\mathfrak{R}| \geqslant 1.$$

We represent $\mathfrak{R}$ as a product of two factors: $\mathfrak{R} = \mathfrak{A}\cdot\mathfrak{B}$, where $\mathfrak{A}$ is the product in the left-hand side of (2.4). The product $\mathfrak{B}$ is bounded from above by Lemma 2.1,

$$(2.7)\qquad |\mathfrak{B}| \leqslant \left(\prod_{i=1}^{n} A_i \prod_{\substack{\vec{x}\in\mathfrak{S}(I),\\ \vec{x}\notin\{\vec{x}^j:\, j=1,\ldots,l\}\\ R(\vec{x})\neq 0}} \max(1,|(\vec{x})_i|)^{m(\vec{x})}\right)^{d(R)}$$
$$\times \exp\left(2t(R)\prod_{i=1}^{n} d(P_i)\right)$$
$$\leqslant \exp\left(c_4\left\{t(R)\prod_{i=1}^{n} d(P_i) + d(R)\sum_{i=1}^{n} t(P_i)\prod_{s\neq i} d(P_s)\right\}\right).$$

Combining (2.6) and (2.7) one gets (2.4).

In order to prove our results in a straightforward way, we make some agreements on the notation that will simplify our symbols.

If we start, in the general case, with the ideal $I = (P_1,\ldots,P_n)$ in $\mathbf{Z}[x_1,\ldots,x_n]$ of dimension zero, then the set

$$\mathfrak{S}(I) = \{\vec{x}_0 \in \mathbf{C}^n\colon P_i(\vec{x}_0) = 0\colon i = 1,\ldots,n\}$$

is a set of vectors in $\mathbf{C}^n$ with algebraic coordinates. The set $\mathfrak{S}(I)$ is naturally divided into components $\mathfrak{S}_\alpha$ closed under the conjugation in $\overline{\mathbf{Q}}$,

$$(2.8)\qquad \mathfrak{S}(I) = \bigcup_{\alpha\in A} \mathfrak{S}_\alpha.$$

The partition of $\mathbb{S}(I)$ into sets $\mathbb{S}_\alpha$ can be made in such a way that all elements of $\mathbb{S}_\alpha$ have the same multiplicity m_α of occurrence in $\mathbb{S}(I)$. We call $\mathbb{S}_\alpha$ an irreducible component of $\mathbb{S}(I)$ and we have

$$\sum_{\alpha\in A} m_\alpha |\mathbb{S}_\alpha| \leq d(P_1)\cdots d(P_n). \tag{2.9}$$

We can define a type and degree of the component $\mathbb{S}_\alpha$ in (2.8). The degree is naturally $|\mathbb{S}_\alpha|$ itself and the type is defined using the sizes of the coordinates of elements of $\mathbb{S}(I)$.

For this we recall Lemma 2.1, where we had nonzero rational integers A_i: $i = 1,\dots,n$, such that

$$A_i \cdot \prod_{\vec{x}\in S'} (\vec{x})_i^{n(\vec{x})}$$

is an algebraic integer for any $S' \subseteq \mathbb{S}(I)$ and $n(\vec{x}) \leq m(\vec{x})$: $\vec{x} \in S'$, and we have a bound

$$|A_i| \cdot \prod_{\vec{x}\in\mathbb{S}(I)} \max\{1,|(\vec{x})_i|^{m(\vec{x})}\} \leq \exp\left\{C_1 \cdot \sum_{j=1}^{n} t(P_j) \cdot \prod_{s\neq j} d(P_s)\right\}: \tag{2.10}$$

$$i = 1,\dots,n.$$

Naturally, the quantity

$$\log\left\{\prod_{i=1}^{n} |A_i| \cdot \prod_{\vec{x}\in\mathbb{S}(I)} \max\{1,|(\vec{x})_i|\}^{m(\vec{x})}\right\} \tag{2.11}$$

can be called *the size* of $\mathbb{S}(I)$. We can define in a similar way the size of the component $\mathbb{S}_\alpha$ as

$$\log\left\{\prod_{i=1}^{n} |a_i^\alpha| \cdot \prod_{\vec{x}\in\mathbb{S}_\alpha} \max\{1,|(\vec{x})_i|\}\right\}, \tag{2.12}$$

where a_i^α are the smallest nonzero rational integers such that $a_i^\alpha \cdot \prod_{\vec{x}\in S_1}(\vec{x})_i$ are algebraic integers for any $S_1 \subseteq S_\alpha$: $i = 1,\dots,n$.

By the *type* of the set $\mathbb{S}_\alpha$ we understand the sum of the degree $|\mathbb{S}_\alpha|$ and its size. We denote the type of $\mathbb{S}_\alpha$ by $t(\mathbb{S}_\alpha)$. Similarly one can define a type of $\mathbb{S}(I)$ as a sum of its degree and size. The type of $\mathbb{S}(I)$ is also denoted by $t(\mathbb{S}(I))$. We note that the degree of $\mathbb{S}(I)$ is not $|\mathbb{S}(I)|$ but rather $\sum_{\vec{x}\in\mathbb{S}(I)} m(\vec{x})$, when elements of $\mathbb{S}(I)$ are counted with multiplicities.

By the definition of types we have

$$\sum_{\alpha\in A} t(\mathbb{S}_\alpha)\cdot m_\alpha \leq t(\mathbb{S}(I)) \leq \exp\left\{C_2 \sum_{j=1}^{n} t(P_j) \prod_{s\neq j} d(P_s)\right\}. \tag{2.13}$$

We want to remark that our decomposition of $\mathbb{S}(I)$ into the union of (disjoint) irreducible components is not unique. It is easier to work with our definition; one can define a canonical decomposition of $\mathbb{S}(I)$ into the maximal union of irreducible components over **Z**.

The case $n = 2$ is very easy to understand using resultants. In this case we have two relatively prime polynomials $P(x, y)$, $Q(x, y) \in \mathbf{Z}[x, y]$ and the set S of their common zeroes

$$S = \{(x, y) \in \mathbf{C}^2: P(x, y) = Q(x, y) = 0\}.$$

Their coordinates can be determined using the resultant of $P(x, y)$, $Q(x, y)$. We make a change of the coordinates to a "normal" [**1a**] form and get new polynomials $P(x', y')$, $Q(x', y')$. In "normal" coordinates defined in [**1**] distinct elements of S have both their coordinates distinct.

Then resultants $R_1(x')$ and $R_2(y')$ of $P'(x', y')$, $Q'(x', y')$ (in "normal" coordintes [**1a**]) with respect to y' and x', respectively, can be writtren as

$$(2.14) \quad R_1(x') = a_1 \prod_{i=1}^{k} (x' - \zeta'_{1,i})^{m_i}, \qquad R_2(y') = a_2 \prod_{i=1}^{k} (y' - \zeta'_{2,i})^{m_i},$$

where $(\zeta'_{1,i}, \zeta'_{2,i})$ is an element of S of multiplicity m_i. In particular, one can represent $R_1(x')$, $R_2(y')$ in terms of the powers of irreducible polynomials,

$$R_1(x') = \prod_{\alpha \in A} P^1_\alpha(x')^{m_\alpha}, \qquad R_2(y') = \prod_{\alpha \in A} P^2_\alpha(y')^{m_\alpha},$$

where $P^1_\alpha(x') = a^\alpha_1 \prod_j (x' - \zeta^\alpha_{1,j})$ and $P^2_\alpha(y') = a^\alpha_2 \prod_j (y' - \zeta^\alpha_{2,j})$ and $(\zeta^\alpha_{1,j}, \zeta^\alpha_{2,j})$ are elements of S. We naturally define

$$S_\alpha = \{(\zeta^\alpha_{1,j}, \zeta^\alpha_{2,j})\},$$

so that $\bigcup_{\alpha \in A} S_\alpha = S$. We can now look on the types of S and S_α: $\alpha \in A$, from the point of view of resultants written in (2.14).

E.g. one can define a size of S as the sum of sizes of R_1 and R_2. This will be equivalent to the previous definition of the size in (2.11) and (2.12).

Consequently we can say

$$t(S) \leq t(R_1) + t(R_2),$$

while for $\alpha \in A$,

$$t(S_\alpha) \leq t(P^1_\alpha) + t(P^2_\alpha).$$

This definition of the type, not expressed in the coordinate form, is more useful in the higher-dimensional case, when we are working with mixed ideals, nonnormal intersections, etc.

The introduced notion of component of a zero-dimensional set is used for solving the main difficulty in the proof. Namely, we are dealing with the situation when $S = S(I)$ is a zero-dimensional set for an ideal $I = (P, Q)$ in $\mathbf{Z}[x, y]$ and the algebraic manifold I is "close" to the point $\bar{\theta} \in \mathbf{C}^2$ in the sense that $|P(\bar{\theta})| < \varepsilon$, $|Q(\bar{\theta})| < \varepsilon$. For the proof it is necessary to replace this notion of "closeness" with an estimate of the distance from $\bar{\theta}$ to elements of S (in the usual, say l^1, metric) in terms of ε and $t(P)$, $t(Q)$. The best possibility would be to have a single element of S close to $\bar{\theta}$ (as it is usual for one-dimensional proofs, see [**15**]).

However the rigorous proof of the existence of a single element of S, approximateing $\bar{\theta}$ is nonelementary and requires the use of "generalized Jacobians" (for $\mathbf{Z}[x_1, \ldots, x_n]$ at $n = 2$) or resolutions of singularities (for $n > 2$). Instead we use the elementary method and we are able to prove only the existence of a single component $\mathbb{S}_\alpha$ of S approximating $\bar{\theta}$ in the sense that $\Pi_{\bar{\zeta} \in \mathbb{S}_\alpha} \|\bar{\theta} - \bar{\zeta}\| \sim \varepsilon$ (see the discussion in [1a]). Surprisingly with this weak statement we are able to finish the proof with the same length of exposition as with the sharper assertion (cf. [16]). The main tool is the Liouvill theorem in the form of Lemma 2.3 above.

Hence we devote special attention to a precise formulation of the result of an approximation of $\bar{\theta}$ by a component $\mathbb{S}_\alpha$ of $S(I)$. We follow arguments from [1a], [1b] and give the corresponding result for zero-dimensional sets in $\mathbf{C}^2$. The same proof works without any changes for arbitrary zero-dimensional sets in $\mathbf{C}^n$. In the proof we use two elementary lemmas from [1] that we included in the proof to make it self-contained. The reader should excuse us for the lengthy formulation of Proposition 2.15 but this way we have a single auxiliary result for further references.

The main auxiliary result is the following.

PROPOSITION 2.15. *Let* $\bar{\theta} = (\theta_1, \theta_2) \in \mathbf{C}^2$ *(with the* l^1*-norm) and* $P(x, y)$, $Q(x, y)$ *be two relatively prime polynomials from* $\mathbf{Z}[x, y]$.

Let $S = S(P, Q)$ *be the set of the zeroes of an ideal* (P, Q) *and* $S = \bigcup'_{\alpha \in A} S_\alpha$ *its representation through irreducible components. If* m_α *is a multiplicity of* S_α, *then one has*

$$\sum_{\alpha \in A} m_\alpha t(S_\alpha) \leq t(S) \leq 4(d(P)t(Q) + d(Q)t(P)) \leq 8t(P)t(Q).$$

Let us assume now that

$$(2.16) \qquad \max\{|P(\bar{\theta})|, |Q(\bar{\theta})|\} \leq \exp(-E)$$

for $E > 0$. One can define the logarithmic distance $E(\bar{\zeta})$ of $\bar{\zeta} \in S$ to $\bar{\theta}$ as

$$(2.17) \qquad \|\bar{\theta} - \bar{\zeta}\| \leq \exp(-E(\bar{\zeta})).$$

Let us put

$$(2.18) \quad \mathcal{E}(S, \bar{\theta}) = \sum_{\bar{\zeta} \in S, E(\bar{\zeta}) > 0} E(\bar{\zeta}) \quad \text{and} \quad \mathcal{E}(S_\alpha, \bar{\theta}) = \sum_{\bar{\zeta} \in S_\alpha, E(\bar{\zeta}) > 0} E(\bar{\zeta})$$

as the definition of (minus logarithm of) distance from $\bar{\theta}$ to S or S_α.

In order to express relations between E, $\mathcal{E}(S, \bar{\theta})$ and $\mathcal{E}(S_\alpha, \bar{\theta})$ we put

$$(2.19) \quad T = \gamma_0(d(P)t(Q) + d(Q)t(P) + d(P)d(Q)\log(d(P)d(Q) + 2))$$

for an absolute constant $\gamma_0 > 0$ ($\gamma_0 \leq 4$) such that $T(S) \leq T$.

Similarly for every $\alpha \in A$ there is a bound T_α of $t(S_\alpha)$ of the form

$$(2.20) \qquad t(S_\alpha) \leq T_\alpha \leq t(S_\alpha) + d(S_\alpha)\log(d(P)d(Q) + 2).$$

In terms of T_α we can formulate results on $\mathfrak{S}(S,\bar\theta)$ and $\mathfrak{S}(S_\alpha,\bar\theta)$. If we assume $E \geqslant 4T$, then

$$\mathfrak{S}(S,\bar\theta) = \sum_{\alpha\in A} m_\alpha \mathfrak{S}(S_\alpha,\bar\theta) \geqslant E - 2T. \tag{2.21}$$

We denote the element of S_α nearest to $\bar\theta$ by $\bar\xi_\alpha$, $E(\bar\xi_\alpha) = \min_{\bar\zeta\in S_\alpha} E(\bar\zeta)$. Then we have

$$E(\bar\xi_\alpha) \geqslant \min\left\{c_5\mathfrak{S}(S_\alpha,\bar\theta),\, c_5\frac{\mathfrak{S}(S_\alpha,\bar\theta)^2}{d(S_\alpha)T_\alpha}\right\} \tag{2.22}$$

and, for other $\bar\zeta \in S_\alpha$ close to $\bar\theta$ we have

$$\sum_{\bar\xi\in S_\alpha,\, E(\bar\xi)\geqslant B_\alpha} E(\bar\xi) \geqslant \frac{\mathfrak{S}(S_\alpha,\bar\theta)}{4} \tag{2.23}$$

where

$$B_\alpha = c_6\mathfrak{S}(S_\alpha,\bar\theta)^{3/2}\cdot\left(d(S_\alpha)T_\alpha^{1/2}\right)^{-1} \quad \text{for } \alpha\in A. \tag{2.24}$$

As an application of these bounds we have the following result, where $d(S_\alpha)$ is replaced by its upper bound T_α. We remark, that in the addition to (2.20),

$$\sum_{\alpha\in A} m_\alpha T_\alpha \leqslant T. \tag{2.25}$$

In particular, there exists $\alpha_0 \in A$ such that

$$\frac{\mathfrak{S}(S_{\alpha_0},\bar\theta)}{T_{\alpha_0}} \geqslant \frac{\mathfrak{S}(S,\bar\theta)}{T} \geqslant \frac{E}{T} - 2. \tag{2.26}$$

Under the conditions (2.26) one has as a corollary of (2.22)–(2.24):

$$E(\bar\xi_{\alpha_0}) \geqslant \min\left\{c_5\mathfrak{S}(S_{\alpha_0},\bar\theta),\, c_5\frac{\mathfrak{S}(S_{\alpha_0},\bar\theta)^2}{T_{\alpha_0}^2}\right\}; \tag{2.27}$$

$$\sum\left\{E(\bar\xi) : \bar\xi\in S_{\alpha_0},\, E(\bar\xi) \geqslant c_6\left(\frac{\mathfrak{S}(S_{\alpha_0},\bar\theta)}{T_{\alpha_0}}\right)^{3/2}\right\} \geqslant \mathfrak{S}(S_{\alpha_0},\bar\theta)/4.$$

PROOF OF THE PROPOSITION 2.15. First of all we must change the system of the coordinates to a "normal" one with respect to a system of polynomials $P(x, y)$, $Q(x, y)$. We use for this Lemma 2.1 **[1b]** and 3.2 **[1a]**. According to this lemma there is a *nonsingular* transformation

$$(\pi)\qquad \begin{cases} x = x'a + y'c, \\ y = x'b + y'd \end{cases}$$

for rational integers a, b, c, d such that

$$\max(|a|,|b|,|c|,|d|) \leqslant M \leqslant \gamma_1 d(P)\, d(Q), \tag{2.28}$$

and which is normal with respect to $P(x, y)$, $Q(x, y)$. It means, from our point of view, first of all that for $\bar{\theta}$ written in new coordinates (x', y') as $\bar{\theta}' = (\theta_1', \theta_2')$ and for *any* common zero $\bar{\zeta}$ of $P(x, y) = 0$, $Q(x, y) = 0$, i.e. element of S, we have in new coordintes $\bar{\zeta}' = (\zeta_1', \zeta_2')$

$$(2.29) \qquad \|\bar{\theta}' - \bar{\zeta}'\|_{l^1} \leqslant 4M^2 \min\{|\theta_1' - \zeta_1'|, |\theta_2' - \zeta_2'|\}.$$

Of all the "normality" properties, the property (2.29) is, certainly, the central one that we use.

In new, "normal" variables, we consider the resultant $R(x')$ of $P'(x', y')$ $(\equiv P(x, y))$ and $Q'(x', y')$ $(\equiv Q(x, y))$ taken with respect to y'. The polynomial $R(x')$ is a polynomial of degree $\leqslant d(P)\,d(Q)$, but the type of $R(x')$ might be slightly higher than that of $R(x)$. This explains why we change the type $t(S)$ by a slightly higher quantity T defined in (2.19). Namely, we take S written in new variables (x', y') as a set S'. Our initial definition of the type was certainly "coordinate-depending" and $t(S')$ may be different from $t(S)$. In order to get the bound of $t(S')$ one can take together with $R(x')$, the polynomial $R_2(y')$, being the resultant of $P'(x', y')$, $Q'(x', y')$ with respect to x'. Then one can define $T = t(S')$ as $t(S') = t(R) + t(R_2)$.

This quantity is certainly bounded by the number T from (2.19):

$$\begin{aligned} t(R) + t(R_2) &\leqslant \gamma_0\{d(P)t(Q) + d(Q)t(P) \\ &\qquad + d(P)\,d(Q)\log(d(P)\,d(Q) + 2)\} \\ &= T. \end{aligned}$$

The term $\log(d(P)\,d(Q) + 2)$ appears in (2.19) since we make the transformation (π) which changes the type of the polynomials à la (2.28).

This factor can be removed, however, from all the statements of 2.15, by more tedious computations. E.g. it is rather easy to get the property (2.22) without studying the transformed resultants, working only with $\mathrm{res}_y(P, Q)$ and $\mathrm{res}_x(P, Q)$.

We prefer, however, to work with T and T_α because the proof is straightforward, though some estimates are suffering.

Now irreducible components $S_\alpha' = \pi^{-1}(S_\alpha)$ (we are writing now in transformed coordinates) are connected with irreducible components of $R(x')$.

Namely, we have

$$R(x') = \prod_{\alpha \in A} P_\alpha(x')^{m_\alpha}$$

where $P_\alpha(x')$ is an irreducible polynomial from $\mathbf{Z}[x']$. Here zeroes of $P_\alpha(x')$ are exactly x'-projection of the set S_α': $\alpha \in A$.

Hence we can work with $P_\alpha(x')$ rather than with $S_\alpha' = \pi^{-1}(S_\alpha)$. According to (2.29) it is enough to bound $|\zeta_1' - \theta_1'|$ from above in order to bound $\|\bar{\zeta}' - \bar{\theta}'\|$ with $\bar{\zeta}' = (\zeta_1', \zeta_2') \in S'$. Here and everywhere in the proof of Proposition 2.15, $\bar{\zeta}' = \pi^{-1}(\bar{\zeta})$, $\bar{\theta}' = \pi^{-1}(\bar{\theta})$.

First of all we can evaluate $\log|R(\theta_1')|$ in terms of

$$E \leqslant -\min\{\log|P(\bar{\theta})|, \log|Q(\bar{\theta})|\}.$$

For this we use the property of the resultants in the classical form.

Namely, we can use the formula from [8]:

$$|\operatorname{res}_x(p,q)| \leqslant \{d(p)H^+(q)|q(x_0)| + d(q)H^+(p)|p(x_0)|\} \times H^+(q)^{d(p)-1} \cdot H^+(p)^{d(q)-1}$$

for arbitrary polynomials $p(x)$, $q(x) \in \mathbf{C}[x]$ and $H^+(p) = \max\{1, H(p)\}$. We put $p(x') = P'(\theta_1', x')$, $q(x') = Q'(\theta_1', x')$ and $x_0 = \theta_2'$.

This way we obtain the inequality

$$\log|R(\theta_1')| \leqslant -E + T. \tag{2.30}$$

In particular, writing $R(x')$ in the form

$$R(x') = a \prod_{\alpha \in A} (x' - \zeta_1'^{\alpha})^{m_\alpha} = a \prod_{\bar{\zeta}' \in S'} (x' - (\bar{\zeta})_1)$$

for $a \in \mathbf{Z}$, and using (2.29) one immediately obtains (2.21).

In order to get other statements (2.22)–(2.24) we simply use the irreducible polynomial $P_\alpha(x')$, whose zeroes are the x'-projection of S' and apply to the polynomial $P_\alpha(x')$ the statement of the following lemma taken from [**1a**, **1b**]:

LEMMA 2.31 [**1a**, **1b**]. *Let $P(x) \in \mathbf{Z}[x]$ be an irreducible polynomial over $\mathbf{Q}$ and*

$$P(x) = a \prod_{i=1}^{d(P)} (x - \eta_i).$$

Let us assume that $\theta \in \mathbf{C}$ and

$$|P(\theta)| \leqslant \exp(-E). \tag{2.32}$$

We arrange zeroes in an order such that

$$|\theta - \eta_i| < 1 \quad \textit{for } i = 1, \dots, r,$$

and put

$$|\theta - \eta_i| = \exp(-e_i): \qquad i = 1, \dots, r, \tag{2.33}$$

and we assume $e_1 \geqslant \cdots \geqslant e_r > 0$. We then have

$$\sum_{i=1}^{r} e_i \geqslant E, \qquad \sum_{i=1}^{r} (i-1)e_i \leqslant 8d(P)t(P). \tag{2.34}$$

If now, $E \geqslant c_7 d(P)t(P)$, then

$$e_1 \geqslant c_8 E \quad \textit{for some } c_8 = c_8(c_7) > 0. \tag{2.35}$$

If, however, $E < c_7 d(P)t(P)$ for a sufficiently small $c_7 > 0$, then we have

$$E > c_9 E^2/d(P)t(P): \qquad c_9 > 0, \tag{2.36}$$

and moreover,

$$\sum_{\substack{j=1\\ e_j \geqslant B}}^{r} e_j \geqslant E/4 \quad \textit{for } B = c_{10}E^{3/2}\left(d(P)t(P)^{1/2}\right)^{-1}. \tag{2.37}$$

REMARK. Here, in order to exclude trivial cases, we assume $E \geqslant c_{11}t(P)$.

PROOF. We define a resultant $\Delta = \Delta(P)$ of $P(x)$,

$$\Delta = a^{2d(P)} \prod_{\substack{i,j=1\\ i\neq j}}^{d(P)} (\eta_i - \eta_j). \tag{2.38}$$

Then $\Delta \in \mathbf{Z}$ and $\Delta \neq 0$, so that $|\Delta| \geqslant 1$.

From (2.32) we immediately obtain

$$\sum_{i=1}^{r} e_i \geqslant E. \tag{2.39}$$

Now keeping in mind the bound

$$|a| \prod_{i=1}^{d(P)} \max\{1, |\eta_i|\} \leqslant (d(P)+1)H(P),$$

from the definition of the resultant $\Delta(P)$ we obtain

$$-\frac{\log|\Delta|}{2} + \sum_{i=1}^{r} e_i(i-1) \leqslant T_1 = 8d(P)t(P). \tag{2.40}$$

We deduce (2.34) from (2.39) and (2.40). The bound (2.36) follows from (2.39)–(2.40) as in Lemma 2.3 [**1a**].

In order to prove (2.37) we put

$$J_C = \{j = 1,\ldots,r\colon e_j < C\}.$$

Let $A = E/8\mathrm{d}(P)$. If

$$\sum_{\substack{j=1\\ j\notin J_A}}^{r} e_j \geqslant 3E/4 \tag{2.41}$$

is not true, then

$$\sum_{\substack{j=1\\ j\in J_A}}^{r} e_j \geqslant E/4.$$

Then

$$|J_A| \cdot A \geqslant \sum_{\substack{j=1\\ j\in J_A}}^{r} e_j \geqslant E/4.$$

But

$$|J_A| \leqslant r \leqslant d(P),$$

i.e. $A \geqslant E_1/4d(P)$, which is impossible. We define B as in (2.37),

$$B = c_{10} E^{3/2}\left(d(P)t(P)^{1/2}\right)^{-1}$$

for some small constant $c_{10} > 0$. Then $B \geqslant A$ according to the previous remark.

Now we want to show (2.37). If (2.37) is not true, then by (2.41),

$$\sum_{\substack{j=1 \\ j \in J_B \setminus J_A}}^{r} e_j \geqslant E/2. \tag{2.42}$$

But according to (2.34) we have

$$|J_B \setminus J_A|^2 \cdot A \leqslant c_{12}\, d(P)t(P).$$

On the other hand, by (2.42),

$$|J_B \setminus J_A| \cdot B \geqslant E/2.$$

Thus

$$B \geqslant \frac{E^{3/2}}{16 d(P)t(P)},$$

which contradicts the choice of B with $c_{10} < 1/16$. Q.E.D.

At last, we derive the statements (2.26)–(2.27). Indeed, if (2.26) is false for every $\alpha_0 \in A$, we get

$$\sum_{\alpha \in A} \mathfrak{S}(S_\alpha, \bar{\theta})m_\alpha \cdot T < \sum_{\alpha \in A} T_\alpha \cdot m_\alpha \cdot \mathfrak{S}(S, \bar{\theta}),$$

which contradicts (2.25). Results (2.27) are a consequence of (2.22)–(2.24). Proposition 2.5 is proved.

REMARK 2.43. In (2.27), the quantity $E(\bar{\xi}_{\alpha_0}) \leqslant -\log\|\bar{\theta} - \bar{\xi}_{\alpha_0}\|$ is bounded below by $c_5 \min\{\mathfrak{S}(S_{\alpha_0}, \bar{\theta}), (\mathfrak{S}(S_{\alpha_0}, \bar{\theta})/T_{\alpha_0})^2\}$. From these two terms, "usually" (in practice), $(\mathfrak{S}(S_{\alpha_0}, \bar{\theta})/T_{\alpha_0})^2$ is a smaller one. Indeed, if $\mathfrak{S}(S_{\alpha_0}, \bar{\theta})$ is large, then (2.27) shows simply that "the bulk" of the measure $\mathfrak{S}(S_{\alpha_0}, \bar{\theta})$ is concentrated in a single term of it, $E(\bar{\xi}_{\alpha_0})$: $E(\bar{\xi}_{\alpha_0}) \geqslant c_5 \mathfrak{S}(S_{\alpha_0}, \bar{\theta})$.

However, one must take even this possibility into consideration.

3. We return now to the set S of common zeroes of $P(x, y)$, $R(x, y)$ constructed in Lemma 1.26 of §1. First of all we state the bound for the degree and size of the set S as defined in (1.31) and then we use Proposition 2.15 to obtain an irreducible component S_0 of S approximating $\bar{\theta}$.

Now the set S has the degree $d(S)$ (the number of elements counted with their multiplicities) bounded by

$$d(S) \leqslant c_1' \cdot d^{4/3} \varphi(d)^{1/3} \log^{1/3} d, \tag{3.1}$$

and its type bounded by

$$t(S) \leqslant c_2' \cdot t\, d^{1/3} \varphi(d)^{1/3} \log^{1/3} d. \tag{3.2}$$

Now we can get an irreducible component S_0 of S approximating $\bar{\theta}$.

LEMMA 3.3. *Let us denote*

$$\mathfrak{T}(S) = \gamma_0(t(S) + d(S)\log(d(S) + 2)) \tag{3.4}$$

for $\gamma_0 \leqslant 2$ *and denote for* $\bar{\xi} \in S$,

$$\|\bar{\theta} - \bar{\xi}\| = \exp(-E(\bar{\xi}))$$

if $\bar{\xi} \in S$. *There exists an irreducible component* S_0 *of* S *such that the following properties are satisfied. If*

$$\mathfrak{T}(S_0) = \gamma_0(t(S_0) + d(S_0)\log(d(S) + 2)),$$

then for

$$\sum\{E(\bar{\xi}) : \bar{\xi} \in S_0,\ E(\bar{\xi}) > 0\} \overset{\text{def}}{=} \mathcal{E}_0 \tag{3.5}$$

and for $\bar{\xi}_0 \in S$ *closest to* $\bar{\theta}$: $\|\bar{\xi}_0 - \bar{\theta}\| \leqslant \|\bar{\xi} - \bar{\theta}\|$, $\bar{\xi} \in S_0$, *we have*

$$\sum\left\{E(\bar{\xi}) : \bar{\xi} \in S_0,\ E(\bar{\xi}) \geqslant c_3' \cdot \frac{\mathcal{E}_0^{3/2}}{d(S_0)\mathfrak{T}(S_0)^{1/2}}\right\} \geqslant \frac{\mathcal{E}_0}{4}; \tag{3.6}$$

$$E(\bar{\xi}_0) \geqslant c_4' \cdot \mathcal{E}_0^2/d(S_0)\mathfrak{T}(S_0).$$

Moreover, the component S_0 *is chosen in a way such that*

$$\mathcal{E}_0/\mathfrak{T}(S_0) \geqslant c_5' \cdot \frac{t \cdot \varphi(d)}{\mathfrak{T}(S)}. \tag{3.7}$$

After the component S_0 is chosen, our strategy is very simple. We take some (L_1, L_2) and use the auxiliary function $F_{L_1,L_2}(z)$ in order to show that the approximation of $\bar{\theta}$ by elements of S_0, given by Lemma 3.3, is too good, and contradicts, to a Small Value Lemma 1.10.

The choice of (L_1, L_2) is straightforward. We demand

$$L_1^2 L_2 = \lambda E_0 \overset{\text{def}}{=} \lambda E(\bar{\xi}_0) \tag{3.8}$$

for a sufficiently small λ, $\lambda > 0$.

The definition of L_1, L_2 can be given as follows:

DEFINITION 3.9. *We put*

$$L_1 = \min\left\{d(S_0), \left[\left(\frac{\lambda E_0\, d(S_0)\log d(S_0)}{\mathfrak{T}(S_0)}\right)^{1/3}\right]\right\}$$

and

$$L_2 = \left[\frac{\lambda E_0}{L_1^2}\right] + 1.$$

LEMMA 3.10. *We have* $L_2 \geqslant L_1$ *and the following important inequalities hold. For a sufficiently large* $C > 0$ *we have*

$$\mathcal{E}_0\lambda > C \cdot \mathfrak{T}(S_0)L_1 \geqslant C \cdot t(S_0)L_1 \tag{3.11}$$

and

$$\mathfrak{E}_0\lambda > C\cdot d(S_0)L_2\log L_1. \tag{3.12}$$

PROOF OF LEMMA 3.10. First of all we must show that $L_2 \geqslant L_1$, i.e. that the definition of the pair (L_1, L_2) is legitimate. Indeed, if $L_1 \leqslant d(S_0)$, then this is the consequence of Definition 3.9 and $\mathfrak{T}(S_0) > d(S_0)\log d(S_0)$. Now let $L_1 \geqslant d(S_0)$, i.e. $L_1 = d(S_0)$. We have in this case $\lambda E_0 d(S_0)\log d(S_0) \geqslant \mathfrak{T}(S_0)\,d(S_0)^3$, which implies $L_2 \geqslant L_1$.

Let us show other statements (3.11)–(3.12) of Lemma 3.10. First of all, if $L_1 \geqslant d(S_0)$, then (3.11)–(3.12) are obvious; the first follows from the definition of L_1; the second is obvious if $d(S_0)$ (or d) is sufficiently large.

Thus we can assume $L_1 \leqslant d(S_0)$. By the definition of $\mathfrak{E}_0$ we have

$$\mathfrak{E}_0/\mathfrak{T}(S_0) \geqslant c_6'\cdot\mathfrak{E}/\mathfrak{T}(S),$$

where $\mathfrak{E} = t\varphi(d)$. We have also $\mathfrak{T}(S_0) \leqslant \mathfrak{T}(S)$, $d(S_0) \leqslant d(S)$. Consequently, by the bounds of $\mathfrak{T}(S)$, $d(S)$ in terms of t, d, we get

$$\mathfrak{E}_0^2 \geqslant c_7'\cdot\mathfrak{T}(S_0)^2\,d(S_0)\log d(S_0)\cdot\varphi(d)/d^2\log^2 d. \tag{3.13}$$

Since $\varphi(d) > C\cdot d^2\log^2 d$ for a sufficiently large $C > 0$, we get (3.11) from (3.13), if one bounds L_1 by a larger quantity, replacing E_0 by $\mathfrak{E}_0$.

At the same time, (3.13) implies (3.12) as the definition of L_2 shows.

4. Let us consider now the auxiliary function $F(z)$ that corresponds to the choice (L_1, L_2) from Definition 3.9 above. There are several possible cases to consider, depending on whether $C_{\lambda_1,\lambda_2}(\bar{\xi}_0) = 0$ for all λ_1, λ_2: $0 \leqslant \lambda_1 \leqslant L_1$, $0 \leqslant \lambda_2 \leqslant L_2$, or not.

Instead of considering several possibilities, we consider at once the general situation. We work now with the polynomials $C_{\lambda_1,\lambda_2}(x, y)$ ($0 \leqslant \lambda_1 \leqslant L_1$, $0 \leqslant \lambda_2 \leqslant L_2$), meaning that they correspond to the pair (L_1, L_2) from Definition 3.9 only.

Since not all polynomials $C_{\lambda_1,\lambda_2}(x, y)$ are zeroes (and this is the *only* property of $C_{\lambda_1,\lambda_2}(x, y)$ we use so far), there is always a partial derivative $\partial C_{\lambda_1,\lambda_2}(x, y)$ of some $C_{\lambda_1,\lambda_2}(x, y)$, which is nonzero at $\bar{\xi}_0$. Since the degree of $C_{\lambda_1,\lambda_2}(x, y)$ is bounded by L_1, let us denote by k_0, $0 \leqslant k_0 < L_1$, the largest integer such that

$$\frac{\partial^{k_1+k_2}}{\partial x^{k_1}\partial y^{k_2}} C_{\lambda_1,\lambda_2}(x, y)\,|_{(x,y)=\bar{\xi}_0} = 0 \tag{4.1}$$

for all rational integers k_1, k_2: $0 \leqslant k_1$, $0 \leqslant k_2$, such that $k_1 + k_2 \leqslant k_0$ and for all λ_1, λ_2: $0 \leqslant \lambda_1 \leqslant L_1$, $0 \leqslant \lambda_2 \leqslant L_2$.

In other words, there exists $\vec{k^0} \in \mathbf{N}^2$ such that $\|\vec{k^0}\|_{l^1} = k_0 + 1$ and

$$\partial^{\vec{k^0}}_{(x,y)} C_{\lambda_1^0,\lambda_2^0}(x, y)\,|_{(x,y)=\bar{\xi}_0} \neq 0 \tag{4.2}$$

for some λ_1^0, λ_2^0: $0 \leqslant \lambda_1^0 \leqslant L_1$, $0 \leqslant \lambda_2^0 \leqslant L_2$, while (4.1) is satisfied. Now, instead of the coefficients $C_{\lambda_1,\lambda_2}(\pi/\omega, \eta/\omega)$ of $F(z)$ we use different coefficients, changing, consequently, the auxiliary function $F(z)$. Namely, we define new polynomials $\tilde{C}_{\lambda_1,\lambda_2}(x, y)$ instead of $C_{\lambda_1,\lambda_2}(x, y)$,

$$\tilde{C}_{\lambda_1,\lambda_2}(x, y) = \frac{1}{\vec{k}^0!} \frac{\partial^{\|\vec{k}^0\|}}{\partial x^{k_1^0} \partial y^{k_2^0}} C_{\lambda_1,\lambda_2}(x, y), \tag{4.3}$$

where $\vec{k}^0 = (k_1^0, k_2^0)$ and, as usual, $\vec{k}^0! = k_1^0! k_2^0!$. Consequently we define new coefficients as $\tilde{C}_{\lambda_1,\lambda_2}(\bar{\xi}_0)$ and a new function $\tilde{F}(z)$ as

$$\tilde{F}(z) = \sum_{\lambda_1=0}^{L_1} \sum_{\lambda_2=0}^{L_2} \tilde{C}\lambda_1,\lambda_2(\bar{\xi}_0)\left(\zeta(z) - \frac{\eta}{\omega} z\right)^{\lambda_1} \wp(z)^{\lambda_2}. \tag{4.4}$$

This function becomes our new auxiliary function corresponding to a pair $(L_1, L_2) \in \mathfrak{L}$.

Let us list some of the properties of $\tilde{F}(z)$ and $\tilde{C}_{\lambda_1,\lambda_2}(x, y)$.

CLAIM 4.5. Polynomials $\tilde{C}_{\lambda_1,\lambda_2}(x, y)$ are polynomials from $\mathbf{Z}[x, y]$, not all zeroes, of degrees $\leqslant L_1$ and of type $\leqslant c_2 \cdot L_2 \log L_1 + L_1 \log 4$. In particular,

$$\log |\tilde{C}_{\lambda_1,\lambda_2}(\bar{\xi}_0)| \leqslant c_8' \cdot L_2 \log L_1 \tag{4.6}$$

for all $0 \leqslant \lambda_1 \leqslant L_1$, $0 \leqslant \lambda_2 \leqslant L_2$. In addition to (4.6), $\tilde{C}_{\lambda_1^0,\lambda_2^0}(\bar{\xi}_0) \neq 0$ according to (4.2).

The statement of Claim 4.5 follows from the definition (4.3) and Lemma 4.5, defining polynomials $C_{\lambda_1,\lambda_2}(x, y)$.

The function $\tilde{F}(z)$ has a lot of interesting features as well. First of all we look at the equations defining $C_{\lambda_1,\lambda_2}(x, y)$ in (4.7),

$$\sum_{\lambda_1=0}^{L_1} \sum_{\lambda_2=0}^{L_2} C_{\lambda_1,\lambda_2}(x, y) \mathfrak{F}_{\lambda_1,\lambda_2,k,m}(x, y) = 0 \tag{4.7}$$

for $k = 0, 1, \ldots, [c_3 L_2]$, $m = 0, 1, \ldots, [c_4 L_1]$. The system of equations (4.7) defines $C_{\lambda_1,\lambda_2}(x, y)$ for a "generic" (x, y); $\tilde{C}_{\lambda_1,\lambda_2}$ is the specialization of C_{λ_1,λ_2}, adjusted to the "nongeneric" point $\bar{\xi}_0$.

Indeed, we have

CLAIM 4.8. For the system $\tilde{C}_{\lambda_1,\lambda_2}(\bar{\xi}_0)$ of complex numbers, not all of which are zeroes, the system of linear equations (4.7) at $(x, y) = \bar{\xi}_0$ is satisfied

$$\sum_{\lambda_1=0}^{L_1} \sum_{\lambda_2=0}^{L_2} C_{\lambda_1,\lambda_2}(\bar{\xi}_0) \cdot \mathfrak{F}_{\lambda_1,\lambda_2,k,m}(\bar{\xi}_0) = 0 \tag{4.9}$$

for $k = 0, 1, \ldots, [c_3 L_2]$, $m = 0, 1, \ldots, [c_4 L_1]$.

To prove (4.9) we apply the derivative $\partial^{\|\vec{k}^0\|}/\partial x^{k_1^0}\partial y^{k_2^0}$ to the system of equations (4.7) and then specialized (x, y) by $(x, y) = \bar{\xi}_0$. Using the definition (4.2) of $\vec{k}^0$ we get

$$\sum_{\lambda_1=0}^{L_1}\sum_{\lambda_2=0}^{L_2}\frac{\partial^{\|\vec{k}^0\|}}{\partial x^{k_1^0}\partial y^{k_2^0}}C_{\lambda_1,\lambda_2}(x, y)\cdot\mathcal{F}_{\lambda_1,\lambda_2,k,m}(x, y)\big|_{(x, y)=\bar{\xi}_0} = 0:$$
$$k = 0, 1, \ldots, [c_3L_2], m = 0, 1, \ldots, [c_4L_1].$$

This means, according to (4.3), exactly the statement (4.9). The statement (4.9) does not mean, however, that the function $\tilde{F}(z)$ has zeroes at $z = \omega/4 + m\omega'$, because by (4.2)

$$\tilde{F}^{(k)}\left(\frac{\omega}{4} + m\omega'\right) = \sum_{\lambda_1=0}^{L_1}\sum_{\lambda_2=0}^{L_2}\tilde{C}_{\lambda_1,\lambda_2}(\bar{\xi}_0)\cdot\mathcal{F}_{\lambda_1,\lambda_2,k,m}\left(\frac{\pi}{\omega}, \frac{\eta}{\omega}\right) \tag{4.10}$$

and, $\bar{\xi}_0 \neq (\pi/\omega, \eta/\omega) = \bar{\theta}$. However, by the choice of $\bar{\xi}_0$, $\bar{\xi}_0$ is rather close to $\bar{\theta} = (\pi/\omega, \eta/\omega)$, which shows that $\tilde{F}(z)$ has *small* values at $\omega/4 + m\omega'$. Namely, we can make a

CLAIM 4.11. For the function $\tilde{F}(z)$ defined in (4.4) we have $\tilde{F}(z) \not\equiv 0$ and for $k = 0, 1, \ldots, [c_3L_2]$, $m = 0, 1, \ldots, [c_4L_1]$ we have

$$\log\left|\frac{1}{k!}\tilde{F}^{(k)}\left(\frac{\omega}{4} + m\omega'\right)\right| < -E_0 + c_9'L_2\log L_1. \tag{4.12}$$

For the proof of Claim 4.11 it is enough to remark that for every polynomial $Q(x, y) \in \mathbf{C}[x, y]$ of the type $\leqslant T$ we have

$$|Q(\bar{\theta})| \leqslant |Q(\bar{\xi}_0)| + \|\bar{\theta} - \bar{\xi}_0\|\exp(\gamma_3 T), \tag{4.13}$$

where $\gamma_3 > 0$.

Now we can apply this remark (4.13) to (4.9) and (4.10).

However in view of our care to the degree in the estimates, this may not be quite satisfactory; because the type of polynomials $\mathcal{F}_{\lambda_1,\lambda_2,k,m}(x, y)$ is bounded by (4.4) by a quantity $c_{10}'L_2 + c_3L_2\log L_2$ rather than by $c_{11}' \cdot L_2\log L_1$. We must use the change of variables and (G, C)-functions. According to this change of variables, $C_{\lambda_1,\lambda_2}(x, y)$ satisfy another system of equations, equivalent to (4.7),

$$\sum_{\lambda_1=0}^{L_1}\sum_{\lambda_2=0}^{L_2}C_{\lambda_1,\lambda_2}(x, y)\cdot\mathcal{K}_{\lambda_1,\lambda_2,k,m}(x, y) = 0 \tag{4.14}$$

for all $k = 0, 1, \ldots, [c_3L_2]$, $m = 0, 1, \ldots, [c_4L_1]$.

By the definition of $\vec{k}^0$, applying $\partial^{\|\vec{k}^0\|}/\partial x^{k_1^0}\partial y^{k_2^0}$ to (4.14) and specializing it at $(x, y) = \bar{\xi}_0$, we get the system of equations

$$\sum_{\lambda_1=0}^{L_1}\sum_{\lambda_2=0}^{L_2}\tilde{C}_{\lambda_1,\lambda_2}(\bar{\xi}_0)\mathcal{K}_{\lambda_1,\lambda_2,k,m}(\bar{\xi}_0) = 0 \tag{4.15}$$

for all $k = 0, 1, \ldots, [c_3 L_2]$, $m = 0, 1, \ldots, [c_4 L_1]$. Then, again in the new variables $w = \wp(z)$, $f(w) = \zeta(z) - \eta z/\omega$, we get the expression for $\tilde{F}^{(k)}(w)$ (for a given $m = 0, 1, \ldots, [c_4 L_1]$)

$$(4.16) \qquad \frac{1}{k!}\tilde{F}^{(k)}(w_0) = \sum_{\lambda_1=0}^{L_1} \sum_{\lambda_2=0}^{L_2} \tilde{C}_{\lambda_1,\lambda_2}(\bar{\xi}_0) \cdot \mathcal{K}_{\lambda_1,\lambda_2,k,m}\left(\frac{\pi}{\omega}, \frac{\eta}{\omega}\right)$$

if $w_0 = \wp(\omega/4)$, $f(w_0) = m \cdot (\omega\eta' - \omega'\eta)/\omega + \zeta(\omega/4) - \eta/4$.

However now the polynomials $\mathcal{K}_{\lambda_1,\lambda_2,k,m}(x, y)$ have the better bound of the type; the type of $\mathcal{K}_{\lambda_1,\lambda_2,k,m}(x, y)$ is bounded by $\gamma_6(k\log(L_1 + 1) + L_2 + L_1 \log(m + 1))$. Hence, returning to an old variable z, we get, using the remark (4.13),

$$(4.17) \qquad \log\left|\frac{1}{k!}\tilde{F}^{(k)}\left(\frac{\omega}{4} + m\omega'\right)\right| < -E_0 + c'_{12} \cdot L_2 \log L_1, \quad k = 0, 1, \ldots, [c_3 L_2],$$

where $\log\|\bar{\theta} - \bar{\xi}_0\| < -E_0$. In other words, (4.12) is proved.

The main algebraic statement that enables us to handle the presence of the "gas" of zeroes $\bar{\xi} \in S_0$ around $\bar{\theta}$ is the following very convenient

LEMMA 4.18. *Let $Q(x, y) \in \mathbf{K}[x, y]$. We assume that*

$$(4.19) \qquad \log|Q(\bar{\xi}_0)| < -\lambda \cdot E_0,$$

$E_0 = E(\bar{\xi}_0)$ for some λ, $0 < \lambda \leqslant 1$. Then for all $\bar{\xi} \in S_0$ with $\|\bar{\theta} - \bar{\xi}\| < 1$, we have

$$(4.20) \qquad \log|Q(\bar{\xi})| < -\lambda E(\bar{\xi}) + \gamma_4 t(Q).$$

If now, in addition to (4.19), *$Q(\bar{\xi}_0) \neq 0$, then*

$$(4.21) \qquad \Sigma^+\{\lambda E(\bar{\xi}) - \gamma_4 t(Q) : \bar{\xi} \in S_0\} \leqslant c'_{13} \cdot \{t(Q) \cdot d(S_0) + d(Q) \cdot t(S_0)\},$$

where Σ^+ means that in the sum only positive elements are counted.

PROOF OF LEMMA 4.18. We have

$$|Q(\bar{\xi})| \leqslant |Q(\bar{\xi}_0)| + \|\bar{\xi} - \bar{\xi}_0\| \exp(\gamma_3 t(Q)).$$

Now for $\bar{\xi} \in S_0$, we have, using the definition of $\bar{\xi}_0$,

$$\|\bar{\xi} - \bar{\xi}_0\| \leqslant \|\bar{\xi} - \bar{\theta}\| + \|\bar{\xi}_0 - \bar{\theta}\| \leqslant 2\|\bar{\xi} - \bar{\theta}\| \leqslant \exp(-E(\bar{\xi}) + \log 2).$$

Since $E_0 \geqslant E(\bar{\xi})$ for $\bar{\xi} \in S_0$ we get (4.20).

Let $Q(\bar{\xi}_0) \neq 0$. Then, since S_0 is an irreducible component, $Q(\bar{\xi}) \neq 0$ for all $\bar{\xi} \in S$. We apply now the Liouville theorem to

$$\left|\prod_{\bar{\xi} \in S'} Q(\bar{\xi})\right|,$$

where $S' = \{\bar{\xi} \in S_0 : \lambda E(\bar{\xi}) - \gamma_4 t(Q) > 0\}$. We have, naturally,

$$\left| \prod_{\bar{\xi} \in S'} Q(\bar{\xi}) \right| > \exp(-c'_{14} \cdot \{t(Q)\, d(S_0) + d(Q) t(S_0)\}). \tag{4.22}$$

Now we get an upper bound for the same product, using (4.20),

$$\left| \prod_{\bar{\xi} \in S'} Q(\bar{\xi}) \right| < \exp\{-\sum (\lambda E(\bar{\xi}) - \gamma_4 t(Q) : \bar{\xi} \in S')\},$$

which establishes (4.21) with the combination of (4.22). The Lemma 4.18 is proved.

Let us apply this lemma to the polynomials arising from the function $\tilde{F}(z)$. As usual in the Gelfond-Schneider method, we have two alternatives: either some $\tilde{F}(\omega/4 + m\omega')$ is bounded below and above by quantities of the same order $\exp(-O(L_1^2 L_2))$, or all $\tilde{C}_{\lambda_1,\lambda_2}(\pi/\omega, \eta/\omega) = \tilde{C}_{\lambda_1,\lambda_2}(\bar{\theta})$ are bounded above by a quantity like $\exp(-O(L_1^2 L_2))$. This is exactly the statement of Lemma 1.10, as applied to $\tilde{F}(z)$.

We will now use Lemma 4.18 for the pair (L_1, L_2) constructed in (3.8) and 3.9. We can simplify the proof using another auxiliary

LEMMA 4.23. *Let* $Q(x, y) \in \mathbf{K}[x, y]$ *and*

$$t(Q) \leqslant c'_{15} \cdot L_2 \log L_1, \qquad d(Q) \leqslant c'_{16} \cdot L_1$$

together with

$$\log | Q(\bar{\theta}) | < -c'_{17} \cdot L_1^2 L_2$$

for some constants $c'_{15}, c'_{16}, c'_{17} > 0$. *If* λ *in* (3.8) *is sufficiently small, then we have* $Q(\bar{\xi}_0) = 0$.

PROOF OF LEMMA 4.23. Indeed, if $Q(\bar{\xi}_0) \neq 0$, then $Q(\bar{\xi}) \neq 0$ for all $\bar{\xi} \in S_0$. By the choice of (L_1, L_2) in 3.9 we can write

$$c'_{18} \cdot L_1^2 L_2 = \lambda E_0$$

for some sufficiently small parameter λ in (3.8), $0 < \lambda < 1$. Then we use Lemma 4.18 and obtain

$$\begin{aligned} \Sigma^+ \{\lambda E(\bar{\xi}) - c'_{19} \cdot t(Q) : \bar{\xi} \in S\} &\leqslant c'_{20}\{t(Q)\, d(S_0) + d(Q) t(S_0)\} \\ &\leqslant c'_{21}\{L_2 \log L_1 \cdot d(S_0) + L_1 t(S_0)\}. \end{aligned} \tag{4.24}$$

Now we use the property of S_0. Namely, we have

$$\Sigma \left\{ E(\bar{\xi}) : \bar{\xi} \in S_0,\ E(\bar{\xi}) \geqslant c'_3 \cdot \frac{\mathcal{E}_0^{3/2}}{d(S_0) \mathcal{T}(S_0)^{1/2}} \right\} \geqslant \frac{\mathcal{E}_0}{4}.$$

However, by the choice of $\mathcal{E}_0$, $\mathcal{T}_0$ in (3.7) one gets

$$\mathcal{E}_0 / \mathcal{T}(S_0) \geqslant (c'_{22})^2.$$

Thus $\mathcal{E}_0^{3/2} / d(S_0) \mathcal{T}(S_0)^{1/2} \geqslant c'_{22} \cdot \mathcal{E}_0 / d(S_0)$.

Thus by Lemma 3.10, we have

$$\lambda c'_{23} \cdot \frac{\mathcal{E}_0^{3/2}}{d(S_0)\mathfrak{T}(S_0)^{1/2}} > C' \cdot t(Q)$$

for a sufficiently large $C' > 0$. In particular, we get

$$\lambda \mathcal{E}_0 \leqslant c'_{24}\{L_2 \log L_1\, d(S_0) + L_1 t(S_0)\},$$

which contradicts the choice of L_1, L_2 and Lemma 3.10. Lemma 4.23 is proved.

We are going to apply Lemma 4.23 in the following way.

COROLLARY 4.25. *Let all the conditions of Lemma* 4.23 *be satisfied. Then, in addition to* $Q(\bar{\xi}_0) = 0$ *we have*

$$\log|Q(\bar{\theta}\,)| < -\tilde{C} \cdot L_1^2 L_2 \tag{4.26}$$

for a sufficiently large constant $\tilde{C} > 0$ (*depending only on* C).

PROOF OF COROLLARY 4.25. Since $Q(\bar{\xi}_0) = 0$, one obtains

$$|Q(\bar{\theta}\,)| \leqslant |Q(\bar{\xi}_0)| + \|\bar{\theta} - \bar{\xi}_0\| \exp(\gamma_3 t(Q)) \leqslant \|\bar{\theta} - \bar{\xi}_0\| \exp(\gamma_3 t(Q)).$$

Now, by the choice of (L_1, L_2) in 3.9 we get

$$\|\bar{\theta} - \bar{\xi}_0\| \leqslant \exp(-E_0) \leqslant \exp(-2\tilde{C} \cdot L_1^2 L_2).$$

This, together with a bound on $t(Q)$ establishes (4.26). Corollary 4.25 is proved.

Corollary 4.25 is applied now to $\tilde{F}^{(k)}(\omega/4 + m\omega')$. First of all, by Lemma 1.10, the following is true:

$$\log\left|\frac{1}{k!}\tilde{F}^{(k)}\left(\frac{\omega}{4} + m\omega'\right)\right| < -c'_{25} \cdot L_1^2 L_2 \tag{4.27}$$

for $m = 0, 1, \ldots, [c_{10} L_1]$ and $k = 0, 1, \ldots, [c_{10} L_2]$. In order to get a better bound on type, we make a change of variables, $\wp(z) = w$, $\zeta(z) - \eta z/\omega = f(w)$. Let us denote by $Q_{k,m}$ a number

$$Q_{k,m} = \frac{1}{k!}\tilde{F}^{(k)}(w)\big|_{w=\wp(\omega/4), f(w) = m(\omega\eta' - \omega'\eta)/\omega + \zeta(\omega/4) - \eta/4} \tag{4.28}$$

for $k = 0, 1, \ldots, [c_{10} L_2]$, $m = 0, 1, \ldots, [c_{10} L_1]$. Then we have

$$Q_{k,m} = \sum_{\lambda_1 = 0}^{L_1} \sum_{\lambda_2 = 0}^{L_2} \tilde{C}_{\lambda_1, \lambda_2}(\bar{\xi}_0) \cdot \mathcal{K}_{\lambda_1, \lambda_2, k, m}(\bar{\theta}\,). \tag{4.29}$$

If one replaces $\bar{\theta}$ by $\bar{\xi}_0$ in (4.29) one gets a polynomial $Q'_{k,m}(\bar{\xi}_0)$ in $\bar{\xi}_0$ of the degree at most $2L_1$, with the coefficients from $\mathbf{K} = \mathbf{Q}(g_2, g_3, \wp(\omega/4), \wp'(\omega/4), i)$ of the type at most $c'_{26} \cdot L_2 \log L_1$. Also, by the choice of $L_1^2 L_2$ and (4.27)–(4.28) we have

$$\log|Q'_{k,m}| < -c'_{27} \cdot L_1^2 L_2. \tag{4.30}$$

Hence, by Lemma 4.23,

$$Q'_{k,m}(\bar{\xi}_0) = 0. \tag{4.31}$$

Now, in (4.29), we replace $\bar{\xi}_0$ by $\bar{\theta}$ in $\mathcal{K}_{\lambda_1,\lambda_2,k,m}$. We have

$$\log\|\bar{\theta} - \bar{\xi}_0\| < \lambda^{-1}L_1^2L_2,$$

and, so (4.31) implies by (4.26),

$$\log|Q_{k,m}| < -\tilde{C}\cdot L_1^2L_2 \tag{4.32}$$

for a sufficiently large constant $C > 0$.

Returning now to the old z-coordinates, we deduce from (4.32), (4.28) that

$$\log\left|\frac{1}{k!}\tilde{F}^{(k)}\left(\frac{\omega}{4} + m\omega'\right)\right| < -\frac{\tilde{C}}{2}\cdot L_1^2L_2 \tag{4.33}$$

if $k = 0, 1, \ldots, [c_{10}L_2]$, $m = 0, 1, \ldots, [c_{10}L_1]$. We can use now the Small Value Lemma 1.10, since $\tilde{C}$ is sufficiently large (with respect to c_{10}, c_{11}).

Consequently, there is an upper bound on the values of the coefficients $\tilde{C}_{\lambda_1,\lambda_2}(\bar{\xi}_0)$,

$$\log|\tilde{C}_{\lambda_1,\lambda_2}(\bar{\xi}_0)| < -\frac{\tilde{C}}{4}\cdot L_1^2L_2. \tag{4.34}$$

Here $\tilde{C}$ is sufficiently large and the comparison with the Lemma 4.23 gives us at once $\tilde{C}_{\lambda_1,\lambda_2}(\bar{\xi}_0) = 0$ for all $\lambda_1 = 0, 1, \ldots, L_1$, $\lambda_2 = 0, 1, \ldots, L_2$. This contradicts the choice of $\tilde{C}_{\lambda_1,\lambda_2}(\bar{\xi}_0)$ in 4.8. Theorem 1.1 is proved.

The method of the proof of Theorem 1.1, with slight modifications for a different auxiliary function only, is applied to other pairs of algebraically independent numbers. For example, in case (i) of §0, for u being an algebraic point of $\wp(x)$, we have in the notation of Theorem 1.1,

$$|R(\zeta(u) - \eta u/\omega, \eta/\omega)| > H(R)^{-c'_{28}d(R)^8\log^9(d(R)+1)}, \tag{4.35}$$

for $\log H(R) > c'_{29}\,d(R)$ for $c'_{28}, c'_{29} > 0$ depending on $\wp(z)$ and u. The proof uses the same auxiliary function as in the proof of Theorem 1.1, but with different choice of L_1, L_2.

Theorem 1.1 implies, in particular, the bound

$$|R(\pi, \omega)| > H(R)^{-C'_0(R)^2\log^2(d(R)+1)}$$

for $R(x, y) \in \mathbf{Z}[x, y]$, $H(R) \geqslant d(R)^{d(R)}$ and a period ω of an elliptic curve with a complex multiplication.

REFERENCES

1. G. V. Chudnovsky, (a) *Algebraic grounds for the proof of algebraic independence*. I. *Elementary algebra*, Comm. Pure Appl. Math. **34** (1981), 1–28.

(b). *Algebraic grounds for the proof of algebraic independence*. II (to appear).

2. ______, *Algebraic independence of constants connected with exponential and elliptic functions*, Dokl. Akad. Nauk Ukrain. SSR Ser. A **1976**, 698–701, 767. (Russian. English summary)

3. ______, *Algebric independence of values of exponential and elliptic functions*, Proc. Internat. Congr. Math. (Helsinki, 1978), vol. 1, Acad. Sci. Fenn., Helsinki, 1980, pp. 339–350; and see Introduction, this volume.

4. ______, Chapter 1, this volume.

5. ______, *Algebraic independence of the values of elliptic function at algebraic point. Elliptic analogue of Lindemann-Weierstrass theorem*, Invent. Math. **61** (1980), 267–280.

6. A. Baker, *Transcendental number theory*, Cambridge Univ. Press, Cambridge, 1975.

7. *Transcendence theory*, Proc. Cambridge Conf. (1976), (A. Baker and D. Masser (eds.)), Academic Press, New York, 1977.

8. A. O. Gelfond, *Transcendental and algebraic numbers*, Dover, New York, 1960.

9. G. V. Chudnovsky, *Measures of irrationality, transcendence and algebraic independence. Recent progress*, Journées Arithmétiques 1980 (J. Armitage, ed.), Cambridge Univ. Press, Cambridge, 1982, pp. 11–82.

10. A. I. Galochkin, *Lower bounds for polynomials of the values of a class of analytic functions*, Math. Sb. **137** (1974), 396–417.

11. G. V. Chudnovsky, *Padé approximation and the Riemann monodromy problem*, Cargèse Lectures, Bifurcation Phenomena in Mathematical Physics and Related Topics, Reidel, Boston, 1980, pp. 448–510.

12. I. R. Shafarevich, *Basic algebraic geometry*, Springer, New York, 1974.

13. W. V. D. Hodge and D. Pedoe, *Methods of algebraic geometry*, Cambridge Univ. Press, Cambridge, 1952.

14. B. L. Van der Waerden, *Modern algebra*, Ungar, New York, 1951.

15. S. Lang, *Introduction to transcendental numbers*, Addison-Wesley, Reading, Mass., 1966.

CHAPTER 9

ANOTHER METHOD FOR INVESTIGATING THE ARITHMETIC NATURE OF VALUES OF FUNCTIONS OF A COMPLEX VARIABLE

Introduction. The purpose of this paper is to investigate a series of questions arising around the problem of determination of properties of analytic functions having, together with their derivatives, rational integer values at a fixed set of points. Our interest in the problem originated from the following statement called the Straus-Schneider theorem:

THEOREM 0. *Let* $\mathbf{K}$ *be an algebraic number field and* $f(z)$ *be a transcendental meromorphic function of finite order of growth* ρ. *Then the set of numbers* $z \in \mathbf{K}$ *such that* $\partial^k f(z)/\partial z^k \in \mathbf{Z}$ *for all* $k \geqslant 0$, *has cardinality at most* $\rho \cdot [\mathbf{K} : \mathbf{Q}]$.

This theorem was proved by Straus for entire functions **[6]** and by Schneider **[5]** for the general case, but with a slightly different upper bound of the cardinality. There appeared a lot of different questions connected with the Straus-Schneider theorem and we refer readers to **[12]** for a comprehensive reference on this subject.

Here we review our progress in this direction. The result directly generalizing the Straus-Schneider theorem is the following proved in **[6]**:

THEOREM 1. *Suppose* $f(z)$ *is a transcendental meromorphic function of finite order* ρ. *Then the set of algebraic numbers* z *such that* $\partial^k f(z)/\partial z^k \in \mathbf{Z}$ *for all* $k \geqslant 0$, *has cardinality at most* ρ.

This estimate cannot be essentially improved. For example, one can consider $f_n(z) = \exp(z \cdots (z - n))$. Neither can the statement of Theorem 1 be improved, since $\mathbf{Z}$ cannot be replaced by $\mathbf{Q}$ or by any other ring of algebraic integers different from the ring of integers in the quadratic imaginary field.

0076-5376/84 $1.00 + $.25 per page

Results, similar to that of Theorem 1, hold in the multidimensional case [**14**]. For example, as we see below one can prove the following:

THEOREM 2. *Suppose* $f(z_1,\ldots,z_n)$ *is a transcendental meromorphic function of finite order of growth* ρ *and suppose that* ω *is an imaginary quadratic number. Then the set of points* $(z_1,\ldots,z_n)\in\overline{\mathbf{Q}}^n$ *such that* $D^{\bar{k}}f(z_1,\ldots,z_n)\in\mathbf{Z}[\omega]$ *for all* $\bar{k}=(k_1,\ldots,k_n)\in\mathbf{N}^n$ *is contained in an algebraic hypersurface in* $\mathbf{C}^n$ *of degree at most* $n\cdot\rho$.

The bound of the degree of the hypersurface can be, actually, improved (see [**14**]).

In §0 we describe the main idea of the proof of Theorems 1 and 2 by looking at Theorem 1 in the particular (but nontrivial) case of $\rho<2$. Then in §§1–3 we present a detailed proof of Theorem 1. In §4 Theorem 2 is dealt with together with the proper reformulation of the Schwarz lemma and definitions of singular and very singular degrees of finite sets of points in $\mathbf{C}^n$. Finally, in §5 we take a new look at Theorems 0 and 1. We study entire functions not only from the point of view of their order of growth, but also from the point of view of their type. New results are proved that show relations between the type of functions and the arithmetic nature of numbers, at which functions admit rational integer values. §5 also contains a discussion on the possible generalization of Theorem 1.

0. The method of conjugate auxiliary functions. Consider an arbitrary transcendental function $f(z)$ of one variable z, meromorphic and of finite order $\leqslant\rho$. We are interested in the structure (cardinality) of the set

$$A_f=\{z\in\overline{\mathbf{Q}}\colon f^{(k)}(z)\in\mathbf{Z}\text{ for all }k\geqslant 0\}$$

or the set $\{z\in\overline{\mathbf{Q}}\colon f^{(k)}(z)\in\mathbf{Z}[i]\text{ for all }k\geqslant 0\}$. The first result in this direction is due to Straus [**6**] (1949) (see also [**5**]):

The set $\{z\in\mathbf{K}\colon f^{(k)}(z)\in\mathbf{Z}\text{ for all }k\geqslant 0\}$ has cardinality $\leqslant\rho[\mathbf{K}:\mathbf{Q}]$, where $\mathbf{K}$ is an algebraic number field.

Moreover, $A_f=\varnothing$ if $\rho<1$ and, by a result of Bertrand [**7**], A_f is finite (and even contains at most one element) if $\rho=1$. We will show that $|A_f|\leqslant\rho$ for any entire (even meromorphic) function or order ρ.

We first indicate the standard procedure, due to M. Waldschmidt [**8, 3, 4**], for constructing an auxiliary function.

Step 1. We construct a polynomial $P(x,y)\in\mathbf{Z}[x,y]$ such that $P(x,y)\not\equiv 0$ and

$$P(x,y)=\sum_{l_1=0}^{L_1}\sum_{l_2=0}^{L_2}p(l_1,l_2)x^{l_1}y^{l_2},$$

where $L_1\to\infty$, $L_2=o(L_1)$, and $P(z,f(z))$ has at each point $w\in S_0\subset A_f$ a zero of order $O(L_1L_2/(\Lambda\,|\,S_0\,|))$.

Step 2. If $f(z) = g(z)/h(z)$, where $g(z)$ and $h(z)$ are entire functions of orders $\leqslant \rho$, we consider the entire function $h(z)^{L_2}P(z, f(z))$, also of order $\leqslant \rho$. Let

$$s_0 = \min\left\{s\colon (P(z, f(z))^{(s)}_{z=w} \neq 0 \text{ for some } w \in S_0\right\}.$$

We apply the maximum modulus principle to the function

$$F(z) = h(z)^{L_2}P(z, f(z))\cdot \prod_{w\in S_0} (z - w_0)^{-s_0}.$$

Then there exists $w_0 \in S_0$ such that

$$P(z, f(z))^{(s_0)}|_{z=w_0} \neq 0$$

and

$$(P(z, f(z)))^{(s_0)}|_{z=w_0} = s_0!F(w_0)\cdot h(w_0)^{-L_2} \prod_{w\neq w_0, w\in S_0} (w - w_0)^{s_0}.$$

Thus, assuming that w_0 is not a zero of $h(z)$ (not a pole of $f(z)$), we obtain

$$|P(x, f(z))^{(s_0)}|_{z=w_0} \leqslant \exp(s_0 \log s_0 + s_0c_1 + L_2c_2)\,|F(w_0)|\,.$$

Now for $R = s_0^{1/\rho}$ we have, as $s_0 \to \infty$,

$$\begin{aligned}|F(z)|_R &\leqslant |h(x)^{L_2}\cdot P(z, f(z))|_R \cdot \left|\prod_{w\in S_0} (z - w)^{-s_0}\right|_R \\ &\leqslant (L_1 + 1)(L_2 + 1)H(P)\exp(c_3L_2R^\rho) \cdot (R/2)^{-s_0|S_0|} \times R^{L_1} \\ &\leqslant \exp\left(c_4L_2s_0 + 2\log(L_1L_2H(P)) - |S_0|\frac{s_0}{\rho}\log s_0 + L_1\log R\right).\end{aligned}$$

Step 3. Consider $\gamma = P(z, f(z))^{(s_0)}|_{z=w_0}$. Since $w_0 \in S_0 \subset A_f$, it follows that w_0 is an algebraic number. Let $\mathbf{Q}(S_0) = \mathbf{K}$ and $[\mathbf{K}:\mathbf{Q}] = n$. Then we have

$$H(f^{(s)}(w_0)) \leqslant s\log s + O(s)$$

as $s \to \infty$ for a meromorphic function $f(z)$ and even

$$H(f^{(s)}(w_0)) \leqslant \left(1 - \frac{1}{\rho}\right)s\log s + O(s)$$

as $s \to \infty$ for an entire function $f(z)$, since $f^{(i)}(w_0) \in \mathbf{Z}$ for all $i \geqslant 0$. Consequently, for $\gamma \in \mathbf{K}$ we have

$$|\overline{\gamma}| \leqslant (L_1 + 1)(L_2 + 1)H(P)\cdot C_5^{L_1}C_6^{L_2}\cdot (L_1)^{L_1}e^{O(s_0+L_2\log s_0)}\exp\left(t_*s_0\log s_0\right),$$

where $t_* = 1$ for a meromorphic $f(z)$ and $t_* = 1 - 1/\rho$ for an entire $f(z)$.

Step 4. We combine the results of Steps 2 and 3, choosing, finally, L_1, L_2, and $H(P)$. Using Siegel's lemma and the fact that $S_0 \subseteq \mathbf{K}$, we can choose $P(x, y) \in \mathbf{Z}[x, y]$, $P(x, y) \not\equiv 0$, so that

$$(P(z, f(z))^{(k)}|_{z=w} = 0$$

for $k \leqslant S_\infty$ and $w \in S_0$, where

$$L_2 = \left[(\log L_1)^{3/4}\right], \qquad S_\infty = \left[L_1 \cdot (\log L_1)^{1/4}\right]$$

as $L_1 \to \infty$ (i.e. $L_1 \gg n, |S_0|, \ldots$). In this case, by Siegel's lemma, $P(x, y)$ can be chosen so that

$$H(P) \leqslant \exp\left(c_7 L_1 (\log L_1)^{3/4}\right)$$

as $L_1 \to \infty$. Now $s_0 \geqslant S_\infty + 1$, i.e.

$$H(P) \leqslant \exp\left(c_8 s_0 (\log s_0)^{1/2}\right).$$

Thus,

$$|\bar{\gamma}| \leqslant \exp\left(c_9 L_1 \log L_1 + t_* s_0 \log s_0\right) \leqslant \exp\left(c_{10} s_0 (\log s_0)^{3/4} + t_* s_0 \log s_0\right).$$

By Step 2,

$$|\gamma| \leqslant \exp\left(c_{11} s_0 (\log s_0)^{3/4} + s_0 (1 - |S_0|/\rho) \log s_0\right),$$

and therefore, since

$$(n-1)\log|\bar{\gamma}| + \log|\gamma| + n \log \operatorname{den}(\gamma) \geqslant 0,$$

we obtain

$$(n-1)t_* + (1 - |S_0|/\rho) \geqslant 0.$$

If $t_* = 1$, then $|S_0| \leqslant n\rho$, and if $t_* = 1 - 1/\rho$, then

$$n - n/\rho - 1 + 1/\rho + 1 - |S_0|/\rho \geqslant 0; \qquad |S_0| \leqslant \rho n - n + 1.$$

From this we obtain the above-mentioned results of Straus ($t_* = 1$) and Bertrand ($t_* = 1 - 1/\rho$ and $\rho \leqslant 1$). However, this does not yield the desired estimate $|A_f| \leqslant \rho$ for $\rho > 1$.

The method requires an essential improvement through a more careful consideration of the conjugate numbers $\gamma \to \gamma^{(i)}$. Let us indicate the fundamentally new feature of our method. It is obvious that the conjugate of the number

$$\gamma = (P(z, f(z)))^{(k)}|_{z=w}$$

where $w \in A_f$, can be obtained by replacing w in $z^{\lambda_1 - s'}|_{z=w}$ by its conjugates $w^{(i)}$, since all of the derivatives $(f^{\lambda_2}(z))^{(s-s')}|_{z=w}$ belong to $\mathbf{Z}$ if $w \in A_f$. In other words, the function $P(z + w^{(i)} - w, f(z))$ has the property that an algebraic number $\gamma = (P(z, f(z)))^{(k)}|_{z=w}$ has as a conjugate $(P(z + w^{(i)} - w, f(z)))^{(k)}|_{z=w}$. This enables us to replace the trivial inequality

$$|\bar{\gamma}|^{(n-1)} \cdot |\gamma| \cdot \operatorname{den}(\gamma)^n \geqslant 1$$

by a more precise one utilizing all of the conjugates $\gamma^{(i)}$.

For a detailed consideration of the conjugates of γ we will use certain auxiliary functions of the form

$$F_j(z) = P\left(z + \lambda_j, f(z)\right)$$

where $P(x, y) \in \mathbf{Z}[x, y]$ and $\{\lambda_j : j \in \mathbf{I}\}$ are chosen so that the conjugates of the numbers $w_i + \lambda_j$ also have the form $w_i + \lambda_{j_1}$, $j_1 \in \mathbf{I}$, for many $j \in \mathbf{I}$. To clarify this idea we will analyze the case $\rho < 2$ and $n = 2$.

As an illustration of the proposed method we will give a very simple proof of the first nontrivial case, where the values of the meromorphic function of order < 2 lie in a real quadratic extension of an imaginary quadratic field.

THEOREM 0.1. *Suppose $f(z)$ is a transcendental meromorphic function in $\mathbf{C}$ of order < 2 and $\mathbf{K}$ is a quadratic field over $\mathbf{Q}$ (or is a quadratic extension of an imaginary quadratic field). If $f^{(k)}(w_i) \in \mathbf{Z}$ for all $k \geqslant 0$, where $w_1, w_2 \in \mathbf{K}$, then $w_1 = w_2$.*

PROOF OF THEOREM 0.1. Suppose $f(z)$ is a transcendental meromorphic function in $\mathbf{C}$ of order $\leqslant \rho_0 < \rho$ (< 2), i.e. there exists an entire function $h(z)$ in $\mathbf{C}$ of order $\leqslant \rho_0$ such that $h(z)f(z)$ is also entire of order $\leqslant \rho_0$. Suppose $\mathbf{K}$ is a quadratic field and $w_1, w_2 \in \mathbf{K}$, $w_1 \neq w_2$, are such that

$$f^{(k)}(w_i) \in \mathbf{Z} \quad \text{for all } k \geqslant 0 \text{ and } i = 1, 2. \tag{0.1}$$

We may assume without loss of generality that $w_1 = 0$. Indeed, we can put $f_1(z) = f(z + w_1)$ and then $f_1(z)$ has order $\leqslant \rho_0$ and $f_1^{(k)}(0) \in \mathbf{Z}$, $f_1^{(k)}(w_2 - w_1) \in \mathbf{Z}$ for all $k \geqslant 0$. Consequently, we assume that

$$w_1 = 0; \qquad w_2 = \zeta \in \mathbf{K}, \quad \zeta \neq 0. \tag{0.2}$$

Let $G = \mathrm{Gal}(\mathbf{K}/\mathbf{Q}) = \{1, \sigma\}$ and denote the conjugate to $\xi \in \mathbf{K}$ by ξ^σ. We will choose a sequence $\{\lambda_i : i \in \mathbf{I}\}$ in $\mathbf{K}$ and work with auxiliary functions of the form

$$F_j(z) = P\big(z + \lambda_j, f(z)\big) \quad \text{for } P(x, y) \in \mathbf{Z}[x, y].$$

The sequence $\{\lambda_j : j = 0, 1, 2, \dots\}$ in $\mathbf{K}$ is defined as follows:

DEFINITION 0.2. For $j = 0, 1, 2, \dots$ put

$$\lambda_j = \begin{cases} 0, & \text{if } j = 0; \\ (k+1)(\zeta^\sigma - \zeta), & \text{if } j = 2k+1,\ k \geqslant 0; \\ (k+1)(\zeta - \zeta^\sigma), & \text{if } j = 2k+2,\ k \geqslant 0; \end{cases} \tag{0.3}$$

$$\lambda_0 = 0, \qquad \lambda_1 = \zeta^\sigma - \zeta, \qquad \lambda_2 = \zeta - \zeta^\sigma, \qquad \lambda_3 = 2(\zeta^6 - \zeta), \dots.$$

In view of the definition of the λ_j, we can describe the conjugates of the numbers $w_i + \lambda_j$. We have

LEMMA 0.3. *If $k = 0, 1, 2, \dots$, then*

$$(w_1 + \lambda_{2k+2})^\sigma = w_1 + \lambda_{2k+1}; \tag{0.4}$$

$$(w_2 + \lambda_{2k+1})^\sigma = w_2 + \lambda_{2k}. \tag{0.5}$$

PROOF OF LEMMA 0.3. According to (0.2), $w_1 = 0$ and $\lambda_{2k+2}^\sigma = \lambda_{2k+1}$. Also, $(w_2 + \lambda_{2k+1})^\sigma = \zeta^\sigma + (k+1)(\zeta - \zeta^\sigma) = (k+1)\zeta - k\zeta^\sigma = \zeta + k(\zeta - \zeta^\sigma) = w_2 + \lambda_{2k}$. Lemma 0.3 is proved.

An analogous assertion holds for the auxiliary "conjugate" functions.

LEMMA 0.4. *Suppose* $P(x, y) \in \mathbf{Z}[x, y]$ *and* $F_j(z) = P(z + \lambda_j, f(z))$, $j = 0, 1, 2, \ldots$. *Then for any* $s \geqslant 0$ *and* $k \geqslant 0$,

$$(0.6) \qquad \left(F_{2k+2}^{(s)}(w_1)\right)^\sigma = F_{2k+1}^{(s)}(w_1),$$

$$(0.7) \qquad \left(F_{2k+1}^{(s)}(w_2)\right)^\sigma = F_{2k}^{(s)}(w_2).$$

PROOF OF LEMMA 0.4. Suppose $P(x, y) = \Sigma_{l_1} \Sigma_{l_2} p(l_1, l_2) x^{l_1} y^{l_2}$ and $p(l_1, l_2) \in \mathbf{Z}$. We have $F_j(z) = \Sigma_{l_1} \Sigma_{l_2} p(l_1, l_2)(z + \lambda_j)^{l_1} f(z)^{l_2}$. Then

$$(0.8) \qquad F_j^{(s)}(w_i) = \sum_{l_1} \sum_{l_2} p(l_1, l_2) \sum_{s=0}^{s} C_s^{s_1} l_1 \cdots (l_1 - s_1 + 1) \times \left(w_i + \lambda_j\right)^{l_1 - s_1} \cdot \left\{f(w_i)^{l_2}\right\}^{(s - s_1)}$$

Since $p(l_1, l_2) \in \mathbf{Z}$ and (0.1) holds, we have

$$(0.9) \qquad \left(F_j^{(s)}(w_i)\right)^\sigma = \sum_{l_1} \sum_{l_2} p(l_1, l_2) \sum_{s_1=0}^{s} C_s^{s_1} l_1 \times \cdots \times (l_1 - s_1 + 1)\left(w_i^\sigma + \lambda_j^\sigma\right)^{l_1 - s_1} \cdot \left\{f(w_i)^{l_2}\right\}^{(s - s_1)}.$$

Therefore, in view of (0.4), (0.5), (0.8), and (0.9),

$$\left(F_{2k+2}^{(s)}(w_1)\right)^\sigma = F_{2k+1}^{(s)}(w_1); \qquad \left(F_{2k+1}^{(s)}(w_2)\right)^\sigma = F_{2k}^{(s)}(w_2).$$

Lemma 0.4 is proved.

The following table illustrates the conjugacy of values of the auxiliary functions.

$i \backslash j$	$\lambda_0 = 0$	$\lambda_1 = \zeta^\sigma - \zeta$	$\lambda_2 = \zeta - \zeta^\sigma$	$\lambda_3 = 2\zeta^\sigma - 2\zeta$	$\lambda_4 = 2\zeta - 2\zeta^\sigma$
$w_1 = 0$	$(0;\ f(0))$	$(\zeta^\sigma - \zeta;\ f(0))$	$(\zeta - \zeta^\sigma;\ f(0))$	$(2\zeta^\sigma - 2\zeta;\ f(0))$	$(2\zeta - 2\zeta^\sigma;\ f(0))$
$w_2 = \zeta$	$(\zeta: f(\zeta))$	$(\zeta^\sigma;\ f(\zeta))$	$(2\zeta - \zeta^\sigma;\ f(\zeta))$	$(2\zeta^\sigma - \zeta;\ f(\zeta))$	$(3\zeta - 2\zeta^\sigma;\ f(\zeta))$

FIGURE 1

Here $(\lambda_j;\ f(0))$ means that we are considering $F_j^{(s)}(0) = \partial^s P(z + \lambda_j, f(z))\big|_{z=0}$, and $(\zeta + \lambda_j;\ f(\zeta))$ means that we are considering $F_j^{(s)}(\zeta) = \partial^s P(z + \lambda_j, f(z))\big|_{z=\zeta}$. An arrow between $(w_i + \lambda_j;\ f(w_i))$ and $(w_i + \lambda_{j+1};\ f(w_i))$ means that $F_j^{(s)}(w_i)$ and $F_{j+1}^{(s)}(w_i)$ are conjugate for all $s \geqslant 0$.

We fix $\varepsilon > 0$ and a sufficiently large natural number N such that

$$1/N - 1 < \varepsilon/2; \qquad \varepsilon < 1/8 \quad \text{Q.E.D.}$$

The main parameter is a natural number L, sufficiently large in comparison with N, ε^{-1}, and $H(\zeta)$. Applying Siegel's lemma and well-known estimates, we obtain the existence of auxiliary functions (see Bertrand [7]):

LEMMA 0.5. *There exists a nonzero polynomial* $P_L(x, y) \in \mathbf{Z}[x, y]$,

$$(0.10)\qquad P(x, y) \equiv P_L(x, y) = \sum_{l_1=0}^{L_1} \sum_{l_2=0}^{L_2} p(l_1, l_2) x^{l_1} y^{l_2},$$

$$L_1 = \left[L(\log L)^{-1/4}\right], \qquad L_2 = \left[(\log L)^{3/4}\right]$$

with rational integral coefficients $p(l_1, l_2)$ *of height*

$$(0.11)\qquad H(P_L) \leqslant c_0 L(\log L)^{1/2},$$

such that the functions

$$(0.12)\qquad F_j(z) = P\left(z + \lambda_j, f(z)\right), \qquad j = 0, 1, \ldots, N,$$

satisfy

$$(0.13)\quad F_j^{(s)}(w_i) = 0 \quad \textit{for all } j = 0, 1, \ldots, N, \textit{ all } s = 0, \ldots, L - 1 \textit{ and } i = 1, 2.$$

Indeed, the number of unknowns $p(l_1, l_2)$ is at least $L(\log L)^{1/2}$, and the number of equations in (0.13) is at most $2(N + 1)L$. Moreover, as is well known,

$$(0.14)\qquad \log\left|\frac{1}{k!}\left\{f(w_i)^l\right\}^{(k)}\right| \leqslant c_1(k + l) + l\log(k + 1).$$

Lemma 0.5 now follows easily.

For $i = 1, 2$ and any $j = 0, 1, \ldots, N$ we denote by u_i^j the smallest $u \geqslant 0$ such that

$$F_j^{(u)}(w_i) \neq 0.$$

It follows immediately from Lemma 0.5 that

$$(0.15)\qquad u_i^j \geqslant L \quad \text{as } i = 1, 2; j = 0, 1, \ldots, N.$$

LEMMA 0.6. *If* $i = 1, 2$ *and* $j = 0, 1, \ldots, N$, *then the number* $F_j^{(u_i^j)}(w_i)$ *belongs to* $\mathbf{K}$, *and for its size* $\left|\overline{}\right|$ *and denominator we have*

$$(0.16)\qquad \begin{aligned} \log\left|\overline{F_j^{(u_i^j)}(w_i)}\right| &\leqslant c_2 u_i^j\left(\log u_i^j\right)^{1/2} + u_i^j \log u_i^j; \\ \log \operatorname{den} F_j^{(u_i^j)}(w_i) &\leqslant c_3 u_i^j\left(\log u_i^j\right)^{-1/4}. \end{aligned}$$

Indeed, it suffices to consider (0.15), (0.11), (0.14), and the explicit form of $F_j^{(s)}(w_i)$ (see (0.8) and (0.9) above).

An upper bound for $|F_j^{(u_i^j)}(w_i)|$ can be obtained by using Schwarz's lemma in the version of Bertrand [7]. By the definition of $h(z)$, the function

$$G_j(z) = h(z)^{L_2} F_j(z)$$

is entire and u_i^j is the smallest $u \geqslant 0$ such that

$$G_j^{(u)}(w_i) \neq 0.$$

Applying Jensen's principle to $G_j(z)$ (cf. [7]), we obtain for any $i = 1, 2$ and $j = 0, 1, \ldots, N$

(0.17)

$$u_1^j + u_2^j \leqslant \frac{1}{\log \Lambda}\left(\log|h^{L_2}F_j(z - w_i)|_{\Lambda r} - \log|h^{L_2}(w_i)| - \log\frac{1}{u_1^j!}|F_j^{(u_i^j)}(w_i)|\right)$$

where $r = 1 + |\zeta|$ and $\Lambda = (u_i^j)^{1/\rho}$. From (0.17), (1.10), and (0.11) we at once obtain

LEMMA 0.7. *For any $j = 0, 1, \ldots, N$ and $i = 1, 2$ we have*

$$u_1^j + u_2^j \leqslant \frac{\rho}{\log u_i^j}\left[c_4 u_i^j (\log u_i^j)^{3/4} + u_i^j \log u_i^j - \log|F_j^{(u_i^j)}(w_i)|\right]. \tag{0.18}$$

From Lemmas 0.6 and 0.7 we obtain

LEMMA 0.8. *For any $j = 0, 1, \ldots, N$ and $i = 1, 2$ we have*

$$u_1^j + u_2^j \leqslant 2\rho\left(u_i^j + c_5 u_i^j (\log u_i^j)^{-1/4}\right). \tag{0.19}$$

PROOF OF LEMMA 0.8. According to Lemma 0.6,

$$-\log|F_j^{(u_i^j)}(w_i)| \leqslant c_6 u_i^j (\log u_i^j)^{3/4} + u_i^j \log u_i^j.$$

Substituting this into (0.18), we obtain (0.19).

It follows from (0.19) that for $j = 0, 1, \ldots, N$ and $i = 1, 2$

$$\log(u_1^j + u_2^j) - \log(2\rho + 1) \leqslant \log u_i^j. \tag{0.20}$$

In view of (0.20), we can put

$$\log u_i^j = \log u^j \quad \text{as } i = 1, 2 \text{ and } j = 0, \ldots, N. \tag{0.21}$$

From Lemmas 0.5 and 0.4 we obtain

LEMMA 0.9. *For all $s \geqslant 0$ and $k = 0, 1, 2, \ldots, [(N - 2)/2]$ we have*

$$\left(F_{2k+2}^{(s)}(w_1)\right)^\sigma = F_{2k+1}^{(s)}(w_1); \qquad \left(F_{2k+1}^{(s)}(w_2)\right)^\sigma = F_{2k}^{(s)}(w_2). \tag{0.22}$$

In particular, by definition of u_i^j,

$$u_1^{2k+2} = u_1^{2k+1}; \qquad u_2^{2k+1} = u_2^{2k}. \tag{0.23}$$

In view of (0.21) and (0.23), we obtain

$$\log u_i^j = \log u \quad \text{for } i = 1, 2 \text{ and } j = 0, \ldots, N. \tag{0.24}$$

In view of Lemma 0.7,

$$\log|F_j^{(u_i^j)}(w_i)| \leqslant -\frac{1}{\rho}\left(u_1^j \log u + u_2^j \log u\right) + u_i^j \log u(1 + o(1)). \tag{0.25}$$

Also, for $i = 1, 2$ and $j = 0, 1, \ldots, N$,

$$(0.26)\qquad 2\log \operatorname{den} F_j^{(u_i^j)}(w_i) + \log|F_j^{(u_i^j)}(w_i)| + \log\left|\left(F_j^{(u_i^j)}(w_i)\right)^{\sigma}\right| \geqslant 0.$$

LEMMA 0.10. *For any $k = 0, 1, \ldots, [(N-3)/2]$ we have*

$$(0.27)\qquad 2(\rho-1)u_1^{2k+1}(1+o(1)) \geqslant u_2^{2k} + u_2^{2k+2};$$

$$(0.28)\qquad 2(\rho-1)u_2^{2k+2}(1+o(1)) \geqslant u_1^{2k+1} + u_1^{2k+3};$$

$$(0.29)\qquad 2(\rho-1)u_2^{0}(1+o(1)) \geqslant u_1^{0} + u_1^{1}.$$

PROOF OF LEMMA 0.10. In view of (0.26), (0.22), and (0.16),

$$c_7 u_1^{2k+1}(\log u)^{3/4} + \log|F_{2k+1}^{(u_1^{2k+1})}(w_1)| + \log|F_{2k+2}^{(u_1^{2k+2})}(w_1)| \geqslant 0.$$

Now using (0.25), we obtain

$$(0.30)\qquad \left(u_1^{2k+1}\log u + u_1^{2k+2}\log u\right)(1+o(1)) - \frac{1}{\rho}\left(u_1^{2k+1} + u_2^{2k+1}\right)\log u - \frac{1}{\rho}\left(u_1^{2k+2} + u_2^{2k+2}\right)\log u \geqslant 0$$

But, by (0.23), $u_1^{2k+2} = u_1^{2k+1}$ and $u_2^{2k+1} = u_2^{2k}$. Then (0.30) implies (0.27):

$$2(\rho-1)u_1^{2k+1}(1+o(1)) \geqslant u_2^{2k} + u_2^{2k+2}.$$

Similarly, from (0.26), (0.22), and (0.16) we obtain

$$\left(u_2^{2k+2} + u_2^{2k+3}\right)\log u(1+o(1)) - \frac{1}{\rho}\left(u_1^{2k+2} + u_2^{2k+2}\right)\log u - \frac{1}{\rho}\left(u_1^{2k+3} + u_2^{2k+3}\right)\log u \geqslant 0.$$

From this we obtain (0.28). Finally, from (0.26) and (0.25) with $j = 0$ and from Lemmas 0.5 and 0.9 we obtain

$$c_8 u_2^0(\log u)^{3/4} + u_2^0\log u(1+o(1)) - \frac{1}{\rho}\left(u_1^0 + u_2^0\right)\log u + u_2^1\log u(1+o(1)) - \frac{1}{\rho}\left(u_1^1 + u_2^1\right)\log u \geqslant 0,$$

i.e. (0.29) holds in view of (0.23).

We define a sequence v_j: $j = 0, 1, \ldots, N$, as follows:

$$(0.31)\qquad v_{2k+1} = u_1^{2k+1}, \qquad v_{2k} = u_2^{2k}.$$

Taking into account (0.27) and (0.28), we see that for $n = 0, 1, \ldots, N-2$ we have

$$(0.32)\qquad 2(\rho-1)v_{n+1}(1+o(1)) \geqslant v_n + v_{n+2}.$$

We will make use of the Čebyšev polynomials (of the second kind) $U_n(x)$: $U_0(x) = 1$, $U_1(x) = 2x$, $U_{n+1}(x) = 2xU_n(x) - U_{n-1}(x)$.

LEMMA 0.11. *If $x_0 = \rho - 1$, then for $n + k \leqslant N$ we have*

$$v_{n+k} \leqslant U_k(x_0)\cdot v_n(1 + o(1)) - U_{k-1}(x_0)v_{n-1}. \tag{0.33}$$

PROOF OF LEMMA 0.11. Assume that (0.33) holds for a given k and any n such that $n + k \leqslant N$. From (0.32) we obtain

$$\begin{aligned} v_{n+k+1} &\leqslant U_k(x_0)v_{n+1}(1 + o(1)) - U_{k-1}(x_0)\cdot v_n \\ &\leqslant 2xU_k(x_0)\cdot v_n(1 + o(1)) - U_k(x_0)v_{n-1} - U_{k-1}(x_0)v_n \\ &\leqslant U_{k+1}(x_0)v_n(1 + o(1)) - U_k(x_0)v_{n-1}. \end{aligned}$$

Lemma 0.11 is proved. Consequently, if $x_0 = \rho - 1 \leqslant 1$, we have

$$0 < L \leqslant U_{N-1}(1)v_1(1 + o(1)) - U_{N-2}(1)v_0. \tag{0.34}$$

Since $U_n(1) = n + 1$, it follows from (0.34) that

$$\frac{v_0}{v_1} \leqslant \frac{N}{N-1} + \frac{\varepsilon}{2} < 1 + \frac{1}{N-1} + \frac{\varepsilon}{2} < 1 + \varepsilon \tag{0.35}$$

for N sufficiently large in comparison with ε^{-1}, if we take into account the choice of N. Thus, in view of (0.31),

$$u_2^0 < (1 + \varepsilon)u_1^1.$$

It follows from (0.29) that

(0.36)

$$2(\rho - 1)u_2^0(1 + o(1)) \geqslant u_1^0 + (1 - \varepsilon)u_2^0 \quad \text{or} \quad (2\rho - 3 + \varepsilon)u_2^0(1 + o(1)) \geqslant u_1^0.$$

Since $w_1 = 0 \in \mathbf{Z}$, we have $F_0^{(u_1^0)}(w_1) \in \mathbf{Z}$ and it follows from (0.18) that

$$\frac{u_1^0 + u_2^0}{\rho} \leqslant u_1^0(1 + o(1)), \quad \text{i.e. } u_2^0 \cdot \frac{1}{\rho - 1} \leqslant u_1^0(1 + o(1)). \tag{0.37}$$

From (0.36) and (0.37) we obtain

$$2\rho - 3 + \varepsilon \geqslant \frac{1}{\rho - 1}.$$

For small ε this implies that $\rho > 2 - \varepsilon$. Letting $\varepsilon \to 0$, we see that $\rho \geqslant 2$. Theorem 0.1 is proved.

The main idea of §0 is realized in the general case below. The proof of Theorem 1 is divided into four parts. In §1 we prove certain auxiliary lemmas. In §2 we construct sets $\{\lambda_j : j \in \mathbf{I}\}$ corresponding to auxiliary functions $F_j(z) = P(z + \lambda_j, f(x))$. The properties of these functions are studied in §3, where we develop a scheme for a random walk in a $(d-1)n$-dimensional lattice. In §4, by analyzing this scheme for a random walk, we obtain $n \leqslant \rho$.

1. Auxiliary lemmas. We will prove a series of auxiliary lemmas in the n-dimensional case.

LEMMA 1.1 (LEMMA 3.4 OF [4]). *Suppose $f(z)$ is a function analytic in a neighborhood of a point $w \in \mathbf{C}^n$. Then there exists a constant $C_1^0 > 0$ such that for* *all* $l \geqslant 1$ *and* $k \in \mathbf{N}^n$, $|k| = K$, *we have*

$$\log\left|\frac{1}{k!}D^k f^l(w)\right| \leqslant C_1^0(K+l) + nl\log(K+1).$$

Furthermore, if $f(z)$ *is entire in* $\mathbf{C}^n$ *of (strict) order* $\leqslant \rho$, *then there exists a constant* $C_2^0 > 0$ *such that*

$$\log\left|\frac{1}{k!}D^k f^l(w)\right| + \frac{K}{\rho}\log\left(\frac{K}{l}+1\right) \leqslant C_2^0(K+l) + nl\log(K+1).$$

Here, as usual, an entire function f in $\mathbf{C}^n$ has (strict) order at most ρ if

$$\log|f|_R \ll R^\rho \quad \text{as } R \to \infty.$$

For $k = (k_1,\ldots,k_n) \in \mathbf{N}^n$ we denote by D^k the operator $\partial^{k_1}/\partial z^{k_1} \cdots \partial^{k_n}/\partial z^{k_n}$ and we put $|k| = k_1 + \cdots + k_n$.

The following lemmas from **[1]** are of a more general nature than the previous ones.

LEMMA 1.2. *Suppose* $f_1,\ldots,f_\nu$ *are analytic functions in a neighborhood of a point* $w \in \mathbf{C}^n$ *such that all of the numbers*

$$D^k f_j(w), \qquad k \in \mathbf{N}^n, \qquad j = 1,\ldots,\nu,$$

are algebraic. Suppose also there exist constants $c[0]$, $c[j]\colon j = 1,\ldots,\nu$, *such that*

$$\log\left|\overline{D^k f_j(w)}\right| \leqslant c[j]\,|k|\log(|k|+1) + c[0]$$

for all $k \in \mathbf{N}^n$, $1 \leqslant j \leqslant \nu$. *If* $P(x_1,\ldots,x_\nu) \in \mathbf{Z}[x_1,\ldots,x_\nu]$ *and* $\deg_{x_i}(P) \leqslant \Lambda_i$: $i = 1,\ldots,\nu$, *then the numbers*

$$\beta_k = D^k P(f_1,\ldots,f_\nu)(w)\colon \qquad k \in \mathbf{N}^n,$$

are algebraic. Furthermore, for any natural number K *and real* τ, $0 < \tau < 1$, *we have*

$$\max_{|k|=K}\log|\overline{\beta_k}| \leqslant \log H(P) + K + \left(\sum_{i=1}^{\nu}\Lambda_i\right)\left((n\nu+1)\log(K+1) + c[0] + 1\right)$$

$$+\begin{cases}\left(\max\limits_{i=1,\ldots,\nu} c[i]\right)K\log(K+1), & \text{if } \max\limits_{i=1,\ldots,\nu} c[i] \geqslant 1,\\ K\log(K+1) - \min\limits_{i=1,\ldots,\nu}(1-\tau)(1-c[i])K\log\left(\dfrac{\tau K}{\nu\Lambda_i}+1\right), & \text{otherwise.}\end{cases}$$

As usual, for a polynomial $P(x_1,\ldots,x_\nu)$ we denote by $H(P)$ the height of the maximum of the moduli of the coefficients. If ξ is an algebraic number, $\xi \in \overline{\mathbf{Q}}$, d is its degree, and $\xi^{(1)} = \xi,\ldots,\xi^{(d)}$ are all of the conjugates of ξ, then $|\overline{\xi}| = \max_{i=1,\ldots,d}|\xi^{(i)}|$.

In the algebraic part of the proof we use Siegel's lemma (see, e.g., Lemma 1.3.1 of **[8]**).

LEMMA 1.3. *Suppose* $\mathbf{K}$ *is an algebraic number field of degree* d. *Also, suppose* $a_{i,j} \in \mathbf{K}$: $i = 1,\ldots,n$, $j = 1,\ldots,m$, *and* δ *is a common denominator of all* $a_{i,j}$ $(i = 1,\ldots,n, j = 1,\ldots,m)$, *and*

$$A \geqslant \max_{1 \leqslant j \leqslant m, 1 \leqslant i \leqslant n} |\overline{a_{i,j}}| n\delta.$$

If $n > dm$, *then the system of equations*

$$\sum_{i=1}^{n} a_{i,j} x_j = 0: \qquad 1 \leqslant j \leqslant m,$$

has a nonzero solution in integers $(x_1,\ldots,x_n) \in \mathbf{Z}^n$ *such that*

$$\max_{i=1,\ldots,n} |x_i| \leqslant \left(\sqrt{2}\, A\right)^{md/(n-md)}.$$

We will also need certain auxiliary analytic results. For a finite $S \subset \mathbf{C}^n$ we denote by $\Omega(S)$ the minimum degree of a hypersurface containing S:

$$\Omega(S) = \min\{\deg(P): P(\bar{x}) \in \mathbf{C}[\bar{x}], P(\bar{x}) \not\equiv 0; P(w) = 0 \text{ for all } w \in S\}.$$

A multidimensional analogue of Schwarz's lemma is Lemma 3.1 of **[1]** (cf. **[2]**).

LEMMA 1.4. *Suppose* $S \subset \mathbf{C}^n$ *is finite and* $\varepsilon > 0$. *Then there exists* $r_0 = r_0(S, \varepsilon) > 0$ *such that for any natural number* M *and any nonzero entire function* $F(\bar{z})$ *in* $\mathbf{C}^n$ *satisfying*

$$D^k F(w) = 0 \quad \text{as } w \in S, |k| < M, k \in \mathbf{N}^n,$$

we have for $R > r > r_0$ *the estimate*

$$\log|F|_r \leqslant \log|F|_R - (1-\varepsilon)\left(\frac{\Omega(S)}{n} - \varepsilon\right) M \log\frac{\varepsilon R}{6nr}.$$

Finally, a consequence of the fact that $\mathbf{C}$ is algebraically closed is the following lemma, which reduces all subsets of $\mathbf{C}^n$ to finite ones.

LEMMA 1.5. *If* $S \subseteq \mathbf{C}^n$ *and for each finite* $S_0 \subseteq S$ *we have* $\Omega(S_0) \leqslant n_0$, *then* S *is contained in an algebraic hypersurface of degree* $\leqslant n_0$.

We present several more lemmas of an auxiliary nature.

LEMMA 1.6. *Suppose* $P(x, y) \in \mathbf{Z}[x, y]$,

$$P(x, y) = \sum_{l_1=0}^{L_1-1} \sum_{l_2=0}^{L_2-1} p(l_1, l_2) x^{l_1} y^{l_2},$$

and $f(z)$ *is an analytic function in a neighborhood of* z_0 *such that* $f^{(k)}(z_0) \in \mathbf{Z}$ *for all* $k \geqslant 0$. *Assume that* z_0 *is an algebraic number. Then for the function*

(1.1) $$F(z) = P(z, f(z))$$

we have

1. $F^{(k)}(z_0)$ *is an algebraic number, and* $F^{(k)}(z_0)$ *belongs to the ring* $\mathbf{Z}[z_0]$ *for all* $k \geqslant 0$.

2. *There exist positive constants* c_1 *and* c_2, *depending only on* z_0 *and* $f(z)$, *such that for the size* $|\overline{F^{(k)}(z_0)}|$ *we have*

$$\begin{aligned}(1.2)\ \log|\overline{F^{(k)}(z_0)}| &\leqslant k\log(k+1)+\log(L_1L_2)\\ &\quad +c_1(k+L_1+L_2)+c_1L_2\log(k+1)+\log H(P)\\ &\leqslant k\log(k+1)+\log H(P)+c_2(k+L_1)+c_2L_2\log(k+2),\end{aligned}$$

where $H(P)$ *is the height of the polynomial* $P(x,y)$.

3. *If* $\delta(z_0)$ *is the denominator of* $z_0\in\overline{\mathbf{Q}}$, *then* $\delta(z_0)^{L_1-k}$ *is a denominator of the algebraic number* $F^{(k)}(z_0)$ *for any* $k\geqslant 0$.

PROOF OF LEMMA 1.6. We have

$$F(z)=\sum_{l_1=0}^{L_1-1}\sum_{l_2=0}^{L_2-1}p(l_1,l_2)z^{l_1}f(z)^{l_2}.$$

Thus, for $k\geqslant 0$,

$$(1.3)\qquad F^{(k)}(z_0)=\sum_{l_1=0}^{L_1-1}\sum_{l_2=0}^{L_2-1}p(l_1,l_2)\cdot\sum_{s=0}^{k}C_k^s l_1\cdots(l_1-s+1)z_0^{l_1-s}\{f(z_0)^{l_2}\}^{(k-s)}.$$

Since $f^{(r)}(z_0)\in\mathbf{Z}$ for all $r\geqslant 0$, we have $\{f(z_0)^l\}^{(m)}\in\mathbf{Z}$ for all $m\geqslant 0$. Moreover, $p(l_1,l_2)\in\mathbf{Z}$ for all $l_1\leqslant L_1-1$, $l_2\leqslant L_2-1$, and $z_0^n\in\mathbf{Z}[z_0]$ for $n\geqslant 0$. Thus, for any $k\geqslant 0$,

$$F^{(k)}(z_0)\in\mathbf{Z}[z_0].$$

Similarly, it follows from (1.3) that $\delta(z_0)^{L_1-1}$ is a denominator of $F^{(k)}(z_0)$ for any $k\geqslant 0$. Thus, assertions 1 and 3 are proved. Let us now estimate the size of $F^{(k)}(z_0)$. In view of (1.3), if we put $|\overline{z_0^*}|=\max\{1,|\overline{z_0}|\}$, we have

$$(1.4)\qquad |\overline{F^{(k)}(z_0)}|\leqslant H(P)\cdot|\overline{z_0^*}|^{L_1}\times\sum_{l_1=0}^{L_1-1}\sum_{l_2=0}^{L_2-1}\sum_{s=0,l_1\geqslant s}^{k}C_k^s l_1\cdots(l_1-s+1)\{f(z_0)^{l_2}\}^{(k-s)}.$$

By Lemma 1.1,

$$\log\left|\frac{1}{k_1!}\{f(z_0)^l\}^{(k_1)}\right|\leqslant c_0(k_1+l)+l\cdot\log(k_1+1)$$

for any $k_1 \geqslant 0$ and $l \geqslant 0$. Substituting into (1.4), we obtain

$$
\begin{aligned}
|\overline{F^{(k)}(z_0)}| &\leqslant H(P)\cdot|\overline{z_0^*}|^{L_1} \\
&\quad\times \sum_{l_1=0}^{L_1-1}\sum_{l_2=0}^{L_2-1}\sum_{s=0,l_1\geqslant s}^{k} \frac{k!}{s!(k-s)!} l_1 \cdots (l_1-s+1)\{f(z_0)^{l_2}\}^{(k-s)} \\
&\leqslant H(P)\cdot|\overline{z_0^*}|^{L_1}\cdot \sum_{l_1=0}^{L_1-1}\sum_{l_2=0}^{L_2-1}\sum_{s=0,l_1\geqslant s}^{k} \frac{k!}{s!(k-s)!} \\
&\quad\times \frac{l_1!}{(l_1-s)!}(k-s)!e^{c_0(k-s+L_2)}\cdot e^{L_2\log(k+1)} \\
&\leqslant H(P)\cdot|\overline{z_0^*}|^{L_1} e^{c_0(k+L_2)+L_2\log(k+1)} \\
&\quad\times \sum_{l_1=0}^{L_1-1}\sum_{l_2=0}^{L_2-1}\sum_{s=0,l_1\geqslant s}^{k} \frac{k!l_1!}{s!(l_1-s)!} \\
&\leqslant H(P)\cdot|\overline{z_0^*}|^{L_1}\cdot e^{c_0(k+L_2)+L_2\log(k+1)} L_1L_2\cdot 2^{L_1}\cdot k!.
\end{aligned}
$$

Finally, for $c_1 = c_1(z_0, c_0) > 0$,

$$
\begin{aligned}
\log|\overline{F^{(k)}(z_0)}| &\leqslant \log(L_1L_2) + \log H(P) + k\log(k+1) \\
&\quad + c_1(k + L_1 + L_2) + c_1L_2\log(k+1).
\end{aligned}
$$

Thus, Lemma 1.6 is completely proved. We will use the estimates of this lemma later on.

In applying the main idea concerning the use of conjugate functions we will need the following lemma.

LEMMA 1.7. *Suppose $P(x, y) \in \mathbf{Z}[x, y]$ and $f(z)$ is an analytic function. Assume that w, λ_1, λ_2 are algebraic numbers such that*

(a) *$f^{(k)}(w) \in \mathbf{Z}$ for all $k \geqslant 0$;*

(b) *$w + \lambda_1$ is conjugate to $w + \lambda_2$, i.e. for some isomorphism $\sigma\colon \mathbf{Q}(w, \lambda_1, \lambda_2) \to \mathbf{C}$ we have $\sigma(w + \lambda_1) = w + \lambda_2$.*

Then for the two auxiliary functions

$$F_{\lambda_1}(z) = P(z + \lambda_1, f(z)) \quad \text{and} \quad F_{\lambda_2}(z) = P(z + \lambda_2, f(z)) \tag{1.5}$$

we have

1. *For all $k \geqslant 0$, the numbers $F_{\lambda_1}^{(k)}(w)$ and $F_{\lambda_2}^{(k)}(w)$ are algebraic and belong to the field $\mathbf{Q}(w, \lambda_1, \lambda_2)$.*

2. *For all $k \geqslant 0$, the algebraic numbers $F_{\lambda_1}^{(k)}(w)$ and $F_{\lambda_2}^{(k)}(w)$ are conjugate and, moreover,*

$$\sigma\left(F_{\lambda_1}^{(k)}(w)\right) = F_{\lambda_2}^{(k)}(w).$$

PROOF OF LEMMA 1.7. Suppose

$$P(x, y) = \sum_{l_1=0}^{L_1-1}\sum_{l_2=0}^{L_2-1} p(l_1, l_2)x^{l_1}y^{l_2}, \tag{1.6}$$

where $p(l_1, l_2) \in \mathbf{Z}$ by hypothesis. Then for any $k \geqslant 0$ we have, in view of (1.6),

$$(1.7) \qquad F_{\lambda_1}^{(k)}(z) = \sum_{l_1=0}^{L_1-1} \sum_{l_2=0}^{L_2-1} p(l_1, l_2) \sum_{s=0}^{k} C_k^s \times l_1 \cdots (l_1 - s + 1)(z + \lambda_1)^{l_1 - s}\{f(z)^{l_2}\}^{(k-s)};$$

$$F_{\lambda_2}^{(k)}(z) = \sum_{l_1=0}^{L_1-1} \sum_{l_2=0}^{L_2-1} p(l_1, l_2) \sum_{s=0}^{k} C_k^s \times l_1 \cdots (l_1 - s + 1)(z + \lambda_2)^{l_1 - s}\{f(z)^{l_2}\}^{(k-s)}.$$

Now $p(l_1, l_2) \in \mathbf{Z}$, and $\{f(w)^l\}^{(m)} \in \mathbf{Z}$ for all $l \geqslant 0$, $m \geqslant 0$ by assumption (a). Consequently, for any $k \geqslant 0$ the numbers $F_{\lambda_1}^{(k)}(w)$ and $F_{\lambda_2}^{(k)}(w)$ belong to the field $\mathbf{Q}(w, \lambda_1, \lambda_2)$. Assertion 1 is proved. Let us show that $F_{\lambda_1}^{(k)}(w)$ and $F_{\lambda_2}^{(k)}(w)$ are conjugate. From (1.7) we obtain

$$(1.8) \qquad \sigma\left(F_{\lambda_1}^{(k)}(w)\right) = \sum_{l_1=0}^{L_1-1} \sum_{l_2=0}^{L_2-1} p(l_1, l_2) \sum_{s=0}^{k} C_k^s \cdot l_1 \cdots (l_1 - s + 1) \times \{\sigma(w + \lambda_1)\}^{l_1 - s} \cdot \{f(w)^{l_2}\}^{(k-s)},$$

since $p(l_1, l_2) \in \mathbf{Z}$ and $\{f(w)^{l_2}\}^{(k-s)} \in \mathbf{Z}$. Applying (b), we obtain from (1.8) that

$$(1.9) \qquad \sigma\left(F_{\lambda_1}^{(k)}(w)\right) = \sum_{l_1=0}^{L_1-1} \sum_{l_2=0}^{L_2-1} p(l_1, l_2) \sum_{s=0}^{k} C_k^s l_1 \cdots (l_1 - s + 1) \times \{w + \lambda_2\}^{l_1 - s} \cdot \{f(w)^{l_2}\}^{(k-s)}.$$

It follows from (1.9) and (1.7) that

$$\sigma\left(F_{\lambda_1}^{(k)}(w)\right) = F_{\lambda_2}^{(k)}(w)$$

for all $k \geqslant 0$. Consequently, the numbers $F_{\lambda_1}^{(k)}(w)$ and $F_{\lambda_2}^{(k)}(w)$ are conjugate for all $k \geqslant 0$ and Lemma 1.7 is completely proved.

We can also write Lemma 1.6 in a more convenient form.

LEMMA 1.8. *Suppose $P(x, y) \in \mathbf{Z}[x, y]$ and $f(z)$ is an analytic function in a neighborhood of w such that $f^{(k)}(w) \in \mathbf{Z}$ for all $k \geqslant 0$. Assume that w and λ are algebraic numbers. Then for the function*

$$F_\lambda(z) = P(z + \lambda, f(z))$$

we have

1. *For $k \geqslant 0$, the number $F_\lambda^{(k)}(w)$ is algebraic and belongs to the ring $\mathbf{Z}[w, \lambda]$.*
2. *There exists a constant $c_3 = c_3(w, \lambda) > 0$ such that for all $k \geqslant 0$ we have for the size $\overline{|F_\lambda^{(k)}(w)|}$ the estimate*

$$\log \overline{|F_\lambda^{(k)}(w)|} \leqslant k \log(k+1) + \log H(P) + c_3(k + \deg_x(P)) + c_3 \deg_y(P) \log(k+2).$$

3. *For the denominator* $\partial(F_\lambda^{(k)}(w))$ *of the number* $F_\lambda^{(k)}(w)$ *we have the estimate*

$$\partial\big(F_\lambda^{(k)}(w)\big) \leqslant \deg_x(P)\cdot\partial(w+\lambda),$$

where $\partial(w+\lambda)$ *is the denominator of* $w+\lambda$.

PROOF OF LEMMA 1.8. Consider the function $g(z) = f(z-\lambda)$ and put

$$F_1(z) = P(z, g(z)).$$

Then $F_1(w+\lambda) = F_\lambda(w)$ and in general, for any $k \geqslant 0$,

$$\frac{d^k}{dz^k}F_1(z)\Big|_{z=w+\lambda} = F_\lambda^{(k)}(w).$$

However, $w+\lambda$ is algebraic, $g(z)$ is analytic in a neighborhood of $w+\lambda$, and $g^{(k)}(w+\lambda) \in \mathbf{Z}$ for all $k \geqslant 0$. Consequently, Lemma 1.8 follows directly from Lemma 1.6. Incidentally, all estimates in Lemma 1.8 can be established directly.

The above auxiliary results will enable us to construct the sequence of auxiliary "conjugate" functions needed for the proof of Theorem 1.

2. Construction of a family of auxiliary functions (beginning of the proof of Theorem 1). Suppose $f(z)$ is a meromorphic function in $\mathbf{C}$ of finite order $< \rho_0$. Then there exists an entire function $h(z)$ in $\mathbf{C}$ of order $< \rho_0$ such that $h(z)f(z)$ is also entire of order $< \rho_0$.

Consider the set A_f of algebraic points of $f(z)$, in the sense that

$$A_f = \big\{z_0 \in \overline{\mathbf{Q}} : f(z) \text{ is analytic in a neighborhood of } z_0, \\ \text{and } f^{(k)}(z_0) \in \mathbf{Z} \text{ for all } k \geqslant 0\big\}.$$

Let $S = \{w_1,\dots,w_n\}$ be some subset of A_f. Then

(2.1) $f(z)$ is analytic in a neighborhood of $w_1,\dots,w_n$,

(2.2) the w_i: $i = 1,\dots,n$ are distinct algebraic numbers,

(2.3) $f^{(k)}(w_i) \in \mathbf{Z}$ for all $k \geqslant 0$ and $i = 1,\dots,n$.

Let $\mathbf{K}$ denote a normal algebraic number field of finite degree containing $\{w_1,\dots,w_n\}$; for example, $\mathbf{K}$ can be the Galois splitting field of $\mathbf{Q}(w_1,\dots,w_n)$. Let G denote the Galois group of $\mathbf{K}$, $G = \mathrm{Gal}(\mathbf{K}/\mathbf{Q})$. For the degree of $\mathbf{K}$ we have

$$[\mathbf{K}:\mathbf{Q}] = |G| = d.$$

Our initial goal is to construct an *infinite* set $\{\lambda_j : j \in \mathbf{I}\}$ of elements of $\mathbf{K}$ such that for any $j \in \mathbf{I}$, $i = 1,\dots,n$, and $g \in G$ there exists $j_1 = F(i,j,g) \in \mathbf{I}$ such that

$$g(w_i + \lambda_j) = w_i + \lambda_{j_1} = w_i + \lambda_{F(i,j,g)}. \tag{2.4}$$

Here and below, the action of an automorphism $g\colon \mathbf{K} \to \mathbf{K}$ on an element ξ of $\mathbf{K}$ will be denoted by $g(\xi)$ or $\xi^{(g)}$.

It is easy to see that for any $w_i, \lambda_j \in \mathbf{K}$ and $g \in G$ we can choose $\lambda_{j_1} \in \mathbf{K}$ such that (2.4) holds, namely,

$$\lambda_{j_1} = w_i^{(g)} + \lambda_j^{(g)} - w_i.$$

Thus, a set $\{\lambda_j: j \in \mathbf{I}\}$ satisfying (2.4) exists by Zorn's lemma. We will *effectively* construct $\{\lambda_j: j \in \mathbf{I}\}$, since in constructing the auxiliary functions *we must distinguish a finite subset* $\{\lambda_j: j \in \mathbf{I}'\}, |\mathbf{I}'| < \infty$.

We will use the *group ring* $\mathbf{Z}[G]$. The elements α of $\mathbf{Z}[G]$ have the form

(2.5) $$\alpha = \sum_{g \in G} n_g \cdot g$$

where $n_g \in \mathbf{Z}$: $g \in G$. Addition in $\mathbf{Z}[G]$ is componentwise and multiplication is induced by G:

$$\Big(\sum_{g\in G} n_g \cdot g\Big)\Big(\sum_{g\in G} m_g \cdot g\Big) = \sum_{g_1, g_2 \in G} n_{g_1} m_{g_2} \cdot g_1 g_2.$$

Finally, $\sum_{g\in G} n_g \cdot g = \sum_{g \in G} m_g \cdot g$ if and only if $n_g = m_g$ for all $g \in G$. We define on the elements α of $\mathbf{Z}[G]$ the trace function $\mathrm{Tr}(\alpha)$:

$$\text{if } \alpha = \sum_{g\in G} n_g \cdot g, \text{ then } \mathrm{Tr}(\alpha) = \sum_{g \in G} n_g \in \mathbf{Z}.$$

Let $\mathbf{Z}_0[G]$ denote the set of elements of trace zero:

(2.6) $$\mathbf{Z}_0[G] = \Big\{\alpha \in \mathbf{Z}[G], \alpha = \sum_{g\in G} n_g \cdot g \colon \sum_{g\in G} n_g = 0\Big\}.$$

We will denote elements of $\mathbf{Z}_0[G]$ by $\mathfrak{s}$. With each element $\mathfrak{s}$ of $\mathbf{Z}_0[G]$ we can associate a vector $v(\mathfrak{s})$ with integral coordinates, indexed by the elements of $G\backslash\{1\}$. Namely,

(2.7) $$\text{if } \mathfrak{s} = \sum_{g\in G} n_g \cdot g, \text{ we put } v(\mathfrak{s}) = (n_g \colon g \in G\backslash\{1\}).$$

The mapping $v\colon \mathbf{Z}_0[G] \to \mathbf{Z}^{G\backslash\{1\}}$ is one-to-one, since an element $\mathfrak{s} = \sum_{g\in G} n_g \cdot g$ of $\mathbf{Z}_0[G]$, having trace zero, is uniquely determined by those n_g for which $g \neq 1$.

The set $\mathbf{Z}_0[G]$ is an ideal of the ring $\mathbf{Z}[G]$. Indeed, the function $\mathrm{Tr}\colon \mathbf{Z}[G] \to \mathbf{Z}$ possesses the following properties:

(2.8)
$$\begin{aligned} &\mathrm{Tr}(\alpha + \beta) = \mathrm{Tr}(\alpha) + \mathrm{Tr}(\beta) && \text{for } \alpha, \beta \in \mathbf{Z}[G];\\ &\mathrm{Tr}(\alpha \cdot n) = \mathrm{Tr}(\alpha) \cdot n && \text{for } n \in \mathbf{Z}, \alpha \in \mathbf{Z}[G];\\ &\mathrm{Tr}(\alpha \cdot g) = \mathrm{Tr}(\alpha) && \text{for } g = 1 \cdot g \in \mathbf{Z}[G], \alpha \in \mathbf{Z}[G]. \end{aligned}$$

It follows from (2.8) that

(2.9) $$\mathrm{Tr}(\alpha \cdot \beta) = \mathrm{Tr}(\alpha) \cdot \mathrm{Tr}(\beta) \quad \text{for } \alpha, \beta \in \mathbf{Z}[G]$$

and that $\mathbf{Z}_0[G]$ is an ideal of $\mathbf{Z}[G]$.

Our main interest is the ring $\mathbf{Z}_0[G]^n$. Elements of $\mathbf{Z}_0[G]^n$ will be denoted by $\mathfrak{A} = (\mathfrak{s}_1, \ldots, \mathfrak{s}_n)$, where $\mathfrak{s}_i \in \mathbf{Z}_0[G]$. For brevity, we put

(2.10) $$\mathcal{G}_0 = \mathbf{Z}_0[G]^n \quad \text{and} \quad \mathrm{C} = G\backslash\{1\}.$$

It will be convenient to represent $\mathcal{G}_0$ as a lattice in $(d-1)n$-dimensional space. For this purpose we use the mapping (2.7). Let $J = \mathrm{C} \times \{1, \ldots, n\}$ and $J_i = \mathrm{C} \times \{i\}$ for $i = 1, \ldots, n$, i.e. $J = \bigcup_{i=1}^n J_i$ and $\mathbf{Z}^J = \prod_{i=1}^n \mathbf{Z}^{J_i}$. In view of (2.7), there exists an isomorphism v_i: $\mathbf{Z}_0[G] \cong \mathbf{Z}^{J_i}$, i.e. there exists an isomorphism $\bar{\bar{v}} = (v_1,\ldots,v_n)$ between $\mathbf{Z}_0[G]^n$ and $\mathbf{Z}^J$:

$$\bar{\bar{v}}\colon \mathbf{Z}_0[G]^n \to \mathbf{Z}^J.$$

We define $\bar{\bar{v}}$ explicitly. If $\mathfrak{A} = (\mathfrak{s}_1,\ldots,\mathfrak{s}_n) \in \mathcal{G}_0 = \mathbf{Z}_0[G]^n$ and $\mathfrak{s}_i = \sum_{g\in G} n_{(g,i)}\cdot g \in \mathbf{Z}_0[G]$: $i = 1,\ldots,n$, put

$$\text{(2.11)} \qquad \begin{aligned} \bar{\bar{v}}(\mathfrak{A}) &= \left(n_{(g,i)}\colon g \in G\backslash\{1\}, i = 1,\ldots,n\right) \\ &= \left(n_{(g,i)}\colon (g, i) \in \mathrm{C}\times\{1,\ldots,n\}\right) = \left(n_j\colon j \in J\right). \end{aligned}$$

Obviously, $\bar{\bar{v}}$ is an isomorphism of $\mathbf{Z}_0[G]^n$ and $\mathbf{Z}^J$ as Abelian groups. Since all $\mathfrak{s}_i$: $i = 1,\ldots,n$, have trace zero, it follows that $\bar{\bar{v}}$ is one-to-one:

$$\text{(2.12)} \qquad \bar{\bar{v}}\colon \mathcal{G}_0 \cong \mathbf{Z}^{\mathrm{C}\times n}.$$

We now define a set of numbers $\{\lambda_{\mathfrak{A}}\colon \mathfrak{A} \in \mathcal{G}_0\}$ in $\mathbf{K}$ satisfying (2.4). Suppose $\mathfrak{A} = (\mathfrak{s}_1,\ldots,\mathfrak{s}_n) \in \mathcal{G}_0 = \mathbf{Z}_0[G]^n$ and $\mathfrak{s}_i = \sum_{g\in G} n_{(g,i)}\cdot g \in \mathbf{Z}_0[G]$: $i = 1,\ldots,n$. Put

$$\text{(2.13)} \qquad \lambda_{\mathfrak{A}} = \lambda(\mathfrak{A}) = \sum_{i=1}^{n} \sum_{g\in G} n_{(g,i)} w_i^{(g)},$$

where $\{w_1,\ldots,w_n\}$ is the subset of $\mathbf{K}$ satisfying (2.1)–(2.3) in which we are interested. Since $\mathbf{K}$ is a normal field, we have

$$\lambda_{\mathfrak{A}} = \lambda(\mathfrak{A}) \in K \quad \text{for } \mathfrak{A} \in \mathcal{G}_0.$$

The isomorphism $\bar{\bar{v}}$: $\mathcal{G}_0 \cong \mathbf{Z}^J$ (see (2.12)), which was defined in (2.11), enables us to define $\lambda(\vec{n})$ for all elements $\vec{n} \in \mathbf{Z}^J$:

$$\text{(2.14)} \qquad \lambda(\vec{n}) = \lambda\left(\bar{\bar{v}}^{-1}(\vec{n})\right) \in \mathbf{K} \quad \text{for } \vec{n} \in \mathbf{Z}^J.$$

Let us verify fulfillment of (2.4). Suppose $i = 1,\ldots,n$, $g \in G$, and $\mathfrak{A} = (\mathfrak{s}_1,\ldots,\mathfrak{s}_n) \in \mathcal{G}_0$, where $\mathfrak{s}_i \in \mathbf{Z}_0[G]$: $i = 1,\ldots,n$. If $g = 1$, then

$$(w_i + \lambda(\mathfrak{A}))^{(g)} = w_i + \lambda(\mathfrak{A}).$$

Suppose $g \neq 1$, i.e. $g \in \mathrm{C}$. If $\mathfrak{s}_i = \sum_{g\in G} n_{(g,i)}\cdot g$: $i = 1,\ldots,n$, then it follows from (2.13) that

(2.15)

$$\begin{aligned} (w_i + \lambda(\mathfrak{A}))^{(g)} &= \left(w_i + \sum_{j=1}^{n} \sum_{g_1\in G} n_{(g_1,j)} w_j^{(g_1)}\right)^{(g)} = w_i^{(g)} + \sum_{j=1}^{n} \sum_{g_1\in G} n_{(g_1,j)} w_j^{(g_1 g)} \\ &= \sum_{j=1, j\neq i}^{n} \sum_{g_1\in G} n_{(g_1,j)} w_j^{(g_1 g)} + w_i^{(g)} + \sum_{g_1\in G, g_1\neq n_{(g^{-1},i)}} n_{(g_1,i)} w_i^{(g_1 g)} \\ &\quad + n_{(1,i)} w_i^{(g)} + n_{(g^{-1},i)} w_i. \end{aligned}$$

We now define an element $\mathfrak{A}_1 = \mathfrak{A}_1(\mathfrak{A}, i, g) \in \mathcal{G}_0$, $\mathfrak{A}_1 = (\mathfrak{s}_1^1, \ldots, \mathfrak{s}_n^1) \in \mathbf{Z}_0[G]^n$, where

$$(2.16)\qquad \mathfrak{s}_j^1 = \sum_{g_1 \in G} n_{(g_1, j)} \cdot (g_1 g) = \mathfrak{s}_j g: \qquad j = 1, \ldots, n, \quad j \neq i,$$

(2.17)

$$\mathfrak{s}_i^1 = \sum_{\substack{g_1 \in G, \\ g_1 \neq 1, g^{-1}}} n_{(g_1, i)} \cdot (g_1 g) + n_{(1, i)} \cdot (1 \cdot g) + n_{(g^{-1}, i)} \cdot (g^{-1} g) + 1 \cdot g - 1 \cdot 1$$
$$= \mathfrak{s}_i g + 1 \cdot g - 1 \cdot 1.$$

Thus,

$$(2.18)\qquad \mathfrak{s}_j^1 = \mathfrak{s}_j g: \qquad j = 1, \ldots, n; \quad j \neq i,$$

$$(2.19)\qquad \mathfrak{s}_i^1 = \mathfrak{s}_i g + g - 1.$$

In view of (2.8) and (2.9), we have

$$\mathfrak{s}_j^i \in \mathbf{Z}_0[G]: \qquad j = 1, \ldots, n,$$

and

$$(2.20)\qquad \mathfrak{A}_1 = (\mathfrak{s}_1^1, \ldots, \mathfrak{s}_n^1) \in \mathcal{G}_0.$$

It follows from (2.16)–(2.19) and (2.13) that

$$(2.21)\qquad w_i + \lambda(\mathfrak{A}_1) = w_i + \sum_{j \neq 1, j \neq i}^{n} \sum_{g_1 \in G} n_{(g_1, j)} w_j^{(g_1 g)}$$
$$+ \sum_{g_1 \in G, g_1 \neq 1, g^{-1}} n_{(g_1, i)} w_i^{(g_1 g)} + (n_{(1, i)} + 1) w_i^{(g)}$$
$$+ (n_{(g^{-1}, i)} - 1) w_i$$
$$= \sum_{j = 1, j \neq i}^{n} \sum_{g_1 \in G} n_{(g_1, j)} w_j^{(g_1 g)}$$
$$+ \sum_{g_1 \in G, g_1 \neq 1, g^{-1}} n_{(g_1, i)} w_i^{(g_1 g)} + (n_{(1, i)} + 1) w_i^{(g)}$$
$$+ n_{(g^{-1}, i)} w_i.$$

Comparing (2.21) and (2.15), we obtain

$$(2.22)\qquad (w_i + \lambda(\mathfrak{A}))^{(g)} = w_i + \lambda(\mathfrak{A}_1),$$

where the $\mathfrak{A}_1 = \mathfrak{A}_1(\mathfrak{A}, i, g)$ in (2.20) is defined in accordance with (2.18) and (2.19). In view of (2.20), $\mathfrak{A}_1$ can be briefly written in the form

$$(2.23)\qquad \mathfrak{A}_1 = \mathfrak{A} g + \Big(0, \ldots, \underset{i-1}{0}, \underset{i}{g - 1}, \underset{i+1}{0}, \ldots, 0\Big),$$

where $(0, \ldots, 0, g - 1, 0, \ldots, 0)$ is the element of $\mathcal{G}_0 = \mathbf{Z}_0[G]^n$ whose ith component is equal to $g - 1 \in \mathbf{Z}_0[G]$ and the others to zero. For brevity, we denote by

$\bar{f}_i$ the vector $(0,\ldots,0,1,0,\ldots,0)$ of length n whose ith coordinate is equal to unity and the others to zero. Then (2.23) can be briefly rewritten in the form

$$\mathfrak{A}_1 = \mathfrak{A}g + \bar{f}_i(g-1) \in \mathcal{G}_0, \tag{2.24}$$

and (2.22) in the form

$$\left(w_i + \lambda(\mathfrak{A})\right)^{(g)} = w_i + \lambda\left(\mathfrak{A}g + \bar{f}_i(g-1)\right). \tag{2.25}$$

Analogues of (2.22) or (2.25) hold for the numbers $\lambda(\vec{n}) \in \mathbf{K}$: $\vec{n} \in \mathbf{Z}^J$, defined by means of (2.14) and (2.11). However, it is more convenient to consider this below in the course of the subsequent argument. For the time being, we choose basis vectors $\vec{e}(j)$ in $\mathbf{Z}^J$:

$$\vec{e}(j)[j_1] = \delta_{j,j_1}: \quad j_1 \in J, \tag{2.26}$$

i.e. the vector $\vec{e}(j) = \vec{e}(g,i)$ in $\mathbf{Z}^J = \mathbf{Z}^{C\times n}$ has a unit coordinate (g,i) and the other coordinates zero for $(g,i) \in J = C \times n$, i.e. $g \in G\setminus\{1\}$, $i = 1,\ldots,n$.

We now introduce notation for those bounded regions in $\mathbf{Z}^J$ inside which we will work with the function $\lambda(\vec{n})$. For a natural number $N \geq 1$, put

$$\begin{aligned} D(N) = \Big\{\vec{n} = \left(n_{(g,i)}: (g,i) \in J\right) \in \mathbf{Z}^J: &\text{ if } i = 1,\ldots,n \text{ and} \\ J_i^+ = \{g \in C: n_{(g,i)} > 0\} &\text{ and } J_i^- = \{g \in C: n_{(g,i)} \leq 0\}, \\ \text{then for } i = 1,\ldots,n, \ \sum_{g\in J_i^+} n_{(g,i)} \leq N, &\ \sum_{g \in J_i^-} -n_{(g,i)} \leq N\Big\}. \end{aligned} \tag{2.27}$$

In particular, for the $(d-1)n$-dimensional cube $C(M)$ in $\mathbf{Z}^J$:

$$C(M) = (\mathbf{Z} \cap [-M, M])^J, \tag{2.28}$$

we have

$$C([N/(d-1)n] \subseteq D(N) \subseteq C(N)). \tag{2.29}$$

Such a specific choice for $D(N)$ makes it more convenient to determine for $\mathfrak{A} \in \mathcal{G}_0$ when $\bar{\bar{v}}(\mathfrak{A}) \in D(N)$. If $\mathfrak{A} = (\mathfrak{s}_1,\ldots,\mathfrak{s}_n) \in \mathcal{G}_0$ and $\mathfrak{s}_i = \Sigma_{g\in G}\, n_{(g,i)} \cdot g$, we put $R_i(\mathfrak{A}) = \{g \in G: n_{(g,i)} > 0\}$, $L_i(\mathfrak{A}) = \{g \in G: n_{(g,i)} \leq 0\}$: $i = 1,\ldots,n$. Then

$$\sum_{g\in R_i(\mathfrak{A})} n_{(g,i)} = -\sum_{g \in L_i(\mathfrak{A})} n_{(g,i)}: \quad i = 1,\ldots,n.$$

It follows from (2.27) and (2.11) that

$$\begin{aligned} \bar{\bar{v}}(\mathfrak{A}) \in D(N) &\Leftrightarrow \sum_{g \in R_i(\mathfrak{A})} n_{(g,i)} \leq N: \quad i = 1,\ldots,n, \\ &\Leftrightarrow \sum_{g\in L_i(\mathfrak{A})} -n_{(g,i)} \leq N, \quad i = 1,\ldots,n. \end{aligned} \tag{2.30}$$

From (2.30) we obtain the important

REMARK 2.1. If $\vec{n} = \bar{\bar{v}}(\mathfrak{A}) \in D(N)$, then $\bar{\bar{v}}(\mathfrak{A}\cdot g) \in D(N)$ for $g \in G$. If $\vec{n} = \bar{\bar{v}}(\mathfrak{A}) \in D(N)$, then for any $g \in G$ and $i = 1,\ldots,n$ we have

$$\bar{\bar{v}}(\mathfrak{A}_1) = \bar{\bar{v}}\left(\mathfrak{A}g + \bar{f}_i(g-1)\right) \in D(N+1).$$

Consequently, the desired family of auxiliary functions $\{F_j(z): j \in \mathbf{I}\}$ can be chosen in the form

$$F_{\vec{n}}(z) = P(z + \lambda(\vec{n}), f(z))$$

for a suitable polynomial $P(x, y) \in \mathbf{Z}[x, y]$ and those $\vec{n} \in \mathbf{Z}^J$ such that $\vec{n} \in D(N)$ for a fixed N, sufficiently large in comparison with d, ρ, and n.

3. Investigation of the family of auxiliary functions (continuation of the proof of Theorem 1). We preserve all of the notation of §2 pertaining to the algebraic points $w_1, \ldots, w_n$ of $f(z)$ and to the family $\{\lambda(\mathfrak{A}): \mathfrak{A} \in \mathcal{G}_0\} = \{\lambda(\vec{n}): \vec{n} \in \mathbf{Z}^J\}$ of elements of $\mathbf{K}$. To this family there will correspond the family of auxiliary functions.

Assume, contrary to Theorem 1, that $\rho < n$. We choose and fix some natural number N, sufficiently large in comparison with $d, n, (n-\rho)^{-1}$,

$$\max_{i=1,\ldots,n} H(w_i), \ldots.$$

Consider the set

$$\{\lambda(\vec{n}): \vec{n} \in D(N)\} = \{\lambda(\mathfrak{A}): \bar{\bar{v}}(\mathfrak{A}) \in D(N)\}, \tag{3.1}$$

where the region $D(N)$ in $\mathbf{Z}^J$ is defined in (2.27), the isomorphism $\bar{\bar{v}}: \mathcal{G}_0 \to \mathbf{Z}^J$ is defined in (2.11), and the elements $\lambda(\mathfrak{A})$: $\mathfrak{A} \in \mathcal{G}_0 = \mathbf{Z}_0[G]^n$, are defined in (2.13).

In addition to the fixed N we consider a natural parameter L, sufficiently large in comparison with N (hence also in comparison with d, n, $(n-\rho)^{-1}$, $\max_{i=1,\ldots,n} H(w_i)$). The auxiliary functions in which we are interested have the form

$$F_{\vec{n}}(z) = P(z + \lambda(\vec{n}), f(z)): \quad \vec{n} \in D(N)$$

or

$$F_{\mathfrak{A}}(z) = P(z + \lambda(\mathfrak{A}), f(z)): \quad \bar{\bar{v}}(\mathfrak{A}) \in D(N),$$

where $P(x, y) \in \mathbf{Z}[x, y]$ and $\deg_x(P) \leq L_1$, $\deg_y(P) \leq L_2$. The parameters L_1 and L_2 are chosen so that

$$L_1 = \left[L(\log L)^{-1/4}\right], \quad L_2 = \left[(\log L)^{3/4}\right]. \tag{3.2}$$

We denote by (l_1, l_2) a pair of integers such that $0 \leq l_1 \leq L_1$, $0 \leq l_2 \leq L_2$. Also, Σ_{l_1, l_2} stands for the sum $\Sigma_{l_1=0}^{L_1} \Sigma_{l_2=0}^{L_2}$, and we put

$$\varphi^{\vec{n}}_{(l_1,l_2)}(z) = (z + \lambda(\vec{n}))^{l_1} f(z)^{l_2}: \quad \vec{n} \in \mathbf{Z}^J. \tag{3.3}$$

From Lemma 1.8 we obtain

LEMMA 3.1. *There exists a constant $c_4 = c_4(N) > 0$ such that for any $i = 1, \ldots, n$, $\vec{n} \in D(N)$, and $k \geq 0$ we have*

$$\varphi^{\vec{n}}_{(l_1,l_2)}(w_i)^{(k)} \in \mathbf{K} \quad \textit{for all } (l_1, l_2).$$

The common denominator $\mathfrak{D}$ of the numbers

$$\left\{\left(\varphi^{\vec{n}}_{(l_1,l_2)}(w_i)\right)^{(k)}: (l_1, l_2); k \geqslant 0, \vec{n} \in D(N), i = 1,\dots,n\right\} \tag{3.4}$$

satisfies the inequality

$$\log \mathfrak{D} \leqslant c_4 L(\log L)^{-1/4}. \tag{3.5}$$

For each $k \geqslant 0$, $\vec{n} \in D(N)$, and $i = 1,\dots,n$ we have the following estimate for the size of $(\varphi^{\vec{n}}_{(l_1,l_2)}(w_i))^{(k)}$:

$$\log\left|\overline{\left(\varphi^{\vec{n}}_{(l_1,l_2)}(w_i)\right)^{(k)}}\right| \leqslant K \log K + c_4 \cdot K. \tag{3.6}$$

where $K = \max\{k, L\}$.

PROOF OF LEMMA 3.1. We have

(3.7)

$$\left(\varphi^{\vec{n}}_{(l_1,l_2)}(w_i)\right)^{(k)} = \sum_{s=0}^{k} C^s_k l_1 \cdots (l_1 - s + 1)(w_i + \lambda(\vec{n}))^{l_1 - s} \cdot \left\{f(w_i)^{l_2}\right\}^{(k-s)}.$$

Therefore, since $w_i \in \mathbf{K}$ and $\lambda(\vec{n}) \in \mathbf{K}$, it follows from (2.3) that

$$\left(\varphi^{\vec{n}}_{(l_1,l_2)}(w_i)\right)^{(k)} \in \mathbf{K} \quad \text{for all } k \geqslant 0, \quad \vec{n} \in \mathbf{Z}^J, \quad i = 1,\dots,n.$$

Let $\delta_0 = \delta_0(N)$ be the common denominator of the numbers

$$w_i + \lambda(\vec{n}): \qquad i = 1,\dots,n; \quad \vec{n} \in D(N).$$

It then follows from (3.7) that $\mathfrak{D} = \delta_0^{L_1}$ is a common denominator of all numbers of the form (3.4). Estimate (3.5) follows from (3.2) with $C_4 \geqslant \ln \delta_0$.

It remains to estimate the size of $(\varphi^{\vec{n}}_{(l_1,l_2)}(w_i))^{(k)}$. For this we use Lemma 1.8. It follows from Lemma 1.8, assertion 2 that

$$\begin{aligned}\log\left|\overline{\left(\varphi^{\vec{n}}_{(l_1,l_2)}(w_i)\right)^{(k)}}\right| &\leqslant k\log(k+1) + c_3(k + L_1) + c_3 L_2 \log(k+2)\\ &\leqslant k\log(k+1) + c_3(k+L) + c_3(\log L)^{3/4}\log(k+2)\\ &\leqslant K\log K + c_4 K,\end{aligned}$$

where $K = \max\{k, L\}$. This implies (3.6), and Lemma 3.1 is proved.

We apply Lemma 1.3 to construct the auxiliary functions $F_{\vec{n}}(z)$.

LEMMA 3.2. *There exists a nonzero polynomial in* $\mathbf{Z}[x, y]$,

$$P(x, y) = \sum_{l_1=0}^{L_1} \sum_{l_2=0}^{L_2} p(l_1, l_2) x^{l_1} y^{l_2}, \tag{3.8}$$

such that for $c_5 = c_5(N) > 0$ *we have*

$$H(P) = \max_{(l_1,l_2)} |p(l_1, l_2)| \leqslant \exp\left(c_5 L(\log L)^{1/2}\right) \tag{3.9}$$

and for the functions

$$F_{\vec{n}}(z) = P(z + \lambda(\vec{n}), f(z)): \qquad \vec{n} \in D(N), \tag{3.10}$$

we have

$$F_{\vec{n}}^{(k)}(w_i) = 0 \tag{3.11}$$

for all $k = 0, \ldots, L-1$, $i = 1, \ldots, n$, *and* $\vec{n} \in D(N)$.

REMARK 3.3. Similarly, we can put

$$F_{\mathfrak{A}}(z) = P(z + \lambda(\mathfrak{A}), f(z))$$

and $F_{\mathfrak{A}}(z) \equiv F_{\vec{n}}(z)$ if $\bar{\vec{v}}(\mathfrak{A}) = \vec{n}$. In the previous notation,

$$F_{\vec{n}}(z) = \sum_{(l_1, l_2)} p(l_1, l_2) \varphi^{\vec{n}}_{(l_1, l_2)}(z).$$

PROOF OF LEMMA 3.2. We apply Lemma 1.3. We can rewrite the system of equations (3.11) defining $p(l_1, l_2)$ in the form

$$\sum_{(l_1, l_2)} p(l_1, l_2) \varphi^{\vec{n}}_{(l_1, l_2)}(w_i)^{(k)} = 0 \tag{3.12}$$

where $k = 0, 1, \ldots, L-1$, $i = 1, \ldots, n$, and $\vec{n} \in D(N)$. This is a system with

$$m_* \leqslant L \cdot n \cdot |D(N)| \tag{3.13}$$

equations and

$$n_* = (L_1 + 1)(L_2 + 1) \geqslant L(\log L)^{1/2} \tag{3.14}$$

unknowns.

It follows from (2.27) and (2.28) that

$$D(N) \subseteq C(N) \quad \text{and} \quad |D(N)| \leqslant |C(N)| \leqslant (2N+1)^{(d-1)n}$$

and

$$m_* \leqslant L \cdot n \cdot (2N+1)^{(d-1)n} \tag{3.15}$$

Using the notation of Lemma 1.3, we see from Lemma 3.1 that

$$\begin{aligned} \log A &\leqslant L \log L + c_4\left(L + (\log L)^{7/4}\right) + c_4 L + \log n_* \\ &\leqslant L \log L + c_6 L. \end{aligned} \tag{3.16}$$

For sufficiently large L we can find nontrivial rational integers $p(l_1, l_2)$ satisfying system (3.12), and

$$\begin{aligned} \max_{(l_1, l_2)} \log |p(l_1, l_2)| &\leqslant \frac{m_* d}{n_* - m_* d}\left(\frac{1}{2} \log 2 + \log A\right) \\ &\leqslant c_7 (\log L)^{-1/2}\left(\tfrac{1}{2} \log 2 + L \log L + c_6 L\right) \\ &\leqslant c_8 L (\log L)^{1/2} \end{aligned}$$

for $c_7 = c_7(n, d, N) > 0$, $c_8 = c_8(n, d, N) > 0$. Thus, (3.9) holds and Lemma 3.2 is proved.

DEFINITION 3.4. For each $\vec{n} \in D(N)$ and $i = 1, \dots, n$ we denote by $u_i^{\vec{n}}$ the smallest natural number $u \geqslant 0$ such that

$$F_{\vec{n}}^{(u)}(w_i) \neq 0.$$

Similarly, $u_i^{\mathfrak{A}}$ is the smallest $u \geqslant 0$ such that

$$F_{\mathfrak{A}}^{(u)}(w_i) \neq 0.$$

Since $P(x, y) \not\equiv 0$ by Lemma 3.2 and $f(z)$ is transcendental, it follows that $F_{\vec{n}}(z) \not\equiv 0$ for any $\vec{n} \in D(N)$ and $u_i^{\vec{n}} < \infty$ exists. In view of assertion (3.11) of Lemma 3.2,

$$u_i^{\vec{n}} \geqslant L \quad \text{for } \vec{n} \in D(N) \text{ and } i = 1, \dots, n. \tag{3.17}$$

Let us estimate the sizes of the numbers $F_{\vec{n}}^{(u_i^{\vec{n}})}(w_i) \neq 0$.

LEMMA 3.5. *For any $\vec{n} \in D(N)$ and $i = 1, \dots, n$,*

$$F_{\vec{n}}^{(u_i^{\vec{n}})}(w_i) \in \mathbf{K},$$

and for its size and denominator we have

$$\log \left|\overline{F_{\vec{n}}^{(u_i^{\vec{n}})}(w_i)}\right| \leqslant u_i^{\vec{n}} \log u_i^{\vec{n}} + c_9 u_i^{\vec{n}} (\log u_i^{\vec{n}})^{1/2} \tag{3.18}$$

and

$$\log \operatorname{den}\left(F_{\vec{n}}^{(u_i^{\vec{n}})}(w_i)\right) \leqslant c_9 L (\log L)^{-1/4} \leqslant c_9 u_i^{\vec{n}} (\log u_i^{\vec{n}})^{-1/4} \tag{3.19}$$

where $c_9 = c_9(N) > 0$.

PROOF OF LEMMA 3.5. We apply Lemma 1.8, taking into account Lemma 3.2 and the fact that $u_i^{\vec{n}} \geqslant L$: $\vec{n} \in D(N)$, $i = 1, \dots, n$. We obtain at once that

$$F_{\vec{n}}^{(u_i^{\vec{n}})}(w_i) \in \mathbf{K}$$

for any $i = 1, \dots, n$, since $w_i \in \mathbf{K}$ and $\lambda(\vec{n}) \in \mathbf{K}$. By Lemma 1.8,

$$\log \operatorname{den}\left(F_{\vec{n}}^{(u_i^{\vec{n}})}(w_i)\right) \leqslant \partial(w_i + \lambda(\vec{n})) L (\log L)^{-1/4}$$

and, in view of (3.17), inequality (3.19) is proved. Then

$$\begin{aligned} \log \left|\overline{F_{\vec{n}}^{(u_i^{\vec{n}})}(w_i)}\right| &\leqslant u_i^{\vec{n}} \log u_i^{\vec{n}} + c_5 L (\log L)^{1/2} + c_4 (u_i^{\vec{n}} + L) \\ &\leqslant u_i^{\vec{n}} \log u_i^{\vec{n}} + c_9 u_i^{\vec{n}} (\log u_i^{\vec{n}})^{1/2}. \end{aligned}$$

Lemma 3.5 is proved.

There remains the analytic part of §3. We use a variant of Schwarz's lemma. For $\vec{n} \in D(N)$ we put

$$G_{\vec{n}}(z) = h(z)^{L_2} \cdot F_{\vec{n}}(z). \tag{3.20}$$

Since $f(z)$ is analytic in a neighborhood of w_i: $i = 1,\ldots,n$, it follows that $h(w_i) \neq 0$ for $i = 1,\ldots,n$. Consequently,

$$G_{\vec{n}}^{(u_i^{\vec{n}})}(w_i) = h(w_i)^{L_2} \cdot F_{\vec{n}}^{(u_i^{\vec{n}})}(w_i) \tag{3.21}$$

and $u_i^{\vec{n}}$ is the smallest $u \geqslant 0$ such that $G_{\vec{n}}^{(u)}(w_i) \neq 0$ for $\vec{n} \in D(N)$ and $i = 1,\ldots,n$.
Applying Jensen's principle to $G_{\vec{n}}(z)$, we obtain the basic analytic inequality:

LEMMA 3.6. *Suppose $\vec{n} \in D(N)$ and $i = 1,\ldots,n$. Then there exists $c_{10} = c_{10}(N) > 0$ such that*

$$\sum_{j=1}^{n} u_j^{\vec{n}} \leqslant \frac{\rho}{\log u_i^{\vec{n}}}\left(c_{10} u_i^{\vec{n}} (\log u_i^{\vec{n}})^{3/4} + u_i^{\vec{n}} \log u_i^{\vec{n}} - \log\left| F_{\vec{n}}^{(u_i^{\vec{n}})}(w_i) \right| \right). \tag{3.22}$$

PROOF OF LEMMA 3.6. Let $\mathrm{no}(g; r)$ denote the number of zeros of the function $g(z)$ in a disc $B(0; r)$ of radius r in which $g(z)$ is analytic. Put $r_i = 1 + \max_{j=1,\ldots,n} |w_i - w_j|$. Then by Definition 3.4, (3.20), and (3.21),

$$\mathrm{no}(G_{\vec{n}}(z - w_i); r_i) \geqslant \sum_{j=1}^{n} u_j^{\vec{n}} \tag{3.23}$$

for $\vec{n} \in D(N)$ and $i = 1,\ldots,n$. Also, it follows from Jensen's formula (see D. Bertrand [7]) that for $\Lambda > 1$ we have

$$\begin{aligned}\mathrm{no}(G_{\vec{n}}(z - w_i); r_i) \\ \leqslant \frac{1}{\log \Lambda}\left(\log |G_{\vec{n}}(z - w_i)|_{\Lambda r_i} - \log \frac{1}{u_i^{\vec{n}}!} G_{\vec{n}}^{(u_i^{\vec{n}})}(w_i) - u_i^{\vec{n}} \log r_i \right).\end{aligned} \tag{3.24}$$

Consider (3.23) and (3.24) with $\Lambda = (u_i^{\vec{n}})^{1/\rho}$. We have

$$\begin{aligned}\sum_{j=1}^{n} u_j^{\vec{n}} \leqslant \frac{\rho}{\log u_i^{\vec{n}}} & \left(\log |h^{L_2} F_{\vec{n}}(z - w_i)|_{\Lambda r_i} \right. \\ & \left. - \log |h^{L_2}(w_i)| - \log \cdot \frac{1}{u_i^{\vec{n}}} \cdot \left| F_{\vec{n}}^{(u_i^{\vec{n}})}(w_i) \right| \right).\end{aligned} \tag{3.25}$$

Since $h(z) \cdot f(z)$ has order $\leqslant \rho_0 < \rho$, it follows from Lemma 3.2 that

$$\begin{aligned}\log |h^{L_2} F_{\vec{n}}(z - w_i)|_{\Lambda r_i} & \leqslant \log(L_1 L_2) + c_5 L (\log L)^{1/2} \\ & + c_{11} L_1 \log(\Lambda r_i) + c_{12} L_2 (\Lambda r_i)^{\rho} \leqslant c_{13} u_i^{\vec{n}} (\log u_i^{\vec{n}})^{3/4},\end{aligned} \tag{3.26}$$

since $u_i^{\vec{n}} \geqslant L$ and $\Lambda = (u_i^{\vec{n}})^{1/\rho}$ for $\vec{n} \in D(N)$ and $i = 1,\ldots,n$.
From (3.25) and (3.26) we immediately obtain (3.22), and Lemma 3.6 is proved.
We mention one more weak, but useful, auxiliary assertion.

LEMMA 3.7. *There exists $c_{14} > 0$ (depending on N, d, w_i) such that for all $\vec{n} \in D(N)$ and $i = 1,\ldots,n$ we have*

$$\sum_{j=1}^{n} \vec{u}_j \leqslant d\rho\left(u_i^{\vec{n}} + c_{14} u_i^{\vec{n}} (\log u_i^{\vec{n}})^{-1/4} \right). \tag{3.27}$$

PROOF OF LEMMA 3.7. In view of Lemma 3.5,

$$-\log|F_{\vec{n}}^{(u_i^{\vec{n}})}(w_i)| \leqslant (d-1)\log|\overline{F_{\vec{n}}^{(u_i^{\vec{n}})}(w_i)}| + d\log \operatorname{den} F_{\vec{n}}^{(u_i^{\vec{n}})}(w_i)$$
$$\leqslant (d-1)u_i^{\vec{n}}\log u_i^{\vec{n}} + c_{15}u_i^{\vec{n}}(\log u_i^{\vec{n}})^{1/2}.$$

Substituting this into assertion (3.22) of Lemma 3.6, we obtain

$$\sum_{j=1}^{n} u_j^{\vec{n}} \leqslant \frac{\rho}{\log u_i^{\vec{n}}}\Big(c_{10}u_i^{\vec{n}}(\log u_i^{\vec{n}} + u_i^{\vec{n}}\log u_i^{\vec{n}}$$
$$+ (d-1)u_i^{\vec{n}}\log u_i^{\vec{n}} + c_{15}u_i^{\vec{n}}(\log u_i^{\vec{n}})^{1/2})\Big).$$

This is (3.27) with $c_{14} \geqslant c_{10} + c_{15}$. Lemma 3.7 is proved.

From Lemma 3.7 follows

REMARK 3.8. As (3.27) shows, for L sufficiently large (in comparison with c_{14}) and for $\vec{n} \in D(N)$ and $i = 1, \ldots, n$ we have

$$\log\left(\sum_{j=1}^{n} u_j^{\vec{n}}\right) - \log(d\rho + 1) \leqslant \log(u_i^{\vec{n}}).$$

Consequently, without loss of generality, we can put

$$\log u_i^{\vec{n}} = \log u^{\vec{n}}: \qquad i = 1, \ldots, n, \tag{3.28}$$

for $\vec{n} \in D(N)$.

We now turn to the main lemma, which connects the conjugates of the numbers of the form $F_{\vec{n}}^{(U_i^{\vec{n}})}(w_i)$ and the other numbers $F_{\vec{m}}^{(u_i^{\vec{m}})}(w_i)$. We will use Lemma 1.7 and the basic relation (2.22) or (2.25), enabling us to pass from $(w_i + \lambda(\vec{n}))^{(g)}$ to $w_i + \lambda(\vec{m})$. Thus, we have the

MAIN LEMMA 3.9. *Suppose $\mathfrak{A} \in \mathcal{G}_0$. If $g \in G$ and $i = 1, \ldots, n$, then the two elements*

$$F_{\mathfrak{A}}^{(k)}(w_i) \quad \text{and} \quad F_{\mathfrak{A}_1}^{(k)}(w_i)$$

where $\mathfrak{A}_1 = \mathfrak{A}_g + \bar{f}_i(g-1)$ are conjugate for any $k \geqslant 0$. Namely, for any $k \geqslant 0$,

$$\left(F_{\mathfrak{A}}^{(k)}(w_i)\right)^{(g)} = F_{\mathfrak{A}g+\bar{f}_i(g-1)}^{(k)}(w_i). \tag{3.29}$$

PROOF OF LEMMA 3.9. In view of (2.22) or (2.25), for any $\mathfrak{A} \in \mathcal{G}_0$, $i = 1, \ldots, n$, and $g \in G$ we have

$$(w_i + \lambda(\mathfrak{A}))^{(g)} = w_i + \lambda(\mathfrak{A}g + \bar{f}_i(g-1)). \tag{3.30}$$

Now $F_{\mathfrak{A}}(z) \equiv P(z + \lambda(\mathfrak{A}), f(z))$, $w_i \in \mathbf{K}$: $i = 1, \ldots, n$, and $\lambda(\mathfrak{A}) \in \mathbf{K}$ for all $\mathfrak{A} \in \mathcal{G}_0$. Since $P(x, y) \in \mathbf{Z}[x, y]$, assertion (3.29) follows from (3.30) and assertion 2 of Lemma 1.7. Let us verify (3.29) directly. In view of (3.8), for $\mathfrak{A} \in \mathcal{G}_0$ and

$i = 1,\dots,n$ we have

$$F_{\mathfrak{A}}^{(k)}(w_i) = \sum_{l_1,l_2} p(l_1, l_2) \sum_{s=0}^{k} C_k^s l_1 \cdots (l_1 - s + 1) \times \left(w_i + \lambda(\mathfrak{A})\right)^{l_1 - s} \cdot \left\{f(w_i)^{l_2}\right\}^{(k-s)}. \tag{3.31}$$

Since $p(l_1, l_2) \in \mathbf{Z}$ and, in view of (2.3), $\{f(w_i)^{l_2}\}^{(k-s)} \in \mathbf{Z}$, it follows from (3.30) with $\mathfrak{A}_1 = \mathfrak{A} g + \bar{f}_i(g-1)$ and from (3.31) that for $g \in G$

$$\left(F_{\mathfrak{A}}^{(k)}(w_i)\right)^{(g)} = \sum_{l_1,l_2} p(l_1, l_2) \sum_{s=0}^{k} C_k^s l_1 \cdots (l_1 - s + 1) \times \left(w_i + \lambda(\mathfrak{A}_1)\right)^{(l_1 - s)} \left\{f(w_i)^{l_2}\right\}^{(k-s)}. \tag{3.32}$$

Comparing (3.32) and (3.31), for $\mathfrak{A}_1$ instead of $\mathfrak{A}$, we obtain (3.29), Lemma 3.9 is proved.

COROLLARY 3.10. *If $\mathfrak{A} \in \mathcal{G}_0$ and $\bar{\bar{v}}(\mathfrak{A}) \in D(N-1)$, then for any $g \in G$ and $i = 1,\dots,n$ we have, for $\mathfrak{A}_1 = \mathfrak{A} g + \bar{f}_i(g-1)$,*

$$\bar{\bar{v}}(\mathfrak{A}_1) = \bar{\bar{v}}\left(\mathfrak{A} g + \bar{f}_i(g-1)\right) \in D(N) \tag{3.33}$$

and

$$u_i^{\mathfrak{A}} = u_i^{\mathfrak{A} g + \bar{f}_i(g-1)}. \tag{3.34}$$

PROOF OF COROLLARY 3.10. $\bar{\bar{v}}(\mathfrak{A}) \in D(N-1)$, then, by Remark 2.1,

$$\bar{\bar{v}}\left(\mathfrak{A} g + \bar{f}_i(g-1)\right) \in D(N).$$

Therefore, $u_i^{\mathfrak{A}_1}$ is defined for $\mathfrak{A}_1 = \mathfrak{A} g + \bar{f}_i(g-1)$. In view of Definition 3.4 and equality (3.29), we conclude that

$$u_i^{\mathfrak{A}} = u_i^{\mathfrak{A} g + \bar{f}_i(g-1)} \quad \text{for } \bar{\bar{v}}(\mathfrak{A}) \in D(N-1),\ i = 1,\dots,n;\ g \in G.$$

Corollary 3.10 is proved.

It follows from (3.29) and (3.34) that for $\bar{\bar{v}}(\mathfrak{A}) \in D(N-1)$, $i = 1,\dots,n$, and $g \in G$ we have

$$\left(F_{\mathfrak{A}}^{(u_i^{\mathfrak{A}})}(w_i)\right)^{(g)} = F_{\mathfrak{A}_1}^{(u_i^{\mathfrak{A}_1})}(w_1), \quad \text{where } \mathfrak{A}_1 = \mathfrak{A} g + \bar{f}_i(g-1). \tag{3.35}$$

We now put together a system of linear inequalities connecting the various $u_i^{\mathfrak{A}}$. We use the natural formula

$$d \cdot \log \operatorname{den} F_{\mathfrak{A}}^{(u_i^{\mathfrak{A}})}(w_i) + \log\left|F_{\mathfrak{A}}^{(u_i^{\mathfrak{A}})}(w_i)\right| + \sum_{g \in G, g \neq 1} \log\left|\left(F_{\mathfrak{A}}^{(u_i^{\mathfrak{A}})}(w_i)\right)^{(g)}\right| \geqslant 0. \tag{3.36}$$

In view of (3.35), we can rewrite (3.36) in the form

$$(3.37)\qquad d\cdot\log\operatorname{den}\left(F_{\mathfrak{A}}^{(u_i^{\mathfrak{A}})}(w_i)\right)+\log|F_{\mathfrak{A}}^{(u_i^{\mathfrak{A}})}(w_i)| + \sum_{g\in G,g\neq 1}\log|F_{\mathfrak{A}g+\vec{f}_i(g-1)}^{(u_i^{\mathfrak{A}g+\vec{f}_i(g-1)})}(w_i)|\geqslant 0$$

for $\bar{\bar{v}}(\mathfrak{A})\in D(N-1)$ and $i=1,\dots,n$. We now take into account (3.22), which, in view of (3.28), we can rewrite in the following form for $\mathfrak{A}_0\in\mathcal{G}_0$, $\bar{\bar{v}}(\mathfrak{A}_0)\in D(N)$, and $i=1,\dots,n$:

$$(3.38)\qquad \log|F_{\mathfrak{A}_0}^{(u_i^{\mathfrak{A}_0})}(w_i)|\leqslant u_i^{\mathfrak{A}_0}\log u^{\mathfrak{A}_0} - \frac{1}{\rho}\sum_{j=1}^{n}u_j^{\mathfrak{A}_0}\log u^{\mathfrak{A}_0}+c_{10}u_i^{\mathfrak{A}_0}(\log u^{\mathfrak{A}_0})^{3/4}.$$

LEMMA 3.11. *For any* $\mathfrak{A}\in\mathcal{G}_0$, $\bar{\bar{v}}(\mathfrak{A})\in D(N-1)$, *and* $i=1,\dots,n$,

$$(3.39)\qquad d(\rho-1)u_i^{\mathfrak{A}}\log u^{\mathfrak{A}}+c_{16}u_i^{\mathfrak{A}}(\log u^{\mathfrak{A}})^{3/4} \geqslant\sum_{g\in G}\sum_{j=1,j\neq i}^{n}u_j^{\mathfrak{A}g+\vec{f}_i(g-1)}\log u^{\mathfrak{A}g+\vec{f}_i(g-1)}.$$

PROOF OF LEMMA 3.11. Substituting (3.38) into (3.37) and using (3.19), we obtain for $\bar{\bar{v}}(\mathfrak{A})\in D(N-1)$

$$(3.40)\qquad \begin{aligned} &u_i^{\mathfrak{A}}\log u^{\mathfrak{A}}-\frac{1}{\rho}\sum_{j=1}^{n}u_j^{\mathfrak{A}}\log u^{\mathfrak{A}}+c_{17}u_i^{\mathfrak{A}}(\log u^{\mathfrak{A}})^{3/4}\\ &+\sum_{g\in G,g\neq 1}u_i^{\mathfrak{A}g+\vec{f}_i(g-1)}\log u^{\mathfrak{A}g+\vec{f}_i(g-1)}\\ &+\sum_{g\in G,g\neq 1}c_{18}u_i^{\mathfrak{A}g+\vec{f}_i(g-1)}\left(\log u_i^{\mathfrak{A}g+\vec{f}_i(g-1)}\right)^{3/4}\\ &-\frac{1}{\rho}\sum_{j=1}^{n}\sum_{g\in G,g\neq 1}u_j^{\mathfrak{A}g+\vec{f}_i(g-1)}\cdot\log u^{\mathfrak{A}g+\vec{f}_i(g-1)}\geqslant 0.\end{aligned}$$

In view of (3.34), we have

$$(d-d/\rho)u_i^{\mathfrak{A}}\log u^{\mathfrak{A}}+c_{19}u_i^{\mathfrak{A}}(\log u_i^{\mathfrak{A}})^{3/4} -\frac{1}{\rho}\sum_{j=1,j\neq i}^{n}\sum_{g\in G}u_j^{\mathfrak{A}g+\vec{f}_i(g-1)}\log u^{\mathfrak{A}g+\vec{f}_i(g-1)}\geqslant 0$$

and Lemma 3.11 is proved.

For any $\mathfrak{A}\in\mathcal{G}_0$, $\vec{n}\in\mathbf{Z}^J$, and $i=1,\dots,n$, if $\bar{\bar{v}}(\mathfrak{A})\in D(N-1)$ or $\vec{n}\in D(N-1)$, we put

$$(3.41)\qquad v_i^{\mathfrak{A}}=u_i^{\mathfrak{A}}\log u^{\mathfrak{A}};\qquad v_i^{\vec{n}}=u_i^{\vec{n}}\log u^{\vec{n}}.$$

Now suppose $i = 1,\dots,n$ and $j = 1,\dots,n$, $j \neq i$. If $g \in G$ and $\bar{\bar{v}}(\mathfrak{A}) \in D(N-1)$, it follows from (3.34) that

$$(3.42)\qquad u_j^{\mathfrak{A}g+\bar{f}_i(g-1)} = u_j^{(\mathfrak{A}g+\bar{f}_i(g-1))g^{-1}+\bar{f}_j(g^{-1}-1)} = u_j^{\mathfrak{A}+\bar{f}_i-\bar{f}_j-\bar{f}_ig^{-1}+\bar{f}_jg^{-1}}.$$

In view of (3.42) and (3.41), we can rewrite (3.39) in the form

$$(3.43)\quad d(\rho-1)v_i^{\mathfrak{A}} + c_{20}v_i^{\mathfrak{A}}\left(\log v_i^{\mathfrak{A}}\right)^{-1/4} \geqslant \sum_{j=1, j\neq i}^{n} \sum_{g\in G} v_j^{\mathfrak{A}+\bar{f}_i-\bar{f}_j-\bar{f}_ig+\bar{f}_jg},$$

for any $\mathfrak{A} \in \mathcal{G}_0$, $\bar{\bar{v}}(\mathfrak{A}) \in D(N-1)$, and $i = 1,\dots,n$. We now apply the mapping $\bar{\bar{v}}$ from (2.11). Obviously,

$$\bar{\bar{v}}\left(\mathfrak{A} + \bar{f}_i - \bar{f}_j - \bar{f}_ig + \bar{f}_jg\right) = \bar{\bar{v}}(\mathfrak{A}) - \vec{e}(g,i) + \vec{e}(g,j)$$

for $g \in G$, $g \neq 1$, i.e. $g \in \mathrm{C} = G\backslash\{1\}$, and for $g = 1$ we obtain simply $\bar{\bar{v}}(\mathfrak{A})$. Replacing $\bar{\bar{v}}(\mathfrak{A})$ by $\vec{n} \in D(N-1)$ in (3.43), we obtain

COROLLARY 3.12. *If $\vec{n} \in D(N-1)$ and $i = 1,\dots,n$, then*
(3.44)

$$d(\rho-1)v_i^{\vec{n}} + c_{20}v_i^{\vec{n}}\left(\log v_i^{\vec{n}}\right)^{-1/4} \geqslant \sum_{j=1, j\neq 1}^{n} v_j^{\vec{n}} + \sum_{j=1, j\neq i}^{n} \sum_{g\in \mathrm{C}} v_j^{\vec{n}+\vec{e}(g,j)-\vec{e}(g,i)}.$$

The vectors $\vec{e}(g,j) \in D(1) \subset \mathbf{Z}^J = \mathbf{Z}^{\mathrm{C}\times n}$ are defined above in (2.26).

The system (3.44) in which we are interested describes, in particular, the scheme of a random walk for n types of particles in $(d-1)n$-dimensional space.

This interpretation is due to D. V. Chudnousky. The essential feature distinguishing (3.44) from the usual random walk schemes is that on the right-hand side of (3.44) there appear $d(n-1)$ summands, and on the left-hand side of (3.44) when $\rho < n$ the constant $d(\rho-1)$ is less than $d(n-1)$. This leads to superharmonic (excessive) functions f satisfying conditions of the type

$$Af(x) \leqslant -\varepsilon f(x)$$

in certain subregions of $\mathbf{Z}^J$ for $\varepsilon > 0$ and a discrete analogue A of the Laplace operator (cf. [**9**]). For $n = 2$, where the problem can be reduced to a walk of one particle in $\mathbf{Z}^{2(d-1)}$, such an approach turns out to be effective (cf. §0, where we examined the case $d = 2$).

Analytic methods from [**9–11**] show that the sytem (3.44) does not have nontrivial solutions for $\rho < n$, hence proving Theorem 1. We are not repeating here all the details, since they are contained completely in [**13**, §4].

4. Multidimensional case. In order to state multidimensional generalization of the Schwarz lemma as a necessary analytic tool for the proof of Theorem 2 we need some definitions. These definitions concern degrees of hypersurfaces in $\mathbf{C}^n$ that contain a finite set of points.

Here and everywhere below S denotes a finite set in $\mathbf{C}^n$.

DEFINITION 4.1. By $\Omega(S)$ we denote the minimal degree of the hypersurface in $\mathbf{C}^n$ containing S. For fixed $K \geqslant 1$ we denote the minimal degree of a hypersurface in $\mathbf{C}^n$ containing each point from S with multiplicity at least K by $\Omega(S, K)$.

Formally we can write this in the form

$$\Omega(S, K) = \min\{\deg P \colon P(\bar{x}) \in \mathbf{C}[\bar{x}] \text{ and } \partial^k P(\bar{w}) = 0 \\ \text{for all } \bar{w} \in S \text{ and } |k| < K \text{ and } k \in \mathbf{N}^n\}.$$

For any $S \subseteq \mathbf{C}^n$ we denote

$$\lim_{K \to \infty} \Omega(S, K)/K = \Omega_0(S).$$

DEFINITION 4.2. Let $P \in \mathbf{C}[x_1, \ldots, x_n]$ be a polynomial of degree $n = \partial(P)$ and let us denote by $\operatorname{ord}_{\bar{x}}(P)$ the order of zero of P at $\bar{x} \in \mathbf{C}^n$. Then for a set S we denote

$$\hat{\Omega}_0(S; P) = \frac{\partial(P) \cdot |S|}{\sum_{\bar{x} \in S} \operatorname{ord}_{\bar{x}}(P)}.$$

(This means how much "proportionally mean" multiplicity is related to $\partial(P)$.)

If $S \subset \mathbf{C}^n$ we denote by $\hat{\Omega}_0(S)$ the "very singular degree of S":

$$\hat{\Omega}_0(S) = \inf\{\hat{\Omega}_0(S; P) \colon P \in \mathbf{C}[x_1, \ldots, x_n]\}.$$

The Schwarz lemma that we consider is important in the following setting. Let $S \subset \mathbf{C}^n$ be fixed and $\varepsilon > 0$ be arbitrary. We consider the entire function $f(\bar{z})$ in $\mathbf{C}^n$ having at any point $\bar{x}_i \in S$ a zero of order $k_i \geqslant 0$: $i = 1, \ldots, |S|$. We are interested in the inequality $|\partial^{\bar{t}} f(\bar{x}_i)| \leqslant k_i! e^{c_1 |\bar{k}|} \cdot |f|_R (R/6nc_e)^{-|\bar{k}| \chi(S)(1-\varepsilon)}$, where $t \in \mathbf{N}^n$, $|\bar{t}| = k_i$, $c_1 = c_1(S, \varepsilon) > 0$, $c_2 = c_2(S, \varepsilon) > 0$ and $\chi(S)$ is called the Schwarz exponent of the form (i) $\hat{\Omega}_0(S)/|S|$ or (ii) $\Omega_0(S)/n|S|$.

One of the examples of such a statement is provided by the following proposition (cf. [**1**]).

PROPOSITION 4.3. *Let S be a finite subset of $\mathbf{C}^n$ and let $0 < \varepsilon \leqslant 1$. There exists a positive number $r_0 = r_0(S, \varepsilon)$ such that for any nonnegative integer M and any nonzero function f entire in $\mathbf{C}^n$ satisfying*

$$\partial^m f(w) = 0 \qquad (w \in S, m \in \mathbf{N}^n, |m| < M)$$

we have for $r > r_0$ and $R > 6nr/\varepsilon$

$$\log|f|_r \leqslant \log|f|_R - (1 - \varepsilon)(\Omega(S)/n - \varepsilon) M \log(\varepsilon R/6nr).$$

In fact, here we can put $\Omega_0(S)$ instead of $\Omega(S)/n$.

For our purpose instead of Proposition 4.3 we use a finer version of the Schwarz lemma.

THEOREM 4.4. *For any finite $S \subset \mathbf{C}^2$, $\varepsilon > 0$, there are $O_0(S, \varepsilon) > 0$, $r_0(S, \varepsilon) > 0$ such that:*

For any entire function $f(\bar{z})$ in $\mathbf{C}^n$ having at points from S not less than $O(f)$ zeroes (counted with multiplicities), we have for $r > r_0$, $O(f) > O_0$, $R > 6nr/\varepsilon$ $|f|_r \leqslant |f|_R \cdot (R\varepsilon/6nr)^{-O(f) \cdot (\hat{\Omega}_0(S)/|S|)(1-\varepsilon)}$.

Instead of Theorem 4.4, using the method of G-invariant system of functions (§§0–3), we prove below a stronger statement.

THEOREM 4.5. *Let $f(z)$ be a meromorphic transcendental function of order $\leqslant \rho$. Then the set $S_{\overline{\mathbf{Q}}}(f)$ of $\overline{w} \in \overline{\mathbf{Q}}^n$ such that*

$$\partial^{\bar{k}} f(\overline{w}) \in \mathbf{Z} \quad \textit{for all } \bar{k} \in \mathbf{N}^n$$

is contained in a hypersurface of very singular degree $\leqslant \rho$, or

$$\hat{\Omega}_0(S) \leqslant \rho, \quad n \geqslant 2 \quad \textit{for any finite } S \subset S_{\overline{\mathbf{Q}}}(f).$$

PROOF. We use our method from §§0–3, [**13**, **14**] on constructing the auxiliary function using group ring of Galois group and random walk associated with this Galois group.

Let S be any finite subset of $S_{\overline{\mathbf{Q}}}(f)$. By algebraic closeness of $\mathbf{C}$ from

$$\Omega(S_1) \leqslant C \quad \text{for any finite } S_1 \subset S_0$$

it follows that $\Omega(S_0) \leqslant C$ with any (infinite) S_0. Thus we take $S \subset S_{\overline{\mathbf{Q}}}(f)$ to be finite and we will prove

$$\hat{\Omega}_0(S) \leqslant \rho.$$

Let us suppose, on the contrary, that

$$\hat{\Omega}_0(S) > \rho. \tag{4.1}$$

We take now the Galois field $\mathbf{K}$ with Galois group G,

$$G = \mathrm{Gal}(\mathbf{K}/\mathbf{Q}), \qquad |G| = [\mathbf{K} \colon \mathbf{Q}] = d, \tag{4.2}$$

such that

$$S \subset \mathbf{K}^n. \tag{4.3}$$

We will enumerate elements of S as

$$\begin{gathered} S = \{\overline{w}_i \colon i = 1, \ldots, |S|\}, \\ \overline{w}_i = (w_{i,1}, \ldots, w_{i,n}) \colon \qquad i = 1, \ldots, |S|, \\ \text{where } w_{i,j} \in k \colon \qquad i = 1, \ldots, |S|, \quad j = 1, \ldots, n. \end{gathered} \tag{4.4}$$

Let $\mathbf{Z}[G]$ be a group ring of G and $\mathbf{Z}_0[G]$ be an ideal in $\mathbf{Z}[G]$ of elements of zero trace:

$$\mathbf{Z}_0[G] = \left\{\alpha = \sum_{g \in G} n_g g \colon \sum_{g \in G} n_g = 0\right\}.$$

Our main objective is $\mathcal{G}_0 = \mathbf{Z}_0[G]^{|S|}$ and we put

$$\mathcal{G}_0 = \mathbf{Z}_0[G]^{|S|}, \qquad C = G/\{1\}, J = C \times \{1, \ldots, |S|\} = C \times |S|.$$

There exists natural isomorphism between $\mathcal{G}_0$ and $\mathbf{Z}^J = \mathbf{Z}^{C \times |S|}$:

$$v \colon \mathbf{Z}_0[G]^{|S|} \to \mathbf{Z}^J, \tag{4.5}$$

where for $\mathcal{A} = (\mathcal{s}_1, \ldots, \mathcal{s}_{|S|}) \in \mathcal{G}_0$, $\mathcal{s}_i = \Sigma_{g \in G} n(g, i)$, $g \in \mathbf{Z}_0[G]$: $i = 1, \ldots, |S|$,

$$v(\mathcal{A}) = (n(g, i) \colon (g, i) \in J = C \times |S|) \in \mathbf{Z}^J.$$

Now for any $\mathcal{A} = (\mathfrak{s}_1,\ldots,\mathfrak{s}_{|S|}) \in \mathcal{G}_0$ and $\mathfrak{s}_i = \sum_{g\in G} n(g,i)\cdot g \in \mathbf{Z}_0[G]$: $i = 1,\ldots,|S|$, we define

$$\lambda_j(\mathcal{A}) = \sum_{i=1}^{|S|} \sum_{g\in G} n(g,i) w_{i,j}^{(g)} \in \mathbf{K}: \qquad i = 1,\ldots,n,$$

and we obtain vector $\bar\lambda(\mathcal{A})$ from $\mathbf{K}^n$ by putting

$$\bar\lambda(\mathcal{A}) = (\lambda_1(\mathcal{A}),\ldots,\lambda_n(\mathcal{A})) \in \mathbf{K}^n,$$

and using v we put for $\vec{n} \in \mathbf{Z}^J$

$$\bar\lambda(\vec{n}) = \bar\lambda(v^{-1}(\vec{n})) \in \mathbf{K}^n \quad \text{for } \vec{n} \in \mathbf{Z}^J. \tag{4.6}$$

Let $\bar{f}_i$: $i = 1,\ldots,|S|$, be basic vector in $\mathbf{Z}^{|S|}$: $\bar{f}_i(\chi) = \delta_{i\chi}$. Then we can define for any $g \in G$,

$$\mathcal{A}_1 = \mathcal{A}\cdot g + \bar{f}_i(g-1) \in \mathcal{G}_0 = \mathbf{Z}_0[G]^{|S|}. \tag{4.7}$$

Now we have the main

CONJUGATE PROPERTY. If $\mathcal{A} \in \mathcal{G}_0$, $\bar{w}_i \in S$ and $g \in G$, then

$$(\bar{w}_i + \lambda(\mathcal{A}))^{(g)} = \bar{w}_i + \bar\lambda(\mathcal{A}g + \bar{f}_i(g-1)): \qquad i = 1,\ldots,n, \tag{4.8}$$

which means

$$(w_{i,j} + \lambda_j(\mathcal{A}))^{(g)} = w_{i,j} + \lambda_j(\mathcal{A}g + \bar{f}_i(g-1))$$

for any $i = 1,\ldots,|S|, j = 1,\ldots,n$.

Now we are ready to define auxiliary functions of the form ($C(\mathfrak{N})$ is a cube $([-\mathfrak{N},\mathfrak{N}] \cap \mathbf{Z})^J$ in $\mathbf{Z}^J$).

$$F_{\vec{n}}(\bar{z}) = P(\bar{z} + \bar\lambda(\vec{n}), f(\bar{z})), \qquad \vec{n} \in C(\mathfrak{N}),$$
$$F_{\mathcal{A}}(\bar{z}) = P(\bar{z} + \bar\lambda(\mathcal{A}), f(\bar{z})), \qquad \mathcal{A} \in \mathcal{G}_0, v(\mathcal{A}) \in C(\mathfrak{N}),$$

where $\bar{z} = (z_1,\ldots,z_n)$ and $\bar\lambda(\mathcal{A})$ are vectors from $\mathbf{C}^n$. We take $\mathfrak{N}$ to be constant, sufficiently large with respect to $\Omega(S)$, n, d, ρ, $(\hat\Omega_0(S) - \rho)^{-1},\ldots$ and take L to be a natural number (parameter) sufficiently large with respect to $\mathfrak{N}$.

We use the Siegel lemma to obtain

EXISTENCE. There exists a nonzero polynomial $P(\bar{x}, y) \in \mathbf{Z}[\bar{x}, y]$, $\deg_{x_i}(P) \leq L_i$, $\deg_Y(P) \leq L_2$,

$$P(\bar{x}, y) = P(x_1,\ldots,x_n, y),$$

where

$$L_1 = \left[L(\log L)^{-1/(2n+2)}\right], \qquad L_2 = \left[(\log L)^{(2n+1)/(2n+2)}\right], \tag{4.9}$$

such that

$$H(P) \leq \exp\left(c_1 L(\log L)^{1/(n+1)}\right), \tag{4.10}$$

$c_1 = c_1(d, n, |S|, \mathfrak{N}) > 0$ and for auxiliary functions

$$F_{\vec{n}}(\bar{z}) = P(\bar{z} + \lambda(\vec{n}), f(\bar{z})): \qquad \vec{n} \in C(\mathfrak{N}) \tag{4.11}$$

we have

$$\partial^{\bar{k}} F_{\vec{n}}(w_i) = 0: \quad \text{for } i = 1,\ldots,|S|,\ \bar{k} \in \mathbf{N}^n, |\bar{k}| < L,$$

and $\vec{n} \in C(\mathfrak{N})$.

For $\vec{n} \in \mathbf{Z}^J$ we denote by $u_i^{\vec{n}}$ the multiplicity of $F_{\vec{n}}(\bar{z})$ at $\bar{w}_i$: $i = 1,\ldots,|S|$,

$$F_{\vec{n}}^{(u_i^{\vec{n}})}(\bar{w}_i) \neq 0. \tag{4.12}$$

Thus

$$u_i^{\vec{n}} \geqslant L \quad \text{for } i = 1,\ldots,|S| \text{ and } \vec{n} \in C(\mathfrak{N}).$$

Now we apply the general Schwarz Lemma 4.4 and obtain any $\vec{n} \in C(\mathfrak{N})$ and $i = 1,\ldots,|S|$ with $c_2 = c_2(\mathfrak{N}) > 0$

$$\sum_{j=1}^{|S|} u^{\vec{n}} \log u^{\vec{n}} \cdot \chi(S) \tag{4.13}$$

$$\leqslant \rho\left(c_2 u_i^{\vec{n}} (\log u_i^{\vec{n}})^{(2n+1)/(2n+2)} + u_i^{\vec{n}} \log u_i^{\vec{n}} - \log|F_{\vec{n}}^{(u_i^{\vec{n}})}(\bar{w}_i)|\right).$$

Here $\chi(S)$ is the Schwarz exponent for S and

$$\chi(S) \geqslant \frac{1}{|S|} \cdot \hat{\Omega}_0(S).$$

Now we apply our "conjugate principle" and obtain

$$\left(\partial^{\bar{k}} F_{\mathcal{A}}(\bar{w}_i)\right)^{(g)} = \partial^{\bar{k}} F_{\mathcal{A}g+\bar{f}_i(g-1)}(\bar{w}_i): \qquad \bar{k} \in \mathbf{N}^n, \tag{4.14}$$

$$u_i^{\mathcal{A}} = u_i^{\mathcal{A}g+f_i(g-1)} \tag{4.15}$$

for $\mathcal{A} \in \mathcal{G}_0$, $i = 1,\ldots,|S|$.

Now using (4.3)–(4.4) and the product formula, we obtain

MARKOVIAN SYSTEM FOR MULTIPLICITIES. For $\vec{n} \in C(\mathfrak{N} - 2)$ we have the following system:

$$d\left(\frac{\rho}{\chi(S)} - 1\right) u_i^{\vec{n}} + c_3 u_i^{\vec{n}} (\log u^{\vec{n}})^{-1/(2n+2)} \tag{4.16}$$

$$\geqslant \sum_{j=1, j\neq i}^{|S|} u_j^{\vec{n}} + \sum_{j=1, j\neq i}^{|S|} \sum_{g\in C} u_j^{\vec{n}+\vec{e}(g,j)-\vec{e}(g,i)}: \ i = 1,\ldots,|S|.$$

Here $\vec{e}(g,i)$ is a basic vector in $\mathbf{Z}^J = \mathbf{Z}^{C\times|S|}$. On the left side of (4.16) we have a constant $d(\rho/\chi(S) - 1)$ and on the right side we have $d(|S| - 1)$ summands, each of which is $\geqslant L$. Then using standard analytic results (see the end of §3 and [**13**]), we obtain

CONCLUSION. If $\mathfrak{N} > \mathfrak{N}(\rho, \chi(S), |S|, (\hat{\Omega}_0(S) - \rho)^{-1})$ and

$$\rho/\chi(S) < |S| \quad \text{or} \quad \rho < \hat{\Omega}_0(S)$$

then (4.16) is impossible for positive $u^{\vec{n}}$: $\vec{n} \in C(\mathfrak{N} - 2)$, $i = 1,\ldots,|S|$.

This proves Theorem 4.4.

The analogous result is valid for the system of functions

$$(f_1(\bar{z}),\ldots,f_{n+1}(\bar{z}))$$

that are algebraically independent and such that $f_1(\bar{z}),\ldots,f_k(\bar{z})$ possess the law of addition over $\mathbf{Q}$. And we consider $\bar{w} \in \mathbf{C}^n$ such that

$$\partial^{\bar{k}} f_i(\bar{w}) \in \overline{\mathbf{Q}} \quad \text{for } i = 1,\ldots,k$$

and

$$\partial^{\bar{k}} f_i(\bar{w}) \in \mathbf{Z} \quad \text{for } i = k+1,\ldots,n+1$$

for $\bar{k} \in \mathbf{N}^n$.

5. Hermite interpolation formula. Let $f(z)$ be an entire function of finite order of growth $\leqslant \rho$. For the investigation of the arithmetic nature of $f^{(k)}(z)$ for an algebraic $z \in \overline{\mathbf{Q}}$ the best approach is to construct an auxiliary function $F(x) = P(x, f(x))$ explicitly using the residue formula.

Let $\omega_1,\ldots,\omega_m$ be distinct complex numbers and we define for an integer $n \geqslant 0$,

$$R_n(x) = \frac{1}{2\pi i}\int_C \frac{f(xz)\,dz}{\prod_{j=1}^m (z-\omega_j)^{n+1}}. \tag{5.1}$$

Then by the Hermite interpolation formula

$$R_n(x) = \sum_{j=1}^m \frac{1}{n!}\partial_z^n\left(\frac{f(xz)}{\prod_{k\neq j}(z-\omega_k)^{n+1}}\right)\Bigg|_{z=\omega_j}. \tag{5.2}$$

By the Newton interpolation series, if $f(z)$ is not a polynomial, then $R_n(x) \neq 0$ for a given $x \neq 0$ and infinitely many $n \geqslant 1$.

The comparison between the formulae (5.1) and (5.2) is what we need to investigate roughly the algebraicity of $f^{(k)}(\omega_j)$: $k \geqslant 0, j = 1,\ldots,m$.

We can let $x = 1$ and C in (5.1) be the circle $|z| = n^{1/\rho}$ for sufficiently large n. Then

$$|R_n(1)| < n^{-mn/\rho} e^{\gamma_5 n}. \tag{5.3}$$

Now let us assume that $\omega_1,\ldots,\omega_m$ are algebraic numbers and

$$f^{(k)}(\omega_j) \text{ are algebraic numbers: } k \geqslant 0; j = 1,\ldots,m, \tag{5.4}$$

such that

$$C_k \cdot f^{(l)}(\omega_j): \qquad l = 0,\ldots,k, j = 1,\ldots,m,$$

are algebraic integers with

$$C_k \in \mathbf{Z}, \qquad |C_k| \leqslant e^{\gamma_6 k}: \qquad k \geqslant 1. \tag{5.5}$$

Then it follows from (5.2) that $n!A_nR_n(1)$ is an algebraic integer for some

$$1 \leqslant A_n \leqslant e^{\gamma_7 n}; \qquad A_n \in \mathbf{Z}.$$

If we know

$$\max_{\substack{k=0,\dots,n\\ j=1,\dots,m}} |\overline{f^{(k)}(\omega_j)}|, \tag{5.6}$$

then we can find $|\overline{R_n(1)}|$. We know (5.6) only in two cases:

(1) when $f^{(k)}(\omega_j) \in \mathbf{Z}$ for all $k \geqslant 0$, then

$$|f^{(n)}(\omega_j)| \leqslant n^{(1-1/\rho)n} e^{\gamma_8 n};$$

(2) when $f(z)$ satisfies an algebraic differential equation $\mathcal{P}(z, f, f', \dots, f^{(q-1)}) = 0$ then

$$|f^{(n)}(\omega_j)| \leqslant n^n e^{\gamma_9 n}.$$

In both cases we obtain immediately effective proofs of two theorems.

THEOREM 5.1 (STRAUS-SCHNEIDER). *Let $f(z)$ be an entire transcendental function of the order $\leqslant \rho$ and let $\mathbf{K}$ be an algebraic number field. If*

$$S_{\mathbf{K}} = \{z \in \mathbf{K}: f^{(k)}(z) \in \mathbf{Z}^{-} \textit{ for all } k \geqslant 0\},$$

then

$$|S_{\mathbf{K}}| \leqslant \rho[\mathbf{K}: \mathbf{Q}]$$

and if $f(z)$ satisfies an algebraic differential equation

$$\mathcal{P}(z, f, \dots, f^{(q-1)}(z)) = 0; \qquad \mathcal{P}(\bar{x}) \in \mathbf{Q}[\bar{x}], \mathcal{P} \not\equiv 0,$$

then for

$$S_{\mathbf{K}}^{*} = \{z \in \mathbf{K}: f^{(k)}(z) \in \mathbf{K} \textit{ for all } k \geqslant 0\},$$

$$|S_{\mathbf{K}}^{*}| \leqslant \rho[\mathbf{K}: \mathbf{Q}].$$

In fact, for the first part of the Theorem 5.1 we have a significant improvement (see Theorem 1).

THEOREM 5.2. *If $f(z)$ is a transcendental meromorphic function of the order of growth $\leqslant \rho$, then*

$$|S_f| = |\{z \in \overline{\mathbf{Q}}: f^{(k)}(z) \in \mathbf{Z} \textit{ for all } k \geqslant 0\}| \leqslant \rho.$$

The method of the proof of Theorem 5.2 is absolutely different from those of the Theorem 5.1. However in the case of an entire function it can be modified to become effective with the same use of the residue formula.

The result of the Theorem 5.2 is already the best possible with respect to ρ; take as an example $f(z) = \exp\{z(z-1)\cdots(z-n)\}$.

However Straus [**6**] proposed the following very fine and nontrivial problem.

PROBLEM (STRAUS). How is the cardinality of the

$$S_f = \{z \in \overline{\mathbf{Q}}: f^{(k)}(z) \in \mathbf{Z} \text{ for all } k \geqslant 0\}$$

connected with the order ρ and the type γ of f,

$$|f|_R < e^{\gamma R^\rho}: R \to \infty.$$

The problem now starts to be connected with the arithmetic properties of the elements of S_f, their divisibility their p-adic properties, etc.

Without loss of generality we can assume

$$0 \in S_f \tag{5.7}$$

(change $f(z)$ to $f(z+w_0)$ for $w_0 \in S_f$, if $S_f \neq \varnothing$).

With given $\rho \geqslant 1$ and γ what numbers can be members of S_f for some transcendental entire $f(z)$ with order ρ and the type γ?

According to Theorem 5.2 we can restrict ourself with the integer value of ρ, $\rho = 2, 3, \ldots$. The most interesting (and already unsolved case) is $\rho = 2$.

If we take a function

$$f(z) = \exp\left\{\frac{z(z \pm 2)}{2}\right\},$$

then we find that the set $\{0, \pm 2\}$ is S_f for f with order 2 and type $\frac{1}{2}$. The number of such examples can be multiplied if we put

$$f(z) = g(z(z-z_1)\cdots(z-z_m)\cdot a)$$

for $z_1, \ldots, z_m$, $a \in \mathbf{Q}$, and take

$$g(z) = \sum_{n=0}^{\infty} \frac{a_n z^n}{n!}$$

as an entire function of order 1 with $a_n \in \mathbf{Z}$ (where $\gamma(g) \geqslant 1$ clearly). See below for such examples.

We can prove rather strong positive results using the explicit expressions for $R_n(z)$ in (5.1)–(5.2). We obtain e.g.

PROPOSITION 5.3. *Let $f(z)$ be a transcendental entire function of order 2 and type γ. Let $S_f = \{0, a\}$ for $a \in \mathbf{Q}$, $a \neq 0$.*

(1) *If $a = p/q$ for $|p|, |q| \geqslant 1$, then*

$$\gamma \geqslant |q|/p^2. \tag{5.8}$$

(2) *If a is a prime number $\geqslant 2$, then*

$$\gamma \geqslant a^{-2+1/(a-1)}. \tag{5.9}$$

For the proof of Proposition 5.3 we can rewrite $R_n(1)$ from (5.2) in a more precise form: take $\omega_1 = 0$, $\omega_2 = a$,

$$n! \cdot R_n(1) = \sum_{j=0}^{n} \frac{(2n-j)!}{(n-j)!j!} a^{-2n+j-1} \times (-1)^{n-j} + \{f^{(j)}(a) + (-1)^{j-1} f^{(j)}(0)\}.$$

Let us suppose that Δ is a rational number > 0,

$$\Delta = c^n, \tag{5.11}$$

such that $n!R_n(1) \cdot \Delta$ is an integer. Then we apply representation (5.1) with a contour C chosen as

$$|z| = n^{1/2} \frac{1}{\gamma^{1/2}} \qquad \text{(here } \rho = 2\text{)}.$$

Then we have

$$|\Delta \cdot n! R_n(1)| \leqslant e^{-\varepsilon n} \cdot \left(c \frac{n \cdot e^{-1} \cdot e}{n \gamma^{-1}} \right)^n = (c\gamma)^n$$

for $n \geqslant n_0(\varepsilon)$. Thus

$$\gamma \geqslant c^{-1}.$$

In case (1) e.g. $c = q^{-1} p^2$ according to (5.10). In case (2) we compute $\nu_a(\cdot)$ in (5.10).

It should be noted that in **[6]** a result, apparently stronger than (1) was announced without proof; but the method of proof of **[6]** provides the result weaker than 5.3. Similarly, results of **[6]** can be improved using representations (5.1), (5.2) for $m > 2$.

We have also examples of the results opposite to this proposition:

EXAMPLE 5.4. (1) For any $\varepsilon > 0$ there exists an integer $n \geqslant 1$ such that $\{0, n\} = S_f$ for some $f(z)$ of order 2 and $\gamma(f) < \varepsilon$.

(2) For a prime p, we have $f_p(z)$ of order 2 and $S_{f_p} = \{0, p\}$ such that

$$\gamma(f_p) = \tfrac{1}{2} \quad \text{for } p = 2$$

and

$$\gamma(f_p) = (2p^{1/p-1})^{-1} \quad \text{if } p \geqslant 3.$$

In particular, our Proposition 5.3 is precise for $a = 2$.

In general case we have such an absolutely new result.

THEOREM 5.5. *Let $f(z)$ be an entire transcendental function of order of growth $n \geqslant 1$. Let $\gamma = 0$, i.e.*

$$|f|_R \ll e^{\gamma(R) \cdot R^n},$$

where $\gamma(R) \to 0$ as $R \to \infty$. Then for

$$S_f = \{z \in \overline{\mathbf{Q}}: f^{(k)}(z) \in \mathbf{Z} \text{ for all } k \geqslant 0\},$$

we have $|S_f| \leqslant n - 1$.

Again we use the representation (5.1) together with the method of "conjugate" functions.

Analogous results can be proved for meromorphic functions and functions in $\mathbf{C}^n$.

Our effective methods give us at last the good tool for the investigation of the

BOMBIERI CONJECTURE. If $f(x)$ is a transcendental meromorphic function of order $\leqslant \rho$ satisfying an algebraic differential equation $P(z, f(z), \ldots, f^{(q-1)}(z)) = 0$, then for the set

$$S_{\overline{\mathbf{Q}}} = \{z \in \overline{\mathbf{Q}}: f^{(k)}(z) \in \overline{\mathbf{Q}} \text{ for all } k \geqslant 0\}$$

we have $|S_{\overline{\mathbf{Q}}}| \leqslant \rho$.

It may happen, however, that the Bombieri conjecture is, in a certain sense, trivial! There are two examples.

(1) (Hermite) If $f(z)$ satisfies $P(f, f') = 0$, then genus$(P) = 0, 1$, i.e. $f(z)$ is reduced to z, e^z or the Weierstrass elliptic function $\wp(z)$.

(2) (Straus) If $f(z)$ is an entire function of the order $\leqslant \rho$ and if

$$|\{z \in \mathbf{C}: f^{(k)}(z) \in \mathbf{Z} \text{ for all } k \geqslant 0\}| > \rho,$$

then $f(z)$ is the sum of exponents.

I think that it would be easier to prove such a type of results:

PROBLEM 1. Let $f(z)$ satisfy all the assumptions of the Bombieri conjecture. If

$$|\{z \in \mathbf{C}: f^{(k)}(z) \in \overline{\mathbf{Q}} \text{ for all } k \geqslant 0\}| \geqslant \rho,$$

then $f(z)$ is reduced to z, e^z or $\wp(z)$.

PROBLEM 2. Let f satisfy $R(f, f', \dots, f^{(q-1)}) = 0$ but not any equation of order $< q - 1$; then for $q \geqslant 3$, $|S_{\overline{\mathbf{Q}}}| \leqslant 1$.

In fact what meromorphic functions f of the finite order of growth can satisfy $R(f, f', f'') = 0$, say? The most known examples are elliptic functions and Painlevé transcendence [**15**].

$$f''' = 12ff' + c_0, \qquad f'' = 6f^2 + c_0 z - g_2/2.$$

REFERENCES

1. M. Waldschmidt, *On functions of several variables having algebraic Taylor coefficients*, Proc. Cambridge Conf., 1976.

2. E. Bombieri, *Algebraic values of meromorphic maps*, Invent. Math. **10** (1970), 267–287.

3. M. Waldschmidt, *Propriétés arithmétiques de fonctions de plusieurs variables*. I, Semênaire P. Lelong, 15° année, 1974/75.

4. ______, *Propriétés arithmétiques de fonctions de plusieurs variables*. II, Semênaire P. Lelong, 16é année, 1975/76.

5. Th. Schneider, *Einführung in die Transcendenten Zahlen*. Springer-Verlag, Berlin, 1957.

6. E.-G. Straus, *On entire functions with algebraic derivatives at certain algebraic points*, Ann. of Math. (2) **52** (1950), 188–198.

7. D. Bertrand, *Equations differentielles algebriques et nombres transcendants dans les domaines complexe et p-adique*, Thèse de 3e cycle, Université Paris VI, 1975.

8. M. Waldschmidt, *Nombres transcendants*, Lecture Notes in Math., Vol. 402, Springer-Verlag, Berlin and New York, 1974.

9. E. B. Dynkin and A. A. Yushkevich, *Markov processes: theorems and problems*, Plenum Press, New York, 1969.

10. B. A. Sevast'ianov, *Verzweigungs prozesse*, R. Oldenburg, Münich, 1975.

11. K. L. Chung, *Markov chains with stationary transition probabilities*, Springer-Verlag, Berlin and New York, 1960.

12. William J. Leveque (ed.), *Reviews in number theory*, Vol. 4, Section Q, Amer. Math. Soc., Providence, R.I., 1974.

13. G. V. Chudnovsky, *A new method for the investigation of arithmetical properties of analytic functions*, Ann. of Math. (2) **109** (1979), 353–376.

14. ______, *Singular points of complex hypersurfaces and multidimensional Schwarz lemma*, Sémenaire Delange-Pisot-Poitou, 19e année, 1977/78, University Publishing House, Paris, 1977, pp. 1–40; reprinted in Progress in Math., Vol. 12, Birkhäuser, Basel, 1981, pp. 29–69.

15. D. V. Chudnovsky, *Riemann monodromy problem, isomonodromy deformation equations and completely integrable systems*, Bifurcation Phenomena in Mathematical Physics and Related Topics, Reidel, Boston, 1980, pp. 385–447.

Translated by G. A. KANDALL

APPENDIX

THE PROOF OF EXTREMALITY OF CERTAIN MANIFOLDS

BY

A. I. VINOGRADOV AND G. V. CHUDNOVSKY

0. Lately one of the most important places in metric number theory is occupied by the problem of investigation of "extremality" for manifolds $\boldsymbol{\Gamma}$ in $\mathbf{R}^n$ (or $\mathbf{C}^n$). In other words, we mean the following problem.

Let us suppose that an m-dimensional manifold $\boldsymbol{\Gamma}$ in $\mathbf{R}^n$ is defined by means of the the system of (smooth enough) equations

$$f_j(x_1,\ldots,x_n)=0: \qquad j=1,\ldots,n-m,$$

and on $\boldsymbol{\Gamma}$ is the naturally induced Lebesgue measure. Whether the manifold $\boldsymbol{\Gamma}$ will be "extremal" in the sense that for any $\varepsilon>0$ and almost all $(x_1,\ldots,x_n)\in\boldsymbol{\Gamma}$ (with respect to the measure) the inequality

$$|a_0+a_1x_1+\cdots+a_nx_n|<\left(\sum_{i=0}^{n}|a_i|\right)^{-m-\varepsilon}$$

has finite number of solutions in integer rational numbers $a_i\in\mathbf{Z}: i=0,\ldots,n$? Besides, it is possible to put a problem about "k-extremality" of $\boldsymbol{\Gamma}$, replacing, in the right side of the inequality, "$-m-\varepsilon$" by "$-k$". In this sense extremality is "k-extremality" for all $k>0$.

The term extremality is introduced, in particular, because "$-m-\varepsilon$" is the best exponent on which one can expect according to Dirichlet's principle. The first result of principal importance in the examination of extremality was the solution by V. G. Sprindzuk (see [**1, 2**]) Mahler's hypothesis on the extremality of the curve $\boldsymbol{\Gamma}_n=(t,\ldots,t^n)$ for any n. Of strong interest is also the result of W. Schmidt [**4**] on the extremality of any "not too flat" curve $\boldsymbol{\Gamma}=(f_1(t),f_2(t))$ in the plane (i.e. with the condition $f_1'f_2''-f_1''f_1'\neq 0$ almost everywhere). Afterwards I. M. Vinogradov's method of trigonometrical sums was successfully applied to the

0076-5376/84 $1.00 + $.25 per page

problems of extremality in the cycle of papers [**2**, **6**, **7**, **13**]. Superposing, principally, analytical conditions, the authors of these papers had proved extremality of many classes of manifolds $\Gamma \subset \mathbf{R}^n$ with $m = \dim \Gamma$ close to n. However the extremality of manifolds like $\Gamma' = (t_1^i t_2^j : i + j \leqslant n)$, $\Gamma'' = (t_1^i t_2^j : i, j \leqslant n)$ that are of the most importance from the arithmetical point of view have not yet been examined for $n > 3$ (for $n = 3$ see [**12**]). But these cases are of particular interest, as the problem of "k-extremality" for Γ', Γ'' simply means the determination of the best k for which for almost all $(x, y) \in \mathbf{R}^2$ (in the sense of Lebesgue measure in the plane) the inquality

$$|P(x, y)| < H(P)^{-k}$$

has a finite number of solutions in nonzero polynomials $P(x, y) \in \mathbf{Z}[x, y]$ of degree $\leqslant n$ and with $H(P)$ being the height of $P(x, y)$. The difficulties arising in this problem are already of algebraic-geometrical character and the final determination of the infimum of such k (presumably $k = n(n+1)/2$) is still difficult. But the estimates that are close to desirable are proved here. For a long time the authors have been examining possible modern approaches to the problem of extremality. They think that the combination of topological and algebraically geometrical methods (resolution of singularities) form a good framework for the solution of this problem. At first we gave the proof for the fist nontrivial case—for the manifolds, connected with the hyperelliptic curve $y^2 = P(x)$, i.e. for $\Gamma_{\mathrm{hy}} = (y^2 x^i : i \leqslant n)$. But then as it turned out the scheme of the proof is transferable also to the more general case of linear combinations of polynomials $P_1(x_1) + \cdots + P_n(x_n)$ i.e. $\Gamma''' = (x_i^j : i \leqslant n, j \leqslant m)$. Simultaneously we succeeded in the simplification of the exposition and made it short and independent of the general methods—intersection theory, u-resultants and topology of real algebraic curves [**8**, **11**]. This Appendix is mainly devoted to the proof of extremality of manifolds connected with linear combinations of polynomials. Let us formulate the result obtained below in the traditional notations of V. G. Sprindzuk [**1**, **7**].

THEOREM III$'$. *Let $n_1, \ldots, n_m$ be natural numbers and $\lambda_1 \neq 0, \ldots, \lambda_m \neq 0$ be real numbers. For the given real numbers $\omega_1, \ldots, \omega_m$ we denote by $w_1 = w_1(\omega_1, \ldots, \omega_m)$, the least upper bound of those $w > 0$ for which there exists an infinite number of systems of polynomials $P_1(x), \ldots, P_m(x)$ from $\mathbf{Z}[x]$ of degree not more than $n_1, \ldots, n_m$ and without constant terms satisfying the inequality*

$$\|\lambda_1 P_1(\omega_1) + \cdots + \lambda_m P_m(\omega_m)\| < H^{-w},$$

$H = \max H(P_j)$, *where $H(P_j)$ is the height of $P_j(x)$, and $\|\cdot\|$ is the distance from the nearest integer. Then*

$$w_1(\omega_1, \ldots, \omega_m) = n_1 + \cdots + n_m$$

for almost all $(\omega_1, \ldots, \omega_m) \in \mathbf{R}^m$.

Theorems I–III, III′ and IV confirm the conjecture suggested in [7, p. 194]. These results in a natural way generalize and supplement the results of Kovalevskaya [**6**] and Kovalevskaya and Bernik [**13**]. They have proved Theorem III′ in the case when $m \gg (\max_{i=1,\dots,m} n_i)^2$. In the proof of the main Theorem I we used Mahler's hypothesis, proved by Sprindzuk, and his method of "essential and nonessential" domains.

For similar results obtained independently see [**17**].

We present also some other results concerning the problems of "k-extremality" of manifolds connected with arbitrary polynomials $P(x_1,\dots,x_n) \in \mathbf{Z}[x_1,\dots,x_n]$. One more result essentially generalizing the theorem from [7] should be noted. Its proof can be obtained using the scheme of proof of Theorem I.

THEOREM IV′. *Let $n_1,\dots,n_m$ be natural numbers, and real numbers $\lambda_1,\dots,\lambda_m$ be such that the inequality*

$$\|\lambda_1 a_1 + \cdots + \lambda_m a_m\| < \left(\prod_{i=1}^{m} (|a_i|+1)\right)^{-1-\varepsilon}$$

has for any fixed $\varepsilon > 0$ a finite number of solutions in integer rational numbers $a_1,\dots,a_m$. Let us denote by $w_2 = w_2(\omega_1,\dots,\omega_m)$ the least upper bound of those $w > 0$, for which there exists an infinite number of systems of polynomials $P_1(x),\dots,P_m(x) \in \mathbf{Z}[x]$ of degree not more than $n_1,\dots,n_m$, correspondingly, satisfying the inequality

$$\|\lambda_1 P_1(\omega_1) + \cdots + \lambda_m P_m(\omega_m)\| < H^{-w}, \qquad H = \max_{i=1,\dots,m} H(P_i).$$

Then for almost all $(\omega_1,\dots,\omega_m) \in \mathbf{R}^m$,

$$w_2(\omega_1,\dots,\omega_m) = n_1 + \cdots + n_m + m.$$

1. This section has auxiliary character. Here we give the exposition of simple but useful lemmas on approximation of points in $\mathbf{C}^n$ in which polynomials from n variables have small values, by the zeroes of these polynomials. In fact in the proof of extremality their results are very important. The proofs of extremality are often based on that for "almost all" polynomials $P(x_1,\dots,x_n) \in \mathbf{Z}[x_1,\dots,x_n]$ instead of the estimate $|P(x_1^1,\dots,x_n^1)| < \varepsilon$ for $\bar{x}^1 = (x_1^1,\dots,x_n^1)$ we can obtain the estimate $|\bar{x}^1 - \bar{x}^0| < O(\varepsilon)$ for $\bar{x}^0 = (x_1^0,\dots,x_n^0)$ being a zero of polynomial P: $P(x_1^0,\dots,x_n^0) = 0$. The other polynomials do not influence the final metric result. Unfortunately this direct scheme is not always working as it is not so simple to determine "almost all" polynomials. For the case $n = 1$ this problem is more difficult than the determination of all the fields with the given discriminant [**1**].

The approximation lemmas given below are used in §2 for the proof of Theorem 1, and afterwards we shall show how by using them as a basis we can obtain essential progress in the problem of k-extremality of smooth multidimensional algebraic manifolds.

We adopt ordinary notation and terminology [**5, 9**]. The symbols **C**, **R**, **Z** and **N** denote, correspondingly, the sets of complex, real, integer rational and natural numbers equipped with the natural algebraic structures. If A is a ring, then $A[x_1,\dots,x_n]$ is a ring of polynomials from n variables over A. For the monomial $R = x_1^{i_1}\cdots x_n^{i_n}$ we put $d(R) = i_1 + \cdots + i_n$, and for $P(x_1,\dots,x_n) \in A[x_1,\dots,x_n]$, $d(P)$ is the maximum of the degrees of monomials entering $P(x_1,\dots,x_n)$. With $P(x_1,\dots,x_n) \in \mathbf{R}[x_1,\dots,x_n]$, $H(P)$ is the maximum of modulus of the coefficients in $P(x_1,\dots,x_n)$.

1.1. We consider at first the problem of approximation of the points of hypersurface $P(x_1,\dots,x_n) = \varepsilon$ by the points of the hypersurface $P(x_1,\dots,x_n)$, being a particular case of the much more general problem mentioned above.

We start from (affine) complex hypersurfaces $P(\vec{x}) = 0 : \vec{x} \in \mathbf{C}^n$. In other words, even real points of the hypersurface $P(\vec{x}) = \varepsilon$ will be approximated by complex zeroes of the polynomial $P(\vec{x}) = P(x_1,\dots,x_n) \in \mathbf{R}[x_1,\dots,x_n]$. An analogous situation has already taken place for $n = 1$ (cf. [**1**]) and this is not an obstable for obtaining the exact metric results. A little later we shall pay special attention to the real case.

Here and afterwards the symbol $P(x_1,\dots,x_n)$ (or $P(x, y)$ for $n = 2$) denotes a polynomial from $\mathbf{R}[x_1,\dots,x_n]$. By $\mathfrak{N}(P)$ we denote the affine complex manifold

$$\mathfrak{N}(P) = \{\vec{x} \in \mathbf{C}^n : P(\vec{x}) = 0\}. \tag{1}$$

It is easy to see that the measure of closeness of the points of the manifold

$$\mathfrak{N}_\varepsilon(P) = \{\vec{x} \in \mathbf{C}^n : P(\vec{x}) = \varepsilon\}$$

for $|\varepsilon| > 0$ to the points of $\mathfrak{N}(P)$ depends essentially on partial derivatives of $P(x_1,\dots,x_n)$. For $\bar{\alpha} = (\alpha_1,\dots,\alpha_n) \in \mathbf{N}^n$ by $P^{(\bar{\alpha})}(x_1,\dots,x_n)$ we shall mean

$$\frac{\partial^{\alpha_1+\cdots+\alpha_n}}{\partial x_1^{\alpha_1}\cdots\partial x_n^{\alpha_n}} P(x_1,\dots,x_n),$$

and $|\bar{\alpha}| = \alpha_1 + \cdots + \alpha_n$. Finally, for $\vec{y} \in \mathbf{C}^n$ we denote the distance from $\vec{y}$ to $\mathfrak{N}(P)$ by $d(\vec{y}, P)$.

The following result is valid for any $\bar{\alpha} \in \mathbf{N}^n$ (cf. L. Hörmander [**14**]).

LEMMA 1.1. *If the degree of the polynomial $P(x_1,\dots,x_n)$ does not exceed m, then for any $\vec{y} = (y_1,\dots,y_n) \in \mathbf{C}^n$ we have*

$$d(\vec{y}, P) \cdot |P^{(\bar{\alpha})}(\vec{y})|^{1/|\bar{\alpha}|} \leqslant 2^m \cdot |\bar{\alpha}|! \cdot |P(\vec{y})|^{1/|\bar{\alpha}|}. \tag{2}$$

PROOF. Let $P(\vec{y}) \neq 0$, i.e. $d = d(\vec{y}, P) \neq 0$. We will show that for any $\vec{z} \in \mathbf{C}^n$ one has

$$|P(\vec{y} + \vec{z})| \leqslant 2^m \cdot |P(\vec{y})| \quad \text{for } |\vec{z}| \leqslant d, \tag{3}$$

where $\vec{y} + \vec{z} = (y_1 + z_1,\dots,y_n + z_n)$. For the proof of (3) we introduce an auxiliary function $p(t) = P(\vec{y} + t\vec{z})$ from $t \in \mathbf{C}^1$. The zeroes of this polynomial $p(t) \in \mathbf{C}[t]$ of degree $\leqslant m - t_i$ satisfy the inequality $|t_i||\vec{z}| \geqslant d \geqslant |\vec{z}|$ according to the definition of d and $|\vec{z}|$. So $|t_i| \geqslant 1$ and

$$\left|\frac{P(\vec{y} + \vec{z})}{P(\vec{y})}\right| = \left|\frac{p(1)}{p(0)}\right| = \left|\prod_{i=1}^{m} \frac{t_i - 1}{t_i}\right| \leqslant 2^m.$$

Consequently, (3) is proved. Now we apply the inequality (3) to the function $F(\vec{z}) = P(\vec{y} + \vec{z})$ analytic in the circle $|\vec{z}| \leqslant d$. Applying the Cauchy formula and the maximum principle to $F(\vec{z})$ we get from (3)

(4) $$|F^{(\bar{\alpha})}(0)| \leqslant |\bar{\alpha}|! \cdot 2^m \cdot |P(\vec{y})/d^{|\bar{\alpha}|}.$$

From (4) we get

$$|P^{(\bar{\alpha})}(\vec{y})| = |F^{(\bar{\alpha})}(0)| \leqslant |\bar{\alpha}|! \cdot 2^m \cdot |P(\vec{y})|/d^{|\bar{\alpha}|},$$

which proves (2).

We deduce from the lemma a corollary for the case $n = 2$ and $|\bar{\alpha}| = 1$ that is of interest to us.

COROLLARY 1.2. *Let*

$$\delta = \min_{\mathbf{R}^2}\left(\max\{|P(x, y)|, |P'_x(x, y)|, |P'_y(x, y)|\}\right),$$

where min *is taken over all the points* $(x, y) \in \mathbf{R}^2$. *Then for any* $(x_1, y_1) \in \mathbf{R}^2$ *there exists* $(x_0, y_0) \in \mathbf{C}^2$ *such that for* $|P(x_1, y_1)| < \delta$ *we have*

$$\sqrt{|x_1 - x_0|^2 + |y_1 - y_0|^2} \leqslant 2^m \cdot |P(x_1, y_1)| \cdot \delta^{-1},$$

where m *is the degree of* $P(x, y)$ *and* $P(x_0, y_0) = 0$.

This follows from (2) for $\bar{\alpha} = (1, 0), (0, 1)$.

Now our main problem is to determine δ at least for smooth curves $\mathfrak{N}(P)$.

We consider that important case, when the point (x_0, y_0) approximating (x_1, y_1) can be chosen real and even in such a way that $(x_1 - x_0)(y_1 - y_0) = 0$.

For this it suffices to consider polynomials of one variable, and we get

LEMMA 1.3. *Let* $P(x) \in \mathbf{R}[x]$ *be a polynomial of degree* $\leqslant m$ *and height* $\leqslant H$. *Denote by* Δ *an arbitrary bounded set of* $\mathbf{R}^1$. *Then there exist constants* $c_0(m, \Delta)$, $c_1(m, \Delta) > 0$, *such that for any* $x \in \Delta$ *there exists a real root* x_0 *of polynomial* $P(x)$, *for which*

$$|P'(x)|^2 > c_0(m, \Delta)|P(x)| \cdot H$$

implies

$$|x - x_0| < c_1(m, \Delta)|P(x)| \cdot |P'(x)|^{-1}.$$

PROOF. Let us denote $P'(x)$ by p', where for simplicity $p' > 0$. Then

(5) $$P'(y) > p'/2 \quad \text{for } |x - y| < c'|P'(x)|/H$$

for $c' = c'(m, \Delta)$. Indeed,

(6) $$P'(y) = p' + (y - x)P''(z)$$

for $z \in [x, y]$. Now $|P'(z)| + |P''(z)| \leqslant c''^m H$, where $c'' = c''(m, \Delta) > 0$. Choosing c' so that $c' < \frac{1}{2}(c'')^{-m}$, (6) immediately implies (5). To prove Lemma 1.3 it suffices to show that $P(x)$ changes the sign in the interval of length $c\,|P(x)| \cdot |P'(x)|^{-1}$ and containing x. Let us put for determinateness $P(x) < 0$. Then we write the Taylor formula

$$P(y) = P(x) + (y - x)P'(z),$$

where $x < y$ and $x \leqslant z \leqslant y$. Now we take (5) into account and obtain

$$P'(z) > \tfrac{1}{2}P'(x) > 0$$

for the case $z - x \leqslant y - x < c' \cdot |P'(x)| \cdot H^{-1}$. Now y is chosen so that

$$(y - x) \cdot P'(z) > |P(x)|,$$

i.e. in order for $y - x > |P(x)| \cdot |P'(z)|^{-1}$. For this, in view of the above, it is necessary that

$$y - x > 2 \cdot |P(x)||P'(x)|^{-1} \quad \text{and} \quad y - x < c'|P'(x)|H^{-1}.$$

The compatibility of these inequalities follows from the assumptions made in the formulation of Lemma 1.3. So, if we choose

$$y = x + |c'''P(x)P'(x)^{-1}|$$

for $c''' < c'$ we obtain Lemma 1.3. The other relations between the signs of $P(x)$ and $P'(x)$ are similarly considered,

$$\operatorname{sgn}(y - x) \cdot \operatorname{sgn} P'(x) = -\operatorname{sgn} P(x).$$

Lemma 1.3 is completely proved.

One can get interesting applications for polynomials in two variables.

LEMMA 1.4. *Let $P(x, y) \in \mathbf{R}[x, y]$ be of degree $\leqslant m$ and height $\leqslant H$ and $\Delta \subset \mathbf{R}^1$ be a bounded subset. As above, we let $\delta = \min_\Delta(\max\{|P(x, y)|, |P_x'(x, y)|, |P_y'(x, y)|\})$, where the minimum is taken over all the points $(x, y) \in \mathbf{R}^2$ (or only over the points $(x, y) \in \Delta$). If for $(x, y) \in \Delta$ one has*

$$\delta^2 > c(\Delta, m)|P(x, y)|H,$$

then there exists $(x_0, y_0) \in \mathbf{R}^2$ such that

$$|x_0 - x| + |y_0 - y| < c(\Delta, m)|P(x, y)|\delta^{-1}, \qquad P(x_0, y_0) = 0$$

and additionally either $x_0 = x$ or $y_0 = y$.

1.2. We need some more lemmas on polynomials and their roots. We cannot use the results of [**1**, **3**] that are usually applied as we are looking not for the nearest root but for the nearest real root. In principle, it is possible to use complex roots, but this leads to complication of the topological picture—to the intersection of complex hypersurfaces with the real ones.

LEMMA 1.5. *We preserve all the notation of Lemma 1.3, $P(x) \in \mathbf{R}[x]$, $d(P) = m$, $H(P) = H$, Δ is a bounded subset of $\mathbf{R}$. If for $x \in \Delta$,*

$$|P'(x)|^2 > c_0(\Delta, m)|P(x)|H, \tag{7}$$

then there exists a real root x_0 *of the polynomial* $P(x)$, *such that*

$$|x_0 - x| < c_1(\Delta, m)|P(x)/P'(x)|. \tag{8}$$

Moreover, in this case

$$|x_0 - x| \cdot |P'(x_0)| \leqslant c_2(\Delta, m)|P(x)|. \tag{9}$$

PROOF. The first part of Lemma 1.5 follows from Lemma 1.3. From the Taylor formula one gets

$$P(x) = P'(x_0)(x - x_0) + \frac{P''(x_0)}{2}(x - x_0)^2 + \cdots.$$

As $|P''(x_0)| \leqslant c_3(\Delta, m)H$, we obtain

$$|P'(x_0)(x - x_0)| \leqslant |P(x)| + c_3 H|x_0 - x|^2 + \cdots. \tag{10}$$

However, by assumption,

$$|x_0 - x| < c_1|P(x)| \cdot |P'(x)|^{-1} \leqslant c_0^{-1/2} \cdot c_1|P(x)|^{1/2}H^{-1/2}.$$

Hence $|x_0 - x|^2 < c_1|P(x)| \cdot H^{-1}$. Thus from (10) we obtain

$$|P'(x_0)(x - x_0)| \leqslant c_4|P(x)|$$

for $c_4 = c_4(m, \Delta)$. The lemma is proved.

This simple lemma is essential as it makes it possible to analyze how close a point is to a hypersurface by *only* taking into account the geometry of this hypersurface (tangents, etc.) and not that of the surrounding domain.

REMARK. As can be seen from the proofs of Lemmas 1.3 and 1.5 the possibility of "good" approximation of the real number x by the root x_0 is changed to the following one:

$$|x_0 - x|^2 < c_1'|P(x)| \cdot H^{-1}. \tag{11}$$

This simpler condition is much more difficult to verify for concrete cases.

We formulate separately some simple statements which are converses to Lemma 1.5.

LEMMA 1.6. *Assume that* $P(x) \in \mathbf{R}[x]$, $d(P) = m$, $H(P) = H$ *and* Δ *is a bounded subset of* $\mathbf{R}^1$. *There is a constant* $c_5(m, \Delta) > 0$ *such that for any* $x \in \Delta$, *the existence of a root* x_0 *of the polynomial* $P(x)$ *such that*

$$|x_0 - x|^2 < c_5 \cdot |P(x)| \cdot H^{-1} \tag{12}$$

implies

$$|P(x)| \leqslant c_6 \cdot |P'(x_0)| \cdot |x_0 - x|, \tag{13}$$

for $c_6 = c_6(m, \Delta)$.

PROOF. Let $|\Delta| = \sup\{|x| : x \in \Delta\} + 2$. Now $|P(x_1)| \leqslant m|\Delta|^m \cdot H$ for any $x_1 \in \Delta$. Analogously for every k, $1 \leqslant k \leqslant m$, we obtain

$$\left|\frac{P^{(k)}(x_1)}{k!}\right| \leqslant \left(C_m^k + \cdots + C_k^k\right) \cdot |\Delta|^{m-k} \cdot H \leqslant m2^m|\Delta|^m H.$$

So we choose $c_5(m, \Delta)$ in such a way that $c_5 m|\Delta|^m < 1$. In this case $|x_0 - x| < 1$ from which, in view of the definition of $|\Delta|$,

$$|P^{(k)}(x_0)/k!| \leqslant m \cdot 2^m \cdot |\Delta|^m \cdot H$$

follows.

Thus according to the Taylor formula,

$$P(x) - \frac{P''(x_0)}{2!}(x - x_0)^2 - \frac{P'''(x_0)}{3!}(x - x_0)^3 + \cdots = P'(x_0)(x - x_0),$$

we obtain

$$(14) \quad |P'(x_0)(x - x_0)| \geqslant |P(x)| - |x_0 - x|^2 \cdot \left(\frac{|P''(x_0)|}{2!} + \cdots + \frac{|P'''(x_0)|}{3!}|x - x_0| + \cdots\right).$$

As c_5 is chosen to be less than $(m|\Delta|^m)^{-1}$, then $|x_0 - x| < 1$ and according to the above

$$(15) \quad |x_0 - x|^2 \cdot \left|\frac{P''(x_0)}{2!} + \frac{P'''(x_0)}{3!}(x - x_0) + \cdots + \frac{P^{(m)}(x_0)}{m!}(x - x_0)^m\right| \leqslant |x_0 - x|^2 m^2 2^m |\Delta|^m H.$$

Now we put $c_5 = (2^{m+1} m^2 |\Delta|^m)^{-1} < (m|\Delta|^m)^{-1}$. So according to (12)

$$(16) \qquad |x_0 - x|^2 m^2 2^m |\Delta|^m H \leqslant |P(x)|/2.$$

Taking (14) — (16) into account we finally obtain

$$|P'(x_0)(x - x_0)| \geqslant |P(x)|/2.$$

Lemma 1.6 is completely proved for $c_6 = 2$.

Unfortunately, as we have already remarked it is difficult to verify the condition (12). That is why we present one more upper bound for $|P(x)|$ in terms of $|x - x_0|$ for the root x_0 of $P(x)$.

LEMMA 1.7. *Let $P(x) \in \mathbf{R}[x]$, $d(P) = m$, $H(P) = H$ and let Δ be a bounded subset of $\mathbf{R}^1$. Then for any $x \in \Delta$ and the root x_0 of $P(x)$ such that $|x_0 - x| < 1$ we have*

$$|P(x)| \leqslant |P'(x_0)| \cdot |x_0 - x| + c_7 H |x_0 - x|^2 \quad \textit{for } c_7 = m^2 2^m |\Delta|^m > 0.$$

PROOF. We use the computations from the previous proof. For $|x_0 - x| < 1$ we get that

$$|P^{(k)}(x_0)/k!| \leqslant m 2^m |\Delta|^m H: \qquad k \leqslant m.$$

But from the Taylor formula

$$|P(x)| \leqslant |P'(x_0)(x - x_0)| + |x_0 - x|^2 \left|\frac{P''(x_0)}{2!} + \cdots\right|$$

$$\leqslant |P'(x_0)(x - x_0)| + m^2 2^m |\Delta|^m H |x - x_0|^2.$$

2. This section is devoted to the proof of a series of theorems on extremality of multidimensional manifolds connected with linear combinations of polynomials.

The general scheme of proof of Theorem 1 has emerged as an elementary and simple version of a general topological and geometrical scheme for investigation of extremality. Without almost any essential changes it is transferred to the case of arbitrary multidimensional manifolds connected with linear combinations of polynomials of type

$$\lambda_1 P_1(x_1) + \cdots + \lambda_n P_n(x_n).$$

The extremality of the manifolds of this type was carried out by V. G. Sprindzuk, E. Kovalevskaya and V. Bernik [**6**, **7**, **13**], making use of the method of trigonometric sums and because of this only the case $n \gg (\max_{i=1,\dots,n} d(P_i))^2$ was considered. We prove extremality without the assumptions of this type. In the proofs of Theorems I and II the separate enumerations of formulae and constants are introduced, as they are not used in the other parts of the text.

Let us consider Euclidean spaces $\mathbf{R}^m : m \geqslant 1$, together with the Lebesgue measure naturally defined on them. For $A \subseteq \mathbf{R}^n$ the symbol mes(A) denotes the measure A in $\mathbf{R}^m$. We do not assume the inclusion $\mathbf{R}^n \subset \mathbf{R}^m : n < m$, considering only the isomorphism $\pi_{n,m} : \mathbf{R}^n \to \mathbf{R}^m$. So there are no ambiguities connected with the use of the same symbol mes$(\cdot)$ for subsets of different $\mathbf{R}^m$. When necessary, we specify on which $\mathbf{R}^m$ the measure is defined.

THEOREM I. *Let m, n be integer rational nonnegative numbers and let $\varepsilon > 0$ be a real number. We denote by A, $A = A(m, n, \varepsilon)$, the set $(x, y) \in \mathbf{R}^2$ for which there exists an infinite number of pairs of polynomials $P(x)$, $Q(x) \in \mathbf{Z}[x]$, with $d(P) \leqslant n$, $d(Q) \leqslant m$; $H(P) \leqslant H$, $H(Q) \leqslant H$ and*

$$|P(x) - Q(y)| \leqslant H^{-n-m-\varepsilon}$$

for $Q(x)$ having zero free coefficient. Then the set A has zero Lebesgue measure.

PROOF. The definition of the set $A = A(m, n, \varepsilon)$ immediately implies its measurability for any $\varepsilon > 0$ and $m, n \geqslant 0$.

The Theorem I for $m + n = 0$ (i.e. $m = n = 0$) and any $\varepsilon > 0$ immediately follows from the Khintchine theorem. It is easy to see that Theorem I is valid for any $\varepsilon > 0$ and also for $m + n = 1$. This follows from the known generalizations of the Khintchine theorem connected with the linear form $ax + by + c$ over $\mathbf{Z}$ [**4**, **5**, **7**].

From this place and up to the final part of the proof it is assumed that Theorem I is true for any $\varepsilon > 0$ and $m + n < I$, $I \geqslant 2$, i.e. the sets $A(m, n, \varepsilon)$ have zero measure, for $m + n < I$. We prove now that the set $A(m, n, \varepsilon)$ has measure 0 for any $\varepsilon > 0$ with $m + n = I$.

Let us assume the contrary; suppose that the set $A_0 = A(m, n, \varepsilon)$ has positive measure for some $\varepsilon = \varepsilon_0 > 0$ and $m + n = I$. By $\pi(n, m, H)$ for $H \in \mathbf{N}$ we denote the set of those pairs of polynomials (P, Q) for which $P(x)$, $Q(x) \in \mathbf{Z}[x]$, $d(P) = n$, $d(Q) = m$, $\max(H(P), H(Q)) = H$ and $Q(x)$ has no free coefficient.

In this notation the set A_0 consists of those $\vec{x} = (x, y) \in \mathbf{R}^2$ for which for an infinite number of $H \in \mathbf{N}$ and an infinite number of pairs $(P, Q) \in \pi(n, m, H)$ we have

$$(1) \qquad |P(x) - Q(y)| < H^{-n-m-\varepsilon} \quad \text{for } \varepsilon = \varepsilon_0 > 0.$$

We suppose further that there is the following relation between n and m:

$$(2) \qquad n \geqslant m.$$

Indeed it does not change the generality. Really, let $\operatorname{mes} A(m, n, \varepsilon) = 0$ be proved for all $m + n = I$ in case of (2) and now let $n < m$. For every $(x, y) \in A(m, n, \varepsilon)$ we have for an infinite number of $H \in \mathbf{N}$ and $(P, Q) \in \pi(n, m, H)$ the inequality (1). Put $Q(z) = zQ_1(z)$ and $Q_1(z) \in \mathbf{Z}[z]$ according to the definition of $\pi(n, m, H)$ and $P(z) = zP_1(z) + a$ with $P_1(z) \in \mathbf{Z}[z]$ and $a \in \mathbf{Z}$. Then (1) can be rewritten in the form

$$|Q_0(y) - P_0(x)| < H^{-n-m-\varepsilon},$$

where $Q_0(z) = Q(z) - a = zQ_1(z) - a$, $P_0(z) = zP_1(z)$ i.e. $d(Q_0) = m$, $d(P_0) = n$, $\max(H(P_0), H(Q_0)) = \max(H(P), H(Q)) = H$ i.e. $(Q_0, P_0) \in \pi(m, n, H)$ where we already have $m > n$. In other words $A(n, m, \varepsilon) \supseteq A(m, n, \varepsilon)$ for any m, n, ε. Thus from $\operatorname{mes} A(m, n, \varepsilon) = 0$ for arbitrary $m + n = I$ and $n > m$ it follows that $\operatorname{mes} A(m, n, \varepsilon) = 0$ for arbitrary $m + n = I$ and $n < m$. Thus the assumption (2) does not restrict the generality of our considerations.

For any $\vec{x} = (x, y) \in \mathbf{R}^2$ we denote by $\mathcal{P}(\vec{x})$ the set of those $(P, Q) \in \pi(n, m, H)$ with some $H \in \mathbf{N}$ for which

$$|P(x) - Q(y)| < H^{-n-m-\varepsilon}.$$

Now let A_0 be the set of those $\vec{x} \in \mathbf{R}^2$ for which $\mathcal{P}(\vec{x})$ contains an infinite number of pairs of polynomials has measure > 0. It is easy to notice that $(P, Q) \in \mathcal{P}(\vec{x})$ implies $(-P, -Q) \in \mathcal{P}(\vec{x})$ for any $\vec{x} \in \mathbf{R}^2$. So without loss of generality we assume further that

(3) if $(P, Q) \in \mathcal{P}(\vec{x})$ for $\vec{x} \in \mathbf{R}^2$ then the highest coefficient of $P(x)$ is positive.

Now for $H, H_0 \in \mathbf{N}$ and $H_0 \leqslant H$ one can put

$$\pi(n, m, H; H_0) = \{(P, Q) \in \pi(n, m, H): \text{the highest coefficient of } P(x) \text{ is equal to } H_0\}.$$

As A_0 has measure > 0, according to the σ-additivity of the measure there exists $N_0 \geqslant 1$ such that the set

$$A_1 = \{(x, y) \in A_0 : |x|, |y| \leqslant N_0\}$$

has measure > 0. As A_0 is invariant also under the transformations $x \to \pm x$, $y \to \pm y$, then the set

$$A_2 = \{(x, y) \in A_0 : N_0^{-1} \leqslant x, y \leqslant N_0\}$$

has measure > 0 too.

Now in view of our approach to the extremality problem it is necessary to change the "smallness" of the difference $P(x) - Q(y)$ to the "closeness" of the point (x, y) to the curve

$$\mathcal{N}(P, Q): P(x) - Q(y) = 0$$

in $\mathbf{A}^2(\mathbf{R})$. We fix the point $\vec{x}_1 = (x_1, y_1) \in A_2$. Let us consider polynomial

$$Q_{(x_1, P)}(z) = Q(z) - P(x_1) \in \mathbf{R}[z] \tag{4}$$

for every pair $(P, Q) \in \mathcal{P}(\vec{x}_1)$. As $(P, Q) \in \mathcal{P}(\vec{x}_1)$ and $\vec{x}_1 \in A_2$, then for $H = \max(H(P), H(Q))$,

$$|Q_{(x_1, P)}(y_1)| < H^{-n-m-\varepsilon}. \tag{5}$$

In order to apply Lemma 1.3, it is necessary to have

$$|Q'_{(x_1, P)}(y_1)|^2 > c_0 H\big(Q_{(x_1, P)}\big) \cdot |Q_{(x_1, P)}(y_1)|,$$

where $c_0(m, \Delta_y) = \gamma_1(m, N_0) = \gamma_1$ and

$$\Delta_y = \{y : N_0^{-1} \leqslant y \leqslant N_0\}.$$

But $H(Q_{(x_1, P)}) \leqslant H(Q) + |P(x_1)| \leqslant H(Q) + nN_0^n H(P) \leqslant \gamma_2 H$ for $\gamma_2 = 1 + nN_0^n$. As

$$Q'_{(x_1, P)}(z) = Q'(z), \qquad H\big(Q_{(x_1, P)}\big) \leqslant \gamma_2 H, \tag{6}$$

for the application of Lemma 1.3 it is necessary to verify the condition

$$|Q'(y_1)|^2 > \gamma_3 H |Q_{(x_1, P)}(y_1)|,$$

where $\gamma_3 = \gamma_1\gamma_2 > 0$. However in view of (4) and (5) it is sufficient to verify the inequality

$$|Q'(y_1)|^2 > \gamma_3 H^{-n-m+1-\varepsilon} \tag{7}$$

for $\gamma_3 = \gamma_3(n, m, N_0) > 0$ and $H = \max(H(P), H(Q))$. We eliminate from A_2 the set of measure zero in order to apply Lemma 1.3. First of all if $m = 0$ then the set $A(m, n, \varepsilon)$ has measure zero according to [**1, 3**]. So we consider the case $m \geqslant 1$.

According to Mahler's hypothesis (the theorem of Sprindzuk [**1, 3**]) there exists a set $\nabla \subseteq \mathbf{R}^1$ of measure zero such that for any $y \notin \nabla$ and polynomial $R(x) \in \mathbf{Z}[x]$ of degree $\leqslant m - 1$ and height $\leqslant mH_1$ from

$$|R(y)| \leqslant \gamma_3^{1/2} H_1^{-m+1-\varepsilon/2} \tag{8}$$

the boundedness of H_1: $H_1 < \psi(y) < \infty$ follows. Now we put

$$A_3 = A_2 \setminus (\mathbf{R}^1 \times \nabla). \tag{9}$$

As mes $A_2 > 0$ and mes $\nabla = 0$, then

$$\text{mes } A_3 > 0. \tag{10}$$

Now let $\vec{x}_1 = (x_1, y_1) \in A_3$ and $(P, Q) \in \mathcal{P}(\vec{x}_1)$ such that $H = \max(H(P), H(Q)) \geqslant \psi(y_1)$. We apply (8) to the polynomial $Q'(y)$ and obtain

$Q'(y) = R(y)$, where $R(y)$ has the degree $\leqslant m-1$ and the height $\leqslant mH(Q) \leqslant mH$ and in view of $y_1 \notin \nabla$ we have

$$|R(y_1)| > \gamma_3^{1/2} H^{-m+1-\varepsilon/2}$$

i.e.

(11) $$|Q'(y_1)| > \gamma_3^{1/2} H^{-m+1-\varepsilon/2}.$$

Raising both sides of (11) to the power two and taking into account the condition (2) that $n \geqslant m$ we obtain

(12) $$|Q'(y_1)|^2 > \gamma_3 H^{-n-m+1-\varepsilon}.$$

Thus (12) proved that the criteria (7) is true

$$|Q'(y_1)|^2 > \gamma_3 H^{-n-m+1-\varepsilon}$$

for arbitrary $\vec{x}_1 = (x_1, y_1) \in A_3$, $(P, Q) \in \mathcal{P}(\vec{x}_1)$ with $H = \max(H(P), H(Q)) \geqslant \psi(y_1)$. Consequently we obtain

LEMMA 2.1. *For any* $\vec{x}_1 = (x_1, y_1) \in A_3$ *and arbitrary* $(P, Q) \in \mathcal{P}(\vec{x}_1)$ *with* $H = \max(H(P), H(Q)) \geqslant \psi(y_1)$ *there exists a real number* y_0 *such that*

(13) $$P(x_1) - Q(y_0) = 0,$$

i.e. $\vec{x}_0 = (x_1, y_0)$ *lies on the curve* $\mathcal{N}(P, Q)$ *and additionally*

(14) $$|y_0 - y_1| \leqslant \gamma_4 |P(x_1) - Q(y_1)| / |Q'(y_1)| \leqslant \gamma_4 |Q'(y_1)| H^{-n-m-\varepsilon}.$$

PROOF. Let us return to the notation (4). According to the results established above and the proven criteria (12) for $\vec{x}_1 \in A_3$ and $H(P) \geqslant \psi(y_1)$ we obtain by (5):

$$|Q'_{(x_1, P)}(y_1)|^2 > c_0 |Q_{(x_1, P)}(y_1)| H(Q_{(x_1, P)}).$$

We now apply Lemma 1.3. We obtain a real number y_0 such that $Q_{(x_1, P)}(y_0) = 0$ and

$$|y_0 - y_1| \leqslant c_1 |Q'_{(x_1, P)}(y_1)|^{-1} |Q_{(x_1, P)}(y_1)|.$$

According to (5) and (6) this means that (14) is satisfied. Lemma 2.1 is proved.

Now for any $\vec{x}_1 = (x_1, y_1) \in A_3$ we put

$$\mathcal{P}_1(\vec{x}_1) = \{(P, Q) \in \mathcal{P}(\vec{x}_1) \colon H = \max(H(P), H(Q)) \geqslant \psi(y_1)\}.$$

As $\psi(y_1) < \infty$, then

(15) the set $\mathcal{P}_1(\vec{x}_1) \subseteq \mathcal{P}(\vec{x}_1)$ is infinite for any $\vec{x}_1 \in A_3$.

Thus we further consider $\mathcal{P}_1(\vec{x})$ instead of $\mathcal{P}(\vec{x}_1)$. We formulate 2.1 in a more convenient form.

COROLLARY 2.2. *For any* $\vec{x}_1 = (x_1, y_1) \in A_3$ *and any* $(P, Q) \in \mathcal{P}_1(\vec{x}_1)$ *there exists a real number* y_0 *such that* $P(x_1) = Q(y_0)$ *and for* $H = \max(H(P), H(Q))$,

$$|y_0 - y_1| < \gamma_4 |Q'(y_1)|^{-1} H^{-n-m-\varepsilon},$$

with

(16) $$|y_0 - y_1| < \gamma_5 H^{-n-1-\varepsilon/2}.$$

PROOF. The first part of the corollary is simply Lemma 2.1. Let us prove (16). For this we use (14) and the inequality (8) with $Q'(y) = R(y)$, $y_1 \notin \nabla$ (as $(x_1, y_1) \in A_3$ see (9)). From (11) one gets

$$|Q'(y_1)| > \gamma_3^{1/2} H^{-m+1-\varepsilon/2}.$$

This together with (14) implies (16) for $\gamma_5 = \gamma_4 \gamma_3^{-1/2}$. Corollary 2.2 is proved.

For the given $\bar{x}_1 = (x_1, y_1) \in A_3$ and fixed $(P, Q) \in \mathcal{P}_1(\bar{x}_1)$ we denote by $\bar{x}_0 = (x_1, y_0)$ the nearest to $\bar{x}_1$ in the direction of the x_1 point of the curve $\mathcal{N}(P, Q)$ (if such a point exists). According to 2.1 and 2.2 such a point $\vec{x}_0 \in \mathbf{R}^2$ exists and moreover the inequalities (14) and (16) are satisfied. It is convenient to present the estimate $|y_0 - y_1|$ is terms of $|Q'(y_0)|$. Let us use Lemma 1.6.

LEMMA 2.3. *If $\bar{x}_1 = (x_1, y_1) \in A_3$, $(P, Q) \in \mathcal{P}_1(\vec{x}_1)$, then there exists a real number y_0 such that $P(x_1) = Q(y_0)$ and for $H = \max(H(P), H(Q))$ we have*

$$|y_0 - y_1| \leq \gamma_4 |Q'(y_1)| H^{-n-m-\varepsilon} \leq \gamma_5 H^{-n-1-\varepsilon/2};$$

$$|y_0 - y_1| \leq \gamma_6 H^{-n-m-\varepsilon} |Q'(y_0)|^{-1}. \tag{17}$$

PROOF. We apply Lemma 1.6 in the case $P(z) = Q_{(x_1, P)}(z)$. Then according to (12), (14) and (16) the conditions of Lemma 1.6 are satisfied. This implies the first part of the lemma and the second part is obtained from the inequalities of Lemma 1.6 and (6).

In the further part of the proof we use the method of the "essential and nonessential domains" of V. G. Sprindzuk in the form given in [**1**, **3**]. We take Lemma 7 from [**3**] in a changed form.

LEMMA 2.4 [**3**]. *For natural k and h let $\mathcal{U}(k, h)$ be the finite collection of real closed intervals. Let us denote by $\mathcal{V}(k, h)$ the subset of $\mathcal{U}(k, h)$ such that for any $I \in \mathcal{V}(k, h)$ there exists $J \in \mathcal{U}(k, h)$, $J \neq I$, for which*

$$\operatorname{mes}(I \cap J) \geq \tfrac{1}{2} \operatorname{mes}(I).$$

We denote by $V(k, h) = \bigcup \mathcal{V}(k, h)$ and $v(k, h)$ the union of intervals of the form $I \cap J$ for $I \in \mathcal{V}(k, h)$ and $J \in \mathcal{U}(k, h)$, $J \neq I$. If the set w of those points of $\mathbf{R}^1$ that are contained in an infinite number of the sets $v(k, h)$ has zero measure then the set W of the elements of $\mathbf{R}^1$ that are contained in an infinite number of the sets $V(k, h)$ also has zero measure.

We need some more notation. For any $A \subseteq \mathbf{R}^2$ and $x_0 \in \mathbf{R}^1$ by $A \upharpoonright x_0$ we denote the one-dimensional section of A by the line $x = x_0$, $A \upharpoonright x_0 = \{y \in \mathbf{R}^1 : (x_0, y) \in A\}$. We recall also that $\pi(H)$ $(= \pi(n, m, H))$ for $H \in \mathbf{N}$ is the set of those pairs $(P, Q) \in \mathbf{Z}[x] \times \mathbf{Z}[x]$, for which $d(P) = n$, $d(Q) = m$, $H = \max(H(P), H(Q))$ for $Q(x)$ being a polynomial without a free coefficient. For any rational integer $k \geq 0$ we put

$$\pi(k) = \bigcup \{\pi(H) : 2^k \leq H < 2^{k+1}\}. \tag{18}$$

For $h, H \in \mathbf{N}$, $h \leq H$, we have introduced above the set $\pi(H; h) = \{(P, Q) \in \pi(H) :$ the leading coefficient of $P(x)$ equals $h\}$. In view of (3) we have

$$\pi(H) = \bigcup \{\pi(H; h) : h \leq H\}. \tag{19}$$

Now we define the set $\pi(k, h)$ for $k \geqslant 0, h \geqslant 1$ as

$$\pi(k, h) = \bigcup \{\pi(H; h) : 2^k \leqslant H < 2^{k+1}\}. \tag{20}$$

We should remark that by definition of $\pi(H; h)$ the set $\pi(k, h)$ is empty for $h \geqslant 2^{k+1}$:

$$\pi(k, h) = \varnothing : \qquad h \geqslant 2^{k+1}. \tag{21}$$

From (18)–(21) we get

$$\pi(k) = \bigcup \{\pi(k, h) : h \leqslant 2^{k+1}\}. \tag{22}$$

For any $H \in \mathbf{N}$ and any pair $(P, Q) \in \pi(H)$ we introduce two types of sets.

$$\begin{aligned} M(P, Q) = \{(x, y) \in \mathbf{R}^2 &: \text{there exists } y_0 \in \mathbf{R}, \\ &\text{such that } (x, y_0) \in \mathfrak{N}(P, Q), \text{ i.e. } P(x) = Q(y_0) \\ &\text{and } |y - y_0| \leqslant \min[\gamma_5 H^{-n-1-\varepsilon/2}, \gamma_6 |Q'(y_0)|^{-1} H^{-n-m-\varepsilon}]\}, \end{aligned} \tag{23}$$

$$\begin{aligned} \Lambda(P, Q) = \{(x, y) \in \mathbf{R}^2 &: \text{there exists a } y_0 \in \mathbf{R}, \\ &\text{such that } (x, y_0) \in \mathfrak{N}(P, Q) \text{ i.e. } P(x) = Q(y_0) \\ &\text{and } |y - y_0| \leqslant \min[\gamma_5 H^{-n-\varepsilon/4}, \gamma_6 \cdot |Q'(y_0)|^{-1} \cdot H^{-n-m+1-3\varepsilon/4}]\}. \end{aligned} \tag{24}$$

As the curve $\mathfrak{N}(P, Q)$ is algebraic, then for any $x \in \mathbf{R}^1$ the set $\Lambda(P, Q) \upharpoonright x$ is a union of a finite number of closed nonintersecting intervals. The family of these intervals is denoted by $\mathfrak{L}(P, Q) \upharpoonright x$ and

$$\Lambda(P, Q) \upharpoonright x = \bigcup \{I : I \in \mathfrak{L}(P, Q) \upharpoonright x\}. \tag{25}$$

We fix $x_0 \in \mathbf{R}^1$ and denote by $\mathfrak{U}(k, h) \upharpoonright x_0$ the set of all the intervals from $\mathfrak{L}(P, Q) \upharpoonright x_0$ for some $(P, Q) \in \pi(k, h)$,

$$\mathfrak{U}(k, h) \upharpoonright x_0 = \bigcup \{\mathfrak{L}(P, Q) \upharpoonright x_0 : (P, Q) \in \pi(k, h)\}. \tag{26}$$

Now we consider a maximal subset $\mathfrak{V}(k, h) \upharpoonright x_0$ of the set that satisfies the property mentioned in Lemma 2.4,

$$\operatorname{mes}(I \cap J) \geqslant \tfrac{1}{2} \operatorname{mes}(I)$$

for any $I \in \mathfrak{V}(k, h) \upharpoonright x_0$ and some $J \in \mathfrak{U}(k, h) \upharpoonright x_0$, $I \neq J$.

We put

$$V(k, h) \upharpoonright x_0 = \bigcup \{I : I \in \mathfrak{V}(k, h) \upharpoonright x_0\}. \tag{27}$$

LEMMA 2.5. *The set $W \upharpoonright x$ of those $y \in \mathbf{R}^1$ that belong to an infinite number of $V(k, h) \upharpoonright x$ for almost all $x \in \mathbf{R}^1$ has measure zero.*

PROOF. Let $W \upharpoonright x_0$ be the set of $y \in \mathbf{R}^1$ which belong to an infinite number of sets $V(k, h) \upharpoonright x_0 : x_0 \in \mathbf{R}^1$.

If $B = \{x \in \mathbf{R}^1 : \operatorname{mes}(W \upharpoonright x) \neq 0\}$, then it is necessary to prove that $\operatorname{mes} B = 0$.

We define by $v(k, h) \upharpoonright x_0$ the union of intervals of the type $I \cap J$ for $I \in \mathfrak{V}(k, h) \upharpoonright x_0$, $J \in \mathfrak{U}(k, h) \upharpoonright x_0$, $I \neq J$. For $x_0 \in B$, $\operatorname{mes}(W \upharpoonright x_0) \neq 0$, and then by Lemma 2.4 and as $\operatorname{mes}(v \upharpoonright x_0) \neq 0$ one gets that the set $v \upharpoonright x_0 \subseteq \mathbf{R}^1$ is composed from those $y \in \mathbf{R}^1$ that belong to an infinite number of $v(k, h) \upharpoonright x_0$.

Let us consider a subset of $\mathbf{R}^2 \mathfrak{B} = \{(x, y) : x \in B, y \in v \upharpoonright x\}$. It is easy to see that for any $x \in \mathbf{R}^1$,

$$\text{(28)} \qquad \mathfrak{B} \upharpoonright x = \varnothing : \ x \notin B \qquad \text{and} \qquad \mathfrak{B} \upharpoonright x = v \upharpoonright x : \ x \in B.$$

Let $2^k \geqslant \gamma_5$ and we choose any $y \in v(k, h) \upharpoonright x$. Then $y \in I \cap J$, where $I, J \in \mathfrak{U}(k, h) \upharpoonright x$, $I \neq J$. According to (25)–(26) we have $y \in \Lambda(P_1, Q_1) \upharpoonright x \cap \Lambda(P_2, Q_2) \upharpoonright x$ for (P_1, Q_1), $(P_2, Q_2) \in \pi(k, h)$ and $(P_1, Q_1) \neq (P_2, Q_2)$ as intervals from $\mathfrak{L}(P, Q) \upharpoonright x$ do not intersect for any (P, Q).

According to (24) we get $y_1, y_2 \in \mathbf{R}^1$ such that for $H_1 = \max(H(P_1), H(Q_1))$, $H_2 = \max(H(P_2), H(Q_2))$: $(x_1, y_1) \in \mathfrak{N}(P_1, Q_1)$, $(x_2, y_2) \in \mathfrak{N}(P_2, Q_2)$ and

$$\text{(29)} \qquad \begin{aligned} |y_1 - y| &\leqslant \min\left[\gamma_5 H_1^{-n-\varepsilon/4}, \gamma_6 |Q_1'(y_1)|^{-1} H_1^{-n-m+1-3\varepsilon/4}\right], \\ |y_2 - y| &\leqslant \min\left[\gamma_5 H_2^{-n-\varepsilon/4}, \gamma_6 |Q_2'(y_2)|^{-1} H_2^{-n-m+1-3\varepsilon/4}\right]. \end{aligned}$$

Now apply Lemma 1.7 to the polynomial $R_i(z) = Q_i(z) - P_i(x) \in \mathbf{R}[z] : i = 1, 2$. As $H_i \geqslant \gamma_5$ and $(x, y_i) \in \mathfrak{N}(P_i, Q_i)$ then in view of 1.7 and (29) we obtain

$$\text{(30)} \qquad |P_i(x) - Q_i(y)| \leqslant |y - y_i| \, |Q_i'(y_i)| + \gamma_7 |y - y_i|^2 H(R_i),$$

$\gamma_7 = \gamma_7(m, N_0) > 0$. Here $H(R_i)$ is the height of the polynomial $Q_i(z) - P_i(x)$, $H(R_i) \leqslant H(Q_i) + H(P_i) \leqslant 2H_i$. Now is view of (2), $n \geqslant m$ and, comparing (29)–(30), we come to

$$\text{(31)} \qquad \begin{aligned} |P_i(x) - Q_i(y)| &\leqslant \gamma_6 H_i^{-n-m+1-3\varepsilon/4} + 2\gamma_5\gamma_7 H_i^{-2n+1-\varepsilon/2} \\ &\leqslant \gamma_8 H_i^{-n-m+1-\varepsilon/2}, \end{aligned}$$

where $\gamma_8 = \gamma_8(n, m, N_0) > 0$ and $i = 1, 2$. As $(P_i, Q_i) \in \pi(k, h)$, then in view of (18), (20) we obtain $H_i \geqslant 2^k : i = 1, 2$. Taking into account inequalities (31) we deduce a new inequality for polynomials $P_3(x) = P_1(x) - P_2(x)$ and $Q_3(x) = Q_1(x) - Q_2(x)$ from $\mathbf{Z}[x]$:

$$\text{(32)} \qquad \begin{aligned} |P_3(x) - Q_3(y)| &\leqslant \gamma_8\left(H_1^{-n-m+1-\varepsilon/2} + H_2^{-n-m+1-\varepsilon/2}\right) \\ &\leqslant 2\gamma_8 (2^k)^{-n-m+1-\varepsilon/2}. \end{aligned}$$

Because $(P_i, Q_i) \in \pi(k, h) : i = 1, 2$, the polynomials $P_3(x)$ and $Q_3(x)$ have the height $\leqslant 2^{k+2}$. Besides, $P_1(x)$ and $P_2(x)$ have, by the definition of $\pi(\mathrm{b}\ h)$, the same leading coefficient h. Thus $P_3(x) = P_1(x) - P_2(x)$ has degree $\leqslant n - 1$. As $Q_i(x) : i = 1, 2$, does not have free coefficient, then $Q_3(x)$ does not have a free coefficient, too. So

$$\text{(33)} \quad d(P_3) \leqslant n - 1, \qquad d(Q_3) \leqslant m, \qquad \max(H(P_3), H(Q_3)) = H_3 \leqslant 2^{k+2},$$

with $Q_3(x)$ having no free coefficient.

According to (33) the inequality (32) can be presented in the form

$$\text{(34)} \qquad |P_3(x) - Q_3(y)| \leqslant \gamma_9 H_3^{-n-m+1-\varepsilon/2}, \qquad \gamma_9 = 2\gamma_8 \cdot 4^{n+m-1+\varepsilon/2}.$$

Thus we proved the inequalities (32)–(34) for arbitrary $x \in \mathbf{R}^1$ and $y \in v(k, h) \upharpoonright x$ for $2^k \geqslant \gamma_5$. If now $(x, y) \in \mathfrak{B}$, then $y \in v \upharpoonright x$ and y belongs to an infinite number of sets $v(k, h) \upharpoonright x$. So for $(x, y) \in \mathfrak{B}$ and an infinite number of pairs of polynomials $(P_3, Q_3) \in \mathbf{Z}[x] \times \mathbf{Z}[x]$, $d(P_3) \leqslant n-1$, $d(Q_3) \leqslant m$, $\max(H(P_3), H(Q_3)) = H_3$, (34) is satisfied, where $Q_3(x)$ does not have a free coefficient. But according to the assumption made in the beginning of the proof of the theorem, all the sets $A(m_1, n_1, \varepsilon_1)$ have zero measure for any $m_1 + n_1 \leqslant m + n - 1$ and $\varepsilon_1 > 0$. Consequently the set $\mathfrak{B}$ is contained in a set of measure 0. Thus

$$\text{mes}\, \mathfrak{B} = 0.$$

From the Fubini theorem it follows that

$$\text{(35)} \qquad \text{mes}(\mathfrak{B} \upharpoonright x) = 0 \quad \text{for almost all } x \in \mathbf{R}^1.$$

By definition of B, $\text{mes}(v \upharpoonright x) \neq 0$ for all $x \in B$. Taking into consideration (28) and (35) we finally obtain $\text{mes}\, B = 0$ and Lemma 2.5 is proved.

LEMMA 2.6. *For almost all $x \in \mathbf{R}^1$ the set $A_3 \upharpoonright x$ has measure zero.*

PROOF OF 2.6. Let us denote by $F \subseteq \mathbf{R}^1$ a set of those $x \in \mathbf{R}^1$, for which a set $W \upharpoonright x$ has measure zero. Then by Lemma 2.5

$$\text{(36)} \qquad \text{mes}(\mathbf{R}^1 \setminus F) = 0.$$

Let us show that for all $x \in F$ (except a set of measure zero) $A \upharpoonright x$ has measure zero. Let $y_0 \in F$ and $\bar{x}_0 = (x_0, y_0) \in A_3$ i.e. $y_0 \in A_3 \upharpoonright x_0$. Then according to Lemma 2.3 and (23) we have for infinite number of pairs of polynomials $(P, Q) \in \mathcal{P}_1(\bar{x}_0)$:

$$\text{(37)} \qquad (x_0, y_0) \in M(P, Q) \quad \text{i.e.}\ y_0 \in M(P, Q) \upharpoonright x_0 : (P, Q) \in \mathcal{P}_1(\bar{x}_0).$$

However $M(P, Q) \upharpoonright x_0$ for any x_0 consists of the union of a finite number of closed intervals, the centers y_i of which are determined as ordinates of points of intersection of a curve $\mathfrak{N}(P, Q): P(x) = Q(y)$, with the line $x = x_0$. Since $m > 0$, for any $x_0 \in F$ the set $M(P, Q) \upharpoonright x_0$ can be represented in the form of the union of $\leqslant m$ intervals

$$M(P, Q) \upharpoonright x_0 = \bigcup \{I : I \in \mathfrak{M}(P, Q) \upharpoonright x_0\}.$$

According to (23)–(25), however, every interval $I \in \mathfrak{M}(P, Q) \upharpoonright x_0$ is contained in some interval $\mathfrak{T}(I) \in \mathfrak{L}(P, Q) \upharpoonright x_0$, and from (23)–(24)

$$\text{(38)} \qquad \text{mes}(I) \leqslant \text{mes}(\mathfrak{T}(I)) \cdot H^{-1-\varepsilon/4}$$

follows, for $(P, Q) \in \pi(H)$. In view of (37), for all $x_0 \in F$ and $y_0 \in A_3 \upharpoonright x_0$ we have

$$\text{(39)} \qquad y_0 \in I \subseteq \mathfrak{T}(I) : \mathfrak{T}(I) \in \mathfrak{L}(P, Q) \upharpoonright x_0,$$

for any $(P, Q) \in \mathcal{P}_1(\bar{x}_0)$ and any infinite set $\mathcal{P}_1(\bar{x}_0)$, $\bar{x}_0 = (x_0, y_0)$. According to the definition of F and Lemma 2.5 y_0 is not contained in an infinite number of sets $V(k, h) \upharpoonright x_0$, i.e. y_0 is not contained in infinite number of intervals from $\mathfrak{V}(k, h) \upharpoonright x_0$ for all possible integers $k \geqslant 0$ and $h \geqslant 1$. According to (39) and the definition of $\mathcal{V}(k, h) \restriction x_0$ (cf. 2.4) for $y_0 \in A_3 \restriction x_0$ this means that

(40) for infinite number (k, h) a point $y_0 \in \mathbf{R}^1$ belongs to an interval of the form $\mathcal{T}(I)$ from $\mathcal{U}(k, h) \restriction x_0$, but not of the form $\mathcal{V}(k, h) \restriction x_0$.

The definition of $\mathcal{V}(k, h) \restriction x_0$ implies that for

$$J_1, J_2 \in (\mathcal{U}(k, h) \restriction x_0) \setminus (\mathcal{V}(k, h) \restriction x_0)$$

and $J_1 \neq J_2$ we have

$$\operatorname{mes}(J_1 \cap J_2) < \tfrac{1}{2}\min(\operatorname{mes}(J_1), \operatorname{mes}(J_2)).$$

Thus any three intervals from $(\mathcal{U}(k, h) \restriction x_0) \setminus (\mathcal{V}(k, h) \restriction x_0)$ do not have common points. Now for $y_0 \in A_3 \restriction x_0$, $y_0 \in [0, N_0]$, and if $y_0 \in I \subseteq \mathcal{T}(I)$, then according to (23)–(24) for $H \geqslant \gamma_5$ we have $\mathcal{T}(I) \subseteq [-2, N_0 + 2]$. So according to the above, the summary length of intervals of the form $\mathcal{T}(I)$ contained in $(\mathcal{U}(k, h) \restriction x_0) \setminus (\mathcal{V}(k, h) \restriction x_0)$ does not exceed $2(N_0 + 4) + \gamma_{10} \leqslant \gamma_{11}$, where $\gamma_{10} = \gamma_{10}(\gamma_5)$. For $I \in \mathfrak{M}(P, Q) \restriction x_0$ and $\mathcal{T}(I) \in \mathfrak{L}(P, Q) \restriction x_0$ for $(P, Q) \in \pi(k, h)$ let us denote by $\mathcal{K}(I)$ the union of intervals $I' \in \mathfrak{M}(P, Q) \restriction x_0$ contained in $\mathcal{T}(I)$. According to (38) and $(P, Q) \in \pi(k, h)$ we have

$$\operatorname{mes}(\mathcal{K}(I)) \leqslant \operatorname{mes}(\mathcal{T}(I))(2^k)^{1-\varepsilon/4}, \qquad \mathcal{K}(I) \subseteq \mathcal{T}(I). \tag{41}$$

Further, let us put

$$\mathfrak{Z}(k, h) \restriction x_0 = \{\mathcal{K}(I) : I \in \mathfrak{M}(P, Q) \restriction x_0, (P, Q) \in \pi(k, h)\}.$$

Taking (26), (39) and (40) into account we obtain

(42) for any $x_0 \in F$ and $y_0 \in A_3 \restriction x_0$, y_0 belongs to infinite numbers of sets $\bigcup\{I : I \in \mathfrak{Z}(k, h) \restriction x_0\}$.

From the foregoing, it follows that for $x_0 \in F$ we have

$$\sum_{\mathcal{K}(I) \in \mathfrak{Z}(k, h) \restriction x_0} \operatorname{mes}(\mathcal{T}(I)) \leqslant \gamma_{11}.$$

By (41) one obtains

$$\sum_{\mathcal{K}(I) \in \mathfrak{Z}(k, h) \restriction x_0} \operatorname{mes}(\mathcal{K}(I)) \leqslant \gamma_{11}(2^k)^{-1-\varepsilon/4}. \tag{43}$$

Now in view of (21) and (43)

$$\sum_{k \geqslant 0} \sum_{h \geqslant 1} \sum_{\mathcal{K}(I) \in \mathfrak{Z}(k, h) \restriction x_0} \operatorname{mes}(\mathcal{K}(I)) \leqslant \sum_{k \geqslant 0} \sum_{1 \leqslant h \leqslant 2^{k+1}} \gamma_{11}(2^k)^{-1-\varepsilon/4}$$

$$\leqslant 2\gamma_{11} \sum_{k \geqslant 0} (2^k)^{-\varepsilon/4} < \infty.$$

According to Borel-Cantelli lemma, a set of $y_0 \in \mathbf{R}^1$ belonging to an infinite number of sets $\bigcup\{I : I \in \mathfrak{Z}(k, h) \restriction x_0\}$ has measure zero for all $x_0 \in F$. By (42)

$$\operatorname{mes}(A_3 \restriction x_0) = 0$$

for $x_0 \in F$, i.e. in view of (36) Lemma 2.6 is proved.

The Fubini Theorem 2.6 implies mes $A_3 = 0$. This contradicts the choice of A_3 (10). Then mes$(A(m, n, \varepsilon)) = 0$ also for $m + n = I$. Theorem I is proved.

Slight changes in the proof of Theorem I lead to the proof of a more general result.

THEOREM II. *Let $n_1,\dots,n_m$ be integers and let $\varepsilon > 0$. Let us denote by $A(n_1,\dots,n_m, \varepsilon)$ the set of those $(x_1,\dots,x_m) \in \mathbf{R}^m$ for which there exists an infinite number of systems of polynomials $P_1(x),\dots,P_m(x) \in \mathbf{Z}[x]$ of degrees $n_1,\dots,n_m$ without free coefficients for which*

$$\|P_1(x_1) + \cdots + P_m(x_m)\| < H^{-n_1 \cdots -n_m-\varepsilon}.$$

Here $H = \max_{i=1,\dots,m} H(P_i)$, and $\|\cdot\|$ is the distance to the nearest integer. Then the set $A(n_1,\dots,n_m, \varepsilon)$ has zero measure in $\mathbf{R}^m$.

PROOF OF THEOREM II. As the main scheme of proof of Theorem I is generalized to this case without substantial changes, we omit wordy considerations. It should be noted however that for $m = 2$ the conditions of Theorems I and II concide, but are formulated differently. Let us assume that Theorem II takes place for all $m_1 < m$ and all $n'_1,\dots,n'_m$, $n'_1 + \cdots + n'_m < I$ and $\varepsilon > 0$, but for $n_1 + \cdots + n_m = I$ and $\varepsilon = \varepsilon_0 > 0$ a set $A_0 = A(n_1,\dots,n_m, \varepsilon) \subseteq \mathbf{R}^m$ has measure > 0. As usual, one can put

$$(1) \qquad n_1 \geqslant \cdots \geqslant n_m, \qquad n_m > 0.$$

Now for $H \in \mathbf{N}$ let us put

$$(2) \qquad \pi(H) = \{(P_1,\dots,P_m) : P_i(x) \in \mathbf{Z}[x],\ d(P_i) = n_i,\ P_i(x) \text{ does not have a free coefficient for } i \geqslant 2 \text{ and } \max_{i=1,\dots,m} H(P_i) = H\}.$$

Let us choose $N_0 \geqslant 1$ such that

$$(3) \qquad \text{mes}(A_1) > 0 \text{ for } A_1 = A_0 \cap [0, N_0] \times [0, N_0],$$

and for $\vec{x} = (x_1,\dots,x_m) \in A_1$ the set $\mathcal{P}(\vec{x})$ of those

$$(P_1,\dots,P_m) \in \cup \{\pi(H) : H \in \mathbf{N}\},$$

for which

$$(4) \quad |P_1(x_1) + \cdots + P_m(x_m)| < H^{-n_1 - \cdots -n_m-\varepsilon}, \qquad (P_1,\dots,P_m) \in \pi(H).$$

is infinite.

According to [1] (cf. proof of Theorem I) there exists a set $\nabla \subset \mathbf{R}^1$ of measure zero such that for $x_0 \notin \nabla$ and any polynomial $R(x) \in \mathbf{Z}[x]$ of degree $\leqslant n_m - 1$ and height $\leqslant n_m H$ satisfying the inequality

$$|R(x_0)| \leqslant c_1^{1/2} H^{-n_m+1-\varepsilon/2}$$

one obtains $H < \psi(x_0) < \infty$. Let us put

$$(5) \qquad A_2 = A_1 \backslash (\mathbf{R}^{m-1} \times \nabla), \qquad \text{mes}(A_2) > 0.$$

For $\vec{x} = (x_1,\ldots,x_m) \in A_2$ we put

$$\mathcal{P}_1(\vec{x}) = \left\{(P_1,\ldots,P_m) \in \mathcal{P}(\vec{x}) : \max_{i=1,\ldots,m} H(P_i) \geqslant \psi(x_m)\right\}.$$

Then for any $\vec{x} = (x_1,\ldots,x_m) \in A_2$ the set $\mathcal{P}_1(\vec{x})$ is infinite. Besides, according to (5) for any $\vec{x} = (x_1,\ldots,x_m) \in A_2$ and $(P_1,\ldots,P_m) \in \mathcal{P}_1(\vec{x})$,

$$H = \max_{i=1,\ldots,m} H(P_i)$$

we have

$$|P_m'(x_m)|^2 > c_1 H^{-2n_m+2-\varepsilon}. \tag{6}$$

Since $m \geqslant 2$, then (2) and (6) give the possibility to apply Lemmas 1.3 and 1.5. We obtain from 1.3 and 1.5

LEMMA 2.7. *For any $\vec{x} = (x_1,\ldots,x_m) \in A_2$ and any $(P_1,\ldots,P_m) \in \mathcal{P}_1(\vec{x})$ there exists $y_m \in \mathbf{R}^1$ such that $(x_1,\ldots,x_{m-1}, y_m) \in \mathfrak{N}(P_1,\ldots,P_m)$, i.e. $P_1(x_1) + \cdots + P_m(y_m) = 0$ and for $H = \max_{i=1,\ldots,m} H(P_i)$,*

$$\begin{aligned} &|x_m - y_m| < c_2 |P_m'(x_m)|^{-1} H^{-n_1-\cdots-n_m-\varepsilon}, \\ &|x_m - y_m| < c_3 H^{-n_1-\cdots-n_{m-1}-1-\varepsilon/2}, \end{aligned} \tag{7}$$

where $c_2, c_3 > 0$ are effectively determined by $n_1,\ldots,n_m$, N_0 and ε.

In view of 1.6 it is possible to have a more convenient form of 2.7.

LEMMA 2.8. *If $\vec{x} = (x_1,\ldots,x_m) \in A_2$ and $(P_1,\ldots,P_m) \in \mathcal{P}_1(\vec{x})$, then there exists $y_m \in \mathbf{R}^1$ for which*

$$P_1(x_1) + \cdots + P_{m-1}(x_{m-1}) + P_m(y_m) = 0, \qquad H = \max_{i=1,\ldots,m} H(P_i)$$

$$|x_m - y_m| < c_2 |P_m'(x_m)|^{-1} \cdot H^{-n_1-\cdots-n_m-\varepsilon} \leqslant c_3 \cdot H^{-n_1-\cdots-n_{m-1}-1-\varepsilon/2},$$

and

$$|x_m - y_m| < c_4 \cdot |P_m'(y_m)|^{-1} \cdot H^{-n_1-\cdots-n_m-\varepsilon}.$$

As before for $A \subseteq \mathbf{R}^m$ and $\vec{x}_0 = (x_1^0,\ldots,x_{m-1}^0) \in \mathbf{R}^{m-1}$ we put $A \upharpoonright \vec{x}_0 = \{y \in \mathbf{R}^1 : (x_1^0,\ldots,x_{m-1}^0, y) \in A\}$.

For h, $H \in \mathbf{N}$, $h \leqslant H$, let us introduce a set $\pi(H, h) = \{(P_1,\ldots,P_m) \in \pi(H)$: the leading coefficient of $P_1(x)$ is equal to $h\}$ and define

$$\begin{aligned} \pi(k) &= \cup \left\{\pi(H) : 2^k \leqslant H < 2^{k+1}\right\}, \\ \pi(k, h) &= \cup \left\{\pi(H, h) : 2^k \leqslant H < 2^{k+1}\right\}. \end{aligned}$$

For $H \in \mathbf{N}$ and $(P_1,\ldots,P_m) \in \pi(H)$ we consider

$$\begin{aligned} M(P_1,\ldots,P_m) = \{&(x_1,\ldots,x_m) \in \mathbf{R}^m : \text{there exists } y_m \in \mathbf{R}^1 \\ &\text{such that } P_1(x_1) + \cdots + P_{m-1}(x_{m-1}) + P_m(y_m) = 0, \\ &|x_m - y_m| \leqslant \min\left[c_3 H^{-n_1-\cdots-n_{m-1}-1-\varepsilon/2}, c_4 \cdot |P_m'(y_m)|^{-1} \cdot H^{-n_1-\cdots-n_m-\varepsilon}\right]\} \end{aligned} \tag{8}$$

(9) $$\Lambda(P_1,\dots,P_m) = \{(x_1,\dots,x_m) \in \mathbf{R}^m : \text{there exists } y_m \in \mathbf{R}^1 \text{ such that } P_1(x_1) + \cdots + P_{m-1}(x_{m-1}) + P_m(y_m) = 0 \text{ and } |x_m - y_m| \leqslant \min[c_3 \cdot H^{-n_1-\cdots-n_{m-1}-\varepsilon/4}, c_4 \cdot |P_m'(y_m)|^{-1} H^{-n_1-\cdots-n_m+1-3\varepsilon/4}]\}.$$

For any $\vec{x} \in \mathbf{R}^{m-1}$ a set $\Lambda(P_1,\dots,P_m)\upharpoonright \vec{x}$ is a union of a finite collection of nonintersecting intervals, the set of which is denoted by $\mathfrak{L}(P_1,\dots,P_m)\upharpoonright \vec{x}$. Further for $\vec{x} \in \mathbf{R}^{m-1}$ and $k \geqslant 0$, $h \geqslant 1$,

(10) $$\mathfrak{U}(k,h)\upharpoonright \vec{x} = \bigcup \{\mathfrak{L}(P_1,\dots,P_m)\upharpoonright \vec{x} : (P_1,\dots,P_m) \in \pi(k,h)\}.$$

Let $\mathfrak{V}(k,h)\upharpoonright \vec{x}$ be a maximal subset of $\mathfrak{U}(k,h)\upharpoonright \vec{x}$ satisfying the conditions of Lemma 2.4 and

(11) $$V(k,h)\upharpoonright \vec{x} = \bigcup \{I : I \in \mathfrak{V}(k,h)\upharpoonright \vec{x}\}.$$

LEMMA 2.9. *For almost all $\vec{x} \in \mathbf{R}^{m-1}$ (in the measure in $\mathbf{R}^{m-1}$) a set $W\upharpoonright \vec{x}$ of those $y \in \mathbf{R}^1$ which belong to an infinite number of sets $V(k,h)\upharpoonright \vec{x}$ has measure zero.*

PROOF OF 2.9. Let $B = \{\vec{x} \in \mathbf{R}^{m-1} : \operatorname{mes}(W\upharpoonright \vec{x}) \neq 0\}$ and $v(k,h)\upharpoonright \vec{x}_0$ be the union of intervals of the form $I \cap J$ for $I \in \mathfrak{V}(k,h)\upharpoonright \vec{x}_0$, $J \in \mathfrak{U}(k,h)\upharpoonright \vec{x}_0$, $I \neq J$. According to 2.4 for $\vec{x}_0 \in B$ one has $\operatorname{mes}(v\upharpoonright \vec{x}_0) \neq 0$, where the set $v\upharpoonright \vec{x}_0 \subseteq \mathbf{R}^1$ is composed of $y \in \mathbf{R}^1$ contained in infinite number of sets $v(k,h)\upharpoonright \vec{x}_0$.

If $y \in v(k,h)\upharpoonright \vec{x}$ with $2^k \geqslant c_3$, then in view of (10) and (11) we have

$$y \in \Lambda(P_1^1,\dots,P_m^1)\upharpoonright \vec{x} \cap \Lambda(P_1^2,\dots,P_m^2)\upharpoonright \vec{x},$$

where $(P_1^1,\dots,P_m^1) \neq (P_1^2,\dots,P_m^2)$ and $(P_1^1,\dots,P_m^1), (P_1^2,\dots,P_m^2) \in \pi(k,h)$.

Taking into account (9) and using Lemma 1.7 we obtain the inequalities

(12) $$|P_1^i(x_1) + \cdots + P_{m-1}^i(x_{m-1}) + P_m^i(x_m)| \leqslant c_5 H_i^{-n_1-\cdots-n_m+1-\varepsilon/2}$$

for $(x_1,\dots,x_m) = (x_1,\dots,x_{m-1}, y)$, where $\vec{x} = (x_1,\dots,x_{m-1}) \in \mathbf{R}^{m-1}$ and $(P_1^i,\dots,P_m^i) \in \pi(k,h) : i = 1,2$. Then polynomials $P_i^3(x) = P_i^1(x) - P_i^2(x)$ have degrees $\leqslant n_i$ and heights $\leqslant H_1 + H_2 \leqslant 2^{k+2}$, with $P_1^3(x)$ having degree $\leqslant n_1 - 1$, as the leading coefficients of $P_1^1(x)$ and $P_1^2(x)$ are equal to h. From (12) it follows for the same k and h that

(13)

$$|P_1^3(x_1) + \cdots + P_m^3(x_m)| \leqslant 2c_5(2^k)^{-n_1-\cdots-n_m+1-\varepsilon/2} \leqslant c_6 \cdot H_3^{-n_1-\cdots-n_m+1-\varepsilon/2}$$

for $c_6 = c_6(n_1,\dots,n_m, N_0, \varepsilon) > 0$ and $H_3 = \max_{i=1,\dots,m} H(P_i^3)$ with $x_m \in v(k,h)\upharpoonright \vec{x}$, $\vec{x} = (x_1,\dots,x_{m-1}) \in \mathbf{R}^{m-1}$. If $\mathfrak{B} = \{(x_1,\dots,x_m) \in \mathbf{R}^m : \vec{x} = (x_1,\dots,x_{m-1}) \in B \text{ and } x_m \in v\upharpoonright \vec{x}\}$, then for $(x_1,\dots,x_m) \in \mathfrak{B}$ one has an infinite number of sequences of polynomials $(P_1^3,\dots,P_m^3)$, satisfying (13). In view of our inductive assumption, the sets $A(n_1 - 1, n_2,\dots,n_m, \varepsilon_1)$ have measure zero (in $\mathbf{R}^m$) for all $\varepsilon_1 > 0$. Thus the measure of $\mathfrak{B}$ (in $\mathbf{R}^m$) is also zero. So $\operatorname{mes}(\mathfrak{B}\upharpoonright \vec{x}) = 0$ (in $\mathbf{R}^1$) for almost all $\vec{x} \in \mathbf{R}^{m-1}$ (with respect to measure in $\mathbf{R}^{m-1}$). However $\operatorname{mes}(v\upharpoonright \vec{x}) \neq 0$ for $\vec{x} \in B$. Consequently $\operatorname{mes} B = 0$ (with respect to measure in $\mathbf{R}^{m-1}$). Lemma 2.9 is proved.

LEMMA 2.10. *For almost all* $\vec{x} \in \mathbf{R}^{m-1}$ *a set* $A_2 \upharpoonright \vec{x}$ *has measure zero.*

PROOF OF 2.10. Let $F \subseteq \mathbf{R}^{m-1}$ be a set of complete measure in $\mathbf{R}^{m-1}$, satisfying conditions of 2.9, i.e. $\operatorname{mes}(\mathbf{R}^{m-1} \setminus F) = 0$. Let us show that for all $\vec{x} \in F$ (except a set of measure zero in $\mathbf{R}^{m-1}$) $A_2 \upharpoonright \vec{x}$ has zero measure. Since $n_m > 0$, i.e. for $(P_1, \dots, P_m) \in \pi(H)$, $P_m(x) \not\equiv 0$, then for all $\vec{x}_0 \in F$, $M(P_1, \dots, P_m) \upharpoonright \vec{x}_0$ is a union of $\leqslant m$ intervals

$$(14) \qquad M(P_1, \dots, P_m) \upharpoonright \vec{x}_0 = \bigcup \{I : I \in \mathfrak{M}(P_1, \dots, P_m) \upharpoonright \vec{x}_0\}.$$

By Lemma 2.8 for $\vec{x}_0 \in F$ and $y_0 \in A_2 \upharpoonright \vec{x}_0$ one has

$$(15)\ y_0 \in I \subseteq \mathcal{T}(I), \quad \text{where } I \in \mathfrak{M}(P_1, \dots, P_m) \upharpoonright \vec{x}_0,\ \mathcal{T}(I) \in \mathfrak{L}(P_1, \dots, P_m) \upharpoonright \vec{x}_0,$$

for any $(P_1, \dots, P_m) \in \mathcal{P}_1(\vec{y}_0)$ and infinite set of $\mathcal{P}_1(\vec{y}_0)$, $\vec{y}_0 = (\vec{x}_0, y_0) \in \mathbf{R}^m$. In view of the choice of F and 2.9, from $\vec{x}_0 \in F$, $y_0 \in A_2 \upharpoonright \vec{x}_0$ it follows that for an infinite number of (k, h) the point y_0 belongs to an interval of the form $\mathcal{T}(I)$ from $\mathcal{U}(k, h) \upharpoonright \vec{x}_0$, but not from $\mathcal{V}(k, h) \upharpoonright \vec{x}_0$.

Finally, for $I \in \mathfrak{M}(P_1, \dots, P_m) \upharpoonright \vec{x}_0$ we denote by $\mathcal{K}(I)$ the union of all $I' \in \mathfrak{M}(P_1, \dots, P_m) \upharpoonright \vec{x}_0$ contained in $\mathcal{T}(I)$. From (8)–(9) it follows that

$$(16) \qquad \operatorname{mes}(\mathcal{K}(I)) \leqslant \operatorname{mes}(\mathcal{T}(I)) \cdot (2^k)^{-1-\varepsilon/4}, \qquad \mathcal{K}(I) \subseteq \mathcal{T}(I)$$

for $I \in \mathfrak{M}(P_1, \dots, P_m) \upharpoonright \vec{x}_0$ and $(P_1, \dots, P_m) \in \pi(k, h)$.

If we put $\mathfrak{Z}(k, h) \upharpoonright \vec{x}_0 = \{\mathcal{K}(I) : I \in \mathfrak{M}(P_1, \dots, P_m) \upharpoonright \vec{x}_0, (P_1, \dots, P_m) \in \pi(k, h)\}$, then from Lemmas 2.4, 2.9 and the definition of F we obtain for $\vec{x}_0 \in F$,

$$\sum_{\mathcal{K}(I) \in \mathfrak{Z}(k, h) \upharpoonright \vec{x}_0} \operatorname{mes}(\mathcal{K}(I)) \leqslant c_7 \quad \text{for } c_7 = c_7(n_1, \dots, n_m, N_0, \varepsilon) > 0.$$

This and (16) imply

$$\sum_{k \geqslant 0} \sum_{h \geqslant 1} \sum_{\mathcal{K}(I) \in \mathfrak{Z}(k, h) \upharpoonright \vec{x}_0} \operatorname{mes}(\mathcal{K}(I)) \leqslant 2c_7 \sum_{k \geqslant 0} (2^k)^{-\varepsilon/4} < \infty$$

for $\vec{x}_0 \in F$. By the Borel-Cantelli lemma $\operatorname{mes}(A_2 \upharpoonright \vec{x}_0) = 0$ for $\vec{x}_0 \in F$. So Lemma 2.10 and thus Theorem II are proved.

It is clear that proofs of Theorems I and II leave a lot of room for generalizations. For example, it presents no difficulty to take into consideration the height H_i of each polynomial $P_i(x)$ separately, since the estimate of $|P_i'(x)|$, which is a key to the proof, can be expressed in terms of $H(P_i)$ and not of $\max_{i=1,\dots,m} H(P_i)$. Besides it is possible to consider a linear combination of polynomials $P_i(x_i)$ with arbitrary but fixed nonzero real coefficients $\lambda_i : i = 1, \dots, m$. For this there is no need at all to make any essential changes in the proof of the theorems, as

$$\partial/\partial x_i(\lambda_1 P_1(x_1) + \cdots + \lambda_m P_m(x_m)) = \lambda_i P_i'(x_i),$$

and the estimate of $|P_i'(x_i)|$ may be obtained from **[1]**. Thus we formulate an immediate generalizations of Theorems I and II, the proofs of which are based on those given above.

THEOREM III. *Let* $n_1,\ldots,n_m$ *be integers,* $\lambda_1 \neq 0,\ldots,\lambda_m \neq 0$ *be real numbers and* $\varepsilon > 0$. *Let* A *be a subset of* $\mathbf{R}^m$ *containing those* $(x_1,\ldots,x_m) \in \mathbf{R}^m$ *for which there exists an infinite number of systems of polynomials* $P_1(x),\ldots,P_m(x)$ *from* $\mathbf{Z}[x]$ *of degrees not higher than* $n_1,\ldots,n_m$ *and without free coefficients satisfying the inequality*

$$\|\lambda_1 P_1(x) + \cdots + \lambda_m P_m(x_m)\| < H^{-n_1-\cdots-n_m-\varepsilon},$$

$H = \max_{i=1,\ldots,m} H(P_i)$. *Then the set* A *has zero measure in* $\mathbf{R}^m$.

Another formulation of Theorems III and III′ is presented in the §0.

If we take the heights $H(P_i)$ into account separately, we arrive at Theorem IV.

THEOREM IV. *Let* $n_1,\ldots,n_m$ *be integers and* $\varepsilon > 0$. *Let* $A = A(n_1,\ldots,n_m, \varepsilon)$ *be a subset of* $\mathbf{R}^m$ *composed of those points* $(x_1,\ldots,x_m) \in \mathbf{R}^m$ *for which there exists an infinite number of systems of polynomials* $P_1(x),\ldots,P_m(x)$ *from* $\mathbf{Z}[x]$ *of degrees not higher than* $n_1,\ldots,n_m$ *and without free coefficients satisfying the inequality*

$$\|P_1(x_1) + \cdots + P_m(x_m)\| < \left(\prod_{i=1}^{m} H(P_i)^{n_i}\right)^{-1-\varepsilon}.$$

Then the sets $A(n_1,\ldots,n_m, \varepsilon)$ *have zero measure in* $\mathbf{R}^m$.

Finally there exists one more version of Theorem III connected with polynomials having free coefficients, but with additional conditions on the approximation properties of the numbers $\lambda_1,\ldots,\lambda_m$ (cf. [**6**], [7]). We mention here

THEOREM V. *Let* $n_1,\ldots,n_m$ *be integers,* $\lambda_1,\ldots,\lambda_m$ *be real and the inequality*

$$\|\lambda_1 a_1 + \cdots + \lambda_m a_m\| < \left(\prod_{i=1}^{m} (|a_i|+1)\right)^{-1-\delta}$$

for any $\delta > 0$ *has only finitely many solutions in integer rational numbers* $a_1,\ldots,a_m$. *If* $\varepsilon > 0$ *and* $A \subseteq \mathbf{R}^m$ *is composed of those* $(x_1,\ldots,x_m) \in \mathbf{R}^m$ *for which there exists an infinite number of systems of polynomials* $P_1(x),\ldots,P_m(x) \in \mathbf{Z}[x]$ *of degrees* $n_1,\ldots,n_m$, *satisfying the inequality*

$$\|\lambda_1 P_1(x_1) + \cdots + \lambda_m P_m(x_m)\| < H^{-n_1-\cdots-n_m-m-\varepsilon},$$

$H = \max_{i=1,\ldots,m} H(P_i)$, *then the set* A *has zero measure in* $\mathbf{R}^m$.

Along the same lines as in Theorem IV it is possible to take into consideration the height of every $P_i(x)$, separately preserving the assumption on $\lambda_1,\ldots,\lambda_m$.

3. In this section we briefly discuss the possibility of the extension of results of §2 to arbitrary algebraic manifolds. The scheme of proof of Theorems I and II shows that the most important point is the transition from "smallness" of the value of the polynomial from $\mathbf{R}[x_1,\ldots,x_n]$ at a point from $\mathbf{R}^n$ to "closeness" of this point to the zero of the polynomial. As is known already, even for $n = 1$ this problem is not easily solvable. When we come to the case $n > 1$ the difficulties increase considerably, especially in view of the appearance of irreducible but not smooth manifolds with singularities of higher multiplicities or nonnormal intersections. The ordinary methods of resolution of singularities are ineffective for the case $n > 1$. Thus we restrict ourselves for the moment to the examination of the problem of k-extremality formulated in §0.

We present results dealing with the problem of estimating functions $w_n(x, y) : (x, y) \in \mathbf{R}^2$. Here $w_n(x, y)$ is the supremum of those $w > 0$ for which there exists an infinite number of polynomials $P(x, y) \in \mathbf{Z}[x, y]$ of degree n satisfying the inequality $|P(x, y)| < H(P)^{-w}$.

We mean at least an estimate $w_n(x, y) \leqslant Cn^2$ for almost all $(x, y) \in \mathbf{R}^2$ and for an absolute constant $C > 0$. The only estimates available up to now were of the form $w_n(x, y) \leqslant O(n^4)$ **[16]** or $w_n(x, y) \leqslant O(n^3)$ for almost all $(x, y) \in \mathbf{R}^2$. Indeed these estimates are trivial consequences of lemmas from §1. We show this below. We also show that the traditional theory of u-resultants **[9]**, which is in fact reducible to an effective version of the Hilbert theorem on the zeros of polynomials (the Nullstellensatz), permits substantial improvement of these results.

To obtain metrical results from theorems on "closeness" of points to algebraic manifolds we need the following simple auxiliary result.

LEMMA 3.1. *Let* $P(x_1, \ldots, x_m) \in \mathbf{R}[x_1, \ldots, x_m]$ *be a polynomial of degree* n, $\Delta \subseteq \mathbf{R}^m$ *be an arbitrary bounded measurable subset of* $\mathbf{R}^m$ *and let* $\mathfrak{N}(P)$ *denote an* $m - 1$*-dimensional manifold in* $\mathbf{R}^m$: $P(x_1, \ldots, x_m) = 0$. *Then for any* $\varepsilon > 0$ *the* $\mathbf{R}^m$ *measure of a set*

$$\{\vec{x} \in \Delta : \text{there exists } \vec{x}_0 \in \mathfrak{N}(P) \text{ and } |\vec{x} - \vec{x}_0| \leqslant \varepsilon\}$$

does not exceed $C(n, m) \cdot \text{mes}(\Delta) \cdot \varepsilon$. *Here* $|\cdot|$ *denotes the norm in* $\mathbf{R}^m$.

Proof of 3.1. is obtained from ordinary formulae of integration taking into consideration a remark to the effect that "$m - 1$-dimensional area" of the surface $P(x_1, \ldots, x_m) = 0$ in Δ does not exceed $C(n, m) \cdot \text{mes}(\Delta)$ for an m-dimensional parallelepiped Δ in $\mathbf{R}^m$.

Using lemmas of §1 and the elimination theory in the classical form **[9]** to consider the estimate of $w(x, y)$ for almost all $(x, y) \in \mathbf{R}^2$. For this it is necessary to consider together with affine manifolds in $\mathbf{A}^2(\mathbf{R})$ also projective manifolds. This way for any polynomial $P(x, y) \in \mathbf{R}[x, y]$ of degree n let us denote by $P(x, y, z) \in \mathbf{R}[x, y, z]$ a homogeneous form of degree n such that $P(x, y, z) = z^n P(x/z, y/z)$. As it is clear from 1.1–1.4 we are mainly interested in

$$\delta = \min_{(x, y) \in \Delta} \left(\max\{|P(x, y)|, |P'_x(x, y)|, |P'_y(x, y)|\} \right),$$

where the min is taken over $(x, y) \in \Delta$, where Δ is a bounded subset of $\mathbf{R}^2$. Naturally, in this case to obtain a nontrivial lower bound of δ the curve $P(x, y) = 0$ should have no singularities in Δ. Thus a natural assumption is smoothness of the curve $\mathfrak{N}(P) : P(x, y) = 0$ in $\mathbf{A}^2(\mathbf{R})$. However, usually it is more convenient to work with projective manifolds. Therefore a more general assumptiion is the smoothness of $\mathfrak{N}(P): P(x, y, z) = 0$ in $\mathbf{P}^2(\mathbf{R})$.

We recall an application of u-resultants [**9**, **10**] which in fact coincides with the Hilbert theorem on the zeroes of polynomials [**8**, **9**].

LEMMA 3.2. *Let* $P_1(x_1,\dots,x_m),\dots,P_m(x_1,\dots,x_m)$ *be homogeneous forms from* $\mathbf{R}[x_1,\dots,x_m]$ *of degrees, respectively* $n_1,\dots,n_m$. *If*

$$P_1(x_1,\dots,x_m),\dots,P_m(x_1,\dots,x_m)$$

do not have common nontrivial roots, then there exists $R = R(P_1,\dots,P_m) \neq 0$, *such that*

$$x_i^\tau R = A_1^i \cdot P_1 + \cdots + A_m^i \cdot P_m$$

for all $i = 1,\dots,m$ *and certain* $A_j^i(x_1,\dots,x_m) \in \mathbf{R}[x_1,\dots,x_m]$ *for a fixed integer* τ. *Here* $R = R(P_1,\dots,P_m)$ *is a homogeneous polynomial over* $\mathbf{Z}$ *of degree* $n_1 \cdots n_{i-1}n_{i+1} \cdots n_m$ *in coefficients of forms* $P_i : i = 1,\dots,m$.

To apply this lemma it is necessary to have also the estimates of the heights and degrees of polynomials $A_j^i(x_1,\dots,x_m)$. Instead of the general examination one can restrict oneself to construction of the basis of inertia forms along the lines [**9**, §§81–82]. We present the corresponding results in the most interesting case $i = m$; when the forms $P_j(x_1,\dots,x_m)$ correspond to the polynomial

$$P_j(x_1,\dots,x_{m-1}) \in \mathbf{R}[x_1,\dots,x_{m-1}]: \qquad j = 1,\dots,m.$$

LEMMA 3.3. *Let polynomials*

$$P_1(x_1,\dots,x_{m-1}),\dots,P_m(x_1,\dots,x_{m-1}) \in \mathbf{R}[x_1,\dots,x_{m-1}]$$

have degrees, respectively $n_1,\dots,n_m$. *Let us suppose that a system of equations*

$$P_1(x_1,\dots,x_{m-1}) = 0,\dots,P_m(x_1,\dots,x_{m-1}) = 0$$

does not have common solutions (including also those at infinity). In other words, a system of homogeneous equations $P_1(x_1,\dots,x_m) = 0,\dots,P_m(x_1,\dots,x_m) = 0$ *does not have nontrivial solutions. Then there exist* $R = R(P_1,\dots,P_m) \in \mathbf{R}$, $R \neq 0$, *and polynomials* $A_1(x_1,\dots,x_{m-1}),\dots,A_m(x_1,\dots,x_{m-1}) \in \mathbf{R}[x_1,\dots,x_{m-1}]$ *such that*

$$R = A_1(x_1,\dots,x_{m-1}) \cdot P_1(x_1,\dots,x_{m-1}) + \cdots + A_m(x_1,\dots,x_{m-1})P_m(x_1,\dots,x_{m-1}).$$

R belongs to a ring generated by coefficients of polynomials

$$P_1(x_1,\dots,x_{m-1}),\dots,P_m(x_1,\dots,x_{m-1})$$

over $\mathbf{Z}$, *and heights and degrees of* $A_i(x_1,\ldots,x_{m-1})$ *are* $d(A_i)$ *and* $H(A_i): i = 1,\ldots,m$, *which can be estimated in terms of* $m, n_1,\ldots,n_m$ *and heights* $H(P_i): i = 1,\ldots,m$ *by*

$$d(A_i) \leqslant \sum_{j=1, j\neq i}^{m} n_j : \qquad i = 1,\ldots,m,$$

$$\ln H(A_i) \leqslant C \cdot \left(\sum_{j=1, j\neq i}^{m} n_1 \cdots n_{j-1} n_{j+1} \cdots n_m \ln H(P_j) \right.$$

$$\left. + (n_1 \cdots n_{i-1} n_{i+1} \cdots n_m - 1) \ln H(P_i) \right)$$

for all $i = 1,\ldots,m$ *and a constant* $C = C(m) > 0$ *effectively determined by* m.

In particular, in the most interesting case, when $P_i(x_1,\ldots,x_{m-1}) \in \mathbf{Z}[x_1,\ldots,x_{m-1}]$ according to 3.2–3.3, $R(P_1,\ldots,P_m) \in \mathbf{Z}$ and as $R \neq 0$, then $|R| \geqslant 1$. In fact this particular estimate is necessary for the following

LEMMA 3.4. *Let us suppose that* $P_1(x_1,\ldots,x_{m-1}),\ldots,P_m(x_1,\ldots,x_{m-1}) \in \mathbf{Z}[x_1,\ldots,x_{m-1}]$ *and a system of homogeneous equations*

$$P_1(x_1,\ldots,x_m) = 0,\ldots,P_m(x_1,\ldots,x_m) = 0$$

does not have nontrivial solutions. For a bounded domain $\Delta \subseteq \mathbf{R}^{m-1}$ *let us put*

$$\delta = \min_{(x_1,\ldots,x_{m-1}) \in \Delta} (\max\{|P_1(x_1,\ldots,x_{m-1})|,\ldots,|P_m(x_1,\ldots,x_{m-1})|\})$$

where the min *is taken over all the elements* $(x_1,\ldots,x_{m-1}) \in \Delta$. *Then for* δ *there exists a nontrivial lower bound expressed in terms of heights and degrees* $H(P_i)$ *and* $d(P_i)$ *of polynomials* $P_i(x_1,\ldots,x_{m-1}): i = 1,\ldots,m$;

$$\delta > C' \cdot |\Delta|^{-\sum_{i=1}^{m} d(P_i)} \times \left(\sum_{i=1}^{m} d(P_i) \right)^{-m+1} \cdot m^{-1} \left(\prod_{i=1}^{m} H(P_i)^{n_1 \cdots n_{i-1} n_{i+1} \cdots n_m} \right)^{-1},$$

where $C' = C'(m, \Delta) > 0$, $|\Delta| = \max_{(x_1,\ldots,x_{m-1}) \in \Delta} \max(|x_1|,\ldots,|x_{m-1}|)$.

It is already sufficient for applications, for example, in the case $m = 3$. We have

PROPOSITION 3.5. *For any* $(x, y) \in \mathbf{R}^2$ *let us denote by* $w_n'(x, y)$ *the supremum of those* $w > 0$, *for which there exists an infinite number of polynomials* $P(x, y) \in \mathbf{Z}[x, y]$ *of degree* n, *defining a smooth curve* $\mathfrak{N}(P)$ *and satisfying*

$$|P(x, y)| < H(P)^{-w}.$$

Then there is an absolute constant $C > 0$ $(C \leqslant 6)$ *such that for almost all* $(x, y) \in \mathbf{R}^2$, $w_n'(x, y) \leqslant C \cdot n^2$.

PROOF. Let Δ be an arbitrary square in $\mathbf{R}^2$ and let us show that for almost all $(x, y) \in \Delta$, $w_n'(x, y) \leqslant C \cdot n^2$.

For any $P(x, y) \in \mathbf{Z}[x, y]$, $d(P) = n$ let us denote by Λ_P a set

$$\Lambda_P = \left\{(x, y) \in \Delta : |P(x, y)| < H(P)^{-Cn^2}\right\}.$$

The totality of those $P(x, y) \in \mathbf{Z}[x, y]$, $d(P) = n$ for which the curve $\mathfrak{N}(P)$ is smooth will be denoted by $\mathfrak{T}$. Then for

$$\delta_P = \min_{(x, y) \in \Delta} \left(\max\{|P(x, y)|, |P_x'(x, y)|, |P_y'(x, y)|\}\right)$$

in view of 3.3 and 3.4 one has

$$\delta_P > C_1 \cdot H(P)^{-(n-1)(3n-1)+1}.$$

Then in view of 1.4 one obtains for $C \geqslant 6$ that for any $(x, y) \in \Lambda_P$ for $P(x, y) \in \mathfrak{T}$ there exists $(x_0, y_0) \in \mathbf{R}$ such that

$$P(x_0, y_0) = 0 \quad \text{and} \quad \left((x - x_0)^2 + (y - y_0)^2\right)^{1/2} < H(P)^{-3n^2-1}.$$

In particular, for any $P(x, y) \in \mathfrak{T}$ one has $\Lambda_P \subseteq M_P$, where $M_P = \{(x, y) \in \Delta_0 :$ there exists $(x_0, y_0) \in \mathfrak{N}(P)$, i.e. $P(x_0, y_0) = 0$ and $|(x, y) - (x_0, y_0)| \leqslant H(P)^{-3n^2-1}\}$, where $|\cdot|$ is the norm in $\mathbf{R}^2$ and Δ_0 is a square in $\mathbf{R}^2$ containing Δ with the following condition, $\rho(\partial\Delta, \partial\Delta_0) \geqslant 2$. Applying 3.1 one concludes

$$\operatorname{mes} M_P \leqslant C_2 \cdot H(P)^{-3n^2-1}$$

for $C_2 = C_2(n; \Delta_0) > 0$. Therefore

$$\sum_{P(x, y) \in \mathfrak{T}} \operatorname{mes} M_P \leqslant \sum_{P(x, y) \in \mathbf{Z}[x, y],\, d(P) = n} \operatorname{mes} M_P$$

$$= \sum_H \sum_{P,\, d(P) = n,\, H(P) = H} C_2 H^{-3n^2-1} \leqslant \sum_H C_3 H^{-n^2-1} < \infty$$

since $n \geqslant 1$. According to the Borel-Cantelli lemma the totality of those $(x, y) \in \Delta$ which are contained in an infinite number of sets M_P, and thus also Λ_P, has zero measure.

It is not difficult to show that C can be chosen $\leqslant 4$. For this one should substitute the application of 1.4. by 1.2, and Lemma 3.1. by its natural generalization—the approximation by complex root of the points from $\mathbf{R}^n$. Besides, when we require the smoothness of $\mathfrak{N}(P)$ in 3.5 it is sufficient to request that the curve be smooth only in affine, but not in projective, space. For this it is only necessary to transform $P_x'(x, y)$ and $P_y'(x, y)$. Indeed the assumption of smoothness of $\mathfrak{N}(P)$ is not necessary in the problem of k-extremality, if we apply the methods of §2 in combination with estimates near singular points along the lines of **[15]**. Thus we obtain

THEOREM 3.6. *For almost all $(x, y) \in \mathbf{R}^2$ we have*

$$w_n(x, y) \leqslant 6n^2.$$

References

1. V. G. Sprindzuk, *Mahler's problem in metric number theory* (Minsk, 1967), Transl. Math. Mono., vol. 25, Amer. Math. Soc., Providence, R. I., 1969.

2. ______, *The method of trigonometric sums in the metric theory of diophantine approximation*, Trudy Mat. Inst. Steklov **128** (1972), part II, 212–228. (Russian)

3. A. Baker, *On a theorem of Sprindzuk*, Proc. Royal Soc. Ser. A **292** (1966), 92–104.

4. W. Schmidt, *Metrische Sätze über simultane approximation abhängiger grössen*, Monatsh. Math. **63** (1964), 154–166.

5. J. W. S. Cassels, *An introduction to diophantine approximations*, Cambridge Univ. Press, Cambridge, 1957.

6. E. I. Kovalevskaya, *Metric theorems on the approximation of zero by a linear combination of polynomials with integral coefficients*, Acta Arith. **25** (1973), 93–104; Dokl. Akad. Nauk BSSR **17** (1973), 1085–1088. (Russian)

7. V. G. Sprindzuk, *Metric theory of diophantine approximations*, Current Problems in the Analytic Theory of Numbers, Minsk, 1974, pp. 178–198. (Russian)

8. I. R. Shafarevic, *Foundation of algebraic geometry*, Springer, New York, 1975.

9. B. L. Van der Waerden, *Modern algebra*, vols. 1, 2, Ungar, New York, 1949–1950.

10. W. V. D. Hodge and D. Pedoe, *Methods of algebraic geometry*, vols. 1–3, Cambridge Univ. Press, Cambridge, 1947.

11. J. P. Serre, *Algèbre locale, multiplicités*, Lecture Notes in Math. vol. 11, Springer-Verlag, Berlin and New York, 1965.

12. R. Slesoraitene, *The Mahler-Sprindzuk theorem for polynomials of the third degree in two variables*. II, Litovski Mat. Sb. **10** (1970), 791–814. (Russian)

13. V. I. Bernik and E. I. Kovalevskaya, *Extremal properties of some surfaces in n-dimensional Euclidean space*, Math. Notes **15** (1974), 247–254.

14. L. Hörmander, *Linear partial differential operators*, Springer, New York, 1963.

15. B. Teissier, *Introduction to equisingularity problems*, Proc. Conf. Algebraic Geometry (Arcata, Calif., 1974), Proc. Sympos. Pure Math., vol. 29, Amer. Math. Soc., Providence, R. I., 1975.

16. A. S. Pyartli, *Diophantine approximations on submanifolds of Euclidean space*, Functional Anal. Appl. **3** (1969), 304–306.

17. V. I. Bernick, *Generated extremal surfaces*, Mat. Sb. **145**(1977), 480–489.

INDEX

ABCDEFGHIJ–CM–8987654